Foreword

This publication, *Composite Materials: Fatigue and Fracture (Third Volume)* contains papers presented at the Third Symposium on Composite Materials: Fatigue and Fracture, which was held in Lake Buena Vista, Florida, 6–7 November 1989. The symposium was sponsored by Committee D30 on High Modulus Fibers and Their Composites. T. Kevin O'Brien, NASA Langley Research Center, presided as symposium chairman and is editor of this publication.

Contents

Overview   1

MATRIX CRACKING AND DELAMINATION

Fracture Mechanics Approaches to Transverse Ply Cracking in Composite
Laminates—LYNN BONIFACE, STEPHEN L. OGIN, AND PAUL A. SMITH   9

Matrix Cracking in Composite Laminates with Resin-Rich Interlaminar Layers—
LARRY B. ILCEWICZ, ERNEST F. DOST, J. W. Mc COOL, AND D. H. GRANDE   30

The Upper Bounds of Reduced Axial and Shear Moduli in Cross-Ply Laminates
with Matrix Cracks—JONG-WON LEE, D. H. ALLEN, AND
C. E. HARRIS   56

Cooling Rate Effects in Carbon Fiber/PEEK Composites—PETER DAVIES,
WESLEY J. CANTWELL, PEAN-YUE JAR, HERVÉ RICHARD, DAVID J. NEVILLE,
AND HANS-HENNING KAUSCH   70

Effects of Moisture Absorption on Edge Delamination, Part I: Analysis of the
Effects of Nonuniform Moisture Distributions on Strain Energy Release
Rate—STEVEN J. HOOPER, RICHARD F. TOUBIA, AND
RAMASWAMY SUBRAMANIAN   89

Effects of Moisture Absorption on Edge Delamination, Part II: An Experimental
Study of Jet Fuel Absorption on Graphite-Epoxy—STEVEN J. HOOPER,
RAMASWAMY SUBRAMANIAN, AND RICHARD F. TOUBIA   107

Effect of Porosity on Flange-Web Corner Strength—HAN-PIN KAN,
NARAIN M. BHATIA, AND MARY A. MAHLER   126

INTERLAMINAR FRACTURE TOUGHNESS

Mixed-Mode Fracture in Fiber-Polymer Composite Laminates—SHAH HASHEMI,
ANTHONY J. KINLOCH, AND GORDON WILLIAMS   143

Effects of T-Tabs and Large Deflections in Double Cantilever Beam Specimen
Tests—RAJIV A. NAIK, JOHN H. CREWS, JR., AND KUNIGAL N. SHIVAKUMAR   169

Experimental Determination of the Mode I Behavior of a Delamination Under
   Mixed-Mode Loading—ANOUSH POURSARTIP AND NARINE CHINATAMBI   187

Stabilized End Notched Flexure Test: Characterization of Mode II Interlaminar
   Crack Growth—KAZURO KAGEYAMA, MASANORI KIKUCHI, AND
   NOBORU YANAGISAWA   210

Initiation and Growth of Mode II Delamination in Toughened Composites—
   ALAN J. RUSSELL   226

Evaluation of the Split Cantilever Beam for Mode III Delamination Testing—
   RODERICK H. MARTIN   243

DELAMINATION ANALYSIS

Fracture Analysis of Transverse Crack-Tip and Free-Edge Delamination in
   Laminated Composites—ERIAN A. ARMANIOS, P. SRIRAM, AND
   ASHRAF M. BADIR   269

Combined Effect of Matrix Cracking and Free Edge Delamination—
   SATISH A. SALPEKAR AND T. KEVIN O'BRIEN   287

Fatigue Delamination Onset Prediction in Unidirectional Tapered Laminates—
   GRETCHEN BOSTAPH MURRI, SATISH A. SALPEKAR AND T. KEVIN O'BRIEN   312

Delamination Analysis of Tapered Laminated Composites Under Tensile
   Loading—ERIAN A. ARMANIOS AND LEVEND PARNAS   340

Analysis of Delamination Growth in Compressively Loaded Composite
   Laminates—MATTHEW D. TRATT   359

A Study of an Implanted Delamination Within a Cylindrical Composite Panel—
   ANTHONY PALAZOTTO AND BRENDAN WILDER   373

STRENGTH AND IMPACT

A Comparison of Experimental Observations and Numerical Predictions for the
   Initiation of Fiber Microbuckling in Notched Composite Laminates—
   E. GAIL GUYNN, WALTER L. BRADLEY, OZDEN O. OCHOA, AND
   JOHN D. WHITCOMB   393

Three-Dimensional Stress Analysis of Plain Weave Composites—
   JOHN D. WHITCOMB   417

Compression Testing of Thick-Section Composite Materials—
   EUGENE T. CAMPONESCHI, JR.   439

Influence of Low-Velocity Impact on Composite Structures—RAM C. MADAN   457

Effects of Stacking Sequence on Impact Damage Resistance and Residual Strength
for Quasi-Isotropic Laminates—ERNEST F. DOST, LARRY B. ILCEWICZ,
WILLIAM B. AVERY, AND BRIAN R. COXON   476

Relevance of Impacter Shape to Nonvisible Damage and Residual Tensile Strength
of a Thick Graphite/Epoxy Laminate—CLARENCE C. POE, JR.   501

Response of Composite Plates to Quasi-Static Impact Events—
RONALD B. BUCINELL, RALPH J. NUISMER, AND JIM L. KOURY   528

Compression of Composite Materials: A Review—EUGENE T. CAMPONESCHI, JR.   550

FATIGUE AND FRACTURE

Fatigue Behavior of Continuous Carbon Fiber-Reinforced PEEK—DON C. CURTIS,
MARK DAVIES, D. ROY MOORE, AND BARBARA SLATER   581

Damage-Based Notched Strength Modeling: A Summary—MARK T. KORTSCHOT
AND PETER W. R. BEAUMONT   596

Fatigue Damage Mechanics of Notched Graphite-Epoxy Laminates—
MARK SPEARING, PETER W. R. BEAUMONT, AND MICHAEL F. ASHBY   617

Hole Effect and Compression Fatigue of T300/N5208 Composite Materials—
DAWEI LAI AND CLAUDE BATHIAS   638

Effect of Interlaminar Normal Stresses on the Uniaxial Zero-to-Tension Fatigue
Behavior of Graphite/Epoxy Tubes—ERHARD KREMPL AND DEUKMAN AN   659

Effects of Notch Geometry and Moisture on Fracture Strength of Carbon/Epoxy
and Carbon/Bismaleimide Laminates—JOHN H. UNDERWOOD,
IAN A. BURCH, AND SRI BANDYOPADHYAY   667

Fatigue Failure Processes in Aligned Carbon-Epoxy Laminates—
MICHAEL R. PIGGOTT AND PATRICK W. K. LAM   686

Fracture of Fibrous Metal MATRIX Composites Containing Discontinuities—
YEHIA A. BAHEI-EL-DIN   696

Fatigue Crack Growth in a Unidirectional SCS-6/Ti-15-3 Composite—
PETER KANTZOS, JACK TELESMAN, AND LOUIS GHOSN   711

Thermomechanical Fatigue of a Quasi-Isotropic Metal Matrix Composite—
BHASKAR S. MAJUMDAR AND GOLAM M. NEWAZ   732

Observations of Fatigue Crack Initiation and Damage Growth in Notched Titanium
Matrix Composites—RAJIV A. NAIK AND W. S. JOHNSON   753

Damage and Performance Characterization of ARALL Laminates Subjected to Tensile Cylic Loading—RICARDO OSIROFF, WAYNE W. STINCHCOMB, AND KENNETH L. REIFSNIDER   772

Effective Crack Lengths by Compliance Measurement for ARALL-2 Laminates—CHRISTOPHER D. WILSON AND DALE D. WILSON   791

An Investigation of the Effects of Temperature on the Impact Behavior and Residual Tensile Strength of an ARamid Aluminum Laminate (ARALL-2 Laminate)—THOMAS C. LEE AND DALE A. WILSON   806

INDEXES

Author Index   823

Subject Index   825

Overview

Although this conference was billed as the third ASTM symposium with the title "Composite Materials: Fatigue and Fracture," it was actually the tenth ASTM symposium addressing these topics. The first was held in Bal Harbour, Florida, in 1973. Since then, the interest in this topic has grown along with the application of advanced composites in primary structures. Furthermore, the interest in these topics has become truly international, as evidenced by the large percentage of papers presented from outside the United States. Of the 40 papers that were presented, 13 had authors or coauthors from other countries, including England, Canada, Switzerland, Japan, France, Australia, and Korea. Furthermore, there was a general balance between papers presented by authors from universities, government, and industry. The conference was organized into nine sessions, with two parallel sessions in the morning, afternoon, and evening of the first day, and single sessions in series on the morning, afternoon, and evening of the second day. The sessions were organized under topics of Matrix Cracking and Delamination, Interlaminar Fracture, Delamination Analyses, Fatigue and Fracture (I, II & III), and Strength and Impact (I & II). One evening session on the first day was devoted to a meeting of the ASTM D30 task group on Interlaminar Fracture Toughness Measurement. The conference sessions were chaired by C. E. Harris, R. H. Martin, W. S. Johnson, and J. H. Crews, Jr. of NASA Langley Research Center, K. L. Reifsnider and W. W. Stinchcomb of Virginia Polytechnic Institute, E. A. Armanios of Georgia Institute of Technology, and A. Russell of DREP, Canada.

Matrix Cracking and Delamination

Boniface, Ogin, and Smith reviewed the growth of matrix cracks in composite laminates. They compared closed-form analyses for strain energy release rates associated with matrix cracking. These models were shown to be consistent with an alternative approach based on an approximate expression for the stress intensity factor at the tip of a growing transverse ply crack.

Ilcewicz, Dost, and McCool analyzed and tested matrix cracking in IM7/8551-7, a graphite epoxy composite reinforced with a toughened interlaminar layer. They found that the in situ strength for thin, angle-ply groups decreases with decreasing interlaminar layer thickness, but this strength is significantly higher than the lower limit exhibited for brittle matrix composites.

Lee, Allen, and Harris proposed a mathematical model utilizing the Internal State Variable (ISV) concept for predicting the upper bounds of the reduced axial and shear stiffness in cross-ply laminates with matrix cracks. Their comparison with experimental data showed the potential applicability of their model to angle-ply laminates subjected to general in-plane loading.

Davies, Cantwell, and Kausch described cooling rate effects on short and long term properties of carbon fiber PEEK composites. Both their short- and long-term results indicated that the internal stresses induced during fast cooling are more detrimental to mechanical properties than the changes in matrix structure observed at slow cooling rates.

Hooper, Subramanian, and Toubia discussed the environmental effects of jet fuel absorption on the delamination of Graphite/PEEK and Graphite/Epoxy composites. A Fickian moisture distribution was used to analyze the effect of absorbed moisture and jet fuel on the strain energy release rate associated with delaminations occurring at straight free edges. The effect of absorption of several fuels and water on the interlaminar fracture toughness was quantified.

Kan, Bhatia, and Mahler analyzed the effect of porosity on flange-web corner strength. They developed a strength of materials model to determine the interlaminar tensile stress, and used this along with experiments on curved flange-web corner elements to establish a relationship between ultrasonic signal attenuation loss due to porosity and the interlaminar tension strength of porous structures.

Interlaminar Fracture Toughness

Hashemi, Kinloch, and Williams interpreted interlaminar fracture tests on fiber composite materials. They showed that even with composites that give linear and reversible loading/unloading curves, correction terms compensating for the fact that these beams do not act as built-in cantilevers are needed when linear beam analyses are employed to deduce values of interlaminar fracture toughness.

Naik, Crews, and Shivakumar documented the effects of large deflections and loading tabs on double cantilever beam (DCB) specimens used to measure interlaminar fracture toughness. They found that large delections and T-tabs cause the loading point to move closer to the delamination tip, leading to an effective crack length that decreases with load. They developed a simple analysis method that accounts for these effects.

Poursartip and Chinatambi discussed their work on experimental determination of the Mode I strain energy release rate in cracked-lap shear (CLS) specimens. They generated a crack opening displacement (COD) profile by measuring the difference in displacement of the two faces of the CLS specimen and allowing for Poisson contraction effects. Results were presented for both thick adhesive joints and delaminations.

Kageyama reported on a stabilized end notched flexure (ENF) test utilizing a special displacement gage for direct measurement of crack shear displacements (CSD). Interlaminar shear fracture toughness and crack length were calculated from the load versus CSD diagram using an analytical relationship between crack length and CSD compliance.

Russell discussed his work on the initiation of Mode II delamination in toughened composites. He found that for tough matrix composites, load-displacement curves generated in ENF tests exhibit a significant amount of nonlinearity below the maximum load corresponding to unstable propagation in shear. The degree of this nonlinearity varied depending on the method used to start the shear delamination. He presented results for IM6/5245C, IM7/8551-7, and AS4/PEEK using either tensile or shear precracks, or starting the delamination directly from the insert. The mechanisms responsible for the nonlinearity were discussed.

Martin described his evaluation of the split cantilever beam (SCB) for Mode III delamination testing. He used a three-dimensional finite element analysis to show that a Mode II component existed at the delamination front near the free edges. This Mode II presence was verified experimentally by observation of shear hackles on the fracture surface. Furthermore, he discovered fiber bridging on the fracture surfaces resulting from the Mode III fracture in the center of the beam. He concluded that the SCB does not represent a pure Mode III delamination.

Delamination Analysis

Armanios, Sriram, and Badir developed an analysis for matrix crack tip and free edge delamination in composite laminates. This paper won the best presented paper award at the conference. They developed a shear deformation model for analyzing the local delaminations originating from transverse cracks in 90° plies. They used their model to predict delamination onset in $(25/-25/90_n)_s$ T300/934 laminates.

Salpekar and O'Brien reported on the combined effect of matrix cracking and free edge on delamination analyses. A three-dimensional finite element analysis was performed on $(0_2/90_4)_s$ cross ply and $(45/-45/90_4)_s$ orthotropic laminates with delaminations growing from 90° matrix cracks. The total G, and the Modes I, II, and III components, calculated along the delamination front, increased from a minimum value calculated in the interior to a maximum value calculated near the free edge. The Mode I component vanished after the delamination had grown a distance of one-ply thickness from the matrix crack. A closed form solution for the total G, calculated from a simple analysis utilizing laminated plated theory, agreed with the asymptotic total G values calculated near the free edge from the three-dimensional analysis.

Murri and O'Brien analyzed delaminations in tapered composite laminates with internal ply drops. A finite element analysis was performed on unidirectional tapered laminates containing internal ply drops to taper the thickness. Strain energy release rates calculated from this model were used with cyclic DCB data to predict the onset of delamination in fatigue due to the tapered geometry.

Armanios and Parmas developed a closed form analysis for the total G associated with delamination in tapered composite laminates under tensile loading. The equilibrium equations were derived using a complementary potential energy formulation on an elastic foundation for the outer belt.

Tratt discussed the effects of delaminations on the compression behavior of composite structures. He used a three-dimensional finite element analysis to predict the onset of midplane and near surface delaminations simulated by embedded Teflon inclusions in compressively loaded AS4/3501-5A carbon fabric-reinforced epoxy laminates.

Palazotto and Wilder described the characteristics of an implanted delamination within a cylindrical composite panel. They found that, for the level of delamination damage investigated, a linear bifurcation gave accurate predictions of global panel strength.

Strength and Impact

Guynn and Bradley compared experimental observations and numerical predictions for the initiation of fiber microbuckling in notched composite laminates. They found that compressive failure of open hole multidirectional thermoplastic composites begins with in-plane fiber microbuckling at the hole boundary. They emphasized that models for compression strength in the literature are conservative because they make no allowance for edge effects. They developed a "fiber on an elastic foundation" model to describe in-phase (shear mode) fiber microbuckling incorporating edge effects.

Whitcomb developed a micromechanics model for woven composite materials. His model showed the effect of several weave parameters on the composite moduli and stresses.

Camponeschi and Kerr evaluated compression testing of thick-section composite materials for use in large Navy structures. They designed a compression test fixture for testing composites greater than one inch in thickness, and tested 48, 96, and 192 ply carbon epoxy and S2-glass epoxy composites.

Madan discussed the influence of low-velocity impact on composite structures. A simple empirical model was proposed incorporating the effects of impact energy, laminate properties, and static influence coefficients. Results indicated that residual strength is a function of the damage present, independent of whether it was inflicted with different impacter masses, velocities, and energies. A semiempirical relation was developed for impact damage area in stitched laminates.

Dost, Ilcewicz, Avery, and Coxon determined the effects of stacking sequence on impact damage resistance and residual strength for quasi-isotropic laminates. They predicted residual compression strength after impact by predicting the stability of sublaminates created by delaminations and matrix cracks. Stacking sequence was varied by changing the thickness of repeating sublaminates or by modifying the sublaminate structure such that it varied through the laminate thickness. They found that ply group thickness affects both the damage resistance and residual strength of the laminate. Laminates with thicker ply groups resulted in greater damage areas following low-velocity impact, but they had increased laminate stability under compression loading for a given damage diameter compared to laminates with thin ply groups.

Poe discussed the relevance of impacter shape to damage of thick graphite epoxy laminates constructed from filament wound rocket motor case cylinders. For a given impacter mass and kinetic energy, the depth and breadth of damage increased with increasing sharpness and decreasing impacter radius. The damage extent was predicted using Hertz's law, Love's solution for hemispherical pressure applied to a semiinfinite body, and a maximum shear stress criterion. Residual tension strength was predicted using fracture mechanics by assuming a surface crack with the same length and depth as the impact damage.

Bucinell, Nuismer, and Koury described the response of filament wound composite plates to large mass impact events. A two degree of freedom lumped mass model was found to predict the effect of mass, velocity, and energy level up to the point of impacter penetration. Detailed response predictions from a distributed mass Rayleigh-Ritz energy model agreed well with experimental data.

Fatigue and Fracture

Curtis, Davies, Moore, and Slater discussed the fatigue behavior of continuous carbon fiber reinforced polyetheretherketone (PEEK). Tension and compression fatigue results were presented for a variety of layup geometries, fiber types, cooling rates, and test temperatures.

Kortschot and Beaumont reviewed their damage-based strength models for notched, cross-ply graphite epoxy laminates. Finite element analysis was used to determine the effect of splitting and delamination on the stress distribution in the zero degree ply. Their model predicted the effect of both notch size and layup on strength, without the need for empirical parameters.

Spearing, Beaumont, and Ashby described fatigue damage mechanisms in notched graphite epoxy laminates. A power law relationship between damage growth and strain energy release rate was developed in which the material constants were shown to be invariant for all the layups investigated. Damage progression was predicted in cross-ply and quasi-isotropic laminates subjected to tension fatigue. Residual strength was calculated as a function of damage dimensions using a stress based criterion with a Weibull distribution in the zero degree ply strength.

Bathias and Lai discussed the hole effect in compression fatigue of T300/5208 composite materials. In order to avoid generalized buckling of the laminated plate, they tested a modified compact specimen. They compared the ratio of the endurance limit to the static strength for various combinations of cyclic tension and compression.

Underwood, Burch, and Bandyopadhyay documented the effects of notch geometry and moisture on fracture strength of carbon/epoxy and carbon/bismaleimide laminates. They performed tension tests on carbon epoxy cross-ply panels with a center notch, of various ply orientations and thickness, to determine translaminar fracture toughness. In addition, they performed bending tests to determine bulk flexural strength. Tests were performed to investigate the effect of notch width, notch length, and ply orientation as well as the effects of moisture and temperature on fracture toughness.

Piggott and Lam reviewed fatigue failure processes in aligned carbon-epoxies. They found that during tensile fatigue, the slope of the S-N curve is increased by plasticizing the polymer and reducing the adhesion between fibers and polymers, and is decreased by reducing the internal microstresses in the composite. They also found that if fiber waviness is present, the antinodes of the wavy fibers create transverse stresses that cause fiber debonding and matrix splitting.

Bahei-El-Din and Dvorak analyzed the fracture of fibrous metal matrix composites containing discontinuities. Unidirectional boron-aluminum specimens with center notches and circular, square, and rectangular holes were tested in tension to failure. The fracture strength was unaffected by the shape of the discontinuity, and was only influenced by the size of the discontinuity. Local stresses were evaluated by a finite element analysis allowing plastic deformation only in specific regions observed in experiments. Evaluation of local stresses at loads corresponding to fracture showed that fracture is controlled by a critical ratio of the average principal stress in a representative volume of the unnotched ligament near the discontinuity to the unnotched tensile strength in the principal stress direction.

Kantzos, Telesman, and Ghosn reviewed the mechanisms of fatigue crack propagation in SiC/Ti-15-3 unidirectional composites. Specimens were tested with fibers parallel or perpendicular to the applied load. Compact tension specimens were tested at room temperature ambient conditions, and single-edge-notched specimens were tested using an in situ loading stage in a scanning electron microscope.

Majumdar and Newaz documented the thermomechanical fatigue (TMF) of metal matrix composites. Isothermal and in-phase thermomechanical fatigue was performed on Ti 15-3/ SCS6 composites. Results showed that for the same stress range, the TMF results in delamination failures and lives that are orders of magnitude less than the isothermal fatigue lives, which are governed by matrix cracking perpendicular to the load direction.

Naik and Johnson discussed fatigue crack initiation and growth in notched titanium matrix composites. Four unnotched laminates of SCS_6/Ti-15-3 had tension fatigue S-N data that could be correlated by a common degradation in the zero degree fiber stress as a function of applied cycles. The double-edge-notched specimen was analyzed using finite elements and was tested to predict the onset of local fiber failure at the notch in fatigue.

Osiroff, Stinchcomb, and Reifsnider discussed their damage and performance characterization of ARALL laminates subjected to tensile cyclic load. They found that at low cyclic stress levels, the fatigue properties of fiber reinforced plies are key factors, whereas at high cyclic stress levels, the fatigue properties of the aluminum plies govern the response of the laminate.

Wilson and Wilson determined effective crack lengths by a compliance method for ARALL-2 laminates as a means for determining stress intensity factor solutions. Fatigue crack growth rate measurements were made on middle-crack tension specimens. An effective crack length was defined to be the length of a through-thickness crack with the same compliance as a crack bridged by the aramid fibers. The variations in compliance and stress intensity factors with effective crack lengths were determined.

Lee and Wilson investigated the effects of temperature on the impact behavior and residual tensile strength of an ARALL-2 laminate. Impact damage was introduced using an instrumented pendulum impact testing system. Damage area and energy absorption were found to vary considerably with temperature. Residual tension strength following impact was measured and was found to increase with increasing test temperature.

Summary

In summary, the editor wishes to thank the authors, session chairmen, and reviewers for working diligently to ensure that the papers included in the symposium, and in this STP, were of high quality. Special thanks is also extended to the ASTM staff for their efforts and perseverance in bringing the publication of this STP to fruition.

T. Kevin O'Brien

U.S. Army Aerostructures Directorate
(AVSCOM), NASA Langley Research Center,
Hampton, Virginia; symposium chairman and
editor.

Matrix Cracking and Delamination

Lynn Boniface,[1] *Stephen L. Ogin,*[1] *and Paul A. Smith*[1]

Fracture Mechanics Approaches to Transverse Ply Cracking in Composite Laminates

REFERENCE: Boniface, L., Ogin, S. L., and Smith, P. A., **"Fracture Mechanics Approaches to Transverse Ply Cracking in Composite Laminates,"** *Composite Materials: Fatigue and Fracture (Third Volume), ASTM STP 1110,* T. K. O'Brien, Ed., American Society for Testing and Materials, Philadelphia, 1991, pp. 9–29.

ABSTRACT: The growth of transverse ply cracks in composite laminates has been investigated both theoretically and experimentally. Some of the closed-form strain energy release rate based analyses of this problem in the literature have been compared and extensions to these approaches are presented. These models have been shown to be consistent with an alternative approach based on an approximate expression for the stress intensity factor at the tip of a growing transverse ply crack. An experimental study of transverse ply crack growth has been carried out using a simple model array of transverse ply cracks in a glass/epoxy laminate. By making the transverse ply sufficiently thick, the specimen compliance was found to change measurably as individual cracks grow. Hence, the strain energy release rate could be determined experimentally (via the compliance relationship) and compared with analytical predictions. Agreement was found to be satisfactory.

KEY WORDS: composite materials, fracture, fatigue (materials), fracture mechanics, composite laminates, ply cracking, strain energy release rate

Composite laminates under fatigue loading exhibit an accumulation of damage, consisting of matrix cracking in the off-axis plies, delamination between layers, splitting (that is, matrix cracking parallel to the fibers in 0° layers) and fiber breakage and debonding. Failure is not due to the initiation and growth of a single crack in a self-similar fashion (as it is in metals for example), which raises the question of the validity of fracture mechanics for composite materials. However, fracture mechanics techniques can be applied to particular types of damage, notably matrix cracking, both between layers (delamination) and within layers (intralaminar cracking). In the present work, we are concerned with a particular class of intralaminar cracking: matrix cracking in the 90° ply of a cross-ply laminate. Various fracture mechanics type approaches have been applied to this problem over the years: the "energetics" approach of Parvizi, Garrett, and Bailey [1]; the finite element based energy release rate approach of Wang and co-workers (for example, Ref 2); the closed-form strain energy release rate approaches of Hahn and Johannesson [3], Poursartip [4], Caslini, Zanotti, and O'Brien [5], Dvorak and Laws [6], and Han, Hahn, and Croman [7]; and the approximate stress intensity based approach of Ogin, Smith, and Beaumont [8] and Ogin and Smith [9].

In the present paper, we compare these approaches, in particular, the closed-form strain energy release rate and stress intensity techniques, and suggest modifications. We also deter-

---

[1] Research fellow, lecturer, and lecturer, respectively, Department of Materials Science and Engineering, University of Surrey, Guildford, Surrey GU2 5XH, United Kingdom.

mine the strain energy release rate associated with the growth of a single transverse-ply crack experimentally and compare the measurements with theory.

Theoretical Background

Stiffness-Crack Spacing Relationship

The geometry of the cross-ply laminate is defined in Fig. 1. The laminate has a central 90° ply of thickness, $2d$, either side of which is a longitudinal ply of thickness, b. The laminate has a width, W, and a gage length, L. The plies have moduli, E_1 and E_2, parallel and perpendicular to the fibers, respectively, and a transverse shear modulus, G_T. For the uncracked laminate, the rule-of-mixtures expression for the modulus parallel to the loading direction is

$$E_0 = (bE_1 + dE_2)/(b + d) \qquad (1)$$

which differs by at most about 5% from the exact expression calculated using laminated plate theory.

Under monotonic loading, an increasing density of cracks appears in the 90° ply with the cracks, in general, spanning the full thickness of the transverse ply and the width of the laminate. The "idealized" cracked laminate is imagined at any stage in loading to have uniformly spaced cracks, separated by a distance, $2s$. This geometry can be analyzed simply using a one-dimensional shear lag analysis (for example, see Refs *10* through *12*). Two results from this analysis that we shall use are first that the variation of the longitudinal stress in the transverse ply is given by

$$\sigma_2 = \sigma \frac{E_2}{E_0} \left[1 - \frac{\cosh(\lambda y)}{\cosh(\lambda s)} \right] \qquad (2)$$

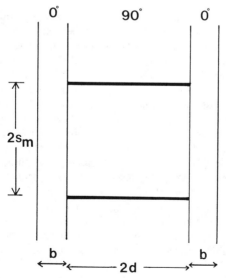

FIG. 1—*Geometry of the cross-ply laminate (edge view).*

and second that the normalized reduction in longitudinal modulus as a result of the cracking is

$$E/E_0 = 1 \bigg/ \left[1 + \frac{dE_2}{bE_1} \frac{1}{\lambda s} \tanh(\lambda s) \right] \qquad (3)$$

The quantity, λ, in Eqs 2 and 3 depends on the assumptions made in the shear-lag analysis. For the approach of Steif used by Ogin, Smith, and Beaumont [12], which assumes a parabolic variation of the longitudinal displacement in the transverse ply, λ is given by

$$\lambda^2 = 3G_T(b + d)E_0/bd^2E_1E_2 \qquad (4)$$

For different forms of the shear-lag analysis, the numerical constant at the front of the expression changes, but the laminate variables entering the expression remain the same (see, for example, Poursartip [4]). In the linear displacement analysis of Garrett and Bailey [10], the three is replaced by a one.

Strain Energy Release Rate Approaches to Cracking

Poursartip [4] calculated the strain energy release rate, G, associated with transverse ply crack growth by making use of the compliance expression

$$G = (P^2/2t) \, dC/da \qquad (5)$$

where

P = load on the laminate,
t = crack thickness (equal to the transverse ply thickness, $2d$), and
dC/da = rate of change of laminate compliance with crack length.

He related the crack length, a, to the idealized crack spacing, $2s$, using the expression

$$a = WL/2s \qquad (6)$$

The compliance, C, is related to the laminate stiffness, E, by

$$C = L/EA \qquad (7)$$

where A is the laminate cross-sectional area.

Combining Eqs 3, 5, 6, and 7 leads to an expression for G

$$G = \left(1 + \frac{b}{d} \right) \frac{dE_2}{bE_1} \frac{1}{E_0} \sigma^2 s \left[\frac{1}{\lambda s} \tanh(\lambda s) - \text{sech}^2(\lambda s) \right] \qquad (8)$$

Caslini, Zanotti, and O'Brien [5], using the same approach as O'Brien [13] had used earlier to analyze delamination, derived an expression for G using the relationship

$$G = -V \frac{\varepsilon^2}{2} \frac{dE}{dA} \qquad (9)$$

where

ε = applied strain,
V = volume of the laminate, and
dE/dA = rate of stiffness change as the flaw extends by an area, dA.

They then made the assumption that all the matrix cracks in the laminate could be treated as a single equivalent flaw. This assumption is inherently the same as that of Poursartip (Eq 6) and consequently the expression for G obtained by Caslini et al. can be rewritten as Eq 8. A similar assumption was made by Han, Hahn, and Croman [7]. They calculated G from the work per unit width needed to close the transverse ply crack. Their final expression for G can be also rewritten as Eq 8.

The assumption in these models that all the matrix cracks in the laminate can be treated as a single equivalent flaw is worthy of further consideration. Later, we derive two further expressions for G that are not based on this premise.

In reality, the crack spacing is not uniform. This means that at any given stage there will be a distribution of crack spacings within the specimen as indicated in Fig. 2a.

If we assume that the transverse ply is homogeneous, then the next transverse ply crack will form in between the two that are currently separated by the largest vertical distance, $2s_m$ (see Fig. 2). We can evaluate the change in laminate compliance as this crack forms as follows.

Suppose the specimen has a "smeared-out" stiffness, E, over that part of the laminate where the crack distribution is not changing. Then let the average stiffness of the laminate in the local region where the crack spacing is $2s_m$ be E_{2s_m} and where it is s_m be E_{s_m}. Before the new crack forms (Fig. 2a), the laminate stiffness is E_a

$$E_a = 1 \left/ \left[\frac{2s_m}{LE_{2s_m}} + \frac{(L - 2s_m)}{LE} \right] \right. \tag{10}$$

and after the crack has formed (Fig. 2c) the stiffness is E_b

$$E_b = 1 \left/ \left[\frac{2s_m}{LE_{s_m}} + \frac{(L - 2s_m)}{LE} \right] \right. \tag{11}$$

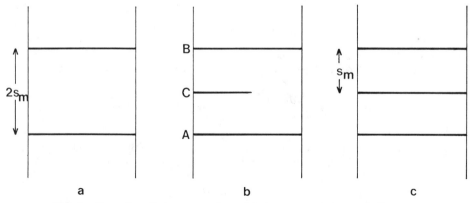

FIG. 2—*Formation of a transverse ply crack between two existing cracks $2s_m$ apart.*

Using this model, the crack length increment, da in Eq 5, is equal to the laminate width, W. So combining Eqs 5, 7, 10, and 11, we obtain

$$G = \frac{P^2}{Wd} \frac{S_m}{2A} \left[\frac{1}{E_{s_m}} - \frac{1}{E_{2s_m}} \right] \tag{12}$$

Using Eq 3, we can substitute for E_{2s_m} and E_{s_m}, so:

$$G = \left(1 + \frac{b}{d}\right) \frac{dE_2}{bE_1} \frac{1}{E_0} \sigma^2 s_m \left[\frac{\tanh(\lambda s_m/2)}{(\lambda s_m/2)} - \frac{\tanh(\lambda s_m)}{\lambda s_m} \right] \tag{13}$$

Note that Eq 13 has the same form as Eq 8 but that there is a different dependence on crack spacing. This is discussed in the next section.

We can attempt a further extension to the strain energy release rate approach by estimating G as a function of (a/W), that is, as a transverse ply crack advances across the width of the laminate (see Fig. 2b).

In this case, suppose that the stiffnesses of the region in which the new crack is growing (bounded by two full-width cracks spaced, $2s_m$) are $E(a)$ and $E(a + da)$ for a crack of lengths a and $a + da$, respectively, then

$$E(a) = \left(\frac{a}{W}\right) E_{s_m} + \left(\frac{W - a}{W}\right) E_{2s_m}$$

and

$$E(a + da) = \left(\frac{a + da}{W}\right) E_{s_m} + \left(\frac{W - (a + da)}{W}\right) E_{2s_m} \tag{14}$$

We can use these equations to find the corresponding expressions for the laminate stiffness. Then using Eqs 5 and 7 we obtain

$$G \approx \left(1 + \frac{b}{d}\right) \sigma^2 s_m \frac{(E_{2s_m} - E_{s_m})}{\left[E_{2s_m} - \frac{a}{W}(E_{2s_m} - E_{s_m}) \right]^2} \tag{15}$$

where we can substitute for E_{2s_m} and E_{s_m} from Eq 3.

Comparison of Strain Energy Release Rate Predictions

It is interesting to compare the values of G calculated using the three different expressions (Eqs 8, 13, and 15). Table 1 shows a comparison of the G values calculated using Eqs 8 and 13 for a range of crack spacings using data for a typical commercial GFRP laminate with a transverse ply thickness of 0.3 mm and the parabolic version of the shear lag analysis. For large crack spacings, the two equations are equivalent. At crack spacings of 1 mm (about three-ply thicknesses), the second approach (Eq 13) starts to indicate some degree of crack interaction that is not shown by the first approach until the cracks are more closely spaced. Moreover, when the crack spacing is of the order of the ply thickness (at which point transverse ply crack-

TABLE 1—*Prediction of G (from Eqs 8 and 13) as a function of crack spacing, 2s, for a commercial GFRP cross-ply laminate (b = d = 0.15 mm).*

Crack Spacing, 2s, mm	5	2	1	0.6	0.3
	$\sigma^2 \times 10^{-3}$ (J/m²)				
G from Eq 8	2.2	2.2	2.2	2.1	1.2
G from Eq 13	2.2	2.2	2.1	1.7	0.7

ing is often observed to saturate), the G value from Eq 13 is about half that from Eq 8, which perhaps suggests that the approach leading to Eq 13 is more realistic.

If we substitute values into Eq 15, then we find that the calculated value of G is not very sensitive to the distance across the width that the crack has traveled. For example, for a crack growing between two existing cracks spaced by $2s = 0.6$ mm, the initial value of G (that is, $a/W \rightarrow 0$) is $1.58 \times 10^{-3}\sigma^2$ which increases linearly to the final value (that is, $a/W \rightarrow 1$) of $1.78 \times 10^{-3}\sigma^2$. This represents an increase of only about 13%, meaning that the crack spacing has a greater influence on G than the crack length.

It is interesting to note that the value of G given by Eq 15 at $a/W = 0.5$ is the same as the value of G given by Eq 13.

Stress Intensity Factor Approach to Transverse Ply Cracking

A stress intensity factor, K, is generally a function of stress and a characteristic length. To estimate K for a transverse ply crack, an idealization of the crack is used, as shown in Fig. 3. The crack is considered to be flat and to extend across the thickness (but not across the width) of the transverse ply. A three-dimensional model (Fig. 3) shows the expected load trajectories around the crack. Away from the crack tip, load is transferred into the 0° plies by shear at the

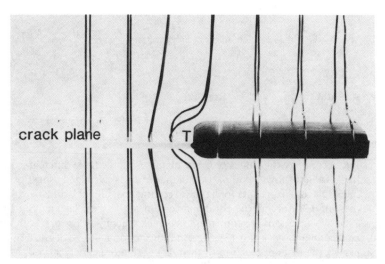

FIG. 3—*Three-dimensional model showing lines that represent the envisaged load trajectories around the tip (T) of a transverse ply crack in a 0/90/0 laminate.*

0/90 interfaces and therefore does not build up at the crack tip. There is, however, a stress intensity due to the localized stress disturbance at the crack tip. With a ply thickness of $2d$, the characteristic length of this disturbance is about d. Since the crack is not a through-thickness crack, but is bounded (and restrained) by the 0° plies, the stress intensity factor is less than $\sigma\sqrt{(\pi d)}$ and is taken to be

$$K = \sigma_t\sqrt{(2d)} \tag{16}$$

where σ_t is the stress in the transverse ply acting on the crack tip. To make the comparison between the G and K approaches, it is appropriate to consider the transverse ply crack growing between two fully formed cracks and estimate σ_t as being equal to the simple maximum stress from the shear-lag (found by putting $y = 0$ into Eq 2). This gives the expression for K as

$$K = \sigma\frac{E_2}{E_0}\left(1 - \frac{1}{\cosh(\lambda s/2)}\right)\sqrt{(2d)} \tag{17}$$

Equations 8 and 13 for G can both be written

$$G = \left(1 + \frac{b}{d}\right)\frac{dE_2}{bE_1}\frac{1}{E_0}\frac{\sigma^2}{\lambda}\cdot f(\lambda s) \tag{18}$$

where $f(\lambda s)$ depends on the assumptions used in deriving an expression for dC/da. Substituting for λ from Eq 4 and using the rule-of-mixtures expression for E_0 (Eq 1), Eq 18 can be rewritten as

$$G = \sqrt{\frac{1 + (dE_2/bE_1)}{12G_TE_2}}\cdot\left(\sigma\frac{E_2}{E_0}\right)^2\cdot 2d\cdot f(\lambda s) \tag{19}$$

For a crack in a laminate consisting of orthotropic plies, we would expect G and K to be related via an expression of the form

$$K^2 = g(\text{moduli})\,G \tag{20}$$

Combining Eqs 19 and 20 yields an expression for K

$$K = g^{1/2}(\text{moduli})\cdot\sqrt[4]{\frac{1 + (dE_2/bE_1)}{12G_TE_2}}\cdot\sigma\frac{E_2}{E_0}\cdot\sqrt{(2d)}\cdot f^{1/2}(\lambda s) \tag{21}$$

This should be compared with Eqs 16 and 17, which can be written

$$K = \sigma\frac{E_2}{E_0}\sqrt{(2d)}\cdot h(\lambda s) \tag{22}$$

where $h(\lambda s)$ depends on the assumptions used in estimating σ_t.

Equations 21 and 22 are similar: in both expressions the ply thickness ($2d$) emerges as the "characteristic length" associated with the stress intensity and both expressions also contain a function of the crack spacing that decreases as the cracks become more closely spaced.

Neither Eq 21 nor Eq 22 suggest any explicit dependence of K on the distance across the width that a transverse ply crack has grown (that is, the value of a/W). However, Eq 15 suggests that this is a minor effect with $K(\alpha\sqrt{G})$ increasing by only about 7% as the crack grows from one edge across the full width. This is in complete contrast to the case of a through-thickness crack in a material where K is proportional to the square root of the crack length.

To investigate the validity of these models, we need experimental measurements of a fracture mechanics parameter. It is possible to measure the strain energy release rate by using the compliance method (Eq 5). The difficulty is that in a conventional cross-ply laminate, where the number of 0° and 90° plies would be comparable, compliance changes would be too small to measure. To overcome this problem, model laminates were made with transverse plies sufficiently thick that the compliance change (and hence G) with crack length could be measured. The experimental method and results are described in the following sections.

Experimental Method

The laminates were fabricated using a simple filament winding technique in which glass rovings ("Silenka" E-glass 600-tex) were wound onto a 300-mm-square steel frame in a 0/90/0 configuration and impregnated with the epoxy matrix (Shell "Epikote" 828 resin cured with nadic methyl anhydride and accelerator benzyldimethylamine in the ratio 100:80:1.5 parts by weight, respectively). The resin was first mixed thoroughly and degassed to remove entrapped air and poured onto the wound frame that was placed inside a large vacuum chamber on a heated plate covered with release film. The chamber was then evacuated and the fibers were wetted as the resin warmed up and was drawn through the rovings under vacuum. When the resin had fully impregnated the fibers, the frame was removed from the chamber and excess resin and air bubbles were expelled from the laminate before placing in an air-circulating oven between thick glass plates. The laminate was cured for 3 h at 100°C under 64 kg weight ($\simeq 7$ kPa pressure) followed by post-curing for 3½ h at 150°C. The laminates were optically transparent since the refractive index of the fiber and the matrix are closely matched. Two nominally identical laminates with the composition (20%0°, 80%90°) were made with fiber volume fractions (determined using a matrix burn-off technique) of approximately 60%. The 0° ply thickness (b) was 0.32 mm with a small variation in 90° ply thickness ($2d$) between the two laminates, that is, 3.02 and 3.20 mm. The material constants for the glass/epoxy system were: $E_0 = 18$ GPa, $E_1 = 43$ GPa, $E_2 = 13$ GPa, $G_T = 4$ GPa, and $\nu_{12} = 0.3$.

Plain rectangular test coupons 220 mm in length and 20 mm wide were cut from the laminates using a water-lubricated diamond saw with a 600-grade grit finish. Aluminum alloy end tabs approximately 50 mm long and 1.5 mm thick were bonded to the coupon using epoxy adhesive.

Longitudinal laminate displacements were measured using an Instron dynamic extensometer with gage lengths of 12.5 and 50 mm, depending on the spacing between cracks. To prevent the extensometer knife-edges from slipping during testing, grooved seats were prepared on the surface of the coupon by resting the knife-edges on a release film in strips of uncured epoxy adhesive that hardens to form the seats.

The change in laminate compliance with increasing crack length was measured as a function of crack spacing for a transverse ply crack growing midway between two fully formed cracks (that is, cracks that span the full width of the ply) spaced at a distance, $2s$, apart (Fig. 4). The crack length was measured at each interval with a traveling microscope and the development of cracks was observed visually and from photographs taken with a 35-mm camera with a macro lens. Testing was carried out on an Instron 1341 50 kN servohydraulic fatigue machine under load control with a tensile sinusoidal waveform at a frequency of 7 Hz and with a stress ratio of $R = 0.1$. The peak cyclic load was chosen to give convenient increments of crack length.

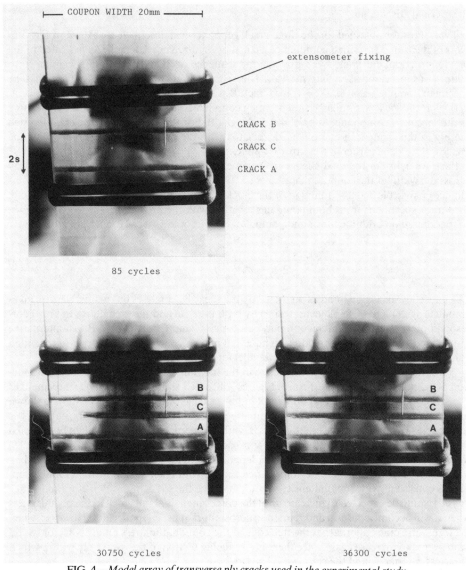

FIG. 4—*Model array of transverse ply cracks used in the experimental study.*

The procedure used in a test was as follows: after measuring the compliance of the uncracked laminate, cracks were initiated at the edge of the coupon. A crack, typically less than about 0.6 mm long, was produced at the required position by notching with a sharp scalpel blade. An increment in crack length was obtained by cyclic loading and the corresponding compliance was measured from the slope of the static load-displacement plot over a constant load range (that is, from zero up to the mean cyclic load). The first two cracks (A and B in Fig. 4) were either grown by fast fracture under static loading or grown gradually under cyclic loading. Compliance was measured as a function of crack length for the third crack, C. The ramp loading was repeated at least three times at each crack length to obtain an average value of compliance, which showed a variation of approximately 0.5%.

Results and Discussion

Typical results obtained for the three-crack geometry just described are shown in Fig. 5 as plots of measured specimen compliance against normalized crack length. As indicated in the previous section, the cracks were grown in the sequence A, B, and C, and the specimen compliance was measured at regular intervals. This sequential growth of the cracks means that the specimen compliance at $a/W = 0$ for Crack B is the same as the specimen compliance at $a/W = 1$ for Crack A. Similarly, the specimen compliance at $a/W = 0$ for Crack C is the same as the specimen compliance at $a/W = 1$ for Crack B. For the present work, we are interested mainly in the changes in compliance associated with the growth of Crack C, for which the analysis developed earlier in the paper is applicable.

First, we consider the total change in specimen compliance as Crack C grows all the way across the width of the coupon, that is, from $a/W = 0$ to $a/W = 1$. Figure 6 shows this compliance change plotted as a function of the crack spacing between Cracks A and B. Data are shown for specimens from both laminates together with the predictions for the compliance change, based on both the linear and parabolic versions of the shear-lag analysis, given by

$$\Delta C = \frac{2s_m}{2(b + d)W} \left[\frac{1}{E_{s_m}} - \frac{1}{E_{2s_m}} \right] \tag{23}$$

The data show the same trend as predicted by the shear-lag analyses but neither version gives complete agreement. The plateaus, shown by both the data and the predictions in Fig. 6 correspond physically to no interaction between Crack C and Cracks A and B. An alternative indication of crack non-interaction is to note that in a plot such as Fig. 5, curves of specimen compliance against crack length will be parallel for non-interacting cracks.

We now consider the progressive change in the measured specimen compliance as a function of crack length for the third crack. Figures 7, 8, and 9 show the compliance increment as a function of crack length for three different initial spacings of Cracks A and B. The data shown in each of these three figures are for nominally identical specimens and the repeatability is reasonable. The overall shape of the curves is, within experimental error, as might be expected on the basis of the simple stress intensity approach. There is an initial regime ($a/W < 0.15$) in which the compliance change with crack length is rapid followed by a linear range and then a final short regime in which the compliance change with crack length is more rapid again. This is consistent with stress intensity arguments since the stress intensity factor (and hence dC/da) would be expected to increase until the crack length is about equal to the ply thickness when it will become constant and remain so across most of the coupon width. However, when the transverse crack approaches the free edge of the coupon, the stress intensity factor would be expected to increase again since there is a reducing section of transverse ply within which the stress diverted round the crack tip is carried.

The predicted curves of compliance increment as a function of crack length for various values of the spacing between Cracks A and B are shown in Fig. 10. These curves are based on the predicted compliance/crack length relationship for the region between Cracks A and B, consistent with Eq 15

$$C = \frac{2s_m}{2(b + d)W} \left[\frac{1}{E_{2s_m} - \frac{a}{W}(E_{2s_m} - E_{s_m})} \right] \tag{24}$$

The trends are similar to those shown by the data, but the analysis is not sufficiently sophisticated to be applicable at the extremes (near $a/W = 0$ and $a/W = 1$).

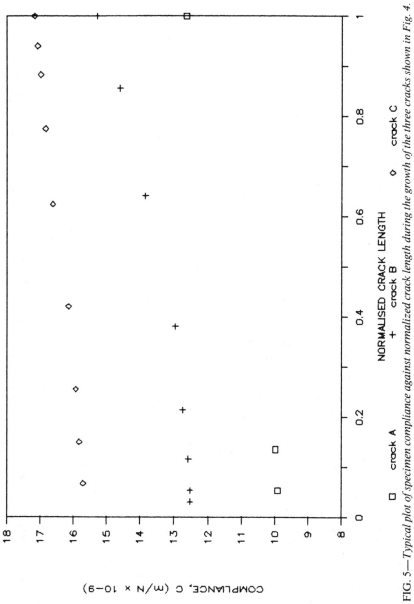

FIG. 5—*Typical plot of specimen compliance against normalized crack length during the growth of the three cracks shown in Fig. 4.*

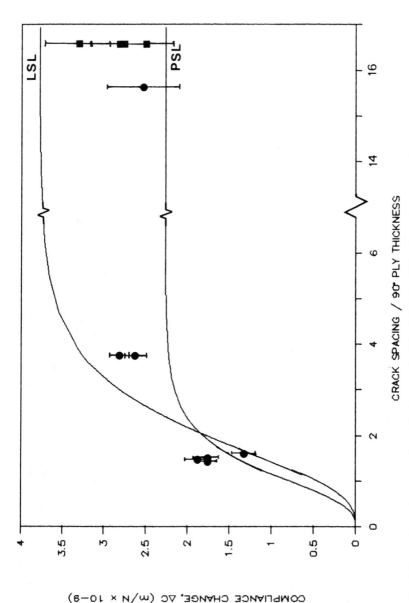

FIG. 6—*Total change in specimen compliance during growth of Crack C from a/W = 0 to a/W = 1 as a function of the spacing between Cracks A and B (normalized with respect to 90° ply thickness):* ● *laminate with 2d = 3.2 mm,* ■ *laminate with 2d = 3.02 mm. Prediction from Eq 23 for linear (LSL) and parabolic shear lag analysis (PSL).*

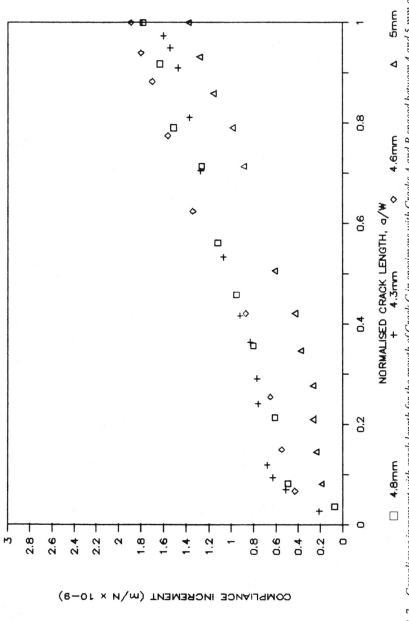

FIG. 7—*Compliance increment with crack length for the growth of Crack C in specimens with Cracks A and B spaced between 4 and 5 mm apart.*

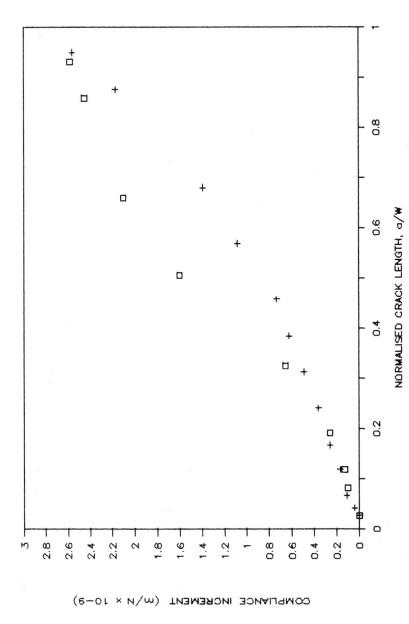

FIG. 8—*Compliance increment with crack length for the growth of Crack C in specimens with Cracks A and B spaced 12 mm apart (data from two nominally identical tests).*

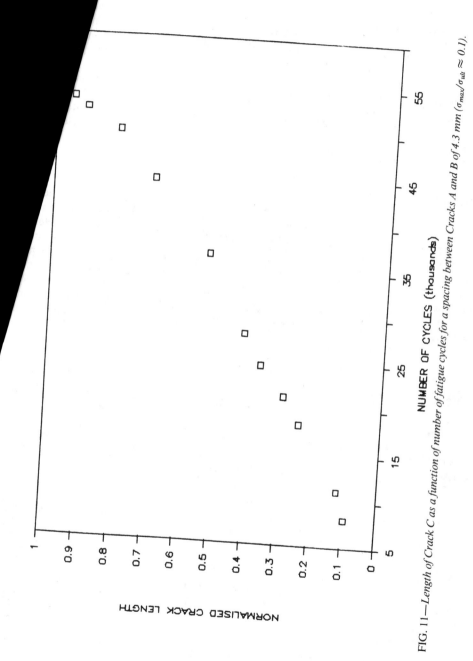

FIG. 11—*Length of Crack C as a function of number of fatigue cycles for a spacing between Cracks A and B of 4.3 mm ($\sigma_{max}/\sigma_{ult} \approx 0.1$).*

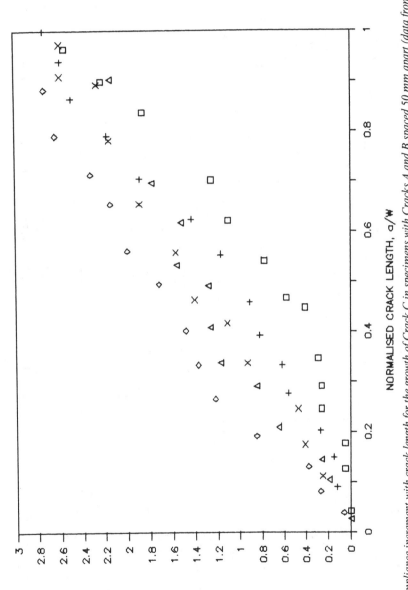

FIG. 9—*Compliance increment with crack length for the growth of Crack C in specimens with Cracks A and B spaced 50 mm apart (data from five nominally identical tests).*

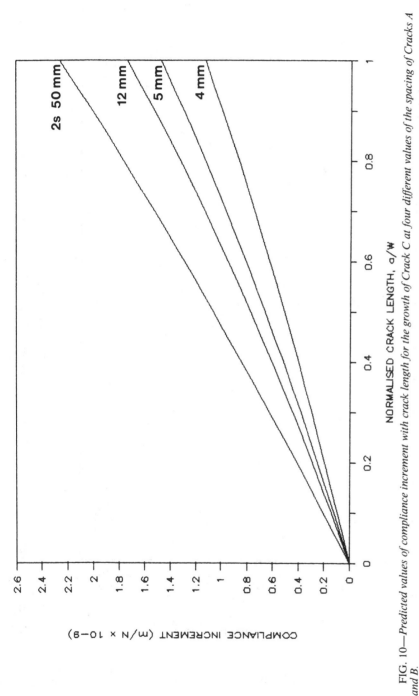

FIG. 10—Predicted values of compliance increment with crack length for the growth of Crack C at four different values of the spacing of Cracks A and B.

Bearing in mind the experimental error, the onl. change that we can determine with confidence is the sl ance/crack length data for each of the three different cra enables experimental values of the strain energy release r shown in Table 2, along with the predicted values of G fro and parabolic shear lag analyses. Agreement is reasonable. To with crack length in more detail experimentally would neces with even thicker transverse plies.

Finally, we look at the crack growth rates. Figures 11, 12, and Crack C as a function of number of cycles for various spacings of growth rates are reasonably constant over most of the specimen wid vious work [14]. A constant growth rate would be expected if the (and the related stress intensity factor) is independent of the len

Concluding Remarks

The experimental results show a strong dependence of the compliance gitudinal spacing between a crack tip and an adjacent crack and, within exp linear dependence on crack length (away from the coupon edges) for all crack the strain energy release rate depends markedly on crack spacing and has a n most, very small, dependence on crack length. This is reflected in the fatigue crac data showing that cracks grow at a roughly constant rate across the specimen wid stant crack spacing. The theoretical predictions of both the strain energy release r stress intensity factor approaches also suggest a strong dependence (of K or G) on c ing. However, our modified strain energy release rate approach predicts a small dep

TABLE 2—Measured and predicted values of strain energy release rate for Crack C as a function the spacing between Cracks A and B.

Crack Spacing, mm	$G_{EXP}{}^a$, $\sigma^2 \times 10^{-2}$ (J/m²)	Predicted G, $\sigma_2 \times 10^{-2}$ (J/m²)	
		$G_{PSL}{}^b$	$G_{LSL}{}^c$
	CP13 LAMINATE, $2d = 3.20$ mm		
	7.1	6.9	5.3
4.8	5.9	7.2	5.7
5.0	5.0	6.0	4.3
4.3	7.2	6.6	4.9
4.6	12.0	10.8	15.8
12	13.0	10.8	15.8
12	13.1	11.0	18.4
50			
	CP14 LAMINATE, $2d = 3.02$ mm		
	12.6	9.8	16.6
50	13.8	"	"
50	11.3	"	"
50	10.1	"	"
50			

^{a} G_{EXP} calculated from Eq 5 using dC/da determined experimentally at $a/W = 0.5$.
^{b} G_{PSL} calculated from Eq 13 using parabolic shear lag analysis.
^{c} G_{LSL} calculated from Eq 13 using linear shear lag analysis.

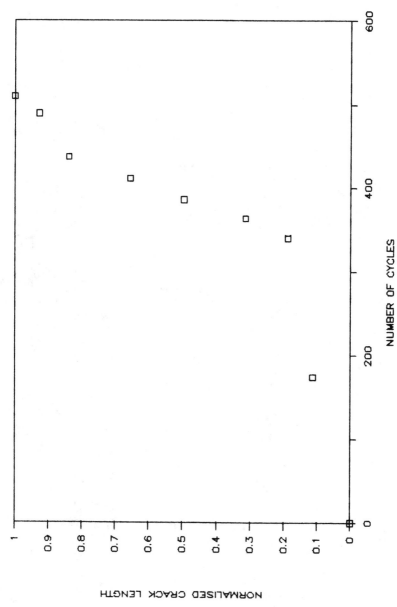

FIG. 12—*Length of Crack C as a function of number of fatigue cycles for a spacing between Cracks A and B of 12 mm ($\sigma_{max}/\sigma_{ult} \approx 0.08$).*

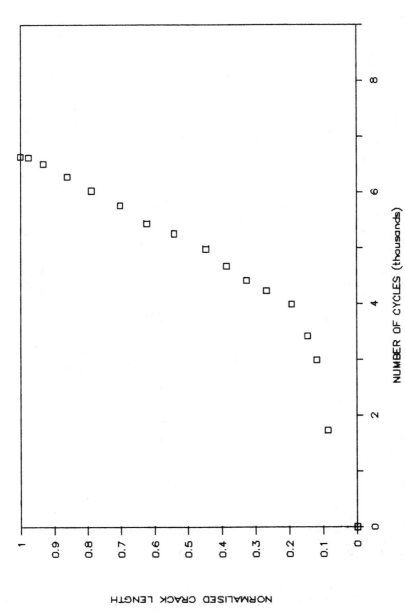

FIG. 13—*Length of Crack C as a function of number of fatigue cycles for a spacing between Cracks A and B of 50 mm ($\sigma_{max}/\sigma_{ult} \approx 0.08$).*

of G on crack length. It is possible that the rule-of-mixtures approximation used to calculate the strain energy release rate/crack length relationship is too crude. An approach that implicitly assumes a constant displacement across the specimen width at the bounding cracks (Cracks A and B in Fig. 4) is unlikely to be strictly correct. It is gratifying to be able to show that the strain energy release rate and stress intensity approaches are consistent and that the transverse ply thickness emerges in both approaches as a "characteristic length" of the problem.

The analyses and experimental data in this paper have been concerned with the (symmetrical) three-crack problem. In commercial laminates with thin transverse plies, many thousands of cracks can grow simultaneously during fatigue cycling, and cracks in all manner of configurations are interacting. In this case, models that implicitly treat an idealized crack array as shuffling to a reduced crack spacing as the overall crack length increases may be a useful approximation. However, a complete model must take into account the crack configuration at any instant and include the number of growing cracks.

Acknowledgments

The authors would like to thank the Science and Engineering Research Council (United Kingdom) for financial support during the course of this work. It is also a pleasure to thank our colleague Mr. R. Whattingham for technical assistance.

References

[1] Parvizi, A., Garrett, K. W., and Bailey, J. E., "Constrained Cracking in Glass Fibre-Reinforced Epoxy Cross-Ply Laminates," *Journal of Materials Science,* Vol. 13, 1978, pp. 195–201.
[2] Wang, A. S. D. and Crossman, F. W., "Initiation and Growth of Transverse Cracks and Edge Delamination in Composite Laminates: I An Energy Method," *Journal of Composite Materials,* Vol. 14, 1980, pp. 71–87.
[3] Hahn, H. T. and Johannesson, T., "Fracture of Unidirectional Composites: Theory and Applications," *Mechanics of Composite Materials,* AMD-Vol. 58, American Society of Mechanical Engineers, 1983, pp. 135–142.
[4] Poursartip, A., "Aspects of Damage Growth in Fatigue of Composites," PhD thesis, Cambridge University Engineering Dept., 1983.
[5] Caslini, M., Zanotti, C., and O'Brien, T. K., "Study of Matrix Cracking and Delamination in Glass/Epoxy Laminates," *Journal of Composite Technology and Research,* Vol. 9, 1987, pp. 121–130.
[6] Dvorak, G. J. and Laws, N., "Analysis of Progressive Matrix Cracking in Composite Laminates: II First Ply Failure," *Journal of Composite Materials,* Vol. 21, 1987, pp. 309–329.
[7] Han, Y. M., Hahn, H. T., and Croman, R. B., "A Simplified Analysis of Transverse Ply Cracking in Cross-Ply Laminates," *Composites Science and Technology,* Vol. 31, 1988, pp. 165–177.
[8] Ogin, S. L., Smith, P. A., and Beaumont, P. W. R., "A Stress Intensity Factor Approach to the Fatigue Growth of Transverse Ply Cracks," *Composites Science and Technology,* Vol. 24, 1985, pp. 47–59.
[9] Ogin, S. L. and Smith, P. A., "Fast Fracture and Fatigue Growth of Transverse Ply Cracks in Composite Laminates," *Scripta Metallurgica,* Vol. 19, 1985, pp. 779–784.
[10] Garrett, K. W. and Bailey, J. E., "Multiple Transverse Fracture in 90° Cross-Ply Laminates of a Glass Fibre-Reinforced Polyester," *Journal of Materials Science,* Vol. 12, 1977, pp. 157–168.
[11] Highsmith, A. L. and Reifsnider, K. L., "Stiffness Reduction Mechanisms in Composites," *Damage in Composite Materials, ASTM STP 775,* K. L. Reifsnider, Ed., American Society for Testing and Materials, Philadelphia, 1982, pp. 103–117.
[12] Ogin, S. L., Smith, P. A., and Beaumont, P. W. R., "Matrix Cracking and Stiffness Reduction During the Fatigue of a $(0/90)_s$ GFRP Laminate," *Composites Science and Technology,* Vol. 22, 1985, pp. 23–31.
[13] O'Brien, T. K., "Characterization of Delamination Onset and Growth in a Composite Laminate," *Damage in Composite Materials, ASTM STP 775,* K. L. Reifsnider, Ed., American Society for Testing and Materials, Philadelphia, 1982, pp. 140–167.
[14] Boniface, L. and Ogin, S. L., "Application of the Paris Equation to the Fatigue Growth of Transverse Ply Cracks," *Journal of Composite Materials,* Vol. 23, 1989, pp. 735–754.

Larry B. Ilcewicz,[1] Ernest F. Dost,[1] J. W. McCool,[1] and D. H. Grande[1]

Matrix Cracking in Composite Laminates with Resin-Rich Interlaminar Layers

REFERENCE: Ilcewicz, L. B., Dost, E. F., McCool, J. W., and Grande, D. H., "**Matrix Cracking in Composite Laminates with Resin-Rich Interlaminar Layers,**" *Composite Materials: Fatigue and Fracture (Third Volume), ASTM STP 1110,* T. K. O'Brien, Ed., American Society for Testing and Materials, Philadelphia, 1991, pp. 30–55.

ABSTRACT: The critical stress or strain causing the onset of matrix cracking in composite laminates has been referred to as in situ transverse lamina strength. A majority of past studies have considered relatively brittle composites exposed to room temperature environments and static load conditions. In the current work, fracture mechanics analysis and test data for a toughened composite material that has a resin-rich interlaminar layer (RIL) were used to investigate in situ strength. Exposure to a range of environmental conditions was considered.

A parametric analysis study was performed to judge the effects of laminate and material variables. A finite thickness effect, indicating an interaction between ply group thickness and effective flaw size, was found dominant. The magnitude of the effect was directly related to RIL stiffness. In situ strength was found to decrease with decreasing RIL stiffness. Experiments with five different laminate layups and six different environmental conditions confirmed the analysis. This work indicates the need to use a fracture mechanics model of actual lamina microstructure and heterogeneous properties to predict in situ strength in materials with RIL.

KEY WORDS: composite materials, fracture mechanics, transverse matrix cracks, graphite/epoxy, environmental effects, toughened matrix, resin-rich interlaminar layers, fatigue (materials), fracture

It is important to understand the mechanics of matrix cracking for composite materials used in aerospace applications. For example, matrix cracks that occur during a low-velocity impact event play a fundamental role in energy dissipation and the generation of delaminations. The increased surface area due to a network of matrix cracks can also alter physical properties such as composite thermal expansion, liquid permeability, and oxidative stability.

Matrix cracks occur in plies of laminated composites due to combined mechanical and environmental stresses. These transverse cracks align with fibers and, when fully formed, span the thickness of individual plies or ply groups stacked together in the same orientation. Matrix cracks redistribute local stress in multidirectional laminates, allowing a crack density to develop in the ply or ply group as a function of load and environmental history.

The critical stress or strain causing the onset of matrix cracking in plies of a laminate has been referred to as in situ transverse lamina strength. This strength is not a material constant. Experiments and analysis have shown that in situ strength increases as the thickness of plies grouped together with the same orientation decreases (for example, Refs *1–6*). These studies have also shown that neighboring plies can impose differing constraint on matrix crack formation, depending on fiber orientation.

Many materials currently used in the aerospace industry have resin-rich interlaminar layers

---

[1] Senior specialist engineer, specialist engineer, senior engineer, and specialist engineer, respectively, The Boeing Company, Seattle, WA 98124.

(RIL). The magnitude of the in situ strengthening effect is expected to decrease if a RIL with significant thickness exists between plies. Relatively soft RIL eliminate some of the constraint imposed by neighboring plies. Recent literature [7,8] has generalized analysis formulations to address some aspects of this problem. The stiffness of RIL depends on environment, loading rate, and stress level. This leads to more complex analysis models to study the mechanics of matrix cracking in materials with RIL microstructure.

Most past studies have verified predictions of matrix cracking in composite laminates using static tests performed in room temperature environments. These tests normally involved model laminates (for example, crossply) and relatively brittle matrix materials. Additional work is needed to verify analysis for a range of environmental conditions, general multidirectional laminates, and toughened matrix materials having RIL.

The objectives of this investigation were twofold. First, finite element analysis and a crack closure technique were used in a parametric study to judge the relative contributions of various factors affecting the strain energy release rate for matrix cracking. These factors included layup, ply group thickness, nearest-neighbor ply orientation, nonlinear lamina properties (dependent on environment), and RIL stiffness. Multidirectional layups and material properties for a graphite/epoxy material with RIL were used in the analysis. The second objective of this study was to compare experimental data with analytical predictions.

Analysis Method

Background

Lamination theory and ply stress or strain failure criteria have generally given conservative predictions of matrix cracking. More successful methods of matrix crack analysis have modeled intrinsic material flaws and either material volume size effects or neighboring ply constraint. A comprehensive review of past analysis methods was given in Ref 9. The following discussion will be limited to those details crucial to predicting matrix cracking in laminates with RIL. This will help justify the analysis method selected for this study.

Statistical methods [3,10–12], based on the principle that probability of failure increases with an increase of material volume, have used a two-parameter Weibull distribution function to model matrix cracking. Such analysis does not account for a change in the effective material volume due to neighboring ply constraint. Results have shown that the statistical parameters determined by curve fitting experimental data change with ply group thickness and neighboring ply orientation, indicating that neighboring ply constraint is important.

Shear lag [1,3,6] and strain field [4] theories apply a fracture mechanics approach to predict matrix cracking. Predictions from these theories compared well with experiments when a critical flaw size is used to impose a lower bound on in situ strength [4,6]. This lower bound corresponds to the transverse strength of $[90]_n$ laminates. Some shear lag models (for example, Refs 1,3, and 6) effectively smear the stiffness of the outer group of plies that constrain matrix crack formation in a laminate. As discussed in Ref 9, this can lead to inaccurate predictions when local stiffness is crucial to material behavior.

Generalized plane strain finite elements, crack closure analysis, and an effective flaw concept have been used in several studies [2,13–17] to predict matrix cracking. An effective flaw is synonymous with the term, critical flaw size, used in other fracture mechanics based approaches. Wang [17] physically defined an effective flaw as a basic property that accounts for the collective interaction of inhomogeneities in the material's real microstructure. As such, it represents the combined effect of the microstructures' response to load (for example, a micro/macro parameter) rather than a measurable defect size. An effective flaw size distribution is assumed to account for variations in the microstructure. Local stiffnesses of constraint plies, including the effects of RIL, can be accurately modeled with the finite element approach.

Each matrix crack model listed previously has yielded results that were verified by experiments; however, there is a fundamental difference between the methods. This difference should become apparent when considering the effects of RIL. Crack density versus strain predictions from statistical methods depend on RIL because of the use of a shear lag analysis to account for stress redistribution near existing matrix cracks. However, in situ strength predictions for statistical based methods are independent of RIL. The finite element [17] and modified shear lag [8] fracture mechanics approaches can both be applied to predict in situ strength for materials with RIL.

The finite element method of Wang [17] was chosen for use in the current study due to several reasons. It was selected over modified shear lag analysis [8] because it was believed to yield a more accurate model of the local constraint of neighboring plies. In addition, finite elements model the interaction between ply group thickness and effective flaw size that will be referred to as a "finite thickness effect." This effect, which is analogous to the finite width effect reported for planar fracture mechanics problems [18], becomes important for problems with RIL. The fracture mechanics prediction of in situ strength for laminates with relatively soft RIL can conceivably decrease with decreasing ply group thickness. Therefore, experimental data from such materials can be used to judge the validity of basic assumptions that form the foundations of statistical and fracture mechanics based approaches.

Current Method

The analysis method used in this study to calculate matrix crack strain energy release rates has been well documented in the literature [2,13–17]. Previous work also documents physical aspects of the analysis method (for example, Refs 9,17). A finite element computer code FCMP, which is described in Ref 16, was used to perform crack closure analysis [19]. This program uses generalized plane strain, linear displacement, non-singular, orthotropic, triangular elements. The finite element mesh had node spacing similar to that shown in Ref 16. The densest mesh occurred near the matrix crack tip where node spacing approached one-twentieth of a ply thickness.

All models assumed Mode I matrix cracking, midplane laminate symmetry, a uniform axial strain distribution, and constant temperature/moisture conditions. The Mode I strain energy release rate, G_I, is given as

$$G_I = f(\delta, F, \Delta a) \tag{1}$$

where δ is the difference in displacements of nodes located at the crack length, a, following Mode I extension of the crack to a length $(a + \Delta a)$. The parameter, F, is the component of nodal force required to close the crack from Length $(a + \Delta a)$ to Length a.

Separate finite element solutions for mechanical and environmental conditions were obtained using the unit load conditions

$$\varepsilon = 1.0 \text{ m/m} \tag{2}$$
$$\Delta T = -1°C$$

where ε is a uniaxial strain and ΔT is a change in temperature from a stress-free condition. These unit solutions can be used to calculate mechanical and thermal energy release rate coefficient functions, denoted C_{eI} and C_{TI}, respectively. The C_{eI} and C_{TI} are defined by

$$C_{eI} = (\delta_e F_e)/(2t \, \Delta a) \tag{3}$$
$$C_{TI} = (\delta_T F_T)/(2t \, \Delta a)$$

where the subscripts $_\varepsilon$ and $_T$ denote strain and temperature. The parameter, t, is a length scale factor relating the actual laminate dimension to those used in the finite element grid (for example, ply thickness). The relationship between G_1, $C_{\varepsilon l}$, and C_{Tl} is given as

$$G_1 = \{(C_{\varepsilon l}\varepsilon^2 t)^{0.5} + (C_{Tl} \Delta T^2 t)^{0.5}\}^2 \tag{4}$$

The failure criteria to predict the onset of matrix cracking is given as

$$\begin{array}{ll} G_1(a_0) = G_{1c} & \text{if } H > a_0 \\ G_{1\max}(a) = G_{1c} & \text{if } H \le a_0 \end{array} \tag{5}$$

where H is thickness of the ply group which cracks, and a_0 is the largest effective flaw size (note that H and a_0 are one-half lengths due to assumed symmetry in the finite element analysis). A method to experimentally determine a_0 will be discussed later. The term, $G_1(a_0)$, is the value of G_1 at $a = a_0$ and $G_{1\max}(a)$ is the maximum value of $G_1(a)$. Equations 4 and 5 can be combined to solve for either a critical ε or ΔT, while holding the other constant.

The two conditions in Eq 5 are interpreted based on a distribution of effective flaw sizes as explained by Wang [16]. An effective flaw that creates a matrix crack must always be less than H. It must also be less than or equal to a_0. In the first condition of Eq 5, H is larger than a_0 and the matrix crack forms from the largest effective flaw for a given material type (that is, a_0). The effective flaw creating a matrix crack in the second condition of Eq 5 depends on $G_{1\max}(a)$. The worst flaws from a distribution in this case change depending on the $G_1(a)$ function for a specific laminate layup. This seems physically acceptable because they are smaller than a_0.

Environmental history can change the residual stress state by effectively shifting the stress free temperature, T_0. As discussed in Ref 9, the relationship

$$T_0 = T_{0\text{dry}} - (\beta_{22}/\alpha_{22})(\text{EMC} - M_0) + T_{\text{vis}} \tag{6}$$

can be used to estimate history dependence. The first term represents an original dry stress-free temperature, $T_{0\text{dry}}$. The need to obtain a unit solution for matrix cracking due to moisture content can be eliminated by assuming that longitudinal thermal (α_{11}) and moisture (β_{11}) coefficients are zero. The effect of moisture swelling can then be simulated by a shift in T_0. The second term estimates this elastic shift in T_0 due to moisture conditioning; where β_{22} is the transverse lamina moisture expansion coefficient, α_{22} is the transverse lamina thermal expansion coefficient, EMC is the current equilibrium moisture content, and M_0 is the moisture content at which lamina swelling begins. The last term in Eq 6, T_{vis}, represents a viscoelastic shift in the stress-free temperature. The value for ΔT used in Eq 4 is given as

$$\Delta T = T - T_0 \tag{7}$$

where T is the current temperature.

Parametric Study

A total of 272 finite element runs were performed to judge material and laminate factors affecting the strain energy release rate for matrix cracking. Multidirectional layups and material properties for a graphite/epoxy material with RIL were used in the analysis. Fractions of ply thicknesses (for example, 0.75) were used in some of the models because of the potential for mixing different grades of tape material in laminate construction for general engineering applications.

TABLE 1—*Lamina moduli for IM7/8551-7.*

| Test Conditions | | | | | | | | | | | | |
Temperature, °C	Moisture[a]	V_f	V_m	E_{l11}, GPa	E_{l22}, GPa	E_{l33}, GPa	G_{l12}, GPa	G_{l13}, GPa	G_{l23}, GPa	v_{l12}	v_{l13}	v_{l23}
21.1	dry	0.57	0.43	151	7.90	7.90	4.55	4.55	2.99	0.32	0.32	0.32
−59.4	dry	0.57	0.43	152	9.18	9.18	5.22	5.22	3.27	0.32	0.32	0.40
82.2	dry	0.57	0.43	151	6.71	6.71	3.98	3.98	2.74	0.33	0.33	0.23
21.1	wet	0.57	0.43	151	7.45	7.45	3.98	3.98	2.74	0.32	0.32	0.36
−59.4	wet	0.57	0.43	152	8.43	8.43	4.73	4.73	3.07	0.32	0.32	0.37
82.2	wet	0.57	0.43	151	6.57	6.57	3.38	3.38	2.44	0.33	0.33	0.35
21.1	dry	0.72	0.28	191	9.77	9.77	6.36	6.36	3.50	0.31	0.31	0.40
−59.4	dry	0.72	0.28	191	10.9	10.9	7.16	7.16	3.73	0.31	0.31	0.47
82.2	dry	0.72	0.28	190	8.60	8.60	5.66	5.66	3.28	0.32	0.32	0.31
21.1	wet	0.72	0.28	190	9.34	9.34	5.66	5.66	3.28	0.31	0.31	0.42
−59.4	wet	0.72	0.28	191	10.3	10.3	6.58	6.58	3.57	0.31	0.31	0.44
82.2	wet	0.72	0.28	190	8.47	8.47	4.90	4.90	3.01	0.32	0.32	0.41

NOTE—Symbols = volume fraction (V), extensional (E) and shear (G) moduli, Poisson's ratio (v), fiber ($_f$), matrix ($_m$), and lamina ($_l$).
[a] Moisture levels correspond to equilibrium conditions obtained in preconditioning chambers; Dry = 71.1°C laboratory oven, and Wet = 60°C/85% relative humidity controlled chamber.

Plots of C_{el} and C_{Tl} as a function of a/t were used to compare the effects of layup, 90° ply group thickness, nearest-neighbor ply orientation, nonlinear lamina properties (dependent on environment), and RIL stiffness. Changes in the values of C_{el} and C_{Tl} physically represent the influence of elastic moduli, expansion coefficients, and laminate construction. Examples will be given in results from the parametric study. The finite element length parameter, 1.0/ply thickness, was used to obtain C_{el} and C_{Tl} appearing in plots. Effects of other factors appearing in the failure criteria for matrix cracking are not evident in plots of C_{el} and C_{Tl}. These factors, which include values of t (that is, actual ply thickness), a_0, ΔT, and G_{Ic}, will receive more attention in the section on experimental results.

Tables 1 and 2 list lamina properties for IM7/8551-7 as a function of environment. Longitudinal and transverse extensional moduli are for lamina subjected to tensile loads. Note that the transverse thermal expansion coefficients for IM7/8551-7 are nearly independent of temperature and; therefore, the assumed unit thermal solution using temperature-dependent moduli for the environment in question is valid for nonlinear analysis. The lamina properties were obtained from a combination of published neat resin data [20], lamina property tests performed at The Boeing Company, and micromechanics analysis [21]. The high fiber-volume fraction was used to model intralaminar regions of laminates with RIL. The RIL modeled in the parametric study was one fifth of a ply thickness to be consistent with materials used in experiments. Neat resin properties used for the RIL are listed in Table 3.

Laminate Layup

The effects of laminate layup, 90° ply group thickness, nearest-neighbor ply orientation, and nonlinear lamina properties were first considered using models that ignore RIL (that is, homogeneous models of each ply in the layup and average lamina properties). Properties used for the lamina in these models are listed with fiber volume fractions of 0.57 in Tables 1 and 2. Models for materials without RIL are expected to be most accurate for untoughened composites with uniform fiber/resin distribution.

TABLE 2—*Lamina expansion coefficients for IM7/8551-7.*

Test Conditions				α_{l11}, μm/ m/°C	α_{l22}, μm/ m/°C	α_{l33}, μm/ m/°C
Temperature, °C	Moisture[a]	V_f	V_m			
21.2	dry	0.57	0.43	−0.48	22.3	22.3
−59.4	dry	0.57	0.43	−0.36	22.3	22.3
82.2	dry	0.57	0.43	−0.58	22.7	22.7
21.1	wet	0.57	0.43	−0.34	30.1	30.1
−59.4	wet	0.57	0.43	−0.20	30.1	30.1
82.2	wet	0.57	0.43	−0.43	30.4	30.4
21.1	dry	0.72	0.28	−0.68	17.2	17.2
−59.4	dry	0.72	0.28	−0.61	17.2	17.2
82.2	dry	0.72	0.28	−0.74	17.4	17.4
21.1	wet	0.72	0.28	−0.61	22.0	22.0
−59.4	wet	0.72	0.28	−0.54	22.0	22.0
82.2	wet	0.72	0.28	−0.65	22.3	22.3

NOTE—$\beta_{l11} = 0.0$ m/m/%MC, $\beta_{l22} = \beta_{l33} = 0.00331$ m/m/%MC for all conditions. Symbols = volume fraction (V), thermal (α) and moisture (β) expansion coefficients, fiber ($_f$), matrix ($_m$), and lamina ($_l$).

[a] Moisture levels correspond to equilibrium conditions obtained in preconditioning chambers; dry = 71.7°C laboratory oven, and wet = 60°C/85% relative humidity controlled chamber.

TABLE 3—*Neat resin properties for 8551-7. Data is from Ref 20 except for the* −59.4°C *test condition.*

Test Conditions		E_m, GPa	ν_m	G_m, GPa	α_m, μm/ m/°C	β_m, m/m/ %MC
Temperature, °C	Moisture[a]					
23.0	dry	3.10	0.36	1.14	46.7	0.00309
−59.4	dry	4.00	0.36	1.47	46.7	0.00309
82.0	dry	2.41	0.39	0.87	46.7	0.00309
23.0	wet	2.83	0.36	1.03	70.0	0.00309
−59.4	wet	3.45	0.36	1.27	70.0	0.00309
82.0	wet	2.34	0.43	0.82	70.0	0.00309

NOTE—Symbols = extensional (E) and shear (G) moduli, Poisson's Ratio (ν), thermal (α) and moisture (β) expansion coefficients, and matrix ($_m$).

[a] Moisture levels correspond to equilibrium conditions obtained in preconditioning chambers; dry = stored in desiccators at ambient temperature, and wet = 74°C/98% relative humidity controlled chamber.

Past literature on matrix cracking has concentrated on $(0_m/90_n)_s$ and $(\pm\theta_m/90_n)_s$ type laminates, with m and n representing variable numbers of plies. Most of the multidirectional laminates analyzed in the current study included a mix of 0, $\pm\theta$, and 90° plies. This class of laminates was considered to be characteristic of many engineering applications. The laminate layup (percentages of 0, $\pm\theta$, and 90 plies) and nearest-neighbor ply orientation were found to have the smallest effects on matrix cracking of the variables studied. As in the past, 90° ply group thickness played a major role in determining the in situ transverse lamina strength of laminates without RIL.

90° Ply Group Thickness

Figure 1 shows typical C_{el} results for laminates with variable 90° ply group thickness. Although laminate layups are distinctly different, each curve has a similar shape that peaks before a/t approaches H/t. As a result, maximum values of C_{el} were found to increase with increasing 90° ply group thickness. Similar behavior was seen in plots of C_{TI}. Although maximum values of C_{el} and C_{TI} continue to increase with 90° ply group thickness greater than shown in Fig. 1, an effective cutoff in values used in matrix crack failure criteria occurs due to the first condition in Eq 5.

As noted in past literature (for example, Refs 15 and 16), C_{el} reaches its maximum at a value for a/t which is less than H/t. The drop in C_{el} for a/t that approach H/t indicates constraint on matrix crack opening in the 90° ply group. This constraint is imposed due to the combined influence of overall laminate layup and nearest-neighbor ply orientation.

Nearest-Neighbor Ply Orientation

For the class of laminate layups studied (that is, those having some 0° plies) the effect of nearest-neighbor ply orientation was found to be small. In order to isolate the effect of nearest-neighbor ply orientation from that of overall laminate layup, several finite element runs were made in which stacking sequence was varied while holding 90° ply group thickness constant. Figure 2 shows some typical results of C_{el} versus a/t for varying nearest-neighbor ply orientation. As expected, higher nearest-neighbor ply group stiffness (that is, 0_3) yields greatest constraint and, hence, the lowest maximum values of C_{el} in the figure. Although the nearest-neigh-

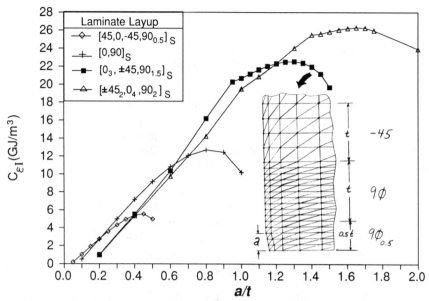

FIG. 1—*Models without RIL for laminate layup and ply group thickness at 21°C/dry.*

bor ply orientation appears significant in Fig. 2, it was found to have a relatively small effect when compared to other variables. Note that the layup considered in Fig. 2 represents two extremes in nearest-neighbor ply group stiffness for typical laminates (that is, two heterogeneous stacking sequences clumping plies of the same orientation).

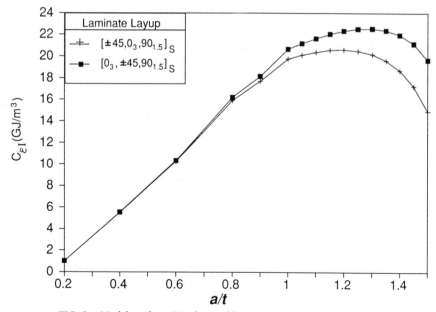

FIG. 2—*Models without RIL for neighboring ply constraint at 21°C/dry.*

Nonlinear Lamina Properties

The nonlinear effect studied in this section pertains to a change in elastic properties as a function of environment. Lamina inplane transverse (E_{22}) and shear (G_{12}) moduli depend on environment (see Table 1). The former is expected to play a significant role in matrix cracking for the 90° plies of a multidirectional laminate subjected to tension. Figure 3 shows the effect of nonlinear material properties on C_{el} for laminates without RIL. The curves align in an order that reflects the corresponding values of E_{22}. Environments yielding the largest and smallest values of E_{22} lead to the highest and lowest values of C_{el} for a given a/t, respectively.

Moisture has a significant effect on lamina in-plane transverse thermal expansion coefficients (α_{22}) (see Table 2). The values of α_{22} and E_{22} both influence C_{TI} for the 90° plies of a multidirectional laminate subjected to a ΔT. Figure 4 shows the effect of nonlinear material properties on C_{TI} for laminates without RIL. The curves separate into two groups (wet and dry). As shown by Coquill and Adams [20], thermal expansion coefficients for 8551-7 increase when the resin is moisture saturated. The higher values of α_{22} for wet environmental conditions lead to relatively high values of C_{TI} for a given a/t. The C_{TI} curves in both the wet and dry groups align in an order that reflect the corresponding values of E_{22}.

Figures 3 and 4 considered how C_{el} and C_{TI} change with nonlinear material properties due to environmental conditions. Since static load conditions (that is, a strain rate of 0.005 m/m/min) were used to obtain moduli, the same was assumed for the analysis. General load functions (for example, creep) and additional nonlinear material behavior (for example, stress) that influence the measured values of E_{22} and G_{12} are expected to yield similar trends in the behavior of C_{el} and C_{TI}; however, the formulation of these parameters is expected to be more complicated than shown in this paper.

The relationship between matrix cracking and environment is not fully apparent when examining Figs. 3 and 4. Values of a_0, ΔT, and G_{Ic} also depend on environment. Since each of these parameters appear with C_{el} and C_{TI} in the failure criteria, the effect of environment on matrix cracking is generally complex. For example, C_{TI} is higher for moisture-saturated lam-

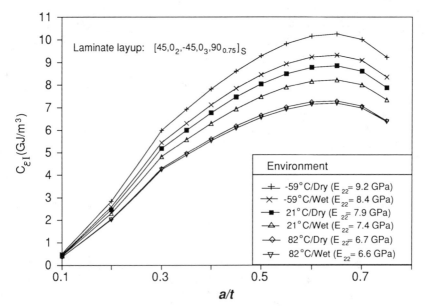

FIG. 3—*Mechanical models without RIL for nonlinear material properties due to environment.*

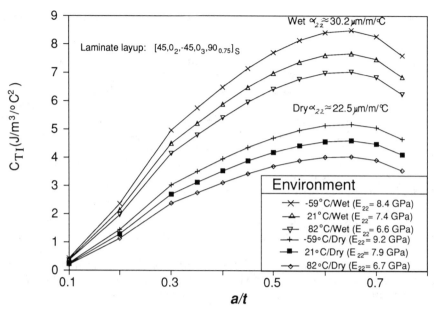

FIG. 4—*Thermal models without RIL for nonlinear material properties due to environment.*

inates in Fig. 4; however, lower values of ΔT will tend to counteract this effect when predicting in situ strength for wet laminates. Complex relationships with environment will be discussed later in the section on experimental results.

Resin-Rich Interlaminar Layers (RIL)

Finite element analyses of materials with RIL were performed using hybrid models (that is, different material properties for the RIL and adjacent lamina). Only one RIL was modeled. It was located between the 90° ply group used for crack growth and the nearest neighbor constraint ply. The RIL thickness was taken as one fifth of a ply thickness and shared between the two adjacent plies (that is, one tenth of each ply). Lamina properties obtained in laminate tests reflect the average fiber volume of materials with RIL (for example, 0.57 in Tables 1 and 2). Values used for intralaminar regions of the two plies located adjacent to the RIL were scaled to a higher fiber volume (see Table 1, $V_f = 0.72$) using neat resin data [20] and micromechanics [21]. This was done in order to more accurately represent the material's microstructure near an RIL. All other plies were modeled using ply properties for the average laminate fiber volume.

Initially, finite element analyses included RILs between each ply in a model. Later it was determined that the "critical location" for an RIL was between the 90° ply group used for crack growth and the nearest-neighbor constraint ply. Any RILs modeled at remote constraint ply interfaces were found to have negligible effect on matrix cracking. The C_{el} and C_{TI} curves were found to have local maxima and minima when RILs were modeled between each ply in the 90° ply group subjected to crack growth. Models that included an RIL only at the critical location tended to follow an average of these local maxima and minima. Predictions of crack growth using models of 90° ply groups with RIL at each ply interface would require individual values of fracture toughness for the RIL and high fiber volume intralaminar regions. Both would be difficult to measure and, therefore, such models currently have limited engineering interest (for example, a micromechanics tool to evaluate trends).

The RIL was simulated in finite element analyses as a layer having properties equal to that of neat resin (see Table 3). The RIL was also modeled with reduced moduli (that is, moduli = $0.25 \times$ neat resin moduli) in order to judge the effects of a range in RIL stiffness.

Figure 5 compares analysis results for laminates with and without RIL. Note that results for models with RIL extend into that portion of the neighboring ply that is RIL. There appears to be a sharp peak in the C_{eI} of laminates modeled with RIL. The peak value of C_{eI} is largest for relatively soft RIL. A peak occurs in laminates with RIL because some of the constraint of neighboring plies is relaxed, causing a "finite thickness" interaction between the effective flaw size and 90° ply group thickness. This interaction has not been previously reported. The effects of 90° ply group thickness and RIL are both shown in Fig. 6. The maximum values of C_{eI} for laminates with RIL are on the order of seven times higher than corresponding laminates without RIL, physically resulting in a greater potential for matrix cracking.

The finite thickness effect appearing in Figs. 5 and 6 for laminates with RIL would also manifest itself in results from $[90]_n$ strength tests with thin laminates (that is, measured strength would drop as laminate thickness approached effective flaw size). Statistically based methods would predict an opposite trend in strength (that is, increased strength with decreased material volume). Figure 7 shows the relationship between unidirectional laminate thickness and a/t. A sharp increase in C_{eI} is evident as a approaches laminate thickness.

Experimental Methods

Two types of specimens were used in experiments for this study, intralaminar fracture toughness and in situ strength tensile coupons. All specimens were fabricated from Hercules

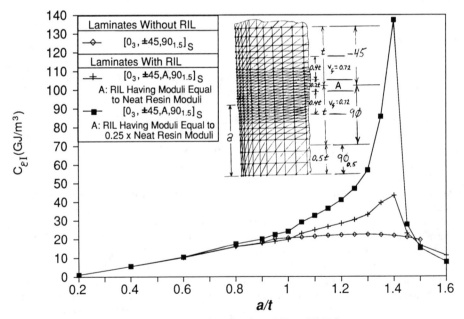

FIG. 5—*Models with and without RIL at 21°C/dry.*

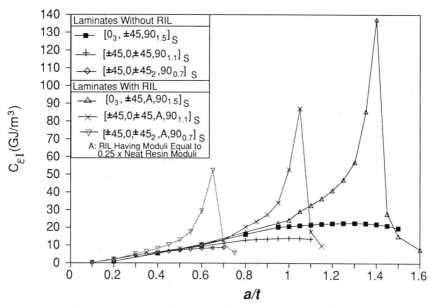

FIG. 6—*Models with and without RIL for three different laminate layups at 21°C/dry.*

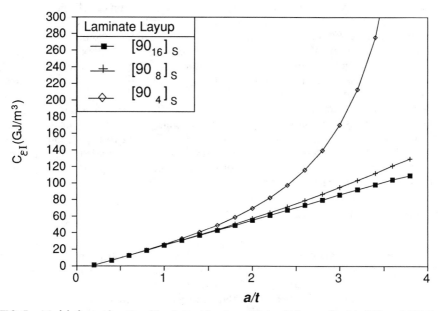

FIG. 7—*Models for unidirectional laminates showing a "finite thickness effect" in* [90]$_4$ *at 21°C/dry.*

IM7G/8551-7, Grade 190 (fiber volume = 0.57 and ply thickness = 0.19 mm), carbon epoxy tape. Six panels were fabricated using a hand-layup process:

1. $[90]_{12}$
2. $[45,90_3,-45,0,90_3,0]_s$
3. $[45_2,90_2,-45_2,0_2,45,-45]_s$
4. $[45,90,-45,45,-45,0,45,-45,45,-45]_s$
5. $[45,90_2,-45,0_4,90_2]_s$
6. $[45,0_2,-45,0_3,90,0_2]_s$

Toughness specimens (3.81 by 25.4 cm) were fabricated from the $[90]_{12}$ panel and in situ strength coupons (2.54 × 15.24 cm) were made from the remaining multidirectional laminates. Both specimen types were untapered.

All panels were cured at 177°C (350°F). Fiberglass tabs were bonded to the in situ strength coupons with a second cure at 121°C (250°F). The tab material was 10 plies of Fiberite MXB7701-7781-Z-6040. The adhesive used to bond tabs was 3M AF163 with an areal density of 293 g/m² (0.06 lb/ft²). Specimens were wet cut using an 80-grit diamond wheel.

An edge notch was machined on each side of fracture toughness specimens at the centerline. Each notch was 0.952 cm long and 0.03 cm wide. Both edges of each in situ strength specimen were polished to get a smooth surface for observing matrix cracking under the microscope. The polishing involved five steps; three passes with aluminum oxide abrasive (120, 320, and 600 grits), and finishing steps from 0.5 to 0.3 μm. The polishing operation typically reduced specimen width by 0.013 cm.

Specimens were preconditioned prior to testing. Coupons denoted "wet" were stored in a chamber with the environment controlled at 60°C and 85% relative humidity for four months prior to testing. The "dry" specimens were stored in an oven maintained at 71°C for one month prior to testing. Tests were performed with wet and dry specimens at three test temperatures; − 59, 21, and 82°C in chambers exposed to laboratory humidity. Each laminate and specimen type was tested for all six environmental conditions.

Intralaminar fracture toughness was measured with the double-edge-notched (DEN) specimen geometry, using a method previously documented [22]. This method resulted in a critical stress intensity factor by incorporating an anisotropic finite width correction factor for DEN specimens. This parameter was converted to G_{Ic} by the orthotropic relationship that used lamina moduli for the corresponding environments.

In situ strength tests consisted of the measurement of matrix crack densities for a 2.54 cm length (centered in the specimen gage length) of the polished edge on both sides of tensile loaded coupons. Crack densities were measured under a light microscope (×50 magnification). Results from both specimen edges were averaged. Specimens were loaded and unloaded in prescribed steps under displacement control at a rate of 0.127 cm/min. After each load step, specimens were removed from the test setup and cracks were counted.

Results and Discussion

Test Results

The results from intralaminar fracture toughness tests for each environmental condition (average of three specimens) are shown in Table 4. Environment was found to have a small influence on values of G_{Ic}. Since $[90]_{12}$ laminates were used in the DEN test procedure, the G_{Ic} values represent an average response of 12 plies in transverse lamina crack growth resistance (that is, combined influence of high fiber volume intralaminar and resin-rich interlaminar regions). Average values are consistent with assumptions used in the finite element models.

TABLE 4—*Intralaminar fracture toughness and effective flaw size properties for IM7/8551-7.*

Test Conditions		G_{Ic}, J/m^2	Y_T, MPa	a_0, mm
Temperature, °C	Moisture[a]			
23.0	dry	513	70.3	0.297
−59.4	dry	496	84.4	0.239
82.0	dry	462	60.9	0.305
23.0	wet	501	51.3	0.488
−59.4	wet	487	61.6	0.384
82.0	wet	452	44.5	0.513

NOTE—Symbols = intralaminar fracture toughness (G_{Ic}), transverse tensile strength (Y_T), and largest effective flaw size (a_0)

[a] Moisture levels correspond to equilibrium conditions obtained in preconditioning chambers; dry = 71.7°C laboratory oven, and wet = 60°C/85% relative humidity controlled chamber.

The values of a_0 were back calculated from finite element analysis of crack closure for a [90]$_{12}$ laminate by matching predicted and measured transverse tensile strengths (Y_T). This was done for each environment. The measured Y_T were obtained using the [90]$_{12}$ fracture toughness specimens (3.81 × 25.4 cm) without notches [22]. Values for Y_T and the calculated a_0 results appear in Table 4. Note that a_0 is very sensitive to environmental conditions. This occurred because Y_T was more strongly affected by environment than notched strength. Values for a_0 are highest for wet conditions. As discussed earlier, a_0 can physically be interpreted as a micro/macro parameter. A higher a_0 for wet conditions seems plausible considering the effects of moisture on the micro stress field at the fiber/matrix scale [for example, Refs 23 and 24].

A total of 164 crack density curves were generated during in situ strength tests (including two replications for each test case and multiple 90° ply groups for most laminates). In addition, a limited number of more detailed microscopy studies were performed to determine trends in matrix crack accumulation/growth. This was done by exposing specimens to a range of peak load levels and performing microscopy on longitudinal cross sections taken from various locations along each specimen width. These studies indicated some influence of specimen free edges. The first matrix cracks appeared in 90° plies on specimen free-edges. Cracks appeared at cross sections taken from the specimen centerline at slightly higher strains.

Cracks commonly occurred in each 90° ply group and the crack density in a group increased with strain. Figure 8 shows a photo montage of the view from a typical specimen edge (Layup 2) after matrix cracks have formed due to tensile loading. Some cracks also appeared in 45° ply groups at load levels approaching specimen failure. Figure 9 shows a view of matrix cracks taken from an axial cross section at the specimen centerline for the same layup and load history as shown in Fig. 8.

Figures 8 and 9 show additional evidence that some free-edge stress effect is present. Figure 8 shows that free-edge matrix cracks in 90° ply groups located near the laminate surfaces tended to tilt at an angle to the through-thickness direction. The angle of tilt changed sign in 90° ply groups located at opposite faces of the laminate and was not evident for cracks appearing in cross sections taken from the specimen centerline (see Fig. 9). This indicated that free-edge interlaminar shear stresses had some influence on matrix cracking. Stronger effects due to interlaminar shear stress were evident in a limited number of tests performed with laminates in which the 45° plies were replaced by 25° plies (that is, matrix crack tilt angle increased).

Test results showed that ply group thickness had little effect on the onset of matrix cracking in IM7/8551-7. Matrix cracks were first observed at applied mechanical strains ranging from 0.004 to 0.008 cm/cm, depending on layup and environment. Note that the highest recorded

FIG. 8—*Specimen free-edge view of matrix cracking in a* $[45,90_3,-45,0,90_3,0]_s$ *laminate.*

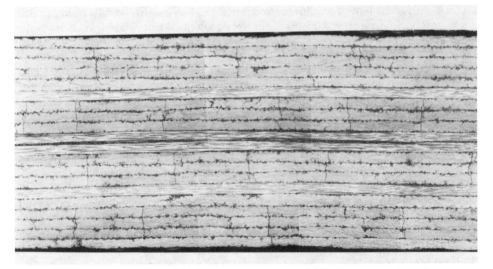

FIG. 9—*Specimen center line view of matrix cracking in a* $[45,90_3,-45,0,90_3,0]_s$ *laminate.*

onset strain was for the thickest 90° ply group studied (four center plies in Layup 5). This is opposite the trends seen with materials studied in the past. Although IM7/8551-7 did not exhibit a significant in situ strengthening effect, the higher toughness of the material lead to relatively high matrix cracking strains. This is particularly true for thick groups of 90° plies that crack at very low mechanical strains in more traditional brittle composites.

Wet preconditioning cycles were expected to relieve thermal stresses (that is, lower ΔT); however, this did not lead to higher applied mechanical strains for the onset of matrix cracking in tests with wet laminates. In fact, many wet laminates tended to crack earlier than their dry counterparts exposed to the same temperatures. This indicates the complex interactions between environmentally dependent parameters that affect matrix cracking.

The ability for IM7/8551-7 to matrix crack under moderate strain levels may actually benefit its resistance to impact damage. For example, results from microscopy work in the literature have indicated that some fiber failure occurs with IM7/8551-7 subjected to impact tests [25]. If IM7/8551-7 exhibited in situ strengthening, coupled with its high toughness, matrix cracks would resist forming during an impact event. This could promote additional fiber failure during impact, potentially degrading post-impact performance.

Comparisons with Analysis

A special modeling scheme was applied to predict the onset of matrix cracking in multidirectional laminates used in tests. This was done to predict cracking in 90° ply groups located away from the test laminate midplane. Results from the parametric study discussed earlier were used to develop the modeling scheme.

The first analysis step was identification of model parameters for a test laminate. These included the percentage of each ply orientation in a laminate, and the candidate 90° ply group thickness. Nearest-neighbor constraint ply orientations were also identified. Finally, the test environment was noted.

The finite element geometry was generated by locating a candidate 90° ply group at the model centerline. An RIL was modeled in only one location; between the candidate 90° ply group and constraint plies. The RIL was shared between the two neighboring plies, each of which was given properties associated with a higher fiber volume (0.72) to simulate actual microstructure near an RIL. All remaining plies in the model had average lamina properties (fiber volume = 0.57). Nearest-neighbor constraint plies were given the same magnitude of orientation as those in test laminates. Plies in the remainder of a model were oriented to yield the same moduli as test laminates. This scheme was thought to minimize the number of plies required for a given model, without compromising accuracy.

Figure 10 shows that the RIL thickness varies dramatically from point to point for a typical cross section of IM7/8551-7. The value of RIL thickness used in finite element models was taken as 0.038 mm (one fifth ply thickness). This represented an average of experimental measurements. The observed growth of matrix cracks across the width of in situ strength specimens may indicate some influence of the varying RIL microstructure of IM7/8551-7. One other feature of note in Fig. 10 is the appearance of a high concentration of toughening agents in the RIL. This is expected to lead to actual RIL moduli that are less than those obtained in neat resin tests for 8551-7 (that is, neat resin has a lower volume fraction of toughening agent). This observation leads to a desire to run two sets of analysis models—one with the RIL having neat resin properties and another with the RIL having reduced properties (0.25 × neat resin moduli). Future work may also consider three-dimensional models of crack growth to account for the free-edge effects, and variable RIL properties (thickness and moduli).

Normalized mechanical and thermal finite element solutions were obtained for each test case. In addition to the two sets of models with RIL, analysis was performed using models without RIL for the purpose of comparison. Material properties corresponding to the candidate test environment are listed in Tables 1 through 4. Equations 6 and 7 were used to obtain ΔT (T_{0dry} = 152°C, and EMC − M_0 = 1.14% for wet laminates). Equations 4 and 5 predicted the "mechanical strain" (that is, ε from Eq 4) required to initiate matrix cracking for a specific laminate and environment. Note that this strain is in addition to the residual ply strains (second term in Eq 4).

A large majority of test results (83%) indicated that the onset of matrix cracking occurred at mechanical strain levels that fell between predictions from the two RIL models. The remain-

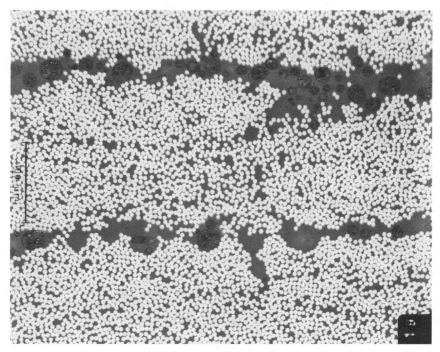

FIG. 10—*Magnified view of ply microstructure, including RIL.*

der of test results fell either slightly above or below this predicted onset range. Typical results comparing experiments and predictions are shown in Figs. 11 through 21.

Figures 11 through 15 demonstrate the effect of varying 90° ply group thickness on matrix cracking for laminates tested at 21°C/dry. The experimental data shows onset strains in a small range between 0.005 and 0.006 cm/cm, essentially independent of ply group thickness. As the strains increased, crack densities did appear to depend on ply group thickness. Larger crack densities were evident with small ply group thicknesses. This relates to redistribution of loads in the neighborhood of matrix cracks.

The RIL models with neat resin moduli generally compared very well with 21°C/dry test data from laminates with 90° ply group thicknesses equal to 2, 3, and 4 plies (for example, Figs. 12 through 15). The same model tended to overpredict 21°C/dry test data for laminates with a single 90° ply (for example, Fig. 11). Two reasons may explain this inconsistency. First, the finite element mesh density used may not have been sufficient to accurately predict the maximum C_{el} for thin ply groups. This could be checked with additional analysis work. A second possible explanation is that a G_{Ic} obtained from notched $[90]_n$ tests represents an average of transverse crack growth resistance within intralaminar and RIL regions of material. This is a reasonable assumption when matrix cracks form in groups of 90° plies, but may be inaccurate when a matrix crack forms in a single ply. The G_{Ic} for crack growth in high fiber volume intralaminar regions is expected to be less than the average, yielding lower matrix crack onset strains in a single 90° ply.

Figures 13 and 15 show crack densities in different 90° ply groups of the same laminate layup tested at 21°C/dry. The only difference is in nearest neighbor constraint plies. Both data and predictions indicate a small increase in in situ strength (<0.001 cm/cm) due to stiffer neighboring constraint plies in Fig. 15.

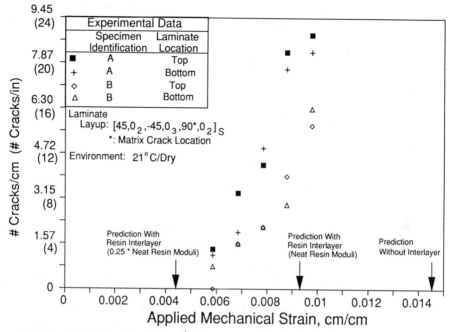

FIG. 11—*Predictions and test data for a laminate with 90° ply group thickness equal to 1 ply.*

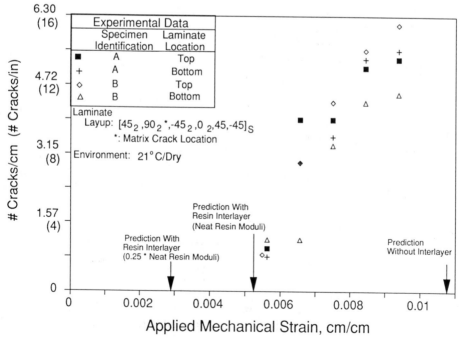

FIG. 12—*Predictions and test data for a laminate with 90° ply group thickness equal to 2 plies.*

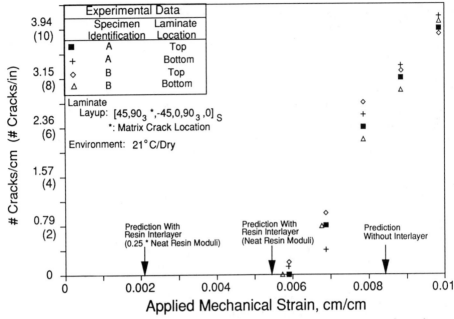

FIG. 13—*Predictions and test data for a laminate with 90° ply group thickness equal to 3 plies.*

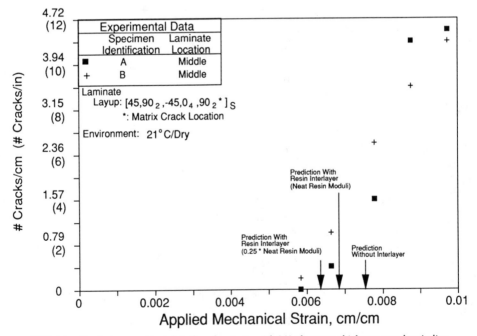

FIG. 14—*Predictions and test data for a laminate with 90° ply group thickness equal to 4 plies.*

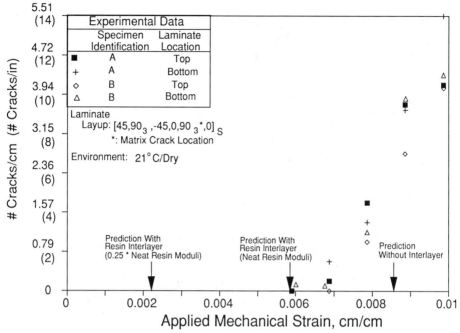

FIG. 15—*Predictions and test data for the second 90° ply group from the laminate shown in Fig. 13.*

The overall effect of environment on matrix cracking in IM7/8551-7 is complex because each key parameter in the failure criterion relates to environment. The values at ΔT decreased as a function of increasing temperature and moisture content (see Eqs 6 and 7). The effect of environment on C_{el} and C_{Tl} was found to be less for the models with RIL than for those without. Values of a_0 increased with increasing temperature and moisture content. This only had an influence on predictions for thick 90° ply groups. Finally, values of G_{lc} were not strongly affected by environment. Although most parameters depended on environment for IM7/8551-7, some tended to neutralize each other when entered in the failure criterion.

Figures 16 through 21 show the effects of environment for one of the laminate layups. Trends in the comparison of experimental data to predictions shown in this sequence of figures are similar to those seen for other laminate layups having a 90° ply group thickness greater than one. Matrix cracks started to appear in −59°C/dry tests at slightly higher strains than the RIL prediction with neat resin moduli (for example, Fig. 16). The onset of matrix cracks in 21°C/dry tests began at or slightly below strains predicted by the RIL model with neat resin moduli (for example, Fig. 17). Matrix cracks first appeared in 82°C/dry and −59°C/wet tests between the two RIL predictions (for example, Figs. 18 and 19). Finally, cracking started in 21°C/wet and 82°C/wet tests at or slightly below strains predicted by the RIL model with reduced moduli (for example, Figs. 20 and 21). These data trends indicate the need to accurately simulate stiffness changes in the RIL as a function of environment.

The relationship between experimental data and RIL predictions suggest that increased moisture and temperature may soften the RIL in IM7/8551-7 more than 8551-7 neat resin. As discussed earlier, a toughening agent used in 8551-7 concentrates in the RIL. The moduli of epoxies have been shown to decrease as a function of increasing concentration of toughening agents (for example, Refs *26* and *27*). The reduced moduli used in the second set of RIL

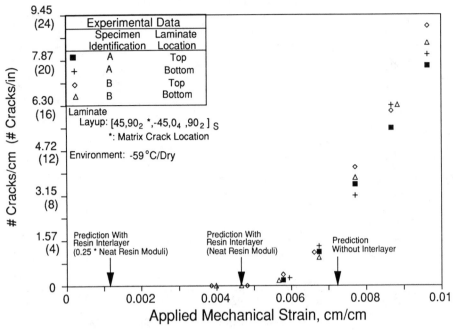

FIG. 16—*Predictions and test data for a laminate exposed to −59°C/dry.*

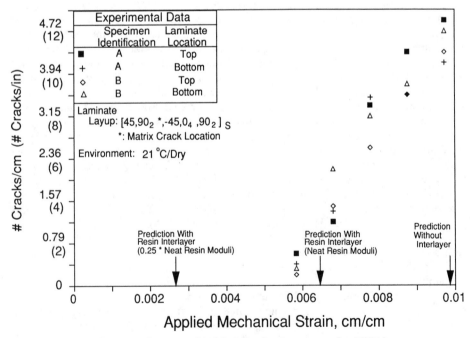

FIG. 17—*Predictions and test data for a laminate exposed to 21°C/dry.*

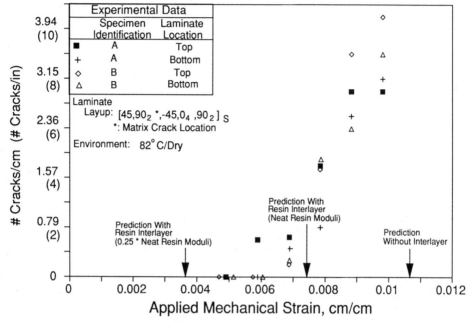

FIG. 18—*Predictions and test data for a laminate exposed to 82°C/dry.*

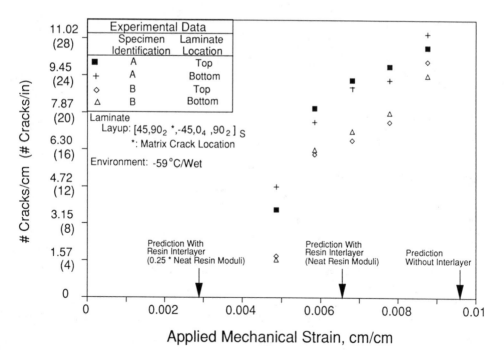

FIG. 19—*Predictions and test data for a laminate exposed to −59°C/wet.*

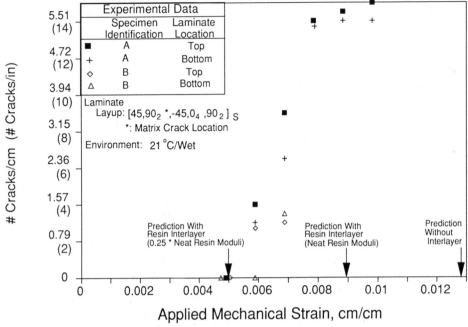

FIG. 20—*Predictions and test data for a laminate exposed to 21°C/wet.*

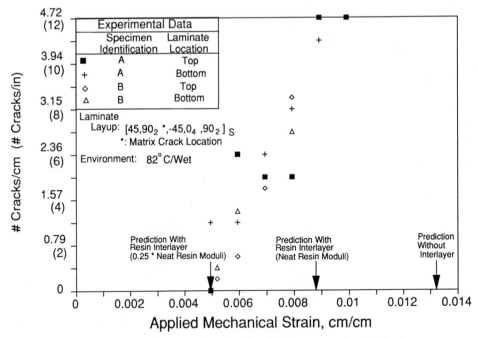

FIG. 21—*Predictions and test data for a laminate exposed to 82°C/wet.*

predictions for this study are within limits expected with the RIL of IM7/8551-7 subjected to high-temperatures and moisture contents. Additional neat resin tests with the increased concentration of toughening agent would be required to confirm this hypothesis.

Viscoelastic shifts in the stress-free temperature (see Eq 6) should be considered when preconditioning for long periods of time in high temperature and moisture environments [9]. The wet preconditioning environment used in this study caused compressive transverse ply residual stresses. If these stresses relax with time, T_0 would increase and result in a small decrease in predictions of ε for $-59°C$/wet and $21°C$/wet conditions. For example, assuming T_0 equals the preconditioning temperature (that is, complete relaxation), the three predictions in Fig. 19 reduce to 0.002, 0.0055, and 0.0084 cm/cm while retaining the same order. Note that a viscoelastic shift in T_0 has no effect on predictions for $82°C$/wet that has a small positive ΔT, which is assumed to be zero in crack closure analysis.

The apparent transverse ply strength of IM7/8551-7 indicated that a finite thickness interaction between an effective flaw and ply group thickness existed. In fact, a large portion of the experimental data indicated the effect was stronger than predicted (that is, single 90° ply groups and hot/wet environments). Despite this evidence, the physical meaning of an effective flaw distribution will probably remain a subject of controversy. Future theories may help solve this controversy; however, Wang's effective flaw concept [17] currently serves as a suitable tool for engineering applications.

Existing shear lag and statistical methods that do not account for RIL generally overpredicted the onset of matrix cracking in IM7/8551-7 by factors up to 400%, depending on environment and layup. An improvement in predictions using these theories may be possible by incorporating an RIL in the model (for example, Refs 7 and 8). A maximum stress failure criterion and lamination theory was also used to predict the onset of matrix cracking. Lamination theory was accurate for $21°C$/dry and $-59°C$/wet conditions; underpredicted the $-59°C$/dry environment; and overpredicted $21°C$/wet, $82°C$/dry, and $82°C$/wet conditions. This trend makes sense considering the relationship between RIL stiffness and environment (that is, the finite thickness effect was strongest for high temperatures and moisture contents).

Conclusions

The fracture mechanics prediction of in situ transverse ply strength for laminates with soft resin-rich interlaminar layers (RIL) can decrease with decreasing ply group thickness. Therefore, experimental data from such materials can be used to judge the validity of basic assumptions that have formed the foundations of past statistical and fracture mechanics based approaches. Finite element analysis was performed to model an interaction between ply group thickness and effective flaw size. The observed behavior was termed a "finite thickness effect."

A parametric analysis study was performed to judge material and laminate factors affecting the strain energy release rate for matrix cracking. The RIL stiffness was found to have the strongest effect of all variables studied. There was a sharp peak in the strain energy release rate of laminates modeled with RIL as crack size approached the 90° ply group thickness. Environment and 90° ply group thickness were also found to have some influence on models.

Intralaminar fracture roughness and in situ strength tensile coupons were used in experiments. A total of six environments and six laminate layups were tested. Test results showed that ply group thickness had a relatively small effect on the onset of matrix cracking for the material studied. Many wet laminates tended to crack earlier than their dry counterparts exposed to the same temperatures. Even without in situ strengthening, the relatively high fracture toughness of the material tested led to moderate resistance to matrix cracking.

A special modeling scheme was applied to predict the onset of matrix cracking in multidirectional laminates used in tests. Experimental results confirmed predictions from analysis, indicating that the RIL must be modeled to predict the onset of matrix cracking. The finite thickness effect was more evident in experimental data than in models for some cases. This suggests a need for some future work to develop more accurate modeling schemes and to measure properties of RIL. Although a tough material was used in tests, a decrease in the in situ strength is expected for other materials with RIL microstructure (for example, poor fiber/resin distribution from a manufacturing process).

Acknowledgments

The authors wish to acknowledge J. Quinlivan, R. Rothschilds, P. Smith, and P. Whalley of The Boeing Company for technical support. Preparation of the manuscript, analysis, and microscopy work were funded by NASA (Contract NAS1-18889, under the technical leadership of W. T. Freeman). Experimental data bases were provided by The Boeing Company.

Use of commercial products or names of manufacturers in this report does not constitute official endorsement of such products or manufacturers, either expressed or implied, by The Boeing Company or The National Aeronautics and Space Administration.

References

[1] Bailey, J. E., Curtis, P. T., and Parvizi, A., *Proceedings,* Royal Society (London), Series A, 366, 1979, pp. 599–623.
[2] Crossman, F. W. and Wang, A. S. D. in *Damage in Composite Materials, ASTM STP 775,* K. L. Reifsnider, Ed., American Society for Testing and Materials, Philadelphia, 1982, pp. 118–139.
[3] Flaggs, D. L. and Kural, M. H., *Journal of Composite Materials,* Vol. 16, March 1982, pp. 103–116.
[4] Morley, J. G. and Pissinou, G., *Journal of Materials Science,* Vol. 21, 1986, pp. 4206–4214.
[5] Narin, J. A. in *Proceedings,* American Society for Composites: 3rd Technical Conference, Technomic Publishing Co., Lancaster, PA, 1988, pp. 472–481.
[6] Flaggs, D. L., *Journal of Composite Materials,* Vol. 19, 1985, pp. 29–50.
[7] Laws, N. and Dvorak, G. J., *Journal of Composite Materials,* Vol. 22, Oct. 1988, pp. 900–916.
[8] Lim, S. G. and Hong, C. S., *Journal of Composite Materials,* Vol. 23, July 1989, pp. 695–713.
[9] Rothschilds, R. J., Ilcewicz, L. B., Nordin, P., and Applegate, S. H., *Journal of Engineering Materials and Technology, Transactions,* American Society of Mechanical Engineers, Lancaster, PA, Vol. 110, April 1988, pp. 158–168.
[10] Fukunaga, H., Chou, T. W., Peters, P. W. M., and Schulte, K., *Journal of Composite Materials,* Vol. 18, July 1984, pp. 339–356.
[11] Peters, P. W. M. in *Composite Materials: Fatigue and Fracture, ASTM STP 907,* H. T. Hahn, Ed., American Society for Testing and Materials, Philadelphia, 1986, pp. 84–99.
[12] Peters, P. W. M. and Chou, T. W., *Composites,* Philadelphia, Vol. 18, No. 1, 1987, pp. 40–46.
[13] Wang, A. S. D. and Crossman, F. W., *Journal of Composite Materials Supplement,* Vol. 14, 1980, pp. 71–87.
[14] Crossman, F. W., Warren, W. J., Wang, A. S. D., and Law, G. E., Jr., *Journal of Composite Materials Supplement,* Vol. 14, 1980, pp. 88–108.
[15] Law, G. E. in *Effects of Defects in Composite Materials, ASTM STP 836,* American Society for Testing and Materials, Philadelphia, 1984, pp. 143–160.
[16] Wang, A. S. D., Chou, P. C., Lei, S. C., and Bucinell, R. B., AFWAL-TR-85-4104, Air Force Materials Laboratory, Dayton, OH, 1985.
[17] Wang, A. S. D. in *Proceedings,* International Symposium on Composite Materials and Structures, Technomic Publishing Co., Lancaster, PA, 1986, pp. 576–584.
[18] Mandell, J. F., McGarry, F. J., Wang, S. S. and Im, J., *Journal of Composite Materials,* Vol. 8, April 1974, pp. 106–116.
[19] Rybicki, E. F. and Kanninen, M. F., *Engineering Fracture Mechanics,* Vol. 9, 1977, pp. 931–938.
[20] Coquill, S. L. and Adams, D. F., "Mechanical Properties of Several Neat Polymer Matrix Materials and Unidirectional Carbon Fiber-Reinforced Composites," NASA CR-181805, National Aeronautics and Space Administration, Hampton, VA, April 1989.
[21] Chamis, C. C., "Simplified Composite Micromechanics Equations for Hygral, Thermal and

Mechanical Properties," NASA TM-83320, National Aeronautics and Space Administration, Cleveland, OH, Feb. 1983.

[22] Coxon, B. R., Walker, T. H., Ilcewicz, L. B., and Seferis, J. C., "Intralaminar Fracture Toughness Characterization for Matrix Cracking in Composites," *Proceedings,* Spring Scanning Electron Microscope Conference, Society for Experimental Mechanics, 1987, pp. 144–151.

[23] Adams, D. F. and Miller, A. K., *Journal of Composite Materials,* Vol. 11, July 1977, pp. 285–299.

[24] Crossman, F. W. and Warren, W. J., "The Influence of Environment on Matrix Dominated Composite Fracture," LMSC-D062004, Lockheed Missiles and Space Co., Palo Alto, CA, Dec. 1985.

[25] Boll, D. J., Bascom, W. D., Weidner, J. C., and Murri, W. J., *Journal of Materials Science,* Vol. 21, 1986, pp. 2667–2678.

[26] Bascom, W. D., Cottington, R. L., Jones, R. L., and Peyser, P., *Journal of Applied Polymer Science,* Vol. 19, 1975, pp. 2545–2562.

[27] Guild, F. J. and Young, R. J., *Journal of Materials Science,* Vol. 24, 1989, pp. 2454–2460.

Jong-Won Lee,[1] *D. H. Allen,*[1] *and C. E. Harris*[2]

The Upper Bounds of Reduced Axial and Shear Moduli in Cross-Ply Laminates with Matrix Cracks

REFERENCE: Lee, J-W., Allen, D. H., and Harris, C. E., **"The Upper Bounds of Reduced Axial and Shear Moduli in Cross-Ply Laminates with Matrix Cracks,"** *Composite Materials: Fatigue and Fracture (Third Volume), ASTM STP 1110,* T. K. O'Brien, Ed., American Society for Testing and Materials, Philadelphia, 1991, pp. 56–69.

ABSTRACT: The present study proposes a mathematical model utilizing the Internal State Variable (ISV) concept for predicting the upper bounds of the reduced axial and shear stiffnesses in cross-ply laminates with matrix cracks. The displacement components at the matrix crack surfaces are explicitly expressed in terms of the observable axial and shear strains and the undamaged material properties. The reduced axial and shear stiffnesses are predicted for glass/epoxy and graphite/epoxy laminates. Comparison of the model with other theoretical and experimental studies is also presented to confirm direct applicability of the model to angle-ply laminates with matrix cracks subjected to general in-plane loading.

KEY WORDS: composite materials, fracture, fatigue (materials), matrix crack, cross-ply laminate, shear, tension, upper bound, stiffness reduction, internal state variable

It has been reported by a number of experimental studies [1–4] that the effective axial stiffness of a cross-ply laminate decreases asymptotically to the solution of the classical ply discount method as the number of matrix cracks approaches infinity. In the classical ply discount method, the out-of-plane stress components are neglected and each cracked layer is assumed to shed its entire load carrying capacity. However, the out-of-plane stress components are much greater than the in-plane stress components near the matrix crack surfaces, because the applied load at far-field is transferred to the cracked layer from the adjacent layers through out-of-plane stresses of the cracked layer near the crack surface that is traction free. These out-of-plane stresses stiffen the entire laminate, resulting in a large discrepancy between experimental observation and the classical ply discount method. A typical example is shown in Fig. 1, in which some of the existing model predictions [2,6,7,9,17] are also plotted.

Among existing models, the study of Hashin based on the complementary strain energy method provides the most effective and straightforward procedure for predicting the lower bounds of the reduced axial and shear stiffnesses of $[0_q/90_r]_s$ type laminates with matrix cracks in 90° layers [7] and the reduced axial stiffness of the same type laminate with matrix cracks in both layers [8]. A brief review of the existing theoretical models [2,5–7,9–12,14–15] is presented in the authors' previous study [17] that provides the upper bound of the reduced axial stiffness.

[1] Research assistant and professor, respectively, Aerospace Engineering Department, Texas A&M University, College Station, TX 77843.

[2] Head, Fatigue and Fracture Branch, NASA Langley Research Center, Hampton, VA 23665-5225.

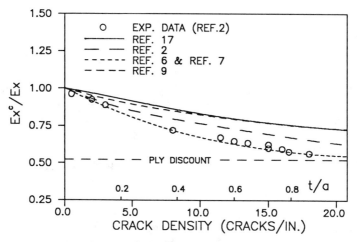

FIG. 1—*Discrepancies between the ply discount method and experimental observation.*

In the following section, after summarizing the authors' previous mathematical model [17], the authors propose an explicit solution for the upper bound of the shear stiffness of a cross-ply laminate.

Model Formulation

The internal state variable (ISV) concept proposed by Vakulenko and Kachanov [13] has been utilized by Allen and other authors [14–16] to predict stiffness reduction of a laminated composite due to matrix cracks and delamination. The internal state variable for a damaged material is defined to be a volume-averaged value of a diadic product between the displacement vector and unit normal vector at the internal traction free surface. The resulting physical quantity becomes an asymmetric second order tensor defined by

$$\alpha_{ij}^{\eta} = \frac{1}{V} \int_{S_2^{\eta}} u_i n_j \, ds \tag{1}$$

where

$\eta = 1, 2, \ldots$ to the number of damage modes in the laminate;
u_i = displacements on the crack surface;
n_j = unit inner normal to the crack surface;
V = local volume over which cracks are averaged; and
S_2^{η} = surface area of cracks in V.

The stress-strain relationships at the ply level are given by

$$\sigma_{ij} = C_{ijkl}\epsilon_{kl} + I_{ijkl}^{\eta}\alpha_{kl}^{\eta} \tag{2a}$$

where

ϵ_{kl} = total strains, and
I_{ijkl}^{η} = damage moduli defined in Refs *14* and *15*.

When the loading direction is perpendicular or parallel to the crack surface, the damage moduli becomes

$$I^{\eta}_{ijkl} = -C_{ijkl} \tag{2b}$$

Lee et al. [17] have demonstrated the usefulness of the ISV concept for cross-ply laminates under uniaxial tension loading. Comparison of their theoretical predictions with available experimental and theoretical studies in the open literature has shown reasonable agreement. The present study extends the ISV approach to a cross-ply laminate under in-plane shear loading. After briefly reviewing the ISV approach for the tension loading case [17], the in-plane shear response of a cross-ply laminate with matrix cracks is presented.

Tension Loading

Consider a laminated composite material with an infinite number of 0° and 90° layers subjected to uniaxial tension loading as shown in Fig. 2. If the interface between the 0 and 90° layers is assumed to remain plane during deformation, the displacement fields in the representative volume element of the cracked layer can be assumed as

$$u = (u_o/a)x + \sum_m \sum_n a_{mn} \sin \alpha_m x \cos \beta_n z \tag{3a}$$

$$v = -(v_o/b)y \tag{3b}$$

$$w = -(w_o/t)z \tag{3c}$$

where

$m, n = 1, 2, 3, \ldots, \infty;$
$\alpha_m = (2m - 1)\pi/2a;$ and
$\beta_n = (2n - 1)\pi/2t.$

The elastic strain components in the volume element are thus given by

$$\epsilon_{xx} = u_o/a + \sum_m \sum_n a_{mn}\alpha \cos \alpha_m x \cos \beta_n z \tag{4a}$$

$$\epsilon_{yy} = -v_o/b \tag{4b}$$

$$\epsilon_{zz} = -w_o/t \tag{4c}$$

$$\gamma_{xz} = -\sum_m \sum_n a_{mn}\beta \sin \alpha_m x \sin \beta_n z \tag{4d}$$

All other components of the strain tensor are zero.
The total potential energy in the volume ($a \times 2t \times 2b$) is given by

$$\Pi = U + V_E = \frac{1}{2} \int_{-b}^{t} \int_{-t}^{b} \int_{0}^{a} C_{ijkl}\epsilon_{ij}\epsilon_{kl} \, dx \, dy \, dz - 2bpu|_{\substack{x=a \\ z=t}} \tag{5}$$

where P is the axial load applied at far-field, and $p \equiv P/2b$.

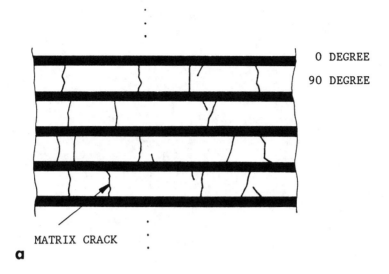

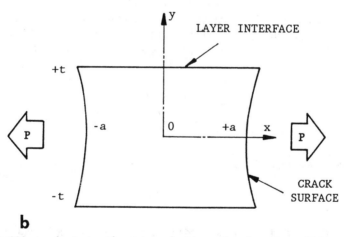

b

FIG. 2—$[0_q/90_r]_\infty$ *type laminate with matrix cracks: (a) overall configuration and (b) one representative 90° element under tension loading.*

Four algebraic equations are obtained by differentiating the total potential energy with respect to unknown constants, u_o, v_o, w_o, and a_{mn}. After a straightforward algebraic manipulation, these constants are determined as follows.

$$v_o/b = (p/2t) \frac{C_{xy}C_{zz} - C_{yz}C_{zx}}{\det[C_{ij}]} \qquad (6a)$$

$$w_o/t = (p/2t) \frac{C_{yy}C_{zx} - C_{xy}C_{yz}}{\det[C_{ij}]} \qquad (6b)$$

$$a_{mn} = \frac{4(-1)^{(m+n)}[-C_{xx}(u_o/a) + C_{xy}(v_o/b) + C_{zx}(w_o/t)]}{at\beta[C_{xx}\alpha_m^2 + G_{xy}\beta_n^2]} \tag{7a}$$

$$u_o/a = (p/2t)\left[\frac{C_{yy}C_{zz} - C_{yz}^2}{\det[C_{ij}]} + \frac{1}{\dfrac{\pi^4}{64\xi} - C_{xx}}\right] \tag{7b}$$

where

$$\xi = \sum_m \sum_n \frac{1}{C_{xx}(2m-1)^2(2n-1)^2 + G_{xy}(a/t)^2(2n-1)^4}. \tag{7c}$$

Utilizing Eqs 1, 6a, 6b, 7a, and 7b, α_{xx} is explicitly given by

$$\alpha_{xx} = \frac{p/2t}{\dfrac{\pi^4}{64\xi} - C_{xx}} \tag{7d}$$

All other components of α_{ij} are assumed to be negligible.

The ISV, α_{xx}, given by Eq 7d represents the contribution of crack opening displacement to the observable axial strain that can be measured from a specimen with matrix cracks under uniaxial tension loading. The α_{xx} can be rewritten in terms of observable strain, u_o/a, using Eqs 7b and 7d.

$$\alpha_{xx} = \frac{(u_o/a)}{1 + \left[\dfrac{C_{yy}C_{zz} - C_{yz}^2}{\det[C_{ij}]}\right]\left[\dfrac{\pi^4}{64\xi} - C_{xx}\right]} \tag{7e}$$

More detailed discussion to the tension loading is to appear in Ref *17*.

Shear Loading

Similarly, if the layer interface is assumed to remain plane during deformation, the displacement fields in the same volume element under in-plane shear loading illustrated in Fig. 3 may be assumed as

$$u = \gamma_o y \tag{8a}$$

$$v = \sum_n a_n f_n(x) \cos \alpha_n z \tag{8c}$$

$$w = 0 \tag{8b}$$

where

$n = 1, 2, 3, \ldots, \infty$; and
$\alpha_n = (2n-1)\pi/2t.$

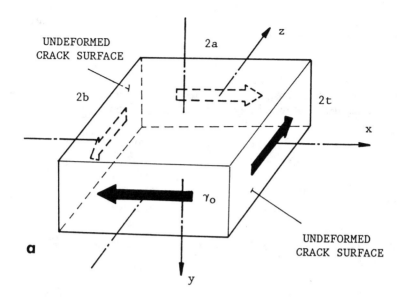

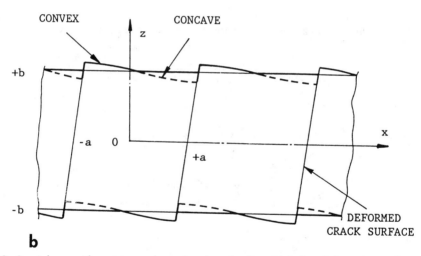

FIG. 3—*A layer with matrix cracks under shear loading: (a) before deformation and (b) after deformation.*

Strain components are then given by

$$\gamma_{xy} = \gamma_o + \sum_n a_n f_{n'}(x) \cos \alpha_n z \qquad (9a)$$

$$\gamma_{yz} = -\sum_n a_n \alpha_n f_n(x) \sin \alpha_n z \qquad (9b)$$

Other strain components are zero.

Since the shear moduli in the x-y and y-z planes are identical, the nontrivial equilibrium equation becomes

$$\gamma_{xy,x} + \gamma_{yz,z} = 0 \tag{10a}$$

In contrast to the uniaxial tension case, the preceding differential equation can be solved directly. Substituting Eqs 9a and 9b into the preceding equation gives

$$\sum_n a_n \cos \alpha_n z [f_{n''}(x) - \alpha_n^2 f_n(x)] = 0 \tag{10b}$$

Thus,

$$f_n(x) = C_{1n} \sinh \alpha_n x + C_{2n} \cosh \alpha_n x \tag{10c}$$

Since v should be antisymmetric with respect to $x = 0$, $C_{2n} = 0$. Along the crack surface, $\gamma_{xy} = 0$. Therefore

$$-\gamma_o = \sum_n a_n C_{1n} \alpha_n \cosh \alpha_n a \cos \alpha_n z \tag{11}$$

Using the Fourier series analysis all unknown constants are determined by

$$a_n C_{1n} \alpha_n = \frac{4\gamma_o(-1)^n}{(2n-1)\pi \cosh \alpha_n a} \tag{12}$$

Therefore

$$v = \sum_n \frac{(-1)^n 8 t \gamma_o \sinh \alpha_n x \cos \alpha_n z}{(2n-1)^2 \pi^2 \cosh \alpha_n a} \tag{13a}$$

$$\gamma_{xy} = \gamma_o \left[1 + \sum_n \frac{(-1)^n 4 \cosh \alpha_n x \cos \alpha_n z}{(2n-1)\pi \cosh \alpha_n a} \right] \tag{13b}$$

$$\gamma_{yz} = -\gamma_o \sum_n \frac{(-1)^n 4 \sinh \alpha_n x \sin \alpha_n z}{(2n-1)\pi \cosh \alpha_n a} \tag{13c}$$

Then the strain energy stored in the unit volume becomes

$$U = \frac{G_{xy}}{2} \int_0^t \int_0^b \int_0^a (\gamma_{xy}^2 + \gamma_{yz}^2) \, dx \, dy \, dz \tag{14}$$

The damaged shear modulus of the volume element is given by

$$G_{xy}^c = \frac{d^2(U/abt)}{d\gamma_o^2} \tag{15}$$

After simple algebraic manipulation, the ratio of the damaged shear modulus to the undamaged shear modulus becomes

$$\frac{G_{xy}^c}{G_{xy}} = 1 - \frac{8}{\pi^2} \sum_n \frac{\tanh \alpha_n a}{(2n - 1)^2 \alpha_n a} \tag{16}$$

It can be shown that the only nonzero internal state variable is given by

$$\alpha_{yx} = \frac{8\gamma_o}{\pi^2} \sum_n \frac{\tanh \alpha_n a}{(2n - 1)^2 \alpha_n a} \tag{17}$$

Note that $\alpha_{xy} = 0$.

If the crack length-distance ratio due to shear loading in a cross-ply laminate is independent of the fiber orientation, the degraded shear modulus of $[0_q/90_r]_\infty$ laminate is given by Eq 16. If $0°$ layers are assumed to be undamaged, Eq 16 should be modified to

$$\frac{G_{LT}^c}{G_{LT}} = \frac{q + rG_{xy}^c/G_{xy}}{q + r} \tag{18}$$

The internal state variables, α_{yx}, given by Eq 17 represent the contribution of crack opening displacement to the observable in-plane shear strain that can be measured from a specimen with matrix cracks under in-plane shear loading.

Modified Constitution

From Eqs (2a), (7e), and 17, modified constitutive equations can be constructed for a single layer containing matrix cracks as functions of the undamaged material constants and the internal state variables, α_{xx} and α_{yx}

$$\begin{Bmatrix} \sigma_{LL} \\ \sigma_{TT} \\ \tau_{LT} \end{Bmatrix} = \begin{bmatrix} Q_{LL} & Q_{LT} & 0 \\ Q_{LT} & Q_{TT} & 0 \\ 0 & 0 & G_{LT} \end{bmatrix} \begin{Bmatrix} \epsilon_{LL} \\ \epsilon_{TT} - \alpha_{xx} \\ \gamma_{LT} - \alpha_{yx} \end{Bmatrix} \tag{19a}$$

or

$$\begin{Bmatrix} \sigma_{LL} \\ \sigma_{TT} \\ \tau_{LT} \end{Bmatrix} = \begin{bmatrix} Q_{LL} & (1 - \zeta)Q_{LT} & 0 \\ Q_{LT} & (1 - \zeta)Q_{TT} & 0 \\ 0 & 0 & (1 - \psi)G_{LT} \end{bmatrix} \begin{Bmatrix} \epsilon_{LL} \\ \epsilon_{TT} \\ \gamma_{LT} \end{Bmatrix} \tag{19b}$$

where

$$\zeta = \frac{\alpha_{xx}}{\epsilon_{TT}}, \tag{20a}$$

$$\psi = \frac{\alpha_{yx}}{\gamma_{LT}}, \tag{20b}$$

$\epsilon_{TT} = u_o/a$,

$\gamma_{LT} = \gamma_o$, and

Q_{ij} = undamaged plane stress stiffness of a layer.

Since ζ and ψ are scalar quantities, the tensor transformation law can be directly applied to Eq 19b. The transformed stiffness tensor of an off-axis angle ply with matrix cracks is then given by

$$Q_{11} = \cos^4\theta Q_{LL} + (1 - \zeta)\sin^4\theta Q_{TT} + (2 - \zeta)\sin^2\theta\cos^2\theta Q_{LT}$$
$$+ 4(1 - \psi)\sin^2\theta\cos^2\theta G_{LT} \quad (21a)$$

$$Q_{22} = \sin^4\theta Q_{LL} + (1 - \zeta)\cos^4\theta Q_{TT} + (2 - \zeta)\sin^2\theta\cos^2\theta Q_{LT}$$
$$+ 4(1 - \psi)\sin^2\theta\cos^2\theta G_{LT} \quad (21b)$$

$$Q_{12} = \sin^2\theta\cos^2\theta[Q_{LL} + (1 - \zeta)Q_{TT}] + [\sin^4\theta + (1 - \zeta)\cos^4\theta]Q_{LT}$$
$$- 4(1 - \psi)\sin^2\theta\cos^2\theta G_{LT} \quad (21c)$$

$$Q_{21} = \sin^2\theta\cos^2\theta[Q_{LL} + (1 - \zeta)Q_{TT}] + [(1 - \zeta)\sin^4\theta + \cos^4\theta]Q_{LT}$$
$$- 4(1 - \psi)\sin^2\theta\cos^2\theta G_{LT} \quad (21d)$$

$$Q_{16} = \sin\theta\cos^3\theta Q_{LL} - (1 - \zeta)\sin^3\theta\cos\theta Q_{TT} + [\sin^3\theta\cos\theta$$
$$- (1 - \zeta)\sin\theta\cos^3\theta]Q_{LT} + 2(1 - \psi)(\sin^3\theta\cos\theta - \sin\theta\cos^3\theta)G_{LT} \quad (21e)$$

$$Q_{61} = \sin\theta\cos^3\theta Q_{LL} - (1 - \zeta)\sin^3\theta\cos\theta Q_{TT} + [(1 - \zeta)\sin^3\theta\cos\theta$$
$$- \sin\theta\cos^3\theta]Q_{LT} + 2(1 - \psi)(\sin^3\theta\cos\theta - \sin\theta\cos^3\theta)G_{LT} \quad (21f)$$

$$Q_{26} = \sin^3\theta\cos\theta Q_{LL} - (1 - \zeta)\sin\theta\cos^3\theta Q_{TT} + [\sin\theta\cos^3\theta$$
$$- (1 - \zeta)\sin^3\theta\cos\theta]Q_{LT} + 2(1 - \psi)(\sin\theta\cos^3\theta - \sin^3\theta\cos\theta)G_{LT} \quad (21g)$$

$$Q_{62} = \sin^3\theta\cos\theta Q_{LL} - (1 - \zeta)\sin\theta\cos^3\theta Q_{TT} + [(1 - \zeta)\sin\theta\cos^3\theta$$
$$- \sin^3\theta\cos\theta]Q_{LT} + 2(1 - \psi)(\sin\theta\cos^3\theta - \sin^3\theta\cos\theta)G_{LT} \quad (21h)$$

$$Q_{66} = \sin^2\theta\cos^2\theta[Q_{LL} + (1 - \zeta)Q_{TT} - (2 - \zeta)Q_{LT}]$$
$$+ (1 - \psi)(\sin^2\theta - \cos^2\theta)^2 G_{LT} \quad (21i)$$

where θ is the off-axis angle.

Note that the conventional transformed stiffness equations result from the preceding equations, if there is no matrix crack in the off-axis angle ply. The modified transformed stiffness tensor just given can be used to construct a laminate constitution using the conventional lamination theory. If the matrix crack density in each layer of a laminate is specified, the crack parameter formulation given by Eq 19b is more convenient than Eq 19a to predict stiffness reduction. However, it is not practical to measure the crack density in each ply of a real laminated structure especially when the structure is subjected to a cyclic loading. In such a case, the internal state variable formulation given by Eq 19a is far more effective than Eq 19b. From a specimen under similar cyclic loading, the internal state variables can be experimentally determined by measuring the stiffness reduction as a function of load history. Then, the experimentally determined internal state variables can be used for predicting stiffness reduction of a laminated structure under cyclic loading without measuring the crack densities in the structure.

Results and Discussion

In the authors' previous work [17], the axial stiffness in cross-ply laminates with matrix cracks were predicted as a function of α_{xx} given by Eq 7e. The comparisons of the predicted axial stiffnesses with experimental data [2,4] presented in Ref 17 are illustrated in Figs. 1, 4, and 5. The mechanical properties of the material in these figures are given in Table 1. The effects of matrix cracks in the ply-level constitution depend on ζ and ψ given by Eqs 20a and 20b, respectively. The numerical values of these two scalar quantities are listed in Table 2 for two material systems in Table 1.

The normalized shear stiffness given by Eq 18 is compared with the lower bound provided by Hashin [7] in Fig. 6. In this figure, the 0° layer is assumed to be undamaged. The normalized shear modulus predicted by the present study is independent of material properties. On the contrary, the result of Hashin depends on the ratio of two shear moduli in the orthotropic and isotropic planes. Also, it should be noted that the present result is for a laminate with infinite number of layers, whereas Hashin's result is for a $[0_q/90_r]$ laminate.

For an off-axis angle ply with matrix cracks, the effective stiffnesses are given by Eq 21a through 21i. However, the observed crack distance, $2a'$, which may be measured from the free edge of a specimen, cannot be directly used for off-axis angle ply. Since the actual crack distance, $2a$, is the value measured in the direction normal to the crack surface, the relationship between the observed and the actual crack distances becomes

$$2a = 2a' \sin \theta \tag{22}$$

By applying the modified stiffness tensor given by Eqs 21a through 21i to the classical lamination theory, the effective stiffness tensor of an angle-ply laminate can be obtained. As a model verification tool, the axial Young's modulus in the loading direction of a $[0/\pm 45]_s$ glass/epoxy laminate is calculated as a function of the undamaged ply properties and the average matrix crack length-distance ratio. The experimental result of Highsmith et al. [1] is extrapolated to obtain the undamaged axial modulus of the laminate in which initial matrix cracks were observed. Then the normalized axial modulus for each data point is compared with the present model prediction as shown in Fig. 7. Since the crack length-distance ratios in +45 and

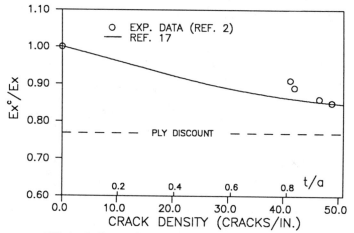

FIG. 4—*Stiffness reduction in $[0/90]_s$ glass/epoxy laminate.*

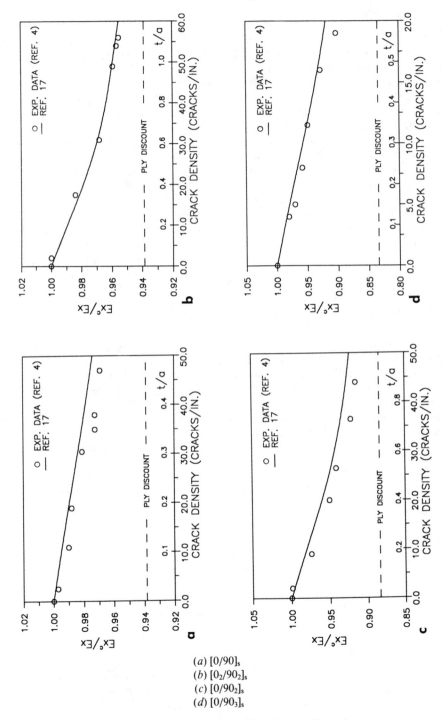

(a) [0/90]$_s$
(b) [0$_2$/90$_2$]$_s$
(c) [0/90$_2$]$_s$
(d) [0/90$_3$]$_s$

FIG. 5—*Stiffness reduction in graphite/epoxy specimens.*

TABLE 1—*Material properties.*

Property, Msi (GPa)	Material	
	Glass/Epoxy, Ref *2*	Graphite/Epoxy, Ref *4*
E_{LL}	6.05 (41.7)	21.0 (144.8)
E_{TT}	1.89 (13.0)	1.39 (9.6)
G_{LT}	0.493 (3.4)	0.694 (4.8)
ν_{LT}	0.3	0.31
ν_{TT}	0.42[a]	0.461[a]
One-ply thickness, in. (mm)	0.008 (0.203)	0.005 (0.127)

[a] Assumed values.

TABLE 2—*Crack parameters.*

t/a	ζ (tension)		ψ (shear) Independent of Material Properties
	Glass/Epoxy	Graphite/Epoxy	
0.0	0.0000	0.0000	0.0000
0.1	0.0812	0.0828	0.0542
0.2	0.1673	0.1704	0.1086
0.3	0.2559	0.2603	0.1628
0.4	0.3415	0.3469	0.2169
0.5	0.4189	0.4247	0.2704
0.6	0.4857	0.4917	0.3224
0.8	0.5891	0.5948	0.4182
1.0	0.6613	0.6664	0.5000
∞	1	1	1

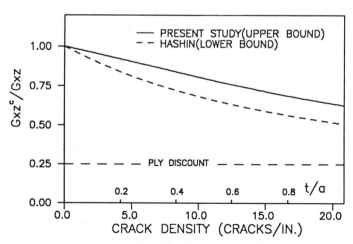

FIG. 6—*Reduced shear modulus of* [0/90₃]ₛ *glass/epoxy laminate.*

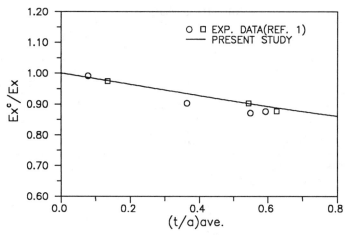

FIG. 7—*Reduced axial stiffness in* $[0/\pm45]_s$ *glass/epoxy laminate.*

$-45°$ layers are not the same, the arithmetic mean value is used for plotting the experimental data and for the model prediction. In this comparison, 0° layers are assumed to be undamaged.

As shown in Fig. 1 and Figs. 4 through 7, the present model gives reasonable upper bounds of axial and shear moduli degraded by matrix cracks. However, the present model encounters difficulty in predicting the effect of matrix cracks in the outermost layers of a laminate. Whenever there exist matrix cracks in the outermost layers, the main assumption that the layer interface is flat during deformation cannot be postulated for the outermost layers. However, the present model may be applicable if the matrix crack densities in the outermost layers are negligibly small compared to those in the interior layers. The analytical solutions to the crack parameters, ζ and ψ, which are the ratios of the internal state variables to the applied far-field strains, include the matrix crack interaction in an explicit form. Numerical values of the crack parameters are also presented for two different material systems as functions of undamaged material properties and the matrix crack length-distance ratio. The internal state variables for tension and shear loadings presented in this study are directly derived from the strain energy loss due to matrix cracks. Thus, by combining the present result with the study of Allen et al. [15], it becomes possible to analytically predict the strain energy release rate of a laminate at a given matrix damage state.

Conclusions

The upper bounds of reduced axial and shear stiffnesses due to matrix cracks are analytically predicted for a cross-ply laminate containing an infinite number of layers. The comparison of the present model predictions with experimental data and Hashin's lower bounds confirms the potential applicability of the present model to general angle-ply laminates even with a finite number of layers.

References

[1] Highsmith, A. L., Stinchomb, W. W., and Reifsnider, K. L., "Stiffness Reduction Resulting from Transverse Cracking in Fiber-Reinforced Composite Laminates," ESM Dept., Virginia Polytechnic Institute and State University, Blacksburg, VA, Nov. 1981.
[2] Highsmith, A. L. and Reifsnider, K. L., "Stiffness-Reduction Mechanisms in Composite Lami-

nates," *Damage in Composite Materials, ASTM STP 775,* K. L. Reifsnider, Ed., American Society for Testing and Materials, Philadelphia, 1982, pp. 103–117.

[3] Ogin, S. L., Smith, P. A., and Beaumont, P. W. R., "Matrix Cracking and Stiffness Reduction during the Fatigue of a [0/90]$_s$ GFRF Laminate," *Composite Science and Technology,* Vol. 22, 1985, pp. 22–31.

[4] Groves, S. E., "A Study of Damage Mechanics in Continuous Fiber Composite Laminates with Matrix Cracking and Interply Delaminations," PhD thesis, Texas A&M University, College Station, TX, 1986.

[5] Laws, N., Dvorak, G. J., and Hejazi, M., "Stiffness Changes in Unidirectional Composites Caused by Crack Systems," *Mechanics of Materials 2,* North Holland, Amsterdam, 1983, pp. 123–137.

[6] Dvorak, G. J., "Analysis of Progressive Matrix Cracking in Composite Laminates," AFOSR-82-0308, Rensselaer Polytechnic Institute, Troy, NY, March 1985.

[7] Hashin, Z., "Analysis of Cracked Laminates: A Variational Approach," *Mechanics of Materials 4,* North Holland, Amsterdam, The Netherlands, 1985, pp. 121–136.

[8] Hashin, Z., "Analysis of Orthogonally Cracked Laminates under Tension," *Journal of Applied Mechanics,* Vol. 54, Dec. 1987, pp. 872–879.

[9] Aboudi, J., "Stiffness Reduction of Cracked Solids," *Engineering Fracture Mechanics,* Vol. 26, No. 5, 1987, pp. 637–650.

[10] Talreja, R., "A Continuum Mechanics Characterization of Damage in Composite Materials," *Proceedings,* Royal Society, London, Vol. 399 A, June 1985.

[11] Talreja, R., "Transverse Cracking and Stiffness Reduction in Composite Laminates," *Journal of Composite Materials,* Vol. 19, 1985, pp. 355–375.

[12] Talreja, R., "Stiffness Properties of Composite Laminates with Matrix Cracking and Interior Delamination," *Engineering Fracture Mechanics,* Vol. 25, Nos. 5/6, 1986, pp. 751–762.

[13] Vakulenko, A. A. and Kachanov, M., "Continual Theory of a Medium with Cracks," Izv. AN SSSR, *Mekhanika Tverdogo Tela,* Vol. 6, 1971, pp. 159–166, in Russian.

[14] Allen, D. H., Harris, C. E., and Groves, S. E., "A Thermomechanical Constitutive Theory for Elastic Composites with Distributed Damage—Part I: Theoretical Development," *International Journal of Solids and Structures,* Vol. 23, No. 9, 1987, pp. 1301–1318.

[15] Allen, D. H., Harris, C. E., and Groves, S. E., "A Thermomechanical Constitutive Theory for Elastic Composites with Distributed Damage—Part II: Application to Matrix Cracking in Laminated Composites," *International Journal of Solids and Structures,* Vol. 23, No. 9, 1987, pp. 1319–1338.

[16] Allen, D. H., Groves, S. E., and Harris, C. E., "A Cumulative Damage Model for Continuous Fiber Composite Laminates with Matrix Cracking and Interply Delaminations," *Composite Materials: Testing and Design (Eighth Conference), ASTM STP 972,* J. D. Whitcomb, Ed., American Society for Testing and Materials, Philadelphia, 1988, pp. 57–80.

[17] Lee, J.-W., Allen, D. H., and Harris, C. E., "Internal State Variable Approach for Predicting Stiffness Reductions in Fibrous Laminated Composites with Matrix Cracks," *Journal of Composite Materials,* Vol. 23, No. 12, 1989, pp. 1273–1291.

Peter Davies,[1] Wesley J. Cantwell,[1] Pean-Yue Jar,[1] Hervé Richard,[1] David J. Neville,[2] and Hans-Henning Kausch[1]

Cooling Rate Effects in Carbon Fiber/PEEK Composites

REFERENCE: Davies, P., Cantwell, W. J., Jar, P-Y., Richard, H., Neville, D. J., and Kausch, H. H., "**Cooling Rate Effects in Carbon Fiber/PEEK Composites,**" *Composite Materials: Fatigue and Fracture (Third Volume), ASTM STP 1110,* T. K. O'Brien, Ed., American Society for Testing and Materials, Philadelphia, 1991, pp. 70–88.

ABSTRACT: This paper examines the influence of cooling rate after molding on short- and long-term properties of continuous carbon fiber reinforced polyetheretherketone (PEEK). First, changes to matrix structure and internal stress levels resulting from cooling panels at different rates are described for composites of PEEK with AS4 and IM6 fibers. The effects of these changes on short-term properties are illustrated by tension, compression, delamination resistance, and drop weight impact tests and results are compared with published data. Long-term properties are then discussed and results from creep loading, tension-tension fatigue, and Mode I delamination fatigue are presented. Both short- and long-term results indicate that the internal stresses induced during fast cooling are more detrimental to the performance of IM6 carbon fiber/PEEK composites than the changes in matrix structure observed at slow cooling rates.

KEY WORDS: carbon fiber, PEEK, cooling rates, crystallinity, internal stresses, delamination, impact, creep (materials), fatigue (materials), composite materials, fracture

Since the introduction of composites based on semicrystalline thermoplastic matrices, doubts have been expressed over the sensitivity of laminate properties to processing conditions. Two aspects in particular have provoked concern, the structure of the matrix and the possibility of the development of high residual stresses. Both of these effects will depend on the rate of cooling of the composite from the molding temperature, and the manufacturers of carbon fiber/PEEK (polyetheretherketone) prepreg have recommended rapid cooling from 380 to 200°C [1].

Slow cooling rates have been shown to increase degree of crystallinity in PEEK [2,3] and AS4/PEEK [4], to promote the growth of larger spherulitic structure in the composite [5], and to result in more extensive transcrystallinity [6]. These changes have been related to a drop in toughness of AS4/PEEK cooled more slowly than 3 K/min. Curtis et al. [7] observed a reduction in G_{Ic} and G_{IIc} between 3 K/min and 1 K/min, and their results are discussed later. Talbott et al. [8] examined faster cooling rates, above 10 K/min, and indicated lower delamination resistance when the rate was decreased. More recently, Vautey presented results showing a drop in G_{Ic} between 20 and 1 K/min, and also a lower impact resistance (larger delaminated area) in AS4/PEEK cooled at 1 K/min than for that cooled at 20 K/min [9]. Little work has been published for other fibers in PEEK nor for other semicrystalline matrix composites, but

[1] Research scientists and professor, respectively, Polymers Laboratory, Ecole Polytechnique Fédérale de Lausanne (EPFL), Switzerland.
[2] Research scientist, Asea Brown Boveri AG (ABB), Baden-Dättwil, Switzerland.

a drop in delamination resistance was recorded for AS4/PPS (poly(phenylene sulfide)) composites after an annealing treatment that increased their degree of crystallinity [10].

The second parameter that may vary with cooling rate is the internal stress level. The high forming temperatures of PEEK composites and the volume shrinkage measurements of Zoller [11] indicate that high stresses may be developed in carbon fiber/PEEK laminates. Even in a unidirectional panel, stresses may be set up due to the difference between fiber and matrix properites and these may be significant at fast cooling rates. Chapman et al. have recently studied residual stress development, modeling skin/core residual stresses in unidirectional AS4/PEEK cooled very rapidly (35 K/s) [12]. However, the residual stresses considered in the present paper are principally those that develop in multidirectional laminates at much slower cooling rates, due to the difference in the properties of adjacent plies. In this case, Jeronimidis and Parkyn have shown that residual stresses up to half the 90° tensile strength may develop in 90° plies in 0/90° AS4/PEEK laminates [13]. These authors reported similar residual curvatures of unbalanced laminates after cooling at rates of 3, 40, and 1000 K/min, but did not discuss cooling rates slower than 3 K/min. The influence of residual stresses on properties of thermoplastic matrix composites has received little attention to date, but such large stresses may cause premature tensile failure, and may also affect fiber alignment and compressive properties.

While some data are available for short-term properties, very little work has been published describing cooling rate effects on long-term behavior, apart from the results of Curtis suggesting a much improved fatigue performance for $((+/-45, -/+45)_2)_s$ AS4/PEEK specimens cooled at 20 K/min than for those cooled more slowly (at 3 K/min) [14]. This indicated an effect of matrix structure, consistent with published fracture toughness trends [7] but occurring at a faster cooling rate than the drop in static G_{Ic} and G_{IIc} values (which was only apparent at 1 K/min). However, more recent work by O'Brien [15] on standard (fast-cooled) AS4/PEEK has highlighted the great influence that residual stress may have on static and dynamic delamination resistance of multidirectional $(35_n/-35_n/0_n/90_n)_s$ layups. One of the aims of the current project was to generate more data to clarify these effects.

The results are presented in four parts. In the first part, cooling rate effects on matrix structure and internal stress levels are described for PEEK composites reinforced with IM6 and AS4 fibers. In the second part, short-term tensile and compression results are given. The third section deals with the influence of cooling rate on fracture, with results from delamination and impact tests. Finally, in the fourth part, results from creep and fatigue tests are presented.

Materials and Methods

All specimens tested in this work were cut from panels molded in-house from ICI prepreg (0.125 mm thick) in a picture frame mold, following the suppliers' recommendations, except for the cooling rate after consolidation at 380°C. This was varied either by transferring the mold to a second heated press or, for slow cooling, by programming the desired cooling rate on a programmable press. The cooling rates of panels were monitored by thermocouples inserted in the mold. The rates quoted in this paper are average rates over the range from 380 to 200°C.

The majority of the tests described here were performed on composites reinforced with IM6 fibers, but as most previously-published work has been concerned with AS4 fiber composites, some tests were also carried out on AS4/PEEK. Fiber volume fraction was nominally 61% throughout.

Degree of crystallinity was estimated by differential scanning calorimetry (DSC), using a Perkin-Elmer DSC4, at a heating rate of 20 K/min. Internal stress levels were compared by molding 8-ply nonsymmetrical $(0_4 90_4)$ panels, cutting them into 20 mm-wide strips and measuring their radii of curvature [16].

Tension testing of 8-ply 1-mm-thick 90° unidirectional, 8-ply $+/-45°$ symmetrical, and 16-ply 2-mm-thick $(0/90/+/-45)_{2s}$ specimens was performed in a temperature chamber on a Zwick 1484 universal test machine, at a crosshead speed of 1 mm/min. Specimen dimensions were 15 mm by 140 mm by thickness. Strains were measured using an extensometer or strain gages. Creep loading at 60, 70, 80, and 90% of the short-term strength was also carried out on this machine. Tension-tension fatigue tests were performed on parallel-sided 2-mm-thick $(+/-30)_{4s}$ specimens, under load control, for a load ratio = 0.1 (minimum load/maximum load). Specimen surface temperature was measured using an infrared thermometer and frequency was chosen to keep surface temperature below 35°C. This required frequencies of 1 to 3 Hz according to the load level applied. Initial tests on $+/-45°$ specimens were not pursued due to significant heating effects even at these frequencies.

Compression tests were performed at 5 mm/min on 5-mm-thick $(0/90/+/-45)_{5s}$ specimens with a 10-mm gage length between the aluminum end tabs, using the supporting fixture described in ASTM Test Method for Compressive Properties of Rigid Plastics (D 695-80).

Double cantilever beam (DCB) and end notch flexure (ENF) specimens were tested to measure Mode I and Mode II delamination resistance, respectively. These 24-ply unidirectional specimens were 20 mm wide, and between four and eight were tested for each condition. Crosshead rates of 2 and 1 mm/min were used for tests on DCB and ENF specimens. Mode I initiation values were measured from a 12-μm-thick aluminum foil starter film-coated with release agent and inserted at mid thickness before molding. G_{Ic} values were determined using an experimental compliance calibration [17] of the form $C = Ka^n$ (where C = compliance; a = measured crack length; and K,n are experimental curve fitting parameters). G_{Ic} was calculated as

$$G_{Ic} = \frac{nP\delta}{2Ba}$$

where P = load, δ = displacement, and B = specimen width.

G_{IIc} was determined using the expression [18]

$$G_{IIc} = \frac{9P\delta a^2}{(2B(2L^3 + 3a^3))}$$

where L is half the distance between supports, 40 mm in this case. Values were measured from short Mode I precracks. The Mode I and Mode II specimen geometries were selected after a preliminary series of tests in which geometry was varied. More details of these tests are available elsewhere [19].

Mode I delamination fatigue tests were performed on specimens of the same geometry, under displacement control at $R = 0.1$ (minimum/maximum displacement), at a frequency of 5 Hz. Initial defect length was 50 mm in all cases. Initiation was detected by the change in compliance as discussed later, as load was continuously followed and a clear compliance increase was apparent. This change was confirmed as corresponding to initiation by sectioning specimens. At this frequency, it was not possible to detect initiation by eye and the compliance change criterion was used both in static and fatigue tests.

Impact damage was introduced into 2-mm-thick $(0/90/+/-45)_{2s}$ panels, using a Ceast instrumented drop weight impact tester with a 12.7-mm-diameter hemispherical indenter. Panels were clamped and the unsupported area was 76 by 126 mm^2. Damage was detected both using Krautkrämer-Branson USIP12 ultrasonic C-scan equipment and by taking sections through damaged specimens. Compression strength after impact was measured using an anti-buckling guide, on 40-mm-wide specimens cut from the impacted panels.

Thin sections ($<10~\mu$m) were prepared by polishing as described elsewhere [5], and examined in transmitted light between crossed polarizers.

Cooling Rate Effects

The two principal cooling rate effects examined here are matrix structure and internal stress level. Characterization of matrix structure is extremely difficult, as within a single specimen a number of morphologies may be found. The structure of PEEK may be described on different scales, from molecular to lamellar up to spherulitic, and the most frequently used means of quantifying the order in the matrix is the degree of crystallinity. In Fig. 1, this parameter, measured by DSC, is shown as a function of cooling rate, and similar increases at slower cooling rates to those reported elsewhere were measured. It may be noted that changing the fiber type does not significantly affect the values, but this does not preclude morphological differences between the two cases. For example, the IM6 fiber is smaller than the AS4 type (5.5 μm diameter rather than 7 μm), which will result in an increased fiber surface for the same volume fraction and may affect crystal nucleation. A large number of polished thin sections have therefore been studied in transmitted polarized light in the optical microscope. Figure 2 indicates some of the features that occur; in this case, for a sample prepared to maximize crystal structure effects. This specimen was held above 400°C for 30 min, cooled in the oven to 320°C at 1 K/min, and held at that temperature for 3 h. Very large spherulites and transcrystalline growth on fibers are apparent, and these are two features that were observed in slow-cooled composite specimens. However, no significant differences in structure have been observed between sections from panels reinforced with the different fibers cooled at the same rate. For composites of both fibers, larger spherulites were seen for cooling rates of 3 K/min and slower, as reported elsewhere [5], and some regions of transcrystalline growth could be found in most samples.

Nonsymmetric $(0_4 90_4)$ laminates were molded in order to visualize the dependence of the

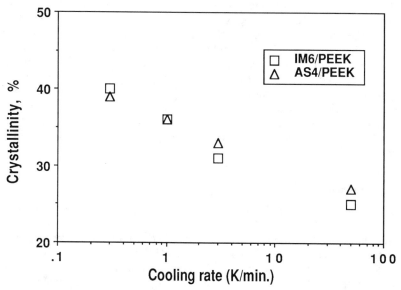

FIG. 1—*Degree of crystallinity (by DSC) as a function of cooling rate for AS4/PEEK and IM6/PEEK.*

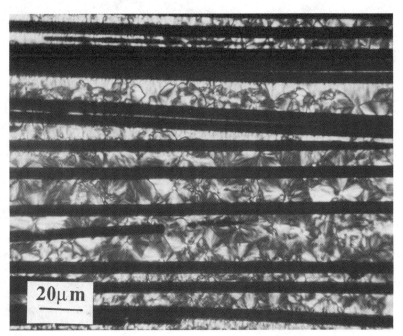

FIG. 2—*Thin section of IM6/PEEK observed in optical microscope between crossed polarizers. Section parallel to fiber taken from sample prepared by heating above 400°C for 30 min, cooling to 320°C at 1 K/ min, and holding for 3 h.*

level of internal stress on cooling rate. Figure 3 shows larger curvature for the slowest-cooled panels, as stresses can relax out during slower cooling. There is a clear difference between the IM6/PEEK panel cooled slowly and those cooled faster than 10 K/min. The results for AS4/ PEEK are also shown but given the high scatter in the data no effect of cooling rate effect could be concluded here for this material. Published values for the same AS4/PEEK layup indicated rather higher internal stresses (smaller radius of curvature) for a panel cooled at 3 K/min [13]. The curvature of nonsymmetric cross-ply laminates has been shown from bimetallic strip theory to depend on the stress-free temperature and on longitudinal and transverse thermal expansion coefficients and moduli [20]. The IM6 fibers are 30% stiffer than AS4 fibers and, according to the fiber suppliers, the thermal expansion coefficients of the two are similar [21]. Therefore, if the longitudinal stiffness is the only difference between the two fibers, then from simple calculations, the radius of curvature of IM6/PEEK would be expected to be larger than that of AS4/PEEK.

Short-Term Properties of IM6/PEEK

Tensile and compressive strengths of different layups, for cooling rates of 50 and either 1 or 0.6 K/min, are presented in Figs. 4 to 6.

For the unidirectional transverse tensile strengths (Fig. 4), no effect is observed except above the glass transition temperature of the PEEK matrix (145°C). Strains were not measured but load-crosshead displacement plots were very similar for both rates.

Similarly, for the +/−45° laminates, failure stresses are not affected by cooling rate (Fig. 5) but, in this case, the failure strains are significantly lower for slower-cooled specimens. Such

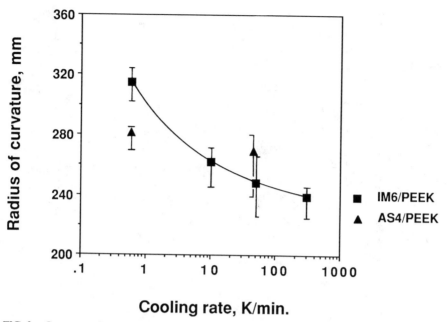

FIG. 3—*Curvature of nonsymmetrical $(0_4 90_4)$ panels after cooling at different rates. Error bars indicate maximum and minimum values.*

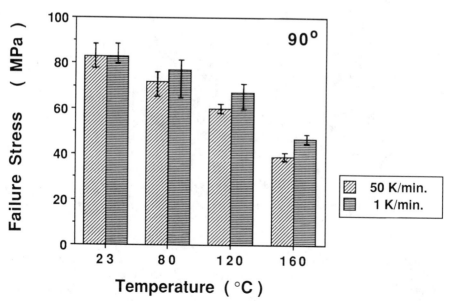

FIG. 4—*Effect of cooling rate on tensile strength of 90° IM6/PEEK specimens as a function of temperature. Error bars indicate maximum and minimum values.*

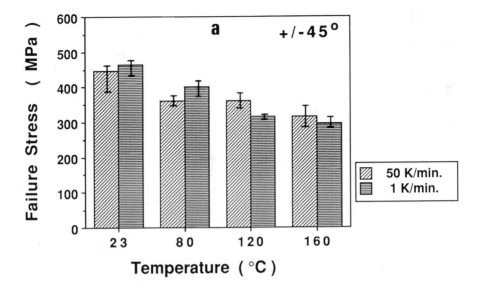

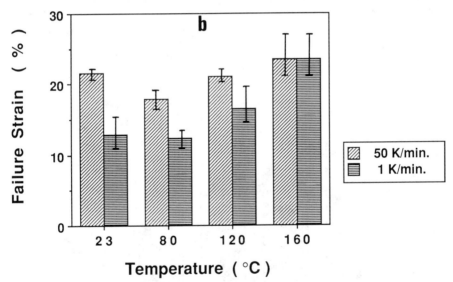

FIG. 5—*Effect of cooling rate on (a) tensile strength and (b) failure strain of +/−45° specimens as a function of temperature. Error bars indicate maximum and minimum values.*

reductions have been reported previously [7,22]. They could be due to matrix structure effects in the slower-cooled material as suggested in Ref 7 for AS4/PEEK, but in the present case much greater volume damage was noted in sections taken from failed fast-cooled specimens, with cracks throughout the specimens [22]. These cracks result in higher strains to failure but the failure stresses are not affected due to the notch insensitivity of this layup. Additional tests

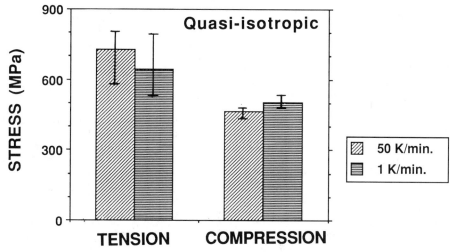

FIG. 6—*Effect of cooling rate on tensile and compression strengths of quasi-isotropic IM6/PEEK specimens. Error bars indicate maximum and minimum values.*

on a panel, of which half was annealed at 270°C for 4 h before testing and half was tested in the as-molded state, suggest that it is the internal stress state that causes the difference in failure strains. The annealed specimens showed lower failure strains than those that had not been annealed, and the difference in degree of crystallinity between the two halves of the panel was small [22].

Results from tension tests on quasi-isotropic specimens show considerable scatter (Fig. 6) but no clear difference between the two cooling rates. In compression there is less scatter in values retained, that is, for those specimens that failed in the gage length, and slower-cooled specimens give slightly higher values (Fig. 6). This may be due to fiber misalignment, as more fiber kinking was observed on the surfaces of faster-cooled panels.

Fracture Behavior of AS4/PEEK and IM6/PEEK

In order to assess whether delamination resistance of AS4/PEEK composites was affected by cooling rate, Curtis et al. used Mode I and Mode II tests [7]. Their Mode I results are replotted in Fig. 7a and show a drop in the value of G_{Ic} corresponding to the onset of instability in passing from material cooled at 3 K/min to 1 K/min. For evaluation, it was necessary to use instability and arrest values as propagation proceeded in unstable jumps. These results were obtained at the Université de Compiègne in France, on 3-mm-thick pre-production material produced at Royal Aircraft Establishment (RAE) Farnborough, United Kingdom, in 1986. Results from other tests performed at Compiègne on 5-mm-thick AS4/PEEK molded a few months later by Imperial Chemical Industries (ICI) were presented recently [9] and these are also plotted on Fig. 7a. These again showed unstable propagation and lower values for specimens cooled at 1 K/min than at 20 K/min. Values are slightly higher than those measured by Curtis et al., and this is believed to be a specimen stiffness effect. Higher G_{Ic} values are frequently obtained on stiffer specimens due to larger fiber bridging and multiple cracking contributions [23].

In 1988, we repeated these tests on 3-mm-thick specimens molded from AS4/PEEK production prepreg, and the results are shown in Fig. 7b. Propagation is now completely stable except at 0.3 K/min.

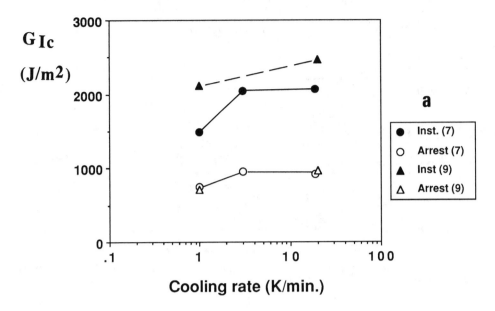

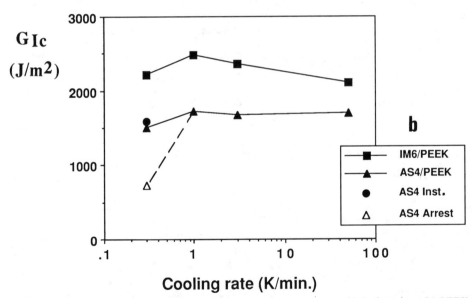

FIG. 7—*Effect of cooling rate on Mode I delamination resistance:* (a) *published results, AS4/PEEK; and* (b) *current work, AS4/PEEK and IM6/PEEK.*

Mode I tests on IM6/PEEK showed no cooling rate dependence (Fig. 7b). It was believed at first that any effect on propagation values might be concealed by an inhomogeneous fiber distribution in this composite [23], but a subsequent study of initiation values also showed no effect of cooling rate. The G_{Ic} values between 1100 and 1500 J/m² were found in both cases. It therefore appears that the matrix in IM6/PEEK composites is even less sensitive to cooling rate than that in the AS4/PEEK material.

Results from Mode II tests are more difficult to interpret as the appearance of nonlinear load-displacement behavior before the maximum load point is open to interpretation [24]. Early results in Fig. 8a were calculated using the maximum load point, whereas for more recent tests, G_{IIc} values corresponding to both the onset of nonlinearity and the maximum load are given. These latter results (Fig. 8b) show no major influence of cooling rate for composites

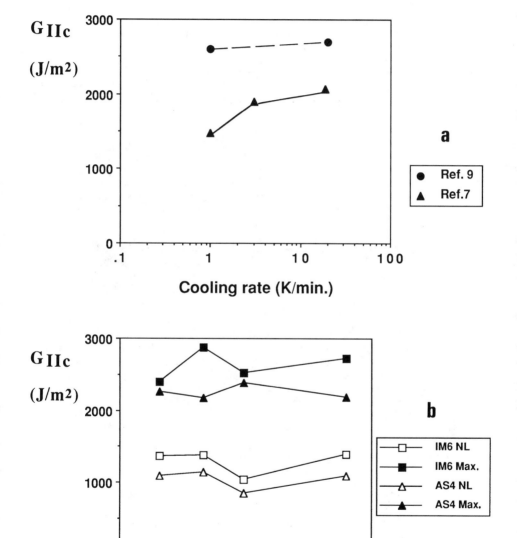

FIG. 8—*Effect of cooling rate on Mode II delamination resistance: (a) published results, AS4/PEEK; and (b) current work, AS4/PEEK and IM6/PEEK.*

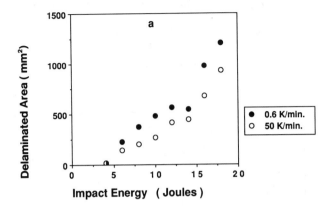

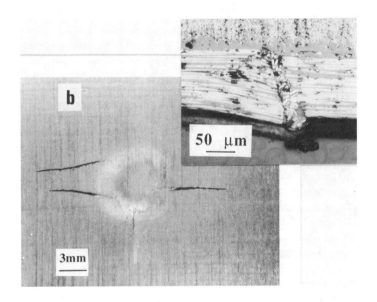

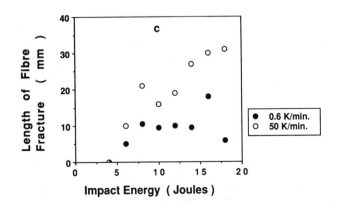

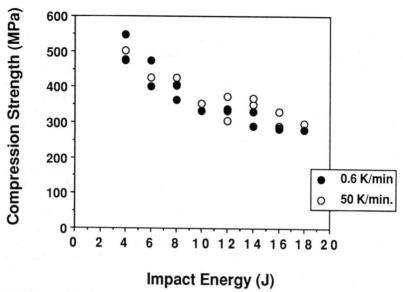

FIG. 10—*Residual compression strength after impact of specimens in Fig. 9, IM6/PEEK.*

of either fiber even at the slowest cooling rate, although a slight trend in the values at the onset of nonlinearity is discussed later.

In order to determine whether the impact resistance of multidirectional layups reflected the insensitivity of the delamination resistance of unidirectional IM6/PEEK to cooling rate, a series of low-velocity impact tests was performed. The results are shown in Fig. 9a where the delaminated area, detected by ultrasonic C-scan, is plotted as a function of incident impact energy. It is apparent that the delaminated area is larger in the slower-cooled panel, and this is in agreement with the results of Vautey [9] on AS4/PEEK. Sections taken from specimens impacted at 10 J showed that damage was confined to delamination at the back surface 0/90° interface and cracking in the corresponding 90° layer. However, this is only part of the story as visual inspection of the impacted faces of panels revealed significant differences. For all impact energy levels, a top surface indentation was apparent but there were also fiber compression failures visible on the impacted surfaces (Fig. 9b). A similar failure mode has been noted by Dorey [25]. In Fig. 9c, the total length of these failures is plotted against impact energy and it is clear that they are far more extensive in the faster-cooled panels. For these panels, a large part of the available impact energy is therefore used in fiber fracture rather than delamination, so the presentation of delaminated area results alone is misleading.

This is underlined by the results from compression after impact tests (Fig. 10) that show no clear difference between residual strengths of fast- and slow-cooled specimens in spite of the larger delaminated areas of the latter.

FIG. 9—*Effect of cooling rate on impact behavior of IM6/PEEK: (a) delaminated area detected by C-scan. Each point is the mean of two values; (b) fiber compression failures on impacted face of fast-cooled panel (50 K/min), 16 J impact. General view of surface and inset micrograph showing fiber buckling failure in top 0° ply; and (c) total length of impacted face fiber fractures, each point mean of two values.*

Long-Term Properties of IM6/PEEK

In a preliminary series of tests to examine the cooling rate dependence of creep behavior of IM6/PEEK, $(90)_{4s}$ and $(+/-45)_{2s}$ specimens were loaded to 60, 70, 80, and 90% of their ultimate tensile strengths, and the strains measures after 1000 and 10 000 s are plotted in Figs. 11 and 12. These figures are consistent with the short-term tensile data (Figs. 4 and 5b) in showing no cooling rate effect at 23°C for the $(90)_{4s}$ materials, but a significantly higher strain in the faster-cooled $(+/-45)_{2s}$ layup.

Delamination fatigue tests were performed on unidirectional specimens in order to investigate the influence of structure on long-term fracture behavior (Fig. 13a). The range of static G_{Ic} values measured on these panels is shown on the y-axis, and similar static and fatigue values were obtained for fast- and slow-cooled specimens. The determination of threshold values for delamination initiation under fatigue conditions was proposed by O'Brien et al. recently [26] to avoid the problems associated with delamination propagation values. However, one problem in comparing data from this type of test is the definition of an initiation criterion. Martin and Murri suggested a "1 to 2% decrease in the load" as corresponding to initiation [27], and in Fig. 13b different criteria are shown on one of the load versus number of cycles plots recorded. The only published Mode I data of this kind available for comparison are those of Martin and Murri [27]. Their results for AS4/PEEK, tested under the same conditions (5 Hz, $R = 0.1$) but with a thicker starter defect, are plotted in Fig. 13c for comparison, and results for IM6/PEEK using the different criteria defined in Fig. 13b are also plotted.

The results from tension-tension fatigue tests on $(+/-30)_{4s}$ specimens are shown in Fig. 14. Four panels were tested and results from all specimens are shown. The results from the faster-cooled panel fall within the scatter band for results from the slower-cooled material, albeit toward the lower limit of this band. These results are discussed later.

Discussion

The effects of matrix structure will be considered first by examining results from tests on unidirectional material, and then residual stress effects and multidirectional laminates will be discussed.

The objective of the first part of this work was to establish what changes are induced in IM6 and AS4/PEEK composites when the cooling rates employed are slower than those recommended. Figure 1 shows that the degree of crystallinity increases, as reported previously, and that this parameter is not affected by fiber type, but the degree of crystallinity is a global parameter and gives no insight into morphological variations. Optical microscopy was therefore employed and indicated a trend of increasing spherulite size at slower cooling rate and distinctly larger spherulites in material reinforced with both types of fiber cooled at rates of 3 K/min and slower. However, despite the preparation of a very large number of thin sections, both perpendicular to and parallel to the fiber direction, no clear difference was observed between AS4/PEEK panels cooled at 1 and 0.3 K/min, nor between the IM6/PEEK panels cooled at these rates. Nevertheless, the Mode I fracture toughness tests (Fig. 7b) indicate a change in propagation mode for the slowest-cooled AS4/PEEK tests, from stable to unstable, which was not observed for the slowest-cooled IM6/PEEK.

Unstable propagation in carbon/PEEK composites was first noted by Carlile and Leach [28], but the causes of the instability are not well understood. Propagation values of G_{Ic} are related to fiber properties and Lang et al. have suggested that values will increase with the minimum radius of curvature of the fiber [29]. Indeed, values presented in Fig. 7 for the smaller-diameter IM6-based composites are higher than those for AS4/PEEK. Assuming that the matrix formulations are the same in the two prepregs then changing to a smaller diameter,

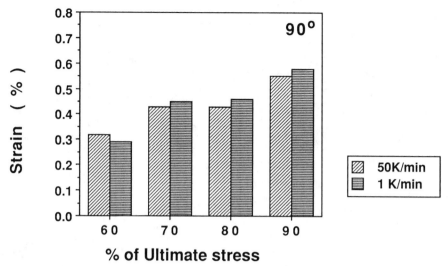

FIG. 11—*Effect of cooling rate on 1000-s creep strains for applied loads corresponding to different percentages of ultimate tensile strength, 90° specimens, IM6/PEEK, 23°C.*

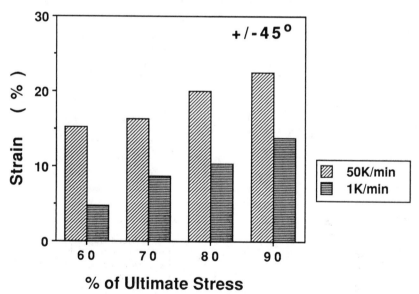

FIG. 12—*Effect of cooling rate on 10 000-s creep strains for applied loads corresponding to different percentages of ultimate tensile strength, +/−45° specimens, IM6/PEEK, 23°C.*

stiffer fiber may also result in changes to the interphase region near the fiber and to the stress field near the crack tip, so that the change from stable to unstable propagation occurs at a cooling rate slower than 0.3 K/min. Unfortunately initiation values are not available for AS4/PEEK as a function of cooling rate, which might have clarified the role of the matrix, as the aluminum foil did not release from the precrack and unstable jumps occurred at initiation in the slower-cooled panels.

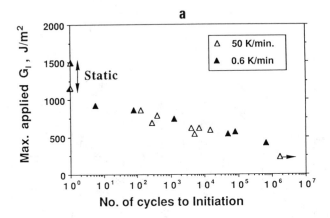

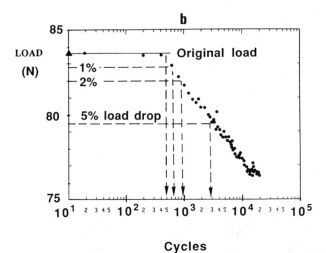

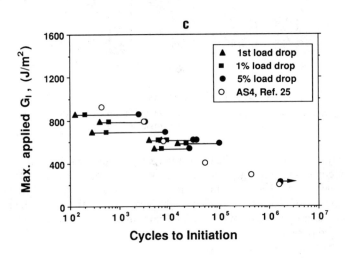

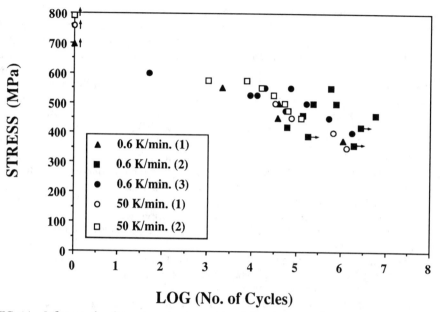

FIG. 14—*Influence of cooling rate on tension-tension fatigue behavior of +/− 30° IM6/PEEK speci-mens, R = 0.1, frequency 1 to 3 Hz; 0.6 K/min specimens taken from three panels; 50 K/min specimens from two panels. Static values on y-axis correspond to failure of tabs, not specimen. Arrows indicate unbro-ken specimens.*

In spite of this anomaly, as far as the short- and long-term tests described here are concerned, it is clear that the Mode I interlaminar fracture behavior of IM6/PEEK is not sensitive to the considerable changes in matrix structure observed to take place over the range from 50 to 0.3 K/min. In addition, for practical purposes the G_{Ic} values for current production AS4/PEEK are still high even at the extremely slow cooling rate of 0.3 K/min.

The Mode II results are also largely independent of cooling rate but it is interesting to note the similarity in the cooling rate dependence of the values at the onset of nonlinearity in Fig. 8b. For both fibers, a drop occurs at 3 K/min that corresponds to the cooling rate at which larger spherulites are observed in both composites. The appearance of nonlinear behavior is believed to correspond to the development of shear microcracks that have been observed during in-situ tests on these materials in the scanning electron microscope [30]. It may be that at this cooling rate the matrix structure is particularly susceptible to microcrack initiation, between spherulites, for example, and in-situ tests are now underway to examine this possibility.

The Mode I delamination fatigue results (Fig. 13a) give an indication of the effect of cooling rate on crack initiation from a thin defect under repeated Mode I loading in a unidirectional laminate. No effect is observed, which confirms the insensitivity of this phenomenon to cool-

←

FIG. 13—*Influence of cooling rate on initiation of a delamination under Mode I fatigue loading, 5 Hz, displacement control ($\delta_{min}/\delta_{max}$ = 0.1): (a) present data, IM6/PEEK, initiation taken as first load drop; (b) recording of load versus number of cycles for G_{max} = 790 J/m², 50 K/min, showing range of lifetimes for different initiation criteria; and (c) comparison of behavior of IM6/PEEK (50 K/min) according to different criteria, with published results for AS4/PEEK (Note that the latter correspond to initiation from a 125-μm-thick defect).*

ing rate noted previously. The threshold value of G_{max} for initiation appears to be higher than that of 120 J/m^2 reported for AS4/PEEK under the same loading conditions elsewhere [27], but the aim here was to compare the two materials over a range of ΔG levels so few of the longer tests that would have enabled the threshold to be accurately determined were performed. Figure 13c shows the influence of the initiation criterion employed, and the results for composites with AS4 fibers fall within the 1 to 5% load drop range for those with IM6 fibers. From the static compliance calibration, using a value of $n = 2.7$, the crack advances, corresponding to load drops of 1 and 2%, can be estimated as 0.2 and 0.4 mm, respectively. A drop of 5% will be roughly equivalent to 1 mm of propagation. However, given the R-curve effect in tough composites, with increasing resistance to crack growth as the crack advances, it seems sensible to employ the first change in compliance as a criterion for initiation. An additional consideration in this type of test is the self-toughening mechanism that has been noted for IM6/PEEK, whose initiation G_{Ic}, increases with time under load [31]. This has not been taken into account, but the most delicate part of these tests is the preparation of the starter defect, which must be carefully opened before the test without precracking the specimen. In spite of these problems, it is clear that initiation under fatigue loading can occur at ΔG values well below half of the static initiation G_{Ic} value.

The short- and long-term 90° tests on unidirectional IM6/PEEK also show no cooling rate dependence (Figs. 4 and 11) so that it may be concluded that the matrix structure does not significantly affect the properties of IM6/PEEK composites.

If stresses set up during cooling are now considered, one manifestation of these stresses is in fiber kinking, which is more evident on the surfaces of fast-cooled panels. These kinks develop due to the large shrinkage of the matrix relative to the fibers that are subjected to residual compressive strains [32]. There is a slight drop in compression strength of faster-cooled quasi-isotropic laminates (Fig. 6) but the main influence of the straightness of fibers can be seen on impact behavior, in Fig. 9. Low-velocity impact has caused both fiber fracture and delamination so that no significant differences in compression strengths after impact were determined for the fast- and slow-cooled materials. This phenomenon is of importance as Mode I and Mode II delamination tests are frequently employed to give an idea of the damage tolerance of a composite. Here, Mode I, Mode II, and compression after impact tests all show no effect of cooling rate, so that a good correlation between the tests is apparent, as has been reported elsewhere for other materials [33]. However, this apparent correlation gives no indication of the difference in failure modes in the fast- and slow-cooled materials, and the more extensive fiber fracture in the former is likely to be more detrimental to residual tensile properties.

In addition to the influence of fast cooling on fiber straightness, residual stresses develop between plies in multidirectional laminates during cooling. The existence of residual stresses in multidirectional composites was demonstrated by the curvature of nonsymmetric laminates in Fig. 3. Stress levels in the IM6/PEEK composites are strongly dependent on cooling rate and these and other published results suggest that stresses in AS4/PEEK are similar or even higher [13]. Their effect on $+/-45°$ IM6/PEEK specimens is shown for short- and long-term loading in Figs. 5 and 12, both of which indicate significantly higher strains in the faster-cooled specimens caused by more volume damage than in those cooled slowly. Few authors have presented creep data for comparison with the results presented here. Comparisons of unreinforced PEEK with an epoxy resin [5] and AS4/PEEK with a carbon/epoxy system [34] have indicated the promising creep behavior of the thermoplastic composite, although the results of Hiel [35] indicate that the properties of the composite at elevated temperature may not be as good as those of commercial PEEK grades. More work is required to fully characterize this effect.

Finally, the tension-tension fatigue test results presented in Fig. 14 are the first results from a comprehensive program investigating the influence of loading conditions and layup on fatigue behavior, and a more detailed study will be reported later. Nevertheless, it is apparent that these results do not agree with the previously-published data on $+/-45°$ laminates [14] where considerably longer fatigue lives were recorded for specimens cooled at 20 rather than 3 K/min. However, those tests were performed on preproduction AS4/PEEK whose properties were shown to be more sensitive to cooling rate. The slightly lower fatigue lives of faster-cooled specimens may reflect the high internal stress levels discussed elsewhere [15], but more tests are clearly necessary.

Overall, the IM6/PEEK tested here showed little sensitivity to cooling rate. The effects that were observed are believed to be primarily due to residual stress effects and the performance of this composite could be improved by cooling slowly after molding. There remains considerable room for optimization of molding cycles and further work is required to look at the effects of different annealing treatments. From the limited comparative tests performed on AS4/PEEK, it is probable that some optimization of the properties of this material is also possible.

Conclusions

Slow cooling after molding of carbon fiber/PEEK laminates has been shown to increase the degree of crystallinity of the matrix and to reduce the level of internal stresses in multidirectional laminates. The influence of cooling rate on properties may be summarized as follows.

Results from matrix dominated tests on unidirectional specimens fabricated from current production prepreg show little influence of cooling rate.

Tests on multidirectional laminates under both static and fatigue loading conditions indicate equivalent or superior performance of specimens from slower-cooled panels.

Based on these tests on current production material, it appears that previous concern over the detrimental effects of the more highly crystalline matrix structure on composite properties is no longer justified. Slower cooling rates or annealing cycles could therefore be used, with beneficial effects on short- and long-term properties.

Acknowledgments

This project was funded by the Swiss Fonds National, under Materials Programme Project NFP19. The authors gratefully acknowledge the assistance of Ciba Geigy Composites Research, Marly, with the impact testing; Brain Senior of the Interdepartmental Microscopy Institute of the EPFL for assistance with microscopy; and Philippe Beguelin of the Polymers Laboratory, EPFL, for help in developing the delamination fatigue test procedure. The AS4/PEEK prepreg was kindly supplied by Dr. D. R. Carlile (ICI Wilton).

References

[1] ICI plc product data sheet.
[2] Velisaris, C. N. and Seferis, J. C., *Polymer Engineering and Science,* Vol. 26, No. 22, Dec. 1986, pp. 1574–1581.
[3] Kumar, S., Anderson, D. P., and Adams, W. W., *Polymer,* Vol. 27, March 1986, pp. 329–336.
[4] Blundell, D. J., Chalmers, J. M., Mackenzie, M. W., and Gaskin, W. F., *SAMPE Quarterly,* Society for the Advancement of Materials and Process Engineering, Vol. 16, No. 4, July 1985, pp. 22–30.
[5] Partridge, I. K., Davies, P., Parker, D. S., and Yee, A. F., "Yield and Fracture in PES and PEEK Matrix Polymers and Their Composites," *Proceedings,* International Conference on Polymers for Composites, Solihull, UK, The Plastics and Rubber Institute, London, 3–4 Dec. 1987, p. 5/1.
[6] Tung, C. M. and Dynes, P. J., *Journal of Applied Polymer Science,* Vol. 33, 1987, pp. 505–520.
[7] Curtis, P. T., Davies, P., Partridge, I. K., and Sainty, J-P., *Proceedings,* Sixth International Conference on Composite Materials (ICCM6/ECCM2), London, July 1987, Elsevier Applied Science Publishers, London, Vol. 4, pp. 401–412.
[8] Talbott, M. F., Springer, G. S., and Berglund, L. A., *Journal of Composite Materials,* Vol. 21, Nov. 1987, pp. 1056–1081.

[9] Vautey, P., *SAMPE Quarterly,* Society for the Advancement of Materials and Process Engineering, Vol. 21, Jan. 1990, pp. 23–28.

[10] Davies, P., Benzeggagh, M. L., and de Charentenay, F. X., *SAMPE Quarterly,* Society for the Advancement of Materials and Process Engineering, Vol. 19, No. 1, Oct. 1987, pp. 19–24.

[11] Zoller, P., *Proceedings,* American Society for Composites Third Technical Conference, Technomic Publishers, Lancaster, PA, Sept. 1988, pp. 439–448.

[12] Chapman, T. J., Gillespie, J. W., Jr., Manson J-A. E., Pipes, R. B., and Seferis, J. C., *Proceedings,* American Society for Composites Third Technical Conference, Technomic Publishers, Lancaster, PA, Sept. 1988, pp. 449–458.

[13] Jeronimidis, G. and Parkyn, A. T., *Journal of Composite Materials,* Vol. 22, May 1988, pp. 404–415.

[14] Curtis, P. T., *Proceedings,* Sixth International Conference on Composite Materials (ICCM6/ECCM2), London, July 1987, Elsevier Applied Science Publishers, Vol. 4, pp. 54–64.

[15] O'Brien, T. K., *Journal of Reinforced Plastics and Composites,* Vol. 7, July 1988, pp. 341–359.

[16] Bailey, J. E., Curtis, P. T., and Parvizi, A., *Proceedings,* Royal Society, London, A 366, 1979, pp. 599–623.

[17] de Charentenay, F. X., Harry, J. M., Prel, Y. J., and Benzeggagh, M. L. in *Effects of Defects in Composite Materials, ASTM STP 836,* American Society for Testing and Materials, Philadelphia, 1984, p. 84.

[18] Russell, A. J. and Street, K. N. in *Progress in Science and Engineering of Composites* (ICCM4), T. Hayashi, K. Kawata, and S. Umekawa, Eds., Japan Society of Composite Materials, Tokyo, 1982, p. 279.

[19] Davies, P., Cantwell, W., Richard, H., Moulin, C., and Kausch, H. H., *Proceedings,* Third European Conference on Composite Materials, Bordeaux France, March 1989, Elsevier Applied Science Publishers, London, pp. 747–755.

[20] Nairn, J. A. and Zoller, P. in *Toughened Composites, ASTM STP 937,* American Society for Testing and Materials, Philadelphia, 1987, pp. 328–341.

[21] Hercules France, S. A., private communication.

[22] Cantwell, W. J., Davies, P., and Kausch, H. H., *Composite Structures,* Vol. 14, 1990, pp. 151–171.

[23] Davies, P., Cantwell, W., Moulin, C., and Kausch, H. H., *Composite Science and Technology,* Vol. 36, 1989, pp. 153–166.

[24] Carlsson, L. A., Gillespie, J. W., and Trethewey, B. R., *Journal of Reinforced Plastics and Composites,* Vol. 5, July 1986, pp. 170–187.

[25] Dorey, G., in *Structural Impact and Crashworthiness,* G. A. O. Davies, Ed., Elsevier Applied Science Publishers, London, Vol. 1, 1984, pp. 155–192.

[26] O'Brien, T. K., Murri, G. B., and Salpekar, S. A. in *Composite Materials: Fatigue and Fracture (Second Volume), ASTM STP 1012,* American Society for Testing and Materials, Philadelphia, 1989, pp. 222–250.

[27] Martin, R. H. and Murri, G. B., "Characterization of Mode I and Mode II Delamination Growth and Thresholds in Graphite/PEEK Composites," NASA Technical Memo 100577, NASA Langley Research Center, Hampton, VA, April 1988.

[28] Carlile, D. R. and Leach, D. C., *Proceedings,* SAMPE Technical Conference, Society for the Advancement of Materials and Process Engineers, Oct. 1983, pp. 82–93.

[29] Lang, R. W., Heym, M., Tesch, H., and Stutz, H., *Proceedings,* Seventh European SAMPE Conference, Society for the Advancement of Materials and Process Engineers, K. Brunsch et al., Eds., Elsevier Science Publishers, Amsterdam, 1986, pp. 261–272.

[30] Senior, B., Davies, P., Cantwell, W., and Kausch, H., "In-Situ Fractography of Fiber Reinforced Composites," presented at EUREM 88, York, England, 1988, Institute of Physics Conference Series No. 93, Vol. 2, IOP Publishing Ltd., pp. 441–442.

[31] Davies, P., Cantwell, W., and Kausch, H. H., "Measurement of Initiation Values of G_{Ic} in IM6/PEEK Composites," *Composite Science and Technology,* Vol. 35, No. 3, 1989, pp. 301–313.

[32] Young, R. J., Day, R. J., Zakikhani, M., and Robinson, I. M., *Composite Science and Technology,* Vol. 34, 1989, pp. 243–258.

[33] Masters, J. E., *Proceedings,* ICCM6/ECCM2, Elsevier Applied Science Publishers, London, Vol. 3, July 1987, pp. 96–107.

[34] Horoschenkoff, A., Brandt, J., Warnecke, J., and Brüller, O. S., *Proceedings,* Ninth European SAMPE Conference, Society for the Advancement of Materials and Process Engineers, Saporiti, F. et al., Eds., Milano, Italy, 1988, pp. 339–349.

[35] Hiel, C., *Proceedings,* Third Technical Conference, American Society for Composites, Sept. 1988, Technomic Publishing, Lancaster, PA, pp. 558–563.

*Steven J. Hooper,[1] Richard F. Toubia,[1] and
Ramaswamy Subramanian[1]*

Effects of Moisture Absorption on Edge Delamination, Part I: Analysis of the Effects of Nonuniform Moisture Distributions on Strain Energy Release Rate

REFERENCE: Hooper, S. J., Toubia, R. F., and Subramanian, R., **"Effects of Moisture Absorption on Edge Delamination, Part I: Analysis of the Effects of Nonuniform Moisture Distributions on Strain Energy Release Rate,"** *Composite Materials: Fatigue and Fracture (Third Volume), ASTM STP 1110,* T. K. O'Brien, Ed., American Society for Testing and Materials, Philadelphia, 1991, pp. 89–106.

ABSTRACT: The effects of a nonuniform moisture distribution on the edge delamination of composite laminates were analyzed. The analysis was performed using a laminated plate theory analysis and a quasi-three-dimensional finite element method. The total strain energy release rate and its mixed mode components were evaluated for both uniform and nonuniform moisture distributions, and the corresponding results were compared. Both of the analyses employed a Fickian diffusion model to determine the hygroscopic concentration profile across the thickness of a laminate.

KEY WORDS: composite materials, fracture, fracture toughness, edge delamination, Fickian moisture diffusion, mixed-mode strain energy release rates, analysis

Delamination has been identified as a significant, and frequently the critical, failure mode in advanced composite materials [1,2]. Delaminations are significant considerations in the design of composite structures since their presence results in reduced laminate stiffness, strength, and fatigue life [3]. The development of free-edge delamination is generally attributed to the existence of singularities near the interfaces of the laminae in the region of a free edge [4–6]. For the case of mechanical loading, the stress concentrations develop due to the mismatch in Poisson's ratios between the adjacent plies [7]. For the case of thermal or hygroscopic loading, the singularities are developed as a result of the mismatch in the coefficients of thermal or hygroscopic expansion or both as well as the mismatch in Poisson's ratio [8–10].

Interlaminar fracture toughness is the accepted measure of delamination resistance under static loading. This quantity is generally evaluated in terms of the critical strain energy release rate, G_c [1–3]. Recently, the effects of residual thermal stresses and uniform moisture absorption on delamination have been studied [10–13]. These studies revealed that residual thermal effects tend to increase G but that moisture effects tend to decrease G. It has been shown that these effects can cancel one another for some graphite-epoxy systems.

Other investigators have shown that the moisture sorption process of advanced composites

[1] Assistant professor of Aerospace Engineering and graduate research assistants, Department of Aerospace Engineering, respectively, The Wichita State University, Institute for Aviation Research, Wichita, KA 67208.

occurs slowly and that an extensive period of time is required for a laminate to reach a steady-state moisture distribution [14–16]. Since these materials develop hygroscopic strains as a result of moisture sorption, it is obvious that the through-the-thickness gradients present during transient moisture diffusion state produce significant interlaminar stresses [9,17]. While many investigators have studied the effects of a uniform moisture distribution on delamination [18–23], little work has been reported on the effects of a nonuniform moisture distribution on delamination.

The present study addressed the question of how a nonuniform moisture distribution affects delamination onset. It also compared these results with results for saturated cases. The effects of a nonuniform moisture distribution on the strain energy release rate for edge delamination of composite laminates were analyzed using a modified laminate theory analysis and a quasi-three-dimensional finite element method. Both analyses employed a Fickian moisture diffusion model [24–27].

Classical Laminate Theory Analysis for an Edge Delamination Tension (EDT) Specimen with a Fickian Moisture Distribution

The total strain energy release rate for edge delamination growth in a composite laminate of finite width is calculated, including the contribution of transient hygroscopic diffusion, based on a classical laminate theory (CLT) analysis. The following derivation closely follows the work of Ref 10. In this analysis, a partially delaminated laminate is modeled as shown in Fig. 1. The strain energy of the entire laminate can be expressed as the sum of the strain energies for each region

$$U = U_{LAM} + U_{90} + U_{SUB1} + U_{SUB2} \tag{1}$$

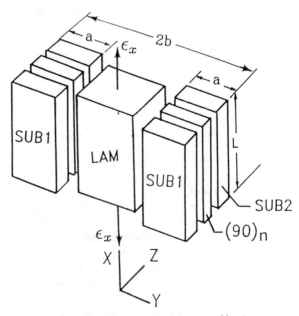

FIG. 1—*Model of a partially delaminated laminate.*

that can be written in terms of the strain energy densities as

$$U = 2(b - a) L t_{LAM}\bar{u}_{LAM} + 2 a L t_{90}\bar{u}_{90} + 2 a L t_{SUB1}\bar{u}_{SUB1} + 2 a L t_{SUB2}\bar{u}_{SUB2} \qquad (2)$$

The strain energy release rate is given by Ref 28

$$G = \frac{dW}{dA} - \frac{dU}{dA} \qquad (3)$$

where $A = 2aL$ is the delamination area. Since the work term, dW/dA, is negligible, substituting Eq 2 into Eq 3 yields the following expression for G.

$$G = t_{LAM}\bar{u}_{LAM} - t_{90}\bar{u}_{90} - t_{SUB1}\bar{u}_{SUB1} - t_{SUB2}\bar{u}_{SUB2} \qquad (4)$$

The specific strain energy for a region can be written as

$$\bar{u} = \frac{1}{2t} \int_{-t/2}^{t/2} \{\epsilon - \alpha \, \Delta T - \beta \, \Delta H\}' \{\sigma\} \, dz \qquad (5)$$

where the prime denotes the transpose operation.

This integral is evaluated on a ply-by-ply basis and thus becomes

$$\bar{u} = \frac{1}{2t} \sum_{i=1}^{N} \int_{z_{i-1}}^{z_i} \{\{\epsilon\} - \{\alpha\}_i \Delta T - \{\beta\}_i \Delta H\}' \{\sigma\} \, dz \qquad (6)$$

For the general case in which both extension and bending occur, the strain at a point is given by

$$\{\epsilon\} = \{\epsilon^\circ\} + z\{k\} \qquad (7)$$

where the midplane strain and curvature vectors can be written as a linear combination of each of the loading conditions

$$\{\epsilon^\circ\} = \{\epsilon^\circ\}^M + \{\epsilon^\circ\}^T + \{\epsilon\}^H \qquad (8)$$

$$\{k\} = \{k\}^M + \{k\}^T + \{k\}^H \qquad (9)$$

The constitutive relationship for the ith ply is

$$\{\sigma\}_i = [Q]_i (\{\epsilon\}_i - \{\alpha\}_i \Delta T - \{\beta\}_i \Delta H) \qquad (10)$$

where $[Q]_i$ represents the transformed reduced stiffness matrix. Substituting Eqs 7 through 10 into Eq 6 yields the following expression for the strain energy density of a laminate

$$\bar{u} = \frac{1}{2t} \sum_{i=1}^{N} \int_{z_{i-1}}^{z_i} \{\{\epsilon^\circ\} + z\{k\} - \{\alpha\}_i \Delta T - \{\beta\}_i \Delta H\}' [Q]_i \{\{\epsilon^\circ\}$$
$$+ z\{k\} - \{\alpha\}_i \Delta T - \{\beta\}_i \Delta H\} \, dz \qquad (11)$$

Finally, the expression for $\bar{u}$ becomes

$$
\begin{aligned}
\bar{u} = \frac{1}{t} \Bigg\{ &\frac{1}{2} [\epsilon^\circ, k] \left[\begin{array}{c|c} A & B \\ \hline B & D \end{array} \right] \left[\begin{array}{c} \epsilon^\circ \\ k \end{array} \right] - [\epsilon^\circ, k] \left[\begin{array}{c} N \\ M \end{array} \right]^T - [\epsilon^\circ, k] \left[\begin{array}{c} N \\ M \end{array} \right]^H \\
&+ \frac{1}{2} \sum_{i=1}^N \{\alpha\}_i' [Q]_i \{\alpha\}_i \Delta T^2 (z_i - z_{i-1}) \\
&+ \sum_{i=1}^N \{\alpha\}_i' [Q]_i \{\beta\}_i \Delta T \int_{z_{i-1}}^{z_i} \Delta H(z) \, dz \\
&+ \frac{1}{2} \sum_{i=1}^N \{\beta\}_i' [Q]_i \{\beta\}_i \int_{z_{i-1}}^{z_i} \Delta H(z)^2 \, dz \Bigg\}
\end{aligned}
\tag{12}
$$

where $[A]$, $[B]$, and $[D]$ represent the extensional, bending-extension coupling, and bending stiffness matrices, respectively, and $\{N\}^T$ and $\{N\}^H$ represent the thermal and hygroscopic stress resultants, respectively. Similarly, $\{M\}^T$ and $\{M\}^H$ are the thermal and hygroscopic moment resultants, respectively.

The local hygroscopic concentration of a partially soaked laminate is in general a function of all three spatial coordinates x, y, and z [24]. The effect of any variation along the x-direction is negligible since only the ends of the specimen would be affected and these regions are not close to the delaminations. Since the specimen is thin, most of the moisture is absorbed through the faces of the specimen that are normal to the z-axis. The simplest prediction of this moisture absorption process is given by the solution to the one-dimensional Fickian diffusion equation as

$$
\Delta H(z,\tau) = C_\infty \left[1 - \frac{4}{\pi} \sum_{m=1}^{\infty} \frac{(-1)^m}{B} \exp\left[- \left\{ \frac{B\pi}{2t} \right\}^2 D_z \tau \right] \cos\left[\frac{B\pi}{2t} z \right] \right]
\tag{13}
$$

where τ denotes time and $B = 2m - 1$.

Moisture sorption will also occur through the edges of the specimens from which the edge delaminations originate. The prediction of this moisture absorption process in this case is given by the solution to Fick's two-dimensional equation of diffusion as

$$
\begin{aligned}
\Delta H(y,z,\tau) = C_\infty \Bigg[1 - \left\{ \frac{4}{\pi} \right\}^2 \sum_{n=1}^{\infty} \sum_{m=1}^{\infty} \Bigg\{ &\frac{(-1)^{n+m+1}}{A B} \\
&\exp\left[\left[- \left\{ \frac{A\pi}{2b} \right\}^2 D_y - \left\{ \frac{B\pi}{2t} \right\}^2 D_z \right] \tau \right] \cos\left[\frac{A\pi}{2b} y \right] \cos\left[\frac{B\pi}{2t} z \right] \Bigg\} \Bigg]
\end{aligned}
\tag{14}
$$

where $A = 2n - 1$ and $B = 2m - 1$.

For the mechanical loading case, ϵ_x is prescribed and the test machine grips impose the following boundary conditions on Eq 12

$$
\epsilon_x = \epsilon_x^\circ
\tag{15}
$$
$$
k_x = k_{xy} = N_y = N_{xy} = M_y = 0
$$

This yields

$$
\begin{bmatrix} \epsilon_x^{\circ} \\ \epsilon_y^{\circ} \\ \epsilon_{xy}^{\circ} \\ 0 \\ k_y \\ 0 \end{bmatrix}^{M} = \begin{bmatrix} A & B \\ B & D \end{bmatrix}^{-1} \begin{bmatrix} N_x \\ 0 \\ 0 \\ M_x \\ 0 \\ M_{xy} \end{bmatrix}^{M}
\tag{16}
$$

that can be solved for ϵ_y, ϵ_{xy}, k_y, N_x, M_x, and M_{xy}.

The thermal and hygroscopic x-direction strains of the sublaminates and $(90)_n$ regions are subject to the following compatibility conditions. The extensional strains ϵ_x are assumed to be the same for all of the sublaminate regions. Thus, the difference between the sublaminate strain and the initial laminate strain is an additional mechanical strain for these regions. It further assumed that the grips of the test machine constrain the curvatures k_x and k_{xy} to zero for all regions. The combination of these effects results in an additional mechanical load acting on each region that is given by

$$
\begin{bmatrix} \epsilon_x^{\circ} \\ k_x \\ k_{xy} \end{bmatrix}^{M}_{ADD} = \begin{bmatrix} \epsilon_x^{\circ} \\ 0 \\ 0 \end{bmatrix}^{T,H}_{LAM} - \begin{bmatrix} \epsilon_x^{\circ} \\ k_x \\ k_{xy} \end{bmatrix}^{T,H}_{REGION}
\tag{17}
$$

These strains, ϵ_x, k_x, and k_{xy}, through an equation similar to Eq 16, lead to additional strains, ϵ_y, ϵ_{xy}, and k_y.

Finite Element Analysis of the Edge Delamination Specimen

The mixed-mode strain energy release rates were determined using a quasi-three-dimensional finite element method (FEM) computer program (Q3DG) [29]. In this analysis, every y-z plane is the same, and because of symmetry, only one quarter of the cross section is modeled as shown in Fig. 2.

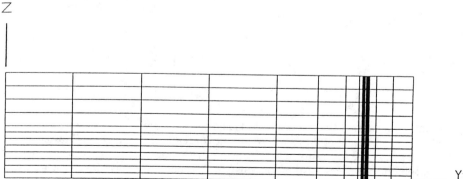

FIG. 2—Finite element model (not to scale).

The Q3DG finite element code was modified to account for the nonuniform moisture state in the partially soaked edge delamination tension (EDT) specimens. Since the basic formulation of this FEM analysis has been well documented in Ref 29, we shall only describe the modifications performed during this research. The finite element stiffness matrix equation for each element can be written as

$$[K]\{u\} + \{F\}_\epsilon + \{F\}_T + \{F\}_H = 0 \tag{18}$$

where

$$
\begin{aligned}
[K] &= \int_{vol} [\tilde{N}_r]'[Q_{rr}][\tilde{N}_r] \, dV \\
\{F\}_\epsilon &= \int_{vol} [\tilde{N}_r]'[Q_{rx}]\{\epsilon^\circ\} \, dV \\
\{F\}_T &= -\int_{vol} ([\tilde{N}_r]'[Q_{xr}]'\{T_x\} + [\tilde{N}_r]'[Q_{rr}]\{T_r\}) \, dV \\
\{F\}_H &= -\int_{vol} ([\tilde{N}_r]'[Q_{xr}]'\{H_x\} + [\tilde{N}_r]'[Q_{rr}]\{H_r\}) \, dV
\end{aligned}
\tag{19}
$$

$[\tilde{N}_r]$ and $[\tilde{N}_x]$ are found from the strain displacement matrix, $[\tilde{N}]$, which is decomposed as shown later since each element has a prescribed mechanical strain, $\epsilon_x = \epsilon^\circ$.

$$
\begin{bmatrix} \dfrac{\epsilon^\circ}{\epsilon_r} \end{bmatrix} = \begin{bmatrix} \tilde{N}_x \\ \hline \tilde{N}_r \end{bmatrix} \{u\} \tag{20}
$$

$[Q_{xx}]$, $[Q_{xr}]$, $[Q_{rx}]$, $[Q_{rr}]$, $\{T_x\}$, $\{T_r\}$, $\{H_x\}$, and $\{H_r\}$ are found from the partitioned form of the stress strain relationships as follows

$$
\begin{bmatrix} \sigma_x \\ \sigma_r \end{bmatrix} = \begin{bmatrix} Q_{xx} & Q_{xr} \\ \hline Q_{rx} & Q_{rr} \end{bmatrix} \left[\begin{bmatrix} \epsilon^\circ \\ \epsilon_r \end{bmatrix} - \begin{bmatrix} T_x \\ T_r \end{bmatrix} - \begin{bmatrix} H_x \\ H_r \end{bmatrix} \right] \tag{21}
$$

where $\{T\}$ and $\{H\}$ are defined as

$$
\{T\} = \begin{bmatrix} \alpha_x \\ \alpha_y \\ \alpha_{xy} \\ 0 \\ 0 \\ 0 \end{bmatrix} \Delta T, \quad \{H\} = \begin{bmatrix} \beta_x \\ \beta_y \\ \beta_{xy} \\ 0 \\ 0 \\ 0 \end{bmatrix} \Delta H \tag{22}
$$

The integrations are performed using the Gauss quadrature technique. Consequently, $\{F\}_H$ is evaluated as a function of the local hygroscopic concentration, ΔH, at the Gaussian point, see Eqs 19 and 22. The ΔH is, in general, a function of location and time and can be calculated from the solution of the Fickian diffusion equation. The solution to a two-dimensional Fickian diffusion equation is plotted as a function of depth coordinate, z, in Fig. 3, and as a function of width coordinate, y, in Fig. 4.

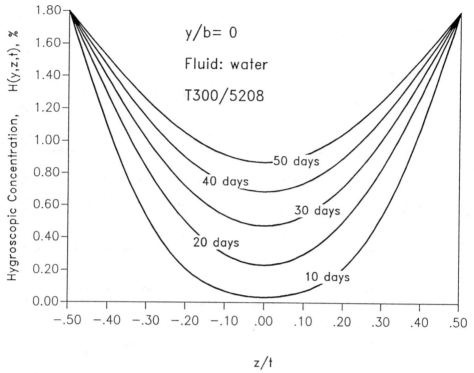

FIG. 3—*Two-dimensional hygroscopic distribution.*

The strain energy release rate components are calculated using the virtual crack closure technique [30] as

$$G_I = [F_{zi} (w_l - w_k) + F_{zj} (w_n - w_m)]/(2d)$$
$$G_{II} = [F_{yi} (v_l - v_k) + F_{yj} (v_n - v_m)]/(2d) \qquad (23)$$
$$G_{III} = [F_{xi} (u_l - u_k) + F_{xj} (u_n - u_m)]/(2d)$$

The forces F_{xi}, F_{yi}, and F_{zi} are the forces in the x-, y-, and z-direction, respectively, acting at Node i as shown in Fig. 5. The forces at Node i are computed from Elements A and B while the forces at Node j are computed from Element A alone. This procedure for calculating the strain energy release rate requires that the mesh be symmetric about the delamination tip.

Results and Discussion

The modified version of Q3DG and the new laminate theory analysis code were verified by analyzing a number of problems, for uniform moisture distributions, described in Ref 10, and comparing the resulting solutions. The same problems were then analyzed for nonuniform moisture distributions, and the solutions of the two analyses were compared. In this comparison study, a one-dimensional Fickian diffusion was employed in both analyses. A two-dimensional Fickian diffusion analysis was also used with the Q3DG finite element analysis. The test

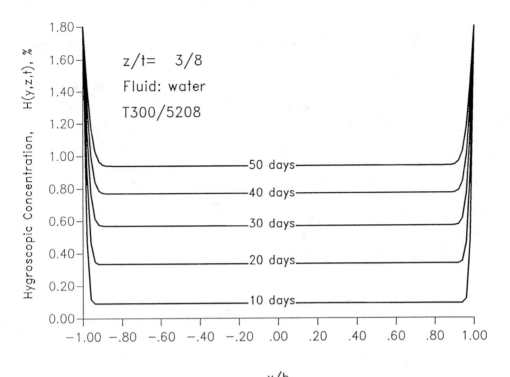

FIG. 4—*Two-dimensional hygroscopic Fickian distribution.*

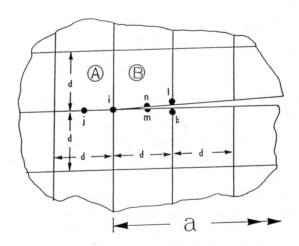

$\bigcirc\!\!A$: Element Number   m : Node Number

FIG. 5—*Elements and nodes near the delamination tip used in the SERR calculations.*

problems addressed the delamination response of T300/5208 graphite-epoxy EDT specimens with the material properties listed here.

$E_1 = 134.4$ GPa (19.5 Mpsi)
$E_2 = 10.2$ GPa (1.48 Mpsi)
$E_{12} = 5.5$ GPa (0.8 Mpsi)
$\nu_{12} = 0.3$
$\alpha_1 = -0.128$ μm/m/K (-0.23 μin./in.°F)
$\alpha_2 = 8.28$ μm/m/K (14.9 μin./in./°F)
$\beta_1 = 0.0$ μm/m/percent weight gain
$\beta_2 = 5560$ μm/m/percent weight gain

Diffusivity coefficients, $D_y = D_z = 1.79 \times 10^{-7}$ m^2/s (1.427×10^{-7} in.^{2}/h) were assumed for this problem.

The analysis was formulated to model a preconditioned EDT specimen that was assumed to be initially dry ($\Delta H = 0$) prior to soaking in water. The saturation weight gain at equilibrium, C_∞, was assumed to be 1.8% and the thermal load, ΔT, was the same for all layups and represents a 450 K (350°F) cure cycle. The environmental condition just described was imposed prior to crack initiation.

The strain energy release rate (SERR), G, for the two methods of analyses was plotted against the average weight gain for both uniform and transient moisture distributions as shown in Figs. 6 through 8. These figures also show that the SERR for uniform moisture distributions, from Ref 10, are in excellent agreement with our results. Similarly, these figures show that the results

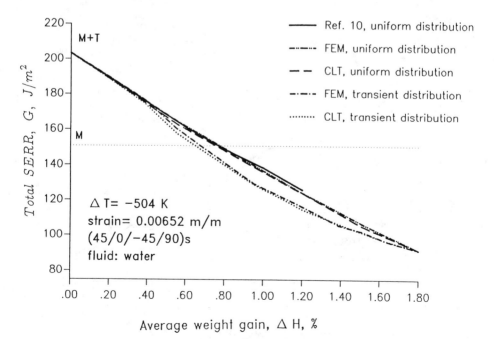

FIG. 6—*Influence of transient moisture diffusion on the strain energy release rate for (45/0/−45/90)$_s$ laminates.*

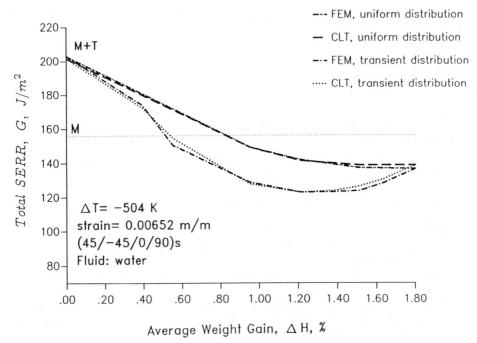

FIG. 7—*Influence of transient moisture diffusion on the strain energy release rate for (45/−45/0/90)s laminates.*

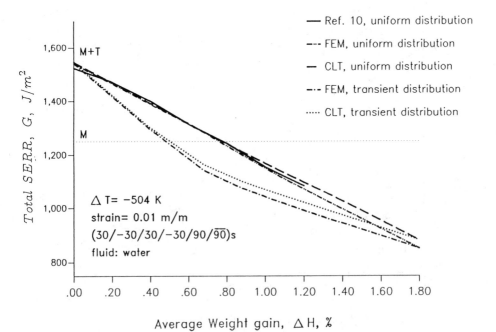

FIG. 8—*Influence of transient moisture diffusion on the SERR for (30/−30/30/−30/90/$\overline{90}$)s laminates.*

predicted by the new CLT and the modified Q3DG finite element code agree for nonuniform moisture distributions.

It is evident from these figures that for a given weight gain the total SERR of a nonuniform moisture distribution is always lower than that of a uniform distribution. It is also apparent that the magnitude of this difference depends on the layup. The difference was smaller for layups with symmetric sublaminates or balanced sublaminates, that is, laminates of the type $(\theta/0/-\theta)$ or $(\theta/\phi/0/-\phi/-\theta)$, than for layups with unsymmetric or unbalanced sublaminates. Clearly, a significant difference can exist between the solutions for uniform and nonuniform moisture distributions.

The SERR due to pure mechanical loads are shown on these plots as dotted horizontal lines. These lines intersect the G versus ΔH curves at the points where the thermal and hygroscopic effects cancel each other [10]. For all layups studied, the hygroscopic effects canceled the thermal effects at a lower average weight gain for Fickian moisture distributions than for uniform distributions. Note that the difference between the pure mechanical SERR, G^M, and the hygro-thermo-mechanical SERR, G^{HTM}, is the contribution of the mechanics effects due to thermal and hygroscopic stresses.

To examine the transient effects of moisture diffusion on delamination, SERR versus elapsed soaking time was plotted in Figs. 9 through 11. The most significant changes in SERR occurred at the initial stage of soaking when the moisture distribution gradients were the highest. Note that for layups having asymmetric sublaminates, the SERR increased near the saturation stage, see Fig. 10, in contrast to layups with symmetric sublaminates for which the SERR decreased monotonically with time.

This study also addressed the effect of crack length on the results predicted by the Q3DG finite element code. The total SERR, G, and the Mode I component, G_I, are plotted versus the

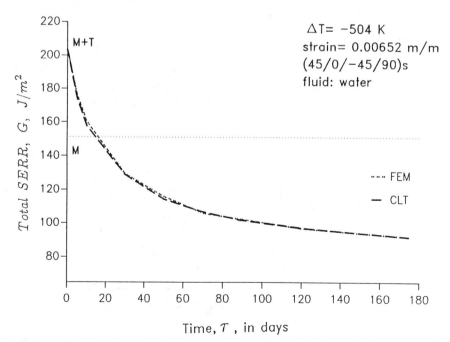

FIG. 9—*Influence of transient moisture diffusion on the SERR for (45/0/−45/90)$_s$ laminates as a function of time.*

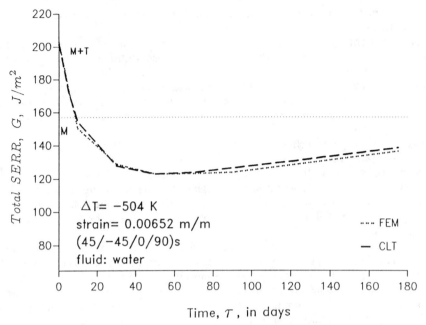

FIG. 10—*Influence of transient moisture diffusion on the SERR for (45/−45/0/90)ₛ laminates as a function of time.*

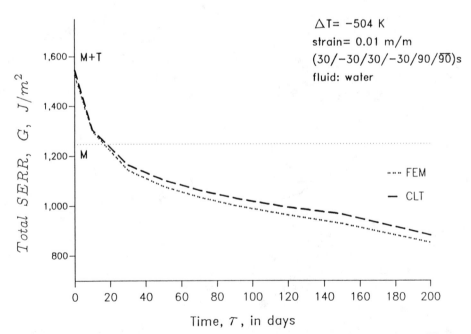

FIG. 11—*Influence of transient moisture diffusion on the SERR for (30/−30/30/−30/90/90̄)ₛ laminates as a function of time.*

crack length, a, for time $\tau = 50$ days as shown in Fig. 12. This figure shows that the strain energy release rate predicted by Q3DG agrees with that predicted by CLT for sufficiently long cracks. Thus, the behavior for a nonuniform moisture distribution is similar to the behavior reported in Ref 3 for a uniform distribution.

The mixed-mode strain energy release rate analysis results are presented in Figs. 13–15 where the percentage of Mode I is plotted versus average weight gain for the various layups. These figures contain data for both uniform and nonuniform moisture distributions. The strain energy release rate mode mix is markedly different for the nonuniform moisture distributions compared to the uniform moisture distributions. These differences can be considerable for some laminate designs. For example, consider the results for the $(35/0/-35/90)_s$ laminates shown in Fig. 15. The percentage of Mode I is a constant 61% for the uniform moisture diffusion case, while it exhibits nearly pure Mode II behavior over a wide range of ΔH for the nonuniform case. Note that the Mode III components are negligible for all cases.

The total strain energy release rates calculated employing one-dimensional and two-dimensional Fickian diffusion models are presented in Figs. 16 and 17. It is apparent that the moisture absorbed through the edges of the specimens does not contribute significantly to the SERR since there are no significant differences between the results obtained using the one-dimensional model and the two-dimensional model. Thus, a one-dimensional diffusion model can be employed without compromising the accuracy of the SERR prediction.

Conclusions

The results of this study support the following conclusions for initially dry T300/5208 graphite-epoxy specimens that were then soaked in water:

1. A nonuniform moisture distribution significantly affects the strain energy release rate as well as the G_I/G and G_{II}/G ratios. Generally, for a given weight gain, the total strain energy

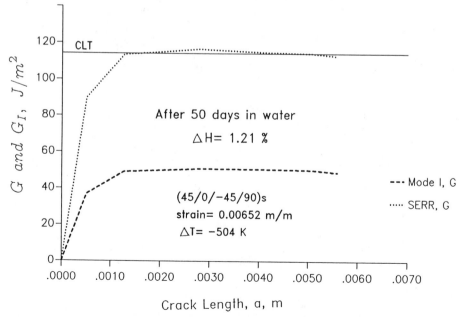

FIG. 12—*Strain energy release rate and its mode mix versus crack length.*

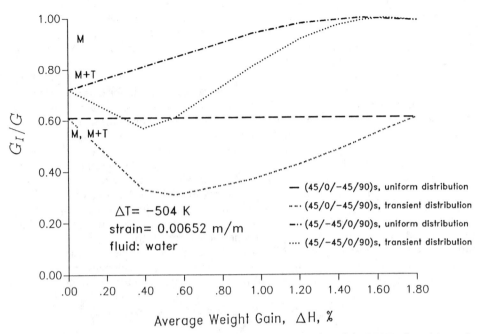

FIG. 13—*Effect of transient moisture diffusion on percentage Mode I of the SERR of quasi-isotropic laminates.*

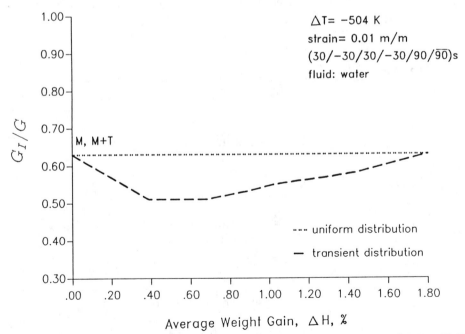

FIG. 14—*Effect of transient moisture diffusion on Mode I percentage of the SERR for (30/−30/30/ −30/90/$\overline{90}$)$_s$ laminates.*

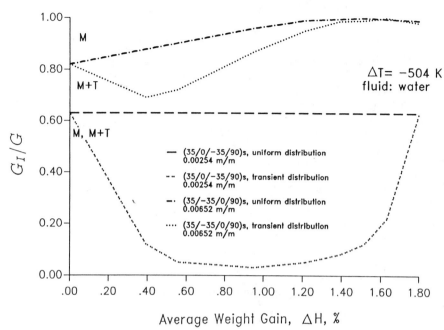

FIG. 15—*Effect of transient moisture diffusion on percentage Mode I for (35/−35/0/90)ₛ and (35/0/−35/90)ₛ laminates.*

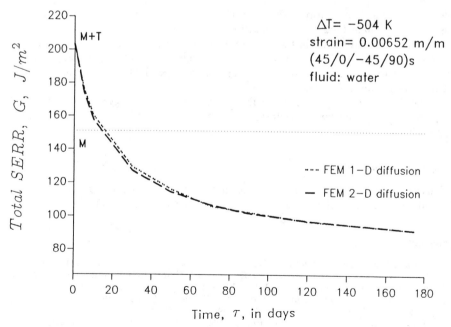

FIG. 16—*Comparison of one-dimensional and two-dimensional Fickian diffusion on the SERR for (45/0/−45/90)ₛ laminates as a function of time.*

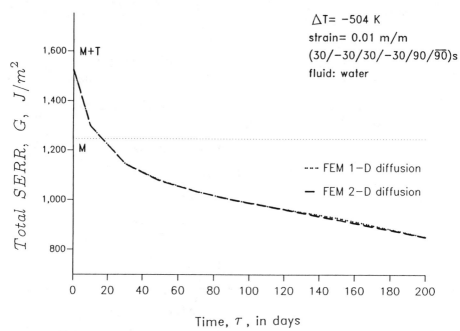

FIG. 17—*Comparison of one-dimensional and two-dimensional Fickian diffusion on the SERR for (30/ −30/30/−30/90/90̄)ₛ laminates as a function of time.*

release rate for a nonuniform moisture distribution is generally less than the strain energy release rate calculated for a uniform moisture distribution. Also, G_I/G ratio for the case of the nonuniform distribution will be less than for the uniform distribution case.

2. Transient moisture conditions significantly affect both the total strain energy release rate and the mixed-mode components. This effect is layup dependent and can dramatically alter the ratios G_I/G and G_{II}/G.

3. The effects of moisture diffusion on the mixed-mode strain energy release rates associated with edge delamination can be accurately modeled using the one-dimensional Fickian diffusion equation.

Acknowledgments

This work was supported by the Army Research Office, Contract No. DAAL 03-87-G-003. The authors would also like to acknowledge the helpful suggestions of Dr. T. Kevin O'Brien.

References

[1] O'Brien, T. K., "Characterization of Delamination Onset and Growth in a Composite Laminate," *Damage in Composite Materials, ASTM STP 775,* K. L. Reifsnider, Ed., American Society for Testing and Materials, Philadelphia, 1982, pp. 140–167.
[2] Garg, A. C., "Fracture Behavior of Carbon Fiber Reinforced Composites—A Review," *Journal of the Mechanical Behavior of Materials,* Vol. 1, No. 1–4, 1988, pp. 101–210.
[3] O'Brien, T. K., "Mixed-Mode Strain Energy Release Rate Effects on Edge Delamination of Composites," *Effects of Defects in Composite Materials, ASTM STP 836,* American Society for Testing and Materials,, Philadelphia, 1984, pp. 125–142.

[4] Wang, A. S. D. and Crossman, F. W., "Some New Results on Edge Effect in Symmetric Composite Laminates," *Journal of Composite Materials,* Vol. 11, Jan. 1977, pp. 92–106.

[5] Herakovich, C. T., Nagarkar, A., and O'Brien, D. A., "Failure Analysis of Composite Laminates with Free Edges," *Modern Developments in Composite Materials and Structures,* J. R. Vinson, Ed., American Society of Mechanical Engineers, 1972, pp. 53–69.

[6] Raju, I. S. and Crews, J. H., "Interlaminar Stress Singularities at a Straight Free Edge in Composite Laminates," *Computers and Structures,* Vol. 14, 1981, pp. 21–28.

[7] Herakovich, C. T., "On the Relationship Between Engineering Properties and Delamination of Composite Laminates," NASA-CR-163956, National Aeronautics and Space Administration, Washington, DC, Feb. 1981.

[8] Wang, A. S. D. and Crossman, F. W., "Edge Effects on Thermally Induced Stresses in Composite Laminates," *Journal of Composite Materials,* Vol. 11, July 1977, pp. 300–312.

[9] Farley, G. L. and Herakovich, C. T., "Influence of Two-Dimensional Hygrothermal Gradients on Interlaminar Stresses Near Free Edges," *Advanced Composite Materials—Environmental Effects, ASTM STP 658,* J. R. Vinson, Ed., American Society for Testing and Materials, Philadelphia, 1978, pp. 143–159.

[10] O'Brien, T. K., Raju, I. S., and Garber, D. P., "Residual Thermal and Moisture Influences on the Strain Energy Release Rate Analysis of Edge Delamination," *Journal of Composites Technology & Research,* Vol. 8, No. 2, 1986, pp. 37–47.

[11] Kriz, R. D. and Stinchcomb, W. W., "Effects of Moisture, Residual Thermal Curing Stresses, and Mechanical Load on the Damage Development in Quasi-Isotropic Laminates," *Damage in Composite Materials, ASTM STP 775,* K. L. Reifsnider, Ed., American Society for Testing and Materials, Philadelphia, 1982, pp. 63–80.

[12] Russell, A. J. and Street, K. N., "Moisture and Temperature Effects on the Mixed Mode Delamination Fracture of Unidirectional Graphite-Epoxy," *Delamination and Debonding of Materials, ASTM STP 876,* W. S. Johnson, Ed., American Society for Testing and Materials, Philadelphia, 1985, pp. 349–370.

[13] Tay, T. E., Williams, J. F., and Jones, R., "Characterization of Pure and Mixed Mode Fracture in Composite Laminates," *Theoretical and Applied Mechanics,* Vol. 7, 1987, pp. 115–123.

[14] Shen, C. H. and Springer, G. S., "Moisture Absorption and Desorption of Composite Materials," *Journal of Composite Materials,* Vol. 10, 1976, pp. 2–20.

[15] Wang, Q. and Springer, G. S., "Moisture Absorption and Fracture Toughness of PEEK Polymer and Graphite Fiber Reinforced PEEK," *Journal of Composite Materials,* Vol. 23, 1989, pp. 434–447.

[16] Yaniv, G., Peimanisis, G., and Daniel, I. M., "Method for Hygromechanical Characterization of Graphite/Epoxy Composite," *Journal of Composites Technology & Research,* Vol. 9, No. 1, Spring 1987, pp. 21–25.

[17] Pipes, R. B., Vinson, J. R., and Chou, T. W., "On the Hygrothermal Response of Laminated Composite Systems," *Journal of Composite Materials,* Vol. 10, April 1976, pp. 129–148.

[18] Garg, A. and Ishai, O., "Hygrothermal Influence on Delamination Behavior of Graphite/Epoxy Laminates," *Engineering Fracture Mechanics,* Vol. 22, No. 3, 1985, pp. 413–427.

[19] Kenig, S., Moshonov, A., Shucrun, A., and Marom, G., "Environmental Effects on Shear Delamination of Fabric-Reinforced Epoxy Composites," *International Journal of Adhesion and Adhesives,* Vol. 9, No. 1, Jan. 1989, pp. 38–45.

[20] Han, K. S. and Koutsky, J., "Effect of Water on the Interlaminar Fracture Behavior of Glass Fibre-Reinforced Polyester Composites," *Composites,* Jan. 1983, pp. 67–70.

[21] O'Brien, T. K., Johnson, N. J., Raju, I. S., Morris, D. H., and Simonds, R. A., "Comparisons of Various Configurations of the Edge Delamination Test for Interlaminar Fracture Toughness," *Toughened Composites, ASTM STP 937,* N. J. Johnson, Ed., American Society for Testing and Materials, Philadelphia, 1987, pp. 199–221.

[22] Chisholm, J. M., Hahn, H. T., and Williams, J. G., "Effect of Seawater on the Fracture Toughness of Pultruded Rods," *Mechanics of Composite Materials—1988,* G. J. Dvorak and N. Laws, Eds., The American Society of Mechanical Engineers, 1988, pp. 117–122.

[23] Wilkins, D. J., "A Comparison of the Delamination and Environmental Resistance of a Graphite-Epoxy and a Graphite-Bismaleimide," NAV-GD-0037, General Dynamics Corp., Fort Worth, TX, Sept. 1981.

[24] Blikstad, M., Sjoblom, P. O. W., and Johannesson, T. R., "Long-Term Moisture Absorption in Graphite/Epoxy Angle-Ply Laminates," *Journal of Composite Materials,* Vol. 18, Jan. 1984, pp. 32–46.

[25] Tsai, S. W. and Hahn, H. T., *Introduction to Composite Materials,* Technomic Publishing Co., Westport, CT, 1980.

[26] Delasi, R. and Whiteside, J. B., "Effect of Moisture on Epoxy Resins and Composites," *Advanced Composite Materials—Environmental Effects, ASTM STP 658,* J. R. Vinson, Ed., American Society for Testing and Materials, Philadelphia, 1978, pp. 2–20.

[27] Shirrell, C. D., "Diffusion of Water Vapor in Graphite/Epoxy Composites," *Advanced Composite Materials—Environmental Effects, ASTM STP 658,* J. R. Vinson, Ed., American Society for Testing and Materials, Philadelphia, 1978, pp. 21–42.

[28] Irwin, G. R., "Fracture," *Handbuch der Physik,* Vol. 6, 1958, p. 551.

[29] Raju, I. S., "Q3DG—A Computer Program for Strain Energy—Release Rates for Delamination Growth in Composite Laminates," NASA CR-178205, National Aeronautics and Space Administration, Washington, DC, Nov. 1986.

[30] Rybicki, E. F. and Kanninen, M. F., "A Finite Element Calculation of Stress Intensity Factors by a Modified Crack Closure Integral," *Engineering Fracture Mechanics,* Vol. 9, 1977, pp. 931–938.

Steven J. Hooper,[1] *Ramaswamy Subramanian,*[1] *and Richard F. Toubia,*[1]

Effects of Moisture Absorption on Edge Delamination, Part II: An Experimental Study of Jet Fuel Absorption on Graphite-Epoxy

REFERENCE: Hooper, S. J., Subramanian, R., and Toubia, R. F., "**Effects of Moisture Absorption on Edge Delamination, Part II: An Experimental Study of Jet Fuel Absorption on Graphite-Epoxy,**" *Composite Materials: Fatigue and Fracture (Third Volume), ASTM STP 1110,* T. K. O'Brien, Ed., American Society for Testing and Materials, Philadelphia, 1991, pp. 107–125.

ABSTRACT: An experimental study was conducted to investigate the environmental effects of jet fuel absorption on the mixed-mode fracture toughness G_c of AS4/3501-6 graphite-epoxy. Edge delamination tension tests of saturated and partially saturated $(30/-30_2/30/90_n)_s$ ($n = 1$, $n = 2$), $(35/-35/0/90)_s$ and $(35/0/-35/90)_s$ layups were conducted. Three types of jet fuels were considered including JP-4 military jet fuel, Jet-A commercial jet fuel, and Type VII rubber swelling test fluid. Additional tests were conducted at the room temperature dry (RTD) and water soaked conditions. The water soaked specimens were also tested in the saturated and partially saturated conditions.

The total and the mixed-mode strain energy release rates were evaluated using a modified quasi-three-dimensional finite element (Q3DG) analysis. Edge delamination data of partially saturated and saturated specimens was used to quantify the relative influence of typical aircraft fluids on interlaminar fracture for composite materials. The effects of hygroscopic strain gradients on the interlaminar fracture toughness was investigated for the fluids considered.

KEY WORDS: composite materials, fracture, environmental effects, delamination, fracture toughness, strain energy release rate, graphite-epoxy, experimental study, jet fuel absorption, partially saturated, edge delamination tension test

Nomenclature

D_z	Diffusion coefficient
E_{Lam}	Longitudinal laminate modulus
E_{11}	Lamina moduli parallel to fiber direction
E_{22}	Lamina moduli perpendicular to fiber direction
G_{12}	Lamina shear moduli
G^M	Strain energy release rate caused by mechanical loading
G^{M+T}	Strain energy release rate caused by mechanical plus residual thermal loading
G^{M+T+H}	Strain energy release rate caused by mechanical plus residual thermal and hygroscopic loading
G_c	Critical strain energy release rate for delamination onset

[1] Assistant professor of Aerospace Engineering and graduate research assistants, Department of Aerospace Engineering, respectively, The Wichita State University, Institute for Aviation Research, Wichita, KA 67208.

G_{Ic}	Critical Mode I strain energy release rate for delamination onset
h	Ply thickness
T	Temperature
t	Laminate thickness
w	Transverse displacement
α_1	Lamina coefficient of thermal expansion in fiber direction
α_2	Lamina coefficient of thermal expansion normal to the fiber direction
β_1	Lamina coefficient of hygroscopic expansion in fiber direction
β_2	Lamina coefficient of hygroscopic expansion normal to fiber direction
$\Delta H(z)$	Percentage moisture weight gain as a function of z
$\overline{\Delta H}$	Average percentage moisture weight gain
ΔT	Temperature differential from cure temperature to test temperature
ϵ	Longitudinal strain
ϵ_c	Delamination onset strain
ν	Poissons ratio
τ	Time

In view of the fact that composite materials are frequently used in the design of primary structure for high-performance aerospace vehicles, it is necessary to closely examine how these materials respond to a typical service environment. Such an environment will expose these materials to fluids such as jet fuel, hydraulic fluid, and deice fluid in addition to water. While some work has addressed the effects of these fluids on material properties such as moduli and strength [1], very little research has been conducted to investigate the effects of a typical service environment on delamination resistance. It is well established that the moisture sorption process of advanced composites occurs slowly and that an extensive period of time is required for a laminate to reach a steady-state saturated moisture distribution [2–6]. Thus, the quantity of moisture absorbed varies as a function of the distance from the exposed surface [7]. Since laminates develop hygroscopic strains as a result of the moisture sorption process, it is obvious that the through-the-thickness gradients present in the transient moisture distribution will produce significant interlaminar stresses [5,8]. Many investigators have studied the effects of a uniform moisture distribution on delamination, but only recently have the effects of a non-uniform moisture distribution on delamination been analyzed [9].

The accepted measure of delamination resistance is the interlaminar fracture toughness, which is generally evaluated in terms of critical strain energy release rate, G_c [10–12]. The residual thermal and moisture stresses have been shown to influence the strain energy release rates for edge delamination of epoxy-matrix composites [8,13–15], but then these studies also were limited to studying the effects of a uniform through-the-thickness distribution of water.

This study addressed two principal problems. The first was to determine how absorption of typical aircraft fluids affects delamination resistance of graphite-epoxy. The second was to determine the effects of partial saturation of typical aircraft fluids on interlaminar fracture toughness. The edge delamination tension test specimen was used in this study. Four layups were tested: $(35/-35/0/90)_s$, $(35/0/-35/90)_s$, and $(30/-30_2/30/90_n)_s$, $n = 1, 2$. These layups were selected based upon a parametric study to optimize the layups for edge delamination tests as reported in Ref 12. All specimens were fabricated without inserts, and thus natural delaminations were obtained. The specimens were exposed to four different fluids including Phillips Jet A commercial jet fuel, JP-4 military jet fuel, Type VII rubber swelling test fluid (jet reference fuel), and water. Type VII rubber swelling test fluid was included in this study since it is a liquid hydrocarbon standard test fluid with a specific chemical composition.

The effects of these fluids on the toughness of graphite-epoxy were analyzed using a modified quasi-three-dimensional finite element analysis employing a Fickian moisture diffusion model as detailed in Ref 9.

Materials

Edge delamination tension (EDT) test specimens were fabricated from ICI Fiberite AS4/3501-6 graphite-epoxy. The specimens were 0.13 m (5.0 in.) long and 0.025 m (1.0 in.) wide. The average cured ply thickness was 0.0002 m (0.008 in.) and the fiber volume fraction was 60%. A modified Fiberite C-9 autoclave cycle was used to cure this material [16]. The autoclave pressure was applied immediately when the temperature reached 394 K (250°F). The subsequent cure temperature was 450 K (350°F).

The specimens were preconditioned in a vacuum drying cycle wherein the specimens were vacuum bagged, placed in an oven and subjected to the temperature cycle illustrated in Fig. 1. A chart recording of the vacuum drying cycle was generated for each batch of specimens. Immediately after the vacuum drying cycle, the specimens were weighed to the nearest 0.0001 g. A total of 50 specimens were prepared for each layup. These were split into five batches. The first batch of seven specimens was tested at room temperature dry (RTD) condition, the second batch of 14 specimens was soaked in water, the third batch of eleven specimens were soaked in Jet A, the fourth batch of eleven specimens was soaked in JP-4 and the fifth batch consisting of the remaining seven specimens was soaked in Type VII rubber swelling test fluid also known as jet reference fuel. Specimens for the partially saturated condition were tested after one to four weeks of soaking in the fluids. The "saturated" specimens were soaked until saturation, which was 190 days. The RTD specimens were tested at room temperature immediately following conditioning in the vacuum drying cycle. The specimens for each batch were chosen from a variety of panel positions.

The average axial modulus only varied by about 2 to 4% for the partially saturated specimens, but the modulus for the saturated specimens showed a significant variation. Hence a

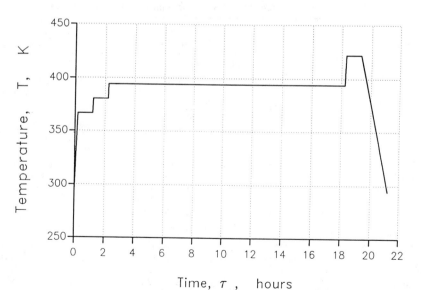

FIG. 1—*Drying cycle.*

single set of lamina properties were used in the analyses for all but the saturated cases. The elastic properties of the material were determined from unidirectional tests at the room temperature dry condition as

E_{11} = 114.17 GPa (16.559 Mpsi)
E_{22} = 7.58 GPa (1.099 Mpsi)
G_{12} = 4.31 GPa (0.625 Mpsi)
ν_{12} = 0.33

The values for the predicted and measured axial laminate moduli for each of the layups and each of the five environments are shown in Table 1. The average measured moduli for the four layups were found to be in close agreement with the moduli calculated from laminated plate theory. In order to simulate the laminate modulus reduction due to moisture absorption, E_{11} was reduced to 108.46 GPa (15.731 Mpsi) and 105.04 Gpa (15.234 Mpsi) for specimens saturated in jet fuel and water, respectively. The values of the thermal expansion coefficients used in this study were

α_1 = -0.094 $\mu\epsilon$/K (-0.17 $\mu\epsilon$/°F)
α_2 = 8.67 $\mu\epsilon$/K (15.6 $\mu\epsilon$/°F)

and ΔT was -504 K (-280°F), which corresponds to a 450 K (350°F) cure cycle. The hygroscopic expansion coefficients used are reported in Table 2. These data were measured using a capacitance type dilatometer. The coefficients of hygroscopic expansion for water reported here closely agree with those reported in Ref 4.

TABLE 1—Comparison of the axial laminate modulus for the saturated states.

Environment	E_{11} Used, GPa (Mpsi)	Calculated E_{Lam}, GPa (Mpsi)	Experimental E_{Lam}, Gpa (Mpsi)
	LAYUP $(35/-35/0/90)_s$		
RTD	114.17(16.559)	54.92(7.965)	54.88(7.960)
Jet A	108.46(15.731)	52.43(7.604)	52.33(7.590)
JP-4	108.46(15.731)	52.43(7.604)	51.48(7.467)
Jet Ref.	108.46(15.731)	52.43(7.604)	53.06(7.696)
Water	105.04(15.234)	50.95(7.389)	50.64(7.344)
	LAYUP $(30/-30_2/30/90)_s$		
RTD	114.17(16.559)	48.31(7.006)	48.87(7.088)
Jet A	108.46(15.731)	46.24(6.706)	46.04(6.677)
JP-4	108.46(15.731)	46.24(6.706)	45.66(6.622)
Jet Ref.	108.46(15.731)	46.24(6.706)	46.56(6.753)
Water	105.04(15.234)	45.00(6.526)	44.89(6.511)
	LAYUP $(30/-30_2/30/90_2)_s$		
RTD	114.559(16.559)	44.12(6.399)	44.37(6.435)
Jet A	108.46(15.731)	42.22(6.123)	41.69(6.046)
JP-4	108.46(15.731)	42.22(6.123)	41.08(5.958)
Jet Ref.	108.46(15.731)	42.22(6.123)	42.67(6.189)
Water	105.04(15.234)	41.07(5.957)	40.71(5.905)

TABLE 2—*Hygral expansion coefficients.*

Fluid	β_1 ($\mu\epsilon$/% weight gain)	β_2 ($\mu\epsilon$/% weight gain)
Water	0	2200
Jet A	0	1900
JP-4	0	1900

Test Procedure

Tension tests of the specimens were performed using the MTS 810 servohydraulic test machine in the stroke control mode at a loading rate of 0.127 cm/min (0.05 in./min). An MTS 632.85 biaxial extensometer with a gage length of 2.54 cm (1.0 in.) was used to measure the longitudinal strain, the transverse displacement, and to detect delamination onset. Strain gages were not used due to concern that the jet fuels might adversely effect the strain gage installations. The biaxial extensometer was mounted along one of the edges across the thickness of the specimen as shown in Fig. 2. The edge of the coupon was simultaneously observed through a microscope of $\times 10$–60 magnifying power. Emery cloth was inserted between the specimen and the machine grips. Load, longitudinal strain, transverse displacement, and longitudinal displacement were continuously monitored and recorded using a personal computer (PC)-based data-acquisition system. The PC-based software was also employed to control the tests. Delamination onset was identified by the abrupt change in the transverse displacement on the transverse displacement-longitudinal strain plots as shown in Fig. 3. This critical point also coincided with the one obtained from the load-strain plots shown in Fig. 4. Delamination onset was further verified by visual observations made through the microscope.

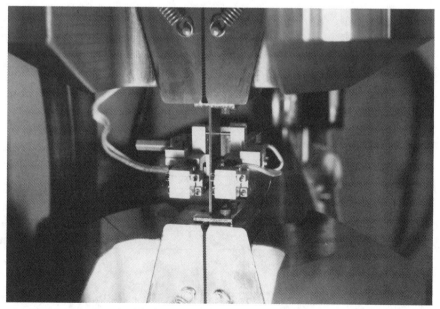

FIG. 2—*Photograph of the biaxial extensometer mounted on an EDT test specimen.*

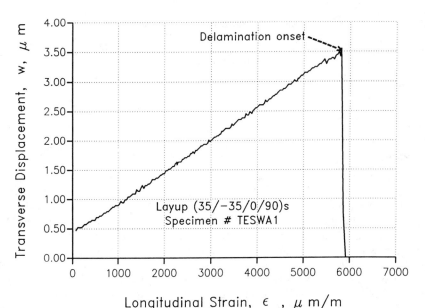

FIG. 3—*Typical transverse displacement strain plot.*

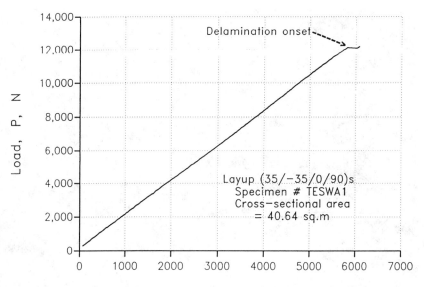

FIG. 4—*Typical load strain plot.*

The thickness of each specimen was recorded at three positions along the length of the specimen and averaged. Mean values of laminate thickness, delamination onset strains, and measured moisture weight gain for each layup and fluid saturated condition were recorded as shown in Table 3. Laminate moduli for all specimens were calculated by a linear regression analysis. These data were averaged for each layup and fluid state.

Data obtained during the tests provided the critical strain, ϵ_c, at delamination onset. This strain value was then used as input to the modified version of Q3DG to determine the critical mixed-mode strain energy release rates. This version of Q3DG is described in Part I of this paper (Ref 9). This analysis was performed to determine the effects of a nonuniform moisture state on a partially saturated EDT specimen. The specimen geometry, lamina material properties, weight gain $\Delta H(z)$, as well as critical strains are included in the input to this computer program. The weight gain, $\Delta H(z)$, was calculated using Fick's law as described in Ref 9 such that the average weight gain $\overline{\Delta H}$ of the mathematical model matched the average weight gain for the specimen being analyzed. The input data for the various fluid saturated cases and the RTD case are given in Table 3.

Experimental Results and Discussions

Results of this study are presented in terms of critical strain, ϵ_c, and strain energy release rates, G^M and G^{M+T+H}, for all cases tested. Test results for specimens saturated in the various fluids and in the RTD condition are presented in Figs. 5 through 8. The average percent weight

TABLE 3—*Data for saturated states.*

Environment	ΔH, % by Weight	Number of Tests	Laminate Thickness, m (in.)	ϵ_c, μm/m or μin./in.
LAYUP $(35/-35/0/90)_s$				
RTD	0.0	5	0.001 63(0.064)	3400
Jet Ref.	0.480	5	0.001 65(0.065)	4040
Jet A	0.789	7	0.001 65(0.065)	4620
JP-4	0.873	6	0.001 68(0.066)	4970
Water	2.24	7	0.001 65(0.065)	6040
LAYUP $(35/0/-35/90)_s$				
RTD	0.0	7	0.001 68(0.066)	3470
Jet Ref.	0.446	5	0.001 73(0.068)	4550
Jet A	0.795	6	0.001 68(0.066)	5180
JP-4	0.877	5	0.001 8(0.071)	5460
Water	2.15	2	0.001 68(0.068)	. . .
LAYUP $(30/-30_2/30/90)_s$				
RTD	0.0	7	0.001 93(0.076)	2420
Jet Ref.	0.459	6	0.002 11(0.083)	3640
Jet A	0.722	7	0.002 01(0.079)	4030
JP-4	0.806	6	0.002 11(0.083)	4250
Water	2.15	6	0.002 16(0.085)	5340
LAYUP $(30/-30_2/30/90_2)_s$				
RTD	0.0	6	0.002 41(0.095)	2060
Jet Ref.	0.421	6	0.002 54(0.100)	3100
Jet A	0.605	7	0.002 54(0.100)	3530
JP-4	0.707	6	0.002 54(0.100)	3760
Water	2.06	5	0.002 62(0.103)	4860

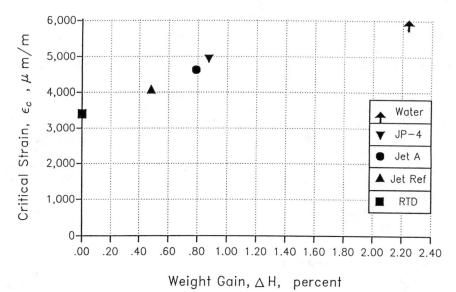

FIG. 5—*Critical strains for specimens saturated in various fluids: layup (35/−35/0/90)_s.*

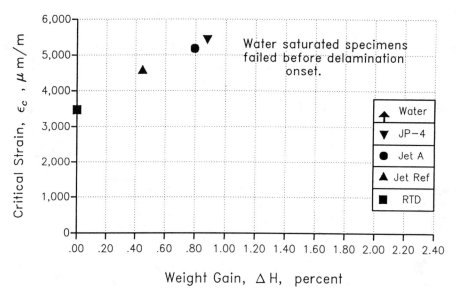

FIG. 6—*Critical strains for specimens saturated in various fluids: layup (35/0/−35/90)_s.*

gain, $\overline{\Delta H}$, at saturation and the corresponding critical strain, ϵ_c, are shown in these figures. The RTD specimens exhibited the lowest ϵ_c and the water saturated specimens the highest ϵ_c. Delamination was observed for all specimens except for $(35/0/−35/90)_s$ specimens saturated in water. These specimens exhibited tensile failure prior to the onset of delamination, thus the water data point is missing in Fig. 6 for this laminate. The total and Mode I critical strain energy release rates are presented in Fig. 9. The solid symbols represent G_c^{M+T+H} and the open

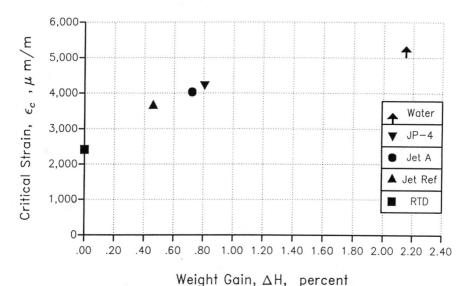

FIG. 7—*Critical strains for specimens saturated in various fluids: layup (30/−30/−30/30/90)ₛ.*

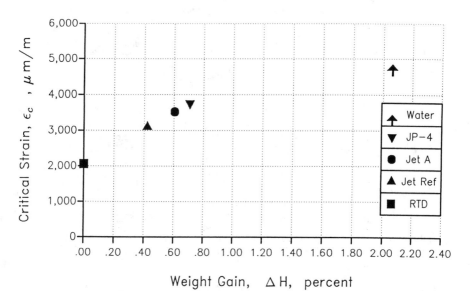

FIG. 8—*Critical strains for specimens saturated in various fluids: layup (30/−30/−30/30/90/90)ₛ.*

symbols are G_{Ic}^{M+T+H}. As seen from this figure and from Tables 3 and 4, G_{Ic}^{M+T+H} for a (35/ −35/0/90)ₛ varied between 72.0 to 94.5% depending upon the fluid absorbed. This is because the sublaminate in the delaminated region for this laminate is unsymmetric and unbalanced. The Mode I percentage was not a function of moisture sorption for the other three laminates. Another result from Fig. 9 is that the fracture toughness, G_c^{M+T+H}, is highest for specimens saturated in water and lowest for RTD specimens. The significance of the residual thermal and

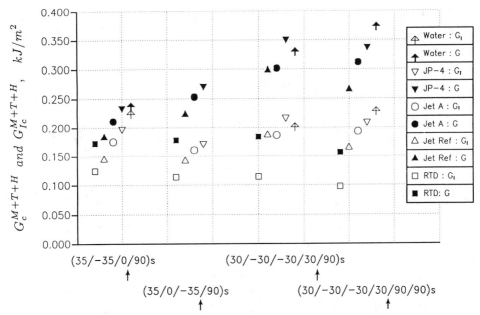

FIG. 9—*Mixed-mode fracture toughness (mechanical, thermal, and hygroscopic).*

TABLE 4—*Comparison of Mode I fracture toughness percentages.*

Environment	$\dfrac{G_{\mathrm{Ic}}^{M}}{G_{c}^{M}}$, %	$\dfrac{G_{\mathrm{Ic}}^{M+T+H}}{G_{c}^{M+T+H}}$, %	$\dfrac{G_{c}^{M}}{G_{c}^{M+T+H}}$, %
LAYUP $(35/-35/0/90)_s$			
RTD	93.57	72.01	53.65
Jet Ref.	93.30	78.61	51.27
Jet A	93.30	83.08	74.60
JP-4	93.30	84.49	77.61
Water	93.16	94.53	104.35
LAYUP $(35/0/-35/90)_s$			
RTD	64.00	63.97	50.25
Jet Ref.	63.51	63.49	63.39
Jet A	63.51	63.50	72.49
JP-4	63.51	63.50	75.06
Water	. . .	. . .	. . .
LAYUP $(30/-30_2/30/90)_s$			
RTD	62.29	62.27	45.43
Jet Ref.	62.66	62.65	62.90
Jet A	61.94	61.93	69.98
JP-4	62.66	62.66	72.66
Water	61.72	61.72	103.48
LAYUP $(30/-30_2/30/90_2)_s$			
RTD	62.35	62.33	45.18
Jet Ref.	62.70	62.69	62.61
Jet A	62.70	62.70	68.93
JP-4	62.70	62.70	72.05
Water	62.48	62.48	99.40

hygroscopic effects are apparent from the data presented in Fig. 10. For the case of specimens saturated in water, G_c^{M+T+H} for all the layups is greater than or at the most equal to G_c^M. The data reported in Table 4 shows that the G_c^M contribution to G_c^{M+T+H} increases with saturation weight gain. This is consistent with the results reported in Ref *13*, where it was pointed out that G_c^T and G_c^H tend to cancel each other.

The results for partially saturated specimens are presented in Figs. 11 through 18 in terms of critical strain, ϵ_c, versus average weight gain, $\overline{\Delta H}$, and ϵ_c versus soaking time. Two points are apparent from these plots. First, the critical strain for a partially saturated laminate does not change over the final 50% of its weight gain. Second, the first 50% of the weight gain occurs within the first one to three weeks of soaking time. Thus G_c^{M+T+H} reaches a plateau after the first 50% of the $\overline{\Delta H}$ at saturation as shown in Figs. 19 through 22.

Damage Development

Observations through the microscope showed that in the $(35/-35/0/90)_s$ layup, the delamination initiates at the 0/90 interfaces. At the critical strain, the delamination extended along the length of the specimen rapidly, typically shifting from one interface, to its symmetric 0/90 counterpart. Few 90° ply cracks were observed. In the $(35/0/-35/90)_s$ layup, delamination initiated at the −35/90 interface. A few 90° ply cracks were observed just before the rapid growth of the delamination. In the $(30/-30_2/30/90)_s$ and $(30/-30_2/30/90_2)_s$ layups, 90° ply crack formation was observed prior to delamination in the −30/90 interface. More cracks were observed for the second of these laminates. At the critical strain, the delamination grew rapidly along the length, shifting from one interface, through the 90° ply cracks, to its symmetric 30/90 counterpart. The delaminations did not shift interfaces at every 90° ply crack. In the case of the moisture saturated specimens, the only difference observed was that each shift in the interface was not limited to just one 90° ply crack, but was spread over at least two or more 90° ply cracks. Thus, in these cases, the shift was not abrupt but more gradual.

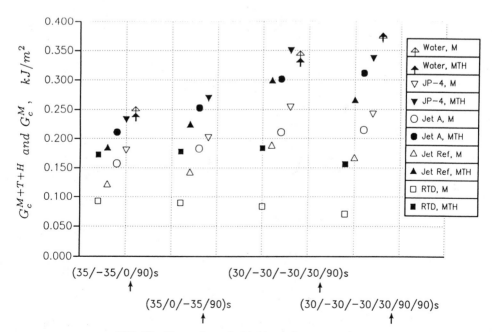

FIG. 10—*Comparison of mixed-mode fracture toughness.*

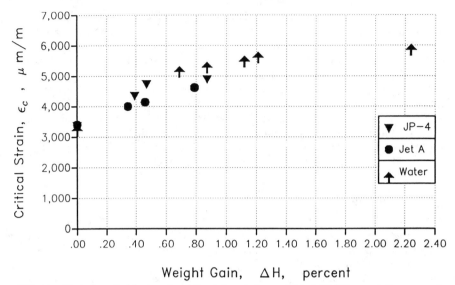

FIG. 11—*Variation of critical strain with weight gain layup (35/−35/0/90)ₛ: partially saturated.*

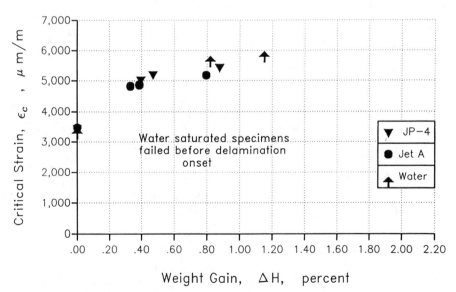

FIG. 12—*Variation of critical strain with weight gain layup (35/0/−35/90)ₛ: partially saturated.*

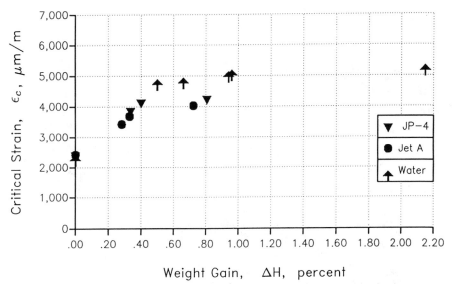

FIG. 13—*Variation of critical strain with weight gain layup (30/−30/−30/ 30/90ₛ: partially saturated.*

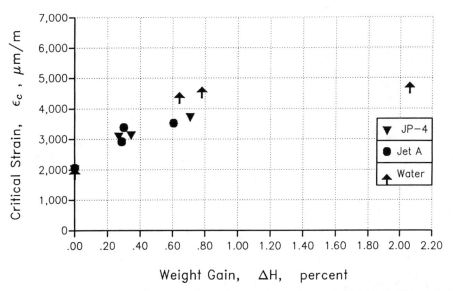

FIG. 14—*Variation of critical strain with weight gain layup (30/−30/−30/30/90/90)ₛ: partially saturated.*

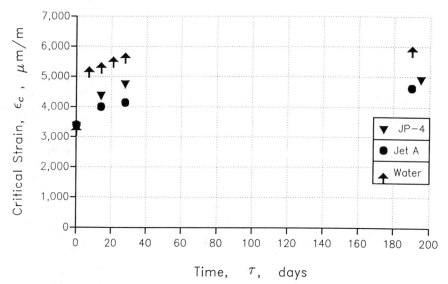

FIG. 15—*Variation of critical strain with time layup (35/−35/0/90)ₛ: partially saturated.*

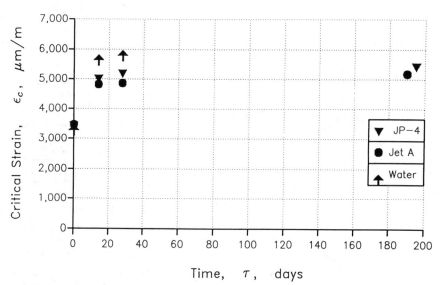

FIG. 16—*Variation of critical strain with time layup (35/0/−35/90)ₛ: partially saturated.*

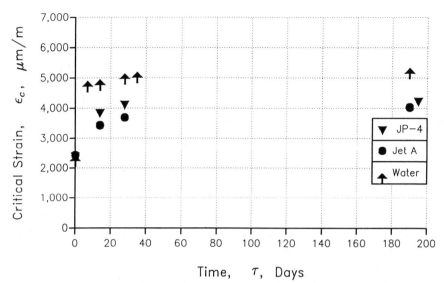

FIG. 17—*Variation of critical strain with time layup (30/−30/−30/30/90)ₛ: partially saturated.*

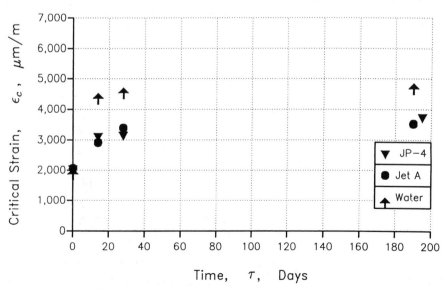

FIG. 18—*Variation of critical strain with time layup (30/−30/−30/30/90/90)ₛ: partially saturated.*

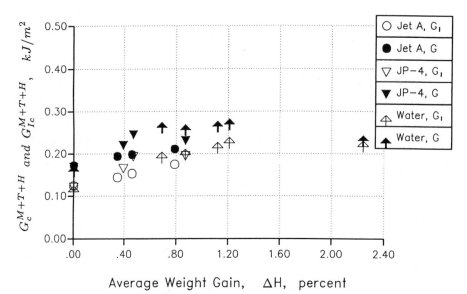

FIG. 19—*Variation of critical SERR* (M+T+H) *with weight gain: layup (35/−35/0/90)ₛ: partially saturated.*

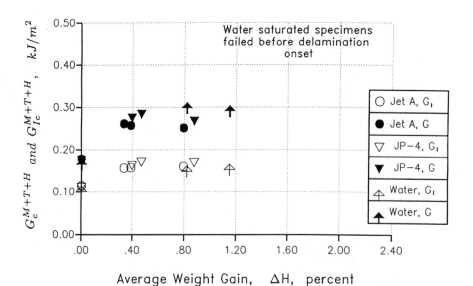

FIG. 20—*Variation of critical SERR* (M+T+H) *with weight gain: layup (+35/0/−35/90)ₛ: partially saturated.*

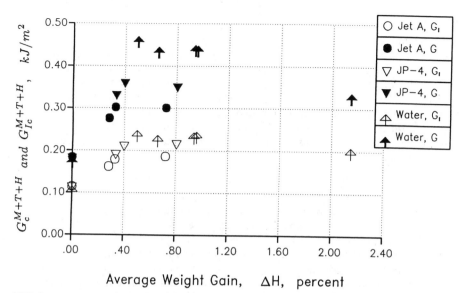

FIG. 21—*Variation of critical SERR (M+T+H) with weight gain: layup (30/−30/−30/30/90)ₛ: partially saturated.*

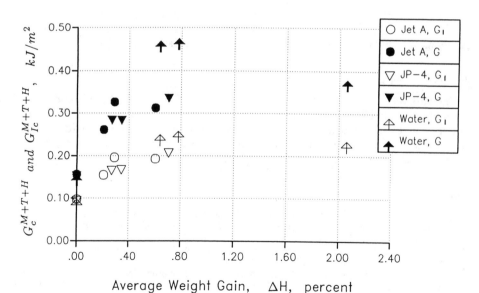

FIG. 22—*Variation of critical SERR (M+T+H) with weight gain: layup (30/−30/−30/30/90/90)ₛ: partially saturated.*

Conclusions

Edge delamination tension (EDT) tests were performed on AS4/3501-6 graphite-epoxy laminates that had been exposed to various aircraft-type fluid environments. Specimens were tested in JP-4 and Jet A jet fuels, water, and Type VII rubber swelling test fluid (jet reference fuel) in both the saturated and partially soaked conditions. Four laminates including (35/ $-35/0/90$)$_s$, (35/0/$-35/90$)$_s$, (30/-30_2/30/90)$_s$, and (30/-30_2/30/90$_2$)$_s$ were tested without inserts during this study. A modified version of Q3DG was used to calculate the mixed-mode critical strain energy release rates, G_{Ic}^{M+T+H} and G_c^{M+T+H} for an edge delamination tension specimen with a through-the-thickness Fickian moisture distribution. Based on the results of this investigation, it is concluded that:

1. Moisture sorption by a composite laminate leads to two effects:
   a. a mechanical effect due to swelling, which significantly influences the mixed-mode strain energy release rates for the edge delamination problem, and
   b. a change in material properties, which significantly influences the mixed-mode fracture toughness of AS4/3501-6 graphite-epoxy laminates.

2. Critical strain energy release rates for partially soaked specimens were nearly equal to the critical strain energy release rates for saturated specimens. Thus this kind of testing can be "accelerated" since the edge delamination toughness reaches an asymptotic value after soaking these specimens for approximately 20 days in a variety of fluids. Further studies have to be conducted to generalize this conclusion to include other composite systems.

3. The room temperature dry (RTD) condition is the most critical condition since this condition results in the lowest critical strain energy release rate.

4. Results have been presented that reflect an increase in the mixed-mode fracture toughness for graphite-epoxy composite material with absorption of jet fuels and water.

5. The Mode I percentage of fracture toughness is affected by moisture sorption only in the case of the sublaminate being unsymmetric and unbalanced.

6. The moisture sorption results in the increase of G_c^M contribution to the fracture toughness, G_c^{M+T+H}. This increase becomes more significant with increasing weight gain because, the contribution of G_c^T and G_c^H tend to cancel each other with increasing weight gain.

Acknowledgments

This work was supported by the Army Research Office, Contract No. DAAL 03-87-G-003. The authors would like to acknowledge the helpful suggestions of Dr. T. Kevin O'Brien. The authors would also like to acknowledge the assistance of Beech Aircraft Corp. in measuring the coefficients of thermal expansion for the material used in this study, and Boeing Military Airplane Company for supplying the JP-4 Military jet fuel.

References

[1] Horn, W. J., Shaikh, F. M., and Soeganto, A., "Degradation of Mechanical Properties of Advanced Composites Exposed to Aircraft Environment," *AIAA Journal,* American Institute of Aeronautics and Astronautics, Vol. 27, No. 10, Oct. 1989, pp. 1399–1405.
[2] Shen, C. H. and Springer, G. S., "Moisture Absorption and Desorption of Composite Materials," *Journal of Composite Materials,* Vol. 10, 1976, pp. 2–20.
[3] Wang, Q. and Springer, G. S., "Moisture Absorption and Fracture Toughness of PEEK Polymer and Graphite Fiber Reinforced PEEK," *Journal of Composite Materials,* Vol. 23, 1989, pp. 434–447.
[4] Yaniv, G., Peimanisis, G., and Daniel, I. M., "Method of Hygrothermal Characterization of Graphite/Epoxy Composites," *Journal of Composites Technology and Research,* Vol. 9, No. 1, Spring 1987, pp. 21–25.

[5] Pipes, R. B., Vinson, J. R., and Chou, T. W., "On Hygrothermal Response of Laminated Composite Systems," *Journal of Composite Materials,* Vol. 10, April 1976, pp. 129–148.

[6] Tsai, S. W. and Hahn, H. T., *Introduction to Composite Materials,* Technomic Publishing Co., Westport, CT, 1980.

[7] Whiteside, J. B., DeIasi, R. J., and Schulte, R. L., "Distribution of Absorbed Moisture in Graphite/ Epoxy after Real-Time Environmental Cycling," *Long-Term Behavior of Composites, ASTM STP 813,* T. K. O'Brien, Ed., American Society for Testing and Materials, Philadelphia, 1983, pp. 192–205.

[8] Farley, G. L. and Herakovich, C. T., "Influence of Two-Dimensional Hygrothermal Gradients on Interlaminar Stresses Near Free Edges," *Advanced Composite Materials—Environmental Effects, ASTM STP 658,* J. R. Vinson, Ed., American Society for Testing and Materials, Philadelphia, 1978, pp. 143–159.

[9] Hooper, S. J., Toubia, R. F., and Subramanian, R., this publication, pp. 89–106.

[10] O'Brien, T. K., "Characterization of Delamination Onset and Growth in a Composite Laminate," *Damage in Composite Materials, ASTM STP 775,* K. L. Reifsnider, Ed., American Society for Testing and Materials, Philadelphia, 1982, pp. 140–167.

[11] Garg, A. C., "Fracture Behaviour of Carbon Fiber Reinforced Composites—A Review," *Journal of the Mechanical Behaviour of Materials,* Vol. 1, No. 1–4, 1988, pp. 101–210.

[12] O'Brien, T. K., "Mixed-Mode Strain Energy Release Rate Effects on Delamination of Composites," *Effects of Defects in Composite Materials, ASTM STP 836,* American Society for Testing and Materials, Philadelphia, 1984, pp. 125–142.

[13] O'Brien, T. K., Raju, I. S., and Garber, D. P., "Residual Thermal and Moisture Influences on the Strain Energy Release Rate Analysis of Edge Delamination," *Journal of Composites Technology and Research,* Vol. 8, No. 2, 1986, pp. 37–47.

[14] Kriz, R. D. and Stinchcomb, W. W., "Effects of Moisture, Residual Thermal Curing Stresses, and Mechanical Load on Damage Development in Quasi-Isotropic Laminates," *Damage in Composite Materials, ASTM STP 775,* K. L. Reifsnider, Ed., American Society for Testing and Materials, Philadelphia, 1982, pp. 63–80.

[15] Russell, A. J. and Street, K. N., "Moisture and Temperature Effects on the Mixed-Mode Delamination Fracture of Unidirectional Graphite/Epoxy," *Delamination and Debonding of Materials, ASTM STP 876,* W. S. Johnson, Ed., American Society for Testing and Materials, Philadelphia, 1985, pp. 349–370.

[16] *ICI Fiberite Materials Handbook,* ICI Fiberite Corp., Tempe, AZ, 15 March 1989.

Han-Pin Kan,[1] Narain M. Bhatia,[1] and Mary A. Mahler[1]

Effect of Porosity on Flange-Web Corner Strength

REFERENCE: Kan, H-P., Bhatia, N. M., and Mahler, M. A., **"Effect of Porosity on Flange-Web Corner Strength,"** *Composite Materials: Fatigue and Fracture (Third Volume), ASTM STP 1110,* T. K. O'Brien, Ed., American Society for Testing and Materials, Philadelphia, 1991, pp. 126–139.

ABSTRACT: A strength of materials based model was developed to determine the interlaminar tensile stress at the corner radii of curved composite structures. The model assumes that the stress state in the local area near the corner is under pure bending. The force in each layer (ply) of the laminate is determined by classical lamination theory and the curved beam analysis method. The interlaminar tensile stress is then obtained by enforcing the force and moment equilibrium in each layer.

To verify the accuracy of this analytical model, a series of corner specimens were fabricated and tested. These specimens were fabricated under controlled conditions and no porosity was detected in the specimens. The corner specimens were tested by pulling apart the sides of the corner to introduce bending moment at the corner. The maximum interlaminar tensile stress computed using the analysis model was compared to the flatwise tensile strength of the material in order to predict the failure moment of the test specimen. Excellent correlations between the measured and predicted failure moment were achieved.

A series of tests were conducted on flange-web corner elements obtained from graphite composite frames that had porosity in their corners. The objectives of these tests were to determine: (1) level of porosity in the corners by using ultrasonic inspections and photomicrographic examinations, and (2) the effect of porosity in the corner on the bending strength and the interlaminar tensile strength. Ultrasonic inspections revealed the locations along the corners where porosity was present. Photomicrographic examinations showed that the porosity occurred in clusters between plies and appeared similar to delaminations. Qualitative levels of porosity, based on photomicrographic observations, were established; these were (1) no porosity, (2) low porosity, and (3) high porosity. Attempts to use acid digestion methods to establish quantitative measures of porosity were not successful because the porosity was localized between only a few plies.

A relationship was established between ultrasonic signal attenuation loss due to porosity and interlaminar tensile strength of porous structures. Strength testing was conducted on the flange-web corner specimens. Post-failure examination of the test specimens showed that interlaminar tension dominated the failure with little evidence of shearing effects. The failure strength data were used with the analytical model to calculate the interlaminar tensile failure stress for each specimen.

KEY WORDS: composite materials, corner radii, interlaminar tensile strength, porosity, nondestructive inspection, fatigue (materials), fracture

Composite materials have been extensively used in aircraft structures in recent years. One area of concern in the application of composites in primary structures is their relatively low interlaminar strength. In addition, interlaminar strength is degraded by manufacturing and processing defects such as porosity. As composite components increase in complexity, there is a higher possibility of porosity existing in the structure. The influence of porosity on static

[1] Principal engineer, engineering specialist, and engineer, respectively, Structural Strength and Life Assurance Research, Northrop Corporation, Aircraft Division, Hawthorne, CA 90250.

strength and fatigue life of composite laminates has been experimentally investigated in Ref *1*. An extensive assessment of these effects on the performance of composite structures was conducted in Ref *2*. These investigations indicate that porosities in composites, either artificially induced or processing caused, can result in significant reduction in structural strength and life.

Porosities tend to concentrate in areas where curing pressure is nonuniform. One typical example is at the flange-web corner of a structure. At these locations, the structure is usually subjected to bending loads that induce high interlaminar tensile stress. To assure adequate strength of the structure, an analysis method is needed to characterize the state of stress at this region and to quantify the influence of porosity on the strength of the structure.

The objective of this study was to experimentally and analytically determine the influence of porosity on the strength of flange-web corners. A strength of materials based analysis was developed to predict the failure load and failure location of a composite corner radius. This analysis method was verified by extensive data correlations. The analysis was then used to quantify the effects of porosity on the flange-web corner strength.

Analysis Development

Analytical prediction of the static strength of curved composite laminates has been discussed in Refs *3* through *9*. Reference *9* presented a review of the analysis methods and failure criteria used for strength prediction of composite corners. A finite element method was used in Refs *3, 4*, and *6* through *8*. In Ref *5*, an elasticity solution was used to determine the stress distribution in a semicircular composite beam. In this paper, a strength of materials based model was used to determine the interlaminar normal stress in a composite laminate at structural corner radii. The stress state in the local area near the corner is assumed to be under pure bending. The governing equations were derived using classical curved beam theory by assuming plane sections in the local region remain plane while deforming. The force in each layer (ply) of the laminate is determined by enforcing the overall equilibrium of the local region. Furthermore, the equilibrium of each layer in the local area is enforced to determine the interlaminar normal stress. The structural configuration, local model, and free body diagram of a layer are shown in Fig. 1.

The tangential strain in the ith layer of the laminate can be expressed as

$$\epsilon_i = \frac{\Delta_i}{R_i} = \frac{P_i}{A_i E_i} \tag{1}$$

where

Δ_i = the elongation of the ith layer in the circumferential direction,
R_i = the radius of the neutral axis of the ith layer,
A_i = the cross-sectional area of the ith layer,
E_i = the Young's modulus of the ith layer in the circumferential direction, and
P_i = the tangential force in the ith layer.

From Eq 1, the force in the ith layer is given by

$$P_i = \frac{\Delta_i A_i E_i}{R_i} \tag{2}$$

Based on the assumption that plane sections in the local region remain plane while deforming, the elongations in different layers are in proportion to the distance from the neutral axis of the

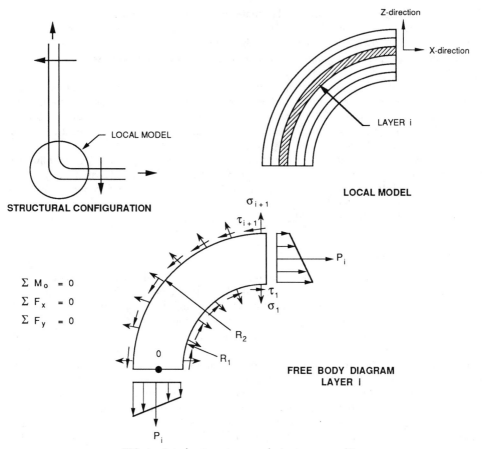

FIG. 1—*Interlaminar stress analysis at corner radii.*

laminate, that is

$$\frac{\Delta_i}{\Delta_k} = \frac{y_i}{y_k} \tag{3}$$

or

$$\Delta_i = \Delta_k \cdot \frac{y_i}{y_k} = \frac{y_i}{y_k} \cdot \left(\frac{P_k R_k}{A_k E_k}\right) \tag{4}$$

The force in the ith layer is then related to the force in the kth layer by substituting Eq 2 into Eq 4, and written as

$$P_i = P_k \left(\frac{R_k}{A_k E_k y_k}\right)\left(\frac{A_i E_i y_i}{R_i}\right) \tag{5}$$

The overall equilibrium of moment requires

$$M = \Sigma P_i y_i \tag{6}$$

Equations 5 and 6 yield the overall equilibrium conditions for the curved laminate under pure applied bending moment. After substituting Eq 5 into Eq 6 and solving for the force in the ith layer, one has

$$P_i = \frac{M}{\dfrac{R_i}{A_i E_i y_i} \Sigma \dfrac{A_k E_k y_k^2}{R_k}} \tag{7}$$

The R_i and y_i terms in Eq 7 are expressed in terms of the neutral axis, R_{NA}, of the laminate, and obtained using the basic definition

$$R_{NA} = \Sigma E_i A_i \bigg/ \int \frac{d(EA)}{r} \tag{8}$$

For laminated sections of uniform width, Eq 8 can be reduced to

$$R_{NA} = \frac{\Sigma E_i t_i}{\Sigma E_i \ln(r_{i+1}/r_i)} \tag{9}$$

where

t_i = the thickness of the ith layer, and
r_i = the radius at the ith interface.

Similarly, the neutral axis of the ith layer can be obtained from Eq 8 and given by

$$R_i = \frac{t_i}{\ln(r_{i+1}/r_i)} \tag{10}$$

The distance, y_i, in Eq 7 is defined as

$$y_i = R_i - R_{NA} \tag{11}$$

The interlaminar normal stress is determined by enforcing the force and moment equilibrium in each layer. The tangential stress in each layer is assumed to be linearly distributed through the thickness of the layer, as shown in Fig. 1. For the pure bending problem considered, a stress-free condition is applied on the inner and outer surfaces of the laminate. A ply-by-ply technique is used to compute the interlaminar normal stresses. Each ply is characterized by the ply thickness and the in-plane Young's modulus (modulus in the circumferential direction of the curved lamina). The interlaminar shear stress is assumed to be negligible [8].

Analytical Verification

To verify the accuracy of this analytical model, a series of corner specimens were fabricated and tested. These specimens were fabricated under controlled conditions and no porosity was detected in the specimens. Figure 2 shows the test specimen and the test setup. The specimens were fabricated from AS4/5250-3 (graphite/BMI) biwoven cloth with a nominal per ply thick-

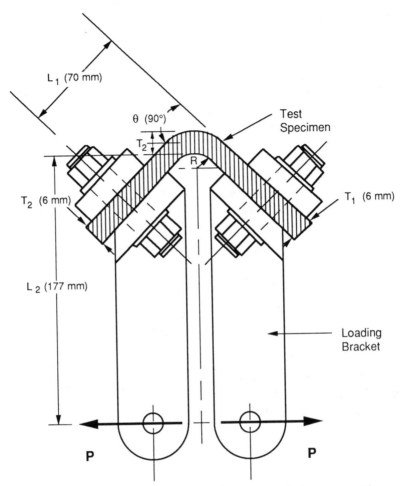

FIG. 2—*Typical specimen geometry and test setup (typical measurements).*

ness of 0.36 mm (0.014 in.). In order to achieve uniform nominal thickness in the corners, the specimens were manufactured using hard tools on both sides. Laminates with three different thicknesses were tested. They were: (1) 5-ply $(45/0/\overline{45})_s$, (2) 11-ply $((45/0)_2/45/\overline{0})_s$, and (3) 21-ply $((45/0)_5/\overline{45})_s$ laminates. The specimens were 50.8 mm (2 in.) wide with a radius of curvature between 1.98 to 9.53 mm (0.078 to 0.375 in.). Cure conditions were similar to those recommended by the material's supplier, Narmco.

Strength testing was conducted using the test setup shown in Fig. 2. The specimens were pulled apart by the force, P, using a displacement-controlled MTS machine to produce pure bending at the corner. A total of 33 specimens were tested under room-temperature ambient and 88°C/1.3% moisture environmental conditions. The bending moment at the corner of each specimen at initiation of failure was determined.

The results of these tests indicated that specimen failure is controlled by normal opening along the ply interfaces. Initial failure occurs by delamination along ply interfaces where interlaminar tension is greatest. Final failure of the specimen is triggered by unstable growth of the delamination. Figure 3 shows the failure sequence of a typical specimen. No visual shear

FIG. 3—*Flange-web corner failure mode under bending load: (a) initial failure—first delamination, and (b) after initial failure—load is continued, more delamination occurs.*

effects were evident in the test specimens. This observation suggested that interlaminar tension failure is the dominant failure mechanism. The analysis was expanded to include shear effects. Results of the very small shear stresses were similar to results found in Ref *1*. In both analyses, corners loaded under a nearly pure bending moment experience a shear stress two orders of magnitude lower than the normal interlaminar stress. This observation tends to justify the assumption used in the analysis that interlaminar shear stresses are negligible.

A maximum interlaminar stress criterion was used to predict the failure moment of the test specimens. The flatwise tension strength of the graphite/BMI was taken as the critical interlaminar stress. Analysis was conducted for each test specimen to obtain the interlaminar stress distribution through the thickness of the laminate at the corner location. Typical distribution of the interlaminar stress is shown in Fig. 4. The figure shows that the maximum interlaminar tension occurs along the fourth interface from the inner surface of the laminate. This location of the maximum interlaminar tension agrees with the location of the delamination observed during the test.

The failure moment was predicted by comparing the maximum interlaminar stress with the flatwise tensile strength of the material. A correlation of the corner bending moment at failure initiation from test data versus the predicted failure moment for each specimen is shown in Fig. 5. The figure shows that the predicted failure moments were within ±5% of measured moments for the two test conditions. These results verify the validity of the analytical model to accurately predict the interlaminar tension stress and the failure moment.

Flange-Web Corner Strength with Porosity

A series of tests were conducted on flange-web corner elements obtained from composite frames that had porosity detected in their corners. Porosity was detected by ultrasonic A-scan and C-scan nondestructive inspection (NDI) methods. The porosity defects were not inten-

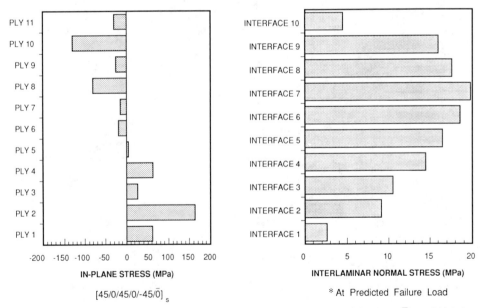

FIG. 4—*Predicted ply stresses for the 11-ply control specimen:* (left) *[45/0/45/0/-45/$\overline{0}$]ₛ and* (right) *at predicted failure load.*

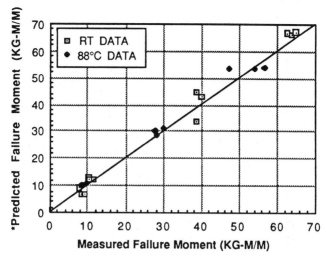

FIG. 5—*Comparison of predicted and measured failure moment (*predicted based on flatwise tensile strength).*

tionally induced, but resulted from manufacturing and processing. Unlike the other corners, these were manufactured with a female tool and pressure bagged on the inner radius. The objectives of these tests were to determine: (1) level of porosity in the corners by using ultrasonic inspections and photomicrographic examinations and (2) effects of porosity on the corner bending strength and the interlaminar tensile strength.

The typical composite Z-frame from which the flange-web corner elements were obtained is shown in Fig. 6. The ultrasonic inspections of the frames revealed the locations in the corners where porosity was present. Test specimens were machined from selected locations along the upper and lower flange-web corners of the frames. The nominal specimen width is shown in Fig. 7. The actual specimen widths varied from 5.6 to 2.2 cm (2.2 to 0.85 in.), and the test data were normalized to unit width.

The specimens from the various locations provided a range of porosity levels as indicated by the ultrasonic signal transmission decibel levels. To determine the physical character of the porosity, initially, specimens from selected locations were subjected to photomicrographic examination. It showed that porosity occurred in clusters between plies appearing similar to a delamination. Based on these observations, three qualitative levels of porosity were established: (1) no porosity, (2) low porosity, and (3) high porosity. Typical photomicrographs of specimens with these three levels of porosity are shown in Fig. 8. The level of porosity in the

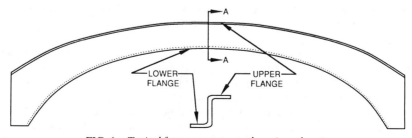

FIG. 6—*Typical frame geometry and specimen location.*

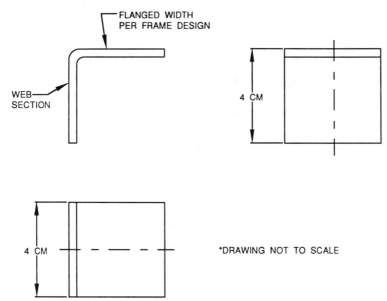

FIG. 7—*Test specimen design for flange-web corner elements (test specimens are machined from a selected frame).*

test specimens was estimated to be the same as that observed in the neighboring sections that were photomicrographed. However, this could not be confirmed by the ultrasonic NDI of the test specimens. Attempts to establish quantitative measures of porosity by using ultrasonic inspection or acid digestion methods were not successful because porosity was very much localized between a few plies.

Testing procedures previously discussed induced a pure bending moment at the corner radii. The failure strength data, for the room temperature and elevated temperature tests, were summarized as applied moment-per-unit width and are presented in Figs. 9 and 10. Analytically predicted moments, based on specific values of interlaminar tensile stresses and nominal laminate thicknesses, are shown for purposes of comparison. Test data showed considerable scatter as compared to analytical prediction. The laminates had porosity and higher than nominal thickness. These two tend to counteract each other with porosity reducing strength and increased thickness increasing strength.

To clarify the effect of porosity on interlaminar tensile strength of this composite material system, the failure strength data were used with the analytical model to calculate interlaminar tensile failure stress for each specimen, based on actual laminate thickness in the corner. The results, plotted in Fig. 11, show the interlaminar tensile failure stress versus temperature for the cases of no porosity, low porosity, and high porosity. These results show that the interlaminar tensile strength is reduced by up to 30% for the low-porosity specimens and up to 50% for the high-porosity specimens.

Additional tests were conducted to establish the lower bounds for the interlaminar tensile strength values and to improve the correlation between ultrasonic NDI signal and laminate failure strength. For these tests, three frames with porosity in their flange-web corners were selected. A total of 62 specimens were tested at five different temperatures. Before testing, each specimen was subjected to a detailed C-scan inspection procedure that measured the loss of attenuation of the ultrasonic signal caused by porosity. This attenuation loss/thickness was numerically read by the scanner. From the specimens measured, the previous identification

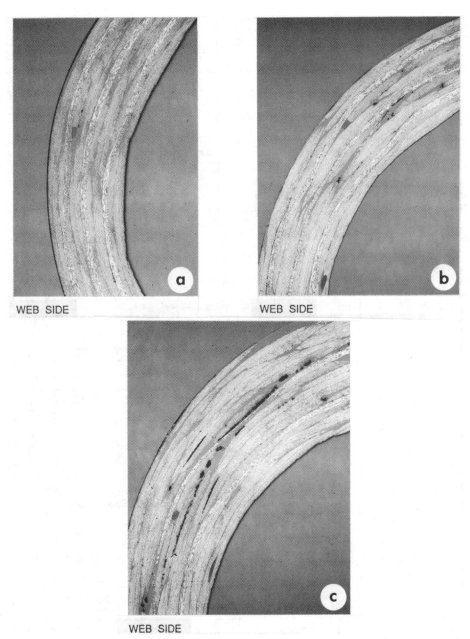

FIG. 8—*Typical photomicrographs of specimens with porosity:* (a) *no porosity,* (b) *low porosity, and* (c) *high porosity.*

of high, low, and no porosity was correlated to the attenuation loss:

1. no porosity = 0 to 0.5 db/mm,
2. low porosity = 0.5 to 3.0 db/mm, and
3. high porosity = 3.0 to 8 db/mm

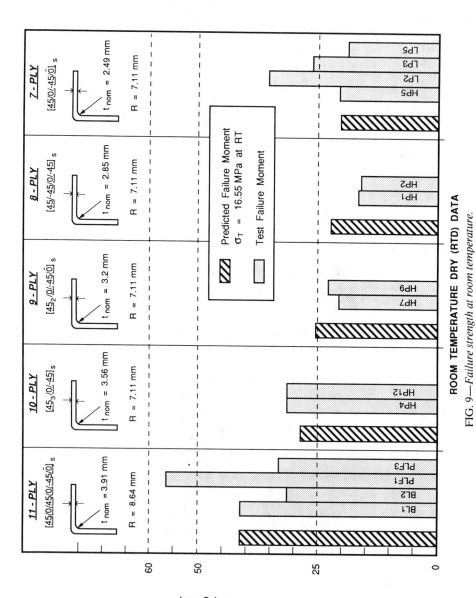

FIG. 9—*Failure strength at room temperature.*

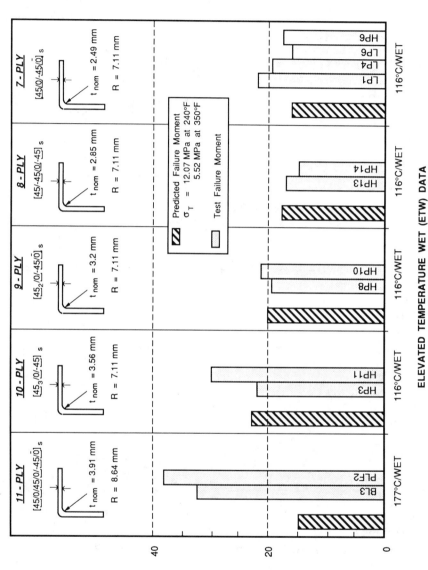

FIG. 10—Failure strength at elevated temperature.

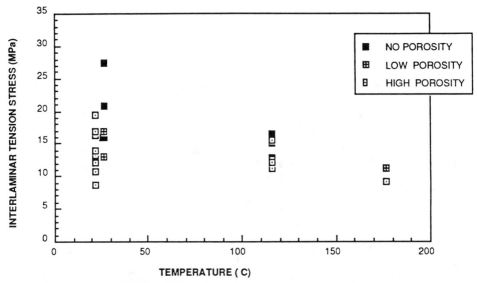

FIG. 11—*Effect of porosity on interlaminar strength.*

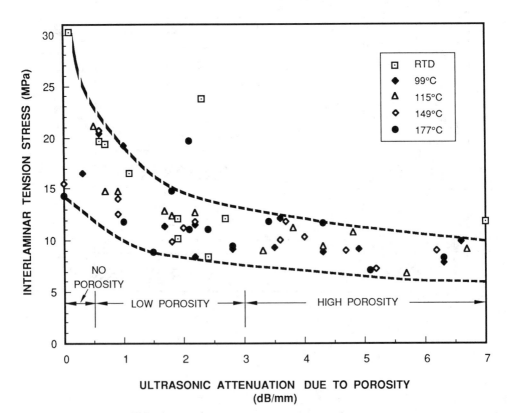

FIG. 12—*Interlaminar stress versus amount of porosity.*

From the strength test data, the interlaminar tensile stress at failure was calculated using the analytical method discussed earlier. The results of interlaminar tensile strength versus ultrasonic signal attenuation loss due to porosity are plotted in Fig. 12. These results show that as attenuation loss increases, indicating higher porosity levels, the interlaminar tensile strength decreases. These results likewise show the lower bound for interlaminar tensile strength for a range of porosity levels based on the C-scan inspection results.

Conclusions

1. Failure of composite corner radii subjected to bending loads is caused by initiation and propagation of delaminations driven by interlaminar tension.

2. Failure load and failure location can be predicted by a simple strength of materials analysis method.

3. Corner specimen tests in conjunction with the analysis method discussed can be used to determine the interlaminar tensile strength of a composite.

4. Tests showed that interlaminar tensile strength is reduced by 30 to 50% for low to high porosity specimens. Therefore, when porosity is observed, structural margins need to be recalculated with reduced interlaminar tensile strength allowables.

5. The relationship demonstrated between ultrasonic signal attenuation loss due to porosity and interlaminar tensile strength can be used as an NDI procedure to assess the strength of structures with observed porosity.

Acknowledgments

This work was conducted under Northrop's Independent Research and Development Projects R-1056 and D-1685.

References

[1] Ramkumar, R. L., Grimes, G. C., Adams, D. F., and Dusablon, E. G., "Effects of Materials and Processes Defects on the Compression Properties of Advanced Composites," Final Technical Report, Contract Nos. N00019-80-C-0490 and N00019-80-C-0484, Naval Air Systems Command, May 1982.

[2] McCarty, J. E. and Ratwani, M. M., "Damage Tolerance of Composites," Interim Report No. 1, Contract No. F33615-82-C-3213, Air Force Wright-Aeronautical Laboratory, Wright-Patterson Air Force Base, OH, March 1983.

[3] Lowry, D. W., Krebs, N. E., and Dobyns, A. L., "Design, Fabrication and Test of Composite Curved Frames for Helicoptor Fuselage Structure," NASA CR 172438, National Aeronautics and Space Administration, Washington, DC, Oct. 1984.

[4] Chang, F. K. and Springer, G. S., "The Strengths of Fiber Reinforced Composite Bends," *Journal of Composite Materials*, Vol. 20, Jan. 1986.

[5] Ko, W. L., "Delamination Analysis of Semicircular Laminated Composite Curved Bars Subjected to Bending," NASA TM 4026, NASA Langley Research Center, Hampton, VA, April 1988.

[6] Sun, C. T. and Kelly, S. R., "Failure in Composite Angle Structures Part I: Initial Failure," *Journal of Reinforced Plastics and Composites*, Vol. 7, May 1988.

[7] Sun, C. T. and Kelly, S. R., "Failure in Composite Angle Structures Part II: Onset of Delamination," *Journal of Reinforced Plastics and Composites*, Vol. 7, May 1988.

[8] Kelly, S. R., "Failure Analysis of Laminated Composite Angles," Ph.D. thesis, School of Aeronautics and Astronautics, Purdue University, West Lafayette, IN, May 1987.

[9] Kedward, K. T., Wilson, R. S., and McLean, S. K., "The Flexure of Simply Curved Composite Shapes," to appear in *Composite Materials: Testing and Design, 9th Volume, ASTM STP 1059*, S. P. Garbo, Ed., American Society for Materials and Testing, Philadelphia.

Interlaminar Fracture Toughness

Shah Hashemi,[1] Anthony J. Kinloch,[1] and Gordon Williams[1]

Mixed-Mode Fracture in Fiber-Polymer Composite Laminates

REFERENCE: Hashemi, S., Kinloch, A. J., and Williams, G., **"Mixed-Mode Fracture in Fiber-Polymer Composite Laminates,"** *Composite Materials: Fatigue and Fracture (Third Volume), ASTM STP 1110,* T. K. O'Brien, Ed., American Society for Testing and Materials, Philadelphia, 1991, pp. 143–168.

ABSTRACT: Mode I, Mode II, and Mixed-Mode I/II interlaminar tests on unidirectional carbon-fiber composites have been conducted using beam specimens. Both a thermosetting-based (an epoxy resin) matrix and a thermoplastic-based (poly(ether-ether ketone)) matrix have been employed. The fracture energy, G_c, has been ascertained and the various correction factors that need to be applied if accurate results are to be obtained are described. Where Mixed-Mode I/II loadings may have been present, the measured value of G_c has been partitioned into the separate G_I and G_{II} components. The experimental results all suggest that the partitioning of G on a global energy basis, as opposed to using local stress-field solutions, is the most appropriate for the laminates. Thus, the differences in energy absorption are a consequence of opening as opposed to sliding, and the symmetry, or otherwise, of these motions is not important. It is thus apparent that a global energy analysis is entirely satisfactory and appropriate for these types of systems. Finally, it is also shown that when the failure locus is plotted in the form of G_I versus G_{II} then it may be theoretically described by a simple interaction parameter model, where the interaction parameter, I_i, has a linear dependence upon the value of G_I/G.

KEY WORDS: composite materials, failure criterion, fracture mechanics, interlaminar failure, local stress-field, mixed-mode tests, fracture, fatigue (materials)

It is now well established that a useful method for characterizing the toughness of composite laminates is to measure the fracture toughness, G_c; the energy per unit area needed to produce the failure [1–4]. Highly anisotropic laminates are unusual in their fracture behavior in that fracture is almost always by delamination and hence can occur by the opening of the crack (Mode I), a sliding mode (Mode II), or mixtures of the two depending on how the laminate is loaded. Isotropic materials almost always fail in a local opening mode since this generally requires less energy. Because the G_c values can be different for various modes, the characterization of the fracture process requires the establishment of some form of failure locus. This is reflected in recent publications [2,5] in which attempts have been made to establish precise procedures for both a pure Mode I test and for pure Mode II. There is also some information on one ratio of mixed-mode loading [6], but much of the published data seems prone to large scatter and uncertainty in the analysis [6,7].

In this paper, we shall add to our previously published results on pure mode loadings [2–4] by describing two forms of mixed-mode beam test and give data on two composites. The two composites are based upon a carbon-fiber thermosetting epoxy-resin material and a carbon-fiber thermoplastic (poly(ether-ether ketone)) material. The basic analysis for G will be given

---

[1] Research fellow, professor, and professor, respectively, Mechanical Engineering Department, Imperial College of Science, Technology, and Medicine, University of London, SW7 2BX, United Kingdom.

(with appropriate correction factors), and the partitioning in the mixed-mode cases will be described in some detail. Such analysis is necessary if mixed-mode tests are to be put on as firm a basis as the current pure mode tests.

Total Energy Release Rate, G

Introduction

All the specimens used here are special cases of the loadings shown in Fig. 1 where the total energy release rate, G, may be derived from moments, M, at the crack root [5,8] and is given by

$$G = \frac{3}{4B^2h^3E_{11}} \left[\left(\frac{2h}{h_1}\right)^3 M_1^2 + \left(\frac{2h}{h_2}\right)^3 M_2^2 - (M_1 + M_2)^2 \right] \qquad (1)$$

where E_{11} is the axial modulus of the laminate.

Pure Mode Loadings

The pure loading modes are achieved by the configurations shown in Figs. 2 and 3, in which $h_1 = h_2 = h$. For pure Mode I loading, the double cantilever beam (DCB) specimen, shown in Fig. 2, has been employed, and here $M_2 = -M_1 = Pa$. For pure Mode II loading the end-loaded split (ELS) specimen, shown in Fig. 3, has been used, and here $M_2 = M_1 = -Pa/2$. Substitution of these values into Eq 1 gives

$$G_I = \frac{12P^2a^2}{B^2h^3E_{11}} \qquad (2)$$

$$G_{II} = \frac{9P^2a^2}{4B^2h^3E_{11}} \qquad (3)$$

The nature of the deformations in these two cases is shown in Fig. 4 where there is *symmetric* motion in both modes with the displacement $v_1 = v_2$ in Mode I and the displacement $u_1 = $

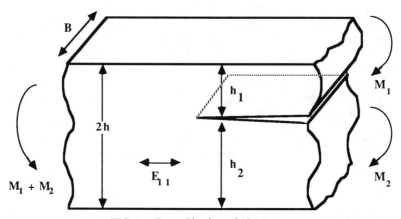

FIG. 1—*General loading of a laminate.*

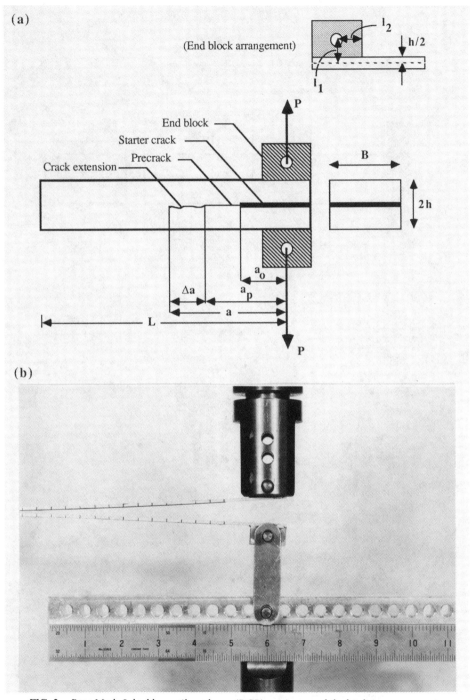

FIG. 2—*Pure Mode I double cantilever beam (DCB) specimen and the loading arrangement.*

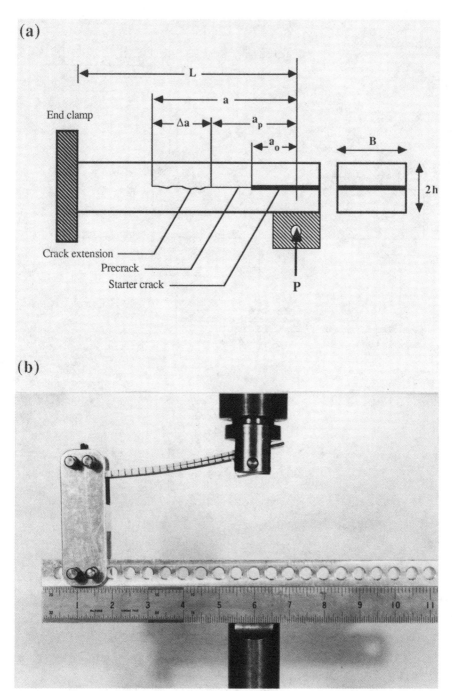

FIG. 3—*Pure Mode II end loaded split (ELS) specimen and the loading arrangement.*

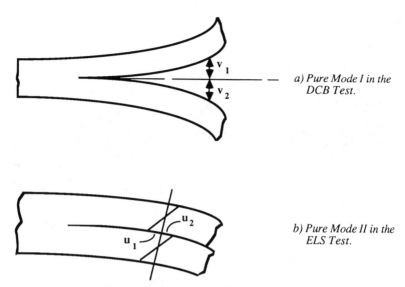

a) *Pure Mode I in the DCB Test.*

b) *Pure Mode II in the ELS Test.*

FIG. 4—*Deformations in the pure mode tests.*

$-u_2$ in Mode II. In the ELS test, the load, P, is equally shared between the two arms giving equal radii of curvature.

Varying Ratio Mixed-Mode (VRMM) Loadings

The test used to give a varying ratio mixed-mode (VRRM) is shown in Fig. 5 [8,9]. For $0 < a < l$, we have pure Mode II loading as in the ELS test, but for $l < a < L$, the loadings result in different moments on each arm given by

$$M_1 = Pl\left(1 - \frac{a}{L}\right) - P_1 a$$

and

$$M_2 = P_1 a$$

where P_1 is the force between the two halves of the beam given by

$$P_1 = \frac{Pl}{2a}\left[\frac{1}{2}\left[1 - \left(\frac{l}{a}\right)^2\right] + \left(1 - \frac{a}{L}\right)\right]$$

The total G is given from Eq 1 as

$$G = \frac{3P^2 l^2}{4B^2 h^3 E_{11}}\left[\left[1 - \left(\frac{l}{a}\right)^2\right]^2 + 3\left[1 - \left(\frac{a}{L}\right)^2\right]\right] \tag{4}$$

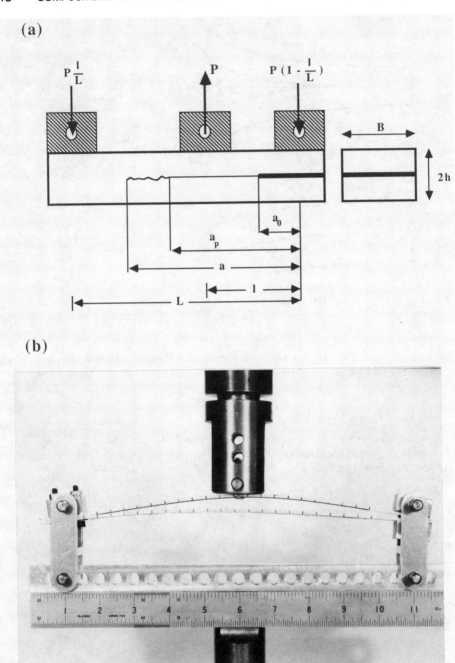

FIG. 5—*Variable ratio mixed-mode specimen (VRMM) and the loading arrangement.*

This may be simply partitioned since the deformations for the two modes can be deduced by dividing M_1 and M_2 as

$$M_1 = M_{II} - M_I$$
$$M_2 = M_{II} + M_I$$

that is, M_{II} bending the arms in the same sense and M_I in an opposing manner. Substituting this into Eq 1 with $h_1 = h_2 = h$ gives

$$G = \frac{9}{B^2 h^3 E_{11}} \left[M_{II}^2 + \frac{4}{3} M_I^2 \right]$$

(5)

where $M_{II} = \frac{1}{2}(M_2 + M_1) = Pl/2(1 - a/L)$ and $M_I = \frac{1}{2}(M_2 - M_1) = Pl/4(1 - l/a)^2)$ giving

$$\frac{G_I}{G} = \cfrac{1}{1 + 3 \left[\cfrac{1 - \left(\dfrac{a}{L} \right)}{1 - \left(\dfrac{l}{a} \right)^2} \right]^2}$$

(6)

Note that at $a = l$ we have pure Mode II and at $a - L$, pure Mode I.

Fixed Ratio Mixed-Mode (FRMM) Loadings

A somewhat simpler test to perform is shown in Fig. 6 in which only one arm of an asymmetric specimen is loaded. This test gives a fixed ratio of Mode I to Mode II loading. Here, G_I/G is independent of crack length, but is a function of $h_1/2h$. In the FRMM test, the loading is $M_1 = -Pa$ and $M_2 = 0$ giving a total G of

$$G = \frac{3P^2 a^2}{4B^2 h^3 E_{11}} \left[\left(\frac{2h}{h_1} \right)^3 - 1 \right]$$

(7)

For the special case of $h_1 = h$, the partitioning may proceed as before with $M_I = M_{II} = -Pa/2$, and using Eq 5, we have

$$\frac{G_I}{G} = \frac{1}{1 + \dfrac{3}{4} \left(\dfrac{M_{II}}{M_I} \right)^2} = \frac{4}{7}$$

(8)

When $h_1 \neq h$, a problem arises because the global deformation can be only partitioned into pure opening and pure sliding but these will not have the symmetries of the conventional definitions of Modes I and II, since $v_1 \neq v_2$ and $u_1 \neq -u_2$. (Although, of course, in pure opening (Mode I) the shear displacements u_1 and u_2 are both zero in value, and in pure sliding (Mode II) the tensile displacements v_1 and v_2 are both zero in value.) This problem has been pointed up recently [10,11] and solutions have been obtained from the local stress-field at the crack

(a)

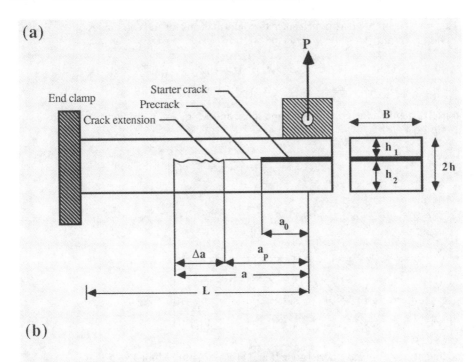

(b)

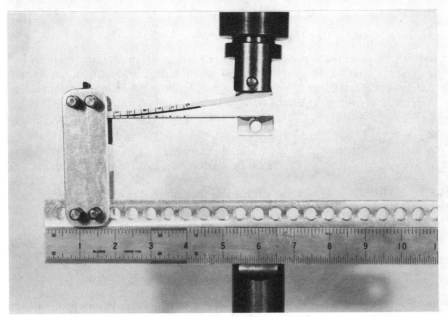

FIG. 6—*Fixed-ratio mixed-mode specimen (FRMM) and the loading arrangement.*

tip using singular field analysis, where the symmetries of the conventional definitions are maintained. However, it is a moot point which division is appropriate in describing a fracture since the singular field can be very local in some geometries. Finite element calculations have shown that they are predominant only over $0.03h$ in a DCB specimen in Mode I and $0.06h$ for the ELS test in Mode II. In an adhesive layer of lower modulus, the size is 2% of the layer thickness [12] since h is small (~ 2 mm) the size of influence is small in absolute terms; say $<$ 0.01 mm. To use the local field solutions, it is necessary that the region of dominance of the field exceeds any local damage or process zone and this is unlikely in the cases considered here. However, the two analyses will now be described.

First, the G will be partitioned into pure opening and pure sliding on a global basis [8]. This may be done by using components of M_{II} that give the same radius of curvature that is achieved when M_1 and M_2 are partitioned as

$$M_1 = M_{II} - M_I$$

$$M_2 = \left(\frac{h_2}{h_1}\right)^3 M_{II} + M_I$$

and on substitution in Eq 1 we have

$$\frac{G_I}{G} = \frac{1}{1 + \dfrac{3h_1^4}{h_2^2(2h)^2}} \tag{9}$$

Note that as $h_1 \to 0$, $G_I/G \to 1$, pure opening and as $h_2 \to 0$, $G_I/G \to 0$, pure sliding.

Second, the solution for the local stress-field partitioning can be deduced from Ref 11 and is tabulated in Table 1 together with the global results. The local-field solution gives the rather surprising result that bending the thin arm ($h_1/2h \to 0$) has 38% of Mode II, which is induced to achieve symmetry.

Asymmetric DCB and ELS Test

Interesting additional tests that emphasize the difference between the global and local-field solutions are the asymmetric DCB (termed the ADCB) and the asymmetric ELS (termed the AELS) test. In the ADCB test, we still have $M_2 = -M_1 = Pa$, as for the symmetric DCB test discussed earlier. Hence, the global solution yields $M_{II} = 0$ and we have pure Mode I loading. In the AELS specimen, we have $M_2 = (h_2/h_1)^3 M_1$ giving $M_1 = 0$, and so we still predict pure Mode II from using the global solution. However, the local stress-field solutions [11] give very different results for the mode of loading in the ADCB and AELS tests. They predict mixed-mode loadings in such tests and the values are given in Table 2. Note that, as mentioned earlier, for the symmetric DCB and ELS test specimens (that is, when $h_1/2h = 0.5$) then the local-

TABLE 1—*Local stress-field and global solution values for the extent of Mode I loading for various $h_1/2h$ ratios in the asymmetric FRMM test.*

$h_1/2h$	0	0.2	0.4	0.5	0.6	0.8	1.0
G_I/G (local-field [11])	0.623	0.611	0.588	0.571	0.549	0.481	0.377
G_I/G (global, Eq 9)	1.00	0.992	0.824	0.571	0.292	0.031	0

TABLE 2—*Local stress-field solution* [11] *values for the extent of Mode I loading as a function of* $h_1/2h$ *for the ADCB and AELS specimens.*

$h_1/2h$	0	0.1	0.2	0.3	0.4	(0.5)
G_1/G (ADCB)	0.623	0.636	0.692	0.798	0.932	(1.00)
G_1/G (AELS)	0.377	0.364	0.308	0.202	0.068	(0)

field and global solutions are in complete agreement and pure Mode I and Mode II loadings, respectively, are observed.

Correction Factors

Theoretical Aspects

The G values computed by beam theory have two major sources of error that must be corrected if accurate results are to be obtained [2,5].

First, the displacements of the beam arms may give rise to pronounced curvature that can cause the effective shortening of the loading arms. In addition, there are errors that arise from tilting and stiffening effects of the loading blocks. The form used to derive G here, Eq 1, uses the applied beam root moments and these may be corrected by a factor, F [5], which may be deduced from large displacement beam theory. Thus any value of G may be corrected by

$$G_{corrected} = G_o \cdot F$$

where G_o is that value computed via the moment form; that is, via load times crack length squared. The various expressions for F are given in Appendix A, together with some typical values computed from the test data where corrections of up to 30% were needed for the FRMM test. No corrections were used in the VRMM test, Fig. 5, although a, l, and L will generally tend to decrease somewhat if large displacements occur. In Eq 4, the ratios l/a and a/L will not be affected generally but the overall value of G_c will be somewhat high because of l^2. However, the general stiff nature of the specimen will mean that the corrections will be small.

Second, and most important, a correction is needed for the rotation of the beam root and shear deformation in the beams that can be incorporated in a correction to the crack length of χh, where χ is a constant and h is the thickness of the cracked section. Theoretically [13], we would expect $\chi = 2.5$ for laminates, but χ may be determined experimentally most conveniently from the DCB test where the compliance from simple beam theory is given by

$$C = \frac{8a^3}{Bh^3 E_{11}}$$

now, if we add the correction factor, χh, to the crack length, then [1–4]

$$\left(\frac{C}{N}\right)^{1/3} = \left(\frac{8}{Bh^3 E_{11}}\right)^{1/3} (a + \chi h)$$

where, by introducing Factor N, we have also corrected the measured displacement (and hence compliance) for the large displacement and end-block effects. The required equations for the correction factor, N, may be deduced from large displacement analysis and are given in

Appendix B. Thus, the preceding expression relating the measured compliance, C, to the crack length via the dimensions and axial modulus of the laminate beam contains all the necessary correction factors.

Experimental Verification

Now, from the preceding expression, by plotting $(C/N)^{1/3}$ versus a, we can obtain χ from the nonzero intercept and E_{11} from the slope. This has been studied in some detail for several materials [2–4] and, for the PEEK composite used here, it was found that $\chi = 2.45$ and $E_{11} = 125$ GPa, and for the epoxy composite, the equivalent values were $\chi = 3.00$ and $E_{11} = 130$ GPa.

These procedures were checked here for the FRMM test for which the compliance expression is

$$C = N\left[\frac{(\gamma - 1)(a + \chi h_2)^3 + (L + 2\chi h)^3}{2Bh^3 E_{11}}\right] \tag{10}$$

where $\gamma = (2h/h_1)^3$ and N is given in Appendix II. Note that both a and L are corrected with χ and that it is difficult to do a direct determination from the compliance data. The scheme that was adopted here was to determine C for various $h_1/2h$ values and then to fix E_{11} and hence find χ from Eq 10. Similarly, the reverse procedure was carried out in which χ was fixed and E_{11} determined. The results are given in Table 3 for the PEEK composite and the data shows remarkable consistency lending confidence to the correction procedure.

In the results that follow, a χh correction is made to all the crack length values used.

Experimental Procedure

Materials

Thermoset Composite—The carbon-fiber epoxy resin composite employed was Fibredux 6376 that was supplied by Ciba Geigy in the form of the prepreg material containing continuous unidirectional carbon fibers of Type T-300 (from Toray Inc., Japan). The prepreg material was pressed in an autoclave using the supplier-recommended cure cycle to obtain a fiber volume fraction of 61%.

Thermoplastic Composite—The thermoplastic composite utilized was PEEK (poly(ether-ether ketone)) that is semicrystalline and was supplied by Imperial Chemical Industries (ICI) in the form of the prepreg containing continuous unidirectional carbon fibers of Type AS4 (from Hercules Inc.). The prepreg material was hot pressed according to the supplier specification to obtain a fiber volume fraction of 61%.

TABLE 3—*Correction factors for the FRMM tests (PEEK composite).*

$h_1/2h$	0.84	0.75	0.69	0.56	0.50	0.44	0.38	0.31	0.25	Mean
χ $E_{11} = 125$ GPa	2.40	2.22	1.98	2.00	2.45	1.95	2.57	2.66	2.86	2.36 ± 0.5
E_{11} (GPa) $\chi = 2.45$	125	127	129	129	125	130	123	123	122	126 ± 4

NOTE—Value of E from independent flexure test = 125 GPa [2].

Specimen Fabrication

Laminate specimens were cut from the fabricated panels with fibers running along the length of the specimen. The starter delamination was produced by placing a double layer of aluminum foil of thickness 10 μm and the length, a_0, across the width of the panel prior to processing. The starter delamination was inserted either between the center plies (symmetric) or, as in the case of FRMM specimens, at different ply interfaces (asymmetric) to obtain different arm thicknesses and therefore different Mode I/II ratios. The length of the starter crack and the specimen dimensions used for all four specimen geometries are given in Table 4.

Mechanical Testing

Test Conditions and Procedure—Delamination toughness tests were performed at ambient temperature (23°C) on laminate specimens using an Instron screw driven machine operated at a constant displacement rate of 2 mm/min. End blocks made of aluminum alloy were bonded to the cracked end of all the specimens as a means of applying the load perpendicular to the interlaminar layer. The surface of the specimens and the end-blocks were prepared for bonding by grit blasting until the surfaces were uniformly dull. The surface of the specimen was then cleaned with acetone and that of the end block with 1,1,1 trichloroethane. The adhesive used for bonding the end blocks was a modified epoxy (Redux 320, supplied by Ciba Geigy), which was cured at 120°C for 60 min. A special jig was designed that allowed all four specimen geometries to be tested that are shown in Figs. 2, 3, 5, and 6. For the Mode II tests, a 1-mm-diameter roller was placed between the crack faces of the split laminate specimens to minimize any frictional effects that might occur by the rubbing of the two surfaces together. This resulted in a small induced Mode I component of a maximum value of 15 J m^{-2}. An extra clip was needed in the VRMM test on the center block to prevent the block breaking away.

Load/Displacement/Crack Length Measurements—Continuous records of load and displacement (P, δ) were obtained for each specimen for the determination of the interlaminar fracture energies. Figure 7 depicts traces for epoxy composite specimens undergoing pure and the combined Mode I/II deformations. Similar traces were obtained for the PEEK composite as has been shown in earlier work [2,4]. The delamination length was measured visually at the surface of the specimen using a traveling microscope. To monitor delamination length white typewriter correction fluid was applied to one edge of all the specimens and fine markings were then put on this edge at about 5 mm intervals (no determination could be made of the interior

TABLE 4—*Nominal specimen dimensions (all dimensions are in mm).*

Geometry	B	2h	L	l	a_0	a_p
			PEEK			
DCB	25	4.30	120	· · ·	20	24
ELS	25	4.30	120	· · ·	23	50
FRMM	25	4.30	120	· · ·	23	50, 60
VRMM	25	4.30	250	114, 76, 64	23	· · ·
			EPOXY			
DCB	25	4.40	120	· · ·	25	30
ELS	25	4.40	120	· · ·	40	50
FRMM	25	4.40	120	· · ·	40	50
VRMM	25	4.40	250	140, 114, 102 90, 76, 64	40	· · ·

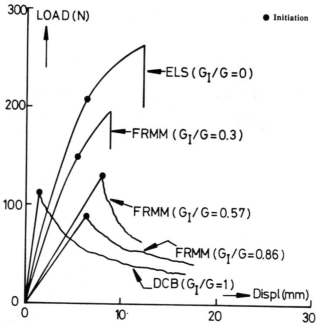

FIG. 7—*Load-displacement traces for epoxy composite specimens undergoing pure and the combined Mode I/II deformations.*

crack tip profile). Delamination lengths were immediately marked on the load versus displacement plot for later identification.

Precracking—Since all the specimens contained aluminum foil to provide a starter crack and the foil introduces a relatively blunt notch, all the specimens were precracked in Mode I at a constant crosshead speed of 2 mm/min to a length, a_p (see Table 4 and Figs. 2, 3, 5, and 6) before the measurements of any critical energy release rate. It is noteworthy that each failure mode produced a fracture surface that appeared different to the naked eye. This change in the fracture surface appearance allowed the length of the initial precrack, a_p, to be measured accurately. Typical fracture surfaces are shown in Figs. 8 and 9.

Results and Discussion

Load-Displacement Traces and Crack Stability

The load-displacement diagrams (see Fig. 7) were generally linear up to initiation though some specimens did show evidence of stiffening (an upward curvature though this is not apparent in Fig. 7). This is mostly due to large displacements and is corrected via F (see preceding and as mentioned in Appendix I) but some of the upward curvature in the P versus δ traces for those tests where some Mode I loading is present may arise from fiber-bridging [2–4]. There was also some evidence of inelastic behavior even in unloading, probably due to the formation of a damage zone giving a decrease in stiffness but the effects were small (less than 5% decrease in compliance).

All the DCB Mode I tests gave stable continuous crack growth while the Mode II ELS tests were stable for the PEEK composite but unstable for the epoxy material. The stability of the configurations used here can be defined in terms of dG/da under fixed displacement condi-

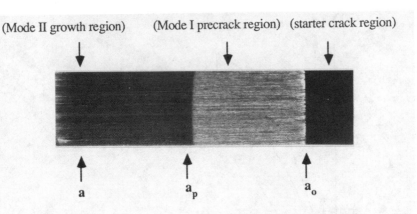

(Mode II growth region)   (Mode I precrack region)   (starter crack region)

a a_p a_o

Mode II ELS fracture surface with usual Mode I precrack.

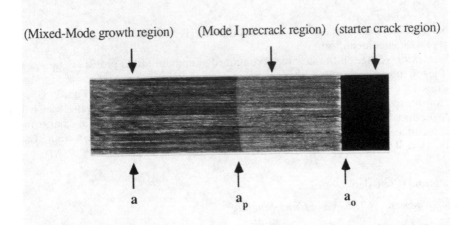

(Mixed-Mode growth region)   (Mode I precrack region)   (starter crack region)

a a_p a_o

FRMM (ratio $G_I/G = 0.57$) fracture surface with usual Mode I precrack.

FIG. 8—*Fracture surface of the PEEK composite.*

(Mode II growth region)   (Mode I precrack region)   (starter crack region)

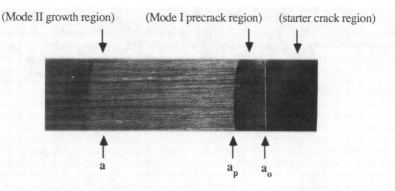

a a_p a_o

Mode II ELS fracture surface with usual Mode I precrack.

(Mixed-Mode growth region)   (Mode I precrack region)   (starter crack region)

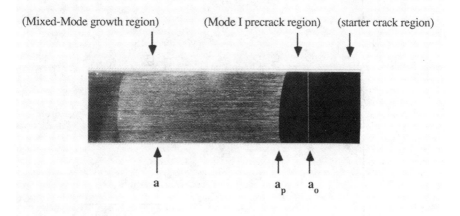

a a_p a_o

FRMM (ratio $G_I /G = 0.57$) fracture surface with usual Mode I precrack.

FIG. 9—*Fracture surface of the epoxy composite.*

tions [8]. If $dG/da > 0$, there will be a tendency toward unstable behavior, but this will depend on dG_c/da since the true condition for instability is $dG/da > dG_c/da$. In the case of the ELS specimen, the a/L ratio used was 0.42, and the theoretical calculations reveal that such a value would lead to unstable crack growth; since unstable crack growth would be predicted [8] to occur when $a/L < 0.55$ if $dG_c/da = 0$. Thus, the stable crack growth associated with the PEEK composite indicates a material effect; that is, $dG_c/da > 0$ and so the resistance to crack growth increases as the crack length increases. The analysis would also suggest that the VRMM tests would be always stable [8] for $a/L < 0.95$, and this was so for both materials. It must be noted, however, that high Mode II content tests could not be employed because of fixing the center block, so all the results were for $G_I/G > 0.2$. Thus, the epoxy composite was not tested in the high Mode II condition where it has a tendency to show unstable crack growth.

The FRMM test is unstable [8] for

$$\left(\frac{h_1}{2h}\right)^3 > \frac{2\left(\dfrac{a}{L}\right)^3}{1 + 2\left(\dfrac{a}{L}\right)^3} \tag{11}$$

All the epoxy composite specimens had initial $a_0/L = 0.42$ and were found to be unstable for $h_1/2h > 0.6$, while Eq 11 predicts instability to occur when, $h_1/2h > 0.5$. For the PEEK composite, initial a/L values of 0.42 and 0.50 were used that gave unstable crack growth at values of $h_1/2h$ of 0.61 and 0.65, respectively. Equation 11 gives values of 0.50 and 0.58, respectively. The analysis, although not precise, is thus a useful guide to behavior.

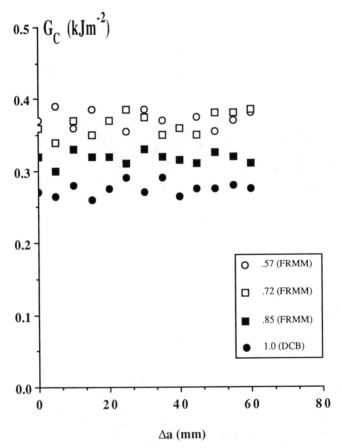

FIG. 10—G_c versus crack growth, Δa, in the FRMM test for the epoxy composite for various G_I/G ratios.

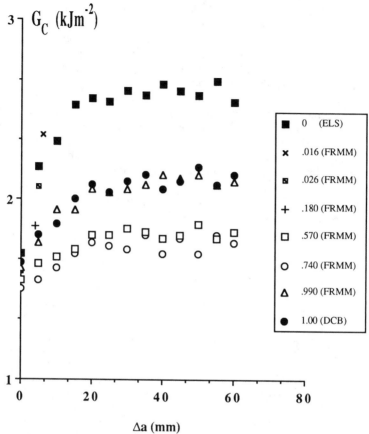

FIG. 11—G_c *versus crack growth,* Δa, *in the FRMM test for the PEEK composite for various* G_1/G *ratios.*

G_c *as a Function of Crack Growth,* Δa

Figure 10 shows G_c versus crack growth, Δa, for the epoxy composite for four conditions in which stable growth occurred. There is no evidence of an R-curve for the epoxy composite. However, it should be noted that, as usual, we have advanced the delamination ahead of the initial foil length, a_0, a small amount before commencing our readings. The first reading then yields the value of G_c (initiation). (This being done to avoid the resin-rich layer just ahead of the starter-crack foil that would lead to unrepresentative values.) In the present tests, this distance (that is, $a_p - a_0$) is about 5 to 10 mm only. From these, and previous results [1–4], we do not consider it likely that an R-curve would have developed and already have attained the steady-state propagation value of G_c within such a short distance. Thus, for the epoxy composite, we conclude that (1) it is highly unlikely that an R-curve develops, and (2) the value of G_c is constant with Δa, and hence G_c (initiation) = G_c (propagation).

Figure 11 shows G_c versus crack growth, Δa, for the PEEK composite. As for the epoxy composite, stable crack growth is observed for $0.5 < G_1/G < 1$ in the FRMM test and in the pure Mode I DCB test. However, the PEEK composite also gives stable crack growth for $G_1/G < 0.5$ in the FRMM test and in the pure Mode II ELS test. As previously reported [2–4], the PEEK composite exhibits a marked R-curve. This was shown to arise from both fiber bridging occurring behind the advancing crack tip and crack-tip damage mechanisms. The presence of an R-curve leads, of course, to the values of G_c (initiation) being less than for steady-state propagation. (As we have previously discussed and demonstrated [3,4], the growing of the delamination ahead of the foil prior to starting our readings will affect that value of G_{Ic} that we define as corresponding to "initiation." However, we have used only a very small extent of such growth (that is, $a_p - a_0$ being a few mm only) so that our definition of the initiation value does represent a consistent and a physically meaningful parameter. Indeed, the definition of "crack initiation" is a common problem to all researchers working in the area of fracture mechanics, whatever the material of interest.) The values of G_c (initiation) appear to be only marginally dependent upon the G_I/G ratio used in the test, while the latter values of G_c (s/s-propagation) are highly dependent upon the G_I/G ratio employed.

As discussed in detail elsewhere [2,3], the presence of the R-curve results from effects such as fiber-debonding, fiber-bridging, and the development of a crack-tip damage zone. These effects will obviously be dependent upon factors like the fiber-matrix adhesion, fiber layup

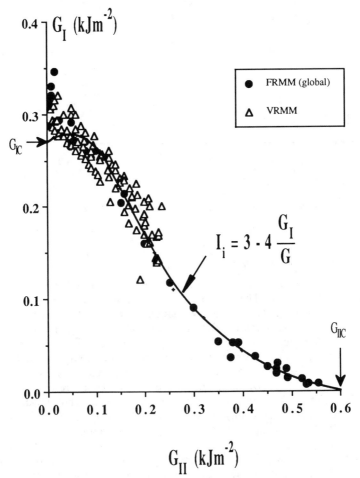

FIG. 12—*The failure locus for the epoxy composite—FRMM data is globally partitioned.*

angle, matrix type, and resin layer thickness. Thus, the exact nature of the R-curve will be a function of the microstructure of the fiber composite, in an analogous manner to the R curve of a ductile metal.

Partitioning of the G_c Data

Figure 12 shows the epoxy composite data for both the VRMM and the FRMM tests plotted as G_I versus G_{II}, together with the pure Mode I and II values. The FRMM data have been partitioned using the global scheme and shows good agreement with the VRMM data. It should be recalled that in the cases of the VRMM test method that there is no difference in the resulting values of G_I and G_{II} from whether the global or local-stress field partitioning methods are employed. Figure 12 also reveals that there is a pronounced difference between the G_{IIc} and G_{Ic} values and that there is some curvature in the locus of failure with a small Mode II content giving a significant increase: 0.28 to 0.33 kJ m^{-2}.

Figure 13 shows the FRMM data partitioned by both schemes (global and the local-stress

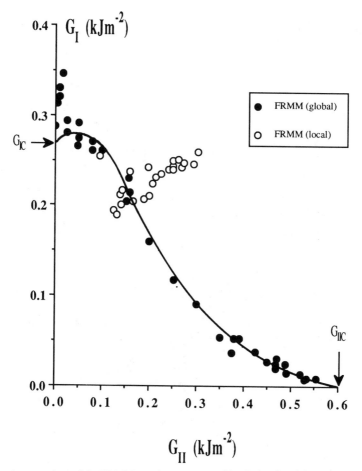

FIG. 13—*A comparison of the FRMM test data partitioned by the local and the global methods for the epoxy composite.*

field) and the marked difference between these two schemes is a reflection of the very significant difference between the values of G_{Ic} and G_{IIc} for the epoxy composite. A major feature of the data shown in Figs. 12 and 13 is the very poor agreement between the values of G_I and G_{II} by partitioning the values obtained from the FRMM tests, using the local-field solutions, and those from the VRMM test.

Figure 14 shows the loci for the PEEK composite for initiation and propagation condition using data from the FRMM test partitioned globally. The initiation data are equivalent to a constant value of total G_c but the propagation values show some difference. Figure 15 shows the propagation data for both the VRMM and FRMM tests compared, the latter partitioned globally, and again there is good agreement. In Fig. 16, the FRMM data are shown partitioned by both schemes and, although it is less pronounced here because of the smaller difference between the Modes I and II values, the global result again gives better agreement with the VRMM data (Fig. 15). The initiation data would simply give a cluster of points on the same line for local partitioning because the total G_c is almost constant.

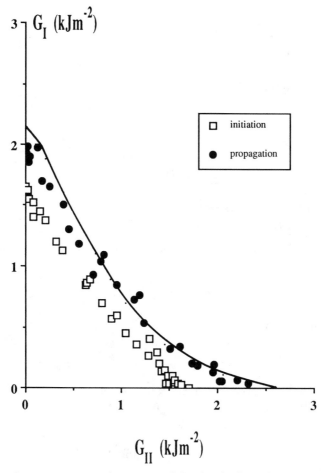

FIG. 14—*Initiation and propagation failure loci for the PEEK composite.*

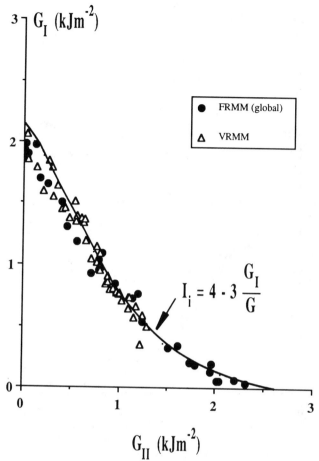

FIG. 15—*A comparison of the FRMM (global) and VRMM test data for propagation for the PEEK composite.*

Asymmetric Double Cantilever Beam (ADCB) and Asymmetric End-Loaded Split (AELS) Tests

A series of ADCB and AELS tests were also performed using the epoxy composite and the dimensions are shown in Table 5. Specimens 1, 2, 3, and 5 were made by machining off one side of a 4.3-mm, centrally cracked laminate while Specimen 4 had an aluminum beam bonded to one side to create asymmetric specimens. The h_2 value given for the latter is an equivalent composite thickness to give the same bending stiffness. Global partitioning would predict pure Mode I for the ADCB tests and pure Mode II for the AELS tests but various G_I/G values arise from local-field partitioning and these values, taken from Table 2, are also shown. The first three specimens were tested in both modes and the ELS values are single instability values. Those values for the ADCB for all four tests are averages for the growing cracks taken over 30 mm and varied from the average by about ± 0.01 kJ m^{-2}.

The symmetric DCB tests gave a value of 0.27 kJ m^{-2} with a standard deviation of ± 0.015,

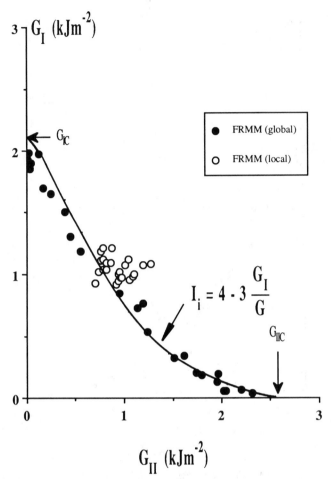

FIG. 16—*A comparison of the FRMM test partitioned by the two methods for the PEEK composite.*

TABLE 5—*ADCB and AELS results for the epoxy composite.*

Specimen Number	h_1, mm	h_2, mm	$h_1/2h$	ADCB G_1/G^b	ADCB G_c, kJ m^{-2}	AELS G_1/G^b	AELS G_c, kJ m^{-2}
1	1.65	2.20	0.43	0.96	0.26	0.04	0.66
2	1.45	2.25	0.39	0.91	0.27	0.09	0.61
3	1.12	2.21	0.34	0.84	0.28	0.16	0.65
4	2.12	7.88^a	0.21	0.70	0.29	0.30	$\cdots$
5	0.56	2.30	0.20	0.69	0.29	0.31	$\cdots$

^{a} Equivalent thickness.
^{b} G_1/G values from local stress-field partitioning [11].
Global partitioning gives the ADCB as pure Mode I ($G_1/G = 1$) and the AELS as pure Mode II ($G_1/G = 0$).

and all the experimental results for the ADCB tests fall within this range. The local-field partitioning scheme predicts up to 30% of Mode II so that, referring to the VRMM data in Fig. 12, we would expect values in the range 0.35 to 0.43 kJ m^{-2} for Specimens 4 and 5 and the results are clearly below this range. The AELS data are all slightly higher than the symmetric ELS data, which gave a G_{IIc} value of 0.60 $\pm$ 0.03 kJ m^{-2}, but the difference between this value and the value from the AELS specimens (0.64 $\pm$ 0.03 kJ m^{-2}) is not statistically significant. Again from Table 2, the local stress-field theory would predict up to 16% Mode I that should give values in the range 0.50 to 0.55 kJ m^{-2}, and again there is no evidence of this.

Criteria for Failure Locus

Both materials give loci that are not far removed from a linear relationship, but there are significant deviations from this. In Fig. 12, for example, a small amount of Mode I gives a rapid decrease in G_c while a small amount of Mode II has very little effect and indeed there is even some evidence of an increase in G_c. Away from these extremes linearity pertains. These effects can be quantified using and interaction parameter, I_i [5] that can be used to describe the loci in the relationship

$$\left(\frac{G_I}{G_{Ic}} - 1\right)\left(\frac{G_{II}}{G_{IIc}} - 1\right) - I_i\left(\frac{G_I}{G_{Ic}} \cdot \frac{G_{II}}{G_{IIc}}\right) = 0$$

For $I_i = 1$, we have the linear case, while for $I_i = 0$, the two cases are completely independent, that is, no interaction. A very strong interaction is a high I_i value and gives rapid changes in each as the other changes. Conversely, for low I_i, values there is little change in either. In Fig. 12, therefore, we find that I_i is high for high G_{II} content and then decreases through unity to zero at low G_{II} content and then becoming negative for lower G_{II} values. This may be modeled approximately by giving I_i a linear dependence on G_I/G, that is

$$I_i = 3 - 4\frac{G_I}{G}$$

and this curve fit is shown in Fig. 12 giving a good representation of the points except at very high G_I/G ratios. A possible mechanism for this type of interaction is surface roughness interlocking that gives rise to the high G_{IIc} value and is more easily overcome with some opening. Similarly, some shear on an opening situation may enhance the interlocking thus increasing G_c. Similar comments apply to the PEEK propagation data shown in Fig. 15 but without the Mode I enhancement so that a suitable fit here would be

$$I_i = 4 - 3\frac{G_I}{G}$$

that is, always positive as shown in Fig. 15. The initiation data here (Fig. 14) is described by I_i = 1; that is, linear.

Conclusions

Mode I, Mode II, and Mixed-Mode I/II interlaminar tests on unidirectional carbon-fiber polymer composites have been conducted using beam specimens. The fracture energy, G_c, has been ascertained and the various correction factors that need to be applied if accurate results are to be obtained have been described and tabulated. The two main sources of error are for (1) large displacements of the beam arms and end-block effects, and (2) the rotation of the beam root and shear deformation in the beams.

Where Mixed-Mode I/II loadings may have been present, the measured value of G_c has been partitioned into the separate G_I and G_{II} components. This has been undertaken using either (1) the global energy analysis, based upon partitioning into pure opening (Mode I) and pure sliding (Mode II), or (2) the local stress-field solutions where the symmetries of the conventional definitions ($v_1 = v_2$ and $u_1 = -u_2$, respectively) are maintained. The experimental results all suggest that the former method of partitioning of G on a global basis is the most appropriate for the laminates. This probably arises because the range of local stress-field dominance is very small and is obscured by local damage. The difference in energy absorption are thus a consequence of opening as opposed to sliding, and the symmetry, or otherwise, of these motions is not important. It is thus apparent that a global energy analysis is entirely appropriate and satisfactory for these types of systems.

Finally, it has been shown that when the failure locus is plotted in the form of G_I versus G_{II}, then it may be theoretically described by a simple interaction parameter model, where the interaction parameter, I_i, has a linear dependence upon the value of G_I/G.

Acknowledgments

The authors are pleased to thank the Science and Engineering Research Council, through the Polymer Engineering Directorate, for financial support and Ciba Geigy plc and ICI plc for provision of materials.

APPENDIX I

Correction Factor, F

The measured crack length, a, may be corrected for the effective shortening of the beam due to the large displacements and the tilting of the end blocks. The correction may be expressed as

$$F = 1 - \Theta_1 \left(\frac{\delta}{L}\right)^2 - \Theta_2 \left(\frac{\delta l_1}{L^2}\right) \tag{12}$$

where Θ_1 and Θ_2 for various geometries are given in Table 6 [5]. Some results for FRMM specimens are presented in Tables 7 and 8.

TABLE 6—Θ_1 and Θ_2 functions for the various specimen geometries.

Mode I DCB
  $\Theta_1 = 0.30$ (Note: $L = a$)
  $\Theta_2 = 1.50$
Mode II ELS
  $\Theta_1 = 0.15(15 + 50(a/L)^2 + 63(a/L)^4)/(1 + 3(a/L)^3)^2$
  $\Theta_2 = -3 \cdot [(1 + 3(a/L)^2)/(1 + 3(a/L)^3)^3] \cdot (L/a)$
Mixed Mode FRMM
  $\Theta_1 = 0.15(15 + (20\gamma - 30)(a/L)^2 + (8\gamma^2 - 20\gamma + 15)(a/L)^4)/(1 + (\gamma - 1)(a/L)^3)^2$
  $\Theta_2 = -3 \cdot [(1 + (\gamma - 1)(a/L)^2)/(1 + (\gamma - 1)(a/L)^3)^3] \cdot (L/a)$
  where $\gamma = [(2h)/h_1]^3$

TABLE 7—*Typical values of* F *for fixed ratio mixed-mode tests (FRMM) on PEEK composite.*

$h_1/2h$	0.84	0.75	0.66	0.56	0.50	0.44	0.38	0.31	0.25	0.22
F	0.73	0.74	0.84	0.85	0.88	0.87	0.86	0.79	0.72	0.69

TABLE 8—*Typical values of* F *for fixed ratio mixed-mode tests (FRMM) on epoxy composite.*

$h_1/2h$	0.84	0.75	0.69	0.56	0.50	0.44	0.38	0.29	0.26	0.21
F	0.88	0.92	0.93	0.95	0.95	0.94	0.95	0.92	0.91	0.89

APPENDIX II

Correction Factor, N

The measured values of the compliance can be corrected by the function, N, given as [5]

$$N = 1 - \Theta_3 \left(\frac{l_2}{L}\right)^3 - \Theta_4 \left(\frac{\delta l_1}{L^2}\right) - \Theta_5 \left(\frac{\delta}{L}\right)^2 \tag{13}$$

For the DCB, ELS, and FRMM specimens, the Θ functions are given in Table 9.

TABLE 9—Θ *functions for the various specimen geometries.*

Mode I DCB
$\Theta_3 = 1.0$ (Note: $L = a$)
$\Theta_4 = 9/8\,(1 - (l_2/a)^2)$
$\Theta_5 = 9/35$

Mode II ELS
$\Theta_3 = 4/(1 + 3(a/L)^3)$
$\Theta_4 = -9/4\{(1 - a/L)(1 + 3(a/L)^3) + 4(1 - (l_2/a)^2) \cdot (a/L)^2 \cdot (1 + 3(a/L)^2)\}/(1 + 3(a/L)^3)^2$
$\Theta_5 = 36/35\{1 + 3/8(a/L)^3[35 + 70(a/L)^2 + 63(a/L)^4]\}/(1 + 3(a/L)^3)^3$

Mixed Mode FRMM
$\Theta_3 = \gamma/(1 + (\gamma - 1)(a/L)^3)$
$\Theta_4 = 9/4\{(1 - a/L)(1 + (\gamma - 1)(a/L)^3) + \gamma(1 - (l_2/a)^2) \cdot (a/L)^2 \cdot (1 + (\gamma - 1)(a/L)^2)\}/(1 + (\gamma - 1)(a/L)^3)^2$
$\Theta_5 = 9/70\{8 + (\gamma - 1)(a/L)^3[35 + 14(2\gamma - 3)(a/L)^2 + (8\gamma^2 - 20\gamma + 15(a/L)^4]\}/(1 + (\gamma - 1)(a/L)^3)^3$
Where $\gamma = [(2h)/h_1]^3$

References

[1] Hashemi, S., Kinloch, A. J., and Williams, J. G., "Corrections Needed in DCB Tests for Assessing the Interlaminar Failure of Fibre-Composites," *Journal of Material Science, Letters,* Vol. 8, 1989, pp. 125–129.
[2] Hashemi, S., Kinloch, A. J., and Williams, J. G., "The Analysis of Interlaminar Fracture in Uniaxial Fibre-Polymer Composites," *Proceedings,* Royal Society, Vol. A427, 1990, pp. 173–199.
[3] Hashemi, S., Kinloch, A. J., and Williams, J. G., "Mechanics and Mechanisms of Delamination in a Poly(Ether Sulphone)—Fibre Composite," *Journal of Composites Science and Technology,* Vol. 37, 1990, pp. 429–462.
[4] Hashemi, S., Kinloch, A. J., and Williams, J. G., "The Effects of Geometry, Rate and Temperature

on the Mode I, Mode II and Mixed-Mode I/II Interlaminar Fracture of Carbon-Fibre/Poly(Ether-Ether Ketone) Composites," *Journal of Composite Materials,* Vol. 24, 1990, pp. 918–956.

[5] Williams, J. G., "Fracture Mechanics of Delamination Tests," *Journal of Strain Analysis,* Vol. 24, 1989, pp. 207–214.

[6] Russell, A. J. and Street, K. N., "The Effects of Matrix Toughness on Delamination," *Toughened Composites, ASTM STP 937,* N. J. Johnston, Ed., American Society for Testing and Materials, Philadelphia, 1987, pp. 275–285.

[7] Wang, S. S., Suemasu, H., and Zahlan, N. M., "Interlaminar Fracture of Random Short Fibre SMC Composite," *Journal of Composite Materials,* Vol. 18, 1984, pp. 574–593.

[8] Williams, J. G., "On the Calculation of Energy Release Rates for Cracked Laminates," *International Journal of Fracture,* Vol. 36, 1988, pp. 101–119.

[9] Hashemi, S., Kinloch, A. J., and Williams, J. G., "Interlaminar Fracture of Composite Materials," *Proceedings,* ICCM Fourth International Conference, Vol. 3, Elsevier, London, 1987, pp. 3.254–3.264.

[10] Thouless, M. D., Evans, A. G., Ashby, M. F., and Hutchinson, J. W., "The Edge Cracking and Spalling of Brittle Plates," *Acta Metallurgica,* Vol. 35, 1987, pp. 1333–1341.

[11] Zhigang, S., "Delamination Specimens for Orthotropic Materials," Harvard University, Division of Applied Science Report, MECH-135, Cambridge, MA, 1988.

[12] Wang, S. S., Mandell, J. F., and McGarry, F. J., "An Analysis of the Crack Tip Stress Field in DCB Adhesive Fracture Specimens," *International Journal of Fracture,* Vol. 14, 1978, pp. 39–58.

[13] Williams, J. G., "End Correction for Orthotropic DCB Specimens," *Journal of Composite Science and Technology,* Vol. 35, 1989, pp. 367–376.

Rajiv A. Naik,[1] *John H. Crews, Jr.,*[2] *and Kunigal N. Shivakumar*[1]

Effects of T-Tabs and Large Deflections in Double Cantilever Beam Specimen Tests

REFERENCE: Naik, R. A., Crews, J. H., Jr., and Shivakumar, K. N., **"Effects of T-Tabs and Large Deflections in Double Cantilever Beam Specimen Tests,"** *Composite Materials: Fatigue and Fracture (Third Volume), ASTM STP 1110,* T. K. O'Brien, Ed., American Society for Testing and Materials, Philadelphia, 1991, 169–186.

ABSTRACT: A simple strength of materials analysis was developed for a double cantilever beam (DCB) specimen to account for geometric nonlinearity effects due to large deflections and T-tabs. A new DCB data analysis procedure was developed to include the effects of these nonlinearities. The results of the analysis were evaluated by performing DCB tests on materials having a wide range of fracture toughnesses. The composite materials used in the present study were T300/5208, IM7/8551-7, and AS4/PEEK.

Based on the present analysis, for a typical load-point deflection/crack length ratio of 0.3 (for AS4/PEEK), T-tabs and large deflections lead to a 19 and 6% effect, respectively, in the computed Mode I strain energy release rate. Design guidelines for DCB specimen thickness and T-tab height were also developed in order to keep effects due to these nonlinearities within 2%.

Based on the test results, for both hinged and tabbed specimens, the effects of large deflection (compared to linear DCB results) on the Mode I fracture toughness (G_{Ic}) were almost negligible (less than 1%) in the case of T300/5208 and IM7/8551-7; however, for AS4/PEEK, the effect was from 2 to 3%. The effects of T-tabs (compared to a linear DCB with tabs) on G_{Ic} were more significant, with T300/5208, IM7/8551-7, and, AS4/PEEK showing a 5, 15, and 20% effect, in that order.

KEY WORDS: double cantilever beam, delamination, fracture toughness, composite materials, large deflection, geometric nonlinearity, strain-energy release rate, loading tabs, fracture, fatigue (materials)

Nomenclature

a	Crack length including end correction
a_{el}	Effective crack length for double cantilever beam (DCB) with large deflections
a_{et}	Effective crack length for DCB with T-tabs
a_{elt}	Effective crack length for DCB with T-tabs and large deflections
a_{exp}, a_i	Experimentally measured crack length
a_0	End correction to account for DCB crack tip rotation
b	DCB specimen width
C_i	Measured initial compliance for DCB specimen
d	Tab height from base to loading point
d_{exp}, δ_i	Measured displacement in a DCB test
E_{11}	Longitudinal composite modulus
E_{22}	Transverse composite modulus

[1] Research scientist and senior scientist, respectively, Analytical Services & Materials, Inc., Hampton, VA 23666.
[2] Senior researcher, NASA Langley Research Center, Hampton, VA 28665-5225.

E_r	Transverse modulus of resin-rich layer
G_{13}	Composite shear modulus
G_1	Strain-energy release rate
G_1^{et}	Strain-energy release rate for DCB with T-tabs
G_1^{elt}	G_1 for DCB with T-tabs and large deflections
G_{Ic}	Mode I fracture toughness
h	Thickness of DCB arm
H	Total tab height from loading point to center of DCB arm
$\overline{H}$	Factor used in calculation of G_{Ic}
I	Moment of inertia
k	Foundation spring constant
M	Bending moment at a section, x, from the crack tip
M_0	Bending moment at crack tip
M_r	Resisting moment due to T-tab rotation
P, P_i	Applied load
t	DCB specimen thickness
t_r	Thickness of resin-rich layer
w	Deflection of the beam at any distance, x, from the crack tip
x	Distance from the crack tip
δ_A	Deflection of DCB arm
Δ_l	Crack length shortening due to large deflections
Δ_t	Crack length shortening due to T-tab rotation
θ	Angle of rotation at DCB loading point

The double cantilever beam (DCB) specimen is a popular specimen for determination of composite Mode I interlaminar fracture toughness (G_{Ic}). The DCB specimens usually consist of several unidirectional plies layed-up with a thin insert at the midplane (Fig. 1) to serve as a starter crack. Load is applied either through metal hinges [1–8] or metal T-tabs [9–16] bonded to the end of the specimen. The crack length is measured along the specimen edges. The measured crack length, applied load, and load-point deflection are then used to compute G_{Ic} by a data analysis method that is usually based on linear beam theory [1,2,5–8,12–18].

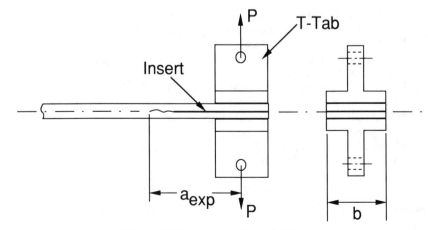

FIG. 1—DCB specimen with bonded T-tabs.

In general, the DCB specimen is designed to limit deflections to the geometrically linear range [1,2]. However, with the advent of tough resin composites, there is an increased possibility of encountering large deflections for DCB specimen thicknesses that have been conventionally used. Large deflections cause an effective shortening of the crack length (Fig. 2a) which leads to errors in the computed G_{Ic} values if DCB data are analyzed using linear beam theory assumptions. In such instances, for a chosen DCB specimen design, an estimate of the effect of large deflections on the computed G_{Ic} values can be made by the analyses of Refs 3 and 4. Alternatively, the design criteria suggested in Refs 1 and 2 can be used to limit DCB deflections to the geometrically linear range.

As mentioned earlier, the DCB specimen is loaded either through bonded hinges or T-tabs. The use of T-tabs shifts the line of action of the load due to the rotation of the DCB arms and leads to an effective crack-length shortening (Fig. 2b). The use of hinged tabs drastically reduces this effect. The effective crack-length shortening in the case of T-tabs increases with load [9–11] and leads to a geometrically nonlinear problem. The analyses in Refs 9 and 11 account for the geometric nonlinearities resulting from both T-tabs and large deflections. Wang et al. [9] and Williams [11] state that the effects of loading tabs may significantly influence the computation of strain-energy release rate in a DCB specimen.

Hinged DCB specimens are, therefore, preferable to T-tabbed specimens. However, based on the authors' personal experience, T-tabs are required for tougher materials to carry the

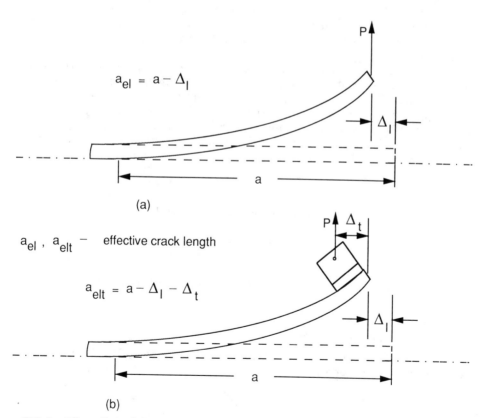

FIG. 2—*Effects of large deflections and T-tabs in a DCB specimen test: (a) effect of large deflections and (b) effect of large deflections and T-tabs.*

higher critical loads. The higher load-carrying capacity of the T-tabs is probably due to the fact that the load gets applied at the center of the thick tab causing a somewhat uniform peel stress at the bondline between the tab and the specimen. In contrast, for hinged specimens, load is applied at one end of the thin hinge leading to a higher peel stress at that end. Thus, for the same load, stresses at the bondline between the hinge tabs and the specimen could be more severe. T-tabs may, therefore, be required for some tough materials to carry the larger critical loads. Although the effect of end rotation on computed G_{Ic} values could be accurately accounted for by the analyses in Refs 9 and 11, these analyses are quite complicated and tedious to use. Conversely, T-tabs could be designed to minimize the effects of end rotation. However, there are no design guidelines available for DCB specimen tabs.

The purpose of this paper is first to present a simple strength of materials analysis to account for geometric nonlinearities resulting from T-tabs and large deflections. Next, DCB design guidelines are presented to minimize the effects of these geometric nonlinearities. Then, a data analysis procedure is developed to compute G_{Ic}. Finally, DCB test results are presented to quantify these effects for commonly used composites over a wide range of toughnesses.

Mathematical Analysis of DCB Specimen with T-Tabs

The analysis of a DCB specimen with T-tabs is presented in the following sections. The DCB arm is first idealized as a beam on an elastic foundation. Equations for the load-deflection behavior are then derived for the case of small deflections followed by derivations for the case of large deflections. Next, expressions for the strain-energy release rate are derived for a DCB with T-tabs. Finally, design guidelines are developed to ensure that DCB deflections are in the geometrically linear range.

The DCB specimen arm is often represented as a cantilever beam [4,8,9]. However, this assumption does not account for the end-rotation at the crack tip. A more suitable assumption is to represent the DCB arm as a beam on an elastic foundation in which end-rotation is possible [18]. The deflection of such a beam with an elastic foundation can be approximated by the deflection of a cantilever beam with an additional end correction length of a_0 (see Appendix). Thus, for the DCB specimen shown in Fig. 3, each arm of the DCB can be idealized as a cantilever beam of length a given by

$$a = a_{exp} + a_0 \tag{1}$$

where a_{exp} is the crack length measured during testing (see Figs. 1, 3, and 4) and a_0 is given by (see Appendix)

$$a_0 = h\sqrt[4]{(E_{11}/6E_{22})} \tag{2}$$

where h is the thickness of each arm and E_{11} and E_{22} are the longitudinal and transverse moduli, respectively.

Small Deflection Analysis

Consider a DCB specimen that has bonded T-tabs of height d (Fig. 3). The application of load P causes an end rotation, θ, that results in a resisting moment, M_r (Fig. 3), given by

$$M_r = PH \sin(\theta) \tag{3}$$

where

$$H = d + (h/2) \tag{4}$$

Double cantilever beam specimen
with end tabs

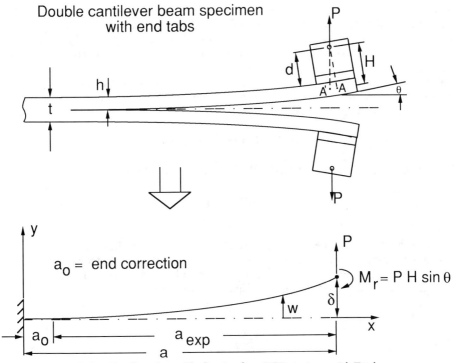

FIG. 3—*Cantilever beam idealization for a DCB specimen with T-tabs.*

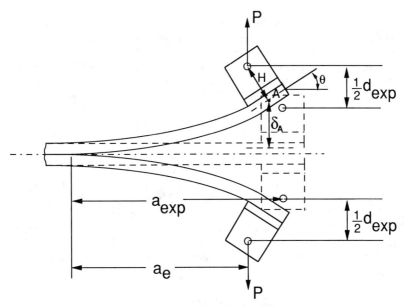

FIG. 4—*Nomenclature for a DCB specimen with T-tabs.*

Note that the tabs are assumed to be rigid in the present analysis. Thus, a DCB specimen with load P applied through end tabs and a measured crack length of a_{exp} can be analyzed as a cantilever beam of length a (Eq 1) with load P and a resisting moment, M_r (see Fig. 3). Note that M_r is a function of θ, which in turn is a function of P leading to a geometrically nonlinear problem.

The cantilever beam in Fig. 3 can be analyzed by first writing the moment at any point, x, along the beam and then using the moment-curvature relationship for a beam. The moment at any point, x, along the beam is,

$$M(x) = P(a - x) - PH \sin(\theta) \tag{5}$$

The moment-curvature relationship for the beam is given by Ref 19,

$$\frac{d^2w}{dx^2} = \frac{M}{E_{11}I} \tag{6}$$

where I is the moment of inertia of the beam cross-section and w is the beam deflection at any point along the beam. Note that the flexural rigidity of the tabs is neglected in Eq 6. Also, the assumption is made that $(dw/dx)^2 \ll 1$. Integrating and applying the appropriate boundary conditions yields expressions for the slope and deflection at Point A (see Fig. 3). The slope is given as

$$\tan(\theta) = \left(\frac{Pa^2}{2E_{11}I}\right)\left(1 - 2\frac{H}{a}\sin(\theta)\right) \tag{7}$$

and the deflection, δ_A, at Point A is given by

$$\delta_A = \left(\frac{Pa^3}{3E_{11}I}\right)\left(1 - 1.5\frac{H}{a}\sin(\theta)\right) \tag{8}$$

Note that Eqs 7 and 8 agree with the slope and deflection of a linear beam (with load P) when H is set equal to zero. Note that the constant slope imposed by the T-tab over the bond region is not accounted for in the preceding equations.

As shown in Fig. 2, the loading tabs cause an effective shortening (Δ_t) of the beam length. This shortening can be easily determined, by inspection of Figs. 3 and 4, as $H \sin(\theta)$. The effective length of a beam with T-tabs can, thus, be approximated by a_{et} where,

$$a_{et} = a\left(1 - \frac{H}{a}\sin(\theta)\right) \tag{9}$$

This effective length, a_{et}, will be used later in the derivation of Mode I strain-energy release rate for a DCB with T-tabs. Note that a_{et} from Eq 9 can be used in the deflection equation of a linear cantilever beam to approximate the deflection of a linear beam with tabs; however, that would give the deflection at Point A' (see Fig. 3) on the beam and not at Point A. Note that Point A' is not a fixed point on the beam and so it does not provide a good reference for the beam deflection. Thus, Point A is used instead as a reference for beam deflection. The deflection at Point A, given by Eq 8, can also be related to the measured (load-point) displacement in a tabbed DCB test. This measured deflection, d_{exp} (see Fig. 4), can be expressed as

$$d_{exp} = 2(\delta_A + H(\cos(\theta) - 1)) \tag{10}$$

The load-deflection behavior of a DCB with tabs can now be examined by using Eqs 7, 8, and 10.

Figure 5 shows a plot of DCB load verses load-point deflection. Both the quantities are presented in a nondimensional form. The dash-dotted curve represents the load-deflection behavior of a linear DCB with tabs for an H/a ratio of 0.3. This will be the case for a 25.4-mm tab and a crack length of 85 mm. Note that the crack lengths used in a DCB test are usually between 50 and 120 mm. For deflection/length ratios of less than 0.05, the DCB with tabs closely follows the simple linear DCB with hinges (dotted line). However, for larger d_{exp}/a ratios, the DCB with tabs departs considerably from linear beam behavior. For a 3-mm-thick DCB specimen made from AS4/PEEK, initial crack extension occurs at a d_{exp}/a ratio of about 0.3 under static loading [15]. At this ratio, there is a difference of 18% between a DCB with tabs and a DCB with hinges.

Large Deflections Analysis

The effects of large deflections on DCB response has been studied in Refs 3, 4, 9, and 11. DCB specimens with hinges were considered in Refs 3 and 4, while tabbed DCB's were analyzed in Refs 9 and 11. The analyses in Refs 9 and 11, however, are very complicated and do not separate the effects of T-tabs and large deflections. The present study uses a simple strength of materials approach, similar to that in Ref 4, to analyze a DCB with tabs undergoing large deflections.

As discussed earlier, large deflections cause an effective shortening of the crack length (see Fig. 2). This shortening can be derived by considering the midplane strain, ϵ_x, given by [20]

$$\epsilon_x = \frac{\partial u}{\partial x} + \left(\frac{1}{2}\right)\left(\frac{\partial w}{\partial x}\right)^2 \tag{11}$$

where u and w are displacements in the x and y directions, respectively. The midplane strain, ϵ_x, is assumed to be zero [4]. Thus, the variation of u with respect to x is given by

$$\partial u/\partial x = -\tfrac{1}{2}(\partial w/\partial x)^2 \tag{12}$$

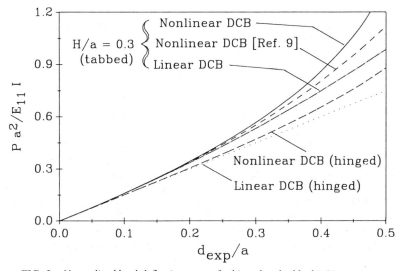

FIG. 5—*Normalized load-deflection curves for hinged and tabbed DCB specimens.*

The term $(\partial w/\partial x)$ can be obtained from Eqs 5 and 6 for a DCB with tabs and is given by

$$\frac{\partial w}{\partial x} = \left(\frac{P}{2E_{11}I}\right)(a^2 - (a - x)^2 - 2 \times H \sin(\theta)) \tag{13}$$

Integrating Eq 12 along the beam length after substituting for $\partial w/\partial x$ from Eq 13 gives the total shortening, Δ_l, of the beam in the x-direction as

$$\Delta_1 = \int_0^a \left(\frac{\partial u}{\partial x}\right) dx \tag{14a}$$

for example,

$$\frac{\Delta_1}{a} = \left(\frac{1}{3}\right)\left(\frac{Pa^2}{E_{11}I}\right)^2 \left(\left(\frac{1}{5}\right) - \left(\frac{5}{8}\right)\left(\frac{H}{a}\right)\sin(\theta) + \left(\frac{1}{2}\right)\left(\frac{H}{a^2}\right)\sin^2(\theta)\right) \tag{14b}$$

The effective crack length (a_{el}) for a beam with large deflections is thus

$$a_{el} = a(1 - (\Delta_1/a)) \tag{15}$$

Using Eqs 9 and 15, it is clear that the effective crack length for a DCB arm with tabs undergoing large deflections can be written as

$$a_{elt} = a\left(1 - \frac{H}{a}\sin(\theta) - (\Delta_1/a)\right) \tag{16}$$

This effective crack length for a DCB arm with T-tabs undergoing large deflections will be used later in the derivation of Mode I strain-energy release rate. Note that, if the effective crack length, a_{elt}, is used in the deflection equation of a linear beam in order to account for the effects of tabs and large deflections, then that would give the deflection at Point A' and not at Point A.

The deflection, at Point A, of a DCB arm with tabs that is undergoing large deflections can be written, using Eqs 8 and 15 as

$$\delta_A = \left(\frac{Pa^3}{3E_{11}I}\right)\left(1 - 1.5\frac{H}{a}\sin(\theta)\right)(1 - (\Delta_1/a))^3 \tag{17}$$

The nonlinear load-deflection response of a DCB with tabs can be plotted using Eqs 7, 10, 14, and 17 and is shown in Fig. 5 (solid curve). At a typical d_{exp}/a ratio of 0.3 [15] and an H/a ratio of 0.3, the nonlinear tabbed DCB (solid curve) is 6% above the linear tabbed DCB (dash-dotted curve) and 25% above the linear hinged DCB (dotted curve). The load-deflection curve for the DCB with tabs and large deflections (solid curve) obtained by the present analysis agrees well with that obtained by the more accurate analysis of Ref 9 (short dashed curve) for d_{exp}/a ratios of less than 0.4. However, for d_{exp}/a ratios of 0.4 and greater there is more than 6% difference between the present analysis and that of Ref 9. The effect of large deflections in the absence of T-tabs can be examined by using Eqs 14 and 17 and substituting $H = 0$. This leads to the long dashed curve in Fig. 5 for a hinged nonlinear DCB. For a typical d_{exp}/a ratio of 0.3, the effect of large deflections in a hinged DCB is only 3%.

Strain-Energy Release Rate Analysis

Based on the deflection equation for a linear DCB with tabs (Eq 8) and its similarity to the linear cantilever beam equation, it is possible to derive a simple equation for the strain-energy release rate. As discussed earlier, the geometric nonlinearities associated with large deflections and T-tabs lead to an effective shortening of the beam. Equations 9, 14, and 15 give a good estimate of this shortening as a function of the applied load.

The strain-energy release rate, G_I, for the DCB is given by [4]

$$G_I = \frac{M_0^2}{bE_{11}I} \tag{18}$$

where b is the width of the specimen and M_0 is the bending moment at the crack tip. For a linear hinged DCB, M_0 is given simply by (Pa). Equation 18 is valid for both linear and nonlinear hinged DCBs. It has been shown in Refs 4 and 11 that it can be used for a DCB with geometric nonlinearities if the moment is calculated by taking into account the shortening of the DCB arm. Thus, if M is given by (Pa_{et}) for a linear tabbed DCB (see Eq 9) and by (Pa_{elt}) for a tabbed DCB with large deflections (see Eq 16) then Eq 18 can also be used for these cases. The strain-energy release rate for a linear DCB with tabs can, therefore, be written as

$$G_I^{et} = \left(\frac{P^2a^2}{bE_{11}I}\right)\left(1 - \frac{H}{a}\sin(\theta)\right)^2 \tag{19}$$

Note that a in Eq 19 is the sum of the measured crack length, a_{exp}, and the end correction, a_0 (see Eq 1). The angle, θ, is a function of the applied load (see Eq 7).

For a DCB with tabs undergoing large deflections, the expression for G_I is given by

$$G_I^{elt} = \left(\frac{P^2a^2}{bE_{11}I}\right)\left(1 - \frac{H}{a}\sin(\theta) - (\Delta_l/a)\right)^2 \tag{20}$$

where Δ_l/a is the shortening in the crack length due to large deflections and is given by Eq 14. The effects of both large deflections and T-tabs on the strain-energy release rate in a DCB can now be studied for a range of d_{exp}/a ratios by using Eqs 7, 8, 10, 14, 17, 19, and 20.

Figure 6 shows the normalized strain-energy release rate $(G_Iba^2/E_{11}I)$ as a function of the normalized DCB load-point deflection (d_{exp}/a) for an H/a ratio of 0.3. For d_{exp}/a ratios of less than 0.3, the short dashed curve for a linear DCB with tabs differs by less than 4% from the results for a linear DCB with tabs, given in Ref 10 (dash-dotted curve). At d_{exp}/a ratios greater than 0.3, the agreement is not very good; but for such high ratios there will be large deflections in the DCB specimen; and, one should use the results for the large deflection case. The solid curve in Fig. 6 corresponds to a DCB with tabs and large deflections (Eq 20) and compares very well with the result from Ref 9. For a typical d_{exp}/a ratio of 0.3 [15], the present solution differs by only 2% from the more accurate solution of Ref 9. Also, for d_{exp}/a ratios greater than 0.3, there is a very good correlation between the present solution and that of Ref 9.

Figure 7 shows the effect of tab height for a range of H/a ratios (based on Eq 20). The solid curve corresponds to a nonlinear hinged DCB and was obtained from Eq 20 by substituting H/a equal to zero. For a typical d_{exp}/a ratio of 0.3, the results for the nonlinear hinged DCB differ by only 3% from those for the linear hinged DCB (dotted curve). The results for the linear hinged DCB were obtained by substituting H/a equal to zero in Eq 19. The difference between the linear hinged DCB and the nonlinear tabbed DCB increases with increasing H/a ratios.

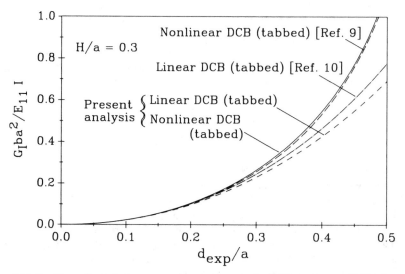

FIG. 6—*Normalized strain-energy release rate curves for DCB specimens with T-tabs.*

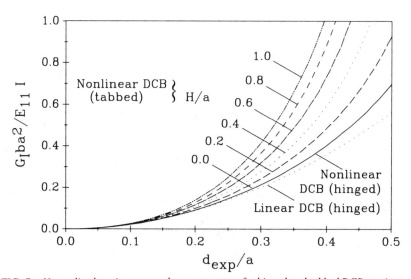

FIG. 7—*Normalized strain-energy release rate curves for hinged and tabbed DCB specimens.*

The percentage variation from the linear beam theory assumptions for DCB specimens, that have T-tab or large deflection effects or both, are summarized in Fig. 8 for a range of d_{exp}/a ratios and an H/a ratio of 0.3. For a nonlinear hinged DCB, the percentage error in using linear beam theory analysis is less than 5% for d_{exp}/a ratios of less than 0.4. However, for a DCB with tabs ($H/a = 0.3$) undergoing large deflections, there could be errors as high as 18% (for a d_{exp}/a of 0.3) if one uses linear beam theory assumptions. For a linear DCB with tabs ($H/a = 0.3$), there would be a 15% error for a d_{exp}/a of 0.3, if one did not account for the effects of the tabs.

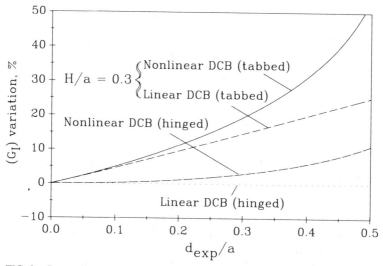

FIG. 8—*Percentage variation in G_I for DCB specimens with geometric nonlinearities.*

Guidelines for Minimizing Geometric Nonlinearity Effects

The G_I expressions in the previous section can also be used to derive design criteria for DCB specimens with T-tabs and large deflections to ensure that DCB deflections are in the geometrically linear range.

DCB Specimens with Large Deflections—Consider a hinged DCB specimen that is undergoing large deflections. The expression for G_I in this case can be derived from Eq 20 by substituting H equal to zero and the corresponding expression for Δ_I from Eq 14

$$G_I = \left(\frac{P^2 a^2}{b E_{11} I}\right)\left(1 - \left(\frac{1}{15}\right)\left(\frac{P a^2}{E_{11} I}\right)^2\right)^2 \tag{21}$$

Notice, that the first term in Eq 21 corresponds to the G_I for a linear DCB and the second term corresponds to a correction attributable to large deflections. If one desires to limit large deflection effects to, say, less than 2%, then the following inequality should hold true

$$\left(1 - \left(\frac{1}{15}\right)\left(\frac{P a^2}{E_{11} I}\right)^2\right)^2 \geq 0.98 \tag{22}$$

Using this inequality and Eq 21 and substituting for I appropriately, an inequality for laminate thickness, t, can be written as

$$t \geq 8.65 \sqrt[3]{(G_{Ic} a^2 / E_{11})} \tag{23}$$

where G_{Ic} is the Mode I fracture toughness for the material being tested and a is the crack length of the DCB specimen. The laminate thickness used for a DCB specimen should satisfy this condition in order to ensure less than 2% effects due to large deflections. The inequality (Eq 23) could also be used to find the crack length, a, for which errors due to large deflection will be less than 2%. If the G_{Ic} value for crack extension from the insert is of interest then, in order to keep large deflection effects below 2%, the insert length should be less than $(0.04\sqrt{(t^3E_{11}/G_{Ic})}$ for a given laminate thickness, t.

DCB Specimens with T-tabs—The expression for G_I, in the case of a DCB specimen with tabs, was derived earlier (see Eq 19). In Eq 19, the first term corresponds to the G_I for a linear DCB and the second term corresponds to a correction attributable to the effects of T-tabs. If one desires to limit T-tab effects to less than 2%, then the following inequality should hold true

$$\left(1 - \frac{H}{a}\sin(\theta)\right)^2 \geq 0.98 \tag{24}$$

Using this inequality and Eqs 7 and 19 and substituting for I appropriately, an inequality for total tab height, H, can be written as

$$H \leq 0.01\sqrt{(0.0434E_{11}t^3/G_{Ic}) + a^2)} \tag{25}$$

Note that H is the total tab height and is given by Eq 4. The tab height used in a DCB specimen with tabs should satisfy the condition (Eq 25) in order to ensure less than 2% effects due to the tabs. For a hinged specimen, H corresponds to the height of the hinge axis of rotation above the centerline of the DCB arm. For DCB specimens that are made thicker in order to minimize large deflection effects, the distance, H, even for hinge tabs could be significant and Eq 25 should be used to check for this tab-like effect.

DCB Testing and Data Analysis

The results of the analysis in earlier sections suggest that the effects of geometric nonlinearity associated with T-tabs on the computed G_I values can be as high as 18%, while the effects of large deflection will usually be less than 5%. In order to illustrate these results, specimens made from unidirectional composites having a wide range of Mode I fracture toughness values were tested both as hinged DCBs and as tabbed DCBs. The DCB data were analyzed using three different procedures. First, a data analysis procedure based on the present analysis was used to account for large deflection and T-tab effects. Next, the well-known Berry procedure [17] based on the compliance method was used. Finally, the area method was used to provide an accurate reference for comparing average G_{Ic} values.

Materials and Specimens

Specimens were about 3 mm thick and were cut from unidirectional, 24-ply panels with a Kapton film (0.0127 mm thick) crack starter at the midplane. Three different panels were made from T300/5208, IM7/8551-7 [21], and AS4/PEEK prepreg according to the manufacturer's instructions. After curing, the panels were cut into 152 by 25 mm DCB specimens. Only two specimens of each material were tested since the focus of the present tests was to illustrate the analysis and compare different data analysis procedures. For each material, one of the specimens was bonded with aluminum alloy hinges and another was bonded with aluminum T-tabs that were 25 mm in height ($d = 25$ mm, see Fig. 3). Edges of the specimens were painted with white water-based typewriter correction fluid and marked at increments of 2.5 mm for visual crack measurements.

Test Procedure

Static tests were performed under displacement control in a screw-driven machine at a constant cross-head rate of 0.0085 mm/s. Load-deflection data was collected through a digital data acquisition system and was also plotted on a *x-y* plotter. Traveling microscopes were used to monitor crack length along both edges of the specimen. Load and deflection were noted for the initial crack extension from the insert and then at every 2.5 mm of crack extension indicated by the marks on the specimen edges. After the crack had extended about 12.7 mm, the specimen was unloaded. The specimen was reloaded and the crack was extended another 12.7 mm while the load, deflection, and crack length were monitored. A third loading-unloading cycle was conducted in a similar manner for another 12.7 mm of crack extension. Typical load-deflection plots are shown schematically in Fig. 9. The geometric nonlinearities associated with T-tabs and large deflections cause an upwardly concave curve with a monotonically decreasing compliance. The initial compliance, denoted by C_i, will be used later in the data analysis procedure.

Data Analysis Procedures

As mentioned earlier, data analysis was performed using three different procedures. This section describes the present analysis to account for the effects of large deflections and T-tabs, Berry's [17] procedure, and the area method.

Present Data Analysis Procedure—The present procedure uses the initial compliance, C_i (Fig. 9), the load, deflection, and crack length data, and the G_I expression given by Eq 20 to determine G_{Ic}. There are, however, two unknowns in Eq 20 that need to be determined before that equation can be used to determine G_{Ic}. These two unknowns are the flexural stiffness, $E_{11}I$, and the angle, θ, at the end of the DCB arms.

The quantity, $E_{11}I$, is directly evaluated by compliance calibration in the present data analysis procedure. However, as indicated in Fig. 9, the compliance of a DCB with geometric nonlinearities changes with load. Based on the analysis, the load-deflection plots in Fig 5 also indicate a compliance that changes with load but for very small deflections ($\delta/a < 0.05$), the curves for the geometrically nonlinear cases coincide with the linear beam curve. Thus, the effects of large deflections and T-tabs are minimal for the initial part of the load-deflection curve and the initial compliance, C_i, can be approximated by linear beam theory assumptions as

$$C_i = \left(\frac{2}{3E_{11}I}\right) a^3 \tag{26}$$

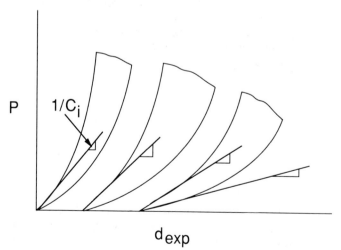

$$d_{exp}$$

FIG. 9—*Typical load-deflection curves for DCB specimens with geometric nonlinearities.*

The quantity in brackets can be determined from a least-squares fit on a log-log plot of C_i versus a. This method was used to determine $E_{11}I$ in the present data analysis procedure.

The preceding procedure eliminates the errors involved in using the values of E_{11} and I in the computation of G_{Ic}. The quantity E_{11} in Eq 20 is actually an effective modulus that, in general, is not the same as the inplane modulus [1,2,13]. Furthermore, the E_{11} in a composite could vary along the laminate thickness due to inhomogeneity caused by the manufacturing process [22]. Also, the moment of inertia I ($= (\frac{1}{12})bh^3$) contains h^3 that could lead to large errors in the computed I due to even small errors in measuring laminate thickness.

The angle, θ, at the end of the DCB arms was determined by using the load-deflection relationship derived earlier (Eq 17) together with the result from Eq 10. The measured deflection can thus be written as

$$d_{exp} = \left(\frac{2Pa^3}{3E_{11}I} \right) \left[1 - 1.5 \frac{H}{a} \sin(\theta) \right] + 2H(\cos(\theta) - 1) \tag{27}$$

A shear correction term [8] given by $(2.4Pa/(bhG_{13}))$ could be added to Eq 27 where G_{13} is the shear modulus and $E_{11}I$ is determined as discussed earlier. The measured load, P, deflection, d_{exp}, and crack length, a, given by Eq 1, are substituted into Eq 27 to determine the angle, θ, iteratively by a numerical scheme. The secant method was used in the present study.

Once the quantities $E_{11}I$ and θ are known, the G_{Ic} for the material can be computed by

$$G_{Ic} = \left(\frac{P^2a^2}{bE_{11}I} \right) \left[1 - \frac{H}{a} \sin(\theta) - (\Delta_l/a) \right]^2 \tag{28}$$

where (Δ_l/a) is given by Eq 14. A shear correction term [8] given by $(1.2P^2/(b^2hG_{13}))$ could be added to Eq 28 to account for shear deformation. In the present study, the effects of the shear correction term were less than 1% and were therefore neglected. However, in some cases this might not be the case and the preceding shear correction should be then used. Note that the term, $(H/a)\sin(\theta)$, corresponds to a correction due to T-tabs and should be used only for a DCB with tabs. Also, the term (Δ_l/a) corresponds to a correction due to large deflections.

Alternatively, the G_{Ic} could be also calculated by first determining $E_{11}I$, as described earlier, and then using Fig. 7 along with the measured d_{exp}/a ratio at crack extension and the H/a ratio to find the corresponding value for $(G_Iba^2/E_{11}I)$. Knowing b, a, and $E_{11}I$, one can find G_I and then calculate G_{Ic} by adding the shear correction of Eq 28 to G_I.

Berry's Method—The Mode I fracture toughness can be determined by Berry's method [17] that uses the load, load-point deflection, and crack length data at the time of crack extension in the following equation [13]

$$G_{Ic} = \frac{n\overline{H}}{2b} \tag{29}$$

where

$$\overline{H} = (1/N) \sum_{i=1}^{N} \left(\frac{P_i \delta_i}{a_i} \right) \tag{30}$$

where P_i, δ_i, and a_i are the measured load, deflection, and crack length, respectively, and N is the total number of data points while n is the slope of the least-squares fit to the compliance-crack length data plotted on a log-log scale. This method has the advantage of not requiring

$E_{11}I$ as the input parameter. Also, since the load, deflection, and crack length are all used in the computation, it might account, at least in part, for the effects of geometric nonlinearity. However, Berry's method is based on linear load-deflection behavior and does not explicitly account for the effects of T-tabs and large deflections.

The Area Method—The load-deflection plots shown in Fig. 9 can be used directly to compute G_{Ic} by accurately measuring the area enclosed by the loading-unloading curve and dividing it by the incremental area created during crack extension. This method implicitly accounts for any geometric nonlinearities since it uses the actual load-displacement curves. However, it can only give average G_{Ic} values and cannot be used to compute the G_{Ic} at the onset of crack growth from the insert.

Results and Discussion

The DCB data for the three different materials was used to compute G_{Ic} for both the hinged and the tabbed specimens. The present data analysis method was used by including all the geometric nonlinearity terms and then also used by neglecting the effects of large deflections and T-tabs. Table 1 shows a comparison of the G_{Ic} values computed using the present method, Berry's method, and the area method. The values shown in the second through fifth columns are for the onset of delamination from the insert while the values shown in the last two columns are average values for the first 12.7 mm of crack extension. The values calculated using Berry's method (BM) should be compared with those from the present method obtained by neglecting the effects due to tabs and large deflections (LB). In general, there is good agreement between these two methods and the slight differences can be attributed to the differences in the two methods of data analysis. The average G_{Ic} values shown in the last two columns are subject to the effects of fiber-bridging and are therefore higher than those computed for the onset of delamination. However, a comparison of these two columns helps validate the results of the present analysis technique.

Table 1 also shows G_{Ic} values computed using the present data analysis method after separately accounting for large deflection (LDT) and T-tab (TB) effects. For both hinged and tabbed specimens, the effects of large deflection (compared to a linear DCB) were less than

TABLE 1—*Comparison of DCB test results.*

Material	$(G_{Ic})_{initial}$ (J/m²)				$(G_{Ic})_{average}$ (J/m²)	
	BM[a]	LB[b]	TB[c]	LDT[d]	Area	LDT[d]
			HINGED[e]			
T300/5208	94	107	· · ·	107	112	119
IM7/8551-7	605	629	· · ·	634	756	670
AS4/PEEK	1564	1593	· · ·	1622	1705	1784
			T-TABBED[f]			
T300/5208	96	102	107	107	85	100
IM7/8551-7	519	536	618	621	858	771
AS4/PEEK	1429	1401	1680	1734	1873	1805

[a] BM = Berry's method [*17*].
[b] LB = Present method neglecting tab and large deflection effects.
[c] TB = Present method considering only tab effects.
[d] LDT = Present method accounting for tab and large deflection effects.
[e] Average initial crack length was 65 mm.
[f] Average initial crack length was 57 mm.

0.8% in the case of T300/5208 and IM7/8551-7, and about 3% for AS4/PEEK. The effects of T-tabs (compared to a linear DCB) were more significant for all the materials with T300/5208 showing a 5% effect, IM7/8551-7 a 15% effect, and, AS4/PEEK a 20% effect.

The effects due to large deflections and T-tabs could have been avoided for the three materials if the DCB thickness and T-tab height were selected according to the guidelines given by Eqs 23 and 25. For T300/5208, the two equations yield (using $G_{Ic} = 87.5$ J/m^2, $E_{11} = 181.3$ GPa, $a = 64$ mm) $t \geq 1.08$ mm and $H \leq 16$ mm. In the present study, a laminate thickness of 3 mm and $H = 25.75$ mm were used, thus, leading to negligible large deflection effects and small T-tab effects. For AS4/PEEK, the guidelines of Eqs 23 and 25 yield (using $G_{Ic} = 1622$ J/m^2, $E_{11} = 136.5$ GPa, $a = 64$ mm) $t \geq 3.14$ mm and $H \leq 4$ mm. The laminate thickness used in this case was 3.23 mm, thus, explaining large deflection errors of about 2% (see Table 1). The tab height, H, used was 25.8 mm that is much higher than the recommended 4 mm leading to the 20% errors due to T-tab effects.

Concluding Remarks

A simple strength of materials analysis was developed for a DCB specimen to account for geometric nonlinearity effects due to large deflections and tabs. A new DCB data analysis procedure was developed to include the effects of these nonlinearities. The results of the analysis were validated by DCB tests performed for materials having a wide range of toughnesses. The materials used in the present study were T300/5208, IM7/8551-7, and AS4/PEEK.

The results of the present simple analysis compared very well with previously developed, more complicated analyses. Based on the analysis, for a typical deflection/crack length ratio of 0.3 (for AS4/PEEK) and an H/a ratio of 0.3, there could be a 19% effect due to T-tabs and a 6% effect due to large deflections on the DCB load-deflection response. The computed strain-energy release rates can be in error by 15% due to tabs and by 3% due to large deflections for the same deflection/crack length ratio of 0.3. In order to keep errors due to these nonlinearities within 2%, the DCB specimen thickness should be greater than $8.65 \sqrt[3]{(G_{Ic}a^2/E_{11})}$ and the total tab height should be less than $0.01 \sqrt{(0.0434E_{11}t^3/G_{Ic}) + a^2}$.

Based on the test results for both hinged and tabbed specimens, the effects of large deflection (compared to a linear DCB) were almost negligible (less than 1%) in the case of T300/5208 and IM7/8551-7; however, AS4/PEEK showed a 2 to 3% effect. The effects of T-tabs (compared to a linear DCB) were more significant for all the materials with T300/5208 showing a 5% effect, IM7/8551-7 a 15% effect, and AS4/PEEK a 20% effect. The average G_{Ic} values computed using the present analysis compared well with those calculated using the area integration method.

APPENDIX

The end deflection of an orthotropic beam on an elastic foundation can be written in a form similar to that given for an isotropic beam in Ref 18 and is given as

$$\delta = \left(\frac{Pa^3}{3E_{11}I}\right)(1 + 3/(\lambda a) + 3/(\lambda a)^2 + 1.5/(\lambda a)^3) \tag{31}$$

where

$$\lambda = \sqrt[4]{((0.25\ k)/(E_{11}I))} \tag{32}$$

and k is the spring constant of the elastic foundation. In the case of a DCB specimen, the elastic foundation consists of a thin resin-rich layer (of thickness t_r) that forms in between two plies and an orthotropic laminate layer (of thickness h) [23]. The combined stiffness of the two layers represents the foundation spring constant, k. By assuming a constant strain in the resin-rich layer and a linearly varying strain distribution in the laminate layer [23], the tranverse stiffnesses are given by ($E_r b/t_r$) and ($2E_{22}b/h$), respectively, where b is DCB width. Since the resin-rich layer and the laminate layer are in series, the foundation spring constant is given by

$$k = \frac{2E_r E_{22}b}{2E_{22}t_r + E_r h} \tag{33}$$

For a very thin resin-rich layer Eq 33 can be simplified as

$$k = 2E_{22}b/h \tag{34}$$

and from Eq 32

$$1/\lambda = h\sqrt[4]{(E_{11}/6E_{22})} \tag{35}$$

Now, Eq 31 can be approximated by replacing 1.5 in the last term with 1.0 as

$$\delta = \left(\frac{P}{3E_{11}I}\right)(a + a_0)^3 \tag{36}$$

where

$$a_0 = 1/\lambda \tag{37}$$

Note that Eq 36 represents the load-point deflection of a cantilever beam of length $(a + a_0)$ and thus a beam on elastic foundation could be approximated by a cantilever beam with an "end correction" a_0 that is given from Eqs 35 and 37 by

$$a_0 = h\sqrt[4]{(E_{11}/6E_{22})} \tag{38}$$

Since the present analysis is primarily for static DCB tests, only the extensional stiffness of the foundation was considered while its rotational stiffness was neglected. It has been shown in Ref 24 that for static cases the rotational stiffness can be neglected but it should be included in analyzing dynamic cases.

References

[1] Ashizawa, M., "Improving Damage Tolerance of Laminated Composites Through the Use of New Tough Resins," Douglas Paper 7250, McDonnell Douglas Corporation, Long Beach, CA, 1983.

[2] Standard Tests for Toughened Resin Composites, Revised Edition, NASA Reference Publication 1092, National Aeronautics and Space Administration, Washington, DC, July 1983.

[3] Devitt, D. F., Schapery, R. A., and Bradley, W. L., "A Method for Determining the Mode I Delamination Fracture Toughness of Elastic and Viscoelastic Composite Materials," Journal of Composite Materials, Vol. 14, Oct. 1980, pp. 270–285.

[4] Whitcomb, J. D., "A Simple Calculation of Strain-Energy Release Rate for a Nonlinear Double Cantilever Beam," Journal of Composites Technology & Research, Sept. 1984, pp. 64–66.

[5] Martin, R. H., "Effect of Initial Delamination on G_{Ic} and G_{Ith} values from Glass/Epoxy Double Cantilever Beam Tests," Proceedings, American Society for Composites, Seattle, WA, Sept. 1988, pp. 688–700.

[6] Johnson, W. S. and Mangalgiri, P. D., "Investigation of Fiber Bridging in Double Cantilever Beam Specimens," Journal of Composites Technology & Research, Vol. 9, 1987, pp. 10–13.

[7] Chai, H., "The Characterization of Mode I Delamination Failure in Non-Woven, Multidirectional Laminates," *Composites,* Vol. 15, No. 4, Oct. 1984, pp. 277–290.

[8] Aliyu, A. A. and Daniel I. M., "Effects of Strain Rate on Delamination Fracture Toughness of Graphite/Epoxy," *Delamination and Debonding of Materials, ASTM STP 876,* W. S. Johnson, Ed., American Society for Testing and Materials, Philadelphia, 1985, pp. 336–348.

[9] Wang, S. S., Suemasu, H., and Zahlan, N. M., "Interlaminar Fracture of Random Short-Fiber SMC Composite," *Journal of Composite Materials,* Vol. 18, Nov. 1984, pp. 574–594.

[10] Wang, S. S. and Miyase, A., "Interlaminar Fatigue Crack Growth in Random Short-Fiber SMC Composite," *Journal of Composite Materials,* Vol. 20, Sept. 1986, pp. 439–456.

[11] Williams, J. G., "Large Displacement and End Block Effects in the 'DCB' Interlaminar Test in Modes I and II," *Journal of Composite Materials,* Vol. 21, April 1987, pp. 330–347.

[12] Hwang, W. and Han, K. S., "Interlaminar Fracture Behavior and Fiber Bridging of Glass-Epoxy Composite under Mode I Static and Cyclic Loadings," *Journal of Composite Materials,* Vol. 23, April 1989, pp. 396–430.

[13] Whitney, J. M., Browning, C. E., and Hoogsteden, W., "A Double Cantilever Beam Test for Characterizing Mode I Delamination of Composite Materials," *Journal of Reinforced Plastics and Composites,* Vol. 1, Oct. 1982, pp. 297–313.

[14] Wilkins, D. J., Eisenmann, J. R., Camin, R. A., Margolis, W. S., and Benson, R. A., "Characterizing Delamination Growth in Graphite-Epoxy," *Damage in Composite Materials, ASTM STP 775,* K. L. Reifsnider, Ed., American Society for Testing and Materials, Philadelphia, 1982, pp. 168–183.

[15] Russell, A. J. and Street, K. N., "The Effect of Matrix Toughness on Delamination: Static and Fatigue Fracture," *Toughened Composites, ASTM STP 937,* N. J. Johnston, Ed., American Society for Testing and Materials, Philadelphia, 1987, pp. 275–294.

[16] Guedra, D., Lang, D., Rouchon, J., Marias, C., and Sigety, P., "Fracture Toughness in Mode I: A Comparison Exercise of Various Test Methods," *Proceedings,* Sixth International Conference on Composite Materials, ICCM & ECCM, London, July 1987, Vol. 3, pp. 346–357.

[17] Berry, J. P., "Determination of Fracture Energies by the Cleavage Technique," *Journal of Applied Physics,* Vol. 34, No. 1, Jan. 1963, pp. 62–68.

[18] Kanninen, M. F., "An Augmented Double Cantilever Beam Model for Studying Crack Propagation and Arrest," *International Journal of Fracture,* Vol. 9, No. 1, March 1973, pp. 83–91.

[19] Timoshenko, S. and Young, D. H., *Elements of Strength of Materials,* 5th ed., Van Nostrand Reinhold Company, New York, 1968.

[20] Timoshenko, S. and Woinowsky-Krieger, S., *Theory of Plates and Shells,* Second ed., McGraw-Hill, New York, 1959.

[21] Gawin, I., "Tough Thermosetting Resins with Superior Damage Tolerance Hercules 8551 Series," *Proceedings,* Thirty-First International SAMPE Symposium and Exhibition, Society for the Advancement of Material and Process Engineering, April 1986, pp. 1205–1213.

[22] O'Brien, T. K., Murri, G. B., and Salpekar, S. A., "Interlaminar Shear Fracture Toughness and Fatigue Thresholds for Composite Materials," *Composite Materials: Fatigue and Fracture, Second Volume, ASTM STP 1012,* P. A. Lagace, Ed., American Society for Testing and Materials, Philadelphia, 1989, pp. 222–250.

[23] Crews, J. H., Jr., Shivakumar, K. N., and Raju, I. S., "Factors Influencing Elastic Stresses in Double Cantilever Beam Specimens," *Adhesively Bonded Joints: Testing, Analysis, and Design, ASTM STP 981,* W. S. Johnson, Ed., American Society for Testing and Materials, Philadelphia, 1988, pp. 119–132.

[24] Kanninen, M. F., "A Dynamic Analysis of Unstable Crack Propagation and Arrest in the DCB Test Specimen," *International Journal of Fracture,* Vol. 10, No. 3, Sept. 1974, pp. 415–430.

Anoush Poursartip[1] and Narine Chinatambi[1]

Experimental Determination of the Mode I Behavior of a Delamination Under Mixed-Mode Loading

REFERENCE: Poursartip, A., and Chinatambi, N., **"Experimental Determination of the Mode I Behavior of a Delamination Under Mixed-Mode Loading,"** *Composite Materials: Fatigue and Fracture (Third Volume), ASTM STP 1110,* T. K. O'Brien, Ed., American Society for Testing and Materials, Philadelphia, 1991, pp. 187–209.

ABSTRACT: Previous work by the authors has shown that it is possible to measure transverse displacements of both faces of a cracked lap shear (CLS) specimen accurately. In this work, this approach is extended so as to determine experimentally the Mode I behavior of the delamination. By considering the difference in the displacements of the two faces, a surface opening displacement (SOD) can be calculated. This SOD is related to the crack opening displacement (COD). It is shown, using the finite element method (FEM) analysis, that the SOD is not identical to the COD, but is uniquely related to it, and is very sensitive to changes in the COD. Experimentally derived SOD profiles at different positions along the width of the crack front are compared with scanning electron microscope examinations of the sectioned specimen, and good agreement is observed. It is shown that there is delayed opening of the crack as the specimen is loaded. The absolute values of the experimental and FEM analysis SOD profiles are in reasonable agreement, and it is shown that G_I values can be calculated from the experimental results. The method is cumbersome and the results are difficult to interpret. However, once the low load profiles are established, the method can be used to monitor subcritical crack growth easily. In all three material systems tested, considerable subcritical crack growth (up to 4 mm) was measured, and the associated G values are 15 to 40% lower than G_c.

KEY WORDS: composite materials, fatigue (materials), fracture, delamination, cracked lap shear, crack opening displacement, surface opening displacement, strain energy release rate, mixed-mode loading, delayed opening, subcritical crack growth

Nomenclature

a	Delamination crack length (mm)
CLS	Cracked lap shear specimen geometry
COD	Crack opening displacement
SOD	Surface opening displacement
E	Young's modulus (GPa)
C, C_0	Compliance, initial compliance
FEM	Finite element method
G	Strain energy release rate (J/m^2)
G_I, G_{II}	Mode I (opening) and Mode II (in-plane shear) components of the strain energy release rate, J/m^2

[1] Assistant professor and senior research technician, respectively, Composites Group, Department of Metals and Materials Engineering, The University of British Columbia, Vancouver, British Columbia, V6T 1Z4 Canada.

G_{Ic}, G_{IIc}	Critical values of energy release rate (J/m^2) for unstable delamination propagation in Mode I and Mode II, respectively
K_1	Mode I (opening) component of stress intensity factor (MPa $\sqrt{m}$)
P	Applied axial load (N)
P^*, G^*	Load and strain energy release rate at onset of subcritical crack growth, as measured from load-SOD curve
P^{**}, G^{**}	Load and strain energy release rate at onset of unstable crack growth, as measured from load-SOD curve
P_c, G_c	Critical load and strain energy release rate, as measured from load-elongation curve
LVDT	Linear variable differential transformer
SEM	Scanning electron microscope
σ_y	Stress normal to the crack plane
ε_y	Strain normal to the crack plane
x	Position along specimen, in the plane of the crack
y	Position in the thickness direction of the specimen
r	Distance from the crack tip
r_{LVDT}	Distance from the LVDTs to the crack tip
θ	Angle to the crack plane
υ	Normal displacement of a crack face
ν	Poisson's ratio
A	Constant of proportionality between G_1 and COD/$\sqrt{r}$
t_0	Thickness of lap and strap
t_2	Thickness of strap
W	Specimen width
L	Specimen length

Delamination growth under mixed-mode loading conditions for both fatigue and static loading is an issue of considerable scientific and practical interest. In order to understand or analyze mixed-mode loading, one needs some method of determining the opening and shear mode components of the strain energy release rate, and then some theory to explain or describe the behavior over a wide range of mixed-mode ratios.

Generally, a finite element method (FEM) solution is used to determine the mixed-mode ratios, and then the experimental results (which are a set of G_c values) are plotted in G_I-G_{II} space, and fitted with an equation of the form

$$\left(\frac{G_I}{G_{Ic}}\right)^n + \left(\frac{G_{II}}{G_{IIc}}\right)^m = 1 \tag{1}$$

It has been reported that $n = m = 1$ gives a good fit to the data [for example, Refs 1 and 2] for many, if not all material systems tested, including AS4-PEEK (APC2). However, tests performed with a new specimen geometry (mixed-mode bending, MMB) capable of a wider range of mixed-mode ratios has shown that, at least for the AS4-PEEK (APC2) system tested there, that a nonlinear form of Eq 1 must be used [3]. Furthermore, there is always some concern that the FEM analyses generally used to separate the Mode I and Mode II components are sensitive to assumptions regarding material properties, homogeneity, boundary conditions, and specimen behavior (that is, nonlinearity). Also, there are cases where FEM analysis cannot be used efficiently because of the complexity of the layup or the geometry.

Therefore, it would be useful and satisfying if there were some fundamental experimental understanding of the crack-driving mechanisms. Furthermore, it is well known that the delamination fatigue growth rate is very sensitive to the strain energy release rate level. Therefore, it has been suggested that a threshold value of G_c be determined and never exceeded in service [4]. An understanding of what is happening at the crack tip would inspire confidence for the use of such an approach.

In this context, this paper describes a method to determine experimentally the Mode I opening behavior of a delamination in a cracked lap shear (CLS) specimen geometry. This work is an extension of preliminary results presented in Ref 5.

Experimental Method

All tests were performed on unidirectional CFRP material. The CLS specimens were approximately 25 mm wide and 211 mm long between the tabs. Three different material systems were used. They were: AS4/3501-6 composite specimens with a strap thickness of 2.11 mm and a lap thickness of 0.87 mm; AS4/3501-6 adhesively bonded specimens with a strap thickness of 2.22 mm, a lap thickness of 1.03 mm, and a FM-300 adhesive bond of thickness 0.28 mm; and AS4/PEEK specimens with a strap thickness of 1.88 mm and a lap thickness of 1.69 mm. Teflon inserts of 25.4 mm length were used as starter cracks. Longer cracks were grown by static loading. All testing was performed on an MTS servohydraulic fatigue machine with hydraulic grips, under computer control.

The CLS specimen and experimental apparatus are shown schematically in Fig. 1. Axial elongation is measured using a linear variable differential transformer (LVDT) attached to the specimen with a 180 mm gage length (not shown in Fig. 1). The axial LVDT is aligned with the center-line of the specimen and is free to rotate around each attachment point. In this manner, any out-of-plane deflections of the specimen are decoupled from the axial deflections.

Out-of-plane deflections of both surfaces of the specimen are measured with two LVDTs mounted horizontally, one against each surface. They are aligned in both the horizontal and vertical planes, and securely attached to a fixture hanging from the top hydraulic grip. The LVDTs can be moved vertically to any position along the length of the specimen, and horizontally to any position across the width of the specimen. Positioning of the LVDT with respect to a fixed origin is always confirmed using an optical traveling microscope. The LVDT plunger tips are replaced with needle points for maximum accuracy.

The resolution of measurement of this arrangement can best be demonstrated by inspecting the results from tests where a unidirectional AS4/3501-6 tensile specimen (of similar dimensions to the CLS specimen) was loaded while monitoring both the in-plane transverse strains (using strain gages) and the out-of-plane transverse deformation (using the LVDTs). A typical result is shown in Fig. 2, with the y-axis representing the transverse deformation in microns. This graph is typical of many similar tests. The strain gage strain data (multiplied by specimen thickness) and the LVDT data are comparable, which should be the case if the laminate is transversely isotropic ($\nu_{12} \approx \nu_{13}$). The noisiness of the LVDT data is very low, much less than 0.1 μm. Errors in the repeatability and hysteresis of the data can lead to larger systematic errors, with the worst typical cases being 0.5 μm. These systematic errors became much worse when we purposely degraded the alignment of the specimen in the grips (leading to an even larger initial apparent dilation on loading) and the seating of the LVDT needle points on the specimen surface (leading to larger apparent hysteresis).

Methodology for Generation of SOD Profiles

The position of the crack tip is determined visually using traveling microscopes, for both the front and back edges. The LVDTs are positioned initially behind the crack tip.

The specimen is then ramped to a maximum load of 13 500 N and the load cell and LVDT

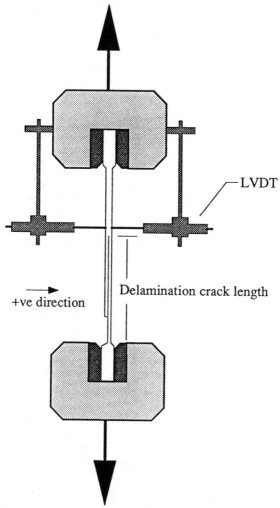

FIG. 1—*CLS specimen geometry and fixture for measuring SOD. (The axial LVDT is not shown.)*

outputs recorded digitally. The LVDTs are then moved up in 0.1-mm increments and the procedure is repeated. In this manner transverse deflection profiles for both surfaces of the specimen can be generated. The specimen surface opening displacement (SOD) profile is the difference between these two deflection profiles.

A typical set of raw results is shown in Fig. 3. The strap face and lap face deflections (Fig. 3a), and the SOD (Fig. 3b) are shown as a function of applied load. Whereas in previous work [5] it was assumed that this value was qualitatively equivalent to the crack opening displacement (COD), we address the issue of their relationship in a more rigorous fashion in a later section.

Inspection of Fig. 3a shows very smooth curves with very low noise levels. However, the strap and lap deflections are similar in magnitude. As a result, inspection of Fig. 3b shows that

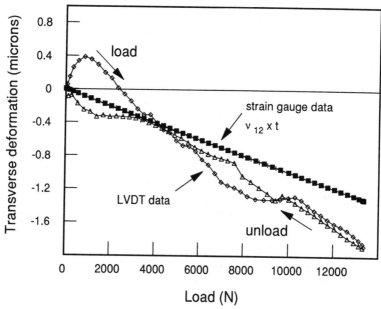

FIG. 2—*Comparison of transverse out-of-plane deformations measured using LVDTs and comparable transverse in-plane deformations measured using strain gages on a tensile AS4/3501-6 composite specimen.*

a plot of the SOD as a function of load is not smooth. There are oscillations of about 0.5 to 1 μm amplitude. However, they are not random noise. Such oscillations were not observed in the validation experiments described previously, and so cannot be due to anything that is common in the setup between the validation tensile geometry and the CLS geometry. Therefore, concerns such as surface roughness, needle-point sticking, and so forth are common to both setups and are not the source of the observed oscillations seen only in the CLS geometry. As the only difference between the specimens is the asymmetrical and cracked nature of the CLS specimen, the observed oscillations most likely represent true specimen behavior.

Since the SOD profiles are built up from a set of tests such as the one in Fig. 3 (each test measuring the SOD as a function of load at one position), a single bad or noisy load cycle will show up as a single bad point on the constructed SOD profile (SOD as a function of position for a given load). Therefore, any systematic feature in a profile is unlikely to be an experimental aberration.

All results presented in this work are unfiltered. Rather than smooth out the data, we have chosen not to draw any conclusions based on changes over small (submillimetre) distances.

SOD profiles were generated for axial traverses along the center line of the specimen, and 1 mm away from the front and back edges, respectively, for the AS4/3501-6 composite specimens. Similar traverses, and an additional vertical traverse 5 mm from the front edge of the specimen were performed on the AS4/3501-6 adhesively bonded specimens. No SOD profiles were generated for the AS4/PEEK material.

For each vertical traverse along the specimen, the experimental procedure described previously was used to generate SOD profiles for applied axial loads ranging from 1000 to 13 500 N in 500 N increments.

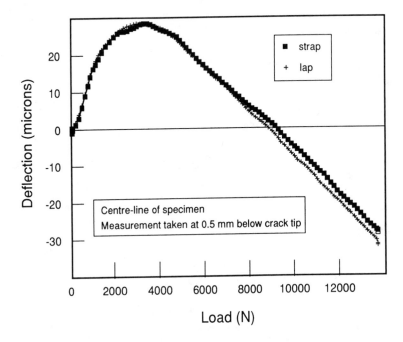

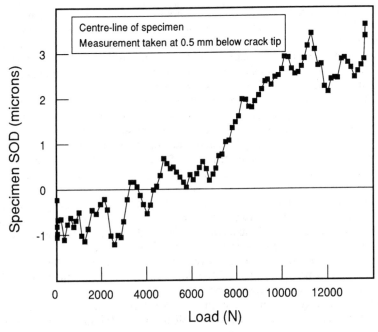

FIG. 3—*AS4/3501-6 composite specimen:* (a) *typical lap face and strap face deflections up to a load of 13 500 N, and* (b) *corresponding SOD.*

Variation of SOD Profile with Applied Load

As will be shown later, in a linear elastic material, the COD is linearly proportional to the stress intensity factor, which is in turn linearly proportional to the local stresses. Furthermore, in a linear elastic material, the SOD is linearly proportional to the COD. In a geometrically linear specimen, the local stresses are linearly proportional to the applied load. Therefore, in a linear system (both material and geometry), the SOD is linearly proportional to the load, P. Thus a normalized profile of SOD divided by load P ($= SOD/P$) as a function of position should be independent of P.

The FEM analysis of the current CLS specimen (described later) predicts that the normalized SOD/P profile is constant over a wide range of loads from 2000 to 25 000 N. The experimental data sets for the center-line traverse for the AS4/3501-6 composite specimen are presented in Figs. 4 and 5. The data are presented in two figures to allow easier viewing, and both figures have identical axes. The profiles become increasingly noisier at low loads as both SOD and P become small. As can be seen, above a load of about 6000 N, the experimental SOD/P profiles superpose. Below 6000 N, the SOD/P profiles are smaller and tend to zero with decreasing load. Note that this is not what is generally meant by geometrically nonlinear behavior in the CLS specimen. What is observed here is some form of delayed crack opening under load.

Similar behavior was seen for the other vertical traverses of the AS4/3501-6 composite specimen and the AS4/3501-6 adhesively bonded specimen. A further example is shown in Fig. 6 for the traverse 1 mm from the front edge of the composite specimen. The data are summarized in Table 1, where the point at which the crack is fully open is recorded. We define a fully open crack to be one where a decrease in load from that condition leads to a decrease in the magnitude of the SOD/P profile.

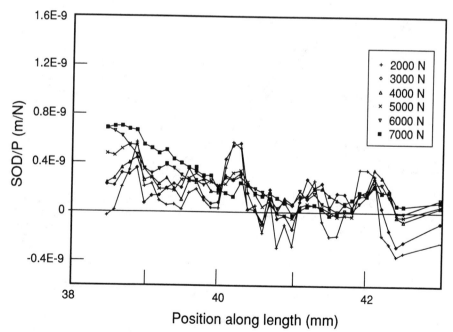

FIG. 4—*Experimental SOD/P profiles (at low loads) measured along the center-line of the AS4/3501-6 composite specimen. The scales on this figure are the same as for Figs. 5 and 6.*

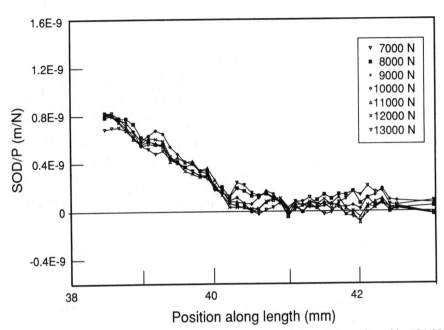

FIG. 5—*Experimental SOD/P profiles (at higher loads) measured along the center line of the AS4/3501-6 composite specimen. The scales on this figure are the same as for Figs. 4 and 6.*

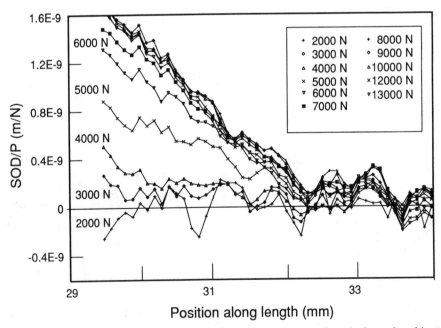

FIG. 6—*Experimental SOD/P profiles measured along a traverse 1 mm from the front edge of the AS4/3501-6 composite specimen. The scales on this figure are the same as for Figs. 4 and 5.*

TABLE 1—*Load required to fully open cracks.*

Position Across Width	Local Crack Length, mm	Opening Load, N
AS4/3501-6 COMPOSITE SPECIMEN		
1 mm from front edge	32.6	5 500
Center line	40.1	6 000
1 mm from back edge	42.5	9 000
AS4/3501-6 ADHESIVELY BONDED SPECIMEN		
1 mm from front edge	97.1	9 000
5 mm from front edge	103.2	10 000
Center line	102.7	8 000
1 mm from back edge	97.5	10 000

Correlation Between SOD Profiles and Visual Crack Observations

In order to determine experimentally the correlation between the SOD profile and the crack tip position, the AS4/3501-6 specimen was sectioned longitudinally, precisely down the center-line and 1 mm from the front and back edges. The sections corresponded as closely as possible to the traverses used to create the SOD/P profiles.

The sectioned pieces were then mounted and inspected in a scanning electron microscope. Great care was taken to set up the pieces inside the SEM, such that the crack tip position observed visually could be located using the same absolute coordinate system used with the LVDTs. In general, it was observed that there is more than one crack tip, with one or more small cracks ahead or overlapping the main crack tip. This is consistent with the evidence of fiber bridging seen with this particular CLS geometry [5].

The SEM observations for the center-line section are shown in Fig. 7. Superimposed on the crack observations are the SOD profiles for $P = 2000$ N and $P = 13\,500$ N. In the region of the main crack tip (located at $x = 40.11$ mm), there is a secondary crack (located from $x = 40.67$ mm to $x = 41.91$ mm). It is interesting that there are no large positive SOD values associated with the secondary crack at the higher load of 13 500 N. On the other hand, there is no Poisson contraction either, which is seen ahead of the crack in Fig. 8.

In similar fashion, the SOD profiles (for $P = 2000$ N and $P = 13\,500$ N) for the traverses close to the front and back edges are superposed on the crack observations in Figs. 8 and 9, respectively. Note that the crack front was not straight and the absolute coordinates for the local crack tip position are different in the three cross sections.

In Fig. 8, we observe that the crack has grown to the current length with considerable fiber bridging. There is good correlation between the SEM observations of the main crack tip and the SOD profile. In Fig. 9, there are two crack tips, with one crack tip about 0.5 mm behind the other. Again, there is good correlation between the SEM observations and SOD profile. Note that the use of radiography or ultrasonics would not provide the necessary submillimetre resolution required to determine the crack front, nor the path of the crack through the thickness, nor the presence of multiple crack tips.

The Relationship Between COD and G_I

Prior to investigating the relationship between COD and SOD, it is helpful to the reader to describe how the COD profile is uniquely related to the Mode I energy release rate.

For simplicity, the following description is for an isotropic material. The conclusions hold for an orthotropic material loaded in the principal directions. The appropriate orthotropic

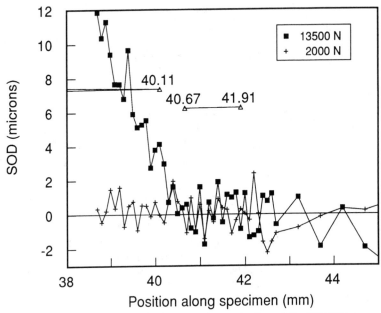

FIG. 7—*Experimental SOD profile measured along the center line of the AS4/3501-6 composite specimen for* P = *2000 and 13 500 N. Superposed on the profile are the crack positions measured in the SEM after sectioning the specimen.*

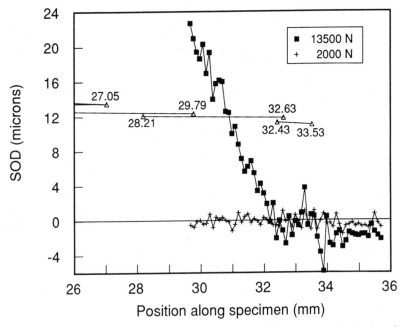

FIG. 8—*Experimental SOD profile measured along a traverse 1 mm from the front edge of the AS4/ 3501-6 composite specimen for* P = *2000 and 13 500 N. Superposed on the profile are the crack positions measured in the SEM after sectioning the specimen.*

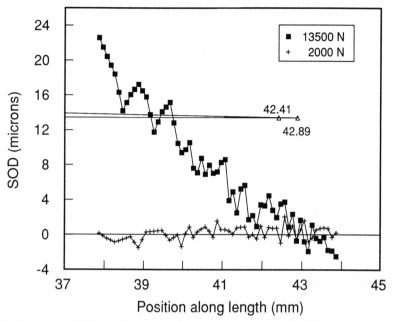

FIG. 9—*Experimental SOD profile measured along a traverse 1 mm from the back edge of the AS4/ 3501-6 composite specimen for* P = *2000 and 13 500 N. Superposed on the profile are the crack positions measured in the SEM after sectioning the specimen.*

solutions are presented in Ref 7. In both cases, the Mode I and Mode II behavior is completely decoupled.

The normal stresses close to the crack tip, for the plane strain condition, are [6]

$$\sigma_y = \frac{K_I}{\sqrt{2\pi r}} \cos\frac{\theta}{2}\left(1 + \sin\frac{\theta}{2}\sin\frac{3\theta}{2}\right) \qquad (2)$$

and the normal displacements close to the crack tip are

$$v = 2(1 + \nu)\frac{K_I}{E}\sqrt{\frac{r}{2\pi}}\sin\frac{\theta}{2}\left[2 - 2\nu + \cos^2\frac{\theta}{2}\right] \qquad (3)$$

It can be seen that both stresses and displacements are affected linearly by the applied stress intensity factor. The only difference is that whereas the stresses scale as $r^{-1/2}$, the displacements scale as $r^{1/2}$. It therefore follows that an accurate knowledge of one allows the determination of the other, and of the stress intensity factor, independent of the boundary conditions.

Along the plane of the crack, we define the COD = $2v$, and the displacement equation reduces to

$$COD = 2v = 8(1 - \nu^2)\frac{K_I}{E}\sqrt{\frac{r}{2\pi}} \qquad (4)$$

In addition, the Mode I strain energy release rate is related to the Mode I stress intensity factor by

$$K_I^2 = \frac{EG_I}{1 - \nu^2} \tag{5}$$

Therefore, substituting Eq 5 into Eq 4 and rearranging leads to

$$\sqrt{G_I} = \left(\frac{1}{8} \sqrt{\frac{2\pi E}{1 - \nu^2}}\right)\left(\frac{COD}{\sqrt{r}}\right)$$
$$= A\left(\frac{COD}{\sqrt{r}}\right) \tag{6}$$

Therefore, if one is able to determine the COD profile as a function of r, then it is possible to calculate G_I. All that is needed is an accurate value of A. For the orthotropic case, this can be derived from Ref 7.

Conceptual Difference Between SOD and COD

The relationship between COD and SOD is shown schematically in Fig. 10. In general, we can state that at any cross section

$$SOD(r) = COD(r) + \int_{x=r} \varepsilon_y \, dy \tag{7}$$

The issue is whether the integrated strains mask the COD. In the case of typical metal specimens, the specimen dimensions are such that it is clearly impossible to infer with any accuracy the COD from the remote SOD values, although there has been work where long holes have been drilled down to near the crack faces [for example, Ref 8].

In the case of relatively thin composite laminates, the SOD values are comparable to the COD values. Effects that cause the SOD values to differ from the COD values are two-fold. First, away from the crack tip, there is a uniform Poisson contraction that is due to the axial loading of the specimen. This is small, consistent, and can be allowed for. Second, in the same zone where the CODs are influenced by the square root singularity, the stresses and strains are highest and the SOD is most likely to differ from the COD. The magnitude of the difference is discussed in the following section.

Another effect in the CLS specimen that must be considered is the shifting of the crack tip with respect to the fixed origins of the LVDTs due to specimen elongation under applied load. This is easily calculated, and for our specimens is of the order of 0.1 mm per load increment of 10 000 N.

Finite Element Method Analysis

It is not possible to calculate accurately the SOD and COD profiles for a CLS specimen using closed-form analyses. Therefore, the SOD and COD profiles were determined using an FEM analysis. As the only package available was the ANSYS Linear program for IBM-PC compatible machines, the following approach developed by Mall (Appendix IX in Ref 9) was used.

The geometrically nonlinear behavior of the CLS specimen was modeled by the beam bending solution of Brussat and Chiu [10], as modified by Mall [9] for finite specimen length. The

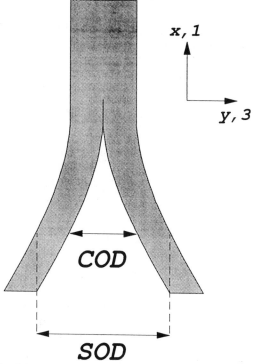

FIG. 10—*Schematic showing the relationship between COD and SOD.*

bending moments and shear forces on a small representative segment of the specimen incorporating the crack tip were then used as input to a linear FEM analysis of this segment. The length of the segment was 8.4 mm, the strap and lap thickness 2.11 and 0.87 mm, respectively. The mesh consisted of 2538 rectangular four-noded isoparametric elements. The zone around the crack tip and the corresponding specimen surfaces above and behind the crack tip had the highest density of elements, with element size of 0.025 by 0.025 mm. Away from the area of interest, the element size increased to 0.4 by 0.1 mm. Plane strain conditions were assumed. The crack length was fixed at 40.11 mm, to correspond to experimental conditions.

Material properties were assumed to be the handbook values for AS4/3501-6 material. In-plane and out-of-plane transverse properties were assumed to be the same. Thus, $E_1 = 138$ GPa, $E_3 = 8.96$ GPa, $\nu_{13} = 0.30$, and $G_{13} = 7.1$ GPa. Only the AS4/3501-6 composite specimen was modeled.

A typical plot of the FEM analysis predictions for COD and SOD are shown in Fig. 11. These results are for an applied load of 10 000 N. The behavior up to 1 mm on either side of the crack tip is shown. The COD values follow the expected square root singularity profile. The SOD values are of the same order, but there is a significant difference between the SOD and COD, especially at the crack tip. The SOD value is nonzero at the crack tip, and is zero about 0.4 mm ahead of the crack tip. Well ahead of the crack tip, the SOD is negative and tends towards the value determined by the Poisson contraction effect.

The normalized values of COD/P and SOD/P, where P is the applied axial load, are plotted in Fig. 12 as a function of P. The range of P is 2000 to 25 000 N. Normalizing the CODs and SODs with respect to the load collapses all the results onto two master curves. This indicates

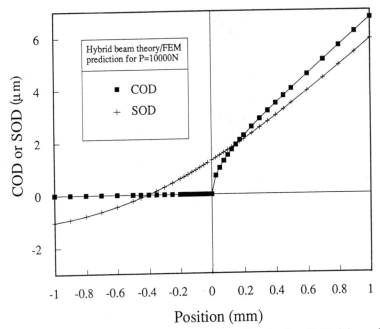

FIG. 11—*FEM analysis solution for the SOD and COD profiles ahead and behind the crack tip in the CLS specimen.*

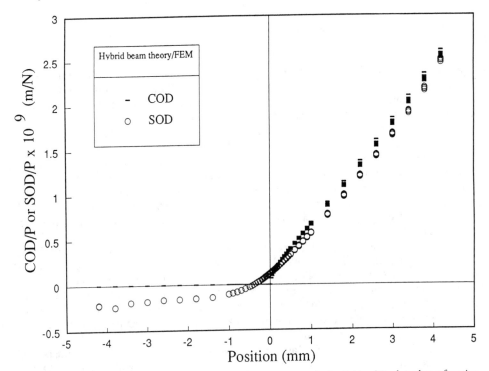

FIG. 12—*COD and SOD profiles normalized with respect to applied axial load P, plotted as a function of distance from crack tip. Results are from the FEM analysis and cover the range P = 2000 to 25 000 N.*

that the specimen is linear with respect to its crack opening behavior, at least for the current geometry according to our numerical analysis. Furthermore, the COD and SOD scale together, with a one-to-one correspondence that is independent of applied load. Thus, a small SOD indicates a small COD, and vice-versa.

In practical terms, it is not simple to convert an SOD profile into a COD profile. The SOD profiles in Figs. 11 and 12 are not linear. Consider Fig. 12 where the term SOD/P normalizes the profiles over a wide range of loads. The slope of the curve over the first 0.6 mm of non-negative values is $\Delta(SOD/P)/\Delta r = 3.4 \times 10^{-7}$ 1/N, and thereafter 4.7×10^{-7} 1/N. To compare with the experimental results (for example, Figs. 4 to 6), one must take care that values over the corresponding positions are being compared.

The FEM results can be used to compute the G_I/G_{II} ratio using the virtual crack closure technique [for example, Ref 9]. The results indicate a very weak load dependence. At a low load of 2000 N, $G_I/G_{II} = 0.37$; at a moderate load of 13 000 N, $G_I/G_{II} = 0.39$; and at a high load of 25 000 N, $G_I/G_{II} = 0.40$. If we assume that the average G_I/G_{II} ratio is 0.40 (as crack growth is observed at loads of about 20 000 N), then this value corresponds to $\Delta(SOD/P)/\Delta r = 3.4 \times 10^{-7}$ 1/N (over the first 0.6 mm from the crack tip). If a lower value of $\Delta(SOD/P)/\Delta r$ is measured experimentally, then G_I must be correspondingly lower. Note that although conceptually the SOD is first related to the COD, and the COD is then related to G_I, in practice we can relate the SOD directly to G_I.

Experimental Determination of G_I

We use the SOD/P profiles to determine the opening mode strain energy release rate experimentally. We concentrate on the fully opened crack behavior. The results are summarized in Table 2. For each specimen, we evaluate $\Delta(SOD/P)/\Delta r$ over the first 0.6-mm range for each experimental traverse. These results are presented in the third column of Table 2.

Next, knowing that G_I scales as the square of the COD (and therefore the SOD), we calculate a ratio of $G_I/G_{I\ center}$. This is tabulated in the fifth column of Table 2. Finally, for the AS4/3501-6 composite specimen, we can use the FEM results to calculate an absolute G_I/G_{total} ratio, provided we make some assumption regarding G_{total}. If we assume that G_{total} remains constant at all points across the width (a local increase in G_I leading to a local decrease in G_{II} such that the local G_{total} is always constant and equal to the global G_{total}), then we can generate the values in the sixth column of Table 2.

TABLE 2—Experimental values for local G_I.

Position Along Width	Crack Length	$\dfrac{\Delta(SOD/P)/\Delta r,^a}{1/N}$	$\dfrac{\Delta(SOD/P)/\Delta r_{exp}}{\Delta(SOD/P)/\Delta r_{FEM}}$	$\dfrac{G_I}{G_{I\ center}}$	$\dfrac{G_I^{\,b}}{G_{total}}$
		AS4/3501-6 COMPOSITE SPECIMEN			
1 mm from front edge	32.6 mm	5.5×10^{-7}	1.6	2.6	0.73
Center	40.1 mm	3.4×10^{-7}	1.0	1.0	0.29
1 mm from back edge	42.5 mm	4.1×10^{-7}	1.2	1.5	0.42
		AS4/3501-6 ADHESIVELY BONDED SPECIMEN			
1 mm from front edge	97.1 mm	6.5×10^{-7}	· · ·	0.4	· · ·
5 mm from front edge	103.2 mm	3.1×10^{-7}	· · ·	0.1	· · ·
Center	102.7 mm	10×10^{-7}	· · ·	1.0	· · ·
1 mm from back edge	97.5 mm	3.1×10^{-7}	· · ·	0.1	· · ·

[a] Over first 0.6 mm.
[b] Assuming FEM result that a $\Delta(SOD)/P)/\Delta r = 3.4 \times 10^{-7}$ corresponds to $G_I/G_{II} = 0.4$ or $G_I/G_{total} = 0.29$, and that G_{total} is constant.

We compare the FEM prediction of $\Delta(SOD/P)/\Delta r = 3.4 \times 10^{-7}$ 1/N for the AS4/3501-6 composite specimen with the measured values in the fifth column of Table 2. The agreement is remarkable for the center-line traverse, but the front and back traverse values are higher. As G_1 is proportional to the square of this value, the G_1 values are significantly higher. Note that the crack front is not perpendicular to the loading direction. The trailing side of the crack (the traverse at 1 mm from the front edge) indicates the highest G_1, as if the crack were being straightened out. This is consistent with the delayed opening loads in Table 1, where the trailing side of the crack opens up first.

In the case of the adhesively bonded specimen, there is considerable variability in the $G_1/G_{1\ center}$ values. The absolute values of $\Delta(SOD/P)/\Delta r$ are close to the values for the composite specimen. This is to be expected as the specimen geometry and material properties are similar (the main difference being the adhesive layer). However, there is no pattern to the G_1 distribution across the crack front. The crack is roughly perpendicular to the loading direction and slightly bowed, with the crack longer in the center of the specimen than at the edges.

It has been suggested [Appendix VIII in Ref 9], on the basis of three-dimensional FEM analysis, that near the free edges of a CLS specimen, the mixed-mode condition is not the I-II condition observed in the center of the specimen, but rather a II-III condition. Near the specimen edges, the material can pull in from the edge in an antiplane shear mode, rather than have to lift off in an opening mode due to symmetry constraints. This is in addition to the change from a plane strain condition in the center of the specimen to a plane stress condition near the edge. Although the specimen analyzed in Ref 9 was very different to the ones tested here, at a distance of 2.5 mm from the edge of the 25-mm-wide specimen, the G_1/G_{total} ratio was predicted to be half that at the center.

Our experimental results for the composite specimen are clearly in disagreement with this numerical prediction. The results for the adhesively bonded specimen are more ambiguous. The center has the highest Mode I content, but the G_1 value 5 mm in from the front edge is lower than just 1 mm from the same edge.

It is not clear whether the variations seen in Table 2 are real, and are due to fiber bridging in the composite specimen or crack wandering through the thickness of the thick adhesive layer in the bonded specimen. Nevertheless, the experimental method is cumbersome, and it is easy to degrade the resolution down to a level inadequate for quantitative work. It is clear that the method is too complicated and probably inaccurate for general use in quantitatively predicting G_1. It may be more suitable for confirming predictions. However, the method's greatest value is that once the groundwork has been done, it can be used easily to detect subcritical crack growth accurately.

Methodology for Crack Growth Experiments

Once an SOD profile has been established for a material, it is possible to position the LVDTs a short distance (at most 2 or 3 mm) behind the crack tip, and then load the specimen to a sufficiently high load to cause delamination growth. Once the crack starts growing, the load is held constant and then the specimen is unloaded. During the test, the output from the load cell, the LVDTs, and the axial elongation of the specimen are recorded.

The results can be then analyzed in the usual fashion. The load-elongation data are plotted, and the onset of nonlinearity in the compliance curve is taken to be the onset of crack growth. The load-SOD data are analyzed in a similar manner. A complete description follows.

Crack Growth Experiments

The common method of determining delamination crack growth in composite laminates is to observe the onset of nonlinear behavior in the load-deflection curve. The compliance of a

CLS specimen is relatively insensitive to crack growth. It is easily shown, for example, Ref *10*, that the compliance of a CLS specimen is

$$C = \frac{(t_2L + (t_0 - t_2)a)}{WEt_0t_2} \quad (8)$$

Differentiating Eq 8 and normalizing with respect to the original compliance yields

$$\frac{1}{C_0}\frac{dC}{da} = \frac{t_0 - t_2}{t_2L + (t_0 - t_2)a} \quad (9)$$

Recognizing that the second term of the denominator is always smaller than the first term

$$\frac{1}{C_0}\frac{dC}{da} \approx \frac{t_0 - t_2}{t_2L} \quad (10)$$

Substituting for our values, we have that for every millimetre of crack growth, the specimen compliance increases by 0.2%. Assume, optimistically, that a 1% change in compliance is detectable, repeatedly and reliably, and can be used to signal the onset of crack growth and hence determine G_c. Even in this case, up to 5 mm of undetected subcritical crack growth prior to unstable crack growth is possible.

In contrast, if we position our LVDTs 1 mm behind the crack tip, then a crack growth of 1 mm will cause a 100% increase in SOD for a constant load, given that the $\Delta(SOD/P)/\Delta r$ is roughly constant near the crack tip.

The sensitivity of the SOD in picking up subcritical crack growth is demonstrated best with the following exhaustive example.

A fresh AS4/3501-6 adhesively bonded specimen was set up, with the LVDTs approximately 1 mm behind the insert tip. The specimen was loaded and unloaded a few times to a load of 13 000 N. From our knowledge of the SOD/P calibration curves for this geometry, we repositioned the LVDTs until we were approximately 0.5 mm behind the insert tip. The load was then ramped to 32 000 N and down again. The SOD is plotted as a function of applied load in Fig. 13. The first load cycle is represented by a solid line. The SOD increases with applied load, such that SOD/P is constant above a load of 10 000 N (indicated by a dotted line). This indicates a fully open crack above 10 000 N, in agreement with previous results (see Table 1). Thereafter, up to a load of 23 000 N, the SOD varies linearly with load. The constant value of SOD/P is used, in conjunction with Fig. 11 or Table 2, to confirm that the LVDTs are placed about 0.5 mm behind the crack tip.

Above 23 000 N, the SOD increases faster with increasing load. There are two possibilities. The first possibility is that there is sufficient nonlinear behavior at the crack tip (that is, plasticity) to give us the observed additional surface displacements. The second possibility is that the crack is growing in a stable manner.

At 32 000 N, the specimen was unloaded. The unloading behavior is shown in Fig. 13 as a dashed line. A permanent set of about 10 μm is measured. Then, without moving the LVDTs, the load is reapplied. The new SOD profile starts at zero, as by definition the SOD is zero at zero load. On loading this time, the SOD is zero for the first 3000 N, and then increases until it varies linearly with load (such that SOD/P is constant) beyond 10 000 N. Again, this is in agreement with the results in Table 1. However, the magnitude of $\Delta(SOD/P)/\Delta r$ is considerably higher than in the first load cycle, indicating that the LVDTs are now further behind the crack tip. Therefore, the nonlinearity beyond 23 000 N in the first load cycle must have been due to crack growth. Using the appropriate value from Table 2, we can establish that the

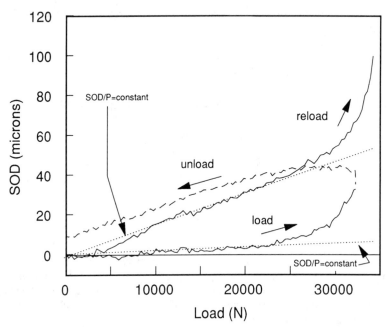

FIG. 13—*Variation in SOD as a function of load for two consecutive loading cycles on a AS4/3501-6 adhesively bonded specimen.*

LVDTs are now about 4 mm behind the crack tip. Therefore, the crack has grown by about 3.5 mm during the first load cycle.

Any uncertainty in the absolute value of $\Delta(SOD/P)/\Delta r$ (as taken from Table 2) will lead to incorrect predictions in the absolute amount of subcritical crack growth, but will not affect the value of the load at the onset of subcritical crack growth.

Inspection of the corresponding load-deflection curves indicates a change in compliance from 1.73×10^{-8} m/N to 1.76×10^{-8} m/N. This is a 1.7% increase in compliance (corresponding to a crack growth of 8.5 mm), and was detected only because the specimen was loaded and unloaded twice. The first load cycle compliance curve is, within the limits of resolution of our system, completely linear. On the basis of the first load-deflection curve alone, we would never detect the subcritical growth. On the other hand, the SOD signal increased by 700% over the same interval.

During the second loading cycle, the SOD increases linearly with P up to about 29 000 N (Fig. 13). The load was increased further until the crack started to grow unstably, and the SOD became unstable, at about 34 000 N. The compliance curve for this second load cycle is shown in Fig. 14. A straight line fit to the first portion of the load cycle is superimposed. There is a gradual decrease in compliance with increasing load. A smaller compliance indicates a shorter crack, and therefore it is clear that other geometrical effects are masking any real increase in compliance due to crack growth. The compliance curve also became unstable at about 34 000 N.

We define the load, P^*, as the load at which the SOD starts to increase nonlinearly with load, that is the point at which SOD/P increases beyond its constant value corresponding to the position of the LVDTs with respect to the crack tip. We define the load, P^{**}, to be the load at which the SOD increase becomes unstable. These values are shown schematically in Fig. 15.

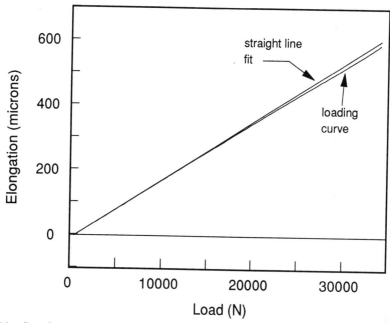

FIG. 14—*Compliance curve for the second loading cycle of the AS4/3501-6 adhesively bonded specimen shown in Fig. 13.*

Before we discuss results for a number of tests, we present in Fig. 16 an SOD versus P plot for a test where the LVDTs are positioned too far behind the crack tip (about 2 mm) and alignment is not good. As can be seen, there is no region where SOD/P is constant (there are no values lying on a straight line through the origin). However, the load values, P^* and P^{**}, are still well defined and are marked on the figure.

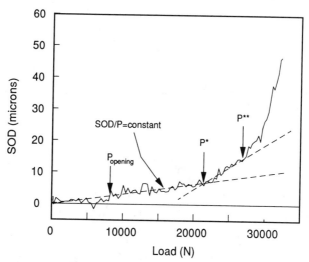

FIG. 15—*Schematic of a typical plot of SOD as a function of P, and definition of P* and P**.*

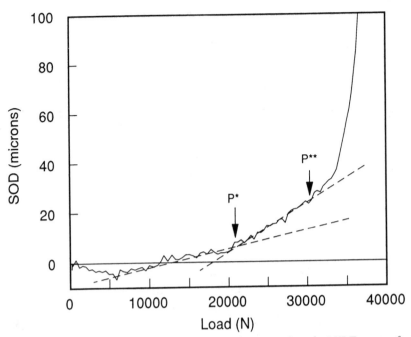

FIG. 16—*Example of a plot of SOD as a function of P, for a case where the LVDTs are too far behind the crack tip, and alignment is not optimum.*

Results for a number of tests on AS4/3501-6 composite specimens and adhesively bonded specimens, and AS4/PEEK composite specimens are presented in Table 3.

The first column in Table 3 is the test identification number. The second column is the nominal crack length. The third and fourth columns are P^* and P^{**}, as defined previously. The fifth column is P_c, the critical load for unstable crack growth as determined by the onset of nonlinearity in the load extension curve. The sixth, seventh, and eighth columns are the corresponding G values calculated using, for example, Ref 10

$$G = \frac{P^2}{2W^2Et_2}\left(1 - \frac{t_2}{t_0}\right) \tag{11}$$

The ninth and tenth columns are the calculated position of the LVDTs relative to the crack tip at Loads P^* and P^{**} according to

$$r_{\text{LVDT}} = \frac{\left(\dfrac{\text{SOD}}{P}\right)_{\text{measured}}}{\Delta(\text{SOD}/P)/\Delta r_{\text{calibration}}} \tag{12}$$

The eleventh column is the difference between the ninth and tenth columns, and is the amount of subcritical crack growth prior to instability. The values of $\Delta(\text{SOD}/P)/\Delta r_{\text{calibration}}$ used are from the SOD/P calibration profiles.

For the AS4/PEEK specimens, no calibration profiles were generated. On the basis of visual measurements of the distance between the LVDTs and the crack tip, and corresponding SOD/P values (Fig. 17), a value of $\Delta(\text{SOD}/P)/\Delta r = 7 \times 10^{-7}$ 1/N was estimated.

TABLE 3—*Subcritical crack growth results.*

Test ID	Nominal Crack Length, mm	P^*, N	P^{**}, N	P_c, N	G^* J/m^2	G^{**} J/m^2	G_c J/m^2	r^* mm	r^{**} mm	Δr, mm
			AS4/3501-6 COMPOSITE SPECIMENS							
F9T4F	21.7	19 655	20 482	22 000	372	404	466	0.7	1.2	0.5
D10T39C	68.5	19 404	21 536	21 750	301	371	379	1.2	1.8	0.6
F3T4F	75.9	19 314	21 337	20 000	354	432	380	0.3	0.8	0.5
F8T5F	94.8	19 357	20 496	20 159	352	395	382	1.6	4.0	2.4
F8T6F	114.1	18 978	20 306	20 000	339	338	376	4.7	6.9	2.2
F8T7F	143.0	18 386	19 685	19 724	318	318	366	4.4	8.2	3.8
Average		19 182	20 640	20 605	339	392	391			
			AS4/3501-6 ADHESIVELY BONDED SPECIMENS							
G7T2F	25.0	27 587	34 482	35 000	725	1132	1167	1.0	2.7	1.7
G8T18C	25.1	28 861	34 233	34 125	834	1173	1166	3.5	5.7	2.2
G6T4F	27.4	30 857	36 857	36 965	985	1405	1413	1.8	4.9	3.1
G9T5C	65.0	28 102	31 423	31 423	703	879	879	1.1	1.5	0.4
Average		28 852	34 249	34 378	811	1147	1156			
			AS4/PEEK SPECIMENS							
A3T1F	79.1	29 731	32 000	34 207	1346	1559	1782	2.3	3.9	1.6
A3T2F	110.2	30 344	33 931	35 310	1402	1753	1899	0.3	1.7	1.4
A3T4F	131.3	27 931	34 138	40 000[a]	1188	1775	2437[a]	0.2	1.8	1.6
Average		29 335	33 356	36 506	1312	1696	1840			

[a] P_c and G_c poorly defined for this specimen as compliance curve showed no abrupt change.

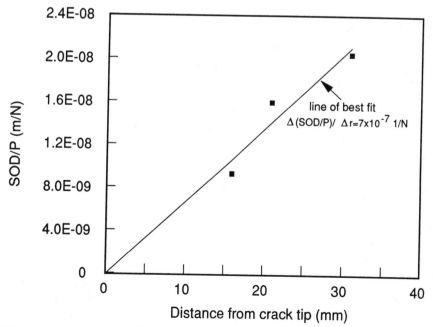

FIG. 17—*Calibration curve of SOD/P as a function of visual measurement of distance behind crack tip (r_{LVDT}) for AS4/PEEK specimens.*

Inspection of Table 3 shows that G^{**} and G_c are very close for the AS4/3501-6 composite and bonded specimens. For the AS4/PEEK specimens, G^{**} is slightly lower than G_c. At longer crack lengths, the AS4/PEEK specimen showed no abrupt changes in the compliance curves, and G_c was indeterminate. However, in all cases G^* is significantly lower than G^{**} or G_c. This indicates that for the CLS specimen, there can be a significant over-estimation of the fracture toughness if the compliance curve is used in isolation, and subcritical crack growth is ignored. Values of G_c are 15 to 40% higher than values of G^*. This is not necessarily the case for the DCB and ENF specimens, where the compliance is much more sensitive to crack growth.

The amount of subcritical crack growth is significant, even when we allow for uncertainty in the absolute values. In the case of the AS4/3501-6 composite specimens, the amount of subcritical crack growth increases monotonically with crack length, up to about 4 mm of growth. In the other systems, subcritical crack growth of up to 3 mm is observed, but there is less of a clear pattern as a function of crack length.

Conclusions

Previously, a method that could give qualitative information on crack tip behavior was presented in Ref 5. In this work, the method is extended and the following conclusions are drawn:

1. It is possible to measure specimen surface opening displacements (SOD) around the crack tip. These values are uniquely related to the crack opening displacements (COD).

2. The SOD is uniquely related to the opening mode strain energy release rate, G_I. It is possible to quantify G_I experimentally, given some finite element method calibration values. However, the method is cumbersome, and the confidence level in the quantitative G_I results is low.

3. Scanning electron microscopy (SEM) shows that often there are secondary cracks, associated with fiber bridging, near the main crack tip. There is good correlation between SOD profiles and SEM observations.

4. The SOD profiles show that the crack remains closed up to a finite and reasonably high load. This delayed opening behavior is unexpected, even if geometric nonlinearity of the specimen is considered.

5. Within the accuracy of the results generated here, there is no indication of a systematic transition from a Mode I component at the center-line to a Mode III component near the edges.

6. There is considerable subcritical crack growth with the CLS specimen for all three material systems tested. The strain energy release rate at the onset of subcritical crack growth was 15 to 40% lower than the fracture toughness calculated from the compliance curves. The amount of subcritical crack growth is a function of the material system, but was in the range of 0.5 to 4 mm. Uncertainties in the experimental measurements may lead to errors in the absolute amount of subcritical crack growth, but not in the associated value of G.

Acknowledgments

This work was carried out under contract with the Canadian Department of National Defence, Defence Research Establishment Pacific (DREP), with Dr. K. N. Street as Scientific Authority. The authors are grateful for useful discussions with Mr. A. J. Russell and Dr. K. N. Street of DREP. The technical help of Mr. L. Gambone, the assistance of Mr. R. Bennett in testing and Mr. E. Jensen (DREP) in making the specimens is much appreciated. The comprehensive and helpful comments and suggestions of the reviewers of this paper led to considerable changes and hopefully improvements.

References

[1] O'Brien, T. K., Murri, G. B., and Salpekar, S. A., "Interlaminar Shear Fracture Toughness and Fatigue Thresholds for Composite Materials," *Composite Materials: Fatigue and Fracture, Second Volume, ASTM STP 1012,* P. A. Lagace, Ed., American Society for Testing and Materials, Philadelphia, 1989, pp. 222–250.
[2] Mall, S., Yun, K. T., and Kochhar, N. K., "Characterization of Matrix Toughness Effect on Cyclic Delamination Growth in Graphite Fiber Composites," *Composite Materials: Fatigue and Fracture, Second Volume, ASTM STP 1012,* P. A. Lagace, Ed., American Society for Testing and Materials, Philadelphia, 1989, pp. 296–310.
[3] Crews, J. H., Jr., and Reeder, J. R., "A Mixed-Mode Bending Apparatus for Delamination Testing," NASA Technical Memorandum 100662, NASA Langley Research Center, Hampton, VA, Aug. 1988.
[4] O'Brien, T. K., "Generic Aspects of Delamination in Fatigue of Composite Materials," *Journal,* American Helicopter Society, Vol. 32, No. 1, Jan. 1987, pp. 13–18.
[5] Poursartip, A. and Chinatambi, N., "Fatigue Growth, Deflections, and Crack Opening Displacements in Cracked Lap Shear Specimens," *Composite Materials: Testing and Design (Ninth Volume), ASTM STP 1059,* S. P. Garbo, Ed., American Society for Testing and Materials, Philadelphia, 1990, pp. 301–323.
[6] Broek, D., *Elementary Engineering Fracture Mechanics,* Third Revision, Martinus Nijhoff Publishers, The Hague, 1982.
[7] Sih, G. C., Paris, P. C., and Irwin, G. R., "On Cracks in Rectilinearly Anisotropic Bodies," *International Journal of Fracture Mechanics,* Vol. 1, 1965, pp. 189–202.
[8] Fleck, N. A., private communication, 1989.
[9] Johnson, W. S., "Stress Analysis of the Cracked Lap Shear Specimen: An ASTM Round Robin," *Journal of Testing and Evaluation,* Vol. 15, No. 6, Nov. 1987, pp. 303–324.
[10] Brussat, T. R. and Chiu, S. T., "Fracture Mechanics for Structural Adhesive Bonds—Final Report," AFML TR-77-163, Air Force Materials Laboratory, Wright Patterson Air Force Base, OH, 1977.

Kazuro Kageyama,[1] *Masanori Kikuchi,*[2] *and Noboru Yanagisawa*[3]

Stabilized End Notched Flexure Test: Characterization of Mode II Interlaminar Crack Growth

REFERENCE: Kageyama, K., Kikuchi, M., and Yanagisawa, N., **"Stabilized End Notched Flexure Test: Characterization of Mode II Interlaminar Crack Growth,"** *Composite Materials: Fatigue and Fracture (Third Volume), ASTM STP 1110,* T. K. O'Brien, Ed., American Society for Testing and Materials, Philadelphia, 1991, pp. 210–225.

ABSTRACT: A stabilized end notched flexure (ENF) test has been proposed for experimental characterization of Mode II interlaminar crack growth. A special displacement gage has been developed for direct measurement of crack shear displacement (CSD) that is the relative shear slip between the upper and lower crack surfaces of the ENF specimen. The test has been carried out under a constant CSD rate, ensuring that the crack growth is always stable. An analytical compliance method has been applied successfully to the stabilized ENF test. Fracture toughness, G_{IIc}, and crack length are calculated from the load versus CSD diagram by using the analytical relationship between crack length and CSD compliance. A computer aided testing (CAT) system has been developed for continuous measurement of the crack length and fracture toughness. Fracture behavior of a unidirectional carbon/epoxy laminate has been examined during crack propagation as well as at the crack initiation by applying the proposed protocol.

KEY WORDS: composite materials, fracture, delamination, crack propagation, laminates, mechanical properties, test method, interlaminar fracture, fiber reinforced composites, epoxy resin, toughness, R-curves, finite element analysis, computers, fatigue (materials)

Nomenclature

a	Crack length
A_i	Coefficients of a polynomial equation that gives a normalized relationship between crack length and crack shear displacement (CSD) compliance
B	Width of end notched flexure (ENF) specimen
C	Load-line compliance
C^{BT}	Load-line compliance derived from the elementary beam theory
E_L	Elastic modulus in the longitudinal direction
E_Z	Elastic modulus in the direction through the thickness
G_{LZ}	Elastic shear modulus
G_{II}	Energy release rate under Mode II loading
G_{II}^{BT}	Mode II energy release rate derived from elementary beam theory
$2H$	Thickness of ENF specimen

[1] Associate professor, Department of Naval Architecture and Ocean Engineering, University of Tokyo, 7-3-1 Hongo, Bunkyo-Ku, Tokyo 113, Japan.

[2] Associate professor, Department of Mechanical Engineering, Science University of Tokyo, 2641 Yamazaki, Noda, Chiba 278, Japan.

[3] Engineer, Aerospace Division, Nissan Motor Co., LTD., 3-5-1 Momoi, Suginami-ku, Tokyo 167, Japan.

I	Moment of inertia
$2L$	Span length between supporting points
M	Bending moment
P	Applied load
v	Load-line displacement
β_1, β_2	Normalized parameters of orthotropy
γ	Normalized CSD compliance
δ	Crack shear displacement (CSD)
δ^{BT}	CSD derived from elementary beam theory
ε_x	Strain in the longitudinal direction
λ	CSD compliance
λ^{BT}	CSD compliance derived from elementary beam theory
ν_{LZ}	Poisson's ratio

The end notched flexure (ENF) test [1] is widely accepted to determine the Mode II interlaminar fracture toughness, G_{IIc}, of fiber reinforced polymer matrix composites. The conventional ENF test is carried out under a constant crosshead rate, and the load versus load-line deflection curve is recorded and used for the data reduction. As a result of the specimen configurations and loading conditions of the ENF specimen, the crack growth is unstable in brittle matrix composite laminates [2]. In the case of recently developed tough matrix composites, on the other hand, subcritical crack growth has been observed. The experimental data indicates that G_{IIc} of the tough matrix composites increases with crack propagation. Optical measurement of crack length of the ENF specimen is not easily carried out because the crack deforms in the shear manner and does not open when the load is applied. The behavior of crack propagation is not monitored during the conventional ENF test. G_{IIc} at the maximum load is usually accepted as the characteristic of the Mode II interlaminar fracture toughness.

For a better understanding of the Mode II interlaminar fracture of composite materials, it might be important to determine the relationship between the fracture toughness and the crack length under controlled crack growth. The R-curve characterizes fracture toughness of the specimen. G_{IIc} at the crack initiation, as well as during the propagation, might be determined reasonably from the R-curve.

A stabilized ENF test has been proposed in the present paper in order to determine the R-curve of Mode II interlaminar fracture. A crack shear displacement (CSD) gage has been developed by the authors (U.S. Patent No. 4914965) for direct measurement of the crack shear displacement, δ, that is the relative shear slip between the upper and lower crack surfaces at the end face of the ENF specimen (see Fig. 1). A new test protocol has been proposed in the present paper based on the CSD measurement technique. The test is carried out under a constant CSD rate, under which condition the crack growth is expected to be stable.

An analytical compliance method has been proposed by one of the authors [3] and successfully applied to the double cantilever beam (DCB) test, in which an analytical relationship between the crack length and the compliance is used for crack length measurement. In the present paper, the technique has been applied to the stabilized ENF test in order to monitor the Mode II interlaminar crack propagation and to determine the R-curve behavior.

Analysis of ENF Specimen

Load-line compliance, C, is defined as the ratio of the displacement, v, under the central loading pin to the applied load, P

$$C = v/P \tag{1}$$

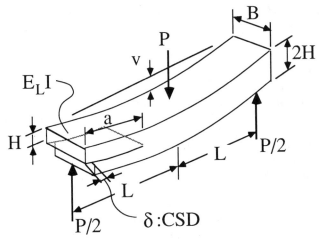

FIG. 1—*Schematic figure of ENF specimen.*

The CSD compliance, λ, is defined as the ratio of the crack shear displacement, δ, to the applied load, P, as shown in Fig. 1

$$\lambda = \delta/P \tag{2}$$

The energy release rate, G_{II}, is given by Eq 3

$$G_{II} = \frac{P^2}{2B} \cdot \frac{dC}{da} \tag{3}$$

where B is the width of the ENF specimen.

At first, elementary beam theory is applied to calculate C and λ, and the results are summarized. Next, a finite element analysis is carried out in order to obtain more accurate solutions, which will be used for the analytical compliance method.

Approximate Analysis Based on Elementary Beam Theory

The load-line compliance, C^{BT}, and the energy release rate, G_{II}^{BT}, has been obtained by Russell and Street [1], based on elementary beam theory.

$$C^{BT} = \frac{2L^3 + 3a^3}{8E_L BH^3} \tag{4}$$

$$G_{II}^{BT} = \frac{9CP^2 a^2}{2B(2L^3 + 3a^3)} \tag{5}$$

where

$2L$ = span length,
a = crack length,
$2H$ = thickness,
B = width, and
E_L = longitudinal modulus of bending as denoted in Fig. 1.

Crack shear displacement, δ^{BT}, is obtained by integrating Eq 6 along the path, Γ, beginning at the lower end corner of the upper beam and ending at the crack tip

$$\delta^{BT} = 2 \int_\Gamma \varepsilon_x \, dx = \frac{H}{E_L I} \int_0^a M \, dx \qquad (6)$$

The bending moment, M, and the moment of inertia, I, of the upper beam are equal to $Px/4$ and $BH^3/12$, respectively. Substituting the values into Eq 6, we obtain Eq 7

$$\delta^{BT} = \frac{3Pa^2}{2E_L BH^2} \qquad (7)$$

where CSD compliance, λ^{BT}, is defined as δ/P and given by Eq 8

$$\lambda^{BT} = \frac{3a^2}{2E_L BH^2} \qquad (8)$$

Equation 8 indicates that λ^{BT} is proportional to the square of crack length, a and that a change of the crack length has a direct effect upon the CSD compliance. From Eqs 4, 5, and 7, a remarkable result is obtained

$$G_{II}^{BT} = \frac{3P\delta^{BT}}{8BH} \qquad (9)$$

Equation 9 indicates that it is not necessary to measure the crack length in order to calculate G_{II}^{BT} when measuring CSD.

Finite Element Analysis

Effects of finite boundaries and stress concentrations at the crack tip and the loading and supporting points are not considered in the framework of elementary beam theory. The effects are considered by using the finite element method for getting more accurate solutions that are applicable to the analytical compliance method. In this paper, the deformation of the ENF specimen has been analyzed numerically by using a two-dimensional finite element method. An isoparametric rectangular element with eight nodes for orthotropic materials is used for the present analysis.

Breakdown of a finite element model is shown in Fig. 2. The L and H are 50 mm and 1.5 mm, respectively. The crack tip Region A in Fig. 2 moves with crack extension. The range of crack length, a, is from 20 to 48 mm, and resultantly that of crack depth ratio, a/L, is from 0.4 to 0.96. The CSD compliance, λ^{FE}, and the load-line compliance, C^{FE}, are calculated as functions of the crack length.

Contact of crack surfaces is considered in Region B in Fig. 2. The distribution of contact stress due to crack closure is assumed to be sinusoidal as shown in Fig. 3. Effect of friction is not considered in the present finite element analysis. The total contact force is equal to $P/4$. The size of the contact area is calculated numerically to satisfy the boundary condition on the crack surfaces, which is that there is no gap on the contact area or no overlap on the free surfaces. The most suitable condition is obtained by assuming a size of $3H$.

A rectangular coordinate system (L, Z) is fixed on the ENF specimen, L being parallel to the longitudinal direction and Z being in the direction through the thickness. The crack plane is normal to the L, Z plane. The material is assumed to be orthotropic with the crack on one

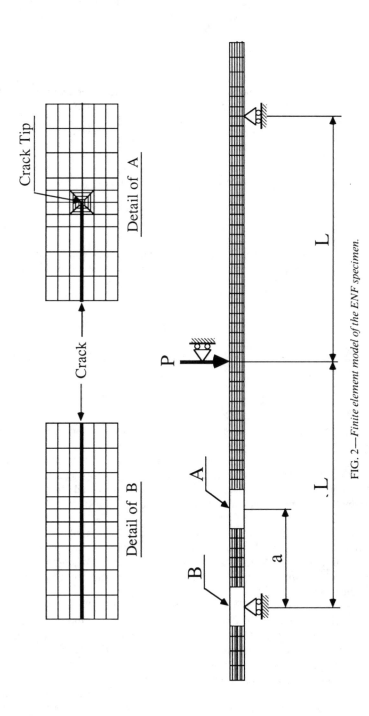

FIG. 2—*Finite element model of the ENF specimen.*

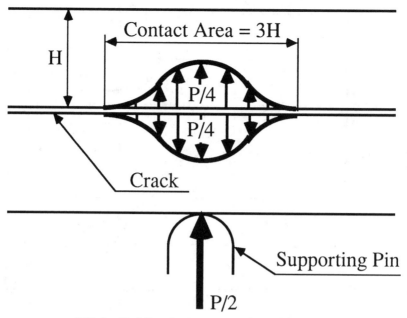

FIG. 3—*Modeling of contact area on the crack surfaces.*

plane of symmetry. The independent elastic moduli in the present problem are E_L, E_Z, G_{LZ}, and ν_{LZ}.

The normalized CSD compliance, γ, is defined as in Eq 10

$$\gamma = \sqrt{BE_L\lambda} \qquad (10)$$

The normalized relationship between crack length and γ is very well approximated by a third order polynomial equation

$$\frac{a}{H} = A_0 + A_1\gamma + A_2\gamma^2 + A_3\gamma^3 \qquad (11)$$

The crack length is evaluated from Eq 11 by applying the analytical compliance method. From elementary beam theory, A_1 is 0.816 and other coefficients are equal to zero. In the present paper, the coefficients, A_i, are obtained from the numerical results of the finite element analysis by applying the least squares method.

According to the same consideration as mentioned in Ref 4, we assume that the orthotropic elasticity can be regulated by using two normalized parameters, β_1 and β_2

$$\beta_1 = \frac{E_Z}{E_L} \quad \text{and} \quad \beta_2 = \frac{E_L}{G_{LZ}} - 2\nu_{LZ} \qquad (12)$$

where β_1 and β_2 are used for the index of orthotropy. The coefficients, A_i, are assumed to be functions of β_1 and β_2, and the results are indicated in Figs. 4 through 7. The coefficients of a unidirectionally reinforced carbon/epoxy laminate, for example, are shown in Table 1 with its elastic moduli. In the case of other composite materials that have different elastic moduli from

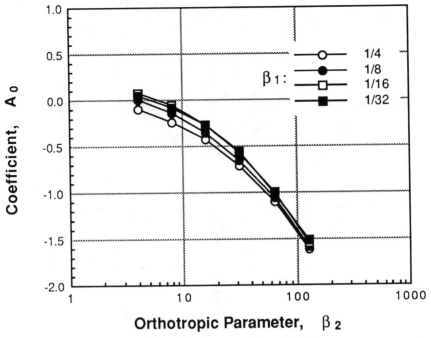

FIG. 4—*Effects of orthotropy on the coefficient* A_0.

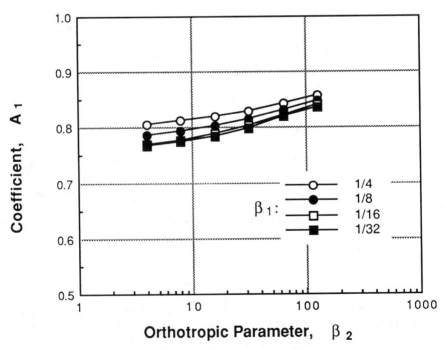

FIG. 5—*Effects of orthotropy on the coefficient* A_1.

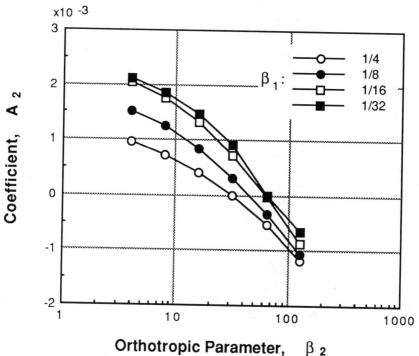

FIG. 6—*Effects of orthotropy on the coefficient* A_2.

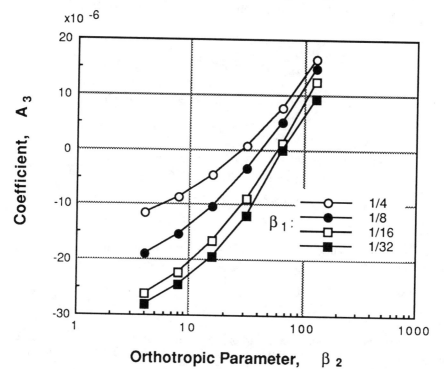

FIG. 7—*Effects of orthotropy on the coefficient* A_3.

TABLE 1—*Elastic moduli and coefficients* A_i *of a unidirectionally reinforced carbon/epoxy laminate tested in the present paper.*

E_L (GPa)	E_Z (GPa)	G_{LZ} (GPa)	ν_{LZ}
113	7.8	4.1	0.34
A_0	A_1	A_2	A_3
-0.520	0.802	0.711×10^{-3}	-0.889×10^{-5}

those listed in Table 1, one would calculate β_1 and β_2, read the values of A_i from Figs. 4 through 7, and apply Eq 11 for calculation of crack length.

The finite element results indicate that the energy release rate, G_{II}, is well approximated by a simple equation that coincides with Eq 9

$$G_{II} = \frac{3P\delta}{8BH} = \frac{3}{8E_LH}\left(\frac{P}{B}\right)^2\gamma^2 \tag{13}$$

The accuracy corresponding to the difference between the value obtained using Eq 13 and the present numerical results is less than 1% for $0.4 \le a/L \le 0.96$, $\frac{1}{32} \le \beta_1 \le \frac{1}{2}$, and $4 \le \beta_2 \le 128$. Equation 13 indicates that G_{IIc} can be calculated from the load versus CSD diagram without direct measurement of the crack length.

Stability of Crack Growth

Stability of crack propagation is discussed based on elementary beam theory. For the ENF specimen under constant load-line displacement (fixed grip) condition, $\partial G_{II}/\partial a$ is obtained as in Ref 2

$$\left(\frac{\partial G_{II}^{BT}}{\partial a}\right)_v = \frac{9v^2a}{8E_LB^2H^3C^2}\left[1 - \frac{9a^3}{2L^3 + 3a^3}\right] \tag{14}$$

Stable crack growth requires $\partial G_{II}/\partial a$ to be less than or equal to zero. This gives

$$a \ge L/\sqrt[3]{3} \approx 0.7L$$

Consequently, for the commonly used $a \approx L/2$, the crack growth is unstable.

Under constant CSD condition, on the other hand, $\partial G_{II}/\partial a$ is obtained from Eq 9

$$\left(\frac{\partial G_{II}^{BT}}{\partial a}\right)_\delta = -\frac{E_LH(\delta^{BT})^2}{2a^3} \tag{15}$$

This quantity is always negative, and thus the crack growth is stable in the case of the CSD-controlled ENF test.

Test Method

Material and Test Specimen

The test material is a unidirectional carbon/epoxy laminate. The nominal thickness of the lamina and laminate are 0.25 and 3.0 mm, respectively. The nominal fiber volume fraction is

60%. Transverse isotropic symmetry is assumed, that is, $E_Z = E_T$, $G_{LT} = G_{LZ}$, and $\nu_{LZ} = \nu_{LT}$. The elastic moduli measured by four-point bending, in-plane shear, and transverse tension tests are listed in Table 1.

The ENF specimens are machined and finished to give dimensions of $2L = 100$ mm, $B = 25$ mm, and $2H = 3$ mm. Twenty millimetres of PTFE (polytetrafluoroethylene) film is inserted as a starter crack. The crack is carefully wedged open and extended about 5 mm beyond the insert in order to introduce a natural crack.

CSD Gage

A special CSD gage has been developed in order to measure the crack shear displacement of the ENF specimen. The relative displacement between the upper and lower crack surfaces is detected by a cantilever of the CSD gage and converted into voltage by using strain gages. Figure 8 illustrates the CSD gage. The gage has a frame with a fixing holder attached to the bottom end, for restraining rotation of the frame with the upper beam of the ENF specimen. The base end of a lever holder, which is a rod-shaped member with a large spring constant, is fixed to the top end of the frame. A pair of adjusting screws are engaged in threaded holes of the frame. The gap between the tip end of the lever holder and the frame can be adjusted by advancing and retracting the lower adjusting screw. A contact is provided on the cantilever so as to be positioned opposite the end face of the lower beam of the specimen. Strain gages are bonded to the cantilever in order to detect the amount of flexing of the cantilever.

The operation of the CSD gage will now be explained. The end portion of the specimen is inserted in the fixing holder and a fixing screw is adjusted to clamp the upper beam of the

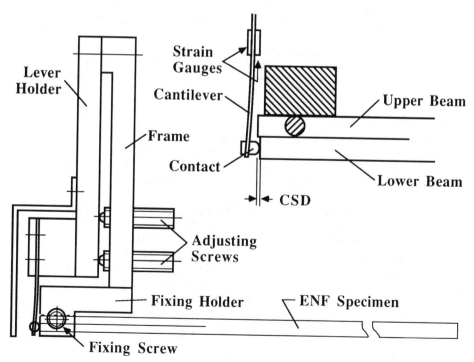

FIG. 8—*Crack shear displacement (CSD) gage.*

specimen. The lower adjusting screw is then turned slowly in the direction that causes the contact of the cantilever to come in contact with the end face of the lower beam of the specimen. After contact has been made, the screw is further turned in the same direction until the reading on the strain gages reaches a prescribed value. The upper adjusting screw acts as a stop for preventing the lever holder coming too close to the frame. While the load is applied to the specimen, the contact follows the shift of the lower beam. The elastic strain of the cantilever is progressively relieved as the contact moves, and the amount of this relief can be read from the strain gages. Therefore, the crack shear displacement can be accurately determined from the output of the strain gages. The bending of the cantilever is greatest just before the measurement begins and progressively decreases as the deformation increases. As a result, there is no risk of the gage being damaged even in the case where an excessive applied load causes the test specimen to break. Work distance of the CSD gage is 1 mm and the gage is calibrated with a micrometer.

Test Procedure

An electrohydraulic testing machine was used for the CSD-controlled ENF test. The output signal of the CSD gage was selected to be the control variable, and the closed-loop control kept the CSD rate constant with crack propagation. The CSD rate was chosen to be 0.03 mm/min in the present paper. Unloading and reloading took place several times during the test to obtain the compliance data. The test was controlled by a computer and the load versus CSD diagram and the CSD compliance were measured automatically. The procedure was performed in much the same way as a DCB test, but it was impossible to measure crack length by using a traveling microscope because the crack propagated in the shear mode. An analytical compliance method based on Eqs 11 and 13 was applied to determine the energy release rate and the crack length. In order to examine the validity of the method, crack length was measured by opening the specimen into two parts after completion of the test. A 5% less secant line method was used in order to determine the fracture toughness at crack initiation. The definition of the onset of crack growth is similar to that described in ASTM Test Method for Plane-Strain Fracture Toughness of Metallic Materials (E 399–83).

Computer Aided Testing System

Control of the testing machine, data acquisition, and analysis of experimental data have been carried out automatically by a computer. Analog signals from the load cell and CSD gage were converted into digital signals and stored in the memory of computer and hard disks. From the digital data of load versus CSD during the unloading and reloading processes, the CSD compliance was determined by using the least squares method. Crack length and fracture toughness were calculated from the compliance by using Eqs 11 and 13. Permanent deflection was estimated by extrapolating the unloading line and the analytical compliance method was extended to the crack propagating processes. The estimated CSD compliance was calculated by assuming that the permanent deflection is equal to that of the previous unloading stage. The energy release rate and crack length were calculated from the estimated CSD compliance during the crack propagation. We could monitor the crack length and fracture toughness almost continuously by applying the computer aided testing technique.

Experimental Results

An example diagram of load versus CSD is shown in Fig. 9. Crack propagation is stable even in the case of the brittle matrix composite. The stable crack growth is an advantage of the CSD-

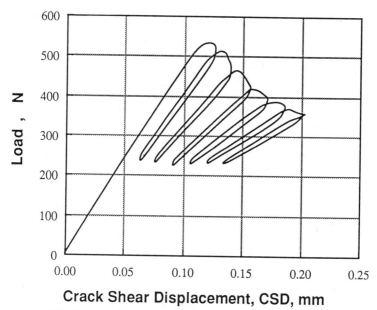

Crack Shear Displacement, CSD, mm

FIG. 9—*Typical load versus CSD curve for stabilized ENF test.*

controlled ENF test. Hystereses in the unloading and reloading processes may be mainly due to the friction on the crack surfaces.

An example diagram of interlaminar fracture toughness, G_{IIc}, versus crack length is shown in Fig. 10. Points marked by ● and ○ represent the results from the CSD compliance during unloading and reloading, respectively. The interpolating curve corresponds to the results during the crack propagation obtained by considering the effect of permanent deflection. Figure 11 represents an experimental relationship between crack length ratio and normalized CSD compliance with the numerical values from Eq 10. Agreement between the experimental and numerical results is excellent. Validity of the analytical compliance method is confirmed.

Interlaminar fracture toughness, G_{IIci}, at the crack initiation is listed in Table 2. The critical load is determined by using 5% less secant line method (ASTM E 399) and the maximum load. G_{IIci} values are calculated from the present analysis (Eq 13) and elementary beam theory (Eq 5).

Figure 12 depicts the change of G_{IIc} with crack propagation, where G_{IIc} is nearly constant during stable crack growth and the average value of the five specimens is 0.650 kJ/m². The scatter of experimental data is small. The effect of crack propagation rate on the fracture toughness during the stable crack growth is shown in Fig. 13. The range of CSD rate is 0.006 to 0.6 mm/min. The CSD rate relates to a crack growth rate of 0.01 to 3.0 mm/s. The crack growth is stable even at the highest CSD rate. The G_{IIc} obtained in the present test is nearly independent of the crack propagation rate.

Discussion

The G_{IIci} determined by applying 5% less secant line method is about 5% smaller than that defined at the maximum load. The result indicates that subcritical crack growth exists even in brittle matrix composites such as a carbon/epoxy system. Accurate estimation of G_{IIci} might be possible for tough matrix composite materials by applying the proposed test procedure.

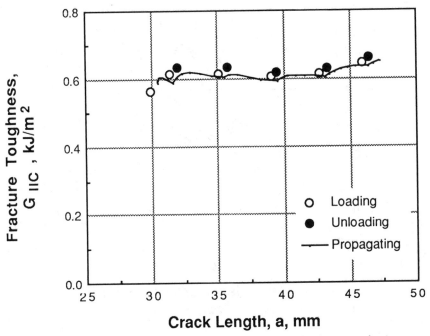

FIG. 10—*An example of* G_{IIc} *versus crack length diagram of carbon/epoxy laminate.*

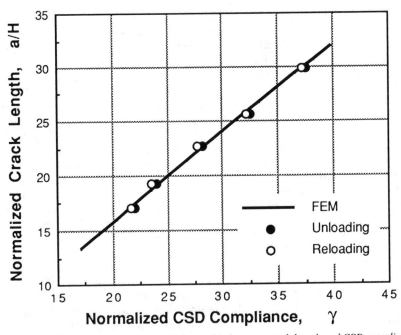

FIG. 11—*Experimental and numerical relationship between crack length and CSD compliance.*

TABLE 2—*Mode II interlaminar fracture toughness, G_{IICi}, at crack initiation.*

	Evaluation of G_{II}		
	Present Analysis	Elementary Beam Theory [1]	
	Estimation of Critical Load		
	5% Less Secant Line	(ASTM E 399)	Maximum Load
Average of G_{IICi} (kJ/m²)($n = 5$)	0.558	0.554	0.585
Standard deviation (kJ/m²)($n = 5$)	0.023	0.025	0.029

Effects of crack growth, Δa, on the CSD compliance and the load-line compliance are examined from Eqs 4 and 8. The results are summarized in Table 3. Crack growth equal to 2.5% of the initial crack length, for example, results in 5.0 and 1.2% increases of the CSD compliance and the load-line compliance, respectively. The crack growth effects on the CSD compliance are about four times as much as on the load-line compliance. More sensitive detection of crack growth can be expected from the CSD compliance than the load-line compliance. A 5% increase of the CSD compliance is nearly equal to the amount of crack growth that the ASTM E 399-83 standard defines as the onset of the crack growth. A 5% increase of load-line compliance, on the other hand, corresponds to 9.6% crack growth.

Conclusions

A CSD gage has been developed for direct measurement of the crack shear displacement of the ENF specimen. A new test protocol has been proposed by using the CSD measurement

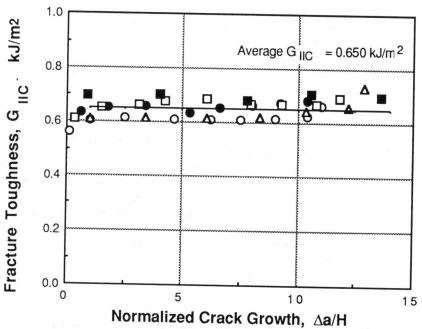

FIG. 12—*Change of G_{IIc} during the stable crack growth for carbon/epoxy laminates.*

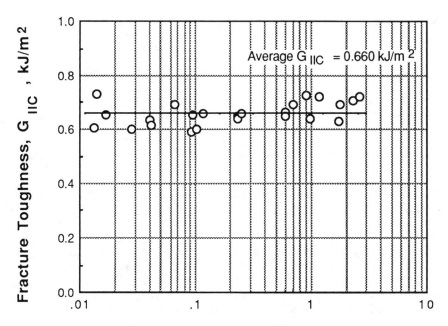

Crack Propagation Rate, da/dt, mm/sec

FIG. 13—*Effect of crack propagation rate, da/dt, on G_{IIc}.*

TABLE 3—*Effects of crack growth on CSD compliance and load-line compliance.*

Crack Growth, $\Delta a/a$ (%)	CSD Compliance, $\Delta\lambda/\lambda$ (%)	Load-Line Compliance, $\Delta C/C$ (%)
2.5	5.0	1.2
5.0	10.3	2.5
9.6	20.0	5.0
10.0	21.0	5.2

technique. The test was carried out under constant CSD rate. The crack growth was always stable under the CSD control. Numerical solutions of the CSD compliance of the ENF specimen are obtained as a function of the crack length. The analytical compliance method has been applied successfully for the determination of G_{IIc} and crack length. The crack initiation was detected with high accuracy from the load versus CSD diagram by using 5% less secant line method similar to the way as proposed in ASTM E 399-83 standard. Interlaminar fracture behavior of unidirectional carbon/epoxy laminate was examined by applying the proposed methods. G_{IIc} was nearly constant during stable crack growth. The average value was 15 to 20% higher than G_{IIci} at crack initiation. The effect of crack growth rate on the fracture toughness was small within the range of the present test.

Acknowledgment

This work was a part of a national program of Basic Technology for Future Industry and financially supported by the Agency of Industrial Science and Technology, Ministry of International Trade and Industry, Japan.

References

[*1*] Russell, A. J. and Street, K. N., "Moisture and Temperature Effects on the Mixed-Mode Delamination Fracture of Unidirectional Graphite/Epoxy," *Delamination and Debonding of Materials, ASTM STP 876,* W. S. Johnson, Ed., American Society for Testing and Materials, Philadelphia, 1985, p. 349.
[*2*] Carlsson, L. A. and Pipes, R. B., *Experimental Characterization of Advanced Composite Materials,* Prentice-Hall, Englewood Cliffs, NJ, 1987.
[*3*] Kageyama, K., Kobayashi, T., and Chou, T.-W., "Analytical Compliance Method for Mode I Interlaminar Fracture Toughness Testing of Composites," *Composites,* Vol. 18, 1987, p. 393.
[*4*] Kageyama, K., Nonaka, K., and Shimamura, S., "A Method of In-Plane Fracture Toughness Test for CFRP," *Proceedings,* Fifth International Conference on Composite Materials, San Diego, CA, 1985, p. 391.

Alan J. Russell[1]

Initiation and Growth of Mode II Delamination in Toughened Composites

REFERENCE: Russell, A. J., "**Initiation and Growth of Mode II Delamination in Toughened Composites,**" *Composite Materials: Fatigue and Fracture (Third Volume), ASTM STP 1110,* T. K. O'Brien, Ed., American Society for Testing and Materials, Philadelphia, 1991, pp. 226–242.

ABSTRACT: The origins of nonlinearity in the Mode II delamination fracture of three organic matrix, carbon fiber composite materials was investigated. This was accomplished by testing specimens with different types of starter cracks and by loading and unloading these specimens several times so that the change in nonlinearity as the delaminations grew could be measured. The load at which crack growth initiated was determined by acoustic emission. Slow crack growth was found to be the principal cause of nonlinearity in the materials tested. The crack velocity obeyed the same power law dependence on G_{II} as has been observed for slow crack growth in visco-elastic polymers. For the first loading from the end of the starter cracks, plastic deformation at the crack tip also contributed to the nonlinearity. Other sources of nonlinearity included an increasing fracture resistance in one of the materials as well as problems associated with producing a clean, sharp starter notch with a straight crack front. The implications of these findings to the use of linear elastic fracture mechanics for predicting delamination behavior of these materials and to the measurement of Mode II interlaminar fracture toughness are also discussed.

KEY WORDS: delamination, fracture, Mode II, nonlinearity, visco-elastic polymers, plastic, *R*-curve, test methods, composite materials, fatigue (materials) fracture

Nomenclature

A, B	Constants relating specimen compliance to crack length
a	Crack length
a_0, a_f	Crack lengths before and after testing
a_i	Crack length at the start of the ith loading step
b	Specimen width
C	Specimen compliance
C_0, C_i, C_f	Compliances corresponding to a_0, a_i, and a_f
C_{init}	Loading compliance prior to deviation from linearity
C_{max}	Compliance at maximum load
ΔC_{max}	Percentage compliance change from C_{init} to C_{max}
C_S	Correction factor for shear compliance
G_I, G_{II}, G_{III}	Mode I, Mode II, and Mode III strain energy release rates
$G_{IIc}(init)$	Value of G_{II} corresponding to P_{init}
$G_{IIc}(max)$	Value of G_{II} corresponding to P_{max}
$G_{IIc}(lin)$	Fracture toughness calculated using a_i and P_{max}
L	Half span of three-point-bend fixture

[1] Leader, composite group, Defence Research Establishment Pacific, FMO Victoria, British Columbia, Canada, VOS 1B0.

m, R	Constants relating crack velocity to G_{II}
P	Load
P_{init}	Load to initiate delamination growth
P_{max}	Maximum load during loading cycle
P_{nl}	Load at which P-δ curve deviates from linearity on first loading
S	Rate of load point displacement
t	Time
V	Crack velocity
δ	Load point displacement
δ_{max}	Value of δ corresponding to P_{max}

The use of fracture mechanics to describe the behavior of delaminations in laminated composite materials is now well established. Two factors have contributed to the ease with which this has been achieved. First, the strain energy release rate, G, can often be calculated quite simply using beam theory or some other strength-of-materials based approach. Indeed, it is possible to calculate not only the total value of G but also the relative magnitude of the individual components, G_I, G_{II}, and G_{III}, in this way [1]. Second, the relatively low fracture toughness of polymer matrix materials such as epoxies and bismaleimides combined with a laminate behavior that is dominated by the linear elastic to fracture properties of the reinforcements has made it possible to use linear elastic fracture mechanics (LEFM) to characterize delamination fracture. A variety of different fracture tests have been developed [2-4] for this purpose, and these have been also used to investigate the various mechanisms involved in interlaminar fracture [5-7].

In an effort to meet the demand for more damage-tolerant composite structures, a number of new composite materials have been developed including those with thermoplastic matrices or those containing a layer of fracture-resistant material between the plies (interleafing). Some of these composite systems have sufficient toughness to bring into question the use of LEFM for predicting delamination behavior. Interlaminar fracture testing of these materials can give rise to load-deflection curves that display significant nonlinearity prior to maximum load. Unlike in metals, where large-scale plastic deformation at the crack tip is responsible for nonlinear behavior, in polymer matrix composite materials a number of other mechanisms may be involved and not all of these would preclude the use of LEFM. For example, Carlsson et al. [8] have proposed that nonlinear behavior could be attributed to either visco-elastic effects or stable crack growth in addition to material yielding.

In the present work, the origins of the nonlinearity observed during the Mode II interlaminar fracture of three different damage-tolerant composite systems were investigated. The influence of the nature of the starter crack on both the nonlinearity and the fracture toughness, G_{IIc}, was also examined. Delamination onset was monitored by acoustic emission and repeated loading and unloading was used to follow the progression of the nonlinearity and to detect any increase in fracture resistance that might occur as the delamination extended. An attempt was also made to evaluate the amount of visco-elastic behavior present by measuring the relationship between crack velocity and strain energy release rate.

Experimental Procedure

The Mode II interlaminar fracture toughness, G_{IIc}, was measured using end notched flexure (ENF) specimens [4,6] loaded as shown in Fig. 1. An extensometer was attached to the three-point-bend fixture to measure the load-point displacement, and a 375-kHz acoustic emission transducer was mounted on the specimen to detect the onset of delamination growth. The transducer output was amplified by a 250 to 500 kHz band-pass preamplifier and the rms volt-

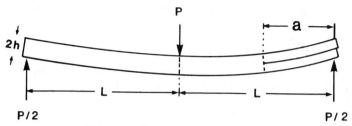

FIG. 1—*End-notched flexure specimen.*

age of this signal was plotted, along with the load, against the load-point displacement on a two-pen *x-y* plotter. The tests were run in the same manner as in previous work [4,9] except that a load-point displacement rate of only 0.3 mm/min was used. This improved the resolution of the acoustic emission signals and allowed ample time to unload the specimen immediately after the maximum load was attained. In some cases, an initial crack length greater than 25 mm was used in order to increase the stability of the fracture process.

Table 1 gives a brief description of each of the three materials tested. All layups were unidirectional and the thicknesses of the specimens and crack starter inserts are also included in Table 1. Three different types of starter cracks were used; the insert itself, a Mode I precrack, and a Mode II fatigue precrack. By looking at the second loading of the nonprecracked specimens, it was also possible to determine the effect of a Mode II static precrack, that is, the crack growth from the first loading could be regarded as the precrack.

Results

Load-Deflection Curves and Nonlinearity

Figure 2 shows two examples of a representative set of load-deflection curves (*P-δ* curves) obtained from the repeated loading and unloading of a single test specimen. During each ramp, the specimen was unloaded as soon as the load was observed to reach a maximum value and begin to drop. In all three test materials, there was a significant amount of nonlinearity in the *P-δ* curves before the maximum load, P_{max}, was reached. Typically four or five loadings were possible on each specimen before the crack reached a length of $\simeq 0.8$ L at which point the test was stopped. The amount of nonlinearity of each loading curve was expressed as

$$\Delta C_{max} = 100 \cdot \frac{(\delta_{max}/P_{max}) - C_{init}}{C_{init}} \tag{1}$$

TABLE 1—*Test specimens.*

Material and Manufacturer	Description	Thickness, mm	Insert
IM6/5245C Narmco/BASF	epoxy/bismaleimide	3.5	0.025-mm TCG[a]
IM7/8551-7 Hercules	epoxy with rubber particles between piles	3.4 and 5.0	0.025-mm TCG
AS4/PEEK I.C.I.	Semi-crystalline thermoplastic	4.0	0.0125-mm aluminum

[a] TCG = Teflon coated glass fabric.

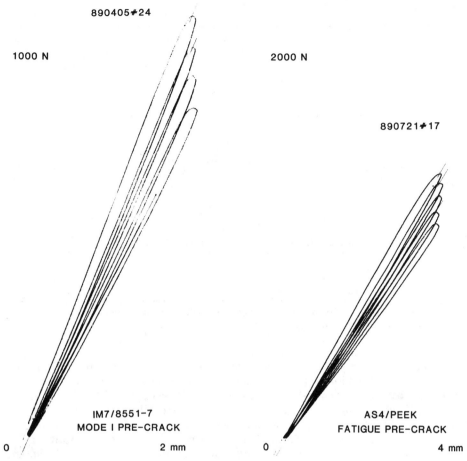

FIG. 2—*Examples of load (Y-axis)—deflection (X-axis) curves for IM7/8551-7 and AS4/PEEK.*

where

δ_{max} = the deflection corresponding to P_{max}, and
C_{init} = the compliances obtained from the slope of the initial linear part of the P-δ curve neglecting the nonlinear portion close to P_{max}.

Table 2 gives the values of ΔC_{max} for each material and starter crack. Several remarks can be made that are valid for all three materials.

1. The nonlinearity drops significantly and becomes relatively constant after the first loading.
2. The nonlinearity is dependent on the type of starter crack for the first loading only.
3. The specimen-to-specimen variation in the amount of nonlinearity was greatest for pre-cracked specimens undergoing their first loading.

Also shown in Table 2 is the permanent deflection resulting from the first loading, expressed as a percentage of δ_{max}. On all subsequent loadings, the P-δ curves returned to the same origin

TABLE 2—*Nonlinearity at maximum load.*

Loading	No Precrack	Mode I Precrack	Fatigue Precrack
		IM6/5245C	
1	1.0	3.7	2.1
2	. . .	2.5	. . .
3	. . .	2.2	. . .
4	. . .	2.3	. . .
P[a]	. . .	0.8	. . .
		IM7/8551-7	
1	3.68 ± 0.43	5.99 ± 1.10	5.41 ± 1.36
2	2.88 ± 0.32	3.18 ± 0.31	2.96 ± 0.26
3	. . .	2.98 ± 0.33	2.90 ± 0.48
4	. . .	3.38 ± 0.32	3.09 ± 0.53
P	2.7	1.9	2.8
		AS4/PEEK	
1	3.80 ± 0.72	8.36 ± 1.41	8.76 ± 1.70
2	1.85	2.55 ± 0.37	2.28 ± 0.33
3	2.1	2.45 ± 0.21	2.33 ± 0.19
4	2.5	2.58 ± 0.33	2.38 ± 0.56
P	1.7	. . .	2.0

[a] P refers to the amount of permanent deflection after the first loading. The mean values are given followed by the standard deviation if more than three tests were run.

without any further permanent deflection occurring (see Fig. 2). In the case of the IM7/8551-7, the permanent deflection of $\simeq 2.5\%$ is sufficient to account for most of the difference in nonlinearity between the first and subsequent loadings. However, for AS4/PEEK the difference is $\simeq 6\%$, which is significantly greater than the permanent deflection of 2%. A possible source of initial nonlinearity in the AS4/PEEK specimens was the nonuniformity of the precracks particularly those grown in fatigue. In Fig. 3, the nonlinearity of the first loading is plotted against the maximum variation in precrack length across the width of the specimen. The increase in nonlinearity that occurs when the crack front is overly curved or skewed is most likely caused by nonuniform crack growth across the width of the specimen.

Acoustic Emission

In order to relate the acoustic emission (AE) output to the P-δ curves, most of the test data were recorded as shown in Figs. 4 and 5. Only the loading curves were recorded and these were offset from each other so that the AE traces (lower curves) did not overlap. Thus, Fig. 4 shows six consecutive loadings of a single specimen and two AE traces corresponding to the first two P-δ curves. The most striking result of the acoustic emission data was the strong, fast rise time, signals that were present early on during the loading of the Mode I precracked specimens but absent in the other specimens. This can be seen by comparing the difference between the AE output from fatigue precracked and Mode I precracked specimens of IM7/8551-7 in Figs. 4 and 5. Similar results were obtained for the other two materials. Since these signals only occurred on the first loading, it seems likely that they resulted from the breaking of the "bridged fibers" that form close to the crack tip during Mode I delamination [4,6].

Two distinct AE signals could be associated with delamination growth. These were short sharp spikes of varying amplitude most likely caused by fiber failures and a more continuous

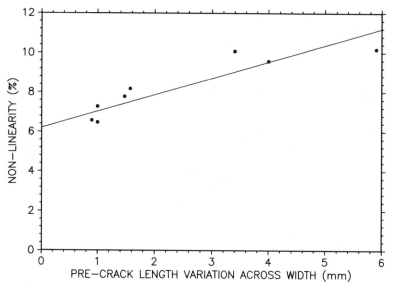

FIG. 3—*Effect of maximum precrack length variation across the width on the first loading nonlinearity of the AS4/PEEK specimens.*

signal that showed up as an increase in the noise level (baseline between spikes) rather than as separately resolved events. This latter type of signal varied in magnitude from material to material and is probably caused by fracture or plastic deformation of the matrix. In general, the increase in AE output coincided with the start of deviation from linearity of the load deflection curve. However, during the first loading, the deviation from linearity occurred before the rise in AE output as is evident from Fig. 4.

Fracture Toughness

Prior to the completion of each test, the loading compliance of the specimen was measured (without allowing any further crack growth) and then each specimen was opened into two parts and the initial and final crack lengths, a_0 and a_f, measured from the fracture surfaces. The crack lengths, a_i, corresponding to each loading compliance, C_i, were then interpolated between a_0 and a_f by means of the following expression

$$C_i = A + Ba_i^3 \qquad (2)$$

where $A = (C_0 a_f^3 - C_f a_0^3)/(a_f^3 - a_0^3)$ and $B = (C_f - C_0)/(a_f^3 - a_0^3)$. Equation 2 can be derived from beam theory [6] and the expressions for A and B are obtained by substituting the initial and final compliances and crack lengths into Eq 2, in turn, and then solving the two resulting equations for A and B. Next, the fracture toughness, G_{IIc}, was calculated using shear modified beam theory [4,10], that is

$$G_{IIc} = \frac{9a^2P^2(C - C_s)}{2b(2L^3 + 3a^3)} \qquad (3)$$

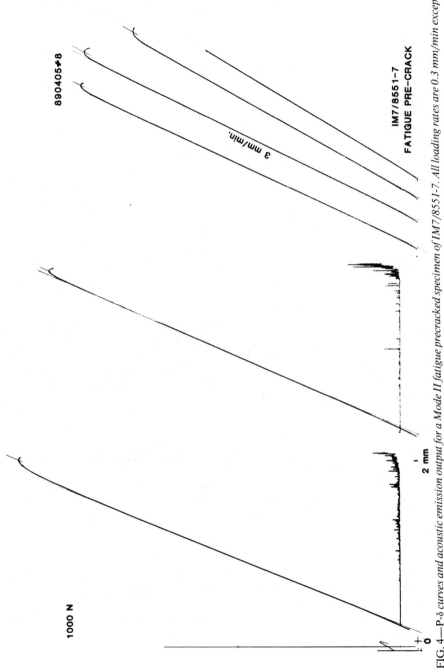

FIG. 4—P-δ curves and acoustic emission output for a Mode II fatigue precracked specimen of IM7/8551-7. All loading rates are 0.3 mm/min except for the fourth loading.

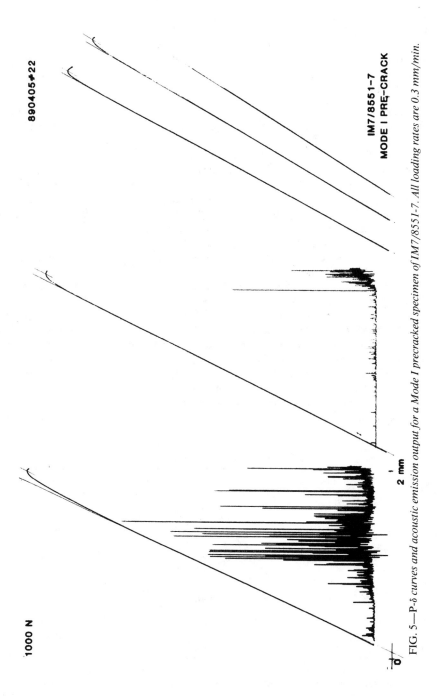

FIG. 5—P-δ curves and acoustic emission output for a Mode I precracked specimen of IM7/8551-7. All loading rates are 0.3 mm/min.

where

a = crack length,
P = load,
C = specimen compliance,
C_s = correction for specimen shear,
b = specimen width, and
L = half span.

For each loading curve, up to three different fracture toughnesses were calculated from Eq 2, namely, $G_{IIc}(\text{lin})$, $G_{IIc}(\text{init})$, and $G_{IIc}(\text{max})$. $G_{IIc}(\text{lin})$ uses the compliance, C_i, and crack length, a_i, corresponding to the linear portion of the load deflection curve together with the maximum load, P_{max} and is the same as the value that is normally calculated for materials that are linear to fracture. $G_{IIc}(\text{init})$ differs from $G_{IIc}(\text{lin})$ in that the load for initiation of delamination growth, P_{init}, is used instead of P_{max}. The load at which delamination growth initiated was determined from the P-δ curve and the AE output. For the first loading only, the load, P_{nl}, at which the P-δ curve became nonlinear was used and the resulting value of G referred to as $G_{IIc}(\text{nl})$. $G_{IIc}(\text{max})$ was calculated using the estimated crack lengths and compliances of the specimen at the maximum load points. This was done by taking $C_{max} = \delta_{max}/P_{max}$ and substituting into Eq 2 as before to obtain a_{max}. (For the first loading only, the permanent deflection was subtracted from δ_{max} prior to calculating C_{max}).

Tables 3 and 4 list the fracture toughnesses obtained for the first and second loadings of each material and starter crack. In all cases, $G_{IIc}(\text{init})$ was less than $G_{IIc}(\text{max})$. Except for the first loading, the fracture resistance increased by 10 to 15% between initiation and maximum load. However, this increase was not cumulative since the $G_{IIc}(\text{init})$ and $G_{IIc}(\text{max})$ values for the third and subsequent loadings were similar to the second. This is shown in Fig. 6 where the fracture resistance is plotted against crack growth for some representative specimens. The slight rise in the IM7/8551-7 curves was typical for that material with the average of all specimens increasing by approximately 3% from loading to loading.

The low values of $G_{IIc}(\text{nl})$ for the first loading reflect the low loads at which the P-δ curves deviated from linearity and may or may not correspond to the onset of delamination growth.

TABLE 3—*Fracture toughnesses—first loading.*

Quantity	No Precrack	Mode I Precrack	Fatigue Precrack
		IM6/5245C	
$G_{IIc}(\text{nl})$	960	554	654
$G_{IIc}(\text{lin})$	971(2)	706(3)	707(2)
$G_{IIc}(\text{max})$	979	770	758
		IM7/8551-7	
$G_{IIc}(\text{nl})$	903 ± 35	796 ± 64	833 ± 60
$G_{IIc}(\text{lin})$	1101 ± 48(5)	1061 ± 68(10)	1047 ± 22(10)
$G_{IIc}(\text{max})$	1185 ± 56	1193 ± 104	1147 ± 48
		AS4/PEEK	
$G_{IIc}(\text{nl})$	2000 ± 302	1278 ± 243	1494 ± 277
$G_{IIc}(\text{lin})$	2981 ± 269(5)	2281 ± 200(5)	2640 ± 133(5)
$G_{IIc}(\text{max})$	3186 ± 305	2752 ± 241	3148 ± 51

NOTE—Mean values ± standard deviation where applicable. Numbers in parentheses refer to the number of specimens tested.

TABLE 4—*Fracture toughnesses—second loading.*

Quantity	No Precrack	Mode I Precrack	Fatigue Precrack
		IM6/5245C	
$G_{IIc}(\text{init})$	. . .	696	. . .
$G_{IIc}(\text{lin})$	. . .	747(1)	. . .
$G_{IIc}(\text{max})$	. . .	798	. . .
		IM7/8551-7	
$G_{IIc}(\text{init})$	1069 ± 46	1059 ± 87	1061 ± 36
$G_{IIc}(\text{lin})$	1122 ± 43(4)	1121 ± 89(8)	1120 ± 31(8)
$G_{IIc}(\text{max})$	1227 ± 43	1220 ± 93	1213 ± 36
		AS4/PEEK	
$G_{IIc}(\text{init})$	2919	2516 ± 250	2921 ± 115
$G_{IIc}(\text{lin})$	3024(2)	2600 ± 258(4)	2997 ± 124(4)
$G_{IIc}(\text{max})$	3223	2770 ± 280	3152 ± 127

NOTE—Mean values ± standard deviation where applicable. Numbers in parentheses refer to the number of specimens tested.

This will be addressed further in the discussion. In general, the Mode I precracks gave the lowest values of $G_{IIc}(\text{nl})$ followed by the fatigue precracks. For IM6/5245C, crack initiation from the end of the insert required 200 J/m² more than the fracture resistance, a clear indication that the resulting "crack tip" was too blunt. However, this was not the case for the other two tougher materials. As was found for the extent of nonlinearity, the type of starter crack only affected the results from the first loading.

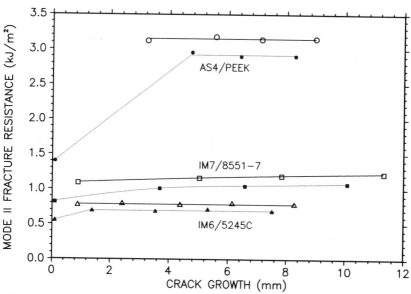

FIG. 6—*Representative R-curves for each material. Solid symbols are $G_{IIc}(init)$ values, open symbols are $G_{IIc}(max)$ values. All three specimens were precracked in tension.*

Time-Dependent Behavior

Two different approaches were taken in order to investigate the time-dependent nature of the fracture process. First, the effect of loading rate was determined by increasing the ram speed by a factor of 10 during one of the reloading cycles. An example of this can be seen in the fourth loading curve in Fig. 4. Table 5 compares the values of ΔC_{max}, G_{IIc}(init), and G_{IIc}(max) obtained at the faster rate with those from the previous slower loading. There was no consistent change in the nonlinearity to maximum load, however, there was a small but significant increase in fracture toughness for both of the materials tested in this way. This method of measuring strain rate effects is particularly sensitive as specimen-to-specimen variations are eliminated.

In other tests, an attempt was made to determine the relationship between G_{II} and crack velocity, V. This was done by stopping the ram just before the load reached P_{max} and then recording the load dropoff against time as shown in Fig. 7. After approximately 45 min, the specimen was unloaded, the final compliance determined as before, and a_0 and a_f measured from the fracture surfaces. Between 10 and 20 mm of crack growth occurred in most tests. The crack length as a function of time, t, was then calculated from Eq 2 as before by taking $C(t) = \delta/P(t)$. A check was made on each specimen to ensure that the load drop-off was the result of crack growth and not shear stress relaxation in the viscoelastic matrix. This was done by comparing the δ/P value at the end of the 45-min hold time with the final reloading compliance, C_f. In all cases, agreement was within 3% indicating that the $\delta/P(t)$ value could indeed be used to calculate the crack length. Finally, G_{II} was determined from Eq 3 and the crack velocity obtained by curve fitting the a versus t data and taking the derivative.

Figure 8 shows examples of the results obtained. Most of the data could be fitted by straight lines on a log-log plot, consistent with the power law relationship often found for slow crack growth in visco-elastic materials [11]. The IM7/8551-7 specimens had the greatest $\log(G_{II})/\log(V)$ slopes followed by AS4/PEEK and then IM6/5245C. However, no unique curve was found to exist for a given material and, in fact, the curve appeared to shift to higher values of G_{II} as the initial crack length increased, as shown in Fig. 9 for IM7/8551-7. The most likely explanation for this behavior is that these tests were not run under equilibrium conditions, that is, G_{II} changed too rapidly to allow time for the plastic zone to reach its steady-state size. Consequently, tests started with different a_0 values experienced different rates of change of G_{II} with crack length and hence different amounts of crack tip blunting.

Discussion

Before attempting to explain the large initial nonlinearity and its dependency on the nature of the starter crack, it is necessary to understand the origin of the constant nonlinearity associated with the reloading curves. In the absence of continued permanent deformation of the

TABLE 5—*Loading rate effects.*

Quantity	Previous Loading	Fast Loading	Change %
		IM7/8551-7	
G_{IIc}(init)	1058	1142	$+7.9 \pm 2.2$
G_{IIc}(max)	1196	1279	$+7.0 \pm 2.5$
ΔC_{max}	3.0	3.1	$+4.7 \pm 13$
		AS4/PEEK	
G_{IIc}(init)	2716	2866	$+5.5 \pm 2.9$
G_{IIc}(max)	2971	3110	$+4.7 \pm 1.1$
ΔC_{max}	2.5	2.3	-6.7 ± 19

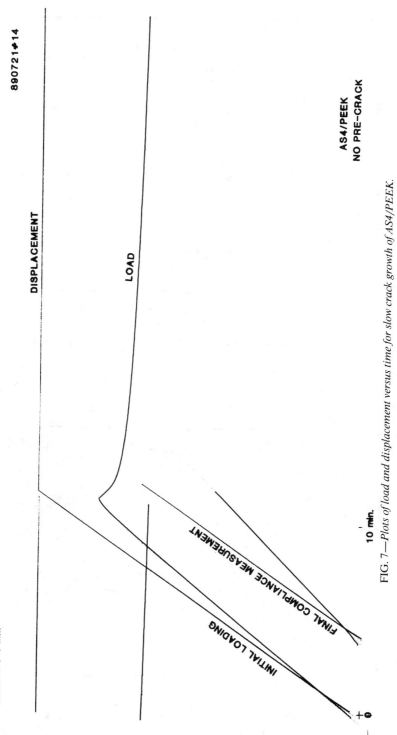

FIG. 7—*Plots of load and displacement versus time for slow crack growth of AS4/PEEK.*

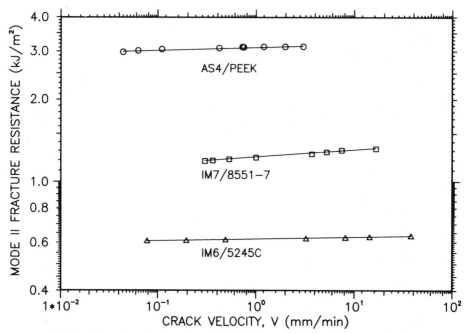

FIG. 8—*Examples of log-log plots of Mode II fracture resistance versus crack velocity.*

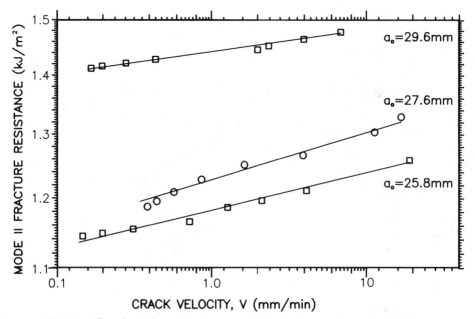

FIG. 9—*Effect of initial crack length on the relationship between G_{II} and crack velocity.*

specimen or of a steeply rising R-curve, the most likely explanation is that the P-δ curve deviates from linearity due to slow crack growth. This is supported by the crack velocity measurements which indicate that growth would be slow at first (<0.01 mm/min) but would increase steadily as both the load and crack length continue to increase. Eventually, a maximum load is reached when the rate of increase in load due to δ increasing is matched by the rate of reduction due to the crack length (and hence compliance) increasing. The equations that govern this behavior are as follows; first

$$\delta = St$$

where S is the loading speed (mm/min). Then, neglecting shear, and rewriting Eq 3 in terms of displacement rather than load

$$G = \frac{9a^2S^2t^2(2L^3 + 3a_0^3)}{2bC_0(2L^3 + 3a^3)^2} \tag{4}$$

And making use of the observed dependence of crack velocity on G_{II}

$$V = RG_{II}^m \tag{5}$$

where R and m are constants obtained by least squares fitting a straight line to the log(V) versus log (G_{II}) data, the increase in crack length can be obtained by integration

$$a = a_0 + \int_0^t V dt \tag{6}$$

and finally the load can be calculated from

$$P = \frac{\delta}{C} = \frac{St(2L^3 + 3a_0^3)}{C_0(2L^3 + 3a^3)} \tag{7}$$

These equations were solved numerically by an iterative procedure for representative values of R and m for each material. Table 6 shows the resulting values of ΔC_{max}, G_{IIc}(init), and G_{IIc}(max). Qualitatively, the predictions agree well with the observed behavior, that is, ΔC_{max} remains constant whereas both G_{IIc}(init) and G_{IIc}(max) increase as the loading rate is raised. Furthermore, the model ranks the materials in the correct order both in terms of the amount of nonlinearity and in the sensitivity to loading rate. However, quantitatively, both the amount of nonlinearity as well as the changes in fracture resistance with loading rate are underestimated. This discrepancy is not surprising given the nonuniqueness of the G_{II} versus V curves (Fig. 9). Clearly, more rigorous tests are required to properly elucidate both the steadystate and transient dependencies of fracture resistance on crack velocity. Nevertheless, the qualitative agreement between the observed effect of loading rate and the simple slow crack growth rate model presented strongly support the contention that most of the nonlinear behavior after the first loading is caused by slow crack growth.

The nonlinearity on the first loading can now be regarded as being made up of two parts, one from slow crack growth and a second contribution from some other mechanism that operates at a lower load. The permanent deformation resulting from the first loading strongly suggests that matrix yielding at the crack tip is responsible. To confirm that the permanent deformation was occurring at the crack tip, one specimen of each material without a delamination was loaded to the same δ_{max} as the fracture specimens, unloaded and then reloaded again. In

TABLE 6—*Predicted Nonlinearity and loading rate effects.*

R	m	Loading Rate, mm/min	ΔC_{max}, %	G_{IIc}(init), J/m^2	G_{IIc}(max), J/m^2
		IM6/5245C			
		0.3	0.28	618	623
4.56E32	155.5	3.0	0.28	628	632
change (%)			0	+1.6	+1.4
		IM7/8551-7			
		0.3	1.2	1193	1278
5.84E-4	36.5	3.0	1.2	1270	1360
change (%)			0	+6.5	+6.4
		AS4/PEEK			
		0.3	0.42	3106	3153
9.31E-52	103.1	3.0	0.42	3176	3224
change (%)			0	+2.3	+2.3

no case was a permanent deformation of more than 0.2% of δ_{max} observed. Figure 10*a* shows schematically why crack tip yielding produces a permanent deflection of the specimen on the first loading only. Although work is constantly required to deform the matrix ahead of the advancing crack tip, the shear deformation can no longer constrain the specimen after the crack has grown through it. If the size of the plastic zone ahead of the arrested crack is smaller than that present at P_{max}, then some additional yielding will occur during each subsequent loading cycle and may contribute to the observed nonlinearities.

The effect of starter crack on the G_{IIc}(nl) values of Table 3 is also consistent with crack tip yielding being responsible for the initial deviation from linearity. The blunt crack tips in the specimens without precracks would produce lower crack tip stresses and require higher values of G_{II} to initiate yielding. The slightly lower values of G_{IIc}(nl) in the Mode I precracked specimens compared to those with fatigue precracks could be caused by the shear hardening already present in the fatigued specimens or it could simply be the result of the bridged fibers breaking prematurely as discussed previously.

The only other likely source of nonlinearity during the first loading is that of a steeply rising *R*-curve. Some further tests were run to determine exactly when crack growth occurred. The edges of some specimens were polished to a metallographic finish and crack extension was detected using high magnification optical microscopy. These tests were run on specimens without precracks as it was easier to tell when a crack had initiated at the end of the insert than whether or not a pre-existing crack had grown. It was observed, for IM7/8551-7, that crack growth began very soon after the start of nonlinearity but, for AS4/PEEK, significant deviation from linearity occurred prior to crack growth. No results were obtained for IM6/5245C due to the unstable crack propagation from the insert. Thus for IM7/8551-7, the fracture resistance increases rapidly at first from <900 J/m^2 (G_{IIc}(nl) for the nonprecracked specimens to $\simeq$1100 J/m^2 and then continues to increase gradually as shown previously in Fig. 6. Based on the microscopic observations, a possible explanation for the increasing fracture resistance is shown schematically in Fig. 10*b*. Whereas the crack can extend with relative ease through the brittle epoxy matrix, the large rubber particles, which are discretely located between the plies, simply deform. As the crack continues to advance, both the number of particles and the resistance of each particle to deformation increases.

In summary then, several different mechanisms may be responsible for nonlinear Mode II interlaminar fracture behavior in polymer matrix composite materials. Three of these mech-

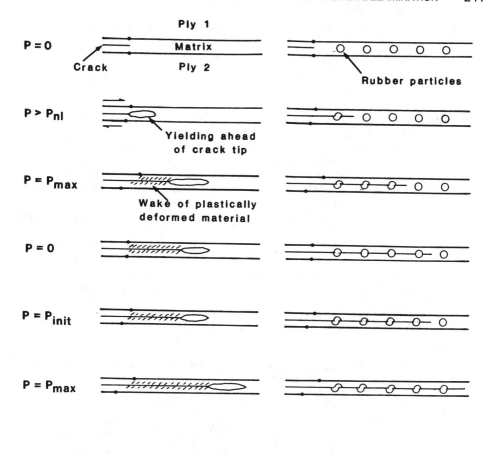

(a) CRACK TIP YIELDING   **(b) IM7/8551-7**

FIG. 10—*Proposed mechanisms to explain (a) crack-tip yielding effects and (b) R-curve behavior of IM7/8551-7.*

anisms (slow stable crack growth, crack tip yielding, and a rising fracture resistance) are natural material responses and are unavoidable in any Mode II interlaminar fracture test. In all three materials, the work of plastic deformation appears to be localized to the crack tip and can therefore be regarded as an inseparable, geometry-independent part of the energy required to advance the crack. As such, it does not appear to be necessary to resort to elastic-plastic fracture mechanics to describe the delamination fracture of these materials, at least at ambient temperature. Other sources of nonlinearity are related to the difficulty in producing an ideal starter crack and include nonuniform crack growth across the specimen width and the failure of bridged fibers resulting from tensile precracks.

Finally, it is worthwhile emphasizing that for visco-elastic materials which undergo slow stable crack growth there is no single fracture parameter which adequately describes this behavior. Rather, there is a need to develop a test method that can properly quantify the relationship between crack velocity, V, and strain energy release rate, G_{II}. Nevertheless, for some purposes such as materials selection or specification, it may still be acceptable (and probably preferable) to define a single fracture toughness, G_{IIc}, that can be measured more simply. Of the four fracture toughnesses evaluated in this investigation, G_{IIc}(init) comes closest to meeting this requirement. However, in order to measure this quantity accurately, it is necessary to frist produce a shear crack that will grow in a self-similar manner.

Conclusions

The following conclusions pertain to the Mode II interlaminar fracture behavior of IM6/5245C, IM7/8551-7, and AS4/PEEK.

1. The principal cause of nonlinear fracture response at ambient temperature is slow stable crack growth. The fracture resistance increases as the crack gradually accelerates.

2. Crack tip yielding results in an initial permanent specimen deflection. Once a zone of shear deformation has been established ahead of the advancing crack tip, plasticity plays only a minor role in any further nonlinearities.

3. In IM7/8551-7, an increase in fracture resistance as the crack extended also contributed to the initial nonlinearity.

4. Nonuniform crack growth across the width of the specimen as well as the tensile failure of bridged fibers in Mode I precracked specimens also affect the initial nonlinearity.

5. Acoustic emission provides a convenient means of monitoring crack initiation once a stable crack has formed in the specimen but is not sufficiently sensitive to define the early stages of shear crack formation.

References

[1] Williams, J. G., "On the Calculation of Energy Release Rates for Cracked Laminates," *International Journal of Fracture,* Vol. 36, 1988, pp. 101–119.

[2] Wilkins, D. J., Eisenmann, J. R., Camin, R. A., Margolis, W. S., and Benson, R. A. in *Damage in Composite Materials,, ASTM STP 775,* K. Reifsnider, Ed., American Society for Testing and Materials, Philadelphia, 1982, pp. 168–183.

[3] O'Brien, T. K., Johnston, N. J., Morris, D. H., and Simonds, R. A., "A Simple Test for the Interlaminar Fracture Toughness of Composites," *SAMPE Journal,* Society for the Advancement of Material and Process Engineering, Vol. 18, No. 4, July/Aug. 1982.

[4] Russell, A. J. and Street, K. N. in *Delamination and Debonding of Materials, ASTM STP 876,* W. S. Johnson, Ed., American Society for Testing and Materials, Philadelphia, 1985, pp. 349–370.

[5] Bradley, W. L. and Cohen, R. N. in *Delamination and Debonding of Materials, ASTM STP 876,* W. S. Johnson, Ed., American Society for Testing and Materials, Philadelphia, 1985, pp. 389–410.

[6] Russell, A. J., "Micromechanisms of Interlaminar Fracture," *Polymer Composites,* Vol. 8, 1987, pp. 342–351.

[7] Chai, H., "Shear Fracture," *International Journal of Fracture,* Vol. 37, 1988, pp. 137–159.

[8] Carlsson, L. A., Gillespie, J. W., and Trethewey, B., "Mode II Interlaminar Fracture Toughness in Graphite/Epoxy and Graphite/PEEK Composites," *Journal of Reinforced Plastics and Composites,* Vol. 5, 1987, p. 170.

[9] Russell, A. J. and Street, K. N. in *Toughened Composites, ASTM STP 937,* N. J. Johnston, Ed., American Society for Testing and Materials, Philadelphia, 1987, pp. 274–294.

[10] Carlsson, L. A., Gillespie, J. W., and Pipes, R. B., "On the Analysis and Design of the End Notched Flexure (ENF) Specimen for Mode II Testing," *Journal of Composite Materials,* Vol. 20, 1986, p. 594.

[11] Kinloch, A. J. and Young, R. J., *Fracture Behaviour of Polymers,* Elsevier Applied Science, London, 1983.

Roderick H. Martin[1]

Evaluation of the Split Cantilever Beam for Mode III Delamination Testing

REFERENCE: Martin, R. H., **"Evaluation of the Split Cantilever Beam for Mode III Delamination Testing,"** *Composite Materials: Fatigue and Fracture (Third Volume), ASTM STP 1110,* T. K. O'Brien, Ed., American Society for Testing and Materials, Philadelphia, 1991, pp. 243–266.

ABSTRACT: A test fixture for testing a thick split cantilever beam for scissoring delamination (Mode III) fracture toughness was developed. A three-dimensional finite element analysis was conducted on the test specimen to determine the strain energy release rate, G, distribution along the delamination front. The virtual crack closure technique was used to calculate the G components resulting from interlaminar tension, G_I, interlaminar sliding shear, G_{II}, and interlaminar tearing shear, G_{III}. The finite element analysis showed that at the delamination front no G_I component existed, but a G_{II} component was present in addition to a G_{III} component. Furthermore, near the free edges, the G_{II} component was significantly higher than the G_{III} component. The G_{II}/G_{III} ratio was found to increase with delamination length but was insensitive to the beam depth. The presence of G_{II} at the delamination front was verified experimentally by examination of the failure surfaces. At the center of the beam, where the failure was in Mode III, there was significant fiber bridging. However, at the edges of the beam where the failure was in Mode II, there was no fiber bridging and Mode II shear hackles were observed. Therefore, it was concluded that the split cantilever beam configuration does not represent a pure Mode III test. The experimental work showed that the Mode II fracture toughness, G_{IIc}, must be less than the Mode III fracture toughness, G_{IIIc}. Therefore, a conservative approach to characterizing Mode III delamination is to equate G_{IIIc} to G_{IIc}.

KEY WORDS: fracture, fatigue (materials), composite materials, delamination, fracture toughness, Mode III testing, split cantilever beam, strain energy release rate

Nomenclature

a	Delamination length
a_0	Initial delamination length
C	Specimen compliance
D	Specimen depth
E_{11}	Tensile longitudinal modulus
F	Force
G	Total strain energy release rate
G_a	Integrated average strain energy release rate
G_b	Strain energy release rate calculated from beam theory
G_c	Critical strain energy release rate
G_g	Global strain energy release rate calculated from compliance variations
G_I	Mode I component of strain energy release rate
G_{II}	Mode II component of strain energy release rate
G_{III}	Mode III component of strain energy release rate

[1] Research scientist, Analytical Services and Materials, Inc., Hampton, VA 23666. Formerly, National Research Council research associate at NASA Langley Research Center, Hampton, VA 23665.

G_{12} Shear modulus in the x-y plane
$G(y)$ Distribution of strain energy release rate along the y-axis
h Beam half thickness
I Second moment of area
P Applied load
u,v,w Displacement in the x-, y-, and z-directions, respectively
x,y,z Axes
δ Beam deflection at loading point
Δ Length of finite element of delamination front
τ Shear stress

With the increased use of laminated fiber reinforced plastics in primary aircraft structural components, the need to understand and predict the failure modes of these components has also increased. There have been many studies over the last decade examining delamination failure of composite materials and structures [1–20]. A delamination may result from high interlaminar stresses causing adjacent plies to come apart. These high stresses are caused by material and geometric discontinuities in the component, and can be tensile, compressive, or shear in nature. Much work has been published on characterizing Mode I (opening or peel) [1–8] and Mode II (sliding or interlaminar shear) [7–15] delamination. Emphasis was initially placed on Mode I fracture testing because it was the most critical mode of fracture with brittle matrix systems [8,15]. Tougher matrix systems resulted in a decreased difference between the Mode I and Mode II fracture toughnesses [14,15]. Mode I and Mode II delamination tests are now sufficiently advanced for the various standards organizations, such as ASTM, to consider. Many delamination problems considered were found to delaminate in a combination of Modes I and II [16–18]. Therefore, Mode III delamination characterization was largely ignored. However, the importance of Mode III delamination is beginning to be appreciated. With the complex loads seen in service, and for certain laminate configurations [19,20], Mode III delamination may occur. Therefore, Mode III delamination needs to be characterized.

In the present literature, there are only a few suggested test methods available for characterizing Mode III delamination in composite materials. Donaldson [21] developed a test using a split cantilever beam (SCB) type arrangement. One arrangement consisted of a unidirectional laminate, adhesively bonded between aluminum bars to give the specimen torsional stiffness as the delamination grew, Fig. 1a. The load was applied by thick metal plates bolted to the aluminum bars. The plates were pinned to the jaw of the test machine. The thick plates helped reduce the Mode I delamination. The test appeared to work successfully for a brittle graphite/ epoxy, but the aluminum bars debonded when a tougher thermoplastic matrix composite was used. Chaouk [20] used a similar split beam configuration using a torsion fixture to introduce the load. Donaldson and Mall also have used the SCB configuration to measure fatigue delamination growth rates [22]. Chai [23] used a modified SCB specimen to test the shear fracture of adhesives. Becht and Gillespie developed a double rail shear test to measure Mode III fracture toughnesses [24]. This test configuration was modified by Gillespie and Becht [25] to a single cracked rail shear test, because of the difficulties in growing two delaminations at one time, Fig. 1b. A Mode III edge delamination test was also investigated by Donaldson [21].

The rail shear configurations have very low compliances and hence accurate values of compliance and change in compliance with delamination growth are difficult to obtain. The SCB, however, is sufficiently compliant to extract specimen compliances from the machine cross head displacements. However, the problem of the adherend debonding prevents the determination of delamination fracture toughness for tougher materials. One solution to the debonding problem is to make the laminates sufficiently thick to provide their own torsional

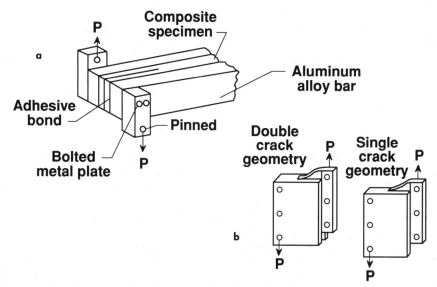

FIG. 1—*Mode III test configurations:* (a) *split cantilever beam specimen and* (b) *crack rail shear specimens.*

stiffness. Also, in Refs *21* and *22,* two bolts were used to transfer the load to the specimen. The data reduction assumed that the load was applied between the center of the two bolts, which may not be entirely accurate. A possible solution to this problem would be to load the laminate edges using a loading nose system, Fig. 2. However, for pure Mode III fracture there must be only τ_{yz} stresses at the delamination front. It is possible that the strain energy release rate along the delamination front of the SCB is not pure Mode III due to the rotation of the beam about

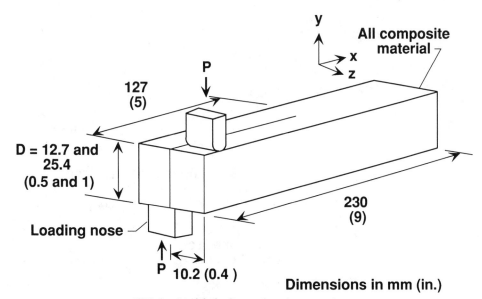

FIG. 2—*Modified split cantilever beam specimen.*

the z-axis causing τ_{xy} stresses at the delamination front. Hence, a G_{II} component may be present at the specimen edges. The presence of a G_{II} component in the SCB specimen has not been previously verified. Therefore, the purpose of this study was to determine if the SCB is suitable for characterizing Mode III delamination, by performing an analysis on this configuration to determine the G distribution along the delamination front. Also, experiments were conducted using the modified SCB configuration. The failure surfaces were examined to determine the mode of failure.

Materials

Unidirectional, 100-ply, glass/epoxy (S2/SP250) panels were manufactured at NASA Langley Research Center according to manufacturer's instructions with the fibers aligned to the x-axis. To simulate a 127-mm (5-in.)-long initial delamination, a folded 0.0127-mm (0.5-mil) Kapton film was inserted between the 50th and 51st ply prior to curing. The average volume fraction for the material used was 64.6%. The volume fraction was determined using ASTM Test Method for Fiber Content of Resin-Matrix Composites by Matrix Digestion (D 3171-76). The specimens were manufactured to the dimensions given in Fig. 2.

The glass/epoxy material properties for use in the finite element analysis and beam theory expressions were obtained from Ref *18* and were $E_{11} = 43.5$ GPa (6.31 Msi), $E_{22} = 17.2$ GPa (2.50 Msi), $G_{12} = 4.14$ GPa (0.60 Msi), and $\nu_{12} = 0.25$. The out-of-plane material properties were equated to the in-plane material properties for use in the three-dimensional finite element analysis, that is, $E_{33} = E_{22}$, $G_{13} = G_{23} = G_{12}$, and $\nu_{13} = \nu_{23} = \nu_{12}$.

Test Procedure

A test fixture for simulating Mode III delamination in the SCB specimens, was manufactured at NASA Langley Research Center. The test fixture is shown schematically in Fig. 3. The

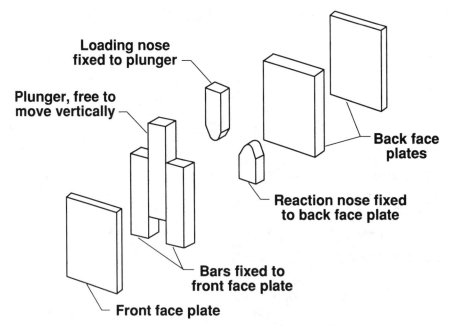

Loading nose fixed to plunger

Plunger, free to move vertically

Back face plates

Reaction nose fixed to back face plate

Bars fixed to front face plate

Front face plate

FIG. 3—*Schematic of Mode III test rig.*

plunger was free to move vertically up and down, but was restrained from movement in any other directions. The lower reaction nose was fixed to the face plate. The face plates aided in restraining any beam rotation about the x-axis in order to suppress any Mode I opening that might occur. Figure 4 shows the test fixture assembled in the testing machine. The specimen was aligned so that the fibers were perpendicular to the loading nose and the face plates. All tests were conducted under ambient conditions of 23°C and 50% humidity.

Several beams were tested at various initial delamination lengths. Initial delamination length was varied by altering the position of the beam in the test fixture prior to testing. The sides of the beam were graduated in 2.5-mm (0.1-in.) intervals to aid in the measurement of delamination length on the edge of the beam. Delamination initiation and propagation were observed visually on both edges using a low powered microscope. The tests were run under displacement control at a cross head displacement rate of 0.5 mm/min (0.02 in./min). The resulting load-displacement plot was recorded on an X-Y plotter. Initiation of delamination

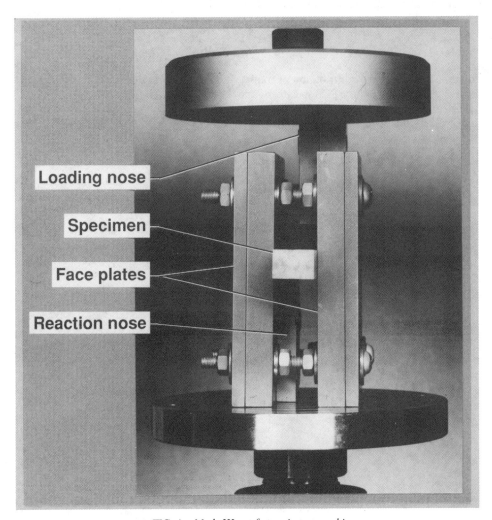

FIG. 4—*Mode III test fixture in test machine.*

from the insert was also observed as a deflection from the initially linear part of the load-displacement plot. Further increments in delamination length were marked on the X-Y plot for subsequent data reduction. Figure 5 shows a typical load-displacement plot. For all tests, on unloading, a sudden drop in load was noticed followed by an unsmooth unloading plot. This unloading path indicates that there was friction present in the test. This friction was probably between the delaminated surfaces, and also between the face plates and the outside edges of the specimen.

Analysis

Beam Theory

The compliance, C, of the SCB specimen can be determined from the deflection of a cantilever beam using elementary beam theory [26] modified for composite materials [27] in a similar way to the double cantilever beam (DCB) specimen [1], thus

$$C = \frac{\delta}{P} = \frac{2a^3}{3E_{11}I} + \frac{D^2a}{4G_{12}I} \tag{1}$$

Equation 1 includes the contribution of transverse shear strain to deflection because of the relatively thick nature of the SCB specimens and the high E_{11}/G_{12} ratio.

The strain energy release rate, G, may be expressed as a function of the derivative of the compliance with respect to delamination length [1], thus

$$G = \frac{P^2}{2D}\frac{dC}{da} \tag{2}$$

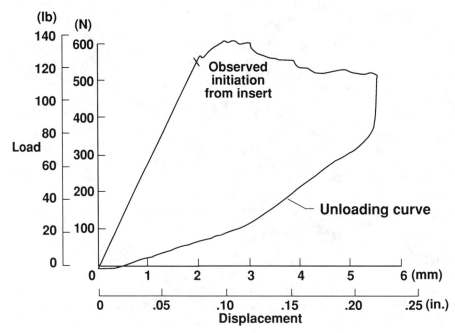

FIG. 5—*Typical Mode III load-displacement curve.*

Therefore, differentiating Eq 1 with respect to delamination length, a, and substituting into Eq 2 yields an expression for strain energy release rate, thus

$$G_b = \frac{P^2}{2D} \left[\frac{2a^2}{E_{11}I} + \frac{D^2}{4G_{12}I} \right] \tag{3}$$

However, at the delamination front of the SCB specimen, the beam theory assumption that the cantilever beam is clamped may not be valid. Any displacement in the x-direction, Fig. 2, at the delamination front will cause a Mode II strain energy release rate. If these displacements are present, the SCB configuration would not yield pure Mode III delamination. Therefore, in order to determine the contribution of the various fracture modes to G_c, a finite element analysis was performed.

Finite Element Analysis

To evaluate the distribution of the different modes of strain energy release rate along the delamination front, a three-dimensional finite element analysis (FEA) was performed using NASTRAN [28]. Two different specimen depths were considered in the analysis, $D = 25.4$ mm (1.0 in.) and $D = 12.7$ mm (0.5 in.). The model consisted of eight-node brick elements (HEXA) and six-node wedge elements (PENTA). NASTRAN's HEXA and PENTA elements are modified isoparametric elements that use selective integration points for different components of strain. For both models, the mesh was refined close to the delamination front in both the x-y plane and the x-z plane. The mesh is shown in Fig. 6 and contained 5603 degrees of freedom.

A unit line load was placed at different delamination lengths between 25.4 and 127 mm (1 and 5 in.). No delamination lengths shorter than 25.4 mm (1 in.) were considered, to prevent any stress concentrations caused by the loading nose from encroaching on the delamination

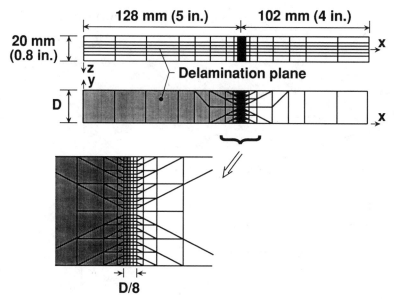

FIG. 6—*Finite element mesh.*

area. These asymmetrical loadings, Fig. 2, caused the model to twist about the x-axis. This rotation was prevented by restraining the outsides of the beam in the z-direction. The restraints ran from the end of the beam to 25.4 mm ahead of the delamination front for all delamination lengths considered. These restraints also prevented any Mode I opening of the SCB.

Strain energy release rate components were calculated using the three-dimensional virtual crack closure technique (VCCT) [29], which assumes that the work done to close the delamination by one element length is equivalent to the strain energy released when the delamination grows by one element length. Therefore, at Node H in Fig. 7 the component strain energy release rates can be evaluated from

$$G_I = \frac{1}{\Delta(y_{i-1} + y_{i+1})} F_z^H (w^B - w^E) \tag{4a}$$

$$G_{II} = \frac{1}{\Delta(y_{i-1} + y_{i+1})} F_x^H (u^B - u^E) \tag{4b}$$

$$G_{III} = \frac{1}{\Delta(y_{i-1} + y_{i+1})} F_y^H (v^B - v^E) \tag{4c}$$

where F^H is the force (in the x-, y-, or z-direction) at Node H, computed from the contribution of the forces of all the elements on one side of the delamination with connectivity at H. The symbols, u, v, and w refer to displacements in the x-, y-, and z-directions, respectively.

The average values of total strain energy release rate, G_a, along the delamination front, were calculated as

$$G_a = \left[\frac{1}{D} \int_0^D G_I(y) \, dy \right] + \left[\frac{1}{D} \int_0^D G_{II}(y) \, dy \right] + \left[\frac{1}{D} \int_0^D G_{III}(y) \, dy \right] \tag{5}$$

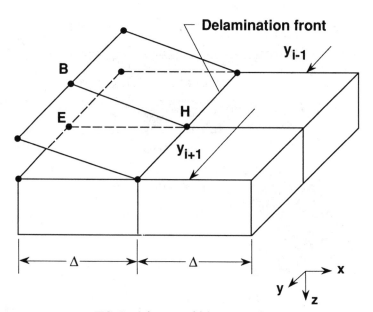

FIG. 7—Schematic of delamination front.

where $G(y)$ is the strain energy release rate distribution along the delamination front calculated using Eq 4. The values of G_a were calculated by numerical integration of the strain energy release rate distributions presented in the next section on results.

In addition, the total strain energy release rate also was calculated globally from the FEA by calculating the change in strain energy from one FEA run at delamination length, a_i, and another at delamination length, a_{i+1}. The global total strain energy release rate, G_g, at $a = (a_i + a_{i+1})/2$ is

$$G_g = \frac{1}{(a_{i+1} - a_i) D} \left[\frac{1}{2} (\Sigma P\delta)_{a_i} - \frac{1}{2} (\Sigma P\delta)_{a_{i+1}} \right] \tag{6}$$

where $(\Sigma P\delta)$ is the sum of the displacements (in the loading direction) of the loaded nodes, multiplied by the applied loads.

Results

Finite Element Analysis

Figures 8a and b show the variation of compliance, C, with delamination length, a, for D = 12.7 and 25.4 mm (0.5 and 1 in.), respectively. The correlation between beam theory (the solid line) and the FEA (open triangles) was good. The beam theory results were consistently below the FEA results because beam theory assumes that the beam is fixed at the clamped end. However, the FEA allows for the y-direction displacement experienced by the beam beyond the delamination front. An analysis where the cantilever beam assumption was replaced by a beam that is partly free and partly supported by an elastic foundation, similar to that conducted for the DCB [30], may yield closer comparison between beam theory and FEA. Also shown in Figs. 8a and b are the compliance values calculated from the experimental tests, open squares. The experimental results are discussed in the next subsection.

Figure 9a shows the total strain energy release rate for $D = 12.7$ mm (0.5 in.) calculated three different ways: (1) by beam theory, Eq 3 (solid line); (2) by the integrated average method, Eq 5 (open squares); and (3) by the global method, Eq 6 (open triangles). Figure 9b shows similar results for $D = 25.4$ mm (1.0 in.). Results using Eqs 5 and 6 yielded good agreement in the values of G/P^2. The beam theory results using Eq 3 were consistently below the FEA results. This difference may again be caused by the differences noted in the determination of compliance.

Figures 10a and b show the distribution of the normalized Mode III component of strain energy release rate, G_{III}/P^2, along the delamination front for $D = 12.7$ and 25.4 mm (0.5 and 1 in.), respectively. Only half the delamination front has been plotted, because the distribution was symmetrical about the x-z plane. For all delamination lengths, the Mode III component was virtually constant along the entire delamination front, but increased at the free edges.

Figures 11a and b show the distribution of the normalized Mode II component of strain energy release rate, G_{II}/P^2, along the delamination front for $D = 12.7$ and 25.4 mm (0.5 and 1 in.), respectively. Again, only half the delamination front has been plotted because the distribution was symmetrical about the x-z plane. The Mode II component increased from zero at the center of the beam to a maximum at the free edge. The Mode I component was zero in all cases because of the restraints set on the model.

Figures 12a and b show the Mode II and Mode III components of strain energy release rate, plotted at a delamination length of 127 mm (5 in.) for $D = 12.7$ and 25.4 mm (0.5 and 1 in.), respectively. Along approximately half the delamination front, G_{II}/P^2 is much larger than G_{III}/P^2. At the free edge, the value of G_{II}/P^2 was approximately six times the value of G_{III}/P^2 for both depths considered.

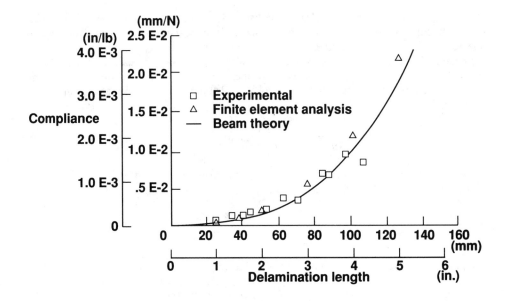

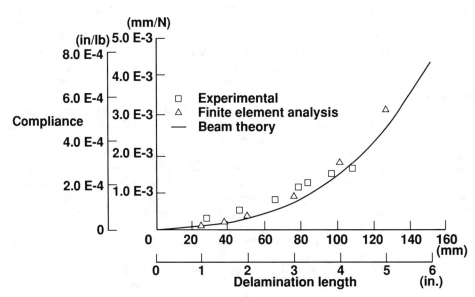

FIG. 8—(a) Compliance versus delamination length, D = 12.7 mm (0.5 in.) and (b) Compliance versus delamination length, D = 25.4 mm (1 in.).

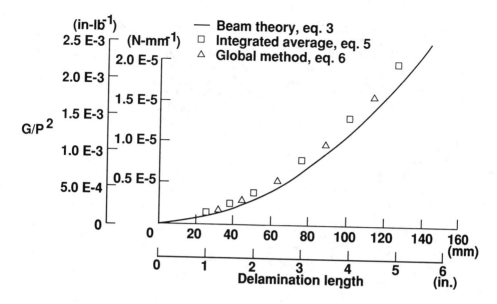

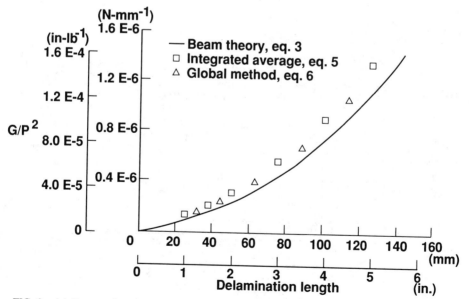

FIG. 9—(a) Computed strain energy release rates, D = 12.7 mm (0.5 in.) and (b) Computed strain energy release rates, D = 25.4 mm (1 in.).

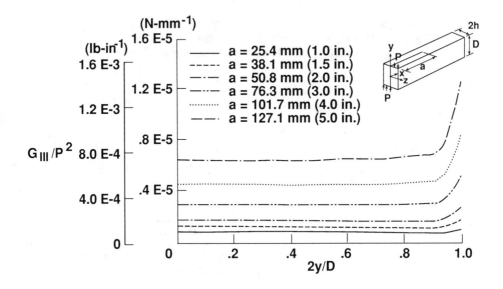

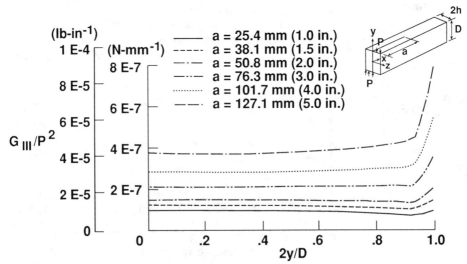

FIG. 10—(a) G_{III} variation along delamination front, $D = 12.7$ mm (0.5 in.) and (b) G_{III} variation along delamination front, $D = 25.4$ mm (1 in.).

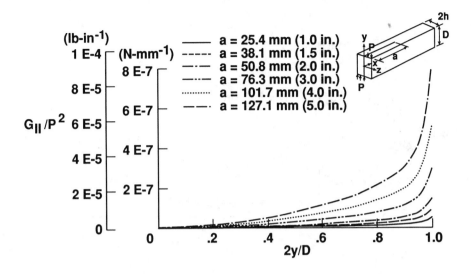

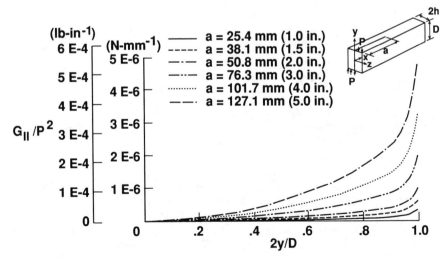

FIG. 11—(a) G_{II} variation along delamination front, $D = 12.7$ mm (0.5 in.) and (b) G_{II} variation along delamination front, $D = 25.4$ mm (1 in.).

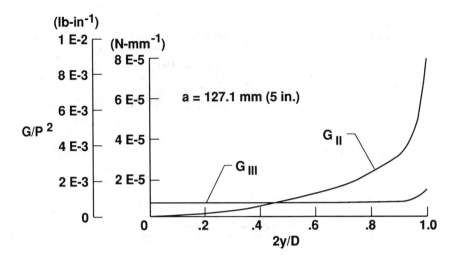

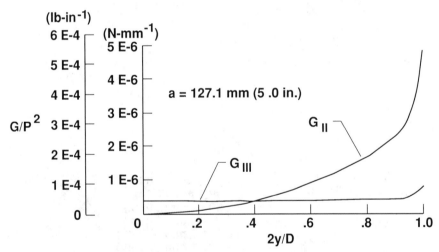

FIG. 12—(a) G_{II} and G_{III} distribution along delamination front, D = 12.7 mm (0.5 in.), a = 127.1 mm (5 in.); and (b) G_{II} and G_{III} distribution along delamination front, D = 25.4 mm (1 in.), a = 127.1 mm (5 in.)

Figures 13a and b show the Mode II and Mode III components together at a delamination length of 25 mm (1 in.) at D = 12.7 and 25.4 mm (0.5 and 1 in.), respectively. For a = 25 mm (1 in.), the Mode II component was only larger than the Mode III component for approximately 15% of the delamination front. At the free edge, the G_{II}/P^2 value was approximately 3.5 times the G_{III}/P^2 value. Figures 14a and b show the G_{II}/G_{III} distribution along the delamination front at various initial delamination lengths for D = 12.7 and 25.4 mm (0.5 and 1 in.) respectively. Figure 14 shows the larger the initial delamination length the greater the proportion of G_{II} present along the delamination front. For all delamination lengths considered, the

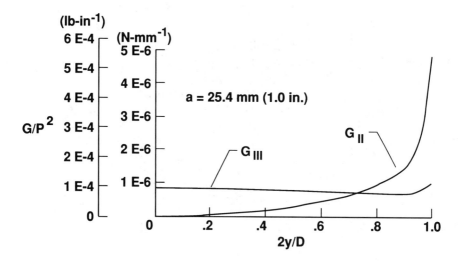

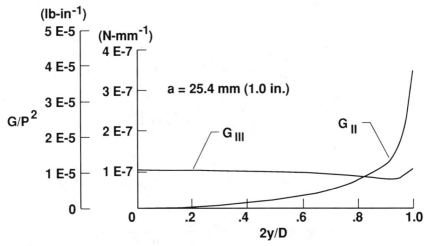

FIG. 13—(a) G_{II} and G_{III} distribution along delamination front, D = 12.7 mm (0.5 in.), a = 25.4 mm (1 in.) and (b) G_{II} and G_{III} distribution along delamination front, D = 25.4 mm (1 in.), a = 25.4 mm (1 in.).

Mode II component was larger than the Mode III component of strain energy release rate at the free edge. The G_{II}/G_{III} distribution was also largely insensitive to beam depth.

Experimental

Figure 15 shows a plot of critical strain energy release rate, G_c, against delamination length for one of the S2/SP250 beams tested where $G_c = (G_I + G_{II} + G_{III})_c$. The term G_c rather than G_{IIIc} has been used, because the results of the finite element analysis showed that delamination

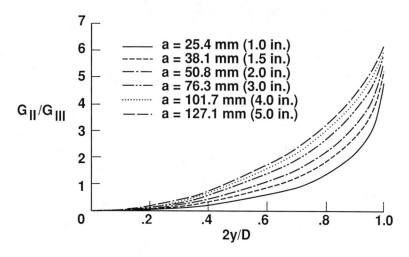

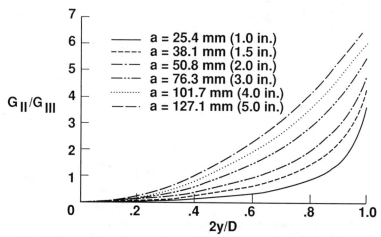

FIG. 14—(a) G_{II}/G_{III} distribution along delamination front, $D = 12.7$ mm (0.5 in.) and (b) G_{II}/G_{III} distribution along delamination front, $D = 25.4$ mm (1 in.).

would not be by pure Mode III alone. The quantity G_c was calculated using the beam theory expression given in Eq 3. The delamination length was taken as that observed at the edge of the beam. In reality, the delamination front was probably not straight, after growth from the insert, but either "U" or "V" shaped due to the variation of G_{II} along the delamination front. No account for the change in shape of the delamination front with delamination extension was taken in Fig. 15. An increase in G_c was observed with an increase in delamination length. This apparent increase in G_c or "R-curve" can be attributed to both fibers bridging the delaminated halves of the beam and a change in the shape of the delamination front caused by the

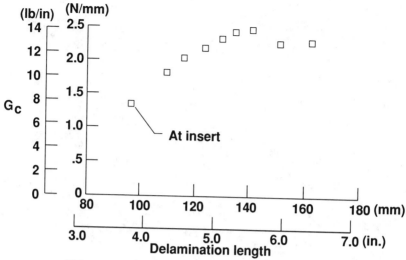

FIG. 15—G_c *versus delamination length*, D = 25.4 mm (1 in.).

G_{II}/G_{III} variation with delamination length. The R-curve effect from fiber bridging is analogous to that seen in the DCB tests using this material [6]. Observation of the delaminated halves of the beams, Fig. 16, shows fiber bridging occurring in the center of the beam only. The longer the initial delamination length, the less widespread the fiber bridging along the delamination front, Fig. 17. Close examination of the failure surface of the specimen, Fig. 18, using a scanning electron microscope, shows the familiar shear hackles at the edge of the specimen caused by Mode II failure of brittle composites [9,10,14,15]. Whereas at the center of the specimen, tangled fibers are visible. These phenomena were consistent with Fig. 14 that showed a large Mode II component near the free edges of the SCB that increased with delamination length. Therefore, Figs. 16 to 18 are further evidence that the SCB test has Mode II failures at the outer edges of the beam.

For material systems that may not experience as much fiber bridging, such as AS4/3501-6, different experimental results to the R-curve shown in Fig. 15 may be expected. With no fiber bridging to increase the apparent G_{IIIc} component, no increase in G_c would be seen as the delamination grew in the specimen. Instead, as the delamination length increased, the Mode II component, along the delamination front, increased. Thus, a decrease in experimental G_c with delamination growth may be observed [21].

A possible cause for fiber bridging observed in the interior of the SCB specimens is the high τ_{yz} stresses in the planes perpendicular to the fibers. These stresses may cause tensile damage in the form of microcracks ahead of the delamination, shown schematically in Fig. 19. The damage ahead of the delamination front could cause the delamination to grow by joining the ends of the microcracks. When the delamination grows and connects different ends of the microcracks then fibers may bridge.

Fiber bridging results in a decrease of experimentally measured compliance. However, at the point of delamination onset at the thin insert, there is no fiber bridging and the delamination front is straight. Therefore, this may be a valid value of compliance to compare with the finite element analysis. Figures 8a and b, show the experimental values of compliance determined at the insert for different initial delamination lengths, a_0, compared with the finite element and beam theory results. The experimental results were higher than both FEA results

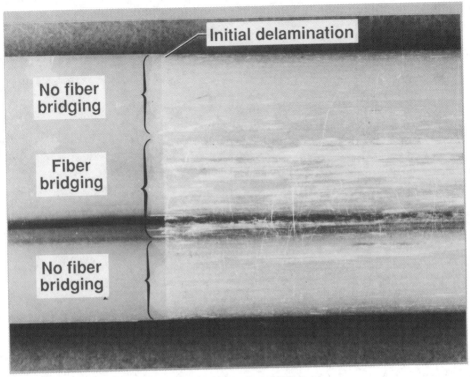

FIG. 16—*Fracture surface of SCB specimen.*

and the beam theory results in most cases. The difference between experimental and theoretical results may possibly be caused by the value of moduli used in the theory. The flexural moduli may be significantly lower than the tensile moduli [31], the latter being used in the analysis presented here. A decrease in the values of E_{11} and G_{12} in the analysis would increase the values of compliance as calculated from Eq 1. Furthermore, compliance was calculated using cross-head displacements, no correction was made for the machine compliance. Any machine deflection would result in increasing the measured compliance. In contrast, the friction observed during the experimental work, would result in reducing the measured compliance. Therefore, the differences between the experimental and analytical values of compliance have not been accounted for at this time.

At initial delamination propagation from the insert, there is no fiber bridging. Therefore, an accurate value of G_c may be determined [6]. Figure 20 shows a plot of experimental G_c versus initial delamination length, a_0, where G_c was calculated from delamination initiation from the insert using Eq 3. A marked decrease in G_c with initial delamination length was observed for both specimen depths considered. Figure 21 shows the variation of G_{II}/G ratio with delamination length, where G_{II} was calculated by averaging the VCCT results over the delamination front. Figure 21 shows the longer the initial delamination length the larger the amount of G_{II} present. Therefore, the greater the quantity of G_{II} at the delamination front the lower the value of G_c. Thus, it can be concluded that $G_{IIc} < G_{IIIc}$. Generally, the 25-mm (1 in.) depth specimens had a higher G_c value than the 12.7 mm (0.5 in.) depth. This result was possibly due to the increased friction caused by the larger surface area.

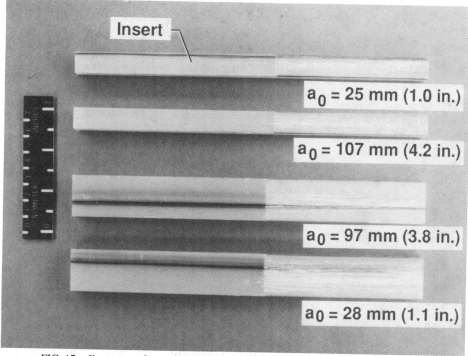

FIG. 17—*Fracture surfaces of SCB specimens with different initial delamination lengths.*

The values of G_c in Fig. 20 were calculated from beam theory. A plot of total G along the delamination front calculated in several ways is shown in Fig. 22. Neither the beam theory nor the integrated average method account for the variation in total G along the front. Hence, there is no valid data reduction technique to obtain a value of G_c from experimental testing of the SCB.

Discussion

If a material characterization test is to be developed, it should take the simplest form possible, to allow testing of the many different material systems for quick quality control screening and material property determination. The SCB test would have represented a simple method to examine Mode III fracture. However, because of the Mode II contribution to failure and the variation in total G along the delamination front invalidating any data reduction technique, it should not be used. Other types of Mode III tests mentioned at the beginning of this paper also have testing effects that may make them unsuitable for Mode III testing. For the case of Mode III delamination, the simplest methods of testing have been attempted with limited success. Other test configurations may exist. However, if the laminate is loaded in pure Mode III, fiber bridging may occur. Thus, only one valid G_{IIIc} value, at the insert, will be obtained from the test. It was shown in this work that G_{IIIc} was larger than G_{IIc}. Hence, in the absence of a pure Mode III test method, a conservative approach to characterizing Mode III delamination, by simply equating G_{IIIc} to G_{IIc}, should be adopted.

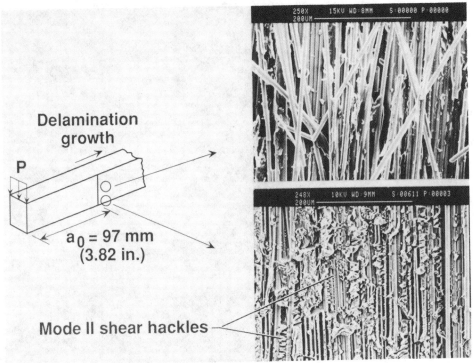

FIG. 18—*Micrographs of fracture surface.*

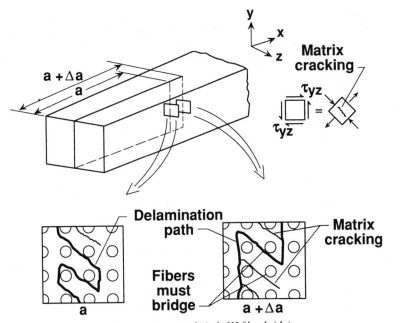

FIG. 19—*Schematic of Mode III fiber bridging.*

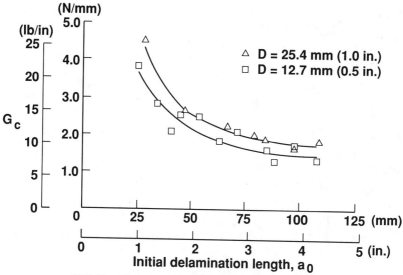

FIG. 20—*Variation of* G_c *with initial delamination length.*

Conclusions

This work investigated Mode III delamination of a glass/epoxy (S2/SP250) composite. A test fixture suitable for testing a thick, split composite beam was developed. A finite element analysis, was conducted on the test specimen to determine the strain energy release rate distribution along the delamination front. The following conclusions were obtained.

1. The finite element analysis showed that at the edge of the delamination front, G_{II} was significantly higher than G_{III} for all beam depths and delamination lengths considered.

2. The distribution of G_{II} and G_{III} along the delamination front was dependent on the delam-

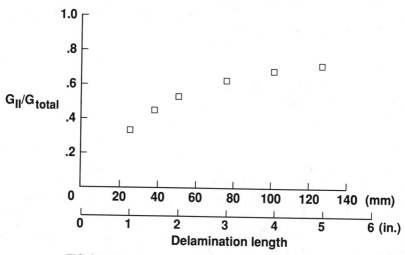

FIG. 21—*Variation of* G_{II}/G *ratio with delamination length.*

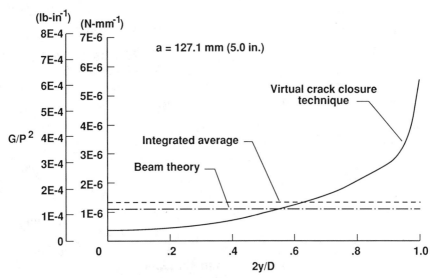

FIG. 22—*Total* G *distribution along delamination front.*

ination length. As the delamination length increased, the ratio of G_{II}/G_{III} along the delamination front increased.

3. The distribution of G_{II} and G_{III} along the delamination front was largely insensitive to beam depth.

4. Total G was nonuniform along the delamination front, thus invalidating any experimental data reduction technique.

5. The distribution of G_{II} and G_{III} along the delamination front was confirmed by examination of the failed surfaces of the test specimens. Where the delamination was Mode II, hackles were present and no fiber bridging was observed. Where the delamination was Mode III, fiber bridging was observed.

6. Plots of G_c as a function of initial delamination length indicated that G_{IIc} was less than G_{IIIc}. Therefore, from the work presented here a conservative approach to characterizing Mode III delamination is to equate G_{IIIc} to G_{IIc}.

Acknowledgments

This work was done while the author held National Research Council Research Associateship at NASA Langley Research Center.

References

[1] Whitney, J. M., Browning, C. E., and Hoogsteden, W., "A Double Cantilever Beam Test for Characterizing Mode I Delamination of Composite Materials," *Journal of Reinforced Plastics and Composite,* Vol. 1, Oct. 1982, pp. 297–313.

[2] Wilkins, D. J., Eisenmann, J. R., Camin, R. A., Margolis, W. S., and Benson, R. A., "Characterizing Delamination Growth in Graphite-Epoxy," *Damage in Composite Materisl, ASTM STP 775,* K. L. Reifsnider, Ed., American Society for Testing and Materials, Philadelphia, 1982, pp. 168–183.

[3] Russell, A. J., "Factors Affecting the Opening Mode Delamination of Graphite/Epoxy Laminates," Defence Research Establishment Pacific (DREP), Victoria, British Columbia, Canada, Materials Report 82-Q, Dec. 1982.

[4] Carlile, D. R. and Leach, D. C., "Damage and Notched Sensitivity of Graphite/PEEK Composites," *Proceedings, Fifteenth National SAMPE Technical Conference, Society for the Advancement of Material and Process Engineering,* Oct. 1983, pp. 82–93.

[5] Keary, P. E., Ilcewicz, L. B., Shaar, C., and Trostle, J., "Mode I Interlaminar Fracture Toughness of Composite Materials Using Slender Double Cantilever Beam Specimens," *Journal of Composite Materials,* Vol. 19, March 1985, pp. 154–177.

[6] Martin, R. H., "Effect of Initial Delamination on G_{Ic} and G_{Ith} Values from Glass/Epoxy Double Cantilever Beam Tests," *Proceedings, American Society for Composites, Third Technical Conference,* Seattle, WA, 25–29 Sept. 1988, pp. 688–700.

[7] Martin, R. H. and Murri, G. B., "Characterization of Mode I and Mode II Delamination Growth and Thresholds in AS4/PEEK Composites," *Composite Materials: Testing and Design, (Ninth Volume), ASTM STP 1059,* S. P. Garbo, Ed., American Society for Testing and Materials, Philadelphia, 1990, pp. 251–270.

[8] Russell, A. J. and Street, K. N., "Factors Affecting the Interlaminar Fracture Energy of Graphite/Epoxy Laminates," *Progress in Science and Engineering of Composites, Proceedings,* Fourth International Conference on Composite Materials (ICCM-IV), Tokyo, 1982, pp. 279–286.

[9] Russell, A. J., "On the Measurement of Mode II Interlaminar Fracture Energies," Defence Research Establishment Pacific (DREP), Victoria, British Columbia, Canada, Materials Report 82-0, Dec. 1982.

[10] Murri, G. B. and O'Brien, T. K., "Interlaminar G_{IIc} Evaluation of Toughened Resin Composites Using the End-notched Flexure Test," AIAA-85-0647, *Proceedings,* Twenty-Sixth AIAA/ASME/ASCE/AHS Conference on Structures, Structural Dynamics and Materials, Orlando, FL, April, 1985, pp. 197–202.

[11] Mall, S. and Kochhar, N. K., "Finite Element Analysis of End Notched Flexure Specimens," *Journal of Composites Technology and Research,* Vol. 8, No. 2, Summer 1986, pp. 54–57.

[12] Salpekar, S. A., Raju, I. S., and O'Brien, T. K., "Strain Energy Release Rate Analysis of the End-Notched Flexure Specimen Using the Finite-Element Method," *Journal of Composites Technology and Research,* Vol. 10, Winter 1988, pp. 133–139.

[13] Carlsson, L. A., Gillespie, J. W., and Pipes, R. B., "On the Analysis and Design of the End Notched Flexure (ENF) Specimen for Mode II Testing," *Journal of Composite Materials,* Vol. 20, Nov. 1986, pp. 594–604.

[14] Carlsson, L. A., Gillespie, J. W., and Tretheway, B. R., "Mode II Interlaminar Fracture of Graphite/Epoxy and Graphite/PEEK," *Journal of Reinforced Plastics and Composites,* Vol. 5, July 1986, pp. 170–187.

[15] O'Brien, T. K., Murri, G. B., and Salpekar, S. A., "Interlaminar Shear Fracture Toughness and Fatigue Thresholds for Composite Materials," *Composite Materials: Fatigue and Fracture, Second Volume, ASTM STP 1012,* P. Lagace, Ed., American Society for Testing and Materials, Philadelphia, 1989, pp. 222–250.

[16] O'Brien, T. K., "Mixed-Mode Strain-Energy-Release-Rate Effects on Edge Delamination of Composites," *Effects of Defects in Composite Materials, ASTM STP 836,* American Society for Testing and Materials, Philadelphia, 1984, pp. 125–142.

[17] Whitcomb, J. D., "Instability-Related Delamination Growth of Embedded and Edge Delaminations," Ph.D. thesis, Virginia Polytechnic Institute and State University, Blockburg, VA, May 1988.

[18] Salpekar, S. A., Raju, I. S., and O'Brien T. K., "Strain Energy Release Rate Analysis of Delamination in a Tapered Laminate Subjected to Tension Load," *Proceedings,* American Society for Composites, Third Technical Conference, Seattle, 25–29 Sept. 1988, pp. 642–654.

[19] O'Brien, T. K. and Raju, I. S., "Strain-Energy-Release Rate Analysis of Delamination Around an Open Hole in Composite Laminates," AIAA-84-0961, *Proceedings,* Twenty-Fifth AIAA/ASME/ASCE/AHS Conference on Structures, Palm Springs, CA, 17–18 May 1984, pp. 526–536.

[20] Chaouk, H., "Edge Delamination Behaviour in Advanced Composite Structures Under Compression Loading," Ph.D. thesis, The University of Sydney, Australia, Feb. 1988.

[21] Donaldson, S. L., "Mode III Interlaminar Fracture Characterization of Composite Materials," *Composites Science and Technology,* Vol. 32, No. 3, 1988, pp. 225–249.

[22] Donaldson, S. L. and Mall, S., "Delamination Growth in Graphite/Epoxy Composites Subjected to Mode III Loading," *Journal of Reinforced Plastics,* Vol. 8, Jan. 1989, pp. 91–103.

[23] Chai, H., "Shear Fracture," *International Journal of Fracture,* Vol. 37, 1988, pp. 137–159.

[24] Becht, G. and Gillespie, J. W., Jr., "Design and Analysis of the Crack Rail Shear Specimen for Mode III Interlaminar Fracture," *Composites Science and Technology,* Vol. 31, 1988, pp. 143–157.

[25] Gillespie, J. W., Jr. and Becht, G., "An Investigation of Interlaminar Fracture of Composite Materials Under Mode III Loading," presented at Composites '88, Boucherville, Quebec, Canada, Nov. 1988.

[26] Timoshenko, S. P., *"Strength of Materials, Part 1,"* 3rd ed., D. Van Nostrand Co., Inc., New York, 1955, p. 170.

[27] Whitney, J. M., Browning, C. E., and Mair, A., "Analysis of the Flexure Test for Laminated Composites," *Composite Materials: Testing and Design (Third Conference), ASTM STP 546,* American Society for Testing and Materials, Philadelphia, 1974, pp. 30–45.

[28] NASTRAN Users Manual, Version 65, Document No. MSR-39, MacNeal-Schwendler Corporation, Nov. 1985.

[29] Shivakumar, K. N., Tan, P. W., and Newman, J. C., Jr., "A Virtual Crack-Closure Technique for Calculating Stress Intensity Factors for Cracked Three Dimensional Bodies," *International Journal of Fracture,* Vol. 36, 1988, pp. R43–R50.

[30] Kanninen, M. F., "An Augmented Double Cantilever Beam Model for Studying Crack Propagation and Arrest," *International Journal of Fracture,* Vol. 9, No. 1, March 1973, pp. 83–91.

[31] Zweben, C., Smith, W. S., and Wardle, M. W., "Test Methods for Fiber Tensile Strength, Composite Flexural Modulus, and Properties of Fabric-Reinforced Laminates," *Composite Materials: Testing and Design (Fifth Conference), ASTM STP 674,* S. W. Tsai, Ed., American Society for Testing and Materials, Philadelphia, 1979, pp. 228–262.

Delamination Analysis

Erian A. Armanios,[1] *P. Sriram,*[1] *and Ashraf M. Badir*[1]

Fracture Analysis of Transverse Crack-Tip and Free-Edge Delamination in Laminated Composites

REFERENCE: Armanios, E. A., Sriram, P., and Badir, A. M., **"Fracture Analysis of Transverse Crack-Tip and Free-Edge Delamination in Laminated Composites,"** *Composite Materials: Fatigue and Fracture (Third Volume), ASTM STP 1110,* T. K. O'Brien, Ed., American Society for Testing and Materials, Philadelphia, 1991, pp. 269–286.

ABSTRACT: Delamination is a predominant failure mode in continuous fiber-reinforced laminated composite structures. Based on the location and direction of growth, there are two distinct types of delamination, namely, free edge delamination and local or transverse crack tip delamination. In many cases, both types occur concurrently with varying levels of interaction. In this paper, a shear deformation model including hygrothermal effects is developed for the analysis of local delaminations originating from transverse cracks in 90° plies located in and around the laminate midplane. A sublaminate approach is used and the model is applied to $(\pm 25/90_n)_s$ T300/934 graphite/epoxy laminates for n values between 0.5 and 8, along with previously developed edge delamination shear deformation models. Critical loads and delamination modes are identified and compared with experimental results. Hygrothermal effects are included in all the models to make the comparisons realistic.

KEY WORDS: graphite composites, graphite/epoxy, delamination, strain energy release rate, fracture mechanics, composite materials, fracture, fatigue (materials)

Nomenclature

a	Crack length
$[A]$	Axial (stretching) stiffness matrix
b	Specimen width
B_j	Coefficients of characteristic equation
$[B]$	Coupling stiffness matrix
c	Specific moisture content; as subscript indicates critical value
$[D]$	Bending stiffness matrix
G_T	Total strain energy release rate
G_j	Strain energy release rate contribution from Mode j
h	Sublaminate thickness
L	Length of modeled portion of specimen
M	Resultant bending moment
n	Number of 90° plies
N	Axial stress resultant
P	Interlaminar normal (or peel) stress
Q	Shear stress resultant

[1] Assistant professor, post doctoral fellow, and graduate research assistant, respectively, School of Aerospace Engineering, Georgia Institute of Technology, Atlanta, GA 30332.

$[\overline{Q}]$ Plane stress sublaminate reduced stiffness matrix
s Eigenvalue
T Test temperature; with subscript, interlaminar shear stress
T_{ref} Stress free temperature
u, U Axial displacement
w, W Transverse (z) displacement
W_e Work done during crack growth
x Axial coordinate
z Transverse (thickness) coordinate
α_{Tj} Coefficient of linear thermal expansion for Sublaminate j
α_{Hj} Moisture swelling coefficient for Sublaminate j
β Shear deformation (rotation)
δ Virtual crack step size
ε Nominal axial strain
ε_z Transverse normal strain
η Mixed-mode fracture law parameter
σ Axial stress
ξ Mixed-mode ratio G_I/G_T

In addition, the following constants have been used for convenience; their definitions are provided in the derivations: a_j, I_j, k_j, χ, φ_j, θ_j, ω_j.

Fiber-reinforced composites are now being used in a wide variety of engineering structures. The concept of directional strength and stiffness has been, for the most part, understood sufficiently to enable efficient load bearing designs. One of the current major issues in composite structures is the understanding and prediction of damage modes and failure mechanisms. A thorough knowledge of the failure mechanisms is bound to lead to the design of efficient and durable structures.

Failures in laminated composite materials often initiate in the form of matrix fractures, namely, transverse matrix cracks and delaminations. Transverse matrix cracks refer to intralaminar failures whereas delaminations refer to interlaminar failures. Transverse cracks usually occur within laminates where the fibers run at an angle to the primary load direction and hence the name. Based on the location and direction of growth, two distinct types of delamination can be discerned. These two types are called edge delamination and local or transverse crack tip delamination. Edge delaminations initiate at the load-free edges of the laminate whereas local delaminations start from a transverse matrix crack. In many cases, both types occur concurrently with varying levels of interaction.

It has been observed in simple tension tests of uniform rectangular cross-section specimen (edge delamination tests) that delaminations initiate along the load-free edges and propagate normal to the load direction. Transverse matrix cracks running parallel to the fibers have been also observed in off-axis and 90° plies. Such transverse cracks extend through the thickness of similarly oriented plies and terminate where the ply orientation changes. Delaminations can also originate at the interfaces where transverse cracks terminate. These transverse crack tip delaminations or local delaminations, grow normal to the transverse crack from which they originate. In the case of 90° plies, the growth direction is parallel to the load.

The growth process of edge delaminations and local delaminations is often modeled using a fracture mechanics approach leading to the calculation of a strain energy release rate. This is because the strain energy release rate can correlate delamination behavior from different loading conditions and can account for geometric dependencies. The strain energy release rate associated with a particular growth configuration is a measure of the driving force behind that failure mode. In combination with appropriate failure criteria, the strain energy release rate provides a means of predicting the failure loads of the structure.

Several methods are available in the literature for analyzing edge delaminations. These include finite element modeling [1-3], complex variable stress potential approach [4], simple classical laminate theory based techniques [5], and higher order laminate theory including shear deformations [6]. Finite element models provide accurate solutions but involve intensive computational effort. Classical laminate theory (CLT) provides simple closed-form solutions and is thus well suited for preliminary design evaluation. However, CLT provides only the total energy release rate, and thus, in a mixed-mode situation, there is insufficient information to completely assess the delamination growth tendency. A higher order laminate theory including shear deformations has the ability to provide the individual contributions of the three fracture modes while retaining the simplicity of a closed-form solution. A shear deformation model is available for off-mid-plane edge delamination and has been shown to agree well with finite element predictions [7].

Crossman and Wang [8] have tested T300/934 graphite/epoxy $[\pm 25/90_n]_s$ specimens in simple tension and reported a range of behavior including transverse cracking, edge delamination, and local delamination. O'Brien [9] has presented classical laminate theory solutions for these specimens, demonstrating reasonable agreement in the case of edge delamination but with some discrepancies in the local delamination predictions. The local delamination model over-estimates the failure strains for $[\pm 25/90_n]_s$ specimens for small n values mainly due to the implicit critical strain energy matching used. A finite element model combining edge and local delaminations has been proposed by Law [10]. His predictions, however, do not fully explain the dependency of the critical strain on the number of 90° plies. A similar three-dimensional finite element analysis including hygrothermal effects has been performed by Wang et al. [11] to determine the delamination onset load for combined delamination, qualitatively demonstrating stable crack growth.

Thermal and moisture effects on the strain energy release rates for interlaminar fracture of unidirectional graphite/epoxy have been investigated by Russell and Street [12]. This investigation also included a study of the effects of shear loading through the use of various test configurations (double cantilever beam, cracked lap shear, etc.). Initiation energies for delamination were found to increase as the proportion of shear loading increased and as the temperature was lowered, but no significant moisture influence was observed. The fracture resistance to crack extension was found to increase under tensile dominated loadings with both temperature and moisture content, but for high shear loading, the resistance was insensitive to the hygrothermal conditions.

O'Brien, Raju, and Garber have presented a CLT-based analysis of mixed-mode edge delamination specimens including hygrothermal effects [13]. They have used finite element modeling to determine the strain energy release rate components. Their results indicate total strain energy release rate increases of as much as 170% due to thermal effects for some T300/ 5208 graphite/epoxy laminates. However, a moisture content of 0.75% has been shown to totally alleviate this increase. According to this analysis, in general, the consideration of thermal effects increases the energy release rate whereas moisture effects have the opposite influence. Further, at high moisture levels, the bending and coupling effects have significant influence on the strain energy. These results have been confirmed using shear deformation models in the case of edge delaminations [14,15]. It was found that the interlaminar stresses follow the same trend as the energy release rate, with increase due to thermal effects and alleviation due to hygroscopic effects.

A higher order plate theory including transverse normal strain and thermal effects has been developed by Whitney [16] for the analysis of midplane edge delaminations. This approach provides the interlaminar stresses also, in addition to the strain energy release rate. A $[0_3/90_3]_s$ graphite/epoxy Mode I specimen was analyzed and the maximum interlaminar normal stress was shown to increase by a factor of 2.7 due to thermal effects, when compared with the pure mechanical strain reference configuration.

In this paper, a shear deformation model is developed for the analysis of local delaminations originating from transverse cracks in 90° plies located in and around the specimen midplane. The sublaminate approach [6] is used to model different regions of the specimen. Such an approach can be used with confidence when the characteristic length of the response is large compared to the individual sublaminate thickness. Comparisons with ply-by-ply finite element models [7] have established the validity of the modeling approach. The closed-form solutions provided by such sublaminate analyses make it simple to identify the parameters that control the response, to understand the behavior, and to select ply stacking sequences that improve the laminate resistance to crack propagation. Plane strain conditions are assumed and thickness strain is neglected. Delaminations are assumed to grow from both ends of the transverse crack tip. Residual hygrothermal stresses are included in the analysis through equivalent applied loads. These equivalent loads are calculated using the laminate hygrothermal and mechanical properties. Inclusion of the residual stresses is essential for comparisons with experimental data since specimen tested under room temperature ambient conditions generally exhibit residual stress effects. The transverse crack is treated as a free boundary and the delamination is considered to be the crack whose growth behavior is to be modeled. The resulting boundary value problem is solved to obtain the interlaminar stresses, total strain energy release rate, and energy release rate components. Critical delamination growth loads are predicted for $[\pm 25/90_n]_s$ T300/934 graphite/epoxy specimens by assuming a simple fracture criterion.

The results are combined with previously developed shear deformation-based edge delamination sublaminate models [14,15] to yield a unified analysis of delaminations and the ability to identify the critical failure mode and loads. Hygrothermal stresses are included in the edge delamination models also. The following sections present details of the local delamination model and a discussion of the results obtained.

Analytical Modeling of Local Delamination

A longitudinal section illustrating the geometry of a generic configuration is shown in Fig. 1. The central region is assumed to be made of 90° plies with an isolated transverse crack in the middle. Delaminations are assumed to grow from both ends of the transverse crack, and towards both specimen ends as shown. From symmetry considerations, only one quarter of the configuration is modeled. The modeled portion of Length L is divided into four sublaminates as shown in Fig. 2. The crack length is denoted by a. The top surface (Sublaminates 1 and 4) is stress free. In order to simplify the analysis, the thickness strain (ε_z) is neglected. The consequence of this, combined with the fact that the transverse displacement (w) is zero along the center line, is that w is zero in Sublaminates 1, 2, and 3. Also, this approximation does not allow for the enforcement of boundary conditions on the shear stress resultants, leading to incorrect estimates of the interlaminar normal stresses. The interlaminar shear stress estimates, however, are reliable [6]. These assumptions lead to considerable simplifications in the analysis. In spite of the simplifications, reliable energy release rate components can be estimated based on the interlaminar shear stress distributions [7].

A generic sublaminate is shown in Fig. 3 along with the notations and sign conventions. The peel and interlaminar shear stresses are denoted by P and T, respectively, with t and b subscripts for the top and bottom surfaces, respectively. The axial stress resultant, shear stress resultant, and bending moment resultant are denoted by N, Q, and M, respectively. A sum-

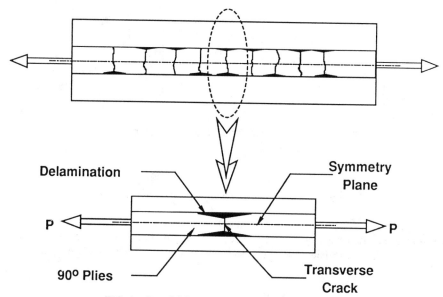

FIG. 1—*Local delamination specimen cross section.*

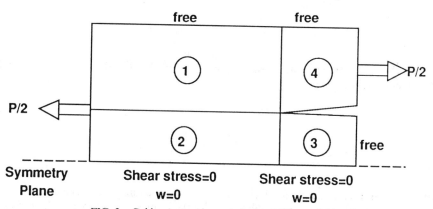

FIG. 2—*Sublaminate schematic for local delamination.*

mary of the governing equations is presented in the following paragraphs for convenience. These equations are derived for a generic sublaminate using the principle of virtual work in Ref 6.

The x and z displacements within the sublaminate are assumed to be of the form

$$u(x,z) = U(x) + z\beta(x) \tag{1}$$

$$w(x,z) = W(x) \tag{2}$$

Here, U represents the axial midplane stretching and W is the transverse displacement. The shear deformation is recognized through the rotation, β. These displacements are the total

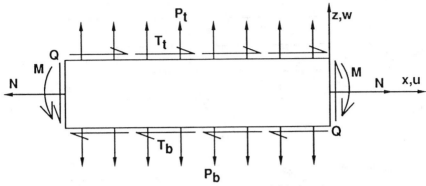

FIG. 3—*Generic sublaminate for local delamination.*

quantities and include the hygrothermal effects. The origin of the coordinate axes for the sublaminates is taken at the delamination tip as shown in Fig. 4. The equilibrium equations take the form

$$N_{,x} + T_t - T_b = 0 \tag{3}$$

$$Q_{,x} + P_t - P_b = 0 \tag{4}$$

$$M_{,x} - Q + \frac{h}{2}(T_t + T_b) = 0 \tag{5}$$

where h is the thickness of the sublaminate. The constitutive relationships in terms of the force and moment resultants are

$$N = A_{11}U_{,x} + B_{11}\beta_{,x} \tag{6}$$

$$Q = A_{55}(\beta + W_{,x}) \tag{7}$$

$$M = B_{11}U_{,x} + D_{11}\beta_{,x} \tag{8}$$

where A_{ij}, B_{ij}, and D_{ij} are the classical laminate theory axial, coupling, and bending stiffnesses, respectively [17]. For a sublaminate thickness of h, the stiffness coefficients are defined as

$$(A_{ij}, B_{ij}, D_{ij}) = \int_{-h/2}^{h/2} \overline{Q}_{ij}(1, z, z^2) dz \tag{9}$$

where the $\overline{Q}_{ij}$ are the transformed plane stress sublaminate reduced stiffnesses. The boundary variables to be prescribed at the sublaminate edges are

N or U
M or β
Q or W

Additionally, at the interfaces between sublaminates, reciprocal traction, and displacement matching boundary conditions have to be specified. The stress resultants in these equations

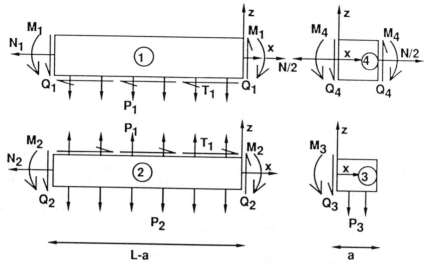

FIG. 4—*Sublaminate forces and coordinate systems.*

include the equivalent hygrothermal loads also. It is convenient to introduce the hygrothermal effects through equivalent axial stress and bending moment resultants defined as follows.

$$N_i^* = \int_{-h/2}^{h/2} \overline{Q}_{ij}[\alpha_{Tj}(T - T_{ref}) + \alpha_{Hj}c] \, dz \tag{10}$$

$$M_i^* = \int_{-h/2}^{h/2} z \, \overline{Q}_{ij}[\alpha_{Tj}(T - T_{ref}) + \alpha_{Hj}c] \, dz \tag{11}$$

Here, T is the test temperature, T_{ref} is the stress free temperature, c is the specific moisture content, and α_{Tj} and α_{Hj} are the transformed coefficients of linear thermal expansion and moisture swelling, respectively. The $\overline{Q}$s are the appropriate transformed stiffnesses. In the constitutive relationships (Eqs 6 through 8), the N and M include the mechanical and hygrothermal contributions. The simpler case where hygrothermal effects are neglected has been presented previously [18].

The solutions in Sublaminates 1 and 2 are coupled by the reciprocal interlaminar stresses denoted T_1 and P_1 and by displacement continuity at the common interface. Assuming exponential solutions for the axial force and bending moment resultants leads to an eigenvalue problem involving the exponential parameters. The characteristic equation is of the form

$$s[B_1 s^4 + B_2 s^2 + B_3] = 0 \tag{12}$$

where s is the eigenvalue parameter, and the B coefficients are given by

$$B_1 = \left(\frac{1}{A_{11(2)}} + \frac{1}{A_{11(1)}} + \frac{h_1^2}{4D_{11(1)}} + \frac{h_2^2}{4D_{11(2)}} \right) \frac{D_{11(1)}}{A_{55(1)}} \frac{D_{11(2)}}{A_{55(2)}} \tag{13}$$

$$B_2 = - \frac{D_{11(2)}}{A_{55(2)}} \left(\frac{1}{A_{11(1)}} + \frac{1}{A_{11(2)}} + \frac{h_2^2}{4D_{11(2)}} \right) - \frac{D_{11(1)}}{A_{55(1)}} \left(\frac{1}{A_{11(1)}} + \frac{1}{A_{11(2)}} + \frac{h_1^2}{4D_{11(1)}} \right) \tag{14}$$

and

$$B_3 = \frac{1}{A_{11(1)}} + \frac{1}{A_{11(2)}} \tag{15}$$

The eigenvalues turn out to be 0 and two nonzero values given by

$$s = \pm \left(\frac{- B_2 \pm (B_2^2 - 4B_1B_3)^{1/2}}{2B_1} \right)^{1/2} \tag{16}$$

For the problem under consideration, all the square roots in this expression lead to real quantities and thus the eigenvalues are real. Since the eigenvalues involve only the stiffness parameters, they are not affected by the inclusion of hygrothermal effects. Further, due to the fact that B_1 has D terms in the numerator, it is much smaller than B_3. This leads to the boundary layer nature of the solution. Since the response (axial forces, moments) have finite values at large distances from the origin, namely, at the ends of the specimen, only the exponentially decaying and constant solutions are used.

Using subscripts to denote the sublaminate of validity, the following boundary conditions from the ends of the modeled region are enforced

$$N_2(0) = 0 \tag{17}$$

$$Q_4(a) = 0 \tag{18}$$

$$\beta_4(a) = 0 \tag{19}$$

$$N_1 + N_2 = \text{applied load} \tag{20}$$

The conditions on N apply only to the mechanical quantities. Further, the following displacement matching conditions are applied.

$$u_1 \left(x, -\frac{h_1}{2} \right) = u_2 \left(x, \frac{h_2}{2} \right) \tag{21}$$

$$U_1(0) = U_4(0) \tag{22}$$

$$U_2(0) = U_3(0) \tag{23}$$

$$\beta_1(0) = \beta_4(0) \tag{24}$$

It should be noted that a β_2 and β_3 matching condition cannot be applied at this level of modeling since it would amount to specifying both W and Q. To eliminate rigid body displacements, U_1 is set to zero at the left end. The following solutions can then be obtained for the stress resultants in Sublaminates 1 and 2

$$N_1 = a_1 e^{s_1 x} + a_2 e^{s_2 x} + \varepsilon A_{11(1)} - N_1^* \tag{25}$$

$$N_2 = - a_1 e^{s_1 x} - a_2 e^{s_2 x} + \varepsilon A_{11(2)} - N_2^* \tag{26}$$

$$M_1 = a_1 k_1 e^{s_1 x} + a_2 k_2 e^{s_2 x} - M_1^* \tag{27}$$

$$M_2 = a_1 k_3 e^{s_1 x} + a_2 k_4 e^{s_2 x} - M_2^* \tag{28}$$

Here k_1 is defined as

$$k_1 = \frac{\dfrac{h_1}{2} s_1^2}{\dfrac{A_{55(1)}}{D_{11(1)}} - s_1^2} \qquad (29)$$

The parameter k_2, is defined in a similar manner using the eigenvalue, s_2. The remaining parameters, k_3 and k_4, are similar to k_1 and k_2 but based on Sublaminate 2 properties. The nominal strain, ε, is defined as

$$\varepsilon = \left(\frac{P}{2b} + N_1^* + N_2^* \right) \frac{1}{A_{11(1)} + A_{11(2)}} \qquad (30)$$

where P is the applied uniform axial force and b is the specimen width. The a_is can be derived from boundary conditions as follows

$$a_1 = \frac{\theta_3 + \theta_4 a}{\theta_d} \frac{1}{A_{11(1)} + A_{11(2)}} \left(\frac{P}{2b} A_{11(2)} + N_1^* A_{11(2)} - N_2^* A_{11(1)} \right) \qquad (31)$$

$$a_2 = -\frac{\theta_1 + \theta_2 a}{\theta_d} \frac{1}{A_{11(1)} + A_{11(2)}} \left(\frac{P}{2b} A_{11(2)} + N_1^* A_{11(2)} - N_2^* A_{11(1)} \right) \qquad (32)$$

with

$$\theta_1 = \frac{s_1}{A_{55(1)}} \left(k_1 + \frac{h_1}{2} \right) \qquad (33)$$

$$\theta_2 = \frac{k_1}{D_{11(1)}} \qquad (34)$$

$$\theta_3 = \frac{s_2}{A_{55(1)}} \left(k_2 + \frac{h_1}{2} \right) \qquad (35)$$

$$\theta_4 = \frac{k_2}{D_{11(1)}} \qquad (36)$$

and

$$\theta_d = \theta_3 - \theta_1 + (\theta_4 - \theta_2)a \qquad (37)$$

The interlaminar shear and peel stresses between Sublaminates 1 and 2 can be obtained using the equilibrium equations (Eqs 3 through 5) as

$$T_1 = a_1 s_1 e^{s_1 x} + a_2 s_2 e^{s_2 x} \qquad (38)$$

$$P_1 = \left(k_1 + \frac{h_1}{2} \right) (a_1 s_1 2 e^{s_1 x}) + \left(k_2 + \frac{h_1}{2} \right) (a_2 s_2^2 e^{s_2 x}) \qquad (39)$$

As mentioned previously, this peel stress estimate is not accurate because of the inability to apply boundary conditions on shear. Recognizing the fact that there are no applied shear forces, it can be concluded that the peel stress distribution should be self-equilibrating. This assumption can be satisfied by including additional exponential terms in the preceding peel stress expression and determining these additional terms by setting the net force and moment due to the peel stress to zero [15]. The peel stress estimated through this correction process is referred to as the modified peel stress.

Proceeding on to Sublaminates 3 and 4, the following solutions can be written

$$N_3 = 0 \tag{40}$$

$$M_3 = \varphi_1 \sinh(\omega_3 x) + \varphi_2 \cosh(\omega_3 x) \tag{41}$$

where

$$\varphi_2 = a_1 k_3 + a_2 k_4 \tag{42}$$

$$\varphi_1 = -\varphi_2 \coth(\omega_3 a) \tag{43}$$

and

$$\omega_3^2 = \frac{A_{55(2)}}{D_{11(2)}} \tag{44}$$

$$N_4 = \frac{P}{2b} \tag{45}$$

$$M_4 = a_1 k_1 + a_2 k_2 \tag{46}$$

The total energy release rate, G_T, is calculated using $G_T = dW_e/da$, where W_e is the work done per unit width by the external (constant) loads on the specimen displacements. For the case where hygrothermal effects are included, there are additional terms due to the work done by the N^*s. In reality, these N^* quantities are not applied loads but correspond to residual stresses. Thus, the additional terms are due to the work done by the applied mechanical strains on these residual stresses. The total energy release including hygrothermal effects is given by

$$G_T = \frac{P}{2b}\left(\frac{P}{b} + N_1^*\right)\left(\frac{1}{A_{11(1)}} - \frac{1}{A_{11(1)} + A_{11(2)}} + I_1 - I_2\right)$$
$$+ \frac{P}{2b} N_2^*\left(-\frac{1}{A_{11(1)} + A_{11(2)}} - I_3 + I_2\right) \tag{47}$$

where the I factors are

$$I_1 = \chi \frac{\theta_2\theta_3 - \theta_1\theta_4}{\theta_d^2}\left(\frac{1 - e^{-s_1(L-a)}}{s_1} - \frac{1 - e^{-s_2(L-a)}}{s_2}\right) \tag{48}$$

$$I_2 = \chi \frac{(\theta_3 + \theta_4 a)e^{-s_1(L-a)} - (\theta_1 + \theta_2 a)e^{-s_2(L-a)}}{\theta_d} \tag{49}$$

with

$$\chi = \frac{1}{A_{11(1)} + A_{11(2)}} \frac{A_{11(2)}}{A_{11(1)}} \tag{50}$$

Parameter I_3, is the same as I_1 but with the ratio $A_{11(1)}/A_{11(2)}$ instead of unity in Eq 48.

Using the virtual crack closure technique, from the relative displacements in the cracked portion and the interlaminar stresses ahead of the crack tip, the Mode I and Mode II energy release rate contributions can be obtained. The Mode III energy release rate is zero from the assumption of plane strain. The Mode II energy release rate is given by

$$G_{II} = \lim_{\delta \to 0} \frac{1}{2\delta} \int_0^\delta T_1(x - \delta)\Delta u(x)dx \tag{51}$$

where δ is the virtual crack step size and Δu is the differential axial displacement across the crack surface. This calculation can be simplified using only the linear part of the differential displacement [7]. In a similar fashion, the Mode I energy release rate can be obtained based on the normal stress (P) and the differential w displacements near the crack front. Since the unmodified peel stress estimate is inaccurate, an alternate approach was used to estimate G_I, the Mode I energy release rate. The total energy release rate for this problem is made up entirely of G_I and G_{II} ($G_{III} = 0$). From an estimate of G_T and G_{II}, an estimate for G_I can be obtained simply as

$$G_I = G_T - G_{II} \tag{52}$$

The critical load for a given specimen can be then evaluated based on an appropriate fracture law. This is illustrated in the results and discussion section.

Analytical Modeling of Edge Delamination

The midplane free-edge delamination configuration is shown in Fig. 5. A uniform axial strain, ε, is applied in the x direction. From symmetry, only one quarter of the cross section of the specimen is considered. The sublaminate scheme and the choice of coordinate axes are illustrated in the cross-sectional view in Fig. 6. Sublaminate 1 represents the uncracked region

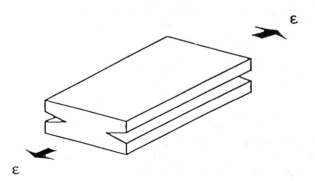

FIG. 5—*Midplane edge delamination.*

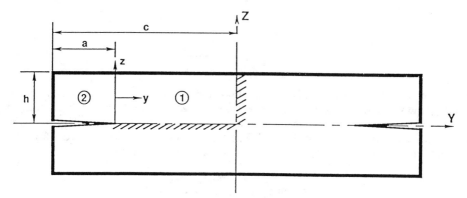

FIG. 6—*Sublaminate schematic for midplane edge delamination.*

while Sublaminate 2 extends only over the cracked (delaminated) region. It can be observed that the sublaminates required for this analysis are plate-like, whereas, in the case of the local delamination analysis, beam-type sublaminates were used. Details of the modeling, solution procedure, and results in the case of $(\theta/-\theta_2/\theta/90_n)$ layup are presented in Ref 15. A similar analysis for off-midplane edge delamination has been presented in Ref 14. Some pertinent edge delamination results in the case of the laminates under consideration will be reproduced here for convenience.

Results and Discussion

The delamination models have been used to study the behavior of $[\pm25/90_n]_s$ T300/934 graphite epoxy specimens for n values of 0.5, 1, 2, 3, 4, 6, and 8. These correspond to the specimens tested by Crossman and Wang [8]. The specimen width and length were fixed at 0.025 and 0.15 m, respectively, as in the tests. In computing the nonmechanical strains and curvatures, the laminate is assumed to be held at the prescribed temperature and moisture levels. In predicting critical strains, the difference between test and stress-free temperatures is assumed to be 155°C and the specimen is assumed to be dry. It is assumed that local delamination occurs under fixed load conditions whereas edge delamination occurs under fixed grip conditions. This difference is a consequence of the modeling approaches used in the analyses. The applied uniform load was 100 MPa axial stress for the local delamination analysis and 0.5% strain for the edge delamination analysis. The solutions were generated using simple computer programs based on the closed-form expressions for the interlaminar stresses and energy release rates.

Local Delamination

An example of the total local delamination energy release rate variation (neglecting hygrothermal effects) with the crack length is presented in Fig. 7. The asymptotic value of G_T is denoted by G_{T0} in the figure. It can be observed that after a certain crack length, the G_T is independent of the crack length. On the basis of curves like the one shown in Fig. 7, the crack length was fixed at 10-ply thicknesses for the remainder of the studies. Typical interlaminar shear stress profiles including the hygrothermal effect are presented in Fig. 8. The corresponding total strain energy release rates appear in Fig. 9. The inclusion of thermal effects increases the stress and the energy release rate while the inclusion of moisture effects has the opposite effect. In

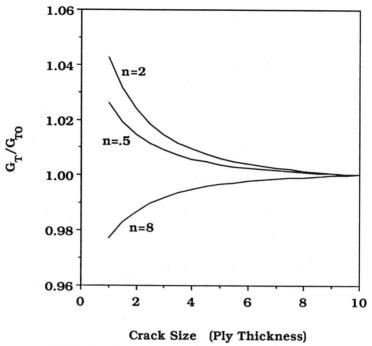

FIG. 7—*Total local delamination energy release rate variation.*

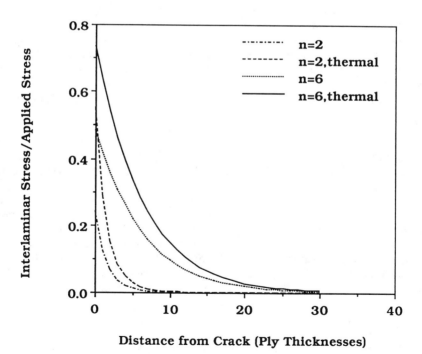

FIG. 8—*Interlaminar shear stress distribution (local delamination).*

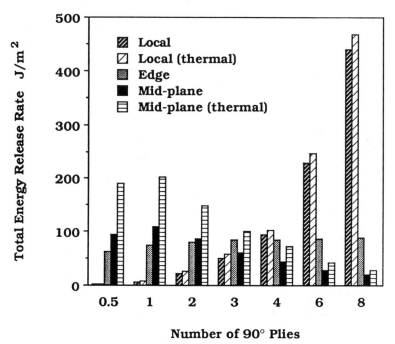

FIG. 9—*Total energy release rate for ($\pm 25/90_n)_s$ graphite epoxy specimen.*

fact a moisture level of about 0.75% almost exactly negates the thermal effects. In Ref *18*, it was reported that after some initial dependence on crack length, the mode mix tended to stabilize to a constant value. Using the model developed here, the asymptotic Mode II component of the local delamination energy release rate was found to be approximately 30% for all n values. In the case of off-midplane edge delamination, the Mode II contribution was less than 10% for the $n = 0.5$ specimen and progressively less for the thicker specimen.

Edge Delamination

As in the case of local delaminations, the interlaminar stress increases with thermal effects and the addition of moisture alleviates this as shown in Fig. 10 for the case of midplane edge delamination. A moisture level of about 0.75% produces a modified peel stress distribution that is indistinguishable from the case of mechanical loading alone. Moreover, the distance at which the modified peel stress reverses its sign is not affected by the residual hygrothermal strains. The hygrothermal influence on midplane delamination strain energy release rate is illustrated in Fig. 11 where the strain energy release rate is plotted versus moisture content for a ($\pm 25/90_2)_s$ laminate. The strain energy release rate follows the trend of increasing with residual thermal stress as in the case of peel stress. Further, residual moisture alleviates the thermal effects, and a moisture level of about 0.75% results in a total alleviation of thermal effects. Similar behavior is observed in the case of off-midplane edge delamination.

Failure Loads and Modes

In order to evaluate the critical loads for local delamination, an appropriate mixed-mode fracture law has to be applied, based on the calculated energy release components. The follow-

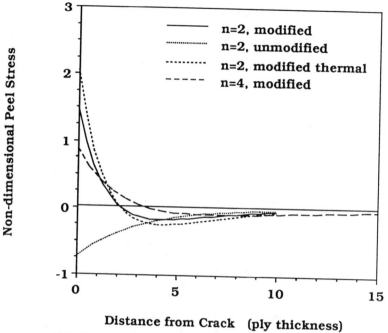

FIG. 10—*Interlaminar normal stress (midplane delamination).*

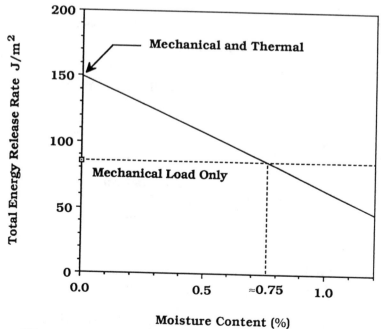

FIG. 11—*Total energy release rate (midplane delamination) for $(\pm 25/90_2)_s$.*

ing mixed-mode criterion [19] has been fitted to the test data of Ref 20 to calculate the mixed-mode G_{Tc} which is then used in the Griffith criterion, $G_T = G_{Tc}$, to obtain the critical delamination growth stress (σ_c) and strain (ε_c) values

$$G_{Tc} = \xi^\eta G_{Ic} + (1 - \xi)^\eta G_{IIc} \qquad (53)$$

Here ξ is the Mode I fraction (G_I/G_T) and G_{Ic} and G_{IIc} are the critical strain energy release rates for the limiting cases of pure Mode I and pure Mode II, respectively. The exponential parameter, η, is a material constant and for the T300/934 system, its value is approximately 0.9. In the case of midplane delamination, since only Mode I is present, G_{Tc} was taken as G_{Ic} (125 J/m^2). Based on the mixed-mode criterion, G_{Tc} was about 400 J/m^2 for the local delamination case ($\xi = 0.7$). The failure loads for edge delamination at the 25/90 interface have been also calculated using the model in Ref 14 and a G_{Tc} value of 150 J/m^2. This G_{Tc} value is different from the value used for midplane delamination due to the limited (less than 10%) presence of Mode II.

In order to consider a worst-case situation, thermal stresses were included and the moisture level was set at zero. Though the thermal stresses had a significant effect on the calculated peak stresses, the effect on the energy release rate was not significant except in the case of midplane edge delamination for the material system and layup considered.

The critical strains are plotted against n, the number of 90° plies in Fig. 12. The experimental results of Ref 8 are also presented in the figure for comparison. The results of the model developed in this paper are represented by the solid and dotted lines while the experimental results are shown as filled squares. The CLT-based model of Ref 9 agrees well with the shear deformation model in terms of the total energy release rate. However, the CLT-based model does

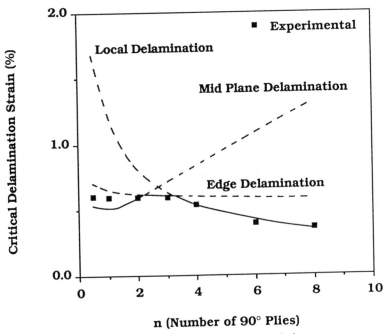

FIG. 12—*Critical delamination strain variation.*

not provide information on the mode split and thus, the value of $G_c(\approx G_{Ic})$ used can lead to bias in the critical strain estimates.

In the experiments, the local delamination phenomenon was observed as the predominant failure mode only for the $n = 4, 6,$ and 8 specimens. The shear deformation model presented in this paper provides good agreement with the experimental data in this range. For $n < 4$, edge delamination either in the midplane or in the 25/90 interface was observed in the tests, in agreement with the edge delamination models. Further, the relative closeness of the calculated critical strains from the midplane and off-midplane edge delamination models implies that, in practice, one could have interaction between these two modes. In such cases, one can expect the delamination to wander around the midplane and the 25/90 interfaces. This is especially so in the case of $n = 0.5$ where midplane delamination is not actually between two distinct layers but in the middle of a single layer. Experimental observations [8] are in agreement with this expectation. Thus, it can be seen that the shear deformation models reproduce the observed behavior with reasonable accuracy and can be used to estimate critical loads for range specimen thicknesses incorporating various delamination modes.

Conclusions

Shear deformation models including hygrothermal effects have been developed to analyze local delaminations growing from transverse cracks in 90° plies and edge delaminations located around the midplane of symmetric laminates. The models have been combined into a unified delamination analysis code in order to predict damage modes and loads in laminated composites. The analytical results of the shear deformation models agree reasonably with critical strain experimental data from $[\pm 25/90_n]_s$ T300/934 graphite epoxy laminates in the range of n from 0.5 to 8. Residual thermal and moisture stresses are found to have only minor effects on the critical strains except in the case of midplane edge delamination for the geometry and material considered. The same failure modes as in the tests are reproduced in the analysis. The integrated delamination code is expected to be of use in design evaluation applications.

Acknowledgments

The authors gratefully acknowledge the financial support provided by NASA under Grant NAG-1-637 and the U.S. Army Research Office under Grant DAAL03-88-C-0003 for performing the research reported in this paper.

References

[1] Wilkins, D. J., Eisemann, J. R., Camin, R. A., Margolis, W. S., and Benson, R. A., "Characterizing Delamination Growth in Graphite-Epoxy," *Damage in Composite Materials, ASTM STP 775*, K. L. Reifsnider, Ed., American Society for Testing and Materials, Philadelphia, 1982, pp. 168–183.
[2] O'Brien, T. K., "Mixed-Mode Strain Energy Release Rate Effects on Edge Delamination of Composites," *Effects of Defects in Composite Materials, ASTM STP 836*, American Society for Testing and Materials, Philadelphia, 1984, pp. 125–142.
[3] Wang, S. S. and Choi, I., "The Mechanics of Delamination in Fiber Reinforced Composite Materials, Part II—Delamination Behavior and Fracture Mechanics Parameters," NASA CR-172270, National Aeronautics and Space Administration, Washington, DC, 1983.
[4] Wang, S. S., "Edge Delamination in Angle Ply Composite Laminates," *Proceedings*, Twenty-Second AIAA/ASME/ASCE/AHS/ASC Structures, Structural Dynamics and Materials Conference, Atlanta, 6–8 April 1981, pp. 473–484.
[5] O'Brien, T. K., "Characterization of Delamination Onset and Growth in a Composite Laminate," *Damage in Composite Materials, ASTM STP 775*, K. L. Reifsnider, Ed., American Society for Testing and Materials, Philadelphia, 1982, pp. 140–167.

[6] Armanios, E. A. and Rehfield, L. W., "Sublaminate Analysis of Interlaminar Fracture in Composites: Part I—Analytical Model," *Journal of Composites Technology and Research*, Vol. 11, No. 4, 1989, pp. 135–146.

[7] Armanios, E. A., Rehfield, L. W., Raju, I. S., and O'Brien, T. K., "Sublaminate Analysis of Interlaminar Fracture in Composites: Part II—Applications," *Journal of Composites Technology and Research*, Vol. 11, No. 4, 1989, pp. 147–153.

[8] Crossman, F. W. and Wang, A. S. D., "The Dependence of Transverse Cracking and Delamination on Ply Thickness in Graphite/Epoxy Laminates," *Damage in Composite Materials, ASTM STP 775*, K. L. Reifsnider, Ed., American Society for Testing and Materials, Philadelphia, 1982, pp. 118–139.

[9] O'Brien, T. K., "Analysis of Local Delaminations and their Influence on Composite Laminate Behavior," *Delamination and Debonding of Materials, ASTM STP 876*, W. S. Johnson, Ed., American Society for Testing and Materials, Philadelphia, 1985, pp. 282–297.

[10] Law, G. E., "A Mixed Mode Fracture Analysis of ($\pm 25/90_n$)$_s$ Graphite/Epoxy Composite Laminates," *Effects of Defects in Composite Materials, ASTM STP 836*, American Society for Testing and Materials, Philadelphia, 1984, pp. 143–160.

[11] Wang, A. S. D., Kishore, N. N., and Li, C. A., "Crack Development in Graphite Epoxy Cross Ply Laminates under Uniaxial Tension," *Composites Science and Technology*, Vol. 24, No. 1, 1985, pp. 1–31.

[12] Russell, A. J. and Street, K. N., "Moisture and Temperature Effects on the Mixed Mode Delamination Fracture of Unidirectional Graphite/Epoxy," *Delamination and Debonding of Materials, ASTM STP 876*, W. S. Johnson, Ed., American Society for Testing and Materials, Philadelphia, 1985, pp. 349–370.

[13] O'Brien, T. K., Raju, I. S., and Garber, D. P., "Residual Thermal and Moisture Influences on the Strain Energy Release Rate Analysis of Edge Delamination," *Journal of Composites Technology and Research*, Vol. 8, No. 2, 1986, pp. 37–47.

[14] Armanios, E. A. and Mahler, M. A., "Residual Thermal and Moisture Influences on the Free-Edge Delamination of Laminated Composites," *Proceedings*, Twenty Ninth AIAA/ASME/ASCE/AHS/ASC Structures, Structural Dynamics and Materials Conference, Williamsburg, AIAA Paper 88-2259, 18–20 April 1988, pp. 371–381.

[15] Armanios, E. A. and Badir, A., "Hygrothermal Influence on Mode I Edge Delamination in Composites," *Composite Structures*, Vol. 15, No. 4, 1990, pp. 323–342.

[16] Whitney, J. M., "Stress Analysis of Mode I Edge Delamination Specimen for Composite Materials," *AIAA Journal*, American Institute of Aeronautics and Astronautics, Vol. 24, No. 7, 1986, pp. 1163–1168.

[17] Jones, R. M., *Mechanics of Composite Materials*, McGraw Hill, New York, 1974.

[18] Sriram, P. and Armanios, E. A., "Fracture Analysis of Local Delaminations in Laminated Composites," *Proceedings*, Thirtieth AIAA/ASME/ASCE/AHS/ASC Structures, Structural Dynamics and Materials Conference, Mobile, AL, AIAA Paper 89-1400, 3–5 April 1989, pp. 2098–2108.

[19] Armanios, E. A., Rehfield, L. W., and Weinstein, F., "Understanding and Predicting Sublaminate Damage Mechanisms in Composite Structures," *Composite Materials: Testing and Design (Ninth Volume), ASTM STP 1059*, S. P. Garbo, Ed., American Society for Testing and Materials, Philadelphia, 1990, pp. 231–249.

[20] Wang, A. S. D., Kishore, N. N., and Feng, W. W., "On Mixed Mode Fracture in Off-Axis Unidirectional Graphite-Epoxy Composites," *Progress in Science and Engineering of Composites*, T. Hayashi, K. Kawata, and S. Umekawa, Eds., International Conference on Composite Materials, ICCM-IV, Tokyo, 1982, pp. 599–606.

Satish A. Salpekar[1] *and T. Kevin O'Brien*[2]

Combined Effect of Matrix Cracking and Free Edge on Delamination

REFERENCE: Salpekar, S. A. and O'Brien, T. K., **"Combined Effect of Matrix Cracking and Free Edge on Delamination,"** *Composite Materials: Fatigue and Fracture (Third Volume),* *ASTM STP 1110,* T. K. O'Brien, Ed., American Society for Testing and Materials, Philadelphia, 1991, pp. 287–311.

ABSTRACT: The effect of the stress-free edge on the growth of local delaminations initiating from a matrix crack in a composite laminate is investigated using a three-dimensional finite element analysis. Two glass epoxy layups, $(0_2/90_4)_s$ and $(\pm 45/90_4)_s$, were modeled with a matrix crack in the central group of eight 90° plies, and delaminations initiating from the matrix cracks in the 0/90 and −45/90 interfaces, respectively. The analysis indicated that high tensile interlaminar normal stresses were present at the intersection (corner) of the matrix crack with the stress-free edge, suggesting that an opening (Mode I) delamination may initiate at these intersections.

In order to analyze the strain energy release rates associated with delaminations that may form at the corners, three different configurations of the local delamination were assumed. One configuration was a uniform through-width strip growing normal to the matrix crack in the direction of the applied load. The other two configurations were triangular shaped delaminations, originating at the intersection of the matrix crack with the free edge, and growing away from the corner. The analysis of the uniform delamination indicated that the magnitude of both the total strain energy release rate, G_T, and its components increased near the free edges. This edge effect was symmetric for the $(0_2/90_4)_s$ layup. However, for the $(\pm 45/90_4)_s$ layup, the G distribution across the front was asymmetric, with the total G and its components having higher values near one free edge than the other. For both layups, the G_I component was large at small delamination lengths, but vanished once the delamination had reached a length of one-ply thickness.

The second and third delamination configurations consisted of triangular-shaped delaminations with straight fronts inclined at angles of 10.6 and 45°, respectively, to the matrix crack. The total G along the delamination front decreased sharply near the matrix crack for both configurations and increased sharply near the free edge for the 10.6° configuration. However, the total G distribution was fairly uniform in the middle of the delamination front for the 10.6° configuration. These inclined models suggest that if the exact geometry of the delamination front could be modeled, a uniform G distribution may be obtained across the entire front. However, because the contour of the delamination front is unknown initially, it may only be practical to model the uniform through-width delamination and use the peak values of G calculated near the free edges to predict delamination onset. For the layups modeled in this study, the total G values near the free edge agreed fairly well with a previously derived closed-form solution. However, a convergence study may need to be conducted to have confidence that peak values of G calculated from three-dimensional finite element analyses near the free edges are quantitatively correct.

KEY WORDS: composite materials, fracture, fatigue (materials), matrix cracking, delamination, finite element analysis, free edge

Nomenclature

a	Delamination length from a matrix crack measured along the x-axis
s	Distance along an inclined delamination front

[1] Research scientist, Analytical Services and Materials, Inc., Hampton, VA 23666.
[2] Senior scientist, Aerostructures Directorate, U.S. Army Aviation Research and Technology Activity (AVSCOM), NASA Langley Research Center, Hampton, VA 23665-5225.

E_{11}, E_{22}, E_{33}	Modulus of elasticity of a unidirectional lamina along and transverse to, the fiber direction
E_{lam}	Axial laminate modulus before delamination
E_{ld}	Modulus of a locally delaminated cross section
E^*	Modulus of an edge delaminated laminate
G	Strain energy release rate
G_T	Total strain energy release rate
G_{av}	Total strain energy release rate averaged along a delamination front
G_I, G_{II}, G_{III}	Mode I, Mode II, and Mode III components of strain energy release rate, respectively
G_{12}, G_{13}, G_{23}	Shear moduli of a unidirectional lamina
h	Ply thickness
m	Number of delaminations growing from a matrix crack
n	Number of plies
P	Axial load
S	Reciprocal of axial laminate modulus
t	Laminate thickness
t_{ld}	Thickness of a locally delaminated cross section
V_i	Relative nodal displacements
F_i	Nodal forces
w	Laminate width
X, Y, Z	Cartesian coordinates
α	Angle between matrix-crack and local delamination front
θ	Fiber angle in degrees, measured counter-clockwise from x-axis
Δ	Delamination increment
ϵ	Axial laminate strain
$\nu_{12}, \nu_{13}, \nu_{23}$	Poisson's ratio of a unidirectional lamina
σ_z	Interlaminar normal stress
σ_0	Remote axial stress

With the increased use of composite materials in primary aircraft structures, there is a need for understanding their failure mechanisms to better predict service life. Composite laminates subjected to monotonic tension and tension fatigue loading undergo a succession of various forms of damage before complete failure. These damage mechanisms include matrix ply cracking [1–3], edge delamination [4–6], and local delaminations arising from matrix ply cracks [7–10]. The progression of damage in a ($\pm 25/90_n$)$_s$ laminate is a good case study to examine the relationship between matrix cracks, edge delamination, and local delamination. This relationship was experimentally demonstrated in Ref 10 for ($\pm 25/90_n$)$_s$ laminates where n varied from ½ to 8. The results are shown schematically in Fig. 1. Under monotonic tension loads, these laminates exhibited 90° matrix cracking for all values of $n \geq 1$ (Fig. 1a). The laminates with $n \leq 3$ developed thumbnail-shaped edge delaminations at the $-25/90$ interfaces, whereas the laminates with $n \geq 4$ developed local delaminations in the vicinity of 90°-ply cracks (Fig. 1b). The delaminated area of the local delaminations was greatest in the region where the matrix crack meets the free edge, indicating that the local delaminations form at these corners. These local delaminations give rise to strain concentrations through the laminate thickness and lead to laminate failure [7–9].

Several researchers have shown that the interlaminar fracture toughness of a composite material may be used to predict the delamination onset and growth [1,8,10,11]. The interlaminar fracture toughness values are expressed in terms of critical values of strain energy release rate. They may be obtained for pure Mode I delamination using the double cantilever beam

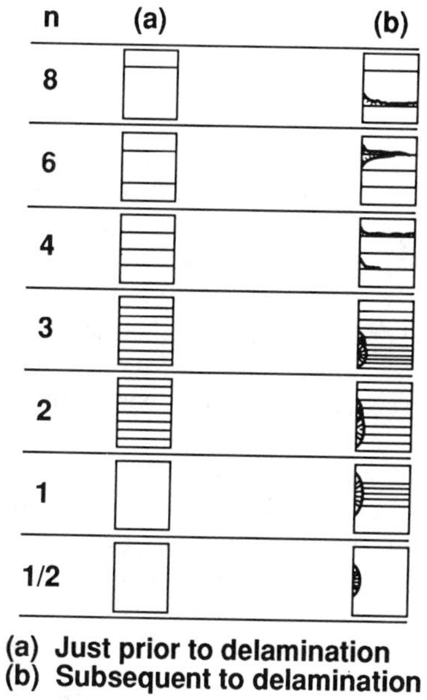

(a) Just prior to delamination
(b) Subsequent to delamination

FIG. 1—*Schematic of the fracture sequence in the* $(\pm 25/90_n)_s$ *laminates:* (a) *just prior to edge delamination, and* (b) *subsequent to edge delamination, Ref 10.*

(DCB) test, and for pure Mode II loading using the end-notched flexure (ENF) test, under monotonic or fatigue loading [12]. The strain energy release rate associated with a delamination under an applied loading is compared with an appropriate value of interlaminar fracture toughness to predict the onset and growth of the delamination. The purpose of this study was to evaluate the strain energy release rate associated with local delaminations, particularly in the proximity of the stress-free edge where the largest delamination area was observed (Fig. 1b), in order to extend this delamination methodology to more realistic cracked-laminate configurations.

Previously, Fish and Lee [13] performed three-dimensional finite element analyses of $(0_2/90_n)_s$ glass epoxy laminates to study the effect of 90° matrix cracks growing from the free edge. The delamination was assumed to grow in from the corner at an angle of 45° to the matrix crack, and was approximated in the form of a saw tooth. The number of 90° plies was varied from one to four, and the effect of changing the matrix crack length and the delamination length was analyzed. However, only the average strain energy release rate across the entire delamination front was evaluated. Their results did not agree well with a simplified closed-form expression for the total G associated with local delamination that was derived earlier by O'Brien [7]. Part of the motivation for this study was to resolve this discrepancy. To this end, the G_T distribution across the front was determined from the three-dimensional finite element analysis, and the maximum value in the distribution was compared to the closed-form equation.

In this study, the $(0_2/90_4)_s$ and $(\pm45/90_4)_s$ glass epoxy laminates were analyzed using the finite element method. The effect of cracking in the 90° plies on the strain energy release rate for local delamination, particularly near the free edge, was evaluated. The G distribution across the delamination front, and the change in G with delamination growth, was obtained for three different orientations of the local delamination with the free edge. The virtual crack closure technique was used to calculate the strain energy release rates. The total G results were also compared with G_T values calculated using the closed-form equation previously derived by O'Brien [7] for local delamination.

Laminate Configuration and Loading

When a $(\pm\theta/90_4)_s$ laminate is subjected to tension load, matrix cracks form in the eight 90° plies at several locations along the length of the laminate, as shown in Fig. 2. Subsequently, either edge delaminations will form between the $-\theta/90$ interface due to Poisson mismatch between the sublaminates (Ref 1 and the Appendix), or local delaminations will form between the $-\theta/90$ interface due to the presence of the matrix crack [7]. In order to study the effect of these damage modes, a representative segment of the laminate containing a single matrix crack, as shown by the shaded area in Fig. 2 will be considered in the analysis.

Figure 3 shows the representative segment of the $(\pm\theta/90_4)_s$ glass epoxy laminate. The segment is $80h$ long, $50h$ wide, and $12h$ thick, where h is the ply thickness. The laminate segment was assumed to be long compared to the local delamination lengths modeled, which were $5h$ long. The laminate was chosen wide enough to capture the effects near the edges and in the interior. The matrix crack is located at the midpoint along the segment length. The origin of the cartesian coordinate system is selected at the point where the matrix crack meets the $-\theta/90$ interface at the free edge. Due to the symmetry of the laminate and loading about its midplane, only the upper half of the laminate was analyzed. The material properties used in the analysis, for the glass epoxy laminate, are given in Table 1. The in-plane properties for a unidirectional ply (for example, E_{11}, E_{22}, G_{12}, ν_{12}) are the same as those in Ref 14. The out-of-plane properties (G_{13}, ν_{13}, G_{23}, ν_{23}) were assumed to be identical to the in-plane properties, and E_{33} was assumed equal to E_{22}. The laminate is subjected to a tension load by applying uniform outward displacements at its ends ($x = -40h, 40h$).

In this study, the three-dimensional finite element analyses of this laminate were performed to obtain the interlaminar stresses and the strain energy release rates associated with a local delamination starting from the matrix crack. Figure 3 shows the first configuration analyzed with a delamination growing uniformly in the x-direction from the 90° matrix crack. Analyses for two other orientations of the local delaminations with the matrix crack were also performed (Fig. 4). The angle of the inclined delamination with the matrix crack was assumed to be 10.6 and 45°, respectively. Quasi-three-dimensional (Q3D) analyses of the laminate were performed to obtain G values in the interior. A closed-form equation for G_T [7] was evaluated and compared with the three-dimensional finite element results.

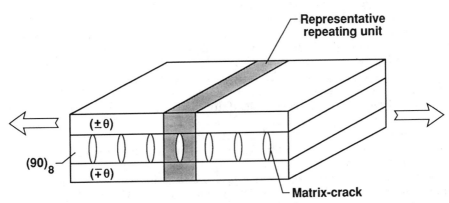

FIG. 2—The $[\pm\theta/90_4]_2$ laminate with matrix cracks.

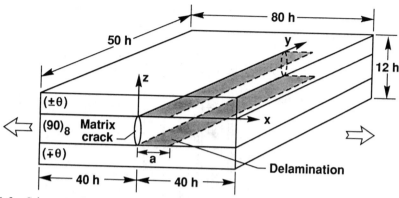

FIG. 3—*Schematic of a uniform through-width delamination, growing from a 90° matrix crack.*

TABLE 1—*Material properties used in the analyses.*

S2/SP250 Glass/Epoxy
$E_{11} = 7.30$ Msi[a]
$E_{22}, E_{33} = 2.10$ Msi
$G_{12}, G_{13}, G_{23} = 0.88$ Msi
$\nu_{12}, \nu_{13}, \nu_{23} = 0.275$

[a] Msi $= 6.895 \times 10^9$ Pa.

Three-Dimensional Finite Element Analysis

Finite Element Models

Uniform Delamination Front—The model consisted of 13 935 nodes and 2808 twenty-node brick elements. The top view of the finite element mesh at the $-\theta/90$ interface is shown in Fig. 5, and the discretization in the thickness direction is shown in Fig. 6. In Fig. 5, a fine mesh in the vicinity of the free edge ($y = 0$) and near the matrix crack ($x = 0$) is used to capture the

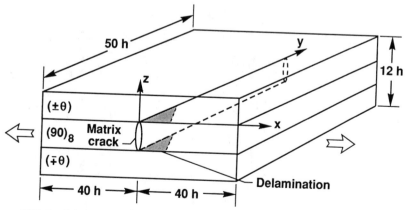

FIG. 4—*Schematic of an inclined delamination, growing from a 90° matrix crack.*

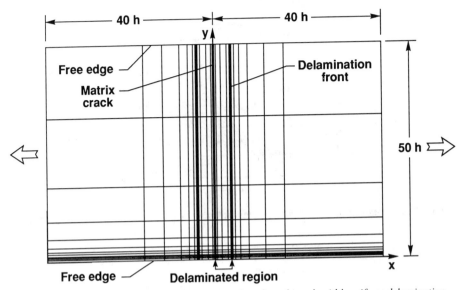

FIG. 5—*Finite element mesh at delamination plane for a through-width uniform delamination.*

presence of high stress gradients. For the same reasons, a fine mesh in the thickness direction near the $-\theta/90$ interface shown in Fig. 6 is used. The smallest element size used, near the edge and also near the crack, was one eighth of a ply thickness. Multipoint constraints were provided at the $-\theta/90$ interface, in the shaded area (on the $z = 0$ plane) shown in Fig. 3. The multipoint constraints can be released independently in the x, y, and z-directions to model delamination growth. The symmetry conditions were applied on the $z = -4h$ plane by imposing zero displacements in the vertical direction, and rigid body motions were constrained in the x and y-directions.

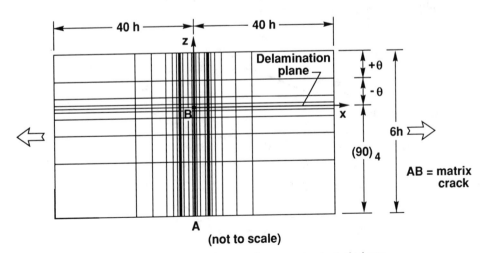

FIG. 6—*Finite element discretization across laminate thickness.*

Inclined Delamination Front—Figure 4 shows the configuration of a delamination that is assumed to grow from the corner of the matrix crack and the free edge. The cartesian coordinate system of the previous analysis is retained. Figure 7 shows the top view of the finite element mesh at the delamination plane. The refinement used near the corner is shown in Fig. 8. The element size along the direction of delamination extension was 0.2 to 0.25 ply thicknesses in the xy plane, and the element size was $0.125h$ in the z (thickness) direction. A total of 3906 twenty-node isoparametric elements, and 18 944 nodes were used.

The analyses were performed for several delamination fronts, inclined at an angle of $\alpha = 10.6°$ with the crack. The delamination distance, a, was measured from the corner, in the x-direction. In order to study the effect on the strain energy release rate, of changing the angle, α, of the delamination front, another finite element model was generated that makes a 45° angle with the matrix crack (y-axis).

Evaluated Parameters

Interlaminar Stresses—The interlaminar normal stresses on the $z = 0$ plane were obtained to identify the location for delamination initiation. The $(\pm 45/90_4)_s$ laminate, similar to that in Fig. 3, but containing only the matrix crack with no delamination, was analyzed. The variation of σ_z along laminate length and laminate width will be discussed.

Strain Energy Release Rates—The strain energy release rate, G, was calculated using the virtual crack closure technique (VCCT) [15,16] and explained here with reference to Fig. 9. The values V_1 through V_5 are the relative displacements between the corresponding top and bottom nodes behind the crack front, and F_1 through F_5 are the forces in the element ahead

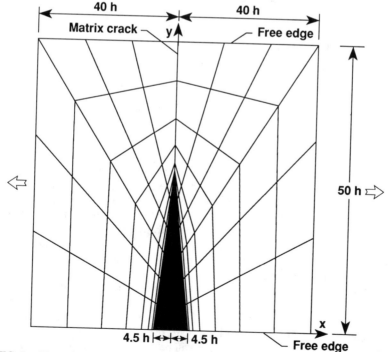

FIG. 7—*Finite element mesh at delamination plane, for a delamination angle, $\alpha = 10.6°$.*

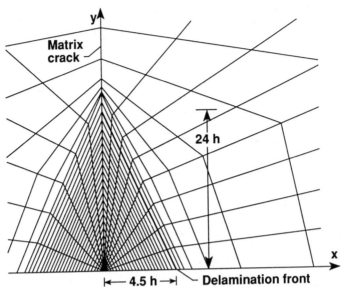

FIG. 8—*Finite element mesh in the vicinity of delamination front, for a delamination angle, $\alpha = 10.6°$.*

of the crack front, above the delamination plane. The three modes of the strain energy release rate and the value of G_T are calculated using the following equations

$$G_I = -\frac{1}{2w\Delta} \sum_{i=1}^{5} (V_i \times F_i)_{\text{opening}} \tag{1}$$

$$G_{II} = -\frac{1}{2w\Delta} \sum_{i=1}^{5} (V_i \times F_i)_{\text{shearing}} \tag{2}$$

$$G_{III} = -\frac{1}{2w\Delta} \sum_{i=1}^{5} (V_i \times F_i)_{\text{tearing}} \tag{3}$$

$$G_T = G_I + G_{II} + G_{III} \tag{4}$$

In the Eqs 1 through 4, Δ is the increment of delamination growth normal to the delamination front and w is the element width along the front. The values of Δ assumed in this analysis are in the range of $0.125h$ to $0.25h$. The delamination length, a, was increased by releasing the multipoint constraints up to the delamination front. The multipoint constraint nodes in the assumed delamination area were completely released from each other in the first analysis. The nodes of the delaminated planes crossing into the opposite plane were constrained in the vertical (z) direction in the next analysis. The quantity G_{av} is calculated as an average of the G_T values along the entire delamination front.

In the uniform delamination analysis (Fig. 5), the mesh is more refined near one free edge ($y = 0$) than the other free edge ($y = 50h$). Thus, it is possible to use a fine mesh near the $y = 0$ edge for a given size of the finite element model. Using this model, the results near both the edges can be obtained. For the $(0_2/90_4)_s$ laminate, the G values will be symmetric about the $y = 25h$ line and the results at the $y = 0$ edge will be the same as the results at the $y = 50h$. In the case of the $(\pm45/90_4)_s$ laminate, two analyses need to be carried out to obtain the effects

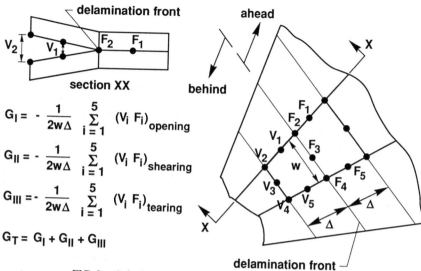

FIG. 9—*Calculation of* G *using virtual crack closure technique.*

near the two edges. The analysis of the $(+45/-45/90_4)_s$ layup yields results at the $y = 0$ edge. The second analysis was carried out for a $(-45/+45/90_4)_s$ layup. The results of this second analysis near $y = 0$ edge are the same as those of the original $(\pm 45/90_4)_s$ layup at $y = 50h$.

In the inclined delamination analyses, the VCCT was used to calculate the Mode I and total strain energy release rate. Small inaccuracies may be present in the results of this analysis because the delamination front does not grow self-similarly. However, a fine mesh is used in the analysis to minimize the inaccuracies, and the results will give a valuable qualitative insight into the variation of G_I and total G along the delamination front.

Quasi-Three-Dimensional Analysis

Quasi-three-dimensional (Q3D) finite element analysis [17] of a y = constant cross section of the laminate was also performed. The finite element discretization used in the Q3D analysis was the same as that used for the three-dimensional model in the XZ plane in Fig. 6. The Q3D formulation assumes an infinitely wide laminate and analyzes a two-dimensional cross section with three degrees of freedom at each node. The results are representative of the laminate interior, (for example, $y = 25h$) away from the stress-free edges. The G_I and G_T values were calculated for three different values of delamination length, a, using the Q3D analysis.

Closed-Form Solution

The total G associated with local delaminations originating at matrix cracks may be also calculated from a previously derived closed-form equation [7] (see Appendix) as

$$G = \frac{P^2}{2mw^2} \left(\frac{1}{t_{ld}E_{ld}} - \frac{1}{tE_{lam}} \right) \tag{5}$$

where the terms in Eq 5 are defined in the Appendix.

Results

Interlaminar Normal Stress

Figure 10 shows a plot of the normalized interlaminar normal stress, σ_z/σ_0, where σ_0 is the remote axial stress on the laminate, in the $-45/90$ interface of the $(45/-45/90_4)_s$ laminate along the laminate width at $x = 0.0625h$ and $x = 0.625h$. The normalized σ_z stresses are tensile. For the plot at $x/h = 0.0625$, that is, very near the crack, there is a sharp increase in the σ_z value near the free edge. The corresponding values of σ_z are much smaller for the $x/h = 0.625$ plot, that is, away from the influence of the matrix crack.

The variation of σ_z/σ_0 along a line in the x-direction, one half of a ply thickness inside the edge ($y/h = 0.5$), is shown in Fig. 11. The σ_z increases sharply near the matrix crack.

Figures 10 and 11 indicate that steep gradients in the interlaminar normal stress exist near the corner of the matrix crack and the free edge on the $z = 0$ plane. Therefore, this corner is a probable site for local delamination onset.

Results of the Uniform Delamination Analysis

In the following section, the results of the Q3D analysis are first compared with interior solutions obtained from the full three-dimensional analysis. Next, the G_T and G_I distributions along the uniform delamination front will be presented for the $(0_2/90_4)_s$ and $(\pm 45/90_4)_s$ layups, also referred to as Layups A and B, respectively. Finally, the variation of G_T with delamination length, a, from the three-dimensional analysis will be compared to the closed-form solution [7]. The values of G_I, G_{II}, G_{III}, and G_T, will be normalized by ϵ^2/h in presenting the results, where ϵ is the remote axial strain and h is the ply thickness. Distances, such as delamination length, will be normalized by the ply thickness, h.

Comparison of Q3D and Three-Dimensional Analyses—The Q3D model analyzed represents a cross section in the interior ($y = 25h$) of the laminate. Figure 12 shows the variation

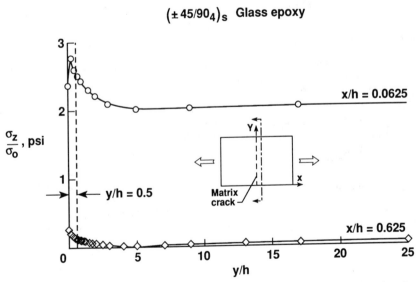

FIG. 10—*Variation of normalized interlaminar normal stress in the* $-45/90$ *interface along laminate width.*

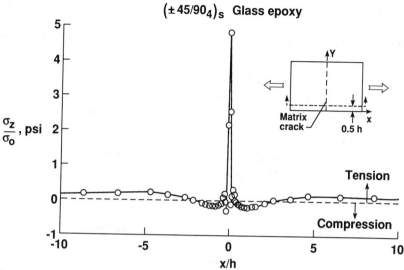

FIG. 11—*Variation of normalized interlaminar normal stress in the* −45/90 *interface along laminate length.*

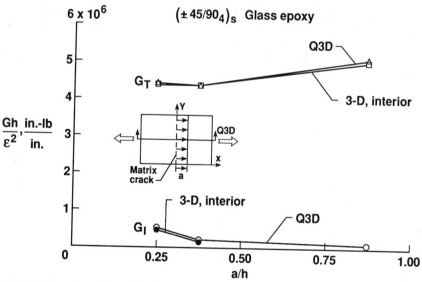

FIG. 12—*Comparison of normalized* G_I *and* G_T *in laminate interior, using Q3D and three-dimensional finite element analyses.*

of normalized G_I and G_T with change in normalized delamination length, for $a/h < 1$. There is a close agreement between the results of the Q3D analysis and the interior ($10 < y/h < 40$) solutions of the three-dimensional analysis.

Normalized G-Distributions Along a Uniform Delamination Front—The distribution of normalized G_I, G_{II}, G_{III}, and G_T along the delamination front, for the $(0_2/90_4)_s$ laminate are shown in Fig. 13 for a delamination length of four ply thicknesses. The plots are assumed to be symmetric about $y/h = 25$, due to the symmetry of the laminate configuration and loading. The opening mode is zero along the front. The G_{II} component (shearing in the x-direction) is constant in the interior, to within about seven ply thicknesses of the free edge, and rises quickly near the edge. The Mode III (shearing in the y-direction) is zero in the interior, and increases in magnitude as y/h approaches either one of the free edges. This plot shows that the strain energy release rates near the free edge are higher than in the interior of the laminate.

The distribution of normalized G_I, G_{II}, G_{III}, and G_T along the delamination front, for the $(0_2/90_4)_s$ laminate are shown in Fig. 14 for a delamination length of 0.375-ply thicknesses. These results are similar to those at $a/h = 4.0$; however, at this smaller delamination length, a Mode I component is also present. This Mode I component is constant in the interior and increases near the free edge. The G_I value is small relative to the corresponding G_{II} value, but is higher than G_{III}.

The normalized G distributions for the $(\pm 45/90_4)_s$ layup at $a/h = 4.0$ are shown in Fig. 15. The results are higher than the similar values in Fig. 13 for Layup A. Furthermore, these distributions are asymmetric, with higher normalized G values near the $y/h = 0$ free edge than near the $y/h = 50$ free edge.

Figure 16 shows normalized G distributions for $a/h = 0.375$, and indicates the presence of Mode I. The G_I rises more steeply near the $y/h = 0$ edge than near the $y/h = 50$ edge.

Mode I at Small Delamination Length—The presence of the opening mode at small delamination length was observed in Figs. 14 and 16. The opening mode does not exist at higher delamination length (Figs. 13 and 15). Figure 17 shows that the normalized G_I decreases with increasing delamination length in Layup A, and vanishes before the delamination has grown as far as one-ply thickness. The G_I is higher close to the free edge ($y/h = 0.375$) compared to the interior ($y/h = 25$). A similar variation is observed for Layup B and is given in Fig. 18. The edge effect is more pronounced in Layup B than in A due to the presence of off-axis plies, which results in a higher Poisson mismatch (see Appendix).

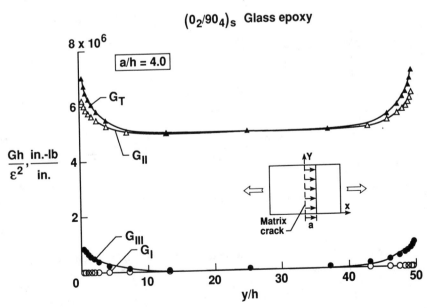

FIG. 13—*Variation of normalized* G *along a uniform delamination front in Layup A, at* a/h = 4.0.

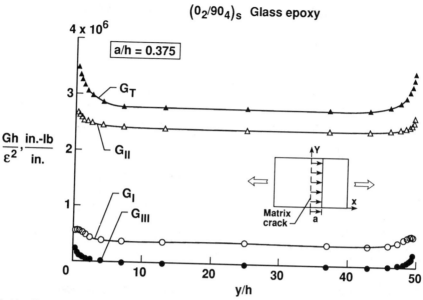

FIG. 14—*Variation of normalized* G *along a uniform delamination front in Layup A, at* a/h = 0.375.

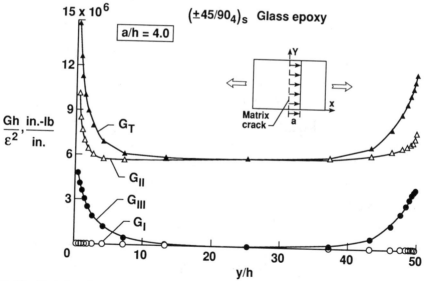

FIG. 15—*Variation of normalized* G *along a uniform delamination front in Layup B, at* a/h = 4.0.

Because G_I decreases with increasing local delamination length, delaminations that form from matrix cracks may initiate in Mode I and then arrest shortly thereafter. This arrest has been observed for local delaminations that form when the laminate is subjected to monotonic loading [10].

Variation of G_T *with Delamination Length*—Figure 19 shows the plots of normalized G ver-

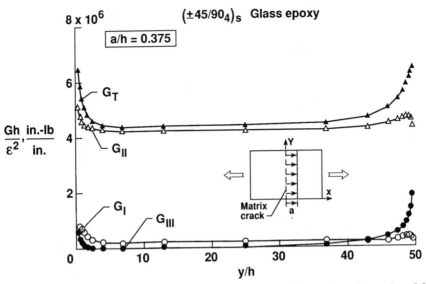

FIG. 16—*Variation of normalized G along a uniform delamination front in Layup B, at* a/h = *0.375.*

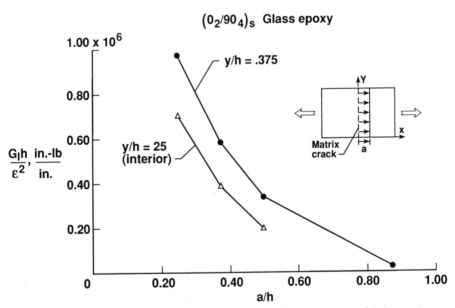

FIG. 17—*Change in normalized G_I due to uniform delamination growth in Layup A.*

sus a/h at $y = 3h/8$, $5h/8$, and at $y = 25h$ (interior) for the $(0_2/90_4)_s$ layup. Normalized total G increases with increasing delamination length, but eventually approaches a constant asymptotic value for each plot. The G_T values closer to the edge are higher than those in the interior. The average normalized G is slightly higher than the interior value due to the influence of the free edge. The value of G_T from Eq 5, normalized by ϵ^2/h where $\epsilon = P/wtE_{lam}$, is greater than

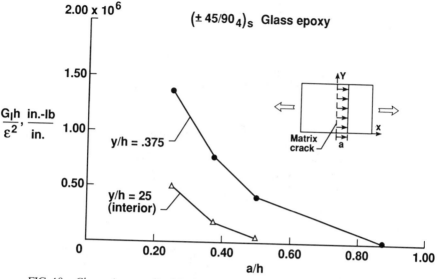

FIG. 18—*Change in normalized G_I due to uniform delamination growth in Layup B.*

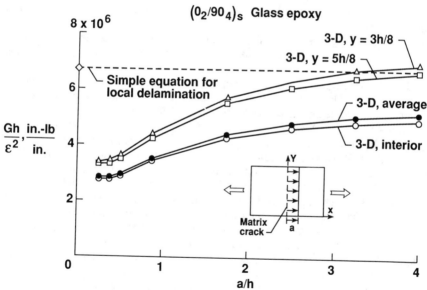

FIG. 19—*Change in normalized G due to uniform delamination growth in Layup A.*

the interior value, but is similar to G_T values calculated very close to the edge using the three-dimensional analysis (Fig. 19).

Figure 20 shows the plots of normalized G versus a/h at $y = 3h/8$, $5h/8$, and at $y = 25h$ (interior) for the $(\pm 45/90_4)_s$ layup. The preceding discussion of the variation of total G with delamination length for the $(0_2/90_4)_s$ layup is also valid in case of $(\pm 45/90_4)_s$ layup. However,

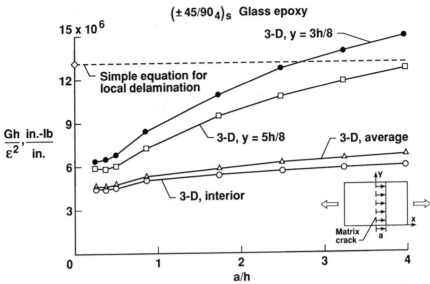

FIG. 20—*Change in normalized* G *due to uniform delamination growth in Layup B.*

the closed-form solution cannot sense the asymmetry in the G distribution for Layup B (Figs. 15 and 16). Hence, it may be calculating the average of the peak G values near the two edges. In any event, a convergence study, using several mesh refinements, is needed to make quantitative comparisons of the closed-form solution to the three-dimensional finite element model.

Based on the occurrence of relatively high G_T and G_I values near the $y/h = 0$ edge in the uniform through-width delamination, the subsequent analyses will focus on the variation of strain energy release rate near that edge for the inclined delamination.

Delamination Contour Near the Free-Edge—Experimental evidence shows that the local delamination has a curved contour near the corner where the matrix crack intersects the free edge (Fig. 1b, $n \geq 4$). This curved contour was approximated as a straight delamination front growing from the corner and inclined at an angle to the matrix crack. In order to obtain the approximate shape of the contour in the vicinity of the corner, the uniform delamination analysis was used. As discussed in an earlier section, the delamination faces are open across the entire width at small delamination lengths. If we assume that the region where the delamination faces are open represents a delamination resulting from Mode I, then the border of the crossover region may represent a first approximation of the actual crack front geometry. When the delamination length, a, was $0.875h$ as shown in Fig. 21, the delaminated faces crossed into one another in some part of the delamination area. Figure 21 also shows the curved line of demarcation between the crossed faces and the open faces. This line is obtained by joining the nodes bordering the crossover region by straight-line segments. The changing angles, α, of these line segments, measured counter-clockwise from the matrix crack, are shown in Fig. 21. The average of the two values of α close to the free edge (10.6°), was used in modeling the inclined delamination.

Results of the Inclined Delamination Analysis ($\alpha = 10.6°$)

The analysis was performed for both the layups analyzed in the earlier section, that is, Layup A, $(0_2/90_4)_s$, and Layup B, $(\pm45/90_4)_s$. Figure 22 shows the variation of normalized G_T along

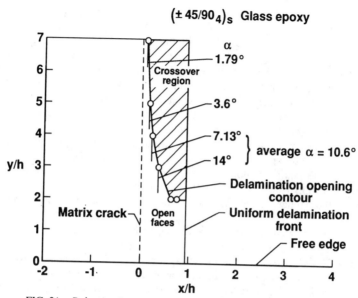

FIG. 21—*Delamination contour near free edge, due to opening mode.*

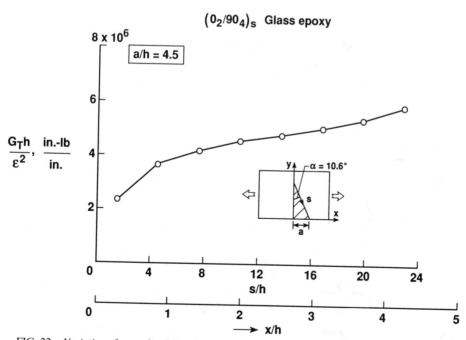

FIG. 22—*Variation of normalized G_T along inclined delamination front for Layup A, (α = 10.6° and a/h = 4.5).*

the delamination front for Layup A, when $a/h = 4.5$. The value $x/h = 0$, which also corresponds to $s/h = 0$, is the point where the front meets the matrix crack. The value of $x/h = 4.5$ corresponds to the intersection of the front with the free edge. The G_T value increases steadily while approaching the free edge ($x/h = 4.5$). The normalized G_T distribution for Layup B in Fig. 23 increases gradually towards the edge in the central portion of the front, but rises steeply near the free edge due to the high Poisson mismatch between the sublaminates (see Appendix). The G_I was zero for both of the layups at these large delamination lengths.

A second analysis was performed for a much smaller delamination length with $a/h = 0.6$. The variation of normalized G_I along the front for Layup B is shown in Fig. 24. The gradual increase in G_I, as x/h approaches zero, is due to the angle, α, being higher than that required to match the delamination profile near the corner (Fig. 21, $y/h > 3$).

Figure 25 shows a similar variation of normalized G_I along the delamination front for Layup B, at $a/h = 0.4$. Comparing Figs. 24 and 25 shows that G_I has a higher value for $a/h = 0.4$ than for $a/h = 0.6$. Hence, as was the case for the uniform delamination, G_I increases with decreasing a/h.

The change in the total G, averaged along the delamination front is plotted as a function of a/h in Fig. 26. The average total G approaches a constant value at large delamination lengths, for both layups. The G_T is lower for Layup A than for Layup B. The increase in total G at small delamination length in Layup B is due to the increase in G_I, discussed in the previous section. A further analysis is needed to determine whether the total G at very small delamination lengths is higher than the constant value attained at larger delamination lengths.

Results of the Inclined Delamination Analysis (45°)

The normalized G_T along a 45° front, shown in Fig. 27 for Layup A, reaches a peak value at x/h near 1 and then decreases with increasing values of x/h (or s/h). This is in contrast to the

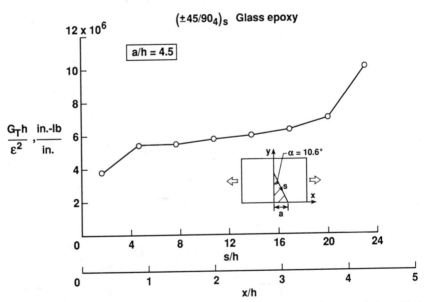

FIG. 23—*Variation of normalized* G_T *along inclined delamination front for Layup B, (*$\alpha = 10.6°$ *and* $a/h = 4.5$).

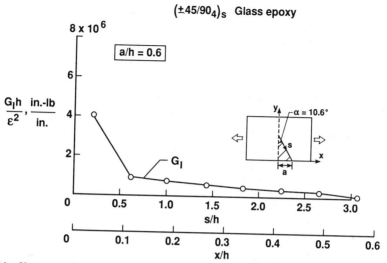

FIG. 24—*Variation of normalized G_I along inclined delamination front for Layup B, ($\alpha = 10.6°$ and a/h = 0.6).*

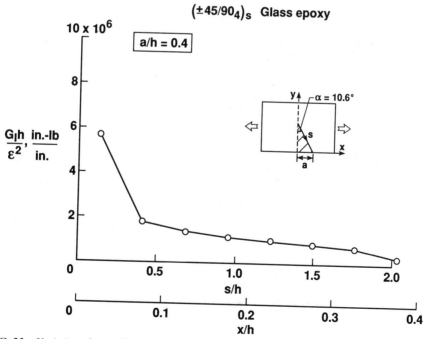

FIG. 25—*Variation of normalized G_I along inclined delamination front for Layup B, ($\alpha = 10.6°$ and a/h = 0.4).*

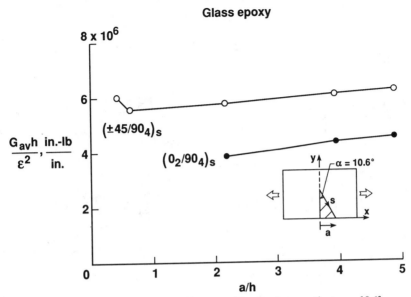

FIG. 26—*Change in normalized* G_{av} *due to delamination growth at* $\alpha = 10.6°$.

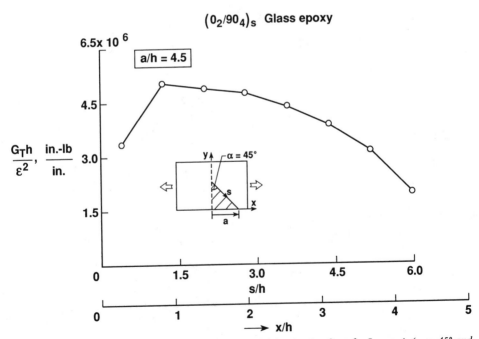

FIG. 27—*Variation of normalized* G_T *along inclined delamination front for Layup A, (* $\alpha = 45°$ *and* $a/h = 4.5$*).*

variation for a similar case at $\alpha = 10.6°$ (Fig. 22), where normalized G_T gradually increases with increasing x/h. The normalized G_T along the front, for Layup B, plotted in Fig. 28, illustrates the same trend as does Layup A. In contrast, the normalized G_T distribution for $\alpha = 10.6°$ shown in Fig. 23 was nearly uniform in the central region, but gradually increased with increasing x/h. Thus, for Layup B, the assumed angle, $\alpha = 10.6°$, yields a more uniform G_T distribution along the laminate front than $\alpha = 45°$. The experimentally observed angle, α, may be higher near the free edge, and lower near the matrix crack, as observed for the delamination profile (Fig. 21). This suggests that the exact contour for the delamination front may need to be modeled to obtain a uniform G distribution.

As shown in Fig. 29, the variation of total G, averaged along the front, approaches a constant at higher a/h, for both layups.

Concluding Remarks

Composite laminates containing a matrix crack and subjected to axial tension were analyzed using the three-dimensional finite element analysis. Glass epoxy $(0_2/90_4)_s$ and $(\pm 45/90_4)_s$ layups were modeled. A matrix crack was assumed to be present in the group of eight 90° plies. Local delaminations were assumed to initiate from the matrix crack at the 0/90 and $-45/90$ interfaces, respectively. Three-dimensional finite element analyses were carried out to evaluate the effect of the free edge on the growth of the local delaminations.

The interlaminar normal stress increased sharply in the vicinity of the corner where the matrix crack meets the free edge, thus indicating that this corner is a potential site for delamination onset. In order to analyze the strain energy release rates associated with delaminations that may form at the corners, three different configurations of the local delamination were assumed. One configuration was a uniform through-width strip growing normal to the matrix crack in the direction of the applied load. The other two configurations were triangular shaped

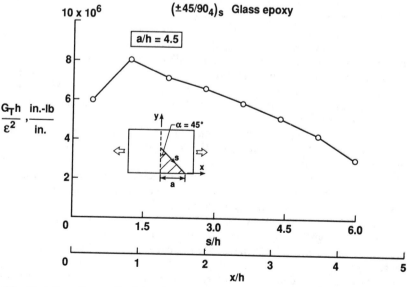

FIG. 28—*Variation of normalized* G_T *along inclined delamination front for Layup B,* ($\alpha = 45°$ *and* $a/h = 4.5$).

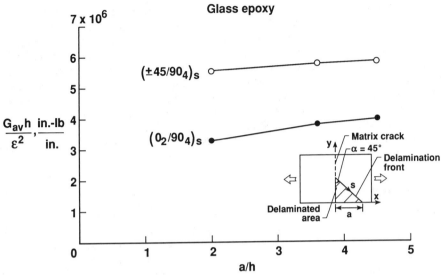

FIG. 29—*Change in normalized* G_{av} *due to delamination growth at* $\alpha = 45°$.

delaminations, originating at the intersection of the matrix crack with the free edge, and growing away from the corner.

For both layups, the total G and its Mode I component, calculated in the interior of the laminate width for the uniform through-width local delamination, matched G values obtained from a quasi-three-dimensional finite element analysis. However, the strain energy release rate and its components, calculated from the three-dimensional finite element model, had higher values near the free edge than in the interior for both layups. For the $(\pm 45/90_2)_s$ layup, the G distributions across the width of the delamination front were asymmetric, with higher G values at one corner than at the other.

For both layups, G_I was nonzero at small delamination lengths, but rapidly decreased with delamination growth, and vanished when the delamination reached a length of one-ply thickness. The total G consisted primarily of G_{II} at delamination lengths greater than one-ply thickness. For all delamination lengths, G_{III} was nonzero only near the free edges. The total G reached a constant value at delamination lengths of about four-ply thicknesses in the interior as well as near the edges.

The second and third delamination configurations consisted of triangular shaped delaminations with straight fronts inclined at angles of 10.6° and 45°, respectively, to the matrix crack. The total G across the delamination front deviated near the matrix crack and the free edge for both configurations. However, the total G distribution was fairly uniform in the middle of the delamination front for the 10.6° configuration. These inclined models suggest that if the exact geometry of the delamination front could be modeled, a uniform G distribution may be obtained across the entire front. However, because the contour of the delamination front is unknown initially, it may only be practical to model the uniform through-width delamination and use the peak values of G calculated near the free edges to predict delamination onset. For the through-width model of the layups in this study, total G values near the free edges agreed fairly well with a previously derived closed-form solution. However, a convergence study may need to be conducted to have confidence that peak values of G calculated from three-dimensional finite element analyses near the free edges are quantitatively correct.

Acknowledgment

This work was performed under NASA contract NAS1-18599. Dr. Salpekar would like to thank Dr. K. N. Shivakumar for helpful discussions, and Dr. I. S. Raju for the use of his three-dimensional finite element computer program.

APPENDIX

Poisson Mismatch and Influence of Delamination on Stiffness

In Ref 7, the effect of local delamination versus edge delamination on laminate stiffness was compared. The stiffness loss due to edge delamination was derived in Ref 16 as

$$E^* = \sum_{i=1}^{m} \frac{E_i t_i}{t} \qquad (6)$$

where m is the number of sublaminates formed by the delamination, E_i and t_i are the modulus (calculated from laminated plate theory) and thickness of the ith sublaminate formed, and t is the original laminate thickness. The difference between E_{lam} and E^* will be greatest for laminates with the largest "Poisson mismatch." The term, Poisson mismatch, refers to the difference between the Poisson's ratio of the original laminate and the Poisson's ratios of the 90° plies and the sublaminates that are formed by the delamination. For example, Table 2 *(top)* shows the Poisson's ratio of the $(0_2/90_4)_s$ and $(\pm45/90_4)_s$ layups, the 0_2 and (±45) sublaminates formed by the delamination, and the 90° plies. The difference between the laminate Poisson's ratio and that of the sublaminates is greatest for the $(\pm45/90_4)_s$ laminate. Hence, the $(\pm45/90_4)_s$ layup has a larger Poisson mismatch than the $(0_2/90_4)_s$ layup.

This difference is also reflected in the difference between the laminated and delaminated moduli, E_{lam} and E^*, calculated from laminated plate theory. Table 2 *(bottom)* lists the moduli for the laminates, sublaminates, and 90° plies, as well as E^* calculated from Eq 6. Also listed is the stiffness loss, $E_{lam} - E^*$, for the two layups. The $(0_2/90_4)_s$ layup has only a 0.6% loss in stiffness due to a delamination in the 0/90 interface, whereas the $(\pm45/90_4)_s$ layup experiences a 7.2% loss in stiffness due to a delamination in the $-45/90$ interface. These stiffness losses reflect the loss in the constraint that the 90° plies originally had on the sublaminates when they were laminated. This stiffness loss is reflected in the strain energy release rate associated with edge delamination [16] given by

$$G = \frac{\epsilon^2 t}{2} (E_{lam} - E^*) \qquad (7)$$

TABLE 2—*Stiffness and Poisson's ratios of laminates and sublaminates.*

Poisson's Ratio (in./in.)[a]					
Layup	ν_{lam}	ν_{90}	ν_{sub}	$\nu_{sub} - \nu_{lam}$	$\nu_{lam} - \nu_{90}$
$(0_2/90_4)_s$	0.104	0.079	0.275	0.171	0.025
$(\pm45/90_4)_s$	0.162	0.079	0.508	0.346	0.083

Stiffness ($\times 10^6$ psi)[b]					
Layup	E_{lam}	E_{90}	E_{sub}	E^*	$E_{lam} - E^*$
$(0_2/90_4)_s$	3.857	2.100	7.300	3.833	0.024
$(\pm45/90_4)_s$	2.462	2.100	2.654	2.285	0.177

[a] in./in. = cm/cm.
[b] psi = 6.895×10^3 Pa or 6.895 kPa.

TABLE 3—*Normalized strain energy release rate values for local delamination.*

Layup	Gh/ϵ^2 (Eq 8) $\times$ 10^6	Gh/ϵ^2 (Eq 9) $\times$ 10^6	(Eq 8 and Eq 9) $\times$ 10^6
$(0_2/90_4)_s$	6.77	6.61	0.16
$(\pm45/90_4)_s$	13.16	10.84	2.32

Because the $(0_2/90_4)_s$ layup has a lower value of $(E_{lam} - E^*)$ than the $(\pm45/90_4)_s$ layup, it will have a lower strain energy release rate for the same applied strain, ϵ.

G_T Due to Local Delamination

Similar to the case of edge delamination, the local delaminations that form at the intersection of matrix cracks and the free edge will relax the constraint between the 90° plies and the sublaminates. However, these local delaminations will also reduce stiffness because of the lost load carrying capability of the cracked ply [7]. Hence, the stiffness loss, and corresponding strain energy release rate, associated with local delamination will be different than for edge delamination. An equation for the total G associated with local delaminations originating at matrix cracks was derived by O'Brien [7]

$$G = \frac{P^2}{2mw^2}\left(\frac{1}{t_{ld}E_{ld}} - \frac{1}{tE_{lam}}\right) \tag{8}$$

where P is the axial load, m is the number of delaminations growing from the matrix crack, and w is the laminate width. This equation assumes that the cracked 90° plies do not carry any load in the delaminated region. The thickness of the locally delaminated region carrying the load (that is, the thickness of the uncracked plies) is denoted as t_{ld}, and has a modulus, E_{ld}, that is calculated from laminated plate theory. The thickness of the entire laminate is t, and the laminate modulus E_{lam} is also calculated using laminated plate theory. Because the stiffness of the original laminate, E_{lam}, and the locally delaminated region, E_{ld}, are calculated using the two-dimensional laminated plate theory, G in Eq 8 accounts for both the loss in constraint as well as the loss in load carrying capability of the cracked off-axis plies [7].

Using E^* from Eq 6 instead of E_{lam} in Eq 8 excludes the effect of Poisson mismatch and isolates the effect of losing the load carrying capability of the cracked plies. Hence, Eq 8 becomes

$$G = \frac{P^2}{2mw^2}\left(\frac{1}{t_{ld}E_{ld}} - \frac{1}{tE^*}\right) \tag{9}$$

Total G values from Eqs 8 and 9, normalized by ϵ^2/h where $\epsilon = P/wtE_{lam}$, are listed in Table 3. Comparing the results of Eqs 8 and 9 indicates that the contribution of Poisson mismatch to the strain energy release rate for local delamination is relatively small for Layup A, but is significantly greater for Layup B.

References

[1] Reifsnider, K. L. and Talug, A., "Analysis of Fatigue Damage in Composite Laminates," *International Journal of Fatigue*, Vol. 3, No. 1, Jan. 1980, pp. 3–11.

[2] Masters, J. E. and Reifsnider, K. L., "An Investigation of Cumulative Damage Development in Quasi-Isotropic Graphite/Epoxy Laminates," *Damage in Composite Materials, ASTM STP 775*, K. L. Reifsnider, Ed., American Society for Testing and Materials, Philadelphia, 1982, pp. 40–62.

[3] Reifsnider, K. L., "Some Fundamental Aspects of the Fatigue and Fracture Response of Composite Materials," *Proceedings*, Fourteenth Annual Meeting of Society of Engineering Science, Lehigh University, Bethlehem, PA, 14–16 Nov. 1977.

[4] O'Brien, T. K., "Mixed-Mode Strain Energy Release Rate Effects on Edge Delamination of Composites," *Effects of Defects in Composite Materials, ASTM STP 836*, American Society for Testing and Materials, Philadelphia, 1984, pp. 125–142.

[5] Adams, D. F., Zimmerman, R. S., and Odom, E. M., "Frequency and Load Ratio Effects on Critical Strain Energy Release Rate G_c Thresholds of Graphite/Epoxy Composites," *Toughened Compos-*

ites, ASTM STP 937, N. J. Johnston, Ed., American Society for Testing and Materials, Philadelphia, 1987, pp. 242–259.

[6] Whitney, J. M. and Knight, M., "A Modified Free-Edge Delamination Specimen," *Delamination and Debonding of Materials, ASTM STP 876,* W. S. Johnson, Ed., American Society for Testing and Materials, Philadelphia, 1985, pp. 298–314.

[7] O'Brien, T. K., "Analysis of Local Delaminations and Their Influence on Composite Laminate Behavior," *Delamination and Debonding of Materials, ASTM STP 876,* W. S. Johnson, Ed., American Society for Testing and Materials, Philadelphia, 1985, pp. 282–297.

[8] O'Brien, T. K., Rigamonti, M., and Zanotti, C., "Tension Fatigue Analysis and Life Prediction for Composite Laminates," NASA TM 100549, NASA Langley Research Center, Hampton, VA, Oct. 1988.

[9] O'Brien, T. K., "Towards a Damage Tolerance Philosophy for Composite Materials and Structures," *Composite Materials: Testing and Design, ASTM STP 1059,* S. P. Garbo, Ed., American Society for Testing and Materials, Philadelphia, 1990, pp. 7–33.

[10] Crossman, F. W. and Wang, A. S. D., "The Dependence of Transverse Cracking and Delamination on Ply Thickness in Graphite/Epoxy Laminates," *Damage in Composite Materials, ASTM STP 775,* K. L. Reifsnider, Ed., American Society for Testing and Materials, Philadelphia, 1982, pp. 118–139.

[11] Murri, G. B., Salpekar, S. A., and O'Brien, T. K., "Fatigue Delamination Onset in Tapered Composite Laminates," NASA TM 101673, NASA Langley Research Center, Hampton, VA, 1989.

[12] Martin, R. H. and Murri, G. B., "Characterization of Mode I and Mode II Delamination Growth and Thresholds in Graphite/PEEK Composites," *Composite Materials: Testing and Design, ASTM STP 1059,* S. P. Garbo, Ed., American Society for Testing and Materials, Philadelphia, 1990, pp. 251–270.

[13] Fish, J. C. and Lee, S. W., "Three-Dimensional Analysis of Combined Free-Edge and Transverse-Crack-Tip Delamination," *Composite Materials: Testing and Design, ASTM STP 1059,* S. P. Garbo, Ed., American Society for Testing and Materials, Philadelphia, 1990, pp. 271–286.

[14] Chan, W. S., Rogers, C., and Aker, S., "Improvement of Edge Delamination Strength of Composite Laminates Using Adhesive Layers," *Composite Materials: Testing and Design (Seventh Conference), ASTM STP 893,* J. M. Whitney, Ed., American Society for Testing and Materials, Philadelphia, 1986, p. 266.

[15] Shivakumar, K. N., Tan, P. W., and Newman, J. C., Jr., "A Virtual Crack Closure Technique for Calculating Stress-Intensity Factors for Cracked Three-Dimensional Bodies," *International Journal of Fracture,* Vol. 36, 1988, pp. R43–R50.

[16] Raju, I. S., Shivakumar, K. N., and Crews, J. C., Jr., "Three-Dimensional Elastic Analysis of a Composite Double Cantilever Beam Specimen," *AIAA Journal,* American Institute of Aeronautics and Astronautics, Vol. 26, No. 12, Dec. 1988, pp. 1493–1498.

[17] Raju, I. S., "Q3DG—A Computer Program for Strain Energy Release Rates for Delamination Growth in Composite Laminates," NASA CR-178205, NASA Langley Research Center, Hampton, VA, Nov. 1986.

Gretchen Bostaph Murri,[1] Satish A. Salpekar,[2] and T. Kevin O'Brien[1]

Fatigue Delamination Onset Prediction in Unidirectional Tapered Laminates

REFERENCE: Murri, G. B., Salpekar, S. A., and O'Brien, T. K., "**Fatigue Delamination Onset Prediction in Unidirectional Tapered Laminates,**" *Composite Materials: Fatigue and Fracture (Third Volume), ASTM STP 1110,* T. K. O'Brien, Ed., American Society for Testing and Materials, Philadelphia, 1991, pp. 312–339.

ABSTRACT: Tapered [0°] laminates of S2/CE9000 and S2/SP250 glass/epoxies, and two different specimen types of IM6/1827I, a graphite/epoxy with a toughened interleaf, were tested. Specimens were subjected to cyclic tension in a hydraulic load frame. The specimens usually showed some initial stable delaminations in the tapered region, but these did not affect the stiffness of the specimens, and loading was continued until the specimens either delaminated unstably, or reached 10^6 to 2×10^7 cycles with no unstable delamination. The final unstable delamination originated at the junction of the thin and tapered regions, and extended into both the tapered and thin regions.

A finite element (FE) model was developed for the tapered laminate, both for the laminate with no initial delamination, and for the tapered laminate with the initial stable delaminations observed in the tests. The analysis showed that for both cases the most likely place for an opening (Mode I) delamination to originate is at the junction of the tapered and thin regions. For each material type, the models were used to calculate the strain energy release rate, G, associated with delaminations originating at that junction and growing either into the thin region between the belt and core plies, or into the tapered region, between the belt and dropped plies. For delamination growth in either direction, calculated values of G reached a peak at a delamination length equal to few ply thicknesses, and then decreased. The highest values of G were calculated for the laminate with an initial stable delamination in the tapered region and an opening mode delamination growing from the junction into the tapered region.

For the materials tested, cyclic G_{Imax} values from double cantilever beam (DCB) tests were used with the maximum G values calculated from the FE analysis for delamination growth in the tapered and thin regions, to predict the onset of unstable delamination at the junction as a function of fatigue cycles. The predictions were compared to experimental values of maximum cyclic load as a function of cycles to unstable delamination from fatigue tests in tapered laminates. The predictions, assuming delamination initiated in the tapered region, agreed reasonably well with the test data, although the correlation was slightly conservative for all except the S2/SP250 material. The measured and calculated results, assuming delamination initiated in the thin region, showed good agreement, although the predictions were slightly unconservative for the S2/SP250 laminates

KEY WORDS: strain energy release rate, finite element analysis, interleaf, dropped ply, delamination, composite materials, double cantilever beam tests, fracture, fatigue (materials)

Nomenclature

a Delamination length along taper
b Delamination length in thin region

[1] Research engineer and senior research engineer, respectively, U.S. Army Aerostructures Directorate, NASA Langley Research Center, Hampton, VA 23665-5225.
[2] Research scientist, Analytical Services and Materials, Inc., NASA Langley Research Center, Hampton, VA 23665-5225.

E_{11}, E_{22} Young's moduli parallel and transverse to the fiber direction
h Thickness of one ply
G Total strain energy release rate
G_I Mode I strain energy release rate
G_{II} Mode II strain energy release rate
G_{Imax} Mode I cyclic strain energy release rate
G_{12} Shear modulus
N Number of loading cycles to delamination onset
N_x Total load per unit width on symmetric half-laminate
P_{max} Maximum cyclic load on tapered laminates
R Ratio of minimum to maximum applied load
V_f Laminate fiber volume fraction
w Specimen width
β Taper angle
ν_{12} Poisson's ratio
σ_0 Applied tensile stress at laminate thick end
σ_n Interlaminar normal stress
τ_{nt} Interlaminar shear stress

Laminated composite structures with tapered thicknesses are currently being designed as a means of tailoring composite parts for specific performance requirements. The thickness of the laminates is typically reduced by dropping plies internally. However, these ply-drop locations are sources for delamination initiation under bending and tension loads. The low delamination durability of such configurations can result in high costs for repair and replacement of parts. In order to design tapered components with improved resistance to such delamination damage, analyses that model the failure mechanism are necessary.

In this study, the effect of tension fatigue loading on tapered laminates was investigated. Tapered specimens were manufactured from three different material types: S2/SP250 and S2/CE9000 glass/epoxies, and IM6/1827I, a graphite/epoxy that had an interleaf material at the delaminating interface. This interleaf is a toughened thermoset that inhibits delamination growth in laminates under static loads [1]. As Fig. 1 shows, under a tension load the continuous belt plies in the tapered region will try to straighten out. The test specimens were loaded in tension fatigue until they either delaminated unstably (as shown in Fig. 1), or reached between 10^6 and 2×10^7 cycles with no unstable delamination. In order to isolate the effect of the geometric discontinuity at the taper, and avoid the influences of matrix cracks and Pois-

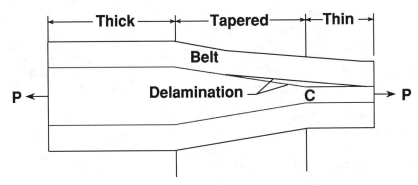

FIG. 1—*Delamination growth in tapered laminate with tension load.*

son's ratio mismatch at the free edge that can occur in many multi-angle laminates, all test specimens were fabricated with unidirectional [0°] plies only. The fatigue tests produced a curve relating the maximum cyclic load to the log of the number of loading cycles at the onset of sudden, unstable delamination.

In addition, a two-dimensional finite element (FE) analysis was used to calculate interlaminar stresses and strain energy release rates associated with delaminations of the type observed in the experiments. In Ref 2, this finite element analysis was used to model a multi-angle tapered layup under a static tension load. The analysis showed that the most likely location for delaminations to initiate was at the junction of the tapered and thin sections (Point C in Fig. 1). The analysis also showed that a delamination growing from this location initially had a strong Mode I component. This failure mode was also expected in the [0°] tapered laminates because the taper geometry was identical.

In Ref 3, double cantilever beam (DCB) tests were conducted in fatigue to characterize the delamination fatigue behavior of the test materials. These DCB tests produced a curve relating the maximum Mode I cyclic strain energy release rate, G_{Imax}, to the corresponding number of loading cycles, N, at which delamination growth begins. In this study, the data from Ref 3 were used with the calculated total strain energy release rates from the finite element analysis to predict the sudden unstable delamination in 0° tapered layups subjected to constant amplitude cyclic loads.

Experiments

Materials

Panels were made of three different materials: S2/SP250, a 121°C (250°F) cure glass/epoxy; 9176°C (350°F) cure glass/epoxy; and IM6/1827I, a 176°C (350°F) cure graphite/epoxy with a toughened thermoset adhesive layer, or interleaf, on one side of the prepreg. The S2/SP250 was manufactured by the 3M Company, the S2/CE9000 was manufactured by Ferro Corporation, and the IM6/1827I was manufactured by American Cyanamid. Eight-ply thick, flat panels were laid up in 0, 90, and ±45° orientations, cured, cut into coupons, and then were tested to measure lamina moduli and Poisson's ratio. Tapered panels were laid up at NASA Langley, in a tool supplied by Bell Helicopter Textron, such that the center of the panel was thicker than the two ends. All panels were cured according to the manufacturers' recommended curing cycles.

Figure 2 shows a drawing of a laminate cut from a cross section of the tapered panel. The laminates had 38 plies in the center thick region, and 26 plies in the thin region near either

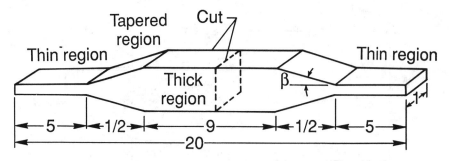

Note: Drawing not to scale

FIG. 2—*Laminate cut from tapered panel.*

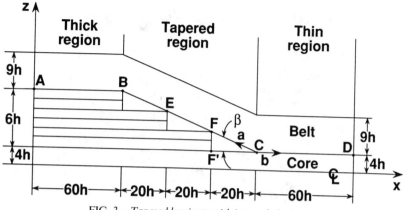

FIG. 3—*Tapered laminate with internal ply drops.*

end. The transition between the thick and thin regions was accomplished by dropping the internal plies two at a time, forming tapered regions that matched the angle, β, of the tool. Strips 25.4 mm (1 in.) wide were cut from each panel, and then cut in the middle of the thick region (Fig. 2) to yield two test coupons, each 254 mm (10 in.) long.

Figure 3 shows a schematic of the upper half of the tapered laminate, with the thick, tapered, and thin regions indicated. Each region had a length of 60-ply thicknesses, $60h$. The interior plies that run the entire length of the laminate are referred to as the core region, and the exterior plies, which also run the entire length of the laminate, enclosing the dropped and core plies, are called the belt plies, as indicated in Fig. 3. Figure 3 also shows the details in the tapered region. The internal plies were dropped at three locations, forming three regions in the taper, each having a length of $20h$, and yielding a taper angle, β, of 5.71°. Small resin pockets, with triangular cross sections, were assumed to form at the end of each pair of dropped plies. Figure 4 shows photographs of the edge of the tapered region for laminates of the three materials. In all cases, resin pockets are visible at the end of the terminated plies. However, depending on the accuracy of the layup, the two plies of a dropped pair often were not aligned exactly with each other, or the dropped plies were not aligned with the corresponding pair on the other side of the midplane. Furthermore, during the curing process, some of the belt plies and core plies either expanded to accommodate the resin flow, or assumed a curvature to facilitate a smoother transition from the thick to the thin regions (see Fig. 5). Hence, the tapered geometry in Fig. 3 was an idealized approximation of the actual detail in the tapered region of the laminates that were manufactured.

For this study, only 0° unidirectional tapered panels were tested so that only the contribution of the taper geometry to the onset of delamination could be studied. However, the IM6/1827 graphite/epoxy laminates were made using two different sequences of interleaf orientations. This resulted in two types of laminates, designated C and T (Fig. 6).

The C laminates were laid up so that the interleaf on all the prepreg plies faced toward the midplane, that is, toward the center line of the laminate thickness, as shown in Fig. 6a, where the location of the interleaf layer is indicated by the arrowhead. Hence, in the C laminates, there was tough interleaf present at every interface, but not along the bottom of the innermost resin pocket.

The T laminates were laid up such that the interleaf in the belt plies and core plies all faced the laminate midplane, as in the C laminates, but the interleaf in the pairs of dropped plies all faced towards each other, as indicated by the arrows in Fig. 6b. Hence, in the T laminates there

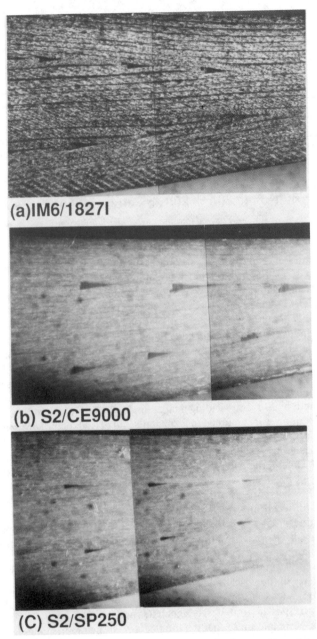

FIG. 4—*Photographs of dropped ply locations for three test materials.*

FIG. 5—*Resin pocket and curved plies in S2/CE9000.*

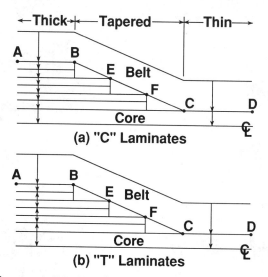

(a) "C" Laminates

(b) "T" Laminates

FIG. 6—*Interleaf locations in IM6/1827I laminates. Arrows indicate direction of interleaf in plies*

was interleaf at the interface between the belt and core plies in the thin region and between the belt and dropped plies in the tapered region; however, there was no interleaf between the core and dropped plies in the tapered region, or along the bottom of the innermost resin pocket.

Static Tests

Static tests were conducted on five flat laminates of each material to determine the basic material properties, and on two tapered laminates of each material to determine the elastic moduli of the thin and thick regions. Specimens were instrumented with extensometers and strain gages and loaded in a servohydraulic load frame. Strips of emery cloth were wrapped longitudinally around the specimen ends, covering the laminate width on both sides in the grips, to protect the laminate from being damaged by the grip teeth. The hydraulic grip pressure was set at 4.83 MPa (700 psi) for the glass/epoxy laminates and at 6.89 MPa (1000 psi) for the graphite/epoxy laminates.

Figure 7a shows the two 25.4-mm (1-in.)-long extensometers that were mounted on the thin and thick regions of the tapered laminates to measure the longitudinal modulus of each region. Specimens were loaded in stroke control at a rate of 0.254 mm/min (0.01 in./min). The load versus displacement plot for the thin and thick regions was recorded on an X-Y-Y' recorder during the loading.

Fatigue Tests

Tapered specimens were instrumented with an extensometer mounted over the tapered region to detect delamination onset (Fig. 7b) and loaded in a servohydraulic load frame in the same manner as in the static tests. Specimens were loaded statically in load control to the mean load, and then cycled sinusoidally at a maximum constant load amplitude corresponding to an R ratio of 0.1 at a frequency of 5 Hz. Specimens were cycled until the onset of unstable delamination was detected visually, audibly, or by monitoring an increase in displacement in

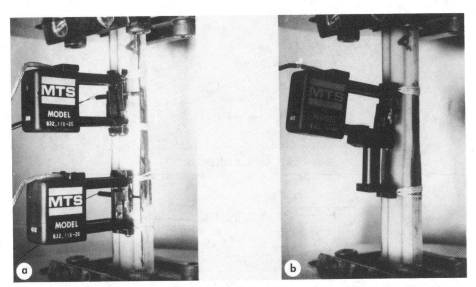

FIG. 7—*Photographs of test specimens in hydraulic load frame with mounted extensometers: (a) static test and (b) fatigue test.*

the tapered region. The latter technique was automated by monitoring a change in the voltage output of the extensometer. If a prescribed displacement change was exceeded, the function generator stopped the loading and the number of loading cycles was recorded. Using this technique, tests could run unattended and the machine would detect the onset of delamination at the junction of the thin and tapered regions. This was always an unstable delamination, and hence, was easily detected. Tests were run for several load levels to determine the number of cycles to unstable delamination onset as a function of the applied maximum cyclic load.

Analysis

In Ref 2 a simple two-dimensional finite element model was developed to determine strain energy release rates associated with delamination growth in laminates of the geometry shown in Fig. 3. Because the laminates considered in this study consisted of only 0° plies, a two-dimensional plane strain analysis was assumed to be sufficiently accurate for studying delamination resulting from the geometric discontinuity in the taper. Also, because of the symmetry of the laminates about the X-axis, it was necessary to model only half of the laminate. The model used eight-noded, isoparametric, parabolic elements with the smallest element size equal to one quarter of the ply thickness, or $h/4$. A refined mesh was used near the plydrop locations to capture the influence of the ply discontinuities on interlaminar stresses. A schematic of the mesh is shown in Fig. 8, where the dimensions in the Z-direction have been exaggerated for clarity. Collapsed eight-noded elements were used at the tips of the resin pockets. The nodes at the end of the thin region were constrained to zero displacement in both the X- and Z-directions and a uniform load per unit width of $N_x = 175.1$ kN/m (1000 lb/in.) was applied to the half-thickness at the thick end.

Because the continuous belt plies in the taper region of the laminate are at an angle, β, to the global coordinate system of the laminate, the moduli of these belt plies must be transformed through the taper angle. This transformation is explained in Appendix A.

Strain energy release rates were calculated using both the virtual crack-closure technique

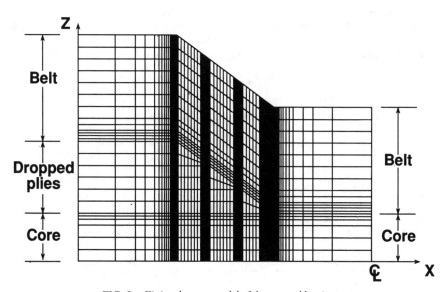

FIG. 8—*Finite element model of the tapered laminate.*

(VCCT) and using a global energy method. The VCCT method uses the local forces ahead of, and the relative displacements behind, the delamination tip to calculate the Mode I and Mode II components of the strain energy release rate, G_I and G_{II}, respectively. The global energy method calculates the total strain energy release rate, G, only. In this method, the difference in the work terms associated with two delamination lengths is divided by twice the difference in the delamination lengths. The result is assumed to be the total strain energy release rate for a delamination midway between these two locations.

Two different laminate configurations were modeled in this study. The first model had 7610 nodes and 2382 elements. It represents a laminate with no initial delamination, in which a delamination was allowed to grow from the junction of the thin and tapered regions (Point C in Fig. 3), either into the thin or tapered regions. The delamination is modeled as growing between the belt and dropped plies a distance, a, as shown in Fig. 3, keeping the thin section completely laminated; or, from Point C into the thin section between the belt and core plies, a distance, b, (Fig. 3) allowing no delamination in the tapered region. Duplicate nodes were created in the model along these interfaces and were constrained to act together. Different delamination lengths were then modeled by releasing the appropriate constraints.

A second, modified model, with 7699 nodes and 2382 elements, was developed to analyze delaminations observed in the test specimens. In this model, duplicate nodes were created along the interface between the dropped plies and core section, to the left of Point F' in Fig. 3, through the tapered region, and along the front edge of the dropped plies, between Points F and F'. Those nodes, as well as those in the tapered section between Points F and B, were released, creating the configuration shown in Fig. 9. As in the first model, different delaminations growing on either side of Point C were modeled by releasing the appropriate constraints. Both models were used to predict strain energy release rates for delamination growth starting at Point C in Fig. 3, for the S2/SP250, S2/CE9000, and IM6/1827I tapered laminates.

Results and Discussion

Experiments

Table 1 lists the basic material properties determined from the flat laminate tests. Table 2 lists the average measured longitudinal moduli of the thin and thick regions of the tapered laminates made from each of the three materials. Also listed in Table 2 are the average measured ply thicknesses and volume fractions, V_f, for the thin and thick regions. The thick regions

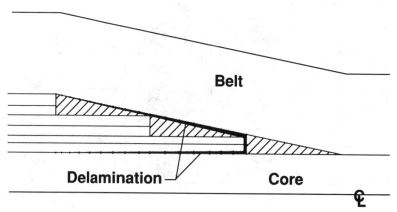

FIG. 9—*Schematic of initial delamination damage.*

TABLE 1—*Material properties.*

	S2/SP250	S2/CE9000	IM6/1827I	Neat Resin [4]
E_{11}	6.600×10^6 psi[a]	7.116×10^6 psi[a]	19.000×10^6 psi[a]	0.595×10^6 psi[a]
E_{22}	2.100×10^6 psi	2.464×10^6 psi	0.856×10^6 psi	0.595×10^6 psi
G_{12}	0.880×10^6 psi	1.100×10^6 psi	0.615×10^6 psi	0.224×10^6 psi
ν_{12}	0.275	0.303	0.361	0.330
V_f, %	57.0	59.0	53.5	$\cdots$

[a] 1 psi = 6.89 kPa.

TABLE 2—*Longitudinal moduli of thin and thick regions in tapered laminates.*

	$E_{11} \times 10^6$ psi[a]	ν_{12}	h, in.[b]	V_f, %
	S2/SP250			
Thin	7.730	0.272	0.0079	57.3
Thick	5.990	0.229	0.0081	55.5
	S2/CE9000			
Thin	7.333	0.271	0.0074	61.2
Thick	6.782	0.223	0.0078	57.6
	IM6/1827I			
Thin	20.85	$\cdots$	0.0079	54.7
Thick	17.39	$\cdots$	0.0082	52.7

[a] 1 psi = 6.89 kPa.
[b] 1 in. = 25.4 mm.

had larger ply thicknesses, and correspondingly lower fiber volume fractions, than the thin regions. Consequently, the moduli measured in the thin region were higher than the moduli measured in the thick region, and the average of the two measurements was similar to the moduli measured on the flat laminates (Table 1). The lamina properties used in the FE analysis were taken from Table 1. The properties given for the IM6/1827I material in Table 1 represent smeared properties for the fiber-reinforced epoxy plus the interleaf. The epoxy plies and interleaf were not discretely modeled in the analysis. The neat resin properties for a typical epoxy from Ref 4 are also given in Table 1.

For all of the IM6/1827I and S2/CE9000 specimens, some initial stable delamination occurred in the specimens before the final, unstable delamination initiated at Point C (Fig. 3). This preliminary delamination was not observed in the S2/SP250 laminates. As shown in Fig. 9, the initial delamination began with a resin crack at the tip of the innermost dropped plies, between the end of the plies and the resin pocket. Under continued loading, delaminations then grew stably from this crack through the tapered region toward the thick region, between the core and dropped plies, and between the belt and dropped plies, extending along the length of the tapered section and sometimes into the thick section a short distance. These delaminations usually formed on both sides of the midplane at a very low number of loading cycles and caused a slight decrease in the stiffness of the test specimens. However, the stiffness loss was not significant in any of the tests, and loading was continued until the final unstable delamination occurred. Reference to Figs. 6a and 6b shows that after this initial delamination occurred, Configurations C and T of the IM6/1827I graphite/epoxy were essentially the same.

Between Points F and D in Fig. 3, both types C and T had interleaf between the belt and resin pocket, and between the belt and core plies in the thin region, but not along the bottom of the resin pocket between C and F′.

The final unstable delamination failure usually occurred along the interface between the belt and dropped plies in the tapered region (with damage sometimes between the core and

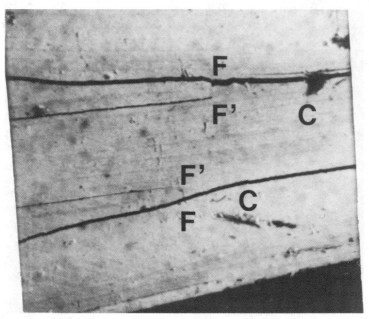

FIG. 10—*Final delamination failure in IM6/1827I.*

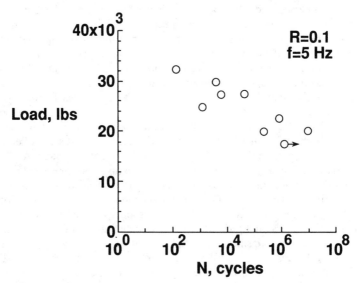

FIG. 11—*Maximum cyclic load as a function of cycles to unstable delamination onset in IM6/1827I, Type T.*

dropped plies also), and between the belt and core plies in the thin region. Figure 10 shows a photo of the damage in the vicinity of Point C for a type T graphite specimen. Unstable delamination was usually evident on both sides of the laminate midplane; however, it was usually more extensive on one side, with the delaminations extending further and occasionally with delaminations at other interfaces. For specimens where the innermost dropped plies were not terminated at locations symmetric about the midplane, the most extensive damage occurred on the side where the dropped plies extend further toward the thin region.

Figures 11 through 14 show the maximum cyclic loads versus the log of the number of loading cycles to unstable delamination onset for each of the four different specimen types tested. Types T and C of the graphite/epoxy specimens showed similar results with moderate scatter (Figs. 11 and 12). Arrows on the data points indicate that the test was terminated before the

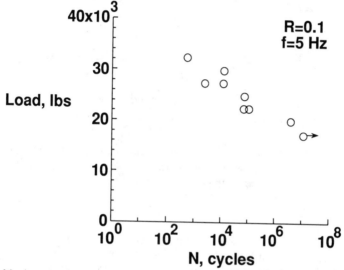

FIG. 12—*Maximum cyclic load as a function of cycles to unstable delamination onset in IM6/18271, Type C.*

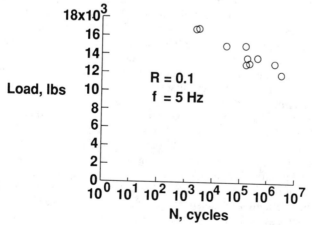

FIG. 13—*Maximum cyclic load as a function of cycles to unstable delamination onset in S2/CE9000.*

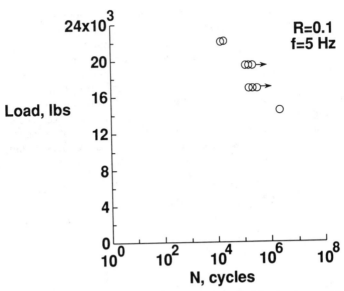

FIG. 14—*Maximum cyclic load as a function of cycles to unstable delamination onset in S2/SP250.*

specimen delaminated. Figures 13 and 14 show the results for the S2/CE9000 and S2/SP250 tests, respectively. The results show little scatter, with the S2/SP250 specimens having slightly longer lives than the S2/CE9000 for the same maximum cyclic load.

Finite Element Model

The finite element model was used to calculate interlaminar normal and shear stress distributions using the three sets of material properties and the epoxy neat resin properties given in Table 1. Results for the laminate without initial stable delamination are shown in Figs. 15 and 16 for the IM6/1827I material. Figure 15 shows the interlaminar normal stress along interface ABCD normalized by the applied tensile stress (σ_n/σ_0). The results suggest that stress singularities may exist at the ply drop locations, as demonstrated by the steep peaks at those locations. However, for the same mesh refinement, the magnitude was highest at Point C, with the interlaminar normal stresses tensile on both sides, indicating that Point C is the most likely place for a Mode I delamination to start. Figure 16 shows the normalized interlaminar shear stresses (τ_{nl}/σ_0) along the same interface. The shear stresses show high peaks at the ply drop locations. These results are consistent with the results of Ref 2 for a glass/epoxy laminate that was not unidirectional. Therefore, it is apparent that these results are due to the taper configuration and not the layup or material modeled.

Interlaminar normal and shear stresses were also calculated for a second model with a resin crack between Points F and F′ (see Fig. 3), that is, between the end of the dropped plies and the adjacent resin pocket, as shown by the inset in Figs. 17 and 18. Figure 17 suggests that singularities may exist at the ply drops. As in Fig. 15, stresses on both sides of Point C are tensile. Therefore, the initial delamination damage does not seem to change the location where the final delamination begins. Comparison of Figs. 16 and 18 shows that for the resin crack configuration, the interlaminar shear stresses again show peaks at the ply drop locations, but in Fig. 18 the shear stress to the right of Point F becomes negative very close to the resin crack.

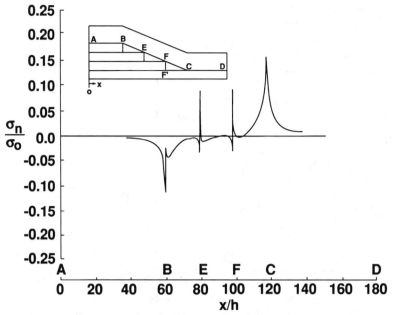

FIG. 15—*Interlaminar normal stresses in IM6/1827I laminate with no initial delamination.*

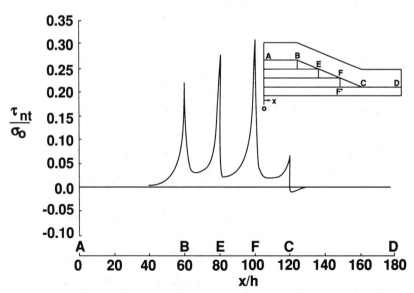

FIG. 16—*Interlaminar shear stresses in IM6/1827I laminate with no initial delamination.*

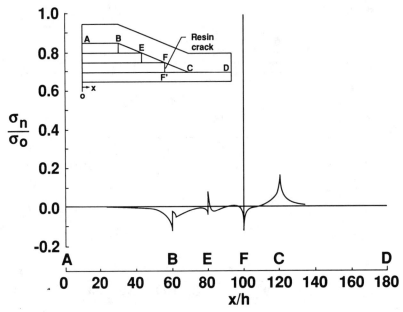

FIG. 17—*Interlaminar normal stresses in IM6/1827I laminate with resin crack.*

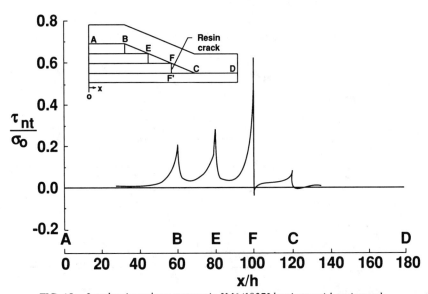

FIG. 18—*Interlaminar shear stresses in IM6/1827I laminate with resin crack.*

The distribution of interlaminar normal and shear stresses for the two glass materials was similar to the graphite.

For each of the three material systems, the finite element model was used to calculate G as a delamination was extended into the taper or thin regions. Delaminations that extended into the taper, along the belt and dropped ply interface from Point C, were designated as Length a, as shown in Fig. 3. Delaminations into the thin section, along the belt and core interface from Point C, were designated as length b, as shown in Fig. 3. First, a delamination was assumed to grow in the tapered section, from Point C between the belt and dropped plies, with no delamination in the thin section ($b = 0$). At each new delamination length a, G was calculated. Similarly, delamination growth into the thin section with no delamination in the tapered section ($a = 0$) was modeled, and values of G were calculated. Calculations were performed for the laminate models with, and without the initial stable delamination. In all cases, the results are given as a normalized value of Gh/N_x^2 in order to permit easier comparison between configurations with different thicknesses or applied loads. Results are shown for the model without the initial stable delamination, with the IM6/1827I material in Fig. 19. Results are given for both the VCCT and energy based methods and show good agreement. Similar good agreement between the VCCT and energy based results was found for all the cases analyzed; therefore, in the remainder of the figures, curves are shown that represent the VCCT results only.

As Fig. 19 shows, G increases rapidly as the delamination grows in either direction, and reaches a peak after the delamination has grown a few ply thicknesses. As the delamination is allowed to grow further, G drops off. The peak value of G was always higher for delamination growth into the taper section than for growth into the thin section. Figure 20 shows the corresponding results for the initially delaminated laminate along with the results for the laminate model with no initial delamination from Fig. 19. As Fig. 20 shows, G is higher in the laminate with initial delaminations at every delamination length, with the highest value of G corresponding to delamination growth in the tapered region. This same pattern was observed for all three material systems modeled. Results for the models with and without the initial delamination, for the S2/CE9000 and S2/SP250 materials are shown in Figs. 21 and 22, respectively. As these two figures show, for the initially delaminated models with delamination growing into the taper, G tends to quickly reach a peak, drops slightly, and then continues to rise as the delamination length increases.

Figure 23 shows the ratio of G_I to total G for the tapered laminate modeled, with delamination growth in the tapered region and in the thin region. The initial delamination growth is primarily Mode I for all three materials. The Mode III component for this unidirectional laminate is assumed to be zero. Hence, in Fig. 23, as the delamination extends into the taper and the Mode I contribution decreases, the Mode II contribution increases until the Mode II com-

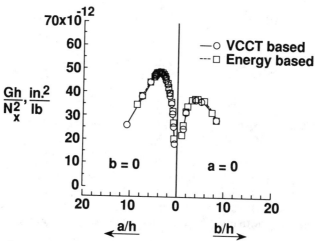

FIG. 19—*Normalized strain energy release rates for IM6/1827I laminate with no initial delamination.*

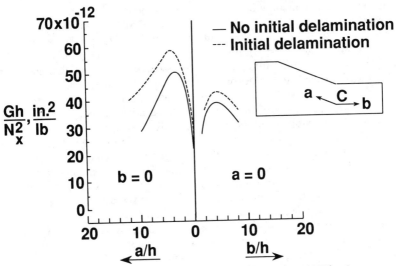

FIG. 20—*Normalized strain energy release rates for IM6/1827I laminate.*

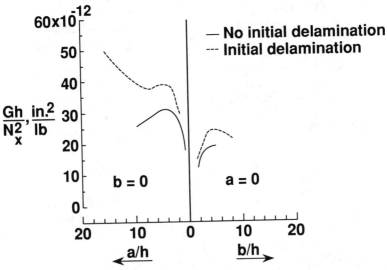

FIG. 21—*Normalized strain energy release rates for S2/CE9000 laminate.*

ponent comprises about 85% of total G near $a = 20h$. For the two glass materials, the percentage of Mode I starts out slightly lower and drops faster, compared to the graphite material, which remains largely Mode I for several ply thicknesses. For all three materials, Fig. 23 shows that the delamination growth into the thin region is almost entirely Mode I for all delamination lengths.

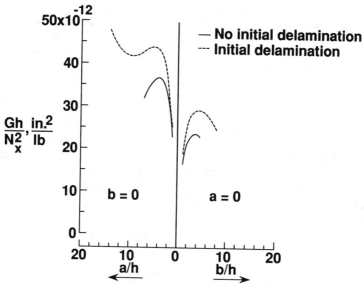

FIG. 22—*Normalized strain energy release rates for S2/SP250 laminate.*

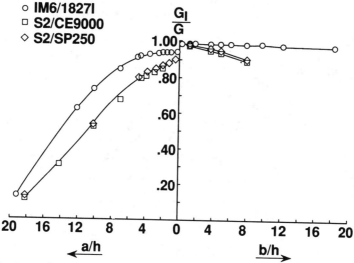

FIG. 23—*Ratio of G_I to total G for delamination growth in laminate with initial delamination.*

Failure Predictions

In Ref *3*, the delamination fatigue behavior of the current test materials under Mode I loading was characterized using DCB tests. The tests were terminated when the initial delamination growth was detected. By testing at various load levels, a curve was generated that related the maximum cyclic strain energy release rate, G_{Imax}, to the log of the number of loading cycles, N, at the onset of delamination. These tests were conducted for the three materials used in the current study. Resulting G-N curves are shown in Figs. 24 through 26 for the IM6/1827I, S2/ CE9000, and S2/SP250 materials, respectively. The data are given in Tables 3 through 5.

An attempt was made to use these G-N data along with the finite element analysis to predict unstable delamination onset during fatigue for the tapered laminates. It is sufficient to use the measured Mode I data from the DCB tests, since the finite element model showed that the initial delamination growth is almost completely Mode I. Since no attempt was made in Ref 3 to produce a statistical curve fit through the DCB data, and since it was not a goal of this study to produce such a curve, the individual DCB data points were used in these predictions. For each of the three materials, the appropriate finite element model was chosen to match the actual failure behavior observed in the tests. The first peak value of Gh/N_x^2 was designated $(Gh/ N_x^2)_{FE}$ (Tables 3 through 5). It was postulated that unstable delamination growth under cyclic loading would initiate in the tapered laminate when the total G value calculated from the FE analysis equaled the cyclic G_{Imax} at which delamination initiated in the DCB tests. For each data point shown in Figs. 24 through 26 (and tabulated in Tables 3 through 5), we can write

$$\left(\frac{Gh}{N_x^2}\right)_{FE} \frac{N_x^2(N)}{h} = G_{Imax}(N) \tag{1}$$

where h is the average measured ply thickness from Table 2 and $N_x(N)$ is the applied maximum cyclic load per unit width on the tapered laminate half-thickness. The maximum cyclic

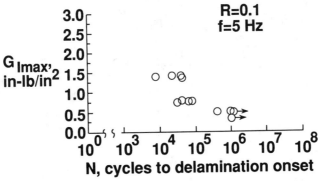

FIG. 24—G_{Imax} as a function of cycles to delamination onset for IM6/1827I.

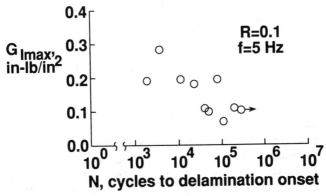

FIG. 25—G_{Imax} as a function of cycles to delamination onset for S2/CE9000.

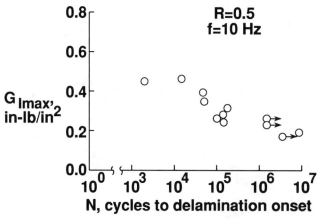

FIG. 26—G_{Imax} as a function of cycles to delamination onset for S2/SP250.

TABLE 3—IM6/1827I graphite/epoxy.

G-N Data (DCB Tests)		Calculated Delamination Onset Loads, 1 lb = 4.95 N	
G_{Imax}, in.-lb/in.^{2}	N, Cycles	Tapered Region[a]	Thin Region[b]
1.417	7 440	27 723	32 858
0.800	41 150	20 830	24 689
0.509	1 015 480	16 615	19 693
0.417	36 840	27 723	32 858
0.794	78 730	20 752	24 596
0.356	1 000 000[c]	13 895	16 469
0.502	1 129 340[c]	16 501	19 557
0.510	415 650	16 632	19 712
0.754	30 260	20 222	23 968
1.420	21 000	27 752	32 892
1.370	38 410	27 259	32 308
0.777	65 570	20 529	24 331

[a] $(Gh/N_x^2)_{FE} = 0.856 \times 10^{-14}$ m^2/N (59.0×10^{-12} in.^{2}/lb).
[b] $(Gh/N_x^2)_{FE} = 0.609 \times 10^{-14}$ m^2/N (42.0×10^{-12} in.^{2}/lb).
[c] Specimen did not reach unstable failure.

load $P_{max}(N)$ on the symmetric tapered laminates is then

$$P_{max}(N) = 2wN_x(N) \qquad (2)$$

where w is the laminate width. Substituting Eq 2 in Eq 1 and solving for $P_{max}(N)$ gives

$$P_{max}(N) = \left[\frac{G_{Imax}(N)h(2w)^2}{\left(\dfrac{Gh}{N_x^2}\right)_{FE}} \right]^{1/2} \qquad (3)$$

TABLE 4—*S2/CE9000 glass/epoxy.*

G-N Data (DCB Tests)		Calculated Delamination Onset Loads, 1 lb = 4.45 N	
G_{Imax}, in.-lb/in.^{2}	N, Cycles	Tapered Region[a]	Thin Region[b]
0.096	55 000	8 651	10 804
0.107	212 300	9 133	11 407
0.195	12 370	12 329	15 399
0.070	120 024	7 387	9 226
0.107	44 510	9 133	11 407
0.196	85 240	12 360	15 438
0.100	284 570	8 829	11 027
0.180	25 000	11 855	14 795
0.192	2 000	12 234	15 280
0.286	4 000	14 931	18 649

[a] $(Gh/N_x^2)_{FE} = 0.566 \times 10^{-14}$ m^2/N (39.0 $\times$ 10^{-12} in.^{2}/lb).
[b] $(Gh/N_x^2)_{FE} = 0.363 \times 10^{-14}$ m^2/N (25.0 $\times$ 10^{-12} in.^{2}/lb).

TABLE 5—*S2/SP250 glass/epoxy.*

G-N Data (DCB Tests)		Calculated Delamination Onset Loads, 1 lb = 4.75 N	
G_{Imax}, in.-lb/in.^{2}	N, Cycles	Tapered Region[a]	Thin Region[b]
0.310	182 000	16 788	20 331
0.223	1 486 000[c]	14 231	17 243
0.347	50 000	17 751	21 510
0.246	1 440 000[c]	14 946	18 111
0.260	100 000	15 366	18 619
0.242	148 000	14 824	17 963
0.277	144 000	15 860	19 218
0.167	3 449 000[c]	12 315	14 922
0.462	15 000	20 483	24 819
0.392	47 000	18 867	22 862
0.187	8 500 000	13 031	15 790

[a] $(Gh/N_x^2)_{FE} = 0.537 \times 10^{-14}$ m^2/N (37.0 $\times$ 10^{-12} in.^{2}/lb).
[b] $(Gh/N_x^2)_{FE} = 0.348 \times 10^{-14}$ m^2/N (24.0 $\times$ 10^{-12} in.^{2}/lb).
[c] Specimen did not reach unstable failure.

For each value of N in Tables 3 through 5, the corresponding G_{Imax} was used in Eq 3 to solve for the predicted maximum cyclic load, P_{max}, assuming delamination grew in the tapered region first. Tables 3 through 5 list the values of $(Gh/N_x^2)_{FE}$ and the predicted values of $P_{max}(N)$ for each set of DCB data. The predicted maximum cyclic loads and corresponding number of cycles to delamination onset are plotted with the tapered laminate test results in Figs. 27 through 30.

Figures 27 and 28 show the predicted failure loads for types T and C of the IM6/1827I material assuming the delamination initiated in the tapered region, compared with the actual test data. The predictions are based on the finite element results for the damaged laminate model. The predictions show good agreement with the test data for both types T and C laminates.

Figure 29 shows the values predicted for the S2/CE9000 laminates using the initially delaminated model, along with the measured failure loads. These predictions are conservative compared to the data. Close examination of the resin pockets of these specimens shows that the plies on both sides of the resin pockets tend to curve in toward the resin, creating a much smaller local taper angle than was modeled (see Fig. 5). A smaller taper angle in the model would result in lower calculated values of Gh/N_x^2, which would in turn yield higher predicted $P_{max}(N)$ values.

For the S2/SP250 laminates, the predictions were calculated using results of the laminate model without the initial delamination, since initial stable delamination growth was not observed in the tapered laminate fatigue tests of this material. Figure 30 shows the predictions and test results. The predictions are very good for this case. However, as indicated in Fig. 26, the DCB data for this material were generated using $R = 0.5$, rather than $R = 0.1$, as for the

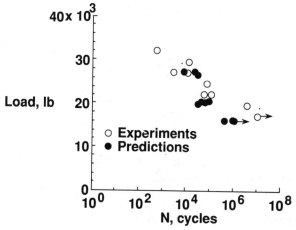

FIG. 27—*Measured and calculated delamination onset in fatigue for IM6/1827I laminates, Type T, assuming delamination onset in tapered region.*

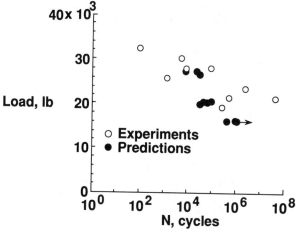

FIG. 28—*Measured and calculated delamination onset in fatigue for IM6/1827I laminates, Type C, assuming delamination onset in tapered region.*

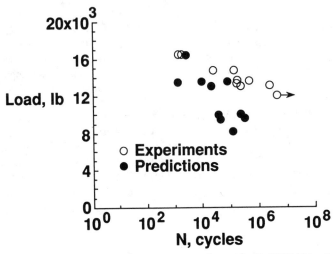

FIG. 29—*Measured and calculated delamination onset in fatigue for S2/CE9000 laminates, assuming delamination onset in tapered region.*

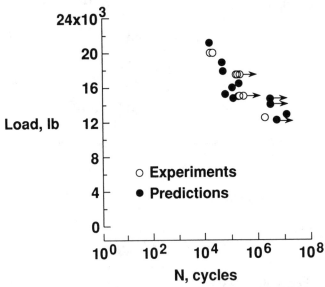

FIG. 30—*Measured and calculated delamination onset in fatigue for S2/SP250 laminates, assuming delamination onset in tapered region.*

other DCB tests and for the tapered laminate tests. Testing at $R = 0.1$ may have the effect of decreasing the cycles to delamination onset, for the same values of G_{Imax}, compared to testing at $R = 0.5$, thereby shifting the data in Fig. 26 down slightly. In that case, $P_{max}(N)$ values calculated using $R = 0.1$ data would be slightly lower than the calculated values (from the $R = 0.5$ data,) and the predictions would be then more conservative compared to the test data. The predictions in Figs. 27 through 30 seem reasonable, however, considering that the geometry of the resin pocket is very idealized. As discussed in the materials section, the resin pocket geometry of the laminates may be quite different from the model due to problems associated with the layup and manufacture of the panels.

Because the peak calculated value of Gh/N_x^2 is higher for delamination onset in the tapered region, it is logical to assume that delamination grows in the tapered region first. However, it is also possible that the thick resin pocket actually inhibits the initiation of delamination growth in that direction, so that delamination grows in the thin region first. Hence, the peak values of Gh/N_x^2 for delamination growth into the thin region were also used with Eq 3 to calculate delamination onset loads assuming the delamination grows into the thin region first. The lower values of $(Gh/N_x^2)_{FE}$ for growth into the thin region have the effect of increasing the predicted delamination onset loads. Figures 31 through 33 show the experimental and calculated results for the IM6/1827 laminates, types T and C, and the S2/CE9000 laminates, respectively. In all cases, the predictions agree fairly well with the experimental results. Results for the S2/SP250 laminates are given in Fig. 34, where the predictions are now slightly higher than the experimental results, and are unconservative; however, the agreement is still reasonable. The calculated values shown in Figs. 31 through 34 are tabulated in Tables 3 through 5.

The fact that the calculated Gh/N_x^2 values decrease after reaching a peak, seems to indicate that stable delamination growth should be expected in the tapered laminates. However, it was shown in Ref 2 that the calculated peak value of G for delamination growth into the taper (or thin) region increases as the fixed delamination length in the thin (or taper) region increases. Therefore, the strain energy release rate increases continually as the delamination grows in either direction from Point C. This results in unstable delamination growth initiating at Point C and growing in both directions.

Conclusions

Tapered [0°] laminates of four different types were tested in cyclic tension until they delaminated. The specimen types consisted of S2/CE9000 glass/epoxy, S2/SP250 glass/epoxy, and

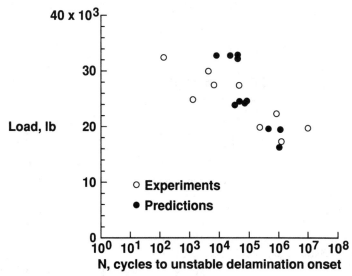

FIG. 31—*Measured and calculated delamination onset in fatigue for IM6/1827I laminates, Type T, assuming delamination onset in thin region.*

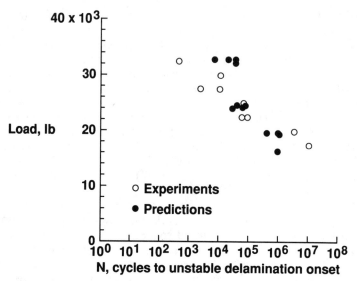

FIG. 32—*Measured and calculated delamination onset in fatigue for IM6/1827I laminates, Type C, assuming delamination onset in thin region.*

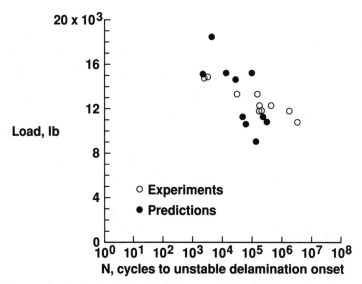

FIG. 33—*Measured and calculated delamination onset in fatigue for S2/CE9000 laminates, assuming delamination onset in thin region.*

two different specimen types of IM6/1827I graphite/epoxy that had a toughened interleaf material at the delaminating interfaces. The delaminations originated at the junction of the tapered and thin sections and grew unstably, both along the taper and into the thin region. A finite element model of the tapered laminate was used to calculate strain energy release rates

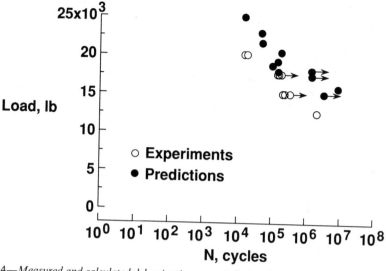

FIG. 34—*Measured and calculated delamination onset in fatigue for S2/SP250 laminates, assuming delamination onset in thin region.*

associated with delamination growth along these interfaces for each of the materials tested. The model showed that the initial delamination is primarily Mode I for all the cases. In addition, double-cantilever beam (DCB) test data were used to characterize the Mode I fatigue delamination behavior of the materials. The DCB results were used along with the finite element calculations to predict delamination onset in the tapered laminates as a function of loading cycles. The following conclusions were reached:

1. Under tensile fatigue loading, the IM6/1827I and S2/CE9000 tapered laminates experienced some initial stable delamination that did not result in significant stiffness loss. For all the test laminates, the final delamination was unstable, initiating at the junction of the thin and tapered regions, and growing in both directions at the interface between the belt and the underlying plies.

2. Finite element calculations showed that the presence of the initial delamination damage increases the strain energy release rate for delamination from the junction point between the thin and tapered regions.

3. The finite element model, when used with fatigue delamination durability data generated from DCB fatigue tests did a reasonable job of predicting unstable delamination onset loads in fatigue for the tapered laminates.

4. Delamination onset predictions, which assumed that the delamination grew first in the tapered region, tended to be conservative. This may be due to differences between the idealized taper angles that were modeled and the less severe taper angles that formed in the specimens during manufacture as a result of the expansion and movement of the plies to accommodate the resin flow at the taper.

5. Although the peak calculated values of Gh/N_x^2 were higher for delaminations initiating in the tapered region, the thick resin pocket may inhibit initial delamination growth in that direction. Calculated values of delamination onset load, assuming initial delamination growth in the thin region, agreed well with the measured delamination onset loads for all cases.

APPENDIX I

Transformation of Stiffness Coefficients

The following transformation was used in Ref 2 where tapered laminates with a multi-angle layup were studied. Since the laminates in the current study were all [0°] unidirectional, there is no rotation θ about the 3-axis. However, the complete transformation is included here for a general case.

The stress-strain relationships for each lamina are transformed from the material coordinate system 1-2-3 (Fig. 35) to the global system X-Y-Z using the following procedure. The three-dimensional stress-strain relationship for a ply in the material coordinate system is

$$\{\sigma\}_{123} = [C]\{\epsilon\}_{123} \tag{4}$$

where

$$\{\sigma\}_{123} = \begin{Bmatrix} \sigma_{11} \\ \sigma_{22} \\ \sigma_{33} \\ \sigma_{12} \\ \sigma_{23} \\ \sigma_{13} \end{Bmatrix} \text{ and } \{\epsilon\}_{123} = \begin{Bmatrix} \epsilon_{11} \\ \epsilon_{22} \\ \epsilon_{33} \\ \epsilon_{12} \\ \epsilon_{23} \\ \epsilon_{13} \end{Bmatrix}.$$

and $[C]_{6\times6}$ is a matrix that can be determined from elastic constants. Following similar notations, the stress-strain relationships for a lamina in the global system can be written as

$$\{\sigma\}_{XYZ} = [C]'\{\epsilon\}_{XYZ} \tag{5}$$

The matrix $[C]'$ is obtained from matrix $[C]$ by rotating the material system 1-2-3 (Fig. 35) to the global coordinate system X-Y-Z through two rotations; that is, a rotation (θ) about the 3 (or Z') axis, and then a rotation (ϕ) about the Y' (or Y) axis [5]. The transformed stiffness coefficient matrix, $[C]'$, is obtained from the material stiffness coefficient matrix, $[C]$, as

$$[C]'_{6\times6} = [T_\phi]_{6\times6}[T_\theta]_{6\times6}[C]_{6\times6}[T_\theta]^T_{6\times6}[T_\phi]^T_{6\times6} \tag{6}$$

where $[T_\theta]$ and $[T_\phi]$ are defined in terms of the appropriate angle as

$$[T_\theta] = \begin{bmatrix} \cos^2\theta & \sin^2\theta & 0 & 2\cos\theta\sin\theta & 0 & 0 \\ \sin^2\theta & \cos^2\theta & 0 & -2\cos\theta\sin\theta & 0 & 0 \\ 0 & 0 & 1.0 & 0 & 0 & 0 \\ -\cos\theta\sin\theta & \cos\theta\sin\theta & 0 & \cos^2\theta-\sin^2\theta & 0 & 0 \\ 0 & 0 & 0 & 0 & \cos\theta & -\sin\theta \\ 0 & 0 & 0 & 0 & \sin\theta & \cos\theta \end{bmatrix}$$

and

$$[T_\phi] = \begin{bmatrix} \cos^2\phi & 0 & \sin^2\phi & 0 & 0 & 2\cos\phi\sin\phi \\ 0 & 1.0 & 0 & 0 & 0 & 0 \\ \sin^2\phi & 0 & \cos^2\phi & 0 & 0 & -2\cos\phi\sin\phi \\ 0 & 0 & 0 & \cos\phi & \sin\phi & 0 \\ 0 & 0 & 0 & -\sin\phi & \cos\phi & 0 \\ -\cos\phi\sin\phi & 0 & \cos\phi\sin\phi & 0 & 0 & \cos^2\phi-\sin^2\phi \end{bmatrix}$$

The superscript, T, in Eq 6 denotes the transpose of the matrix. Furthermore, the plane strain conditions require that $\epsilon_{YY} = \epsilon_{XY} = \epsilon_{YZ} = 0$. Incorporating these conditions in Eq 5 yields the stress-strain relation-

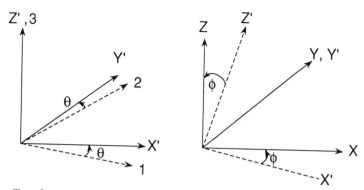

FIG. 35—*Transformation from material coordinate system to global coordinate system:* (left) *1-2-3 material system rotation θ about 3-axis and* (right) *X-Y-Z global coordinate system rotation φ about Y'-axis.*

$$\{\sigma\}_{XZ} = [C]'_{XZ}\{\epsilon\}_{XZ} \tag{7}$$

where

$$\{\sigma\}_{XZ} = \begin{Bmatrix} \sigma_{XX} \\ \sigma_{ZZ} \\ \sigma_{XZ} \end{Bmatrix} \text{ and } \{\epsilon\}_{XZ} = \begin{Bmatrix} \epsilon_{XX} \\ \epsilon_{ZZ} \\ \epsilon_{XZ} \end{Bmatrix}$$

The matrix $[C]'_{XZ}$ is obtained from the global $[C]'$ matrix as

$$[C]'_{XZ} = \begin{bmatrix} C'_{11} & C'_{13} & C'_{16} \\ C'_{31} & C'_{33} & C'_{36} \\ C'_{61} & C'_{63} & C'_{66} \end{bmatrix} \tag{8}$$

References

[1] Chan, W. S., Rogers, C., and Aker, S., "Improvement of Edge Delamination Strength of Composite Laminates Using Adhesive Layers," *Composite Materials: Testing and Design (Seventh Conference) ASTM STP 893,* J. M. Whitney, Ed., American Society for Testing and Materials, Philadelphia, 1986, p. 266.

[2] Salpekar, S. A., Raju, I. S., and O'Brien, T. K., "Strain Energy Release Rate Analysis of Delamination in Tapered Laminate Subjected to Tension Load," *Proceedings,* American Society for Composites, Third Technical Conference, Seattle, WA, Sept. 1988, pp. 642–654.

[3] Martin, R. H. and O'Brien, T. K., "Characterizing Mode I Fatigue Delamination of Composite Materials," *Proceedings,* Fourth Technical Conference of the American Society for Composites, Blacksburg, VA, Oct. 1989, pp. 257–266.

[4] Shivakumar, K. N. and Crews, J. H., Jr., "Bolt Clampup Relaxation in a Graphite/Epoxy Laminate," *Long Term Behavior of Composites, ASTM STP 813,* T. K. O'Brien, Ed., American Society for Testing and Materials, Philadelphia, 1983, pp. 5–22.

[5] Jones, R. M., *Mechanics of Composite Materials,* Scripta, Washington, DC, 1975, pp. 48–51, 325–326.

Erian A. Armanios[1] *and Levend Parnas*[1]

Delamination Analysis of Tapered Laminated Composites Under Tensile Loading

REFERENCE: Armanios, E. A. and Parnas, L., "**Delamination Analysis of Tapered Laminated Composites Under Tensile Loading,**" *Composite Materials: Fatigue and Fracture (Third Volume), ASTM STP 1110*, T. K. O'Brien, Ed., American Society for Testing and Materials, Philadelphia, 1991, pp. 340–358.

ABSTRACT: A study was conducted to analyze tapered composite laminates under tensile loading. A tapered construction made of S2/SP250 glass/epoxy laminate was used to achieve a thickness reduction using three consecutive dropped plies over a distance of 60 ply thicknesses. The principle of minimum complementary potential energy was used to determine interlaminar stresses. The interlaminar peel stress distribution shows a higher tensile intensity at the taper/thin portion juncture. The total strain energy release rate is determined using a simplified membrane model. Results are compared with a finite element simulation.

KEY WORDS: fiber-reinforced composite materials, strain energy release rate, delamination, fracture mechanics, fatigue (materials), fracture, composite materials

The use of composites in primary structures is ever increasing in commercial and military aircraft applications due, mainly, to their potential for creating significant weight savings, damage tolerance, and elastic tailoring. Some of the distinguishing features in the use of composites are substantial reductions in the number of fasteners due to both reduced part count and widespread use of co-curing and secondary bonding.

Modern composite rotor hub designs utilize this concept and introduce hingeless and bearingless structural components to reduce weight, drag, and number of parts. For example, the flapping flexure region of some composite rotor hubs are tapered in order to create an effective hinge for elastic tailoring. The tapered design is achieved by internally dropping a number of plies. However, these ply drops introduce geometric and material discontinuities around their locations. These discontinuities generate large interlaminar stresses that may cause delamination and ultimately lead to premature failure. A knowledge of the interlaminar stresses and energy release rate distribution is essential to understanding the failure mechanisms in tapered composites and designing against them.

Most of the efforts associated with the analysis of delamination in composites have been limited to uniform laminates. However, fewer numbers of studies are concerned with tapered constructions.

An analysis of a single-step ply drop configuration under tensile and compressive loading has been presented in Ref *1*. Interlaminar shear and peel stresses in the vicinity of the ply drop-off were estimated using a finite element simulation. The elements used were either eight-noded quadrilateral, or six-noded triangular, isoparametric elements. The failure predictions did not show good agreement with test results. This may be due to the stress failure criteria

[1] Assistant professor and NATO scholar, respectively, Georgia Institute of Technology, Atlanta, GA 30332.

used and the accuracy of the interlaminar stresses predicted by the finite element analysis in the vicinity of the ply dropoff.

Curry et al. [2] conducted experiments on graphite/epoxy tapered laminates subjected to compression loads. They also presented a three-dimensional finite element analysis to obtain the three-dimensional stress state. Failure was predicted using an interlaminar stress criterion.

Wu and Weber [3] analyzed a laminate with a single ply drop, using a quasi-three-dimensional isoparametric finite element method. This was developed for the stress analysis of variable thickness laminated plates. The plates have an infinite width and are subjected to a uniform in-plane loading. Numerical results for interlaminar stresses were given for a single step plate with various stacking sequence and fiber orientations. The effect of resin filler on the stress distribution was also analyzed.

Fish and Lee [4] analyzed a tapered beam having three ply drops using a finite element formulation. A combination of hybrid- and displacement-based models was used in the analysis. Interlaminar stress values averaged over a ply thickness distance were used in connection with the Tsai-Wu failure criteria to predict delamination onset. The results showed good agreement with experimental failure loads.

Salpekar et al. [5] presented a finite element analysis of a tapered composite laminate subjected to a tensile load. The elements used were an eight-noded isoparametric element. The laminate was made of glass/epoxy material. The tapered geometry consisted of a group of plies dropped at three distinct locations as shown in Fig. 1. Interlaminar stresses, total strain energy release rate, and the distribution of energy release rate components were determined. The variation of strain energy release rate with crack growth indicated a simultaneous growth along the taper and thin section.

In this study, a simple extensional model is used to analyze tapered laminates subjected to tensile loading. A generic configuration similar to the laminate used in Ref 5 is shown in Fig. 1 where a 38-ply-thick laminate with a stacking sequence of $([0_7/\pm45]/[\pm45]_3/[0/\pm45/0])_s$ is reduced to 26-ply laminate of $([0_7/\pm45]/[0/\pm45/0])_s$ by dropping three inner sets of plies. The top set of plies with a $[0_7/\pm45]$ layup, is referred to as the belt section, and the inner plies of $[0/\pm45/0]$ layup constitutes the core section. Due to the symmetry, only one half of it is considered.

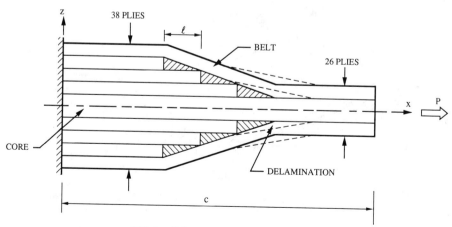

FIG. 1—*Edge view of the tapered structure.*

The basic analysis approach adopted here utilizes two levels of modeling, a global and local scale. The global scale is concerned with overall generalized forces and strains such as axial force and displacements leading to the global energy balance of the laminate. A simple consistent deformation assumption is the foundation of this model. The total strain energy release rate is determined from the work done by the external loads. The work is based on the axial stiffness of different elements in the tapered configuration. The distribution of interlaminar stresses is determined on the basis of the equilibrium conditions of the belt section elastically supported by the dropped inner plies, using the principle of minimum complementary potential energy.

Analysis

The tapered laminate made of S2/SP250 glass/epoxy and shown in Fig. 1 is assumed to be fixed at $x = 0$, and subjected to an axial load at $x = c$. The taper is achieved by dropping three inner sets of plies a distance ℓ apart.

As a result of the tapered geometry, an interlaminar tensile peel stress is created between the belt and core regions at the juncture of the taper and thin uniform sections [5]. These stresses can result in a pop-off of the belt from the core initiating a delamination. This is schematically shown in Fig. 1. The objective of this work is to develop a simple modeling approach for the delamination analysis between the belt and inner plies. The total strain energy release rate associated with a delamination at the interface between the belt and the inner regions is presented first. It is based on the rate of change of work done by the applied load with delamination growth. This is followed by the determination of the interlaminar shear and peel stress distribution based on the equilibrium of the belt region.

Strain Energy Release Rate

The sublaminate modeling of cracked laminate configurations and corresponding sublaminate stiffnesses are shown in Figs. 2 and 3, respectively. The laminate is assumed to be under a plane stress state coinciding with the x-z plane. A delamination between belt and core sections is assumed to grow parallel to the belt direction in the tapered and thin uniform sections.

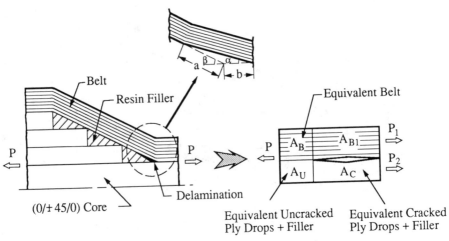

FIG. 2—*Sublaminate modeling approach.*

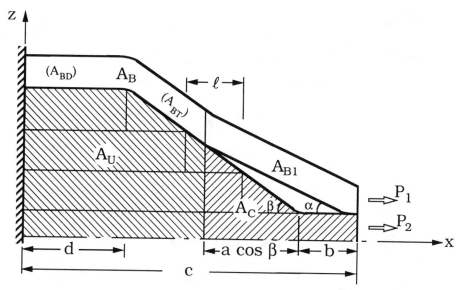

FIG. 3—*Dependency of core and belt stiffnesses on delamination.*

The uncracked region of the laminate is fixed at the left end and subjected to an axial force, P, while the axial loads in the cracked belt and inner regions at the right end are denoted by P_1 and P_2, respectively, as shown in Fig. 3.

The delaminations in the taper and thin uniform sections are denoted by a and b, respectively. The inner plies in the taper portion are modeled by two equivalent sublaminates. The stiffness properties are smeared to obtain the effective cracked and uncracked stiffnesses that are designated by A_C and A_U, respectively. These stiffnesses change from one ply drop group to another with crack growth. The effective uncracked and cracked stiffnesses of the inner section can be represented in the three consecutive ply drop regions as follows.

Region 1: $0 < a' < \ell$

$$A_U = \frac{d + 3\ell - a'}{\dfrac{d}{A_R} + \dfrac{\ell}{A_3} + \dfrac{\ell}{A_2} + \dfrac{\ell - a'}{A_1}} \tag{1}$$

$$A_C = A_1 \tag{2}$$

Region 2: $\ell < a' < 2\ell$

$$A_U = \frac{d + 3\ell - a'}{\dfrac{d}{A_R} + \dfrac{\ell}{A_3} + \dfrac{2\ell - a'}{A_2}} \tag{3}$$

$$A_C = \frac{a' + b}{\dfrac{a' - \ell}{A_2} + \dfrac{\ell + b}{A_1}} \tag{4}$$

Region 3: $2\ell < d' < 3\ell$

$$A_U = \frac{d + 3\ell - a'}{\dfrac{d}{A_R} + \dfrac{3\ell - a'}{A_3}} \tag{5}$$

$$A_C = \frac{a' + b}{\dfrac{a' - 2\ell}{A_3} + \dfrac{\ell}{A_2} + \dfrac{\ell + b}{A_1}} \tag{6}$$

where

$a' = a \cdot \cos \beta,$
$h = $ ply thickness,
$d = $ length of uniform thick portion ($60h$),
$\ell = $ distance between two consecutive ply drop locations,
$A_1 = 2hQ^{45} + 2hQ^0,$
$A_2 = 4hQ^{45} + 2hQ^0,$
$A_3 = 6hQ^{45} + 2hQ^0,$
$A_R = 8hQ^{45} + 2hQ^0,$
$Q^0 = \overline{Q}_{11}$ of a $0°$ ply, and
$Q^{45} = \overline{Q}_{11}$ of a $\pm 45°$ ply.
$$\tag{7}$$

Parameters A_1, A_2, and A_3 represent the effective axial stiffness of the first, second, and third inner ply drop sets, respectively, starting from the thin portion of the tapered laminate. The effective axial stiffness of the thick uniform core region is denoted by A_R and defined in Eq 7, whereas the belt axial stiffness, A_B, is given by

$$A_B = \frac{d + 3\ell - a'}{\dfrac{d}{A_{BD}} + \dfrac{3\ell - a'}{A_{BT}}} \tag{8}$$

where

$$A_{BD} = 7hQ^0 + 2hQ^{45}. \tag{9}$$

The effective belt stiffness in the uniform region is represented by A_{BD}. The stiffness A_{BT} and A_{B1} are the same as A_{BD} but at inclined angles β and α, respectively as shown in Fig. 3. A three-dimensional transformation is required in order to estimate the effective axial stiffnesses of the inclined belt region (A_{BT} and A_{B1}). This is due to the belt layup and the orientation of the different belt portions to the loading axis. The three-dimensional transformation is presented in Appendix I.

For the extensional analysis, a membrane model is used. In this model, the overall bending of the sublaminates is neglected [6] and the response is controlled by their axial stiffnesses expressed in Eqs 1 through 9. Bending and coupling stiffnesses are neglected at this level of modeling. The equilibrium equations for a membrane behavior, reduce to

$$N_{,x} = 0 \tag{10}$$

and the displacement field is assumed to be

$$u(x, z) = U(x) \tag{11}$$

and

$$w = 0 \tag{12}$$

The constitutive relationships are represented by

$$N = A_{11}U_{,x} \tag{13}$$

where the axial stiffness coefficient is denoted by A_{11}.

The stress and displacement fields, are determined based on the effective section stiffnesses given in Eqs 1 through 9. In this model, load is shared by the core and the belt portions according to their respective stiffness ratios at the fixed end

$$P_1 = \frac{PA_B}{A_B + A_U} \tag{14}$$

$$P_2 = \frac{PA_U}{A_B + A_U} \tag{15}$$

where P is half of the total applied axial load at the ends, and P_1 and P_2 denote the resultant axial force in the belt and core regions, respectively.

Apply Eqs 10 and 13 to the cracked and uncracked regions of the laminate and use Eqs 14 and 15 to determine the axial displacements at $x = c$, the cracked section of the laminate as

$$U_1 = \frac{P(a' + b)}{(A_B + A_U)} \left(\frac{A_B}{A_{B1}} - 1 \right) \tag{16}$$

$$U_2 = \frac{P(a' + b)}{(A_B + A_U)} \left(\frac{A_U}{A_C} - 1 \right) \tag{17}$$

where $c = 120h$.

The mismatch between the displacements of the delaminated region given by

$$U_1 - U_2 = \frac{P(a' + b)}{(A_B + A_U)} \left(\frac{A_B}{A_{B1}} - \frac{A_U}{A_C} \right) \tag{18}$$

is controlled by the taper angle, crack length, and the associated pop-off angle.

The external work done can be computed as

$$W_e = \tfrac{1}{2}(P_1U_1 + P_2U_2) \tag{19}$$

Substitute from Eqs 14 through 17 into Eq 19 to get

$$\frac{2W_e}{P^2} = \frac{(a' + b)}{(A_B + A_U)^2} \left(\frac{A_B^2}{A_{B1}} + \frac{A_U^2}{A_C} - A_B - A_U \right) \tag{20}$$

The total strain energy release rate, G_T, is determined from the rate of change of the external work done for a delamination of area $\mathcal{A}$ by

$$G_T = \frac{dW_e}{d\mathcal{A}} \tag{21}$$

where

$\mathcal{A} = w \cdot a$, and
$w = $ width of the laminate.

The expression for G_T given in Eq 21 is associated with a delamination growing along the tapered region.

Interlaminar Stresses

For the taper geometry under consideration, the dropped plies create a stress diffusion/concentration problem that originates from the belt-ply drop locations. The concentrated transverse normal and shear forces at the ply drop locations as well as the shear stress in the resin pocket are determined first. This is followed by a sublaminate modeling of the belt and inner region in order to determine the piecewise exponentially decaying interlaminar stresses at the interface between the belt and dropped plies.

The interlaminar stresses at the interface between the belt and inner plies of the tapered section are determined by considering the equilibrium of the belt region. The tapered belt section isolated in Fig. 4 is subjected to the applied axial end forces, N_1 and N_2. These belt forces are obtained from the global equilibrium approach described in the previous section. The analytical model assumes the belt in the tapered region as a beam elastically supported by the dropped plies. A schematic of the modeling approach and the free body diagram of the tapered belt section appear in Figs. 4 and 5, respectively.

The belt section is connected to the inner plies over three resin pockets. It is also supported by the inner plies at distinct ply drop locations. The resin pockets act as an elastic foundation while the ply drop corners act as concentrated supports and they together constrain the deflec-

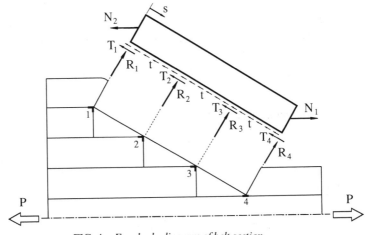

FIG. 4—*Free body diagram of belt section.*

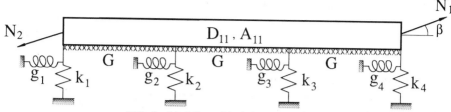

FIG. 5—*Modeling of the belt in taper region.*

tion of the belt section. The resin pockets are assumed to carry primarily shear stress and they are represented by distributed shear springs with a constant foundation stiffness parameter, G. The value of G is assumed to be equal to the shear stiffness of the bulk epoxy resin. The supports at the ply drop locations are modeled as extensional and concentrated shear springs whose stiffness are denoted by k_i and g_i ($i = 1,2,3,4$), respectively.

The dropped plies are assumed to have a contact length of a ply thickness, h. For a unit laminate width, this constitutes the effective area of the linear springs that are assumed to extend from the interface to the laminate centerline. The modeling of the springs is shown in Fig. 6.

The extensional spring stiffness is calculated by

$$k_i = \frac{AE_{33}}{L_i} \tag{22}$$

where A is the effective contact surface area. The distance between the interface and laminate centerline is L_i. Its value varies depending on the particular ply drop location. The out-of-plane extensional modulus of the dropped plies is represented by E_{33}. The stiffness of the concen-

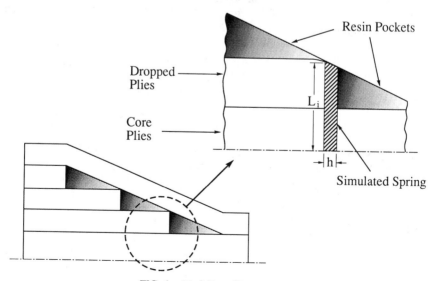

FIG. 6—*Modeling of linear springs.*

trated shear springs are similarly calculated as

$$g_i = \frac{AG_{13}}{L_i} \tag{23}$$

where G_{13} is the out-of-plane shear stiffness of the ply.

A minimum complementary potential energy formulation is used to estimate the interlaminar stresses. The total complementary potential energy consists of the bending and extensional energy contributions

$$\Pi^c = \Pi_b + \Pi_e + \Pi_k \tag{24}$$

where Π_b and Π_e represent the bending and extensional energy contributions, respectively, of the belt and core region while Π_k is the energy stored in the elastic springs. These are given as

$$\Pi_b = \frac{1}{2} \int_0^{3\ell} \frac{M^2(s)}{D_{11}} ds \tag{25}$$

$$\Pi_e = \frac{1}{2} \int_0^{3\ell} \frac{N^2(s)}{A_{11}} ds \tag{26}$$

$$\Pi_k = \frac{1}{2} \int_0^{3\ell} \frac{\tau^2(s)}{G} ds + \frac{R_1^2}{2k_1} + \frac{R_2^2}{2k_2} + \frac{R_3^2}{2k_3} + \frac{R_4^2}{2k_4} + \frac{T_1^2}{2g_1} + \frac{T_2^2}{2g_2} + \frac{T_3^2}{2g_3} + \frac{T_4^2}{2g_4} \tag{27}$$

where s denotes the axial component of the coordinate system in the tapered section. The bending and extensional stiffness coefficients are D_{11} and A_{11}, respectively. They are calculated by using the three-dimensional analysis described in Appendix I. The unknown extensional and shear spring force resultants are denoted by R_i, $T_i (i = 1,2,3,4)$, respectively. The constant resin filler shear stress is an additional unknown.

The total number of unknowns in this formulation is nine. These unknowns are constrained by the following equilibrium equations

$$R_3 = 3R_1 + 2R_2 - 3N_{22} \tag{28}$$

$$R_4 = -2R_1 - R_2 + 2N_{22} + N_{12} \tag{29}$$

$$T_4 = -T_1 - T_2 - T_3 - 3\ell t + N_{11} - N_{21} \tag{30}$$

where N_{11}, N_{12}, N_{21}, and N_{22} denote the components of the extensional resultant stresses at the two ends of the belt section. The resin filler shear stress is denoted by t.

The bending moment and axial force terms in each of the three ply drop regions are written as

Region 1: $0 < s < \ell$

$$M(s) = R_1 s - N_{22} s \tag{31}$$

$$N(s) = T_1 + ts + N_{21} \tag{32}$$

Region 2: $\ell < s < 2\ell$

$$M(s) = R_1 s + R_2(s - \ell) - N_{22} s \tag{33}$$

$$N(s) = T_1 + T_2 + ts + N_{21} \tag{34}$$

Region 3: $2\ell < s < 3\ell$

$$M(s) = R_1 s + R_2 s - R_2 \ell - (3R_1 + 2R_2 - 3N_{22})(s - 2\ell) \tag{35}$$

$$N(s) = T_1 + T_2 + T_3 + ts + N_{21} \tag{36}$$

Using the expressions in Eqs 33 through 36, the bending energy in Eq 25 can be written as

$$
\begin{aligned}
\Pi_b = {} & \frac{1}{2D_{11}} \int_0^\ell [R_1 s - N_{22} s]^2 \, ds \\
& + \frac{1}{2D_{11}} \int_\ell^{2\ell} [R_1 s + R_2(s - \ell) - N_{22} s]^2 \, ds \\
& + \frac{1}{2D_{11}} \int_{2\ell}^{3\ell} [R_1 s + R_2 s - R_2 \ell - (3R_1 + 2R_2 - 3N_{22})(s - 2\ell)]^2 \, ds
\end{aligned}
\tag{37}
$$

Similarly, the energy of extensional loads can be expressed by

$$
\begin{aligned}
\Pi_e = {} & \frac{1}{2A_{11}} \int_0^\ell (T_1 + ts + N_{21})^2 \, ds + \frac{1}{2A_{11}} \int_\ell^{2\ell} (T_1 + T_2 + ts + N_{21})^2 \, ds \\
& + \frac{1}{2A_{11}} \int_{2\ell}^{3\ell} (T_1 + T_2 + T_3 + ts + N_{21})^2 \, ds
\end{aligned}
\tag{38}
$$

The energy stored in the elastic springs is written as

$$
\begin{aligned}
\Pi_k = {} & \frac{3}{2} \frac{c^2}{G} \ell + \frac{1}{2k_3}(3R_1 + 2R_2 - 3N_{22})^2 \\
& + \frac{1}{2k_4}(-2R_1 - R_2 + 2N_{22} + N_{12})^2 \\
& + \frac{R_1^2}{2k_1} + \frac{R_2^2}{2k_2} + \frac{T_1^2}{2g_1} + \frac{T_2^2}{2g_2} + \frac{1}{2g_4}(-T_1 - T_2 - T_3 - 3\ell t + N_{11} - N_{21})^2 \\
& + \frac{T_1^2}{2g_1} + \frac{T_2^2}{2g_2} + \frac{T_3^2}{2g_3}
\end{aligned}
\tag{39}
$$

The complementary potential energy in Eqs 37 through 39 is expressed in terms of six unknowns, namely R_1, R_2, T_i ($i = 1,2,3$), and t. By minimizing the total complementary potential energy with respect to these unknown force resultants, the following linear system of equations is obtained

$$
\begin{aligned}
& \left(\frac{\ell^3}{D_{11}} + \frac{1}{k_1} + \frac{9}{k_3} + \frac{4}{k_4} \right) R_1 + \left(\frac{3}{2} \frac{\ell^3}{D_{11}} + \frac{6}{k_3} + \frac{2}{k_4} \right) R_2 \\
& \qquad\qquad = \frac{5}{3} \frac{\ell^3}{D_{11}} N_{22} + \left(\frac{9}{k_3} + \frac{4}{k_4} \right) N_{22} + \frac{2}{k_4} N_{12}
\end{aligned}
\tag{40}
$$

$$\left(\frac{3}{2}\frac{\ell^3}{D_{11}} + \frac{6}{k_3} + \frac{2}{k_4}\right) R_1 + \left(\frac{2}{3}\frac{\ell^3}{D_{11}} + \frac{1}{k_2} + \frac{4}{k_3} + \frac{1}{k_4}\right) R_2$$
$$= \frac{1}{3}\frac{\ell^3}{D_{11}} N_{22} + \left(\frac{6}{k_3} + \frac{2}{k_4}\right) N_{22} + \frac{1}{k_4} N_{12} \quad (41)$$

$$\left(9\frac{\ell^3}{A_{11}} + \frac{3\ell}{G} + \frac{9\ell^2}{g_4}\right) t + \left(\frac{9}{2}\frac{\ell^2}{A_{11}} + \frac{3\ell}{g_4}\right) T_1 + \left(4\frac{\ell^2}{A_{11}} + \frac{3\ell}{g_4}\right) T_2 + \left(\frac{5}{2}\frac{\ell^2}{A_{11}} + \frac{3\ell}{g_4}\right) T_3$$
$$= -\frac{9\ell^2}{A_{11}} N_{21} + \frac{3\ell}{g_4}(N_{11} - N_{21}) \quad (42)$$

$$\left(\frac{9}{2}\frac{\ell^2}{A_{11}} + \frac{3\ell}{g_4}\right) t + \left(\frac{3\ell}{A_{11}} + \frac{1}{g_1} + \frac{1}{g_4}\right) T_1 + \left(\frac{2\ell}{A_{11}} + \frac{1}{g_4}\right) T_2 + \left(\frac{\ell}{A_{11}} + \frac{1}{g_4}\right) T_3$$
$$= -\frac{3\ell}{A_{11}} N_{21} + \frac{1}{g_4}(N_{11} - N_{21}) \quad (43)$$

$$\left(4\frac{\ell^2}{A_{11}} + \frac{3\ell}{g_4}\right) t + \left(\frac{2\ell}{A_{11}} + \frac{1}{g_4}\right) T_1 + \left(\frac{2\ell}{A_{11}} + \frac{1}{g_2} + \frac{1}{g_4}\right) T_2 + \left(\frac{\ell}{A_{11}} + \frac{1}{g_4}\right) T_3$$
$$= -\frac{2\ell}{A_{11}} N_{21} + \frac{1}{g_4}(N_{11} - N_{21}) \quad (44)$$

$$\left(\frac{5}{2}\frac{\ell^2}{A_{11}} + \frac{3\ell}{g_4}\right) t + \left(\frac{\ell}{A_{11}} + \frac{1}{g_4}\right) T_1 + \left(\frac{\ell}{A_{11}} + \frac{1}{g_4}\right) T_2 + \left(\frac{\ell}{A_{11}} + \frac{1}{g_3} + \frac{1}{g_4}\right) T_3$$
$$= -\frac{\ell}{A_{11}} N_{21} + \frac{1}{g_4}(N_{11} - N_{21}) \quad (45)$$

The concentrated normal and shear forces at the ply drop regions and the interlaminar shear in the resin filler are estimated by solving the simultaneous system of equations in Eqs 40 through 45 and using Eqs 28 through 30.

Equations 40 through 45 produce resultant normal and shear forces concentrated at the ply drop locations in addition to a constant shear stress due to resin-rich pockets. These concentrated forces are then transformed to obtain their averaged stress equivalents in a boundary layer zone around the discontinuities. The boundary layer length is calculated based on a local sublaminate analysis [7] whose details are given in Appendix B.

Results and Discussion

The material properties of S2/SP250 glass/epoxy are given in Table 1. The distance, ℓ, between ply drop locations is taken as $20h$. The strain energy release rate is numerically calculated from the work done by external forces. The variation of the total strain energy release rate parameter, $\overline{G}_T$, defined as $G_T h/P^2$ with a delamination growth along the taper appears in

TABLE 1—*Material properties of S2/SP250 glass-epoxy.[a]*

E_{11}	$E_{22} = E_{33}$	$G_{12} = G_{13}$	G_{23}	$\nu_{12} = \nu_{13}$	ν_{23}
50.3	14.5	6.1	3.4	0.275	0.416

[a] Stiffnesses are in GPa, and ply thickness = 0.216 mm.

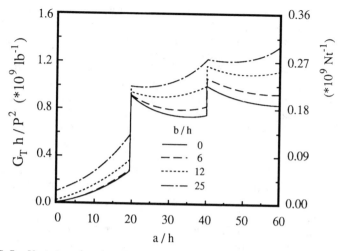

FIG. 7—*Variation of total strain energy release rate parameter with delamination.*

Fig. 7. The distribution of $\overline{G}_T$ for a delamination along the taper section ranging from 0 to the total length of the taper ($60h$) region is shown in the figure. The discrete jumps at a' equal to $20h$ and $40h$ correspond to the ply drop locations, reflecting the effect of the material and geometric discontinuities. The strain energy release rate associated with a delamination along the tapered region increases with the delamination growth, b, along the uniform portion. This is in agreement with the results of Ref 5.

A comparison of the total strain energy release rate parameter per unit width with the results of Ref 5 is provided for the case of $b = 24h$ in Fig. 8. The total strain energy release rate distribution of Ref 5 is based on a two-dimensional finite element simulation. Half of the tapered laminate was modeled using 7610 nodes and 2382 eight-noded, isoparametric, parabolic elements. The distribution predicted by the present approach captures the nonlinear variation of

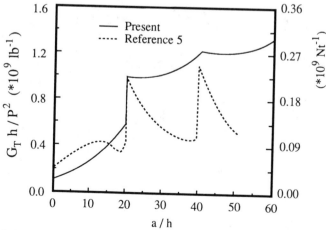

FIG. 8—*Comparison of total strain energy release rate parameters (b/h = 24).*

the strain energy release rate with delamination and agrees qualitatively within the first drop $0 < a' < 20h$ as shown in Fig. 8. The decrease in the total strain energy release rate parameter predicted by the finite element distribution within the second and third ply drop regions is in contrast with the increase shown by the present approach.

The differences depicted in Fig. 8 are primarily due to the smeared stiffness modeling approach adopted in the present model in comparison to the finite element crack-closure where the local variations in the stiffness in the neighborhood of the delamination are considered. A sublaminate analysis [7] where the resin pocket as well as the ply drop stiffnesses are accounted for, is needed in order to capture their local effects on the total strain energy release rate distributions. The present smeared stiffness approach is a global model that provides qualitative results and trend information. The model is effective for investigating delamination onset along the tapered region. This global model was used to determine the axial stress resultants in the belt and inner region. In estimating the interlaminar stresses at the interface between the belt and the ply drops, the local stiffness variations of the resin pockets and the ply drop regions were considered in the model.

The interlaminar transverse normal and shear stress distributions are presented in Figs. 9 and 10, respectively. The interlaminar stress distributions from the finite element simulation of Ref 5 are plotted in dotted lines. While the shear stress is of the same sign, the normal stress distribution changes from a high tensile value in the taper and thin uniform section interface to a compressive value close to the thick uniform section. The interlaminar peel stress distribution shows a compressive stress concentration at the ply drop Locations 1, 2, and 3. It does not, however, exhibit the sudden change of sign predicted by the finite element model [5].

Conclusion

A simple model is developed for the delamination analysis of a tapered laminate under tensile loading. Two levels of modeling are utilized, a global and a local scale. The total strain energy release rate is estimated in terms of axial stress resultants and effective stiffness distributions based on the global energy balance of the laminate. The interlaminar stresses are determined on the basis of equilibrium conditions and local stiffness variations at the ply drop locations.

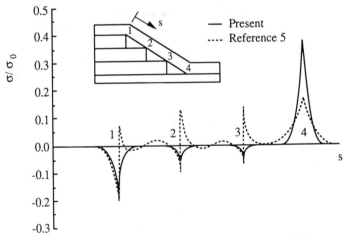

FIG. 9—*Distribution of interlaminar peel stress along the belt.*

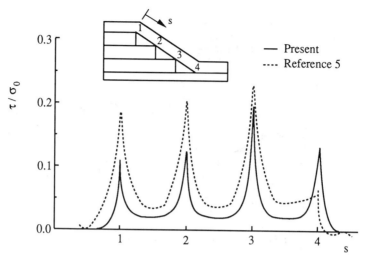

FIG. 10—*Distribution of interlaminar shear stress along the belt.*

Closed-form expressions for the total strain energy release rate distribution and interlaminar stresses are derived. The interlaminar stress and the total strain energy release rate distributions are in qualitative agreement with a finite element solution. The developed analysis shows that a membrane model captures the nonlinear variation of the strain energy release rate with delamination growth.

The present approach is useful in understanding the basic mechanics of the problem and predicting the factors controlling the behavior. It is not intended to compete with large-scale numerical approaches, but rather to serve as the means for selecting and evaluating candidate configurations and providing trend information.

Acknowledgments

The authors gratefully acknowledge the financial support provided by the NASA Langley Research Center under Grant NAG-1-637. Sincere appreciation is extended to NASA Contract Monitor Gretchen Murri and to Drs. T. Kevin O'Brien and Satish Salpekar for their helpful discussions.

APPENDIX I

Transformation of Stiffnesses

A three-dimensional transformation of stiffnesses is required in order to estimate the effective axial stiffness of the belt regions, A_B and A_{B1}. This is due to the belt layup and the orientation of the different belt portions to the loading axis as shown in Fig. 11.

The loading direction is along Axis 1 in the 1, 2, 3 reference system. The principal material coordinates are denoted by $1'$, $2'$, and $3'$. First, a transformation relationship is developed to obtain the equivalent stiffness of the belt layup about the $3'$ axis. This is followed by a transformation relationship associated with the taper angle around the 2-axis.

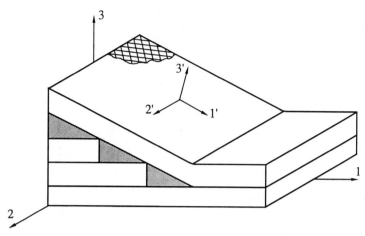

FIG. 11—*Global and local coordinate systems.*

The stress-strain relationships in the principal material coordinates for an orthotropic laminate are given by

$$\{\sigma\}_{6\times1} = [Q]_{6\times6}\{\epsilon\}_{6\times1} \qquad (46)$$

where Q_{ij} are the reduced stiffness coefficients.

The presence of angle plies in the belt region making an angle, θ, in the $1'2'$-plane, results in the following constitutive relationship

$$\{\sigma'\} = [\overline{Q}]\{\epsilon'\} \qquad (47)$$

where the transformed reduced stiffnesses, $\overline{Q}_{ij}$, are given in terms of reduced stiffnesses, Q_{ij}, as

$$
\begin{aligned}
\overline{Q}_{11} &= c^4Q_{11} + 2c^2s^2Q_{12} + s^4Q_{22} + 4c^2s^2Q_{66}, \\
\overline{Q}_{22} &= s^4Q_{11} + 2c^2s^2Q_{12} + c^4Q_{22} + 4c^2s^2Q_{66}, \\
\overline{Q}_{12} &= c^2s^2Q_{11} + (c^4 + s^4)Q_{12} + c^2s^2Q_{22} - 4c^2s^2Q_{66}, \\
\overline{Q}_{66} &= 4c^2s^2Q_{11} - 8c^2s^2Q_{12} + 4c^2s^2Q_{22} + 4(c^2 - s^2)^2Q_{66}, \\
\overline{Q}_{33} &= Q_{33}, \\
\overline{Q}_{13} &= c^2Q_{13} + s^2Q_{23}, \\
\overline{Q}_{23} &= s^2Q_{13} + c^2Q_{23}, \\
\overline{Q}_{44} &= c^2Q_{44} - 2csQ_{45} + s^2Q_{55}, \text{ and} \\
\overline{Q}_{55} &= s^2Q_{44} + 2csQ_{45} + c^2Q_{55}.
\end{aligned}
\qquad (48)
$$

Where $c = \cos\theta$, and $s = \sin\theta$, and the angle, θ, is a rotation around the $3'$-axis.

Any ply in the belt portion of the taper makes an angle, β, with the loading axis if it is in the uncracked belt portion and an angle, α, if it is in the cracked belt portion as shown in Fig. 3. By performing a rotation about the 2-axis, the stiffness along the loading axis, takes the form

$$\{\sigma\}_{6\times1} = [C]_{6\times6}\{\epsilon\}_{6\times1} \qquad (49)$$

where σ_{ij} and ϵ_{ij} are in 1,2,3-axis system and C_{ij} represents the elements of the transformed stiffness matrix in this coordinate system.

For plane stress conditions in the 1–3 plane (that is, $\sigma_{i2} = 0$; $i = 1, 2, 3$), the stress-strain relationships reduce to

$$\sigma_{11} = (C_{11} - C_{12}^2/C_{22})\epsilon_{11} \tag{50}$$

where

$$C_{11} = \bar{c}^4\overline{Q}_{11} + 2\bar{c}^2\bar{s}^2\overline{Q}_{13} + \bar{s}^4\overline{Q}_{33} + \bar{c}^2\bar{s}^2\overline{Q}_{55}, \tag{51}$$
$$C_{12} = \bar{c}^2\overline{Q}_{12} + \bar{s}^2\overline{Q}_{23}, \text{ and} \tag{52}$$
$$C_{22} = \overline{Q}_{22} \tag{53}$$

where $\bar{c}$ and $\bar{s}$ are the cosine and sine of the angles that the cracked or uncracked belt portions makes with the loading axis.

The coefficient of ϵ_{11} in Eq 50 represents the transformed axial stiffness used in the derivation of A_B and A_{B1}.

APPENDIX II

Boundary Layer Analysis of Interlaminar Stresses

The minimization of the complementary potential energy provides estimates of the extensional and shear force resultants at the ply drop locations. A sublaminate analysis approach [7] is proposed to determine the equivalent stress distribution around the ply drops. In this model, individual plies of a laminate or groups of plies—sublaminates—are treated as laminated units represented by their effective stiffnesses. A particular belt-resin combination is singled out and the schematic of the modeling is shown in Fig. 12. The average resin thickness is h over a distance, $20h$, while the belt thickness is $9h$. The interlaminar peel and shear stresses are denoted by P and T, respectively.

The overall equilibrium equations for the membrane model encompass the following governing equations for each sublaminate

$$N_{,x} + n = 0 \tag{54}$$

$$Q_{,x} + q = 0 \tag{55}$$

$$-Q + m = 0 \tag{56}$$

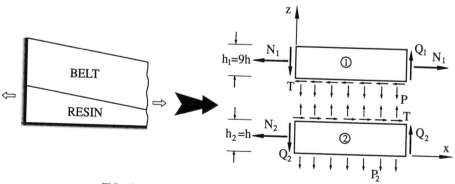

FIG. 12—*Sublaminate model for belt-resin combination.*

where N and Q are the axial and shear stress resultants, respectively. Parameters n, q, and m represent the effective distributed axial force, lateral load, and moment, respectively. They are defined as

$$n = T_2 - T_1 \tag{57}$$

$$q = P_2 - P_1 \tag{58}$$

$$m = \frac{h}{2}(T_2 + T_1) \tag{59}$$

The interlaminar shear and peel stresses at the top of the sublaminate are denoted by T_2 and P_2, respectively, while the corresponding stresses at the bottom of the sublaminate are designated as P_1 and T_1.
The constitutive relationships are given by

$$N = A_{11}U_{,x} + B_{11}\beta_{,x} \tag{60}$$

$$Q = A_{55}(\beta + W_{,x}) \tag{61}$$

$$M = B_{11}U_{,x} + D_{11}\beta_{,x} \tag{62}$$

The present formulation recognizes shear deformation through the rotation, β.
The axial displacement, u, and the transverse displacement, w, are assumed to be of the form

$$u(x, z) = U(x) + z\beta(x) \tag{63}$$

$$w(x) = W(x) \tag{64}$$

The transverse displacement, w, is zero at $z = 0$, hence, the transverse displacement associated with the belt and core region reduce to

$$W_1 = W_2 = 0 \tag{65}$$

Subscripts 1 and 2 refer to the belt and core regions, respectively.
The displacement continuity at the interface of two sublaminates is given as

$$u_1\left(x, -\frac{h_1}{2}\right) = u_2\left(x, \frac{h_2}{2}\right) \tag{66}$$

where $h_1 = 9h$ and $h_2 = h$.
The continuity of displacements and stresses at the interface permits one to express the variables associated with Sublaminates 1 and 2 in terms of axial stress resultants, N_1 and N_2. The solution reduces to two coupled differential equations

$$N_{1,x} + N_{2,x} = 0 \tag{67}$$

$$\frac{N_1}{\left(A_{11} - \dfrac{B_{11}^2}{D_{11}}\right)_1} - \frac{N_2}{(A_{11})_2} - \frac{(9h)^2}{4(A_{55})_1}N_{1,xx} - \frac{h^2}{4(A_{55})_2}N_{2,xx} = 0 \tag{68}$$

for which the solution is assumed to be

$$(N_1, N_2) = (\bar{A}_1, \bar{A}_2)e^{sx} \tag{69}$$

Substitute from Eq 69 into Eqs 67 and 68 to get the following

$$s\left\{s^2\left[\frac{81h^2}{4(A_{55})_1}+\frac{h^2}{4(A_{55})_2}\right]-\left[\left(\frac{D_{11}}{A_{11}D_{11}-B_{11}^2}\right)_1+\frac{1}{(A_{11})_2}\right]\right\}=0 \tag{70}$$

where s is the characteristic root.

The boundary conditions associated with the sublaminate sections are given at $x = 0$

$$U_1 = U_2 = 0 \tag{71}$$

and at $x = \ell = 20h$

$$N_2 = 0, N_1 = \overline{N} \tag{72}$$

The axial forces related to Sublaminates 1 and 2 become

$$\begin{Bmatrix} N_1 \\ \\ N_2 \end{Bmatrix} = \begin{Bmatrix} \dfrac{1}{1+L_{12}}\dfrac{\cosh sx}{\cosh 20sx}+\dfrac{L_{12}}{1+L_{12}} \\ \\ -\dfrac{1}{1+L_{12}}\dfrac{\cosh sx}{\cosh 20sx}+\dfrac{1}{1+L_{12}} \end{Bmatrix}\overline{N} \tag{73}$$

where

$$L_{12} = (A_{11})_1/(A_{11})_2 \tag{74}$$

Finally, the shear and peel stresses at the interface between Sublaminate 1 and 2 are obtained from the equilibrium equations as

$$T = \frac{s\overline{N}}{(1+L_{12})}\frac{\sinh sx}{\cosh 20sh} \tag{75}$$

$$P = \frac{hs^2\overline{N}}{2(1+L_{12})}\frac{\cosh sx}{\cosh 20sh} \tag{76}$$

Equations 75 and 76 can be simplified as

$$T = \alpha_1 \sinh sx \ (i = 1,2,3,4) \tag{77}$$

$$P = \gamma_i \cosh sx \ (i = 1,2,3,4) \tag{78}$$

where α_i and γ_i are coefficients whose values vary for each ply drop location. For both quantities, the boundary layer length is found to be 30% of the total sublaminate length. In this context, the boundary layer length is defined as the distance where the interlaminar shear stress decays to 1% of its maximum value. The concentrated force determined by the minimization of the complementary potential energy can be expressed in terms of the corresponding interlaminar stresses as

$$T_i = 2\int_{0.7\ell}^{\ell} T(x)\, dx \ (i = 1,2,3,4) \tag{79}$$

and

$$R_i = 2\int_{0.7\ell}^{\ell} P(x)\, dx \ (i = 1 - 4) \tag{80}$$

Equations 79 and 80 provide expressions for α_i and γ_i in terms of point forces, T_i and R_i, respectively.

Peak stresses are then calculated by using Eqs 75 and 76 and expressions for α_i and γ_i obtained in Eqs 79 and 80 as

$$(\tau)_{\text{peak}} = T(x = \ell) = \alpha_i \sinh s\ell \tag{81}$$

$$(\sigma)_{\text{peak}} = P(x = \ell) = \gamma_i \cosh s\ell \tag{82}$$

References

[1] Kemp, B. L. and Johnson, E. R., "Response and Failure Analysis of a Graphite-Epoxy Laminate Containing Terminating Internal Plies," *Proceedings,* Twenty-Sixth AIAA/ASME/ASCE/AHS Structures, Structural Dynamics, and Materials (SDM) Conference, AIAA Paper No. 85-0608, April 1985, pp. 13–24.

[2] Curry, J. M., Johnson, E. R., and Starnes, J. H., Jr., "Effect of Dropped Plies on the Strength of Graphite-Epoxy Laminates," *Proceedings,* Twenty-Ninth AIAA/ASME/ASCE/AHS Structures, Structural Dynamics and Materials (SDM) Conference, Monterey, CA, AIAA Paper No. 87-0874, 6–8 April 1987, pp. 737–747.

[3] Wu, C. M. and Webber, P. H., "Analysis of Tapered (in Steps) Laminated Plates Under Uniform Inplane Load," *Composite Structures,* Vol 5, 1986, pp. 87–100.

[4] Fish, J. C. and Lee, S. W., "Delamination of Tapered Composite Structures," *Engineering Fracture Mechanics,* Vol. 34, No. 1, 1989, pp. 43–54.

[5] Salpekar, S. A., Raju, I. S., and O'Brien, T. K., "Strain Energy Release Rate Analysis of Delamination in a Tapered Laminate Subjected to Tension Load," Third Technical Conference, American Society for Composites, Seattle, WA, 26–29 Sept. 1988, pp. 642–654.

[6] Rehfield, L. W., Armanios, E. A., and Weistein, F., "Analytical Modeling of Interlaminar Fracture in Laminated Composites," Composites '86: Recent Advances in Japan and the United States, *Proceedings,* Third Japan-U.S. Conference on Composite Materials, K. Kawata, S. Umekawa, A. Kobayashi, Eds., Japan Society for Composite Materials, Tokyo, 1986, pp. 331–340.

[7] Armanios, E. A. and Rehfield, L. W., "Sublaminate Analysis of Interlaminar Fracture in Composites: Part I—Analytical Model," *Journal of Composites Technology & Research,* Vol. 11, No. 4, Winter 1989, pp. 135–146.

Matthew D. Tratt[1]

Analysis of Delamination Growth in Compressively Loaded Composite Laminates

REFERENCE: Tratt, M. D., "**Analysis of Delamination Growth in Compressively Loaded Composite Laminates,**" *Composite Materials: Fatigue and Fracture (Third Volume), ASTM STP 1110*, T. K. O'Brien, Ed., American Society for Testing and Materials, Philadelphia, 1991, 359–372.

ABSTRACT: An analytical and empirical study was conducted to investigate delamination behavior in composite structures.

The objective of the analysis was to predict the threshold stress for the initiation of delamination growth in compressively loaded composite laminates. MSC/NASTRAN geometrically nonlinear finite element analysis and a virtual crack opening technique were used to compute strain energy release rate distributions around circular delaminations. The distributions were computed for various compressive stress levels and the threshold stress for delamination growth was then determined by applying the strain energy release rates in a mixed-mode fracture criterion.

A number of static compression tests were conducted on specimens made from AS4/3501-5A graphite fabric-reinforced epoxy. The specimens contained circular delaminations of various sizes. The delaminations had been created by implanting Teflon inclusions in the prepreg layup.

Computed predictions of the delamination growth threshold stresses were significantly in error relative to the test data. The analysis did, however, concur with observed trends and provided qualitative insight into the process of quasi-static delamination growth.

KEY WORDS: composite materials, delamination, fracture, graphite-epoxy materials, mixed-mode loading, strain energy release rate

Nomenclature

a	Crack (delamination) length
E_x, E_y, E_z	Material moduli
F_r, F_t, F_z	Radial, circumferential, and normal forces on finite element nodes
G_{xy}, G_{xz}, G_{yz}	Material shear moduli
G_I, G_{II}, G_{III}	Strain energy release rates for Mode I, Mode II, and Mode III fracture
G_{Ic}, G_{IIc}	Critical Mode I and Mode II strain energy release rates for delamination growth
MS	Margin of safety with respect to fracture
$\delta_r, \delta_t, \delta_z$	Relative radial, circumferential, and normal displacements of finite element nodes
$\nu_{xy}, \nu_{xz}, \nu_{yz}$	Material Poisson ratios
σ_{b1}, σ_{b2}	Measured buckling stress of thinner and thicker sublaminates
σ_i	Estimated threshold stress for the initiation of delamination growth $= (\sigma_{b2} + \sigma_{max})/2$
σ_{max}	Measured maximum compressive stress
Θ	Angular position on delamination front

[1] Senior research engineer, Rohr Industries, Inc., Chula Vista, CA 91912-0878.

The ever-expanding role of composites in aircraft structural design has necessitated an enhanced understanding of composite defect and damage tolerance. The deleterious effects of delaminations have made these defects the subject of particularly extensive research.

Delamination effects are known to be most severe in the case of compressively loaded structures. Research has revealed that delamination-related failure initiates after the buckling of the sublaminates adjacent to the delamination. This is followed by interlaminar fracture with the consequence of delamination growth, general instability, and crippling. This failure mode can significantly undermine laminate compressive strength.

It is important for manufacturing quality assurance and product support that methods be developed to assess the severity of delaminations in composite hardware. Empirical research, although effective in a limited application, is costly and cannot address all possible circumstances.

A number of researchers have recognized the practical limitations of empirical methods and have attempted to develop analytical techniques to model delamination effects. A variety of computational methods have been used, the most common being Rayleigh-Ritz energy methods [1–6] and finite element analyses [7–10]. In most cases, the analyses were performed by special purpose computer codes.

This paper presents the results of an analytical study using the MSC/NASTRAN finite element code to model delamination growth. MSC/NASTRAN is a versatile code that is currently in wide use in the aircraft and other industries. The analytical results are compared with data obtained from a test program that investigated delamination behavior in graphite fabric-reinforced epoxy.

Experimental Work

Specimen Preparation

Composite panels were prepared from eight plies of AS4 (6K 5HS)/3501-5A graphite fabric-reinforced epoxy. The thickness of the cured panels was approximately 2.9 mm. The warp fibers in all plies were unidirectionally oriented so as to be aligned with one edge of the panel. Delaminations were created in the panels by embedding circular inclusions in the prepreg layup at a depth of two plies. The inclusions, which were 12.7 to 50.8 mm in diameter, consisted of a layer of 120 glass fabric sandwiched between two layers of Teflon film. Although the inclusions were approximately 0.4 mm thick, they did not result in any externally measurable eccentricity in the graphite-epoxy material. Ultrasonic inspection of the panels revealed that the resulting delaminations were very well defined and that everywhere else the panels were free of significant defects.

The panels were cut such that the delaminations were centered in 102 by 84 mm test specimens. The two longer edges of the specimens were potted into steel fittings that were then precisely ground to ensure flatness and parallelism.

Test Procedure

The test arrangement was that shown in Fig. 1. Compressive load was applied through the potted fittings. The two unloaded edges of the specimen were clamped by the front and back face supports of the test fixture. Sublaminate buckling characteristics were monitored by a pair of back-to-back axial strain gages that were bonded to the specimen at the center of the delamination. Sublaminate buckling was indicated by the onset of strain reversal in the measured load-strain response. Some specimens were fitted with three pairs of gages that were bonded across the width of the specimen. This additional instrumentation served to verify the uniformity of the loading and to detect any overall specimen buckling.

FIG. 1—*Compression test arrangement.*

All tests were conducted under stroke-control (at a rate of 0.02 mm/s) in ambient conditions. Load, strain, and crosshead displacement data were recorded continuously during testing.

Experimental Results

Specimens with 12.7 mm delaminations failed due to overall specimen crippling. This was not the desired failure mode as it was not delamination-related. This was not the case for 25.4 and 38.1 mm delaminations. A representative load-strain plot for these delaminations appears in Fig. 2. The pronounced strain reversal measured on both surfaces of the laminate indicates that the two sublaminates buckled and then deflected in an outward direction. (In this case, for a 38.1 mm delamination, the two-ply sublaminate buckled first at a load of approximately 30 kN. This was followed by the buckling of the remaining six-ply sublaminate at approximately 64 kN.) The sublaminate separation resulted in an opening of the delamination and, assumedly, some degree of Mode I crack loading along the delamination front. Strain reversal subsequently detected at the outer strain gages indicates the onset of delamination growth.

A typical plot of load as a function of crosshead displacement appears in Fig. 3. The yielding-type response evident in the plot confirms the damage growth within the specimen. Post-failure ultrasonic inspection revealed extensive bidirectional delamination growth. The growth was greatest, however, in a direction transverse to the applied load where damage was found to extend across the width of the specimen. Final failure resulted from crippling.

The load-displacement plots did not always provide definitive evidence of the initiation of delamination growth. Audible cracking, generally noted shortly after the buckling of the

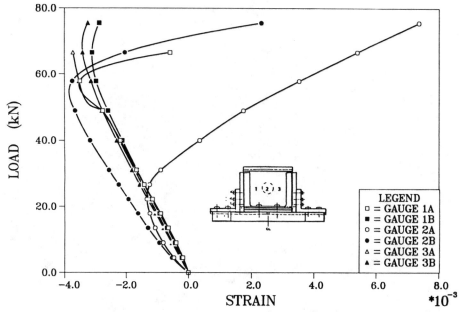

FIG. 2—*Load-strain response for 38.1 mm delamination.*

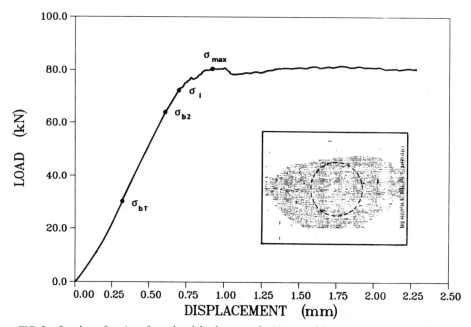

FIG. 3—*Load as a function of crosshead displacement for 38.1 mm delamination and post-failure ultrasonic C-scan.*

thicker sublaminate, was not accompanied by any significant changes in the load-displacement response. There were also large variations between specimens in the point at which audible cracking started. Acoustic emission analysis would have been appropriate here, but the necessary equipment was not available.

The threshold stress for delamination growth was, therefore, assumed to be the average of the sublaminate buckling stress (that of the thicker sublaminate) and the peak compressive stress measured during the test. This point is marked on Fig. 3 and is designated as σ_i. It appears that this may be a good estimate of the delamination growth threshold. At this point in Fig. 3, as was the case in many of the load-displacement plots, the trace shows a departure from linearity and a jaggedness that is characteristic of fracture events. It should be noted that this point also corresponds to that at which the outermost pairs of strain gages indicate the onset of strain reversal in Fig. 2.

The test data are summarized in Table 1. Reported are the stresses at the onset of sublaminate instability, the maximum stresses, and the estimated growth threshold stresses. It can be seen from the table that the compressive strength was reduced by as much as 60% relative to the laminate ultimate compressive strength.

Analysis

The MSC/NASTRAN finite element computer code was used to analyze the delamination compression specimens. Finite element models, such as that in Fig. 4, were constructed for each specimen configuration.

Due to the symmetry of the problem, it was necessary to model only one-quarter of the specimen. Symmetry boundary conditions were enforced on two edges ($u = 0$ on $x = 0$ and $v = 0$ on $y = 0$). Clamped constraints ($w = 0$) were imposed on the remaining two edges. Load was introduced as a uniform displacement along the edge $x = H$.

The model was constructed of six- and eight-node solid isoparametric elements. The material properties assigned to the elements were:

$$E_x = E_Y = 60.3 \text{ GPa}$$
$$E_z = 6.90 \text{ GPa}$$
$$G_{xy} = 5.45 \text{ GPa}$$
$$G_{xz} = G_{yz} = 0.97 \text{ GPa}$$
$$\nu_{xy} = 0.062$$
$$\nu_{xz} = \nu_{yz} = 0.25$$

Contact conditions between the sublaminates were modeled by NASTRAN "CGAP" elements. These are scalar elements with bilinear stiffness characteristics; they have a high com-

TABLE 1—*Summary of delamination compression test data.*

Diameter, mm	σ_{b1}, MPa	σ_{b2}, MPa	σ_{max},[a] MPa	σ_i, MPa Test	σ_i, MPa Analysis
0	360 ± 1%	360 ± 1%	428 ± 3%[b]	. . .	. . .
12.7	383 ± 5%	383 ± 5%	436 ± 6%[b]	. . .	. . .
25.4	211 ± 6%	250 ± 3%	287 ± 2%	268 ± 2%	409
38.1	101 ± 9%	201 ± 7%	283 ± 6%	243 ± 4%	271

[a] The laminate ultimate compressive strength is approximately 720 MPa.
[b] Failure was by overall specimen crippling, so σ_i has no significance.

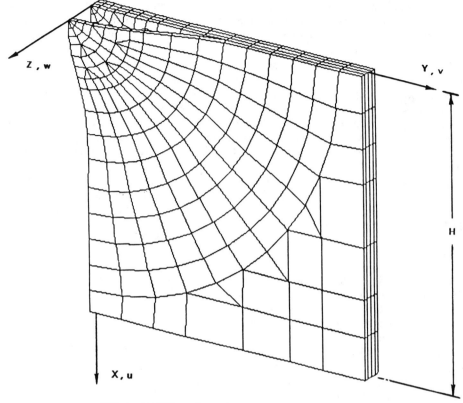

FIG. 4—*NASTRAN finite element model (deformed shape).*

pressive stiffness and a negligible tensile stiffness. By connecting the nodes of the sublaminates with these elements, sublaminate overlapping was prevented while allowing unhindered sublaminate separation.

Model behavior was first verified by performing a NASTRAN bifurcation buckling analysis on each model. The results of the analysis are plotted with the corresponding test data in Fig. 5. The computed sublaminate buckling stresses generally agree to within 10% of measured values.

Having confirmed correct (and reasonably accurate) model behavior, each model was submitted for postbuckling analysis. This was accomplished by executing the NASTRAN solution sequence for geometrically nonlinear (large displacement) analysis. The solution procedure is described in Ref *11*.

Strain energy release rates along the delamination front were calculated for various postbuckled compressive stress levels. This was done by means of a virtual crack opening technique. The nodal forces (F_r, F_t, and F_z) along the delamination front (defined in terms of cylindrical coordinates, as shown in Fig. 6) were retrieved from the NASTRAN output data for each stress level. The finite element model was then modified by disconnecting the elements on the boundary of the delamination, effectively simulating a delamination or crack growth of Δa. The model was then resubmitted for analysis and for each stress level the relative displacements (δ_r, δ_t, and δ_z) were determined for the previously coincident boundary nodes.

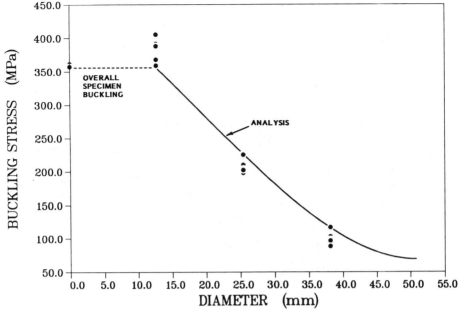

FIG. 5—*Measured and computed sublaminate buckling stress, σ_{bl}, as a function of delamination size.*

Given the nodal forces and displacements from the two NASTRAN solutions, one was able to calculate the Mode I, Mode II, and Mode III strain energy release rates along the delamination front as follows:

$$G_I = F_z \delta_z / 2\Delta A$$
$$G_{II} = F_r \delta_r / 2\Delta A$$
$$G_{III} = F_t \delta_t / 2\Delta A$$

It is clear from these expressions that the Mode I, Mode II, and Mode III strain energy release rates are the normal, radial, and tangential components of energy released per unit area of crack extension. As illustrated in Fig. 6, the change in crack surface area ahead of each node on the delamination front is

$$\Delta A \approx \bar{a}\, \Delta\Theta\, \Delta a$$

where $\bar{a}$ is the average crack length over the increment of growth, Δa; Δa is equal to 3.2 mm (the length of one element); and $\Delta\Theta$ is equal to 0.13 rad (7.5°) for both 25.4 and 38.1 mm delaminations.

The crack opening technique was initially implemented using two approaches. The first approach was to model crack growth at only one node at a time. Only elements adjacent that node were disconnected, leaving the rest of the model intact. The relative nodal displacements at that location were determined and the corresponding strain energy release rates were calculated. This process was repeated for each node along the delamination front to obtain strain energy release rate distributions.

The second approach was to disconnect all the elements along the delamination front simultaneously and determine the strain energy release rates at each node.

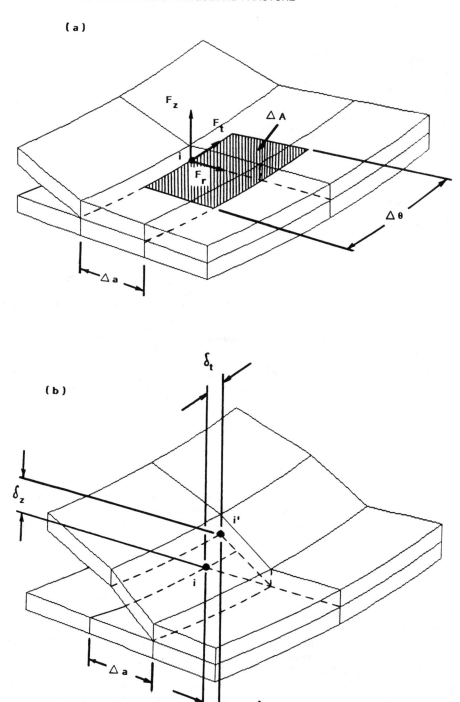

FIG. 6—*Finite element nodal forces and displacements used in virtual crack opening technique:* (a) *forces acting at node* i *and* (b) *relative displacements between nodes* i *and* i'.

The first method is intuitively more satisfying; one would expect delamination growth to initiate at a unique location and not simultaneously along the entire delamination front. This method yielded strain energy release rate values that were too low, however, and produced very poor results in the subsequent fracture analysis. Only the results of the second method are reported here.

The Mode I, Mode II, and Mode III strain energy release rate distributions for the 25.4 and 38.1 mm delaminations appear in Figs. 7 and 8, respectively. The plots show that there is a significant degree of mixed-mode loading on the delamination front. They also reveal large gradients in G_I, G_{II}, and G_{III}. This explains why delamination growth was not uniform in all directions. It is also interesting to note the change in the shape of the strain energy release rate distributions with increasing applied stress. This is an indication of the nonlinear nature of the problem.

For both delamination sizes, it is evident that G_I is the most dominant component of strain energy release rate. Fracture in the specimens was, therefore, most likely driven by Mode I crack loading. Examination of Figs. 7a and 8a reveals that the peak values of G_I occur at $\Theta = 90°$ and that these values increase with delamination size. The former observation explains why delamination growth propagated in a direction transverse to the applied load. The latter explains why delamination growth initiated at a lower stress in the case of larger delaminations.

Mixed-Mode Fracture

The computed strain energy release rate distributions describe the state of elastic energy contained along the delamination front at various compressive stress levels. In order to determine

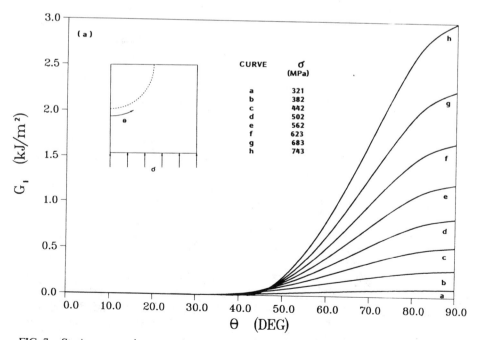

FIG. 7—*Strain energy release rate distributions for 25.4 mm delamination: (a) Mode I, (b) Mode II, and (c) Mode III.*

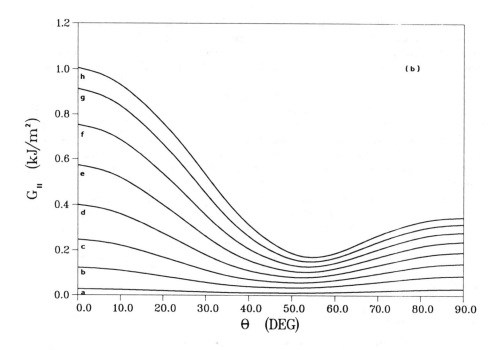

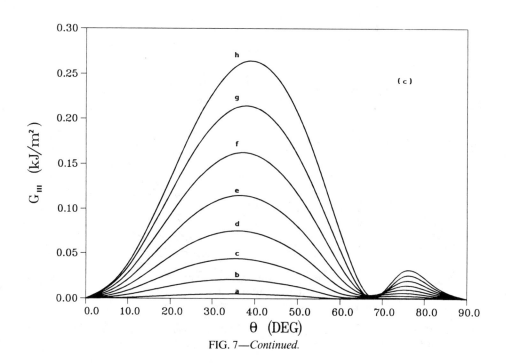

FIG. 7—*Continued.*

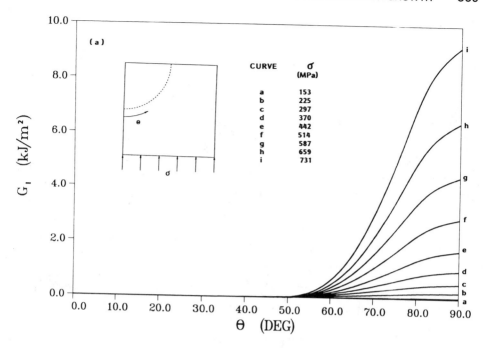

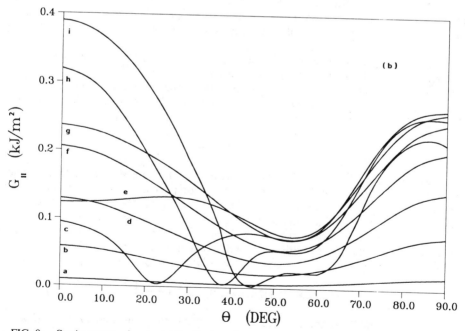

FIG. 8— *Strain energy release rate distributions for 38.1 mm delamination: (a) Mode I, (b) Mode II, and (c) Mode III.*

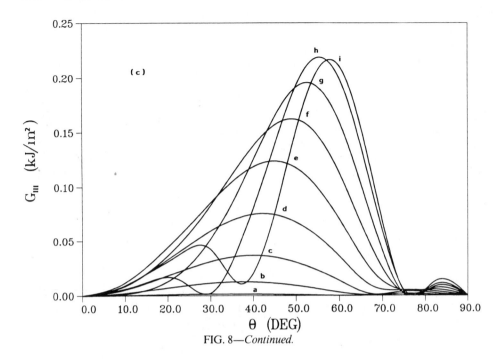

FIG. 8—*Continued.*

the location of fracture initiation and the corresponding stress level, the strain energy release rates were evaluated relative to the crack resistance of the graphite-epoxy material. Due to the mixed-mode nature of the loading along the delamination front, this required a mixed-mode fracture criterion.

A simplifying assumption was made in selecting a fracture criterion. Because G_{III} was shown to be relatively small compared to G_I and G_{II}, it was assumed that Mode III loading did not contribute significantly to fracture. Consequently, the Mode III component of strain energy release rate could be ignored and the fracture criterion could then be expressed in terms of just G_I and G_{II}. The criterion selected was

$$(G_I/G_{Ic}) + (G_{II}/G_{IIc}) = F \geq 1 \tag{1}$$

where G_{Ic} and G_{IIc} are the critical strain energy release rates for pure Mode I and Mode II fracture, respectively. Work done by Johnson and Mangalgiri [12] demonstrated for a number of material systems that this criterion provides the best fit to mixed-mode fracture test data.

Double cantilever beam tests and end-notched cantilever beam tests were conducted on the AS4/3501-5A material to measure G_{Ic} and G_{IIc}. It was found that $G_{Ic} = 0.42$ kJ/m^2 and $G_{IIc} = 1.26$ kJ/m^2. These values were substituted into Eq 1 and the margins of safety for fracture were determined at each point along the delamination front and at each compressive stress level

$$MS(\sigma,\Theta) = [1/F(G_I(\sigma,\Theta), G_{II}(\sigma,\Theta))] - 1$$

For both 25.4 and 38.1 mm delaminations, it was found that the lowest margin of safety occurs at an angle of $\Theta = 90°$. This margin of safety was then plotted as a function of applied stress, as illustrated in Fig. 9. From this plot it was possible to determine the threshold stress for the initiation of fracture and delamination growth. The results of this fracture analysis

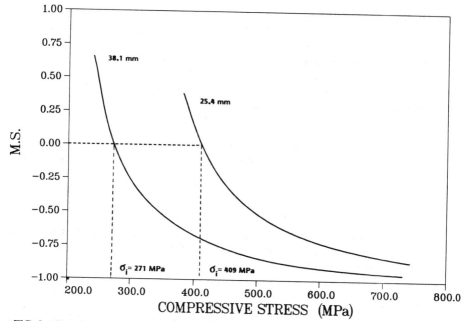

FIG. 9—*Computed margin of safety with respect to the initiation of fracture at* $\Theta = 90°$ *for 25.4 and 38.1 mm delaminations. Fracture initiates at a stress corresponding to MS = 0.*

appear in Table 1. The computed results are as much as 53% in error for a 25.4 mm delamination. The agreement with test data is somewhat better for a 38.1 mm delamination (12%). In both cases, the computed predictions are unconservative relative to the test data. It follows that the computed strain energy release rates were too low.

It was thought that low strain energy release rate values could have resulted from excessive model stiffness. An attempt was made to reduce the model stiffness by enhancing in-plane and through-thickness mesh refinement. The mesh density was doubled in the vicinity of the delamination front. This change did not, however, significantly alter the strain energy release rates. Further efforts in mesh refinement were not feasible due to already considerable computation times.

The source of the discrepancy between computed and empirical results has not yet been identified. The unconservative nature of the predictions suggests that failure mechanisms other than interlaminar fracture (such as fiber buckling and matrix cracking) may have contributed to delamination growth and specimen failure. If this was the case, the analysis would require a more sophisticated failure criterion or set of criteria to account for various failure modes. This is being investigated.

Conclusions

An empirical and analytical study was conducted to characterize the effects of embedded circular delaminations in compressively loaded graphite-epoxy laminates. Experiments demonstrated that delamination-related failure is initiated by the buckling of the sublaminates adjacent to the delamination. Sublaminate buckling is followed by interlaminar fracture with the consequence of delamination growth, general instability, and crippling. This failure mode was shown to reduce laminate compressive strength by as much as 60%.

The objective of the computational analysis was to model delamination growth by predicting the threshold stress for growth initiation. Computed predictions were unconservative and significantly in error relative to the test data.

The analysis provided qualitative insight into the process of quasi-static delamination growth. It was revealed that there is significant mixed-mode crack loading along the delamination front as well as large gradients in G_I, G_{II}, and G_{III}. Fracture analysis also showed that, driven primarily by Mode I crack loading, delamination growth propagates in a direction transverse to the applied compressive load. This finding was consistent with experimental observations.

The analysis was evaluated on the basis of very limited test data. The potential benefits of this analytical method dictate that it should not be dismissed until further verification has been done.

Acknowledgments

The author would like to thank Tan Phung for supervising the fabrication of the test specimens and Rudy Van Dorssen and Dave Woods for performing the tests. This research was funded entirely by Rohr Industries, Inc.

References

[1] Kan, H. P., Deo, R. B., Shah, C., and Kinslow, R., "Resistance Curve Approach to Predicting Residual Strength of Composites," Report No. AFOSR-TR-87-0062, Air Force Office of Scientific Research, Bolling Air Force Base, DC, Sept. 1986.

[2] Webster, J. D., "Flaw Criticality of Circular Disbond Defects in Compressive Laminates," Report No. CCM-81-03, Center for Composite Materials, University of Delaware, Newark, DE, June 1981.

[3] Wang, J. T. and Connolly, J. J., "A Simple Analysis Technique to Evaluate Delaminated Composite Laminate Under Compressive Load," *Proceedings,* Twentieth International SAMPE Technical Conference, Society for the Advancement of Material and Process Engineering, Sept. 1988, pp. 505–516.

[4] Ashizawa, M., "Fast Interlaminar Fracture of a Compressively Loaded Composite Containing a Defect," *Proceedings,* Fifth DoD/NASA Conference on Fibrous Composites in Structural Design, Report No. NADC-81096-60, Naval Air Development Center, Warminster, PA, Jan. 1981, pp. 269–291.

[5] Chai, H. and Babcock, C. D., "Two-Dimensional Modelling of Compressive Failure in Delaminated Laminates," *Journal of Composite Materials,* Vol. 19, Jan. 1985, pp. 67–98.

[6] Bottega, W. J. and Maewal, A., "Delamination Buckling and Growth in Laminates," *Journal of Applied Mechanics,* Vol. 50, March 1983, pp. 184–189.

[7] Whitcomb, J. D., "Finite Element Analysis of Instability Related Delamination Growth," *Journal of Composite Materials,* Vol. 15, Sept. 1981, pp. 403–426.

[8] Whitcomb, J. D., "Strain Energy Release Rate Analysis of Cyclic Delamination Growth in Compressively Loaded Laminates," *Effects of Defects in Composite Materials, ASTM STP 836,* American Society for Testing and Materials, Philadelphia, 1984, pp. 175–193.

[9] Whitcomb, J. D., "Mechanics of Instability-Related Delamination Growth," NASA Technical Memorandum 100622, National Aeronautics and Space Administration, NASA Langley Research Center, Hampton, VA, May 1988.

[10] Whitcomb, J. D., "Three-Dimensional Analysis of a Postbuckled Embedded Delamination," *Journal of Composite Materials,* Vol. 23, Sept. 1989, pp. 862–889.

[11] *MSC/NASTRAN Application Manual,* Vol 1., J. Joseph, Ed., The MacNeal-Schwendler Corporation, Los Angeles, CA, June 1983.

[12] Johnson, W. S. and Mangalgiri, P. D., "Influence of the Resin on Interlaminar Mixed-Mode Fracture," *Toughened Composites, ASTM STP 937,* N. J. Johnston, Ed., American Society for Testing and Materials, Philadelphia, 1987, pp. 295–315.

Anthony Palazotto[1] and Brendan Wilder[1]

A Study of an Implanted Delamination Within a Cylindrical Composite Panel

REFERENCE: Palazotto, A. and Wilder, B., "A Study of an Implanted Delamination Within a Cylindrical Composite Panel," *Composite Materials: Fatigue and Fracture (Third Volume), ASTM STP 1110,* T. K. O'Brien, Ed., American Society for Testing and Materials, Philadelphia, 1991, pp. 373–389.

ABSTRACT: This research studied the effects of an inserted circular delamination on the strength of an eight-ply, quasi-isotropic composite laminate. Cylindrical panels were loaded compressively, and the global buckling load of the delaminated plies were examined. It was found that, for the level of delamination defect investigated, a linear bifurcation analysis gave fairly accurate predictions of the global panel strength. Load values, which caused local instabilities in the region of panel damage, were also recorded and compared to an approximate analysis.

During the experimental phase of this research, certain physical responses of the panels led to a set of simplifying assumptions. It was thought that there existed a value of strain that when applied to a circular delamination would cause the sublaminate to buckle. The value of strain in the sublaminate should be related to a value of strain in the curved panel. Some method of relating these two values of strain was suggested by the results of experimental tests.

KEY WORDS: composite materials, fracture, fatigue (materials), delamination, circular delamination, buckling

Forty curved panels were fabricated and tested for this research. The panels were constructed of Hercules AS4/3501-6 graphite/epoxy, 304.8 mm prepeg tape. All panels were laid up with a $(0/-45/45/90)_s$ quasi-isotropic geometry. The panels were cylindrical in shape having a radius of curvature of 304.8 mm measured to the outside convex surface of the panel. They had a trimmed height of 330.2 mm and an arc length of 317.5 mm (Fig. 1). The material properties were determined to be (from coupon testing)

$$E_1 = 129.93 \text{ GPa}$$
$$E_2 = 10.12 \text{ GPa}$$
$$G_{12} = 6.27 \text{ GPa}$$
$$v_{12} = 0.28$$
$$v_{21} = 0.022$$

Initially, delaminations were introduced by placing two disks of Dupont 300 A, 0.5-mil FEP Teflon film back-to-back between plies during layup. The disks were coated with RAM 225 release agent prior to being placed in the panel to ensure 100% debonding. Originally, delaminations were to be placed between Plies 1-2, 2-3, 6-7, and 7-8 (Fig. 2). The inserts placed between Plies 2-3 and 6-7 formed the desired delamination. However, problems were encoun-

[1] Professor and former graduate student, respectively, Aeronautics and Astronautics Department, Air Force Institute of Technology, Wright-Patterson Air Force Base, OH, 45433.

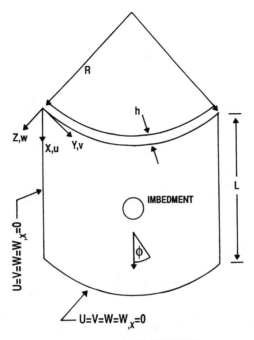

X, Y, Z   SURFACE COORDINATES
u, v, w   DISPLACEMENTS
h   PANEL THICKNESS
R   PANEL RADIUS
L   PANEL HEIGHT
φ   PLY ORIENTATION

FIG. 1—*Panel geometry.*

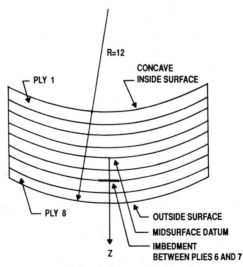

FIG. 2—*Cross section of panel.*

tered when the Teflon disks were placed only one ply in from the exterior surface (between Plies 1-2 and 7-8). After the panels were removed from the autoclave, the delaminations were externally visible. They appeared to be a low, circular blister approximately the size of the Teflon insert with fractures in the matrix running parallel to the 0° fiber orientation of the exterior blistered ply (Fig. 3).

Horban [1] had previously noted this problem when he used 1-mil mylar inserts. He had then tried using 0.5-mil mylar inserts, and found no matrix cracking or blistering effect. As a result of Horban's experience, it was thought that using one 0.5-mil Teflon insert with RAM 225 release agent as opposed to the two inserts placed back-to-back would give acceptable results. Four panels were manufactured in this manner. Two of the panels had 50.8-mm (2-in.) diameter inserts while the other two had 101.6-mm (4-in.) diameter inserts. All four panels exhibited the matrix cracking and blistering seen previously. At this point, it was not known what was causing the blisters. Panels were constructed using one 0.5-mil Teflon disk with no RAM 225 release agent applied. Blisters still occurred. It was then thought that perhaps volatiles were being formed beneath the Teflon disk resulting in a vapor pressure as the panels were heated in the autoclave. Panels were then constructed using one 0.5-mil Teflon disk that had been perforated with 20 small holes. Blisters still occurred. Thus, since the issue of one-ply-deep inserts has not been resolved, no specimens were tested under compressive loading with such a delamination insert.

Panel Layup and Curing

The panels were laid up in curved steel molds with a 304.8 mm radius of curvature. Panels that were to receive Teflon inserts were carefully measured to the center of the intended delamination using a steel scale, and the Teflon was placed using a RAM 225 release agent to ensure total delamination. The laminates were then cured in an autoclave according to cycle B-240-

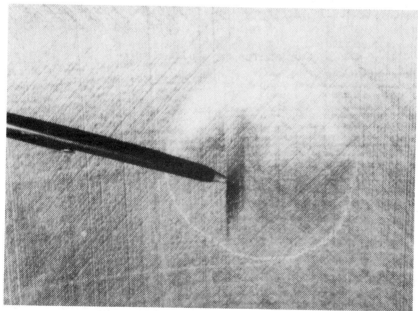

FIG. 3—*Blistered delamination (matrix cracking).*

T, Rev A [2]. After curing, the panels were removed from the bagging and visually inspected and C-scanned using a hand scanner to ensure that no large voids or delaminations were present. During the C-scanning process, the inserted delaminations were marked so that the panels could be accurately trimmed. The panels were then checked at 13 locations to determine the average panel thickness and the average ply thickness. Thickness variation within individual panels was small (on the order of 0.2%); however, the average panel thickness variations from panel to panel, amongst the group of panels tested, was 6.8%.

Next, two test panels were cut from each mold panel using a curved steel fixture to hold the panel and a radial arm saw with a water-lubricated 177.8-mm (7-in.), diamond-tipped blade. The panels were trimmed to a 330.3 mm height by a 317.5 mm arc length to allow for mounting in a compression test fixture. The unsupported dimension of the panel, when in the fixture, was 304.8 mm by 304.8 mm. Strain gages ((M&M) CEA-250-350UW-350) were placed on each panel to determine the uniformity of load introduction to the panel and to determine the load at which strain reversal occurred at delamination. Strain gages 1 and 2 (SG1 and SG2) were placed on the concave and convex sides of the panel, respectively. These gages were at the center of the panel's damaged area so that strain differential could be determined due to the bending in the damaged region of the panel. Gages SG3 and SG4 were placed at the top edge of the panel to track the uniformity at which the top edge displaced during the loading process.

Experimental Setup

The test fixture used in this research is a modification of the original General Dynamics 1974 design that was used by Wilkens [3] in his early work with curved composite panels and modified extensively by Refs 1 and 2. The panel test device consists of a 10 866-kg 25 000-lb capacity MTS compression machine with an Interface Model 1220-BF load cell. The panel is mounted in a fixture (Fig. 4) that clamps the top and bottom edges of the panel and applies a knife edge support with negligible restraining moments to the side supports. These boundary conditions can be thought of as representative of the edge restraints caused by a continuous panel restrained against rotation at one end by a major structural support while being restrained by panel stiffeners in the transverse direction.

The fixture gives the following edge boundary condition:

Top edge: $v = w = w_{,x} = 0$; $u =$ prescribed.
Bottom edge: $u = v = w = w_{,x} = 0$.
Vertical edges: $w = w_{,x} = 0$; $u = v = w_{,y} =$ free.

The fixture's vertical supports allow 6.35 mm of the panel to protrude above the top of the supports, to allow for the loading head displacement. An aluminum flange located at the center of the base plate provides a mount for an linear variable differential transducer (LVDT), which measures the vertical displacement. And, since the bottom test fixture is not permanently attached to the compression machine, this flange also provides a mount to secure the bottom test head in a uniform position for each test.

Test Procedure

After the panel was placed in the test fixture, a 1.33-kN seating load was applied and all of the supports were tightened. The purpose of this seating load was to ensure that the panel's top and bottom edges are clamped parallel to the heads and to reduce the chance of introducing a bending deformation at the supports. The seating load was removed, and all of the channels

FIG. 4—*Fixture set up.*

were set to zero. The panels were then compressively loaded at a constant displacement of 1.27 mm/min displacement with a load cell measuring the applied load. The displacement was introduced at the top edge of the panel. A LVDT measured the relative movement of the top head and bottom heads of the compression machine. Data were recorded three times per second from the LVDT, the load cell, and the four strain gages. The data were saved on a VAX 11/780 computer and later to a magnetic tape backup so that experimental graphs could be plotted. In addition, a CRT was used that allowed the progression of panel loading to be tracked during the test.

Ten of the panels tested had a 13.33-kN seating load applied to determine if there was a difference in strain reversal at the delamination as a result of ply separation brought about by the high initial load. This seating load had the effect of smoothing the introduction of the test load, since there was very little initial displacement of the panel in the supports at initial test loads. However, the load didn't seem to affect the shape of the strain separation curve.

Experimental Results

When compressively loaded, panels with a 50.8-mm circular delamination experienced global panel buckling at a mean load that was 13.2% lower than STAGSC-1 predictions, (The

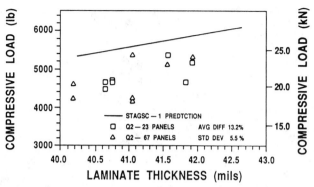

FIG. 5—*STAGSC-1 predictions versus experimental results for delaminations.*

STAGSC-1 program uses the finite element technique [4], and the authors incorporated its bifurcation capabilities, Fig. 5). These experimental buckling loads, for delaminated panels, were only slightly lower than the experimental buckling loads for undamaged panels. However, initial sublaminate instability was detected at a much lower load. When these panels were compressively loaded, the delaminated region (the sublaminate) would reach its local critical load and buckle or "snap." This local snapping effect can be detected visually as a blistered region at the delamination (Fig. 6), or it can be detected experimentally through the use of SG1 and SG2 centered over the delamination. This local sublaminate snapping causes a

FIG. 6—*Snapped sublaminate.*

release of strain energy and a lessening of the compression on the strain gage attached to the sublaminate. A strain reversal effect resulted, an ideal depiction of which can be seen in Fig. 7. In this figure, the strain reversal is seen to occur at approximately 3.61 kN. At local sublaminate buckling, SG1 on the base laminate is in compression, while SG2 on the sublaminate is seen to have its compressive strain relieved. SG2 goes into tension as the delaminated plies snap out. Another effect that can be seen from Fig. 7 is the lessening of compression on the base laminate at approximately 16.46 kN. It is thought that this indicates the onset of global panel buckling. An attempt was made, in this research, to predict this sublaminate instability through the use of the linear superposition of simple finite element models and a plane strain approximation. This method will be presented in the next section.

During the experimental testing of the panels the deformation was symmetrical. Hence, only axial strain was assumed to be present. Also, an axial load was applied to the panel in the plane of the panel. This led to the assumption, for convenience, that a state of plane strain could be approximated for the panel in a strip in the region of the delamination. The delaminated region of the panel was closely observed after experimental testing, and it was found to have undergone a plastic deformation as well as a delamination growth as a result of panel buckling (Fig. 8). When the panel was unloaded and removed from the test fixture, the delaminated area was clearly visible and retained the same general deformed shape it had attained under the buckling load. Another phenomenon that was noted was the shape of the buckled sublaminate (which occurred before any global buckling). Most of the 50.8-mm delamination would snap out into a single blistered region that appeared to be a low bubble on the surface

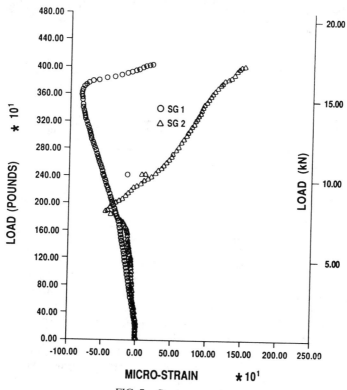

FIG. 7—*Strain reversal.*

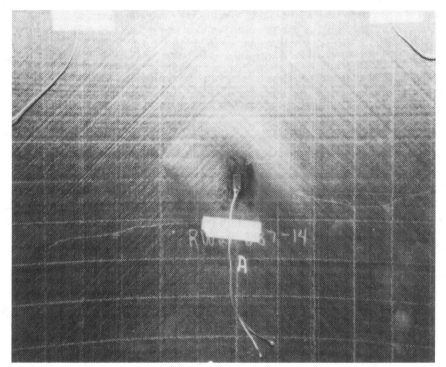

FIG. 8—*Plastic deformation.*

of the panel. The 101.6-mm sublaminates, on the other hand, would snap into a pattern of "ridges and valleys" that were oriented at approximately a 45° angle to the x-direction. This indicates the possibility of shear stresses developing at the sublaminate boundary in the post buckled state. It is also thought that the base laminate, which was in contact with the sublaminate, was laterally bracing the delaminated plies and causing them to buckle at higher modes. This phenomenon would tend to increase the critical load at which larger sublaminates initially became unstable relative to an unrestrained sublaminate.

Analytical Technique

This section is an attempt at calculating the onset of sublaminate buckling. The experimental phase of this study produced observations that indicated the delamination area never grew even after global buckling. Thus, the authors wanted to get some idea why this was so different than the observations when plates were considered.

Based on those observations described in the previous section as well as in Refs *1* and *2*, a strip of the panel in the vicinity of the delaminations was considered. This strip, shown by the dashed lines in Fig. 9, is flat in the longitudinal direction (which is the direction of loading) and slightly curved in the transverse direction. In the transverse, or y-direction, the strip is restrained against deformation by the eight-ply thickness of panel between the strip and the fixture's vertical support, while in the x- and z-directions, the strip is unrestrained against deflection. An edge displacement is applied to the strip in the x-direction, with restraint in the y-direction, and free movement in the z-direction. Hence, a plane strain assumption was assumed for this region of the panel.

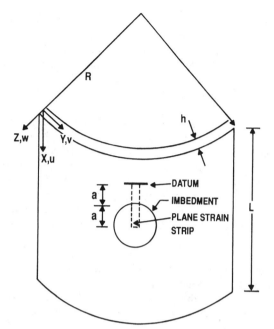

FIG. 9—*Plane strain strip.*

For the panel geometry investigated, a uniform displacement was applied to the top edge of the curved panel. It was found that this displacement caused an apparent constant strain field through the panel in the loading direction, and introduces an effective strain to the delaminated plies via an axial deflection of the sublaminate boundary. The relationship between strain in the curved panel and strain in the delaminated plies was used since strain is an easily measurable quantity for associating the response of the curved panel with the behavior of the sublaminate plies. Since bending is negligible, the delamination is primarily affected by the deflection in the axial direction that is transmitted by a constant strain in the curved panel. To investigate the effect of this deflection on the sublaminate, a finite element model was used to relate this value of N_x to the eight-ply curved panel. This model represented the strip shown in Fig. 9, had eight-plies through the z-direction, and contained a two-ply delaminated region. This model was loaded using the strain that had caused buckling in the finite element model of the two-ply circular sublaminate.

The resulting N_x value, through the cross section of the plane strain model, was calculated at a section of the model one delamination radius away from the delamination tip. This datum was chosen so that the calculated force resultant would be far enough from the discontinuity so that model responses due to the delamination tip could be avoided and thus simulating the major characteristics of the global panel. This force resultant in the eight-ply plane strain model could then be related to the global panel load, and would represent a force in the eight-ply laminate corresponding to the buckling strain in the $(0/-45)$ sublaminate.

In summary the prediction of the compressive load, which will cause the delaminated region of a composite panel to snap, was developed as follows:

1. A model of the curved panel containing no delamination was run using the STAGSC-1 finite element program for a normalized edge displacement. An N_x value at the top edge of the panel was calculated for this displacement, and the strain in the panel associated with this force

was calculated at the datum position shown in Fig. 9. The strain associated with the applied load in this model was then applied to a plane strain model in Step 4.

2. The delamination area of the panel with the asymmetric $(0/-45)$ orientation was modeled as a circular plate using STAGSC-1. This model was given a uniform edge displacement; the bifurcation load was determined; and the strain in the disk and the N_x value at the plate's buckling were determined. The strain obtained at bifurcation for this model was then used to load the plane strain model in Step 3.

3. A finite element program [5] was used to run a plane strain analysis of an eight-ply composite containing a delamination. This model was evaluated using the strain obtained at buckling for the circular plate of Step 2. The strain from Step 2 is used to calculate an end displacement in the plane strain model that yields an equivalent strain at the delamination tip. For this value of strain, the N_x value, through the eight-ply section of the model, is calculated. This calculation is performed at a cross section that is one delamination radius away from the delamination tip so that stress concentrations due to the delamination radius are avoided, and so that a uniform N_x value can be calculated that corresponds to an eight-ply section of the curved panel.

4. A plane strain model was also run for a section of panel that contained no delamination. This model was run using the strain in the curved panel of Step 1 and is used to determine an N_x in a plane strain model that is equivalent to the load applied to the curved panel. The model was constructed such that the strain at the outer edge of the delaminated region calculated in Step 1 was used as the input load for the plane strain model. The N_x value in an eight-ply section of the model located one delamination radius away from the delamination was calculated.

5. Thus, the N_x values have been calculated in two plane strain models. One force resultant corresponds to a uniform load applied to the top edge of a curved panel; and the other force resultant corresponds to the buckling load for a $(0/-45)$ circular plate. Using these force resultants, and the force resultants obtained in the two STAGSC-1 models, a load is computed that when applied to the top edge of the curved panel will cause buckling of the circular delamination. This load is simply a percentage of the load applied to the curved 304.8-mm panel.

The models used in this numerical method and the process of calculating a local sublaminate buckling load will be discussed in more detail in the balance of this paper.

Curve Panel—STAGSC-1 Model

One of the standard geometries that STAGSC-1 is capable of generating is a curved cylindrical panel. For the finite element model used in this analysis, the surface of the shell was discretized into an 18 by 18-element mesh consisting of 16.9-mm-square elements. Previous work done by Siefert [2], with this geometry, had shown that accurate results could be obtained if the element size was between 12.7 and 25.4 mm. The element selected for this analysis was the quadrilateral 410 element referred to as the QUAF 410 in the STAGSC-1 user manual. This element was selected because it works well for the linear analysis and the cylindrical shells, and because it has 12 fewer degrees of freedom than the more accurate QUAF 411 element.

Both linear material properties and bifurcation buckling were used for this model. The model was loaded at its tip edge with a constant end displacement in the positive direction. The boundary conditions used for the model were the same constraints assumed in the experimental work. Using the STAGSC-1 model, the strains and stresses in the panel could be determined for a given value of load, P. Also computed was the force resultant at the top edge of the panel, N_{x1}, associated with this load.

The introduction of a uniform displacement to the top edge of the curved composite panel causes a strain in the central location of the delamination tip. The critical strain at this point, which will cause the sublaminate to buckle, is discussed in the next section.

Sublaminate—STAGCS-1 Model

The authors were interested in modeling the buckling of a circular plate to characterize the plane strain strip assumption previously discussed. The simple, yet fairly accurate condition for a strip under the plane strain assumption was one of complete geometric symmetry within the plate and, therefore, the subsequent convergence study and model was predicted on this strain feature.

The delaminations inserted into the panels, as stated previously, were 50.8 and 101.6 mm in diameter. The STAGSC-1 model discussed in this section was used to determine the force resultant and the strain necessary to cause buckling of the $(0/-45)$ circular delaminated region of the panels. The force resultant at the model boundary, N_{x3}, and the edge displacement of the boundary were determined using a STAGSC-1 annular ring shell geometry and various finite element mesh refinements.

A convergence study was first performed considering isotropic material to determine an optimum finite element mesh and to check the accuracy of the STAGS QUAF 410 element. A mesh was modeled over a quarter of a circle since symmetric loading and boundary conditions could be applied. A closed-form Bessel function solution exists [6] for a uniform axisymmetric compressive load acting on a circular plate. Therefore, a uniform edge displacement was applied radially inward at the outer boundary, and the N_x value at plate buckling was recorded. A refined model was run for a clamped edge support, and the STAGSC-1 buckling load was compared to the Bessel solution of Ref 6 and found to yield acceptable results. It should be pointed out that the authors felt the clamped edge condition, in which $u_x = u_y = w_{,x} = w_{,z}$ are prescribed (free, in this case) and w and $w_{,y}$ are constrained (though approximate), best characterized the actual delaminated boundary with respect to the overall sublaminate restraint (u_x is radial displacement; w is the z displacement; $w_{,x}$ is the rotation about the thickness axis; $w_{,x}$ is the rotation about the tangent line at the boundary; $w_{,y}$ is the rotation about the radial line at the boundary). Thus, this preliminary model investigation became valuable in the subsequent sublaminate investigation.

As a further check on the accuracy of the finite element model used in the composite plate analysis, comparison was made to Shivakumar and Whitcomb's work [7]. Reference 7 had obtained convergence with their model for an elliptical delamination with anisotropic layups, so it was used in this study to check the accuracy of the authors final model. Shivakumar and Whitcomb's model uses the TRINC 320 triangular plate element that has 18 degrees of freedom. Their model had 738 degrees of freedom as opposed to the 377 degrees of freedom in the 10 by 8 model using the QUAF 410 incorporated herein. Consequently, it was a more expensive model to run. Reference 7 used the 320 element that correlated well with the closed-form Bessel solution, but not as close as the model using the QUAF 410 rectangular elements.

Thus, the 10 by 8 QUAF 410 model shown in Fig. 10 was used to obtain N_x at buckling for the $(0/45)$ circular delaminated region, N_{x3}. It is required that the entire plate be modeled due to the ply layup. Also determined using this model was the nominal ε_x strain in the sublaminate at local buckling. The strain was then used in the plane strain model discussed in the next section to determine an equivalent N_x value in an eight-ply laminate, N_{x4}. This force resultant is the force resultant in an eight-ply plane strain model associated with the strain at buckling for a two-ply, $(0/-45)$ circular sublaminate.

Also calculated from this STAGSC-1 model was the average normal stress at the sublaminate boundary at local buckling. The STAGSC-1 output, for a 50.8-mm delamination, yielded an average N_x value for an axisymmetric edge displacement of 0.025 mm, 8.429 N/mm. This

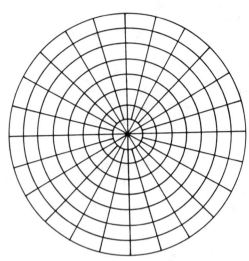

FIG. 10—*360° mesh.*

value was then multiplied by the eigenvalue, 0.095, to determine the average force resultant at the boundary at the onset of sublaminate instability. This force resultant was then divided by the cross-section dimension of the $(0/-45)$ plies, 2×0.122 mm thickness, to determine the average normal stress, 3284 kPa, at the sublaminate boundary at local buckling of the model. This stress will be used in the next section as a comparison to check the validity of the plane strain model.

Plane Strain Finite Element Model

The finite element code used for this analysis was developed by the Air Force Flight/Dynamics Laboratory [5]. The model is used in this analysis to tie the strain at local sublaminate buckling to the strain in a plane strain section of the curved panel. The strain in this eight-ply plane strain model could then be used to calculate an axial load that would cause the sublaminate to buckle.

The eight-ply plane strain delaminated model shown in Figs. 11 and 12 was loaded with an end displacement equivalent to the strain required to buckle the $(0/-45)$ sublaminate. This displacement was applied one delamination radius away from the delamination tip so that a uniform cross-sectional force resultant could be achieved.

The stresses in the vicinity of the delamination tip were recorded and are shown in Fig. 12. The average axial stress in the $(0/-45)$ section of the model was calculated to be 3255 kPa. This value is for a plane strain calculation associated with the strain at buckling in the $(0/-45)$ STAGSC-1 circular disk model. The STAGSC-1 circular disk model discussed previously yielded an average stress value of 3284 kPa for this same value of strain. The difference between these two stresses is only 0.88%. Therefore, it appears that by using a symmetrical loading pattern, a load distribution over the upper two plies is equivalent to a plane strain phenomenon. Also of note in this plane strain model is the magnitude of stress in the z-direction. In the vicinity of the delamination tip, these stresses varied from 4.69 to 13.65 kPa. This level of stress is low, and it can be observed that up to the point of sublaminate buckling, there is no danger of Mode I crack growth due to these stresses. (This was expected since the loading is assumed in the plane of the delamination.) Further crack growth features would have to

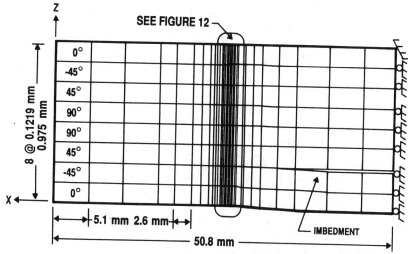

FIG. 11—*Delamination finite element model.*

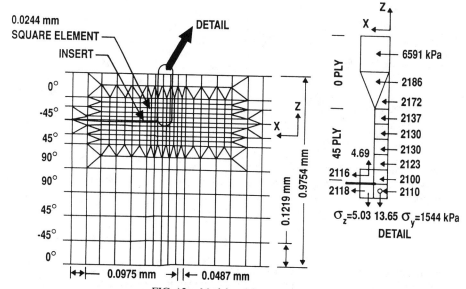

FIG. 12—*Model at delamination tip.*

consider the delaminated region in the post-buckling regimen and was not analytically considered. Yet, even after global buckling occurred, C-scanning did not indicate any delaminate growth.

This plane strain model was designed to determine a force resultant, N_{x4}, through an eight-ply plane strain section of the composite. This force resultant is the N_x value in a quasi-isotropic, eight-ply section associated with the strain at sublaminate instability for a two-ply $(0/-45)$ delamination.

The strain in the curved 25.4-mm panel resulting from a panel load has been related to a force resultant in a plane strain model, and the strain in a circular disk at local buckling has been related to a force resultant in a plane strain model. It is now possible to calculate the load, which when applied to the top edge of the 50.8-mm curved panel, will cause local instability of a delaminated region of the panel. This calculation is explained in the next section. It must be realized that this approach presented is only an estimate of the onset of the sublaminate buckling.

Plane Strain Model—No Delamination

A plane strain analysis was performed for the laminate to correlate the force resultant and strain in a plane strain model with the force resultant and strain in a plane strain strip of the curved composite panel.

The average N_x value distributed throughout the model was calculated for the strain (obtained from the curved panel model) that was associated with a displacement at the top edge of the curved 50.8-mm panel.

First, the Q values were calculated for the material

$$Q_{11} = \frac{E_1}{1 - v_{12}v_{21}} \tag{1}$$

$$Q_{22} = \frac{E_2}{1 - v_{12}v_{21}} \tag{2}$$

$$Q_{66} = G_{12} \tag{3}$$

The $\overline{Q}_{11}$ value was computed in x-direction, where Θ is the angle between the fiber direction and the panel's x-direction.

$$\overline{Q}_{11} = Q_{11} \cos^4 \Theta + 2(Q_{12} + 2Q_{66}) \sin^2 \Theta \cos^2 \Theta + Q_{22} \sin^4 \Theta \tag{4}$$

For plane strain, the σ_x stress can be calculated.

It was observed, as previously mentioned, that the panel yields a constant ε_x strain. One could quickly compute this strain by using the displacement at the panel's top edge, and divide this value by the overall length of the panel, thus yielding a σ_x of

$$\sigma_x = \overline{Q}_{11}\varepsilon^\circ \tag{5}$$

The stress is obtained for each of the plies, and a force resultant, N_{x2}, is calculated. This force resultant is then the value of N_x in the eight-ply plane strain strip associated with a deflection induced through loading of the curved panel without a delamination.

Analytical Prediction

Based on the previous analysis, and using linear superposition, the following equation was developed to predict local buckling of the sublaminates

$$P_f = P \cdot \alpha \cdot \beta \tag{6}$$

where α is the ratio of the force resultant of a curved panel under edge displacement passing into a plane strain region, and β is the ratio of force resultants in the plane strain models that creates buckling of the circular inserts.

$$\alpha = N_{x2}/N_{x1} \tag{7}$$

$$\beta = N_{x3}/N_{x4} \tag{8}$$

where

$P_f \equiv$ predicted load in pounds at which local sublaminate snapping will occur;
$P \equiv$ load in pounds at top edge of the curved panel;
$N_{x1} \equiv N_x$ value related to a nominal edge displacement for the curved panel;
$N_{x2} \equiv N_x$ value in plane strain model with no delamination. This value is calculated for a strain equivalent to the strain in the curved panel due to a load, P;
$N_{x3} \equiv N_x$ value at buckling for a circular disk;
$N_{x4} \equiv N_x$ value for plane strain model containing a delamination. This force resultant is calculated using the strain at buckling for the circular disk model.

Substituting Eqs 7 and 8 into Eq 6, the predicted load applied to the top of the curved 50.8-mm panel that would cause local buckling in the delaminated plies is

$$P_f = P \frac{N_{x2}N_{x3}}{N_{x1}N_{x4}} \tag{11}$$

As the average thickness of the laminate plies increases, the value at which local sublaminate snapping will occur, also increases. This is seen by noting, through an empirical relationship derived in Ref 8, that N_{x1} (the N_x value for the 50.8-mm curved panel) increases by panel thickness, h, raised to the 2.06 power (approximately by h^2). N_{x3} for the circular disk increases by approximately the cube of the panel thickness; and N_{x2} and N_{x4} both increase linearly with thickness. Therefore, the snapping of the sublaminate was expected to increase linearly with panel thickness. This linear variation of predicted snapping load was plotted as a linear prediction of snapping load versus panel thickness in Fig. 13. Using this relationship, the load causing local instability of a sublaminate could be determined based on panel thickness. This relationship appeared to be valid for the range of panel thicknesses used in this research.

The relationships between analytic predictions and experimental snapping values are shown in Fig. 13. The data points shown in Fig. 13 are in pairs. Each pair represents two panels from the same mold with the same panel thickness. These data pairs are shown connected by a line with a bar at midheight representing the average snapping load for the pair. The snapping loads for the 50.8-mm (2-in.) delaminations fell below analytic predictions. This was expected since linear models were used in the analysis, which, in general, increases the load at which predicted failure will occur. For these panels, the experimental snapping loads differed from the analytic predictions by a mean of 31.3% with a standard deviation of 21.4%.

Only three data points (shown in Fig. 14) were obtained for the snapping of the 101.6 mm (4-in.) delaminations. This is a very sparse data set; however, it was noted that these panels tended to snap at higher load values than were predicted. A possible explanation for this is that the 101.6 mm (4-in.) sublaminates were being restrained within their domain by the base laminate, and are therefore buckling at higher modes. These panels differed from the analytic predictions by a mean of 14.6 with a standard deviation of 5.2.

This analytic technique is seen to give predictions of local sublaminate buckling within

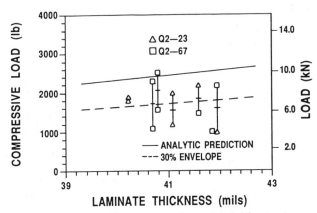

FIG. 13—*Strain reversal load, 50.8 mm (2 in.) delamination.*

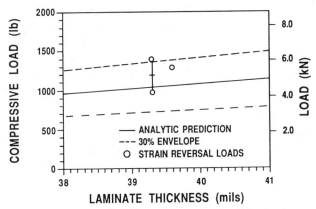

FIG. 14—*Strain reversal load, 101.6 mm (4 in.) delamination.*

approximately 30% of experimental values. These predictions are good engineering approximations obtained using a linear two-dimensional model to predict a nonlinear, three-dimensional local instability.

Conclusions

It is possible to state from the study that:

1. In general, the linear bifurcation predictions obtained using the STAGSC-1 finite element code were in close agreement with experimental results. For undamaged panels, the experimental global buckling load was an average of 12.9% lower than predicted values (as expected) with a standard deviation of 2.8%.

2. For the implanted delamination sizes studied, only slight decreases in global panel buckling were noted.

3. The phenomena of local sublaminate snapping of the delaminated plies can be predicted using a superposition of a series of linear finite element models. The use of these models results in a fairly accurate and inexpensive prediction of local instabilities caused by local buckling. The local instability is generally 40% of the panels' global buckling force.

4. The stresses in the vicinity of delamination crack tip were calculated in a plane strain model. The model was loaded with a strain sufficient to cause buckling of the $(0/-45)$ delaminated plies. At this load, the stresses in the z-direction at the crack tip were calculated. These stresses, normal to the plane of the delamination, were much less than the stress required to initiate crack growth in this material.

References

[1] Horban, B. A. and Palazotto, A. N., "Experimental Buckling of Cylindrical Composite Panels with Eccentrically Located Circular Delaminations," *Journal of Spacecraft and Rockets*, Vol. 24, No. 4, July–Aug., 1987, pp. 349–352.

[2] Siefert, G. R. and Palazotto, A. N., "The Effect of a Centrally Located Midplane Delamination on the Instability of Composite Panels," *Journal of Experimental Mechanics*, Vol. 26, No. 4, Dec. 1986, pp. 330–336.

[3] Wilkins, D. J., "Compression Buckling Tests of Laminated Graphite/Epoxy Curved Panels," AIAA Paper No. 74-32, presented at the AIAA Twelfth Aerospace Sciences Meeting, Washington, DC, American Institute of Aeronautics and Astronautics, 30 Jan. 1974.

[4] Almroth, B. O., Brogan, F. A., and Stanley, G. M., "Structural Analysis of General Shells," *Volume II User Instruction for STAGSC-1*, Lockheed Palo Alto Research Laboratory, CA, Jan. 1981.

[5] Sandhu, R. S., "Alternate Strength Analysis of Symmetric Laminates," Technical Report AFFDL-TR-73-137, AD 779927, Air Force Flight Dynamics Laboratory, Wright Patterson Air Force Base, OH, Feb. 1974.

[6] Dym, C. L. and Shames, I. H., *Solid Mechanics—A Variational Approach*, McGraw-Hill, New York, 1973.

[7] Shivakumar, K. N. and Whitcomb, J. D., "Buckling of a Sublaminate in a Quasi-Isotropic Composite Laminate," *Journal of Composite Materials*, Vol. 19, Jan. 1985, pp. 2–18.

[8] Wilder, B. L., "A Study of Damage Tolerance In Curved Composite Panels," Masters thesis, School of Engineering, The Air Force Institute of Technology, AFIT/GA/AA/88M-3, March 1988.

Strength and Impact

E. Gail Guynn,[1] Walter L. Bradley,[1] Ozden O. Ochoa,[1] and
John D. Whitcomb[1]

A Comparison of Experimental Observations and Numerical Predictions for the Initiation of Fiber Microbuckling in Notched Composite Laminates

REFERENCE: Guynn, E. G., Bradley, W. L., Ochoa, O. O., and Whitcomb, J. D., "A Comparison of Experimental Observations and Numerical Predictions for the Initiation of Fiber Microbuckling in Notched Composite Laminates," *Composite Materials: Fatigue and Fracture (Third Volume), ASTM STP 1110*, T. K. O'Brien, Ed., American Society for Testing and Materials, Philadelphia, 1991, pp. 393–416.

ABSTRACT: Previous experimental observations indicate that the compressive failure of open-hole multidirectional thermoplastic composites begins with in-plane fiber microbuckling at the hole. Growth of this damage requires little additional load, suggesting that compressive strength is controlled by initiation, rather than propagation, of in-plane microbuckling. The significant role of local constraint on the initiation of fiber microbuckling is indicated in this work.

Stacking sequence and resin ductility (varied as a function of the test temperature) are the two primary variables in this investigation, both of which provide systematic variation of support to the load-bearing 0° plies. The material system selected for this investigation is a thermoplastic composite, APC-2: AS4/PEEK.

The compression specimens used are 2.54 cm wide with a 2.54 cm long gage length. Each specimen contains a 0.3175-cm-diameter semicircular notch at each free edge, centered along the gage length. This polished notch facilitates the observation of the initiation of damage in the form of in-plane fiber microbuckling into the notch. High-temperature tests are conducted using a controlled forced-air heat gun rather than an environmental chamber to allow high-magnification observation of the failure processes at the higher temperatures.

Geometric and material nonlinear two-dimensional finite element analysis is used to model the effects of the initial fiber waviness in combination with matrix nonlinearity on the fiber microbuckling initiation strain levels and the resulting shear strain developed in the matrix.

KEY WORDS: composite materials, compression, open-hole/notched composites, fiber microbuckling, fiber shear, finite element analysis, high temperature tests, fracture, fatigue (materials)

One problem impeding the widespread application of composites is their inherent weakness in compressive strength when compared to the tensile properties of the same material. This result is not surprising, given that the composite's tensile and compressive strength comes primarily from long narrow fibers. The effective strength of these fibers in compression is generally much lower than their strength in tension due to fiber microbuckling that leads to mac-

[1] Graduate doctoral student and professors, respectively, Texas A&M University, College Station, TX 77843.

roscopic compressive failure. Nevertheless, it is desirable to develop composite systems with the smallest penalty in compressive strength possible.

Previous work [1–9], including an extensive literature review [5], on tough matrix composite laminates containing center holes indicates that the compressive failure of these composites initiates with microbuckling of the 0° fibers toward the unsupported surface in 0° plies at a free surface. Compressive failure in the laminates that did not contain surface 0s initiated with in-plane fiber microbuckling into the center hole. Growth of this damage to a critical size at which catastrophic failure occurs required little additional load. This result suggests that the compressive strength is controlled by initiation of in-plane fiber microbuckling, particularly for the PEEK system [1–4]. The in- and out-of-plane fiber microbuckling appears to precede shear crippling damage in the systems studied in Ref 1–4. Furthermore, the fiber microbuckling and shear crippling often lead to local delamination when the local strain, necessary to accommodate the large localized interlaminar shear strains, exceeds the resin ductility. These local delaminations apparently do not propagate (to become macroscopic delaminations) until final compressive failure when large scale brooming or delamination or both occurs.

The significant role of local constraint on the initiation of fiber microbuckling is also indicated in Refs 1–5. For example, when ±45° fibers were used as surface plies, in-plane fiber microbuckling initiated at a strain that is much higher than the initiation strain when the 0° fibers were the surface plies and the microbuckling was out-of-plane. These results indicate that one reason most compression strength models overpredict the actual compression strengths observed in experiments is that they make no allowance for free surface effects, that is, these models implicitly assume infinitely large laminates.

Many theories have been developed to predict the compressive strength of unidirectional composite laminates. However, few similar investigations have been conducted for multidirectional composite laminates, even though these laminates are more widely used than unidirectional laminates in composite structural components. Furthermore, an uncertainty in the modes and mechanisms of compression failures in composites remains because of the lack of a clear understanding of all possible modes of failure. In order to more accurately predict the compressive strength of composite laminates, the strain level at which fiber microbuckling initiates must be determined. Additionally, the factors that affect this strain level must be studied systematically. Accurate evaluation of this strain level is important because this process appears to control the strength of ductile composite systems.

The experimental observations [1–4] and an extensive literature review [5] suggest at least four factors that affect the strain level at which fiber microbuckling initiates and thus, partially controls the composite's compressive strength. These factors are as follows: (1) the degree of fiber waviness, (2) the fiber/matrix interfacial bond strength, (3) the effects of the free surfaces, and (4) the nonlinear resin shear constitutive behavior. Two additional factors that may be also important in the determination of the strain level at which either in- or out-of-plane, or both, fiber microbuckling initiates are (5) the orientation of the supporting plies adjacent to the 0° plies through the thickness of a laminate and (6) the thickness of the resin-rich regions between plies.

The objectives of the experimental program presented in this paper are to explore the effects of the free surfaces, the nonlinear constitutive behavior of the resin, the orientation of the supporting plies adjacent to the 0° plies, and the thickness of the resin-rich region between plies on the in-plane or out-of-plane, or both, fiber microbuckling initiation strain levels. The nominal strain at fiber microbuckling initiation in the 0s and the subsequent development of damage are monitored as a function of these independent variables. The objectives of the finite element analysis (FEA) are to show the effects of the initial fiber waviness in combination with matrix nonlinearity on the shear strain developed in the matrix.

Experimental Procedures

Material

The material selected for this investigation is APC-2, an aromatic polymer composite manufactured by Fiberite Corporation—An ICI (Imperial Chemical Industries) Company. APC-2 is an advanced structural composite composed of continuous carbon fibers (AS4) and "Victrex" PEEK (Polyetheretherketone), a semicrystalline thermoplastic matrix material.

The glass transition temperature, T_g, for PEEK is 143°C (290°F), and the melting temperature, T_m, is 335°C (635°F). Resin properties [10] are given in Table 1. The anisotropic fiber properties [10] are given also in Table 1. The diameter of this fiber is 7 μm. The fiber volume fraction is 61% by volume and 68% by weight. This thermoplastic system was selected for this investigation because the compression strength of thermoplastics in general is a primary concern for industry applications. Typically, thermoplastics have compression strength values approximately 50% less than the observed tensile strength, compared to 75 to 80% for graphite/epoxy systems. Lamina material properties [10] for APC-2 are summarized in Table 1.

To evaluate the factors that affect the compressive failure strength, the laminate stacking sequence is varied systematically. A relatively simple baseline stacking sequence $[(\pm 45/0_2)_3/\pm 45/0]_s$, has been selected. Systematic variations of this stacking sequence allow for a detailed study of the effects of supporting fiber orientation, fiber waviness, interfacial bond strength, and resin-rich regions between plies. For consistency, these variations are made through the laminate thickness.

Table 2 details the laminates used to study the effects of the local constraint (supporting fiber orientation) on fiber microbuckling initiation. Five stacking sequences of APC-2 are used to vary the support to the 0° fibers. The $\pm 45°$ plies in the baseline stacking sequence are replaced with either $\pm 15°$ plies, $\pm 75°$ plies, or $90_2°$ plies. Additionally the 45s and 0s are interchanged in one laminate to determine the effect of surface 0s on the initiation of fiber microbuckling.

The effect of the resin-rich regions (Table 3) is evaluated by adding a PEEK resin film at each 45/0 interface through the thickness of the baseline stacking sequence. In the stacking

TABLE 1—*Mechanical properties of constituent materials.*

Material	E_{11}, GPa	E_{22}, GPa	G_{12}, GPa	ν_{12}
PEEK resin	3.60	3.60	1.30	0.42
AS4 fiber	235	14.0	28.0	0.20
APC-2 lamina	134	8.90	5.10	0.30
Isotropic fiber[a]	67.2	67.2	28.0	0.20

[a] Assumed for FEA.

TABLE 2—*Effects of supporting fiber orientation.*

Stacking Sequence	Variable Investigated
$[(0_2/\pm 45)_3/0/\pm 45]_s$	vary support to 0s
$[(\pm 15/0_2)_3/\pm 15/0]_s$	vary support to 0s
$[(\pm 45/0_2)_3/\pm 45/0]_s$	baseline layup
$[(\pm 75/0_2)_3/\pm 75/0]_s$	vary support to 0s
$[(90_2/0_2)_3/90_2/0]_s$	vary support to 0s

TABLE 3—*Effects of resin-rich regions.*

Stacking Sequence	Variable Investigated
$[(\pm45/0_2)_3/\pm45/0]_s$	baseline APC-2, no PEEK film added
$[(\pm45/f/0_2/f)_3/\pm45/f/0]_s$	vary resin rich region: $f = 0.025$ mm, one layer of PEEK film
$[(\pm45/f_3/0_2/f_3)_3/\pm45/f_3/0]_s$	vary resin rich region: $f = 0.025$ mm, one layer of PEEK film

sequences, the addition of resin film is denoted by f. Two different thickness, 0.025 mm (f) and 0.075 mm (f_3) of resin film are added to the laminates.

Methods

Compression Specimen Geometry—The compression specimen is 2.54 cm wide by 10.16 cm long with a semicircular notch (3.175 mm diameter) at each free edge, centered along the gage section. The gage length for this specimen is 2.54 cm. Preliminary experimental results, detailed in Ref 11, indicate that reducing the gage length from 5.08 (used in Refs 1-4) to 2.54 cm does not introduce significant end effects but does minimize specimen bending and the incidence of Euler buckling. These specimens are strain-gaged front and back (at the specimen's center) with longitudinal strain gages to monitor any specimen bending and to provide accurate measurement of the fiber microbuckling initiation strains.

Compression Test Methods—Compression tests are conducted in a specially designed ultra-high axial alignment Material Test System (MTS) machine in the Materials and Structures Laboratory of Texas A&M University. The specimens are loaded in compression to failure in the servocontrolled hydraulic test stand at a relatively slow rate in displacement control to provide more stable growth of the shear crippling zone. The fiber microbuckling process occurring in the radius of the semicircular edge notches is monitored with a Wild M8 Zoom Stereo-microscope. This testing system is shown in Fig. 1. Tests were interrupted at the first indication of fiber microbuckling and subsequently observed in the scanning electron microscope (SEM).

This specially designed MTS contains a collet-type grip arrangement. The collet inserts that grip the specimen are machined for an ideal specimen thickness, for example, 5.334 ± 0.025 mm. To further improve the alignment and gripping support provided during the compression tests, a fixture has been designed to precision cast resin shims symmetrically onto the specimen ends. The thickness of the specimen plus the cast shims is within the thickness tolerances required by the collet inserts, and the shims are uniformly bonded to the specimen ends. The epoxy used as shim material consists of an all-purpose resin, DER 31 [12], and a curing agent or hardener, DEH 24 [12], supplied by The Dow Chemical Company [13], Freeport, Texas.

Some preliminary high-temperature tests are conducted at 77°C (170°F) to provide variation in the constitutive behavior of the PEEK matrix, and, thus, additional variation in the support provided to the 0° fibers. To provide easier observation and access to the specimen, a forced-air heat gun with a controller, rather than an environmental chamber, is used to heat the specimen gage section.

Two types of experimental information are obtained from these notched compression tests. First, the nominal strain associated with the initiation of fiber microbuckling is measured. It should be noted that this value is the nominal strain at which either in- or out-of-plane fiber microbuckling is sufficiently general to be observed with the stereomicroscope. This strain level is defined as fiber microbuckling initiation, ϵ_I. Second, the measured nominal strain is used in conjunction with two-dimensional finite element analysis to determine the local strain at the notch during fiber microbuckling initiation.

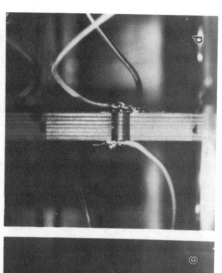
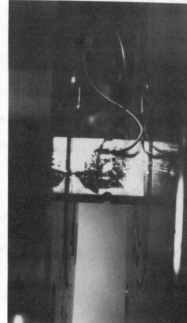

(a) Overview of MTS test stand, controller, and data acquisition.
(b) Closeup showing gripped specimen, stereomicroscope in the background, and video image of notch edge.
(c) Surface view of edge-notched specimen with strain gages, gripped in MTS test stand.
(d) Edge view of specimen with strain gages.

FIG. 1—*Compression test facilities.*

Determination of Lamina Constitutive Behavior—The constitutive behavior for shear loading is determined from tensile and compressive tests of a $[\pm 45]_{ns}$ laminate where n is 2 for tension and n is 8 for compression tests. These tests are conducted at the same temperatures (21°C (70°F), 77°C (170°F), and 132°C (270°F)) as the notched compression tests to show the considerable change in the yield strength and the shape of the shear stress-strain curves at these temperatures.

The specimens for the tension tests are 2.54 cm wide by 22.9 cm long with two 2.81 cm long glass/epoxy tabs bonded to each end of the coupon [14]. The specimens for the compression tests are prepared in a manner similar to the notched compression specimens except without the edge notches. High-strain (5%) longitudinal-transverse strain gages are bonded onto both types of specimens using a high strain (15%) adhesive. The tests are conducted in displacement control at a rate of 1 mm/min.

Two specimens, one compression-loaded and one tension-loaded, are monitored for matrix cracking with dye-penetrant enhanced X-radiography. Zinc iodide is used as the dye-penetrant. Each of these tests is paused at every 1% increment of longitudinal strain. The dye-penetrant was applied, allowed to soak, and then the X-ray was taken.

The nonlinear shear stress-strain relationship for these laminates is calculated using the method described in Refs 15–17. From the remotely applied axial stress, σ_{xx}, the longitudinal strain, ϵ_{xx}, and the transverse strain, ϵ_{yy}, the in-plane shear stress, τ_{12}, and the in-plane shear strain, γ_{12}, in each lamina are computed by the following relationships

$$\tau_{12} = \frac{\sigma_x}{2} \text{ and } \gamma_{12} = \epsilon_{xx} - \epsilon_{yy}. \tag{1}$$

Using Eq 1, the nonlinear shear stress-strain behavior of the lamina is determined.

Finite Element Analysis

Geometric and material nonlinear two-dimensional finite element analysis is used to show the effects of initial fiber waviness in combination with matrix nonlinearity on the fiber microbuckling initiation strains and the subsequent shear strain developed in the matrix. The finite element program ABAQUS [18] was used to perform the analysis.

Once the fibers are no longer considered perfectly straight, as typically observed in composites, the problem to be solved is not the classical Euler problem. Herein, the term "fiber microbuckling" has been defined to refer to large lateral deflections of initially wavy fibers leading to fiber breakage, rather than a bifurcation instability. Instability is defined as the point at which additional applied displacement no longer gives an increase in the load carried by the column. This point defines the initiation of fiber microbuckling in models with an initial fiber waviness.

Model

Two-dimensional FEA is used to model fiber microbuckling of an infinite series of wavy fibers and matrix. The unit cell, shown with an initial waviness in Fig. 2, is used to represent an infinite array of fibers embedded in matrix. The unit cell consists of one fiber (shaded region) with one matrix half width (white regions) along each side. Multipoint constraint boundary conditions are used to make the unit cell behave as an infinite series of fibers and matrix regions. For clarity, an axis, s, is defined to traverse along the column length, following the contour of the unit cell. The origin of this axis is defined at $x = 0$ and $y = 0$, the base of

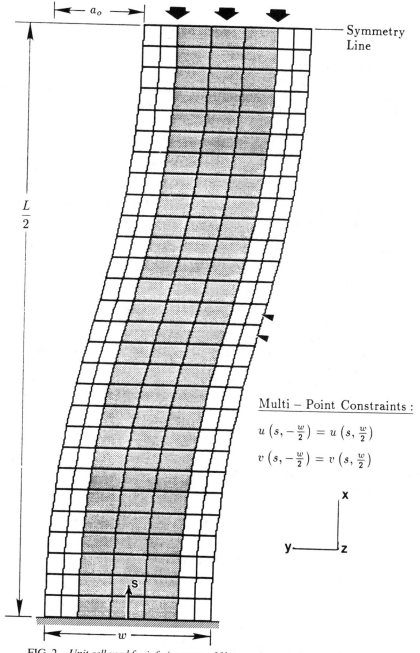

FIG. 2—*Unit cell used for infinite array of fibers and matrix for wavy fiber model.*

the column at the center of the fiber width. Using this axis with $u = u(s,y)$ and $v = v(s,y)$, the multipoint constraint boundary conditions are as follows

$$u\left(s, -\frac{w}{2}\right) = u\left(s, \frac{w}{2}\right) \text{ and } v\left(s, -\frac{w}{2}\right) = v\left(s, \frac{w}{2}\right) \tag{2}$$

where w is the width of the unit cell. Because the fiber length, L, is much larger than the initial amplitude of the waviness a_0, the x and s axes are nearly collinear.

In this model, symmetry is utilized to reduce computer computations. Nodal displacements (indicated by arrows on Fig. 2) are applied at the column symmetry line. The initial fiber waviness is defined as the ratio of the initial amplitude of the sine wave to the original length of the sine wave, a_0/L. Four arbitrary waviness ratios are modeled: (1) $a_0/L = 0.0000$, (2) $a_0/L = 0.0025$, (3) $a_0/L = 0.0050$, and (4) $a_0/L = 0.0075$.

Implementation

The matrix, fibers, and homogeneous regions are modeled with ABAQUS CPS8R elements. These plane stress elements are eight-noded and are evaluated using reduced integration.

Constituent Properties—For the model, the fiber volume fraction, V_f, is 60% and the diameter of the fiber, d_f, is 7.6 μm. The width of the modeled matrix regions are equal and were determined to be 5.1 μm, based on d_f and V_f. The length of the fiber, L, is selected based on measurements from SEM micrographs. The ratio, L/d_f, was measured from the micrographs to be 87.

The equivalent stress-strain behavior of the resin is derived from the lamina τ_{12}-γ_{12} data. For the analysis, it is assumed that the resin is isotropic, homogeneous, and has constitutive behavior that can be expressed in the form of the Ramberg-Osgood equation. Additionally, it was assumed that $G_f \gg G_m$ ($G_m/G_f = 0.046$), and thus, the shear is primarily transferred by the resin. Using these assumptions, $\tau_{12} = \tau_{12f} = \tau_{12m}$ and $\gamma_{12m} = \gamma_{12}/V_m$. The subscripts f and m correspond to the fiber and matrix, respectively. Once the shear constitutive behavior of the resin is derived, the equivalent stress-strain ($\bar{\sigma}$-$\bar{\epsilon}$) behavior of the resin is computed using

$$\bar{\sigma} = \left[\frac{1}{2}[\sigma_x - \sigma_y)^2 + (\sigma_y - \sigma_z)^2 + (\sigma_z - \sigma_x)^2] + 3[\tau_{xy}^2 + \tau_{yz}^2 + \tau_{zx}^2]\right]^{1/2} \tag{3}$$

and, for proportional loading (the components of applied stress remain in constant ratio to one another throughout the straining process) only

$$\bar{\epsilon} = \left[\frac{2}{9}[(\epsilon_x - \epsilon_y)^2 + (\epsilon_y - \epsilon_z)^2 + (\epsilon_z - \epsilon_x)^2] + \frac{1}{3}[\gamma_{xy}^2 + \gamma_{yz}^2 + \gamma_{zx}^2]\right]^{1/2} \tag{4}$$

Proportional loading may be assumed because this datum is derived from a tension test. For the stress state in the $[\pm 45]_{ns}$ specimens, the equivalent stress-strain behavior is reduced to $\bar{\sigma} = \sqrt{3}\tau_{xy}$ and $\bar{\epsilon} = \gamma_{xy}/\sqrt{3}$. The nonlinear constitutive behavior of the resin is fit using the Ramberg-Osgood stress-strain relationship

$$\bar{\epsilon} = \frac{\bar{\sigma}}{E} + \frac{\alpha\bar{\sigma}}{E}\left[\frac{|\bar{\sigma}|}{\sigma_0}\right]^{n-1} \tag{5}$$

where α is the yield offset, σ_0 is the yield stress, E is the Young's modulus, and n is the hardening exponent for the "plastic" term. Assuming $\alpha = 1$, the values determined for E, σ_0, and

TABLE 4—*Rambert-Osgood parameters.*

Temperature, °C (°F)	E, MPa	σ_0, MPa	n
21 (70)	3899	114.2	8.0748
77 (170)	3899	82.65	5.8676
132 (270)	3899	34.11	3.2625

n for the best fit through the data are given in Table 4. The equivalent stress-strain data, using the parameters from Table 4, are plotted in Fig. 3. The axial modulus for the Ramberg-Osgood fit is 8.3% stiffer than the given matrix Young's modulus (Table 1). Figure 3 shows that as the temperature is increased, the resin yield strength decreases. This trend derives from the resin shear constitutive behavior.

Isotropic fiber properties are assumed for this analysis (Table 1). The values for G_{12} and ν_{12} are fixed to the actual values and E_{11} ($E_{22} = E_{11}$) necessary for isotropy is computed.

Results and Discussion

The results from the experimental program and finite element analysis are presented in this section. It should be noted that absolute values for stress, strain, and fiber microbuckling initiation strains are plotted in all graphs and bar charts throughout this paper.

Experimental Results

Lamina Constitutive Behavior—The shear stress-strain curves obtained from both tension and compression testing for the APC-2 laminae, derived from Eq 1, are shown in Fig. 4. Radiographs (not shown) indicate that matrix cracking has initiated in the tension-loaded specimens by the 2% axial strain level while the compression specimen showed no signs of damage at axial strain levels up to 9.7%. The flatter 21°C (70°F) curve in Fig. 5, obtained from tension testing, is attributed to matrix cracking in the laminate. Consequently, compression testing of $[\pm 45]_{8s}$ specimens is used for determination of the lamina constitutive behavior. Lamina shear constitutive behaviors obtained from compressive loading, derived using Eq 1, for 21°C (70°F), 77°C (170°F), and 132°C (270°F) tests are also shown in Fig. 4. The derived resin shear constitutive behavior are shown in Fig. 5. Additionally, an assumed theoretically linear behavior has been added to the figure. From this datum, the elastic shear modulus of the resin, G_m, was measured to be 2.0 GPa, significantly higher than given in Table 1 [10]. The higher PEEK shear modulus is attributed to the fiber contribution and the fiber/matrix interaction. These curves indicate a significant reduction in the resin shear stress yield strength of the resin as the temperature is increased. This trend is reflected in the equivalent stress-strain data shown in Fig. 3. It is anticipated that this reduction decreases the amount of support for the fibers and reduces the strain level at which fiber microbuckling initiates.

Effects of Supporting Ply Orientation—The test matrix used to illustrate the effects of supporting ply orientation on the initiation of fiber microbuckling in 0° plies was given in Table 2. Data showing these effects are presented in this section.

Figure 6 shows a surface view of in-plane and out-of-plane fiber microbuckling of a $[(0_2/\pm 45)_3/0/\pm 45]_s$ laminate tested at 21°C (70°F). Figures 6a and b show the in-plane fiber microbuckling into the semicircular notch and also fiber microbuckling out-of-plane toward the free surface. Higher magnifications of typical fiber damage, both shear and tensile/bending-type fiber breaks, are shown in Figs. 6c and d. Figure 6a was one of the micrographs used for measurement of the column length for the finite element analysis.

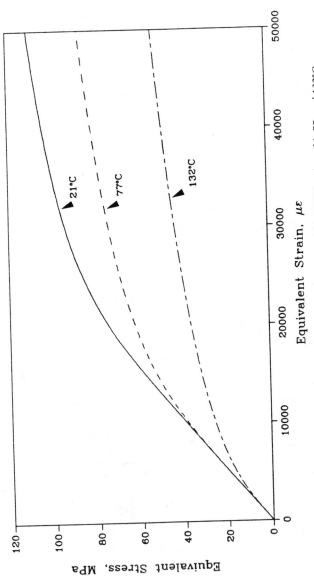

FIG. 3—*Equivalent stress-strain behaviors (used in FEA) of the PEEK resin at 21, 77, and 132°C.*

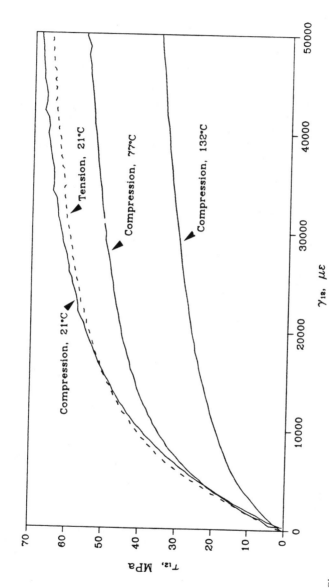

FIG. 4—*Shear stress-strain data obtained from the tension testing of a* [±45]₂ₛ *specimen at 21°C and the compression testing of* [±45]₈ₛ *specimens at 21, 77, and 132°C.*

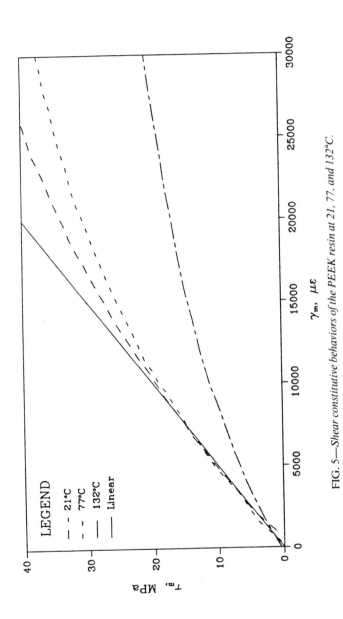

FIG. 5—*Shear constitutive behaviors of the PEEK resin at 21, 77, and 132°C.*

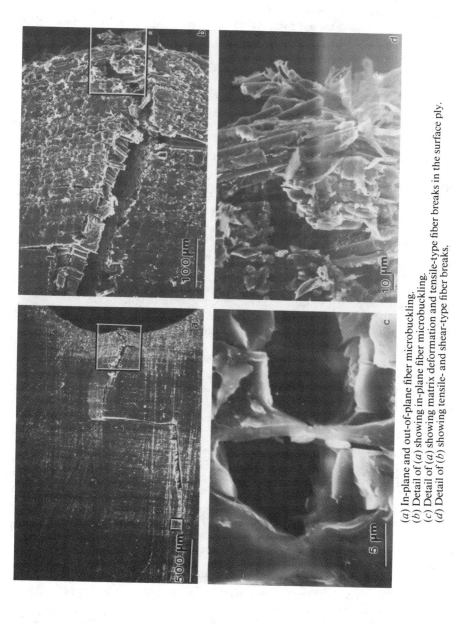

(a) In-plane and out-of-plane fiber microbuckling.
(b) Detail of (a) showing in-plane fiber microbuckling.
(c) Detail of (a) showing matrix deformation and tensile-type fiber breaks in the surface ply.
(d) Detail of (b) showing tensile- and shear-type fiber breaks.

FIG. 6—*Surface view of in-plane and out-of-plane fiber microbuckling of a* $[(0_2/\pm45)_3/0/\pm45]_s$ *laminate tested at 21°C.*

Figure 7 is a bar chart showing the nominal axial strain for fiber microbuckling initiation, ϵ_I, for two layups, one with ± 45 plies as surface plies and one with 0°_2 plies as surface plies. In-plane fiber microbuckling initiated in specimens with $\pm 45^\circ$ plies as surface plies at strain levels of ≈ 7042 $\mu\epsilon$. On the other hand, out-of-plane fiber microbuckling toward the free surface initiated in the specimens with 0°_2 plies as surface plies at strain levels of ≈ 4580 $\mu\epsilon$. These strain levels indicate a 35% reduction in the initiation strain level for specimens with out-of-plane fiber microbuckling, compared to initiation by in-plane fiber microbuckling. This 35% reduction emphasizes the importance of including the effect of the free surface in fiber microbuckling models for accurate predictions. Similar reductions were observed in the analytical model in Ref 19.

The effects of supporting ply orientation on in-plane fiber microbuckling are shown in Fig. 8. This bar chart shows the nominal axial strain level for in-plane fiber microbuckling in specimens with $\pm 15^\circ$, $\pm 45^\circ$, $\pm 75^\circ$, and 90°_2 plies as surface plies. The data indicate the $\pm 75^\circ$ plies provide the most resistance to fiber microbuckling while the $\pm 15^\circ$ plies provide the least resistance. However, this result did not seem intuitively correct to the authors. Consequently, the SCFs were computed using two-dimensional finite element analysis of an orthotropic plate with two semicircular edge notches. Additionally, fringe plots of the stress distributions (not included) were examined to verify that the strain gages were in remote locations, with respect to the notches, on the specimens. The SCFs are listed in Table 5 for these laminates, and the local fiber microbuckling initiation strain levels are shown as a function of the supporting ply orientation in the bar chart in Fig. 9. The trend in Fig. 9 also indicates that an optimum angle ($\approx 75^\circ$) of supporting ply orientation exists. This optimum angle allows the local constraint to maximize the resistance to fiber microbuckling. More indirectly, the various supporting ply orientations cause different residual stresses in the laminate, and, thus, a different initial stress state that may influence the fiber microbuckling process and initiation levels.

Observation of the local strain values in Fig. 9 poses another question; namely, what about fiber failure? Tensile strains to failure for AS4 fibers are reported to be 14 500 $\mu\epsilon$ [20]. However, DeTeresa [21] has observed fiber shear failure strains of $\approx 36\,000$ $\mu\epsilon$ for a single AS4 fiber embedded in matrix and loaded in compression. To investigate this phenomenon, one of the notches of a baseline specimen was observed in the SEM, prior to testing. Then, the specimen was loaded to a local strain level of $\approx 17\,000$ $\mu\epsilon$ at the notch and unloaded. Observation of the specimen under an optical microscope indicated that the discontinuous 0° plies have protruded into the notch, similar to the result in Refs 22 and 23. Subsequent SEM examination of the same notch showed failure by fiber shear and bending. Micrographs showing the multiple fiber failures are shown in Fig. 10. Figure 10a shows an overview of the fiber damage in one of the group of 0s closer to the specimen free surface. Figure 10b is a higher magnification (from Fig. 10a) of multiple failures by fiber shear. Figure 10c shows the fiber damage in the center group of 0s; these plies had the highest apparent density of fiber breaks. Figure 10d is an excellent example of a fiber shear failure. One possible explanation for fiber shear failures is as follows. Assume that each of the fibers in a ply have different initial in-plane wavinesses. Consequently, the straighter fibers support a higher local stress than those with the larger initial wavinesses. These straighter fibers then reach the stress level necessary to cause fiber shear, prior to ply microbuckling and catastrophic damage development. These micrographs indicate that failure by fiber shear may precede fiber microbuckling in these thermoplastic laminates, although fiber microbuckling still appears to be the strength limiting stage of compression strength of composites.

Effects of Resin-Rich Regions Between Plies—The test matrix for the effects of resin-rich regions between plies was given in Table 3. The room temperature data illustrating these effects are shown in Fig. 11. In-plane fiber microbuckling initiates at nominal strain levels of ≈ 7042 $\mu\epsilon$ for the baseline layup, ≈ 6580 $\mu\epsilon$ for the laminate with one layer of neat resin at the 0/45

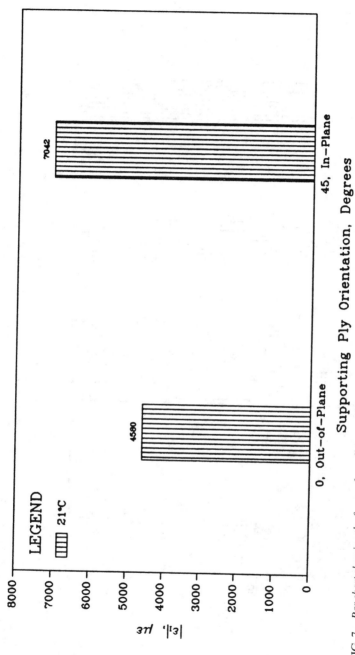

FIG. 7—*Bar chart showing the free surface effects on fiber microbuckling initiation in* [($\pm45/0_2$)$_3$/±45/0]$_s$ *and* [(0_2/±45)$_3$/0/±45]$_s$ *laminates tested as 21°C.*

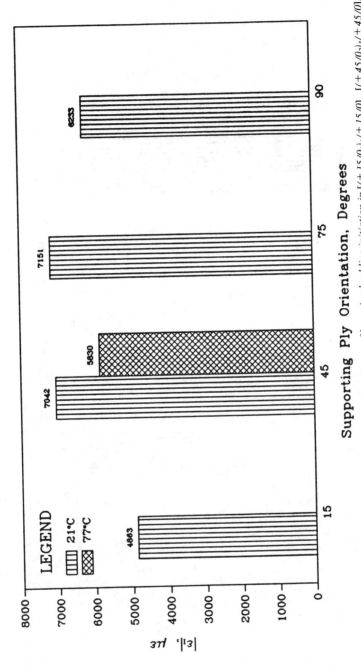

FIG. 8—*Bar chart showing the effects of supporting ply orientation on fiber microbuckling initiation in* $[(\pm 15/0_2)_3/\pm 15/0]_s$, $[(\pm 45/0_2)_3/\pm 45/0]_s$, $[(\pm 75/0_2)_3/\pm 75/0]_s$, *and* $[(90_2/0_2)_3/90_2/0]_s$ *laminates tested at 21 and 77°C.*

TABLE 5—*Strain concentration factors.*

Stacking Sequence	SCF
$[(0_2/\pm45)_3/0/\pm45]_s$	3.28364
$[(\pm15/0_2)_3/\pm15/0]_s$	4.19298
$[(\pm45/0_2)_3/\pm45/0]_s$	3.28364
$[(\pm75/0_2)_3/\pm75/0]_s$	3.41052
$[(90_2/0_2)_3/90_2/0]_s$	3.69487

interfaces, and ≈ 5475 $\mu\epsilon$ for the laminate with three layers of neat resin at the 0/45 interfaces. Compared to the baseline layup, the introduction of one layer of neat resin at the 0/45 interfaces through the laminate thickness causes a 7% reduction in the strain level for in-plane fiber microbuckling initiation. The addition of three layers of neat resin at 0/45 interfaces through the laminate thickness causes a 22% reduction in the initiation strain level, approximately three times the reduction caused by the addition of one layer of resin. These results indicate that the addition of soft resin layers on each side of the 0_2 plies significantly reduces the resistance to in-plane fiber microbuckling. These additions also allow for more out-of-plane fiber microbuckling into the resin-rich regions.

Effects of Resin Constitutive Behavior—The lamina, and thus resin, constitutive behavior was varied by testing at two temperatures 21°C (70°F) and 77°C (170°F). The change in the shear constitutive behavior, attributed to temperature effects, was shown in Fig. 4 for the lamina and in Fig. 5 for the neat resin. Some preliminary results showing the effects of the resin constitutive behavior on the initiation of fiber microbuckling in the baseline laminates were included in Fig. 8. These nominal axial strain results show that for the baseline layup, in-plane fiber microbuckling initiates at ≈ 7042 $\mu\epsilon$ at 21°C (70°F) and at ≈ 5830 $\mu\epsilon$ at 77°C (170°F). These results indicate that for the baseline layup, reducing the resin shear stress yield strength causes a 17% reduction in the strain level for in-plane fiber microbuckling initiation, and, thus, significantly reduces the resistance to in-plane fiber microbuckling.

Finite Element Results

The infinite series model (Fig. 2) is used to show the effects of initial fiber waviness on fiber microbuckling initiation strains and the development of shear strain in the matrix. The matrix constitutive behavior assumed is the 21°C nonlinear data (Fig. 5). Figure 12 is a plot of the average global axial stress, σ_{xx}, as a function of the average applied global strain, ϵ_{xx}, for four different initial fiber wavinesses ($a_0/L = 0.0000$, $a_0/L = 0.0025$, $a_0/L = 0.0050$, and $a_0/L = 0.0075$). The curve for the perfectly straight fiber continues to the critical buckling strain (eigenvalue analysis) of 94 100 $\mu\epsilon$, compared to 80 605 $\mu\epsilon$ predicted using the relationships on page 141 of Ref 24. Fiber microbuckling (instability indicated by arrows on Fig. 12) for the initially wavy infinite series model occurred at 27 046 $\mu\epsilon$ for $a_0/L = 0.0025$, 21 923 $\mu\epsilon$ for $a_0/L = 0.0050$, and 18 369 $\mu\epsilon$ for $a_0/L = 0.0075$. These strains represent 71.2, 76.7, and 80.5% reductions, compared to the perfectly straight fiber, in the fiber microbuckling strain levels for $a_0/L = 0.0025$, 0.0050, and 0.0075, respectively. These trends indicate that a small amount of waviness (for example, 0.25%) causes a significant reduction (for example, 71.2%), compared to those for the perfectly straight fiber, in the initiation strain levels. Additional increases in the initial fiber waviness cause further reduction in the fiber microbuckling initiation strains.

Figure 13 is a plot of the local shear strain, γ_{xy}, as a function of the applied global axial strain, ϵ_{xx}, for the same four models. The local shear strain is plotted for the region of maximum shear, (two elements indicated by arrows on Fig. 2). This graph illustrates that when the fibers are

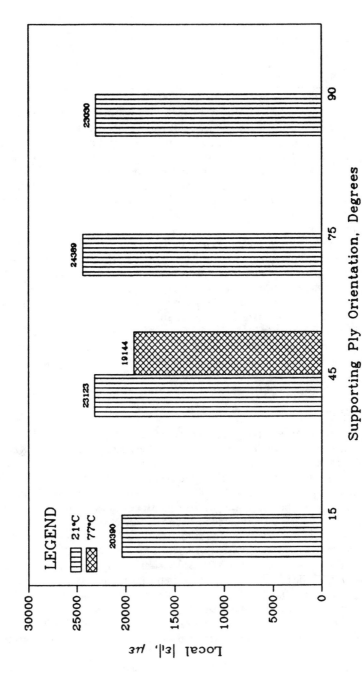

FIG. 9—Bar chart showing the local fiber microbuckling initiation strain levels for the laminates in Fig. 8.

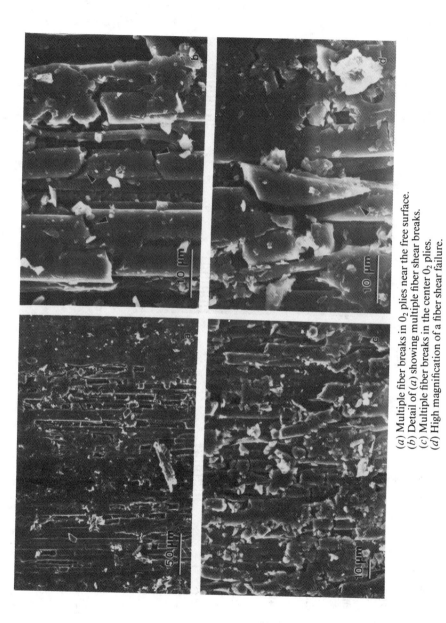

(a) Multiple fiber breaks in 0_2 plies near the free surface.
(b) Detail of (a) showing multiple fiber shear breaks.
(c) Multiple fiber breaks in the center 0_2 plies.
(d) High magnification of a fiber shear failure.

FIG. 10—*Baseline specimen loaded to a local strain level of ≈ 17000 $\mu\epsilon$.*

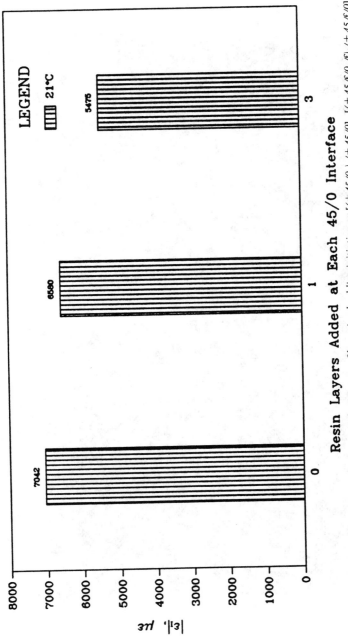

FIG. 11—*Bar chart showing the effects of resin-rich regions on fiber microbuckling initiation on* [(±45/0₂)₃/±45/0]ₛ, [(±45/f/0₂/f)₃/±45/f/0]ₛ, *and* [(±45/f₃/0₂/f₃)₃/±45/f₃/0]ₛ *laminates tested at 21°C.*

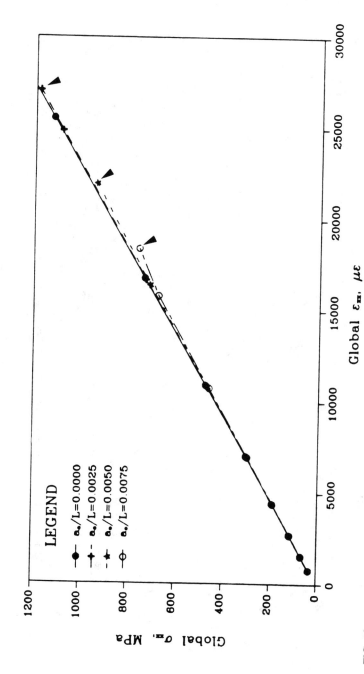

FIG. 12—*Average global axial stress versus average applied axial strain for four initial fiber waviness values. Instability for case $a_0/L = 0.0000$ occurred at 94 000 $\mu\varepsilon$.*

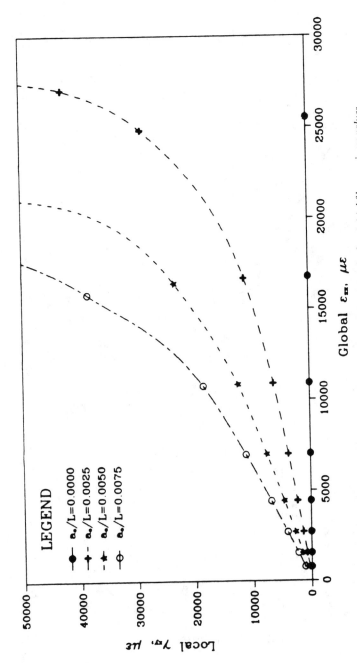

FIG. 13—*Local shear strain versus average applied axial strain for four initial fiber waviness values.*

perfectly straight, zero shear strain is developed in the matrix, as expected. As the initial fiber waviness increases, the amount and rate of shear strain development in the resin for a given applied strain increases. Additionally, the applied strain at which the resin reaches its shear stress yield strength is decreased (asymptote of curves) as the initial fiber waviness increases. The axial strains that correspond to the asymptotic behavior in Fig. 13 correlate well with the fiber microbuckling initiation strains in Fig. 12. This correlation is additional proof that as the tangent modulus decreases (with increasing applied load), the resistance to fiber microbuckling decreases. Similar trends were observed in a parallel study [25] for an infinite model using a minimum total potential energy approach.

Summary

The experimental results indicate that the local constraints (free surfaces, supporting ply orientation, and resin-rich regions) significantly affect the strain level for the initiation of in-plane fiber microbuckling. SEM examination of a damaged specimen indicates that multiple fiber breaks (shear and tensile type) appear to precede fiber microbuckling and occur at stress levels lower than those required for catastrophic damage. Additionally, preliminary results at an elevated temperature, 77°C (170°F), showed the shear stress yield strength of the resin was reduced and consequently, the resistance to fiber microbuckling was also reduced.

The combination of nonlinear matrix constitutive behavior and initial fiber waviness cause significant reductions in the fiber microbuckling initiation strain levels, particularly when compared to the straight fiber model. These two factors cause nonlinear global stress-strain responses of the models (Fig. 12). Furthermore, it was shown that increasing the initial fiber waviness causes a large increase in the rate and amount of shear strain developed in the matrix (Fig. 13). The fiber microbuckling initiation strains correlate well with the applied strains associated with the asymptotic local shear strains. It appears that the combination of these two factors causes the localized shear strains to exceed the resin shear stress yield strength leading to premature fiber microbuckling initiation.

Acknowledgments

The authors gratefully acknowledge the financial and mentor-type support of this project provided by NASA Langley Research Center, Hampton, Virginia, under Grant Number NAG-1-659, monitored by Dr. Charles E. Harris. Additionally, the authors acknowledge with gratitude the material contributions by Fiberite Corporation—An ICI Company, particularly the contact, Dr. David Leach.

References

[1] Guynn, E. G., "Micromechanics of Compressive Failures in Open Hole Composite Laminates," Department of Mechanical Engineering, Texas A&M University, College Station, TX, Masters thesis, Dec. 1987.

[2] Guynn, E. G., Bradley, W. L., and Elber, W., "Micromechanics of Compression Failures in Open Hole Composite Laminates," *Composite Materials: Fatigue and Fracture (Second Symposium), ASTM STP 1012,* American Society for Testing and Materials, Philadelphia, 1989, pp. 118–136.

[3] Guynn, E. G. and Bradley, W. L., "Measurements of the Stress Supported by the Crush Zone in Open Hole Composite Laminates Loaded in Compression," *Journal of Reinforced Plastics and Composites,* Vol. 8, March 1989, pp. 133–149.

[4] Guynn, E. G. and Bradley, W. L., "A Detailed Investigation of the Micromechanisms of Compressive Failure in Open Hole Composite Laminates," *Journal of Composite Materials,* Vol. 23, May 1989, pp. 479–504.

[5] Guynn, E. G. and Bradley, W. L., "Experimental Observations and Finite Element Predictions for

the Initiation of Fiber Microbuckling in Notched Composite Laminates," Annual Progress Report for NASA Research Grant NAG-1-659, NASA Langley Research Center, Hampton, VA, Oct. 1989.

[6] Hahn, H. T. and Williams, J. G., "Compression Failure Mechanisms in Unidirectional Composites," *Composite Materials: Testing and Design (Seventh Conference), ASTM STP 893,* J. M. Whitney, Ed., American Society for Testing and Materials, Philadelphia, 1986, pp. 115–139.

[7] Starnes, J. H. and Williams, J. G., "Failure Characteristics of Graphite/Epoxy Structural Components Loaded in Compression," *Mechanics of Composite Materials—Recent Advances,* Z. Hashin and C. T. Herakovich, Eds., Pergamon Press, New York, 1982, pp. 283–306.

[8] Sohi, M. M., Hahn, H. T., and Williams, J. G., "The Effect of Resin Toughness and Modulus on Compressive Failure Modes of Quasi-Isotropic Graphite/Epoxy Laminates," *Toughened Composites, ASTM STP 937,* N. J. Johnston, Ed., American Society for Testing and Materials, Philadelphia, 1987, pp. 37–60.

[9] Hahn, H. T., "Compressive Failure of Unidirectional Composites," presented at the Thirteenth International Symposium for Testing and Failure Analysis, Los Angeles, Nov. 1987.

[10] Fiberite Corporation—An ICI (Imperial Chemical Industries) Company, Orange, CA, "APC-2: The Product of High Technology," trade name material data sheets, 1988.

[11] Guynn, E. G. and Bradley, W. L., "Micromechanics of Composite Laminate Compression Failures," Annual Progress Report for NASA Research Grant NAG-1-659, NASA Langley Research Center, Hampton, VA, Aug. 1988.

[12] Trademark of The Dow Chemical Company.

[13] Barron, D., private communication, Oct. 1988.

[14] Carlsson, L. A. and Pipes, R. B., *Experimental Characterization of Advanced Composites Materials,* 1st ed., Prentice-Hall, Inc., Englewood Cliffs, NJ, 1987, pp. 54–65.

[15] Petit, P. H., "A Simplified Method of Determining the In-Plane Shear Stress-Strain Response of Unidirectional Composites," *Composite Materials: Testing and Design, ASTM STP 460,* American Society for Testing and Materials, Philadelphia, 1969, pp. 83–93.

[16] Rosen, B. W., "A Simple Procedure for Experimental Determination of the Longitudinal Shear Modulus of Unidirectional Composites," *Journal of Composite Materials,* Vol. 6, Oct. 1972, pp. 552–554.

[17] Hahn, H. T., "A Note on Determination of the Shear Stress-Strain Response of Unidirectional Composites," *Journal of Composite Materials,* Vol. 7, July 1973, pp. 383–386.

[18] *ABAQUS,* Hibbitt, Karlsson, and Sorenson, Inc., Providence, RI, 1987.

[19] Guz, A. N. and Lapusta, Y. N., "Stability of Fibers Near a Free Cylindrical Surface," Institute of Mechanics, Academy of Sciences of the Ukrainian SSR, Kiev, translated from *Prikladnaya Mekhanika,* Vol. 24, No. 10, Oct. 1988, pp. 3–9.

[20] Leeser, D., private communication, June 1989.

[21] DeTeresa, S. J., private communication, Dec. 1989.

[22] Potter, R. T. and Purslow, D., "The Environmental Degradation of Notched CFRP in Compression," *Composites,* July 1983, pp. 206–225.

[23] Purslow, D. and Potter, R. T., "The Effect of Environment on the Compression Strength of Notched CFRP-A Fractographic Investigation," *Composites,* Vol. 15, April 1984, pp. 112–120.

[24] Jones, R. M., *Mechanics of Composite Materials,* 1st ed. McGraw-Hill Book Company, New York, 1975.

[25] Davis, J., "The Effects of Fiber Waviness on the Compressive Response of Fiber-Reinforced Composite Materials," Masters thesis, Department of Aerospace Engineering, Texas A&M University, College Station, TX, Dec. 1989.

John D. Whitcomb[1]

Three-Dimensional Stress Analysis of Plain Weave Composites

REFERENCE: Whitcomb, J. D., "**Three-Dimensional Stress Analysis of Plain Weave Composites,**" *Composite Materials: Fatigue and Fracture (Third Volume), ASTM STP 1110*, T. K. O'Brien, Ed., American Society for Testing and Materials, Philadelphia, 1991, pp. 417–438.

ABSTRACT: Techniques were developed and described for performing three-dimensional finite element analysis of plain weave composites. This paper emphasizes aspects of the analysis that are different from analysis of traditional laminated composites, such as the mesh generation and representative unit cells. The analysis was used to study several different variations of plain weaves that illustrate the effects of tow waviness on composite moduli, Poisson's ratios, and internal strain distributions. In-plane moduli decreased almost linearly with increasing tow waviness. The tow waviness was shown to cause large normal and shear strain concentrations in composites subjected to uniaxial load. These strain concentrations may lead to earlier damage initiation than occurs in traditional cross-ply laminates.

KEY WORDS: composite materials, weaves, stress analysis, three-dimensional composite materials, fracture, fatigue (materials)

Nomenclature

C_{ijkl}	Constitutive coefficients
D	6 by 6 constitutive matrix
E_{11}, E_{22}, E_{33}	Young's moduli for orthotropic tow
$\overline{E}_x, \overline{E}_y, \overline{E}_z$	Average normalized Young's moduli for woven composite
F_x^i, F_y^i, F_z^i	Restraint forces in x-, y-, and z-directions, respectively, for load case, i
G_{12}, G_{23}, G_{13}	Shear moduli for orthotropic tow
$\overline{G}_{xy}, \overline{G}_{yz}, \overline{G}_{xz}$	Average normalized shear moduli for woven composite
H	Half-thickness of finite element model
l_1, l_2	Half-lengths of wavy and straight parts of tow
P	Applied load in the x-direction
u, v, w	Displacements in x-, y-, and z-directions
u_0, v_0, w_0	Specified displacements in x-, y-, and z-directions
W	Half-width of finite element model ($W = l_1 + l_2$)
x, y, z	Cartesian coordinates
$\varepsilon_x, \varepsilon_y, \varepsilon_z$	Normal strains
$\varepsilon_1, \varepsilon_2, \varepsilon_3, \varepsilon_{12}, \varepsilon_{23}, \varepsilon_{13}$	Strains with respect to material coordinate system
$\nu_{12}, \nu_{23}, \nu_{13}$	Poisson's ratios for orthotropic tow
$\overline{\nu}_{xy}, \overline{\nu}_{xz}, \overline{\nu}_{zy}$	Average normalized Poisson's ratios for woven composite

[1] Associate professor, Texas A&M University, Aerospace Engineering Department, College Station, TX, 77843-3141.

Traditionally, advanced composite structures have been fabricated from tape prepreg that was stacked to form a laminate. This type of construction tends to give optimal in-plane stiffness and strength. Since the primary loads usually are in-plane, this fabrication procedure appeared logical. However, there are at least two reasons why the usual laminated construction may not be best. First, secondary loads due to load path eccentricities, impact, or local buckling can sometimes dominate the failure initiation because of the low through-thickness strength of traditional laminates. Second, for thick laminates there are many laminae that must be assembled. This results in tedious labor with many opportunities for mistakes in orienting the laminae.

Weaving is an alternate fabrication technique that has received considerable attention recently. The interlacing of fiber bundles in woven composites increases out-of-plane strength. Woven mats are thicker than a traditional lamina, hence, fabrication of thick composites is less labor intensive and less prone to assembly error. These enhanced properties are obtained at the expense of some in-plane stiffness and strength. How much stiffness and strength is lost depends on the weave architecture. Because of the immense variety of possible weaves, it is not practical to determine optimal weave architecture through tests alone. Analytical models are needed that can predict the effect of various weave parameters on the mechanical properties.

Most of the analytical models that currently exist for woven composites were developed for prediction of moduli (for example, Refs 1,2,3). These models are based on many simplifying assumptions, similar to those found in classical laminate theory that may be appropriate for moduli prediction, but preempt the extension of the model for strength prediction. Models developed for stress analysis have generally been quasi-three-dimensional (Q3D) [4,5]. Since a Q3D analysis only models a single representative plane, little of the three-dimensional character of a woven composite is included. Reference 2 includes some three-dimensional analysis, but the model was far too crude to permit stress analysis. Reference 6 used a more refined three-dimensional finite element model than that in Ref 2, but the model was still probably not sufficiently refined for stress analysis (no stress distributions were reported). There has been no detailed three-dimensional analysis of the stresses or strains in woven composites.

This paper has two objectives. The first objective is to describe a refined finite element based three-dimensional analysis of woven composites. Figure 1 shows a schematic of the repeating unit for a single mat of a plain weave composite, which is the particular woven form considered in this paper. The grid that overlays the tows in Fig. 1 is there to clarify the geometry of the tows. The discussion of the analysis will emphasize aspects of the analysis that would be new to persons who have only analyzed conventional laminated composites. The second objective is to present a few results that illustrate the effect of tow waviness on the effective moduli, Poisson's ratios, and the internal strain distributions.

Analysis

Three-dimensional finite element analysis was used to study the behavior of plain weave composites. The finite element method was selected because of its flexibility in modeling complex shapes, spatial variation of material properties, and arbitrary boundary conditions. For moduli calculations, a crude model can be used; for detailed stress analysis, a much more refined model can be assembled, without any significant increase in professional manhours (compared to a crude model). In the following sections, various aspects of the analysis will be discussed. First, the configurations will be described, followed by sections on the finite element meshes and the material properties.

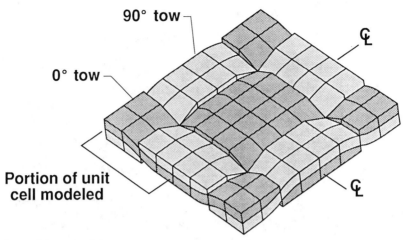

FIG. 1—*Schematic of plain weave unit cell. (Neat resin regions removed to show tow structure.)*

Configurations

The configurations analyzed consisted of mats of plain weave stacked to make a laminate. Figure 2 shows symmetric and unsymmetric stacking of the mats. The symmetry in Fig. 2a is with respect to the interfaces between the mats. The geometry of an actual laminate is expected to be a complicated mix of these stackings. For this initial study, the symmetric idealization in Fig. 2a will be used. Such a laminate can be considered to be an assemblage of unit cells like that indicated in Fig. 1. Because of symmetry within the unit cell, only one-fourth of the unit cell is actually modeled. Admittedly, this is a highly idealized plain weave composite. In actual composites the "synchrony" of the waviness in the mats could be hardly assured. Also, the interface between mats would not be planar. Furthermore, the shape of the tow cross sections is likely to vary much more than is practical to model. In spite of the simplifications intrinsic to the unit cell definition, this simplified cell definition is probably a reasonable starting point for detailed three-dimensional stress analysis. Figure 3a shows a coarse finite element model of a plain weave. This model has a fairly small length of wavy tow (l_1) compared to the length of straight tow (l_2). Several different ratios of wavy to straight tows were considered. The ratio was varied by holding the length, l_1, constant and varying l_2. The tow dimension, H, was also held constant. The term "waviness ratio" will be used as a measure of the fraction of tow that is inclined relative to the load direction. The waviness ratio is defined to be $l_1/(l_1 + l_2)$. Note that tow waviness varies inversely with tow width for a plain weave; narrow tows result in very wavy weaves.

Both extension and shear loadings were used. The boundary conditions for extension and shear were implemented quite differently, so they will be discussed separately. The dimensions and coordinate directions discussed later are defined in Fig. 3a.

The boundary conditions for extension loads involve constraints on both displacements and net normal force on the same plane. Because of symmetry, there are no shear forces on the planes. Imposing a specific displacement (zero or nonzero) is a simple matter in finite element analysis (for example, specifying $u = 0$ on $x = -W$). Imposing a constraint on both displacements and forces on the same plane is much more complicated. For example, imposing the condition that $v = $ constant on $y = W$ and that the net restraint force is zero is not simple.

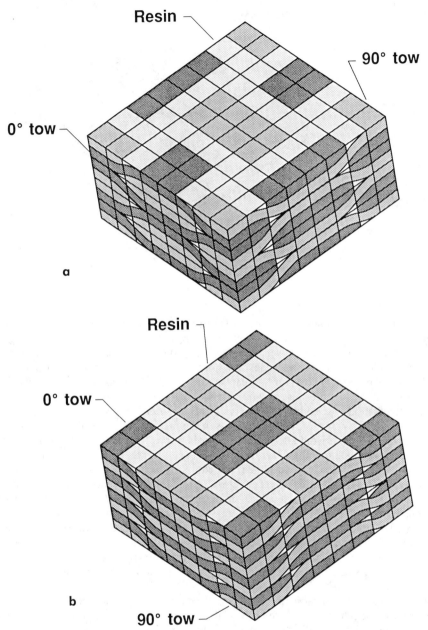

FIG. 2—*Stacking of plain weave mats: (a) symmetric stacking and (b) unsymmetric stacking.*

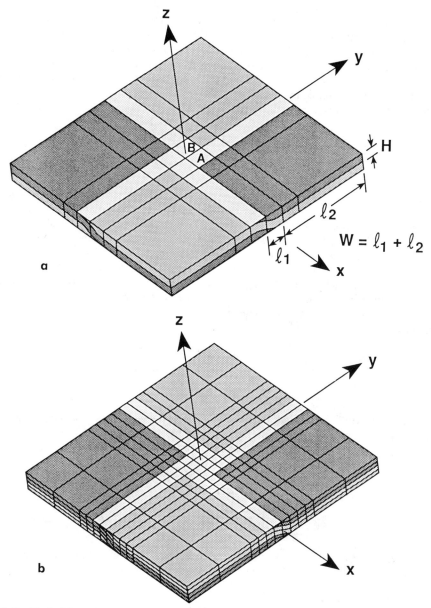

FIG. 3—*Typical finite element meshes: (a) coarse mesh (595 nodes, 96 elements) and (b) refined mesh (3793 nodes, 768 elements).*

Multipoint constraints could be imposed, but this is a complicated programming task and the number of nodes involved would result in a marked increase in the bandwidth of the equations. The approach taken herein is simple and does not affect the bandwidth. This approach, which is based on the principle of superposition, will be explained next by describing the steps required to obtain uniaxial loading of the unit cell in the x-direction. Because superposition is used, this approach is limited to linear analysis.

The first step is to impose normal constraints on all six faces of the finite element model.

$$\begin{aligned}
&\text{constrain } u \text{ on } x = \pm W\\
&\text{constrain } v \text{ on } y = \pm W\\
&\text{constrain } w \text{ on } z = \pm H
\end{aligned} \tag{1}$$

A displacement $(u = u_0)$ is imposed on $x = W$ and the other constrained displacements are set to zero. The normal constraint forces are calculated for the three planes $(x = W, y = W,$ and $z = H)$ and are defined to be F_x^1, F_y^1, and F_z^1, where the superscript indicates that these are forces from Loading Case 1. The subscript indicates the direction of the force (and implicitly the plane on which the force acts). Next, a displacement $(v = v_0)$ is imposed on $y = W$ and the other constrained displacements are set to zero. The corresponding constraint forces $(F_x^2, F_y^2,$ and $F_z^2)$ are determined. Finally, a displacement $(w = w_0)$ is imposed on $z = H$ and the other constrained displacements are set to zero. The corresponding constraint forces $(F_x^3, F_y^3,$ and $F_z^3)$ are determined. These nine constraint forces are used in the following equations

$$\begin{aligned}
F_x^1 + aF_x^2 + bF_x^3 &= P\\
F_y^1 + aF_y^2 + bF_y^3 &= 0\\
F_z^1 + aF_z^2 + bF_z^3 &= 0
\end{aligned} \tag{2}$$

The unknowns are the load in the x-direction, P, and the scaling coefficients (a and b). The last two equations express the condition that the net normal force on the $y = W$ and the $z = H$ planes must be zero. Solving the last two equations yields

$$b = \frac{-F_y^1 F_z^2 + F_z^1 F_y^2}{F_y^3 F_z^2 - F_z^3 F_y^2} \tag{3}$$

$$a = \frac{-F_y^1 - bF_Y^3}{F_Y^2}$$

These values of a and b can be used in Eq 2 to determine the load. The average normal strains and two of the average Poisson's ratios are

$$\varepsilon_x = \frac{u_0}{2W}$$

$$\varepsilon_y = \frac{av_0}{2W}$$

$$\varepsilon_z = \frac{bw_0}{2H} \tag{4}$$

$$\nu_{xy} = -\frac{\varepsilon_y}{\varepsilon_x}$$

$$\nu_{xz} = -\frac{\varepsilon_z}{\varepsilon_x}$$

The effective Young's modulus in the x-direction $(\overline{E}_x)$ is calculated using the following energy balance equation

$$\tfrac{1}{2}Pu_0 = \tfrac{1}{2}\overline{E}_x\varepsilon_x^2 \text{ Vol} \tag{5}$$

For the models considered in this paper, one could have calculated $\overline{E}_x$ based on just the average stress on the plane ($x = W$) and the specified strain. In the more general case, the geometry might be such that there is no simple cross sectional area that could be used to determine average stress. Such would be the case for the configuration in Fig. 1, if the neat resin regions were left out.

The remaining Poisson's ratios and Young's moduli can be determined in a similar manner.

The exact boundary conditions for shear loading of an infinite array of unit cells require many multipoint constraints to impose antisymmetry. Such boundary conditions permit the faces of the model to warp. The multipoint constraints complicate the analysis and significantly increase the computational cost. Hence, the following approximate boundary conditions were used. For G_{xy}

$$u = ay \text{ and } v = ax \text{ on } x = \pm W \text{ and } y = \pm W \tag{6a}$$

For G_{yz}

$$v = az \text{ and } w = ay \text{ on } y = \pm W \text{ and } z = \pm H \tag{6b}$$

For G_{xz}

$$u = az \text{ and } w = ax \text{ on } x = \pm W \text{ and } z = \pm H \tag{6c}$$

These boundary conditions are very simple to impose. Consider the case of shear in the xy plane. Known nonzero displacements are prescribed on four faces. On $z = \pm H$, the displacement (w) is set to zero. This is possible because on the average, shear in the xy plane does not cause normal strain in the z-direction for the plain weave configuration studied herein. This is not generally true for textiles. For other material architectures, a superposition procedure (like that described earlier) might be necessary.

Finite Element Meshes

Figure 3 shows typical coarse and refined meshes for the plain weave composite. The coarse mesh had 595 nodes and 96 elements. The fine mesh had 3793 nodes and 768 elements. The elements were 20-node isoparametric hexahedrons. The meshes exhibit cyclic symmetry. In fact, a mesh is synthesized from the basic unit in Fig. 4 rotated at 0, 90, 180, and 270° about the z-axis.

In order to minimize the number of elements required, compatibility of displacements was not completely maintained at the center of the model. Figure 5 illustrates the incompatibility. Elements A and B are the same as those so labeled in Fig. 3a. They are shown separated in Fig. 5 to show the interface between the two elements and to show Element C. There is some incompatibility in the displacements because Node 9 is not connected to any node in Element A. However, since the resin material (for example, Element A) is much more compliant than the tows, this approximation in the modeling is probably not significant. Even with the various simplifications, the internal shape of the mesh is fairly complicated. Work is needed to determine what refinements to the modeling would yield the greatest improvements in accuracy.

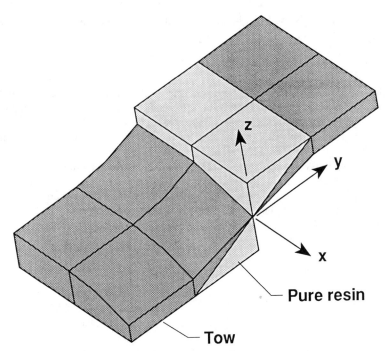

FIG. 4—*Basic element group used to generate mesh.*

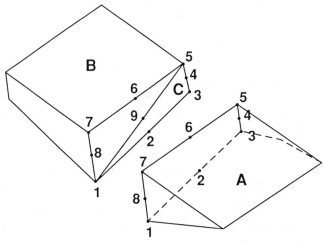

FIG. 5—*Closeup of Elements A and B from Fig. 3a. Elements are separated to show incompatability.*

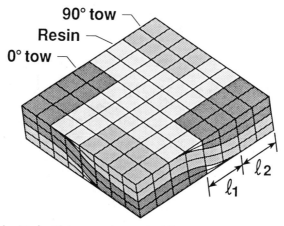

FIG. 6—*Mesh with large waviness ratio. (Waviness ratio = $l_1/(l_1 + l_2)$.)*

Some of the meshes had a larger ratio of wavy tow to straight tow than other meshes. Figure 6 shows a mesh with a waviness ratio of 0.5, compared to a waviness ratio of 0.167 for the mesh in Fig. 3.

Material Properties

The material properties chosen were for a hypothetical linear elastic graphite/epoxy composite. The two properties were selected to be the same as for unidirectional tape prepreg material. The resin was assumed to be isotropic. Table 1 summarizes the assumed material properties.

There are two angles that describe the orientation of a tow at any point. There is a primary rotation about the z-axis of either 0 or 90°. This angle is constant for a particular tow. There is also a secondary rotation about the x- or y-axis due to the waviness. The amount of rotation about the x- and y-axes varies spatially. This second rotation was determined based on the direction of the normal to the surface defined by four of the midside nodes, which are labeled a, b, c, and d in Fig. 7. The surface normal direction was calculated as the cross product $N = ca \times db$. Appendix I describes the material property transformations required because of these rotations. A right-handed coordinate system is assumed. Positive rotations are defined to be

TABLE 1—*Assumed material properties.*

	Generic Graphite/Epoxy [7]	Neat Resin [8]
E_{11}	13.4×10^{10} Pa	0.345×10^{10} Pa
E_{22}	1.02×10^{10} Pa	0.345×10^{10} Pa
E_{33}	1.02×10^{10} Pa	0.345×10^{10} Pa
ν_{12}	0.3	0.35
ν_{23}	0.49	0.35
ν_{13}	0.3	0.35
G_{12}	0.552×10^{10} Pa	0.128×10^{10} Pa
G_{23}	0.343×10^{10} Pa	0.128×10^{10} Pa
G_{13}	0.552×10^{10} Pa	0.128×10^{10} Pa

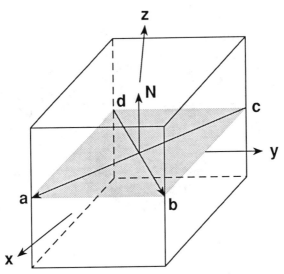

FIG. 7—*Calculation of element orientation.*

counterclockwise from the material axis to the global axis (or clockwise from the global axis to the material axis).

Results and Discussion

Several configurations were analyzed to determine the effect of tow waviness on effective moduli, Poisson's ratios, and internal strain distributions. The results of these analyses will be discussed in this section.

Figures 8a and b show the effect of tow waviness on the effective moduli and three of the Poisson's ratios. The properties are normalized by those for a conventional (0/90)$_s$ laminate fabricated from tape material. The (0/90)$_s$ laminate is the limiting case when the waviness ratio goes to zero. In fact, for a waviness ratio of 0.167 (which was the smallest waviness ratio considered), the moduli and Poisson's ratios are within about 10% of the values for a (0/90)$_s$ laminate. For reference, the properties for a (0/90)$_s$ laminate are listed here.

$$E_x = 7.25 \times 10^{10} \text{ Pa} \qquad E_z = 1.27 \times 10^{10} \text{ Pa}$$
$$G_{xy} = 0.552 \times 10^{10} \text{ Pa} \qquad G_{yz} = 0.426 \times 10^{10} \text{ Pa}$$
$$\nu_{xy} = 0.0424 \qquad \nu_{zy} = 0.0795 \qquad \nu_{xz} = 0.455$$

Most of the results were generated using coarse models. A few results were generated using refined models. The differences in the moduli for the coarse and fine models were insignificant, so only the results for the coarse mesh are shown in Fig. 8a. There was more difference between the coarse and fine models for the Poisson's ratios. Hence, results for both model refinements are shown in Fig. 8b. The solid circles are the results for the coarse model. The solid circles are the results for the fine model. Only two waviness ratios were analyzed using the fine model.

Figure 8a shows that $\overline{E}_x$, $\overline{E}_z$, and $\overline{G}_{xy}$ decrease almost linearly with increased waviness. The waviness has the largest effect on $\overline{E}_x$. Some of the decrease can be attributed to the increase in resin content as waviness increases. The volume of neat resin pockets increases from 3.2% for the case with a waviness ratio of 0.167 to 12.5% for the case with a waviness ratio of 0.5. If one

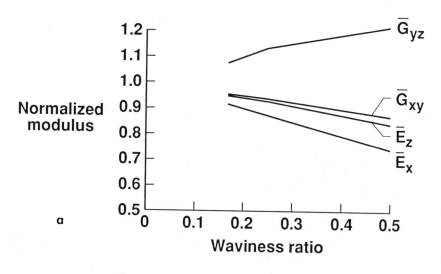

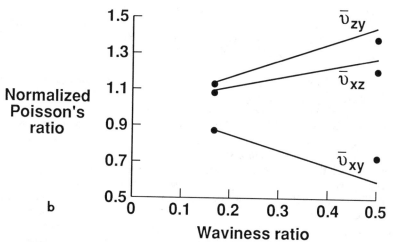

FIG. 8—*Effect of waviness on (a) moduli and (b) Poisson's ratio. The moduli and Poisson's ratios are normalized with respect to those for a (0/90)$_s$ laminate. The curves indicate results for the coarse mesh. The filled circles indicate results for the fine mesh. (Waviness ratio = $l_1/(l_1 + l_2)$.)*

assumes the tape material had a fiber volume ratio of 0.6, then the overall fiber volume ratio for the weave would be 0.58 and 0.52 for waviness ratios of 0.167 and 0.5, respectively. Thus, there can be significant decreases in the fiber volume ratios due to the resin pockets. Conversely, if one attempts to maintain a fiber volume ratio in a woven material comparable to that for a tape material, the result might be incomplete bonding of the fibers. The transverse shear modulus ($\overline{G}_{yz}$) increases with increased waviness. Figure 8b shows that the in-plane Poisson's ratio ($\overline{\nu}_{xy}$) decreases with increased waviness. The other two Poisson's ratios ($\overline{\nu}_{xz}$ and $\overline{\nu}_{zy}$) increase with waviness. The variation is essentially linear. There is more difference between the results for the coarse and refined models for large waviness than for small waviness. Numerical values of the coarse model results in Fig. 8 are tabulated in Table 2.

TABLE 2—*Effect of waviness on moduli and Poisson's ratios (waviness ratio = $l_1/(l_1 + l_2)$) (coarse mesh results).*

	Waviness Ratio		
	0.167	0.25	0.5
$\overline{E}_x$	0.92	0.88	0.75
$\overline{E}_z$	0.95	0.93	0.84
$\overline{G}_{xy}$	0.96	0.94	0.87
$\overline{G}_{yz}$	1.10	1.14	1.22
$\overline{\nu}_{xy}$	0.87	0.81	0.60
$\overline{\nu}_{zy}$	1.14	1.21	1.45
$\overline{\nu}_{xz}$	1.10	1.14	1.28

No experiments were performed as part of this study, but some qualitative comparisons can be made with the results in Ref 9. Reference 9 presents analytical and experimental in-plane modulus results for a plain weave, oxford weave, five harness satin weave, and an eight harness satin weave. No results were presented for a (0/90/90/0) tape laminate, but the eight harness satin results should have close to the same moduli, since it has low waviness. The loss in axial stiffness ($\overline{E}_x$) shown in Fig. 8 agrees qualitatively with Ref 9. Figure 8 shows a decrease in $\overline{\nu}_{xy}$ with increased waviness, but the analytical results in Ref 9 predict no change and the experiments showed an increase. Figure 8 shows a decrease in $\overline{G}_{xy}$ with increased waviness, which agrees with the experimental results in Ref 9, but disagrees with the analytical results. There is obviously a need for further analytical and experimental work to clarify the source of these inconsistencies.

Strain distributions were calculated for weaves with waviness ratios of 0.167 and 0.5. Refined finite element meshes were used. The procedure for smoothing the strains to obtain nodal strains is described in Appendix II. Because of limited space, only a few of the results will be presented here. Also, only uniaxial loading was considered for the results herein. The results will be presented in terms of isostrain contours. The strains are normalized by the magnitude of the applied axial strain. These strains are calculated relative to the material coordinate system, in which the x_1-axis is along the tow direction, the x_2-axis is in-plane and perpendicular to the x_1-axis, and the x_3-axis is perpendicular to the other two axes. For example, ε_1 is the longitudinal strain in the fiber regardless of the tow orientation with respect to the global coordinate system. Since no convergence study was performed, the results should be considered qualitative.

For the case with a waviness ratio of 0.167, Fig. 9 shows the 0° tow for which strain contours will be presented. Figure 10 identifies the analogous area for the case with a waviness ratio of 0.5. The area in Fig. 10 labeled ABCD is identical in geometry to the area labeled ABCD in Fig. 9. Hence, the only difference between the two models is the straight-tow region in the mesh with the smaller waviness ratio.

Figure 11 shows contours of constant (ε_1) for the case with a waviness ratio of 0.167. The largest strain concentration is at $(x,y,z) = (0,0,0)$. There are strong, fairly complicated strain gradients near $(x,y,z) = (0,0,0)$, but a few tow thicknesses away from the $y = 0$ plane, the variation in strain on any $y = y_0$ plane is independent of y_0. A two-dimensional analysis would probably do a reasonable job in the region away from the $y = 0$ plane, but obviously could not predict the peak strains, which occur near the origin.

Figures 12a through f show ε_1, ε_2, ε_3, ε_{12}, ε_{23}, and ε_{13} contours for the region labeled ABCD in Fig. 11. As just mentioned, the largest ε_1 occurs at the origin. The magnitude there is approximately 2.7 times the average strain for the unit cell. This strain concentration would be expected to cause fiber breakage at a lower global strain than would occur in a tape laminate.

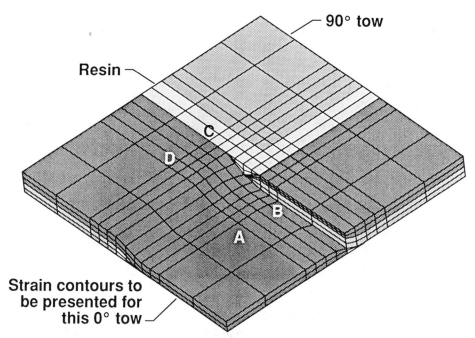

FIG. 9—*Identification of area for which strain contours will be presented. (Waviness ratio = 0.167; waviness ratio = $1_1/(1_1 + 1_2)$.)*

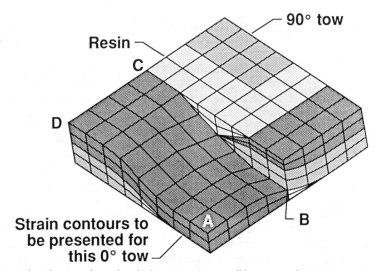

FIG. 10—*Identification of area for which strain contours will be presented. (Waviness ratio = 0.5; waviness ratio = $1_1/(1_1 + 1_2)$.)*

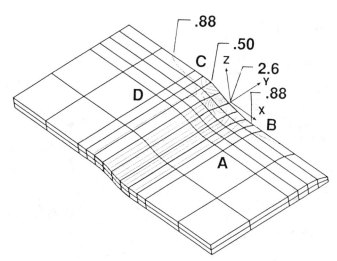

FIG. 11—*Normalized axial strain, $\varepsilon_1/\varepsilon_x^\circ$.*

Much of the strain concentration is likely due to tapering of the tow in both the *x*- and *y*-directions to a zero thickness at the origin (see Fig. 4). Herein lies a particularly sticky modeling problem. Even though the tow cross section may actually vary as indicated by the finite element model, the tapering is due to migration, not termination, of the individual fibers. That is, fibers near the *y* = 0 plane that are aligned parallel to the *x*-axis away from the origin are squeezed further away from the *y* = 0 plane as they approach the origin, resulting in a zero tow thickness at the origin. This fiber migration causes a very complicated variation in the local constitutive properties of the tow. There was no attempt in this paper to account for this. The straightening of the tow under tensile loads causes an increase in ε_1 at Point E and a decrease in ε_1 at Point F.

Figure 12*b* shows that the ε_2 strains are small. The peak magnitude occurs on the *y* = 0 plane, probably because that is the boundary of the tow, where there is an abrupt change in local properties.

Figure 12*c* shows that both the magnitude and variation of ε_3 are large. The presence of large ε_3 strains has been previously reported in Ref *4*. In that study a quasi-three-dimensional analysis was used. Figure 12*c* shows that the strain variation is nearly two-dimensional in character, except where the peak occurs. The large magnitude of ε_3 might cause delamination.

A uniaxially loaded conventional $(0/90)_s$ laminate fabricated from tape prepreg has no shear strains away from free edges. Figures 12*d*, *e*, and *f* show that a plain weave fabric has significant shear strains. The peak ϵ_{12} is about the same magnitide as the average axial strain. Particularly under fatigue loading, the ϵ_{12} could lead to intratow cracking. The peak ε_{23} (Fig. 12*e*) is also nearly as large as the average axial strain. The peak occurs along the tow boundary and might lead to some intertow cracking, particularly under fatigue.

The ε_{13} strain component (Fig. 12*f*) is by far the largest strain component. This strain component is due to the eccentricity of the two ends of the tow. The very large magnitude suggests that delamination initiation might be dominated by this strain component.

The strain contours for the case with a waviness ratio of 0.5 are very similar to those in Fig. 12. The contours for ε_1 and ε_{13} are shown in Fig. 13 to illustrate the similarity. Interestingly, the peak strains tend to be larger for the less wavy weave for the same average axial strain. Table 3 lists the minimum and maximum strains for the two weaves. However, the less wavy weave

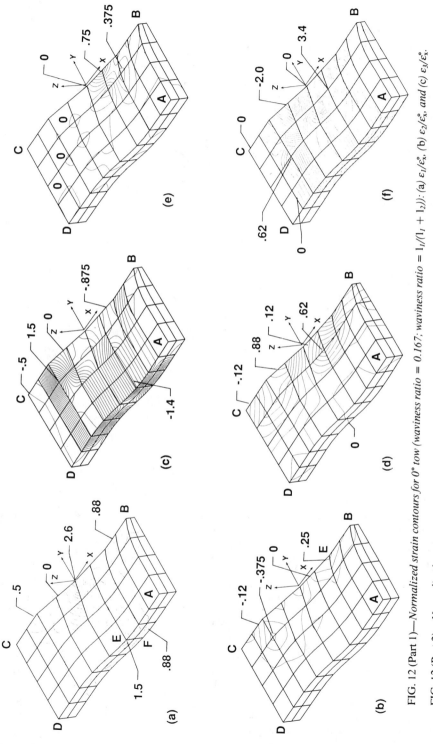

FIG. 12 (Part 1)—*Normalized strain contours for 0° tow (waviness ratio = 0.167; waviness ratio = $l_1/(l_1 + l_2)$): (a) $\varepsilon_1/\varepsilon_x^\circ$, (b) $\varepsilon_2/\varepsilon_x^\circ$, and (c) $\varepsilon_3/\varepsilon_x^\circ$.*

FIG. 12 (Part 2)—*Normalized strain contours for 0° tow (waviness ratio = 0.167; waviness ratio = $l_1/(l_1 + l_2)$): (d) $\varepsilon_{12}/\varepsilon_x^\circ$, (e) $\varepsilon_{23}/\varepsilon_x^\circ$, and (f) $\varepsilon_{13}/\varepsilon_x^\circ$.*

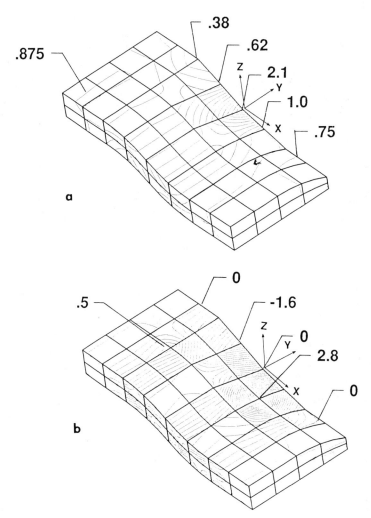

FIG. 13—*Normalized strain contours for 0° tow (waviness ratio = 0.5; waviness ratio = $l_1/(l_1 + l_2)$):* (a) $\varepsilon_1/\varepsilon_x^\circ$ *and* (b) $\varepsilon_{13}/\varepsilon_x^\circ$.

TABLE 3—*Maximum normalized strains for waviness ratios of 0.167 and 0.5 (waviness ratio =* $l_1/(l_1 + l_2)$).

Normalized Strain[a]	Waviness Ratio = 0.167		Waviness Ratio = 0.5	
	Min	Max	Min	Max
$\bar{\varepsilon}_1$	0.44	2.7	0.35	2.2
$\bar{\varepsilon}_2$	−0.42	0.29	−0.32	0.27
$\bar{\varepsilon}_3$	−1.4	1.6	−1.4	1.2
$\bar{\varepsilon}_{12}$	−1.0	1.0	−0.82	0.82
$\bar{\varepsilon}_{23}$	−0.83	0.83	−0.68	0.68
$\bar{\varepsilon}_{13}$	−3.4	3.4	−2.9	2.9

[a] Normalized strain = strain/specified axial strain.

also has a higher axial stiffness. For a given average strain, the average stress is higher for the less wavy weave. If both weaves are subjected to the same average stress, the peak strains in the more wavy weave are about as large or larger than for the less wavy weave, which is more in line with intuition.

Conclusions

Techniques were developed and described for performing three-dimensional finite element analysis of plain weave composites. The discussion of the analysis emphasized aspects of the analysis that are different from analysis of traditional laminated composites, such as mesh generation and representative unit cells.

The analysis was used to study several different weaves to determine the effects of tow waviness on composite moduli, Poisson's ratios, and internal strain distributions. The average normalized composite moduli ($\overline{E}_x$, $\overline{G}_{xy}$, and $\overline{E}_z$) all decreased with increasing waviness. The average normalized out-of-plane shear modulus ($\overline{G}_{xz}$) increased with increasing waviness. As expected, there are significant strain gradients. The magnitude of these strain concentrations suggest that damage initiation will occur at a significantly lower global strain than for a traditional cross-ply laminate. However, some of the most severe strain concentrations occur where the weave geometry is most difficult to model with confidence.

Because a resin pocket was not included at the center of the model, the cyclic symmetry required a zero tow thickness at the center (which is troublesome). This problem should probably be avoided in future studies by including a resin pocket. Also, the tow width was assumed to vary independently of tow thickness and waviness, which was a simplifying assumption resulting from a lack of experimental data. There is an obvious need for experimental work to characterize the variation of the fiber tow geometry and to document the initiation of damage under static and fatigue loads. There is also a need for much more analysis to identify the errors caused by the various simplifying assumptions.

APPENDIX I

Material Property Transformations

This appendix describes the material property transformations that are required between the global and material coordinate systems. Transformations from one coordinate system to another are needed for strains, stresses, and the material properties. Although the finite element analysis calculates the strains and stresses relative to the global coordinate system, it is reasonable to report them relative to the material coordinate system. The material properties are known relative to the material coordinate system, but the finite element formulation requires the properties relative to the global coordinate system. Hence, there are transformations in "both directions".

To obtain the strains with respect to the material coordinate axes requires a second order tensor transformation. The transformation rule between the global and material coordinate systems is given by Ref *10*

$$\varepsilon'_{ij} = a_{im}a_{jn}\varepsilon_{mn} \tag{7}$$

where the ε'_{ij} are the strains in the material coordinate system and a_{ij} = cosine of angle between the x'_i and x_j axes that are the material coordinate axes and the global coordinate axes, respectively. Figure 14 illustrates the definition of the a_{ij} for a rotation about the z-axis only.

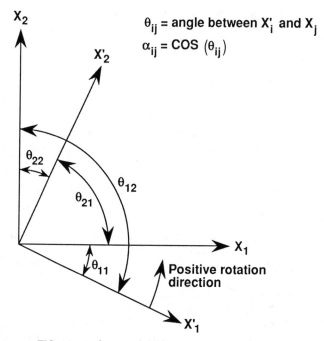

FIG. 14—*Definition of* a_{ij} *for rotation about* x_3 *axis.*

In expanded form, Eq 7 can be written as

$$
\begin{bmatrix}
\varepsilon'_{11} \\
\varepsilon'_{22} \\
\varepsilon'_{33} \\
\varepsilon'_{12} \\
\varepsilon'_{21} \\
\varepsilon'_{23} \\
\varepsilon'_{32} \\
\varepsilon'_{13} \\
\varepsilon'_{31}
\end{bmatrix}
= [T_1]
\begin{bmatrix}
\varepsilon_{11} \\
\varepsilon_{22} \\
\varepsilon_{33} \\
\varepsilon_{12} \\
\varepsilon_{21} \\
\varepsilon_{23} \\
\varepsilon_{32} \\
\varepsilon_{13} \\
\varepsilon_{31}
\end{bmatrix}
\tag{8}
$$

where

$$
[T_1] =
\begin{bmatrix}
a_{11}a_{11} & a_{12}a_{12} & a_{13}a_{13} & a_{11}a_{12} & a_{12}a_{11} & a_{12}a_{13} & a_{13}a_{12} & a_{11}a_{13} & a_{13}a_{11} \\
a_{21}a_{21} & a_{22}a_{22} & a_{23}a_{23} & a_{21}a_{22} & a_{22}a_{21} & a_{22}a_{23} & a_{23}a_{22} & a_{21}a_{23} & a_{23}a_{21} \\
a_{31}a_{31} & a_{32}a_{32} & a_{33}a_{33} & a_{31}a_{32} & a_{32}a_{31} & a_{32}a_{33} & a_{33}a_{32} & a_{31}a_{33} & a_{33}a_{31} \\
a_{11}a_{21} & a_{12}a_{22} & a_{13}a_{23} & a_{11}a_{22} & a_{12}a_{21} & a_{12}a_{23} & a_{13}a_{22} & a_{11}a_{23} & a_{13}a_{21} \\
a_{21}a_{11} & a_{22}a_{12} & a_{23}a_{13} & a_{21}a_{12} & a_{22}a_{11} & a_{22}a_{13} & a_{23}a_{12} & a_{21}a_{13} & a_{23}a_{11} \\
a_{21}a_{31} & a_{22}a_{32} & a_{23}a_{33} & a_{21}a_{32} & a_{22}a_{31} & a_{22}a_{33} & a_{23}a_{32} & a_{21}a_{33} & a_{23}a_{31} \\
a_{31}a_{21} & a_{32}a_{22} & a_{33}a_{23} & a_{31}a_{22} & a_{32}a_{21} & a_{32}a_{23} & a_{33}a_{22} & a_{31}a_{23} & a_{33}a_{21} \\
a_{11}a_{31} & a_{12}a_{32} & a_{13}a_{33} & a_{11}a_{32} & a_{12}a_{31} & a_{12}a_{33} & a_{13}a_{32} & a_{11}a_{33} & a_{13}a_{31} \\
a_{31}a_{11} & a_{32}a_{12} & a_{33}a_{13} & a_{31}a_{12} & a_{32}a_{11} & a_{32}a_{13} & a_{33}a_{12} & a_{31}a_{13} & a_{33}a_{11}
\end{bmatrix}
$$

Assuming the strain tensor is symmetric ($\varepsilon_{ij} = \varepsilon_{ji}$), and performing trivial modification gives the transformation in terms of the engineering shear strains, which are two times the tensor shear strains.

$$
\begin{bmatrix}
\varepsilon'_{11} \\
\varepsilon'_{22} \\
\varepsilon'_{33} \\
2\varepsilon'_{12} \\
2\varepsilon'_{23} \\
2\varepsilon'_{13}
\end{bmatrix}
= [T_2]
\begin{bmatrix}
\varepsilon_{11} \\
\varepsilon_{22} \\
\varepsilon_{33} \\
2\varepsilon_{12} \\
2\varepsilon_{23} \\
2\varepsilon_{13}
\end{bmatrix}
\tag{9}
$$

where

$$
[T_2] =
\begin{bmatrix}
a_{11}a_{11} & a_{12}a_{12} & a_{13}a_{13} & 2a_{11}a_{12} & 2a_{12}a_{13} & 2a_{11}a_{13} \\
a_{21}a_{21} & a_{22}a_{22} & a_{23}a_{23} & 2a_{21}a_{22} & 2a_{22}a_{23} & 2a_{21}a_{23} \\
a_{31}a_{31} & a_{32}a_{32} & a_{33}a_{33} & 2a_{31}a_{32} & 2a_{32}a_{33} & 2a_{31}a_{33} \\
a_{11}a_{21} & a_{12}a_{22} & a_{13}a_{23} & a_{11}a_{22} + a_{12}a_{21} & a_{12}a_{23} + a_{13}a_{22} & a_{11}a_{23} + a_{13}a_{21} \\
a_{21}a_{31} & a_{22}a_{32} & a_{23}a_{33} & a_{21}a_{32} + a_{22}a_{31} & a_{22}a_{33} + a_{23}a_{32} & a_{21}a_{33} + a_{23}a_{31} \\
a_{11}a_{31} & a_{12}a_{32} & a_{13}a_{33} & a_{11}a_{32} + a_{12}a_{31} & a_{12}a_{33} + a_{13}a_{32} & a_{11}a_{33} + a_{13}a_{31}
\end{bmatrix}
$$

Stress is also a second order tensor, so the transformation is the same as for the tensor strains (Eqs 7 and 8). Since there is no distinction between engineering and shear stresses, the further manipulations resulting in Eq 9 are not applicable.

The material properties are ordinarily defined with respect to the material coordinate axes, but the finite element formulation requires that they be defined with respect to the global coordinate axes. The material constitutive coefficients comprise a fourth order tensor. Hence, the form of the transformation rule is [10]

$$
C_{ijkl} = a_{im}a_{jn}a_{ko}a_{lp}C'_{mnop}
\tag{10}
$$

Note that the a_{ij} in Eq 10 are different than those in Eq 7, since the transformation is in the opposite direction (that is, from the material axes to the global axes).

Equation 10 is cumbersome. Alternately, we can derive a matrix form of the transformation based on the invariance of the strain-energy density. Also, this alternate transformation makes use of the already calculated matrix (T_2). The invariance can be expressed as

$$
\underline{\varepsilon}^T D\underline{\varepsilon} = \underline{\varepsilon}'^T D'\underline{\varepsilon}'
\tag{11}
$$

where

$$
\underline{\varepsilon} =
\begin{bmatrix}
\varepsilon_{11} \\
\varepsilon_{22} \\
\varepsilon_{33} \\
2\varepsilon_{12} \\
2\varepsilon_{23} \\
2\varepsilon_{13}
\end{bmatrix}
$$

and D is a 6 by 6 constitutive matrix.

Equations 3 and 5 can be combined to obtain

$$
\underline{\varepsilon}^T D\underline{\varepsilon} = \underline{\varepsilon}^T T_2^T D' T_2 \underline{\varepsilon}
\tag{12}
$$

From Eq 12, it is obvious that the transformation rule is

$$
D = T_2^T D' T_2
\tag{13}
$$

APPENDIX II

Strain Smoothing

This appendix describes the procedure used to smooth the strains. The smoothing procedure calculates nodal strains based on the strains calculated at the quadrature points. The procedure is based on Ref *11*. It is assumed that strains are available at eight quadrature points. The ordering of the terms in the matrices that will be described depends on the numbering sequence for the corner nodes in a single element and the quadrature points. Figure 15 shows the numbering sequence for the nodes and the quadrature points. For clarity in the sketch, some nodes and quadrature points are not shown. However, the numbering patterns for the hidden nodes and points are the same as for those shown.

The smoothing procedure begins with assuming a functional form for the strains within an element. The form assumed herein is

$$\varepsilon(x_1, x_2, x_3) = S^i(x_1, x_2, x_3)C^i \text{ sum on } i \qquad (14)$$

where $[S^i] = [1 \; x_1 \; x_2 \; x_3 \; x_1x_2 \; x_1x_3 \; x_2x_3 \; x_1x_2x_3]$.

Equation 14 can be written for the eight quadrature points in the following form

$$[\varepsilon] = \begin{bmatrix} 1 & -A & -A & -A & B & B & B & -C \\ 1 & -A & -A & A & B & -B & -B & C \\ 1 & -A & A & -A & -B & -B & B & C \\ 1 & -A & A & A & -B & B & -B & -C \\ 1 & A & -A & -A & -B & B & -B & C \\ 1 & A & -A & A & -B & -B & B & -C \\ 1 & A & A & -A & B & -B & -B & -C \\ 1 & A & A & A & B & B & B & C \end{bmatrix} \begin{bmatrix} C^1 \\ C^2 \\ C^3 \\ C^4 \\ C^5 \\ C^6 \\ C^7 \\ C^8 \end{bmatrix} \qquad (15)$$

where $A = 1/\sqrt{3}$, $B = \frac{1}{3}$, and $C = A \times B$

Equation 15 can be solved for the coefficients (C^i).

$$[C^i] = [H]^{-1}[\varepsilon^i] \qquad (16)$$

where H is the coefficient matrix in Eq 15.

With the C^i known, the strains at the nodes can be determined using Eq 14. The smoothing procedure can be summarized in terms of a single matrix multiplication.

$$[\varepsilon_n^i] = [T][\varepsilon_q^i] \qquad (17)$$

where Subscripts n and q indicate nodal or quadrature values, respectively and the matrix $[T]$ is

$$\begin{bmatrix} A & B & B & C & B & C & C & D \\ B & C & A & B & C & D & B & C \\ C & B & B & A & D & C & C & B \\ B & A & C & B & C & B & D & C \\ B & C & C & D & A & B & B & C \\ C & D & B & C & B & C & A & B \\ D & C & C & B & C & B & B & A \\ C & B & D & C & B & A & C & B \end{bmatrix} \qquad (18)$$

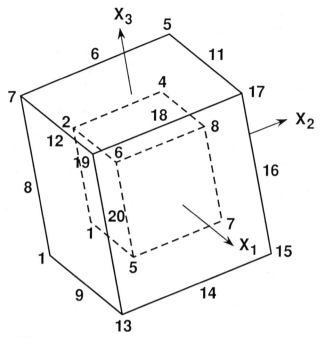

FIG. 15—*Numbering sequence for nodes and quadrature points.*

where

$$A = 5 + 3 \cdot E$$
$$D = 5 - 3 \cdot E$$
$$B = -1 - E$$
$$C = -1 + E$$

where $E = \sqrt{3}$.

This procedure gives the nodal strains for an individual element. Most nodes in a mesh are shared by several elements. A simple averaging procedure was used to smooth discontinuities in the strain field. Since strain discontinuities are physically possible at bimaterial interfaces, the averaging was performed among elements of the same material type. This averaging procedure must be performed before transforming the strains to the material coordinate system.

References

[1] Halpin, J. C., Jerine, K., and Whitney, J. M., "The Laminate Analogy for 2 and 3 Dimensional Composite Materials," *Journal of Composite Materials,* Vol. 5, Jan. 1971, pp. 36–49.
[2] Ishikawa, T., "Anti-Symmetric Elastic Properties of Composite Plates of Satin Weave Cloth," *Fibre Science and Technology,* Vol. 15, 1981, pp. 127–145.
[3] Ishikawa, T. and Chou, T. W., "Stiffness and Strength Behavior of Woven Fabric Composites," *Journal of Material Science,* Vol. 17, 1982, pp. 3211–3220.
[4] Avery, W. B. and Herakovich, C. T., "A Study of the Mechanical Behavior of a 2-D Carbon-Carbon Composite," Virginia Polytechnic Institute and State University Report VPI-E-87-15 or CCMS-87-13, Blacksburg, VA, Aug. 1987.

[5] Kriz, R. D., "Influence of Damage on Mechanical Properties of Woven Composites at Low Temperatures," *Journal of Composite Technology & Research,* Vol. 7, No. 2, Summer 1985, pp. 55–58.

[6] Dow, N. F. and Ramnath, V., "Analysis of Woven Fabrics for Reinforced Composite Materials," NASA Contract Report 178275, NASA Langley Research Center, Hampton, VA, April, 1987.

[7] Wang, S. S. and Choi, I., "The Mechanics of Delamination in Fibre-reinforced Composite Laminates. Part I—Stress Singularities and Solution Structure. Part II—Delamination Behavior and Fracture Mechanics Parameters," NASA CR 172269 and CR 172270, NASA Langley Research Center, Hampton, VA, Nov. 1983.

[8] Halpin, J. C., *Primer on Composite Materials: Analysis,* Technomic Publishing Co., Inc., Lancaster, PA, 1984, p. 166.

[9] Foye, R. L., "The Mechanics of Fabric-Reinforced Composites," *Proceedings,* Fiber-Tex Conference, Sept. 1988, NASA Conference Publication 3038, NASA Langley Research Center, Hampton, VA, pp. 237–247.

[10] Frederick, D. and Chang, T. S., *Continuum Mechanics,* Scientific Publishers, Inc., Cambridge, MA, 1972, p. 29.

[11] Hinton, E., Scott, F. C., and Ricketts, R. E., "Local Least Squares Stress Smoothing for Parabolic Isoparametric Elements," *International Journal for Numerical Methods in Engineering,* Vol. 9, 1975, pp. 235–256.

Eugene T. Camponeschi, Jr.[1]

Compression Testing of Thick-Section Composite Materials

REFERENCE: Camponeschi, E. T., Jr., **"Compression Testing of Thick-Section Composite Materials,"** *Composite Materials: Fatigue and Fracture (Third Volume), ASTM STP 1110,* T. K. O'Brien, Ed., American Society for Testing and Materials, Philadelphia, 1991, pp. 439–456.

ABSTRACT: As composite materials become more attractive for use in large Navy structures, the need to understand the mechanical response of composites greater than 6.4 mm (0.25 in.) in thickness becomes a necessity.

In this program, a compression test fixture that allows the testing of composites up to 25.4 mm (1 in.) in thickness and greater was designed and refined. This fixture was used to evaluate the effects of constituents, fiber orientation, and thickness on the compressive response of composite materials. In addition, the fixture was used to determine if the failure mechanisms observed for thick composites are similar to those that have been observed and reported for composite materials less than 6.4 mm (0.25 in.) thick.

The in-plane moduli, in-plane and through-thickness Poisson's ratios, compression strength, and failure mechanisms of the thick composites were shown to be independent of material thickness. The predominant failure mechanisms for both materials were kink bands and delaminations, and were identical in geometry to those that have been reported for composite materials in the range of 2.54 mm (0.1 in.) thick.

Although unchanged with thickness, the through-thickness Poisson's ratio for the carbon- and glass-reinforced laminates were found to be significantly nonlinear, resulting in changes in this property of up to 58% from the initial region of the strain-strain curve to the final region of the strain-strain curve.

KEY WORDS: composite materials, thick-section materials, compression testing, three-dimensional properties, fatigue (materials), fracture

The high specific compressive strength of composite materials make them highly attractive as candidate materials for Naval applications. In many cases, the material thickness required for these applications is much greater than those that have been demonstrated to date. For example, in considering composite cylinders subjected to external pressure, scale model testing has been conducted on unstiffened cylinders nominally 203 mm (8 in.) in diameter with a wall thickness of 15 mm (0.6 in.) [1].

The results from such tests have indicated that thick-walled carbon-reinforced composite cylinders do not reach collapse pressures expected from a three-dimensional stress analysis of a thick orthotropic shell [2] coupled with allowable strength from thin uniaxial compressive strength tests. For example, a collapse pressure corresponding to a laminate stress of 965 MPa (140 ksi) is expected for $[0/0/90]_{ns}$ carbon/epoxy shells, but wall strengths of 552 to 690 MPa (80 to 100 ksi) are routinely measured ($[0/0/90]_{ns}$ refers to a symmetric laminate with a group of two 0° plies and one 90° ply repeated $2n$ times). In contrast to these findings, comparable tests of fiberglass reinforced cylinders [3,4] have resulted in expected and achieved laminate strengths of 827 MPa (120 ksi).

[1] Project engineer, David Taylor Research Center, Code 2802, Annapolis, MD 21402.

Possible explanations for the unexpected low strength of thick carbon-reinforced cylinders fall into the categories of material issues, stress analysis issues, or manufacturing issues. In terms of materials issues, the elastic constants or strengths determined for thin (less than 3.2 mm [0.125 in.]) materials may not be appropriate for materials that are greater than 6.4 mm (0.25 in.) in thickness. What are the trends for the compressive properties of composite materials with increasing thickness?

Stress analysis requirements that arise for thick composites include the need for fully three-dimensional analysis and the incorporation of nonlinear materials effects into these analyses should the effects be significant. The capability to perform complex three-dimensional stress analysis exists, yet accurate three-dimensional material data properties and three-dimensional failure criteria do not.

The manufacturing issues of concern for thick composite shells include the effect of residual stress, material nonuniformity, the development of layer waviness, and the presence of material property gradients through the thickness of the component.

Certainly all of these issues are interrelated, but they should be investigated independently to identify the relative importance of each parameter with respect to the performance of thick structures. In this investigation, the effect of thickness on material response and the development of three-dimensional compressive properties have been addressed. The elastic constants, strength, and failure mechanisms of carbon and S2 glass-reinforced composites are studied as a function of increasing section thickness.

Test Method Development

A survey of compression test methods to identify one that would be appropriate for testing composites between 6.4 mm and 25.4 mm (0.25 and 1.00 in.) thick reveals a myriad of possible methods for materials less than 6.4 mm (0.25 in.) thick, and none for greater thickness [5]. What was learned from this survey is that an end-loaded test coupon with simple clamping blocks on the ends is the most economical and appropriate for thick composites. The development of a fixture to test thick specimens in compression was undertaken and the following criteria were applied: the fixture must allow thick-section testing capability beginning at 6.4 mm (0.25 in.); must allow further scale up for thicker, wider, and longer specimens; must prevent load eccentricities; must allow an unsupported gage length; and must prevent splitting or brooming failures from occurring near the load introduction points.

A fixture design that met these requirements is similar to one used by Irion and Adams [6] for 2.54-mm (0.1-in.)-thick specimens, and a cross section of the final design is shown in Fig. 1. A photograph showing the size of the 25.4-mm (1.0-in.) David Taylor Research Center (DTRC) fixture compared to the ASTM Test Method for Compression Properties of Unidirectional or Crossply Fiber-Resin Composites (D 3410-87) IITRI fixture can be found in Fig. 2. In the DTRC fixture, load is applied to the ends of the specimen and clamping blocks are used to provide stability and prevent end-brooming at the point of load introduction. A hardened steel plate is inserted between both ends of the specimen and the test machine crosshead platens and act as load bearing surfaces. A self-aligning spherical seat is placed between one end of the specimen and the load machine to assist in aligning the specimen axis and the loading axis.

Preliminary studies on test fixture design showed fixture alignment rods were unnecessary since the specimen thickness and the clamping blocks provided adequate fixture/specimen stability. These studies also showed that the size and number of clamping bolts was critical since significant bolt stresses develop due to through-thickness Poisson displacements. Initial compression tests with 48-ply specimens showed four 6.4-mm (0.25-in.) bolts in each half of the fixture could not withstand the stresses created by the specimen through-thickness Poisson

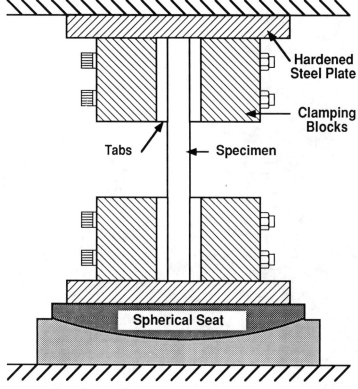

FIG. 1—*Schematic of DTRC thick-section compression fixture.*

effects. The following equation was developed to determine bolt stress as a function of applied longitudinal load

$$\text{SIG}_b = \frac{(\text{NU}_{13})(\text{SIG}_c)(E_b)(E_3)(A_s)(L_c)}{(E_1)(L_b E_3 A_s + E_b A_t L_c)}$$

where

SIG_b = bolt stress,
SIG_c = compression strength of composite sample,
NU_{13} = through-thickness Poisson's ratio of composite sample,
E_1 = longitudinal modulus of elasticity of composite material sample,
E_b = modulus of elasticity of bolts,
E_3 = through-thickness modulus of elasticity of composite material sample,
A_t = total cross-sectional root area of all bolts,
A_s = area of contact between sample and one clamping block,
L_b = length of bolts, and
L_c = thickness of composite sample.

This equation provides the stress in each bolt as a function of specimen properties, specimen geometry, bolt modulus, and bolt length. The final bolt configuration consisted of six 12.7-

FIG. 2—*Photograph of 25.4 mm (1.0 in.) DTRC thick-section compression fixture and IITRI compression fixture.*

mm (0.5-in.) bolts for 48- and 96-ply specimens, and ten 15.9-mm (0.625-in.) bolts for the 192-ply specimens. The bolt torque applied to each fixture prior to testing was 6.8, 20.3, 67.8 $N \cdot m$ (5, 15, and 50 $ft \cdot lb$) for the 48, 96, and 192-ply specimens, respectively.

The crosshead displacement rate used in this investigation was chosen to provide a strain rate of approximately 0.0025 l/s. The equivalent crosshead rates were 25.8 mm/min (0.015 in./min) (48-ply specimen), 30.6 mm/min (0.020 in./min) (96-ply specimen), and 61.2 mm/min (0.040 in./min) (192-ply specimen).

Material Systems and Specimen Geometry

The two material systems evaluated in this investigation were AS4/3501-6 carbon/epoxy and S2/3501-6 fiberglass/epoxy. They were chosen to investigate the effects of carbon and glass fiber reinforcements in a common epoxy matrix in terms of the mechanical response observed when these fibers are used as reinforcements in thick unstiffened cylinders.

The carbon-reinforced prepreg tape was supplied by Hercules Inc. as AS4 fiber with 3501-6 176°C (350°F) epoxy resin (150 g/cm^2, 0.49 oz/ft^2 areal weight). The S2 glass-reinforced prepreg was supplied by Fiberite as S2 glass fiber also with 3501-6 350 F epoxy resin (205 g/cm^2, 0.68 oz/ft^2 areal weight). Both systems were supplied as 304-mm (12-in.) wide prepreg tape and were autoclave cured at the DTRC. An autoclave air temperature schedule that was slightly different than those used for thin (<48 ply) epoxy-based composites was used. This air temperature was determined from test cures on 96- and 192-ply laminates with thermocouples placed within the test panels to monitor temperature through the panel thickness during cure.

Following fabrication, samples from all panels were removed and tested for fiber volume fraction (FVF) and void content (ASTM Test Method for Fiber Content of Resin-Matrix Composites by Matrix Digestion (D 3171-76) and ASTM Test Method for Void Content of Reinforced Plastics (D 2734-70). The following values were determined:

	48 ply	96 ply	192 ply
AS4/3501-6	58.4%/−1.33%	60.0%/0.34%	60.3%/−0.57%
S2/3501-6	57.6%/0.27%	53.8%/0.97%	58.0%/0.64%

Multiple specimens were machined from panels 48, 96, and 192 plies thick, resulting in nominal specimen thicknesses of 6.4, 12.7, and 25.4 mm (0.25, 0.50, and 1.0 in.) The $[0]_n$ and $[0/0/90]_{ns}$ laminate stacking sequences were fabricated for the 48- and 96-ply panels, and $[0/0/90]_{32s}$ were fabricated for the 192-ply panels. Tabs were fabricated separately from the same material as each specimen, with a $[0/90]_{ns}$ orientation, and adhesively bonded to the test specimen panels after both the specimens and tabs were cured. Tabs were bonded to the 48- and 192-ply panels with American Cyanamid FM-123-2 film adhesive, and the 96-ply panels with Hysol 907 two-part epoxy.

After tabbing, individual specimens were then rough cut from the large panel on a diamond saw, and were machined to final tolerance by grinding the specimen ends, sides, and tab surfaces parallel and perpendicular within 0.025 mm (0.001 in.).

Specimen width and tab length were chosen to be comparable to those commonly used when conducting the ASTM D 3410-87 IITRI test. A tab geometry with no taper at the gage section was selected since this geometry is used for thin-section, 2.54-mm (0.1-in.) IITRI compression test specimens.

The maximum allowable specimen gage section length was determined on the basis of a Euler column buckling analysis that assumed the rectangular cross-section specimen acts as a pinned end column. The effects of transverse shear were also included in the analysis, and the expression for the allowable gage length/thickness ratio is [7]

$$\frac{1}{t} = 0.9069 \left[\frac{E_x}{Y_{ult}} \left(1 - 1.2 \frac{Y_{ult}}{G_{xz}} \right) \right]^{1/2}$$

where

l = specimen length,
t = specimen thickness,
E_x = longitudinal modulus,
G_{xz} = through-thickness shear modulus, and
Y_{ult} = ultimate compressive strength.

Since specimen geometry must be prescribed prior to final testing, material properties must be assumed to determine the permissible gage length/thickness ratio. The assumed material properties and the resultant gage length/thickness ratios are listed in Table 1.

The final speciman geometries were designed so that the width was four times the specimen

TABLE 1—*Maximum permissible gage length/thickness ratios.*

	S2 glass/3501-6		AS4/3501-6	
	$[0]_n$	$[0/0/90]_{ns}$	$[0]_n$	$[0/0/90]_{ns}$
Strength, Mpa (ksi)	1207 (175)	827 (120)	1517 (220)	965 (150)
E_x, Gpa (Msi)	51.7 (7.5)	34.5 (5)	117 (17)	75.8 (11)
G_{xz}, Gpa (Msi)	4.83 (0.7)	4.83 (0.7)	4.83 (0.7)	4.83 (0.7)
l/t	5.0	5.2	6.3	7.0

thickness, the gage length was five times the specimen thickness, and the tab length was five times the specimen thickness (with a minimum tab length of 62.5 mm [2.5 in.]). The nominal specimen dimensions are summarized in Table 2 and the specimen geometry is shown in Fig. 3.

Five specimens of each thickness and orientation were evaluated for the 48- and 192-ply thicknesses, and four of each (two with a 3:1 l/t ratio and two with a 5:1 l/t ratio) were evaluated for the 96-ply thickness. Foil backed electrical resistance strain gages were used in this investigation to monitor strain. Single gages or unstacked 0/90 CEA-06 type gages were used with lengths of 3.2 or 6.4 mm (0.125 or 0.250 in.) The 48-ply and one half of the 96-ply specimens were instrumented with a longitudinal gage on each face. The remaining half of the 96-ply and all of the 192-ply specimens were instrumented with strain gages on both faces and both edges as shown in Fig. 3.

Results and Discussion

The results from this program include longitudinal modulus of elasticity, in-plane and through-thickness Poisson's ratio, ultimate compression strength, and ultimate compression strain at failure. These data as well as the observed failure mechanisms are discussed in the next three sections. Tables 3 and 4 summarize the elastic constant, strength, and strain-to-failure data. The data in Table 3 were determined using an initial secant tangent slope calculation between 0.1 and 0.3% strain.

TABLE 2—*Nominal specimen dimensions.*

	48-Ply	96-Ply	192-Ply
Thickness, mm (in.)	6.4 (0.25)	12.7 (0.50)	25.4 (1.0)
Width, mm (in.)	25.4 (1.0)	50.8 (2.0)	101.6 (4.0)
Length, mm (in.)	158.8 (6.25)	190.5 (7.5) and 165.1 (6.5)	381.0 (15.0)
Gage length, mm (in.)	31.8 (1.25)	63.5 (2.5) and 38.1 (1.5)	127.0 (5.0)
Tab length, mm (in.)	63.5 (2.5)	63.5 (2.5)	127.0 (5.0)
Tab thickness, mm (in.)	3.2 (0.125)	3.2 (0.125)	4.4 (0.25)

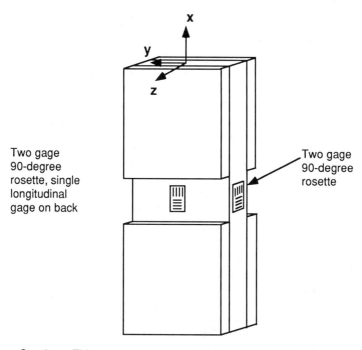

Two gage
90-degree
rosette, single
longitudinal
gage on back

Two gage
90-degree
rosette

Specimen Thickness : t   Tab length: 5t, 2.5 in. min.
Width : 4t   Thickness: .25t, .125 in. min.
Gage Length : 5t, 3t

FIG. 3—*Specimen geometry and strain gage locations.*

Elastic Constants

The longitudinal modulus of elasticity (E_x) was recorded for all three specimen thicknesses and NU_{xy} and NU_{xz} were recorded for the 96- and 192-ply specimens. The E_x data from Table 3 are represented in graphical form in Fig. 4. This plot shows that the longitudinal moduli for these materials is independent of specimen thickness. For purposes of comparison to other published data, the values of E_x can be adjusted for fiber volume fraction effects from 115.8 GPa (16.8 Msi) (60% FVF) for the [0] AS4/3501-6 converts to 125.5 GPa (18.2 Msi) (65% FVF) and from 51.0 GPa (7.4 Msi) (55% FVF) to 60.0 GPa (8.7 Msi) (65% FVF) for the [0] S2/3501-6 specimens.

A comparison of NU_{xy} and NU_{xz} for the unidirectional 92-ply specimens show both the carbon and fiberglass materials to be transversely isotropic. The measured values of NU_{xz} for the $[0/0/90]_{ns}$ laminates were compared to theoretically predicted values in Ref 8 and agree well.

When reducing the strain-strain data used to determine NU_{xz} for the $[0/0/90]_{ns}$ laminates, significant nonlinearities in the curves were observed as shown in Fig. 5. These nonlinearities were quantified by comparing the initial and final slope of the NU_{xz} strain-strain curves. The initial slope was determined by the secant tangent method between 0.1 and 0.3% strain, and the final slope was determined by the same method between strain-at-failure and 0.2% less than strain-at-failure. The results of this comparison are shown in Fig. 6. Since the nonlinearities in NU_{xz} were so significant, a test was conducted on two 192-ply laminates to determine

TABLE 3—*Summary of elastic constant results.*

Number of Plies	[0]	[0/0/90]	[0]	[0/0/90]
	Longitudinal Modulus, GPa (Msi)			
	AS4/3501-6		S2/3501-6	
48	117.1 (16.99)	80.26 (11.64)	52.86 (7.66)	41.37 (6.00)
96	115.0 (16.68)	79.98 (11.60)	49.02 (7.11)	38.27 (5.55)
192	· · ·	81.22 (11.78)	· · ·	40.75 (5.91)
	NU_{xy}			
	AS4/3501-6		S2/3501-6	
48	· · ·	· · ·	· · ·	· · ·
96	0.332	0.067	0.290	0.157
192	· · ·	0.068	· · ·	0.167
	NU_{xz}			
	AS4/3501-6		S2/3501-6	
48	· · ·	· · ·	· · ·	· · ·
96	0.322	0.450	0.306	0.363
192	· · ·	0.472	· · ·	0.357

TABLE 4—*Summary of strength and strain-to-failure results.*

Number of Plies	[0]	[0/0/90]	[0]	[0/0/90]
	Ultimate Compression Strength, MPa (ksi)			
	AS4/3501-6		S2/3501-6	
48	1160 (168.2)	1067 (154.7)	1275 (184.9)	988.7 (143.4)
96	852.2 (123.6)	891.5 (129.3)	976.3 (141.6)	930.1 (134.9)
192	· · ·	841.9 (122.1)	· · ·	797.8 (115.7)
	Longitudinal Strain-to-Failure, %			
	AS4/3501-6		S2/3501-6	
48	1.00	>1.48	2.56	2.46
96	.79	1.26	2.06	2.62
192	· · ·	1.16	· · ·	2.01

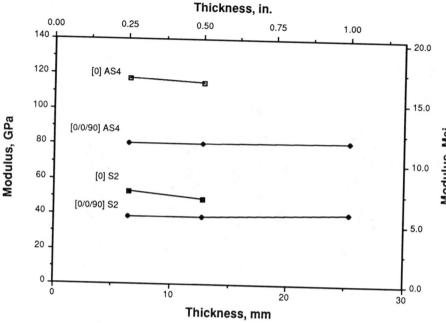

FIG. 4—*Longitudinal modulus as a function of thickness.*

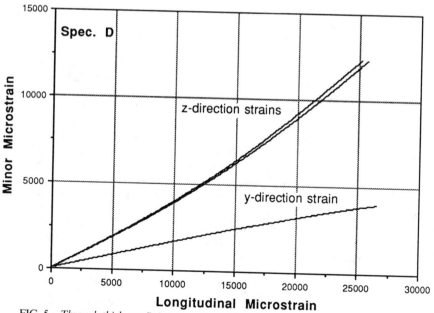

FIG. 5—*Through-thickness Poisson's ratio nonlinearity 96-ply [0/0/90] S2 glass/3501-6.*

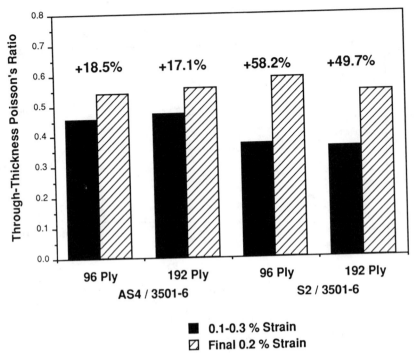

FIG. 6—*The NU$_{xz}$ change for [0/0/90]$_{ns}$ laminates.*

if the nonlinearities were reversible. For one AS4 and one S2 glass-reinforced coupon, the first compression test was conducted to 75% of ultimate stress, the load was slowly reversed, and the specimen was reloaded to failure. The strain-strain data for the test to failure tracked the data for the initial test, attributing this nonlinearity to a reversible phenomena and not damage development.

The strain gages used to determine NU$_{xz}$ were mounted on the edge of the compression test specimens as described earlier, so the effect of free-edge stresses must be considered when analyzing the edge strain data. To estimate the sign and magnitude of the sigma$_z$ stress, a free-edge stress analysis that utilizes a force and moment balance in the free-edge zone as suggested by Pagano and Pipes [9] was performed. This analysis indicated that the sigma$_z$ stresses on the free-edge of the 192-ply carbon and S2 glass laminates is less than 13.8 MPa (2 ksi) and is compressive. Therefore, the effect of these stresses would be to decrease the free-edge Poisson strains compared to strains away from the free-edges, making the edge-measured NU$_{xz}$ nonlinearities more conservative than in the center of the laminates.

A similar comparison for nonlinearities seen in the longitudinal moduli is shown in Fig. 7. These nonlinearities are not as significant as those for NU$_{xz}$, however, a drop in modulus of 20% could significantly effect strength and stability analyses for thick composite shells.

Nonlinearities for the longitudinal modulus and NU$_{13}$ for the unidirectional specimens were also measured and the results follow:

	AS4/3501-6	S2 glass/3501-6
E_x	10% decrease	no change
NU$_{13}$	12.1% increase	12.9% increase

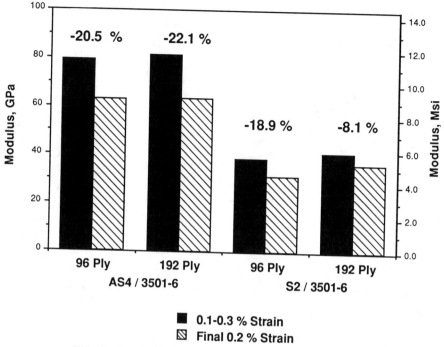

FIG. 7—*Longitudinal modulus change for* [0/0/90]$_{ns}$ *laminates.*

Ultimate Compression Strength

Figure 8 shows the ultimate compressive strength as a function of specimen thickness for both materials and orientations. These curves show a sharp decrease in compression strength with increasing thickness for the unidirectional specimens. Even at a thickness of 6.4 mm (0.25 in.), the strength of the unidirectional carbon and fiberglass coupons was lower than strengths determined using shear loading test techniques such as the IITRI, Celanese, or sandwich beam methods documented in ASTM D 3410-87. Unidirectional compression strengths from end-loaded coupons are typically reported to be lower than from the methods in ASTM D 3410-87 [10,11] and the strengths measured using 6.4-mm (0.25-in.)-thick specimens in this investigation are comparable to those previously reported. Due to the continually decreasing strength in testing 12.7-mm (0.50-in.)-thick unidirectional specimens and the lack of interest in nesting large numbers of unidirectional plies even in thick laminates, no 192-ply unidirectional coupons were fabricated or tested.

The strength of the AS4/epoxy [0/0/90]$_{ns}$ laminates dropped 13.8% in going from 6.4 to 12.7 mm (0.25 to 0.5 in.) and dropped 5.6% in going from 12.7 to 25.4 mm (0.5 to 1.0 in.). The S2 glass/epoxy laminates showed a lower decrease from 6.4 to 12.7 mm (0.25 to 0.5 in.) (9.9%) than from 12.7 to 25.4 mm (0.5 to 1.0 in.) (14.2%). At first look, this trend of decreasing strength with increasing thickness appears significant, but when considering that the failure location for these laminates is where the tabs terminate and the gage section begins, the drop in strength can be explained by the gage section Poisson expansion that occurs in these thick laminates [12]. The difference between the through-thickness expansion of the gage section and the through-thickness expansion of the region within the fixture grips is the effective gage section expansion, and this expansion results in fiber curvature where the coupon exits the

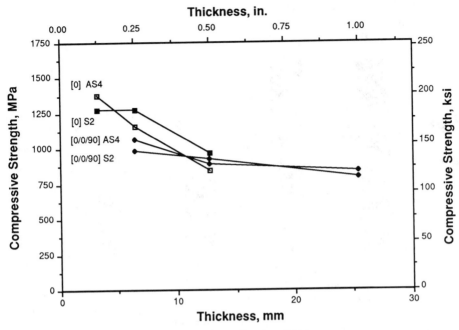

FIG. 8—*Strength as a function of thickness.*

clamping blocks. Since this effective gage section expansion increases with increasing thickness, then so will the corresponding fiber curvature. Fiber curvature has been shown to have a strong influence on compression strength [13,14], and Ref 12 shows this curvature accounts for the drop in compression strength measured for the thick [0/0/90]$_{ns}$ laminates.

One other fact determined in the strength study is that the low compressive strength experienced in thick, unstiffened, carbon-reinforced composite cylinders is not attributable to the effect of thickness on uniaxial compressive strength. That is, the strength of the thick [0/0/90]$_{ns}$ AS4/epoxy coupons did not drop to a level of 552 to 690 MPa (80 to 100 ksi) as seen in testing of thick-walled shells.

Failure Mechanisms

The failure mechanisms observed in the [0] carbon coupons were the same for the 48- to 96-ply specimens. All failures occurred on the ends of the specimens at the load introduction point. The most predominant failure characteristic was a brittle shear plane that initiated at the top corner of the specimen and propagated down and across the width of the specimen.

This failure location is in a region of high stress concentration, and the observed failure mechanism is not typically reported for thin, shear loaded compression coupons that have higher compression strengths than achieved here. Typically, failure for these thinner, shear-loaded specimens occurs in the specimen gage section and failure is a combination of longitudinal splitting, fiber breakage, and kink-band formation. Attempts to change the location and mode of failure with increased fixture preload were unsuccessful.

The failure mechanisms observed for the [0] S2 glass coupons were also the same for the 48- and 96-ply specimens. For the 48-ply specimens, two failed at the end, two failed at the tab-gage section interface, and one failed in the center of the gage section. For the 96-ply coupons,

three failed at the end and one failed simultaneously at the end and the tab-gage length interface. Regardless of failure location, the predominant failure mechanism seen in these specimens were kink-bands that ran through the thickness of the specimens. Longitudinal splitting
also occurred when the failure extended into the gage section, since no support was provided
by the fixture in this region.

The failure mechanisms in the $[0/0/90]_{ns}$ laminates were the same for all three thicknesses
tested. Kink-bands that propagated through the thickness of the specimens at an angle of 20
to 30° were present in all of the failed $[0/0/90]_{ns}$ coupons. For all ten of the 48-ply laminates,
four of the eight 96-ply laminates, and eight of the ten 192-ply laminates, these kink-bands
were located at the gage length tab termination interface; the others failed at the specimen
ends. Figures 9 and 10 are overall photographs of representative 96- and 192-ply $[0/0/90]_{ns}$

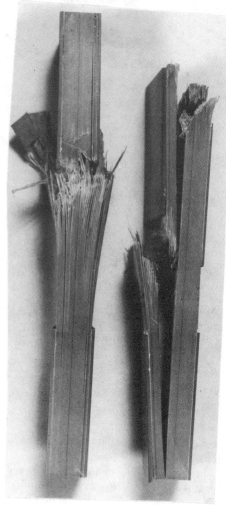

FIG. 9—The 96-ply $[0/0/90]_{16s}$ fractured specimens; (left) AS4/3501-6 and (right) S2/3501-6.

FIG. 10—*The 192-ply [0/0/90]$_{32s}$ fractured specimens; (left) AS4/3501-6 and (right) S2/3501-6.*

specimens, and Figs. 11 and 12 show close-up views of the kink bands that occurred in the 96- and 192-ply [0/0/90]$_{ns}$ coupons.

The number of end failures for the 96-ply laminates could possibly be attributed to the fact that tabs on these specimens were ground so that the final tab thickness-to-specimen thickness ratio (1:6) was less than for the 48- (1:2) and and 192-ply (1:4) laminates. This ratio was intended to be 1:4 for the 96-ply specimens as for the 192-ply specimens. Even so, the strength

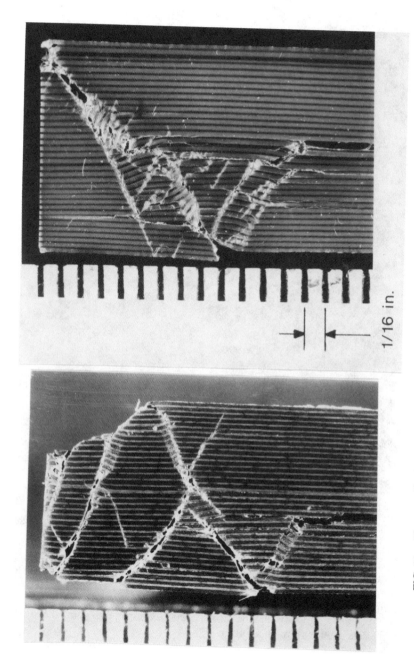

FIG. 11—Close-up of kink-bands in 96-ply [0/0/90]₆ₛ specimens; (left) AS4/3501-6 and (right) S2/3501-6.

FIG. 12—*Close-up of kink-bands in 192-ply* [0/0/90]$_{32s}$ *specimens;* (left) *AS4/3501-6 and* (right) *S2/3501-6.*

of the laminates that failed on the ends was the same as for those that failed in the tab termination region.

The predominant difference in the failure characteristics between the carbon and S2 glass 25.4-mm (1.0-in.)-thick coupons is that the carbon coupons completely separated upon failure whereas the S2 glass coupons remained intact.

Conclusions

The conclusions from this investigation concern the effect of thickness on the compressive response of [0] and [0/0/90]$_s$ carbon- and fiberglass-reinforced composite materials. The longitudinal modulus of these materials was insensitive to sample thickness. The in-plane and through-thickness Poisson's ratios were also independent of thickness, however, large changes in the through-thickness Poisson's ratio with applied load were observed for the [0/0/90]$_s$ laminates. A change in NU$_{xz}$ of 57.4% was recorded for the S2/3501-6 laminates.

The strength of the [0] specimens was very sensitive to thickness in tests conducted on coupons up to 12.7 mm (0.5 in.) thick. The [0/0/90]$_s$ laminates showed a decrease in strength of approximately 22% from 6.4 mm (0.25 in.) to 25.4 mm (1.0 in.). The failure characteristics for both materials in all thicknesses was similar to observations regularly reported for thin composite coupons. The presence of shear planes and kink-bands through the specimen thickness predominated. Delaminations were routinely observed propagating from the kink-bands and resulted in excessive damage development when failures occurred in the vicinity of the gage-section. When failures occurred within the clamping blocks, delamination was suppressed and the resulting failure characteristics were much more preserved.

In reference to the objective of this research as summarized in the beginning of this paper, the trends in material properties, strength, and failure mechanisms of thick composites do not account for the observed response of thick composite shells when subjected to hydrostatic pressure. The strength of thick, carbon-reinforced laminates has been found to be at least as high as for thick fiberglass-reinforced laminates. With the exception of the nonlinearities seen in the through-thickness Poisson's ratio, the material elastic constants reported are equivalent to those found in thin coupons. The effect of the these nonlinearities on shell response will not be known until these effects are incorporated in three-dimensional shell analysis. However, since the nonlinearities reported are greater for the fiberglass than the carbon composites, it appears that they will not provide an explanation for the poor response of thick, unstiffened carbon-reinforced shells compared to fiberglass shells.

Acknowledgments

The author would like to acknowledge the lab support provided by Tom Mixon, Jim Kerr, and Bonnie Paddy, the support of Dave Moran, Bruce Douglas, Joe Crisci from the DTRC IR/IED program office, and the support of Jim Kelly from the DARPA Advanced Submarine Technology program. He would also like to acknowledge the support provided by Dick Wilkins and the other members of his graduate committee at the University of Delaware.

References

[1] Garala, H. J., "Experimental Evaluation of Graphite/Epoxy Composite Cylinders Subjected to External Hydrostatic Compressive Loading," Proceedings, 1987 Spring Conference on Experimental Mechanics, Society for Experimental Mechanics, Bethel, CT, 1987, pp. 948–951.

[2] Leknitskii, S. G., Theory of Elasticity of an Anisotropic Body, Mir Publishers, Moscow, 1981.

[3] Dally, J. W., Nelson, H. R., and Cornish, R. H., "Fatigue and Creep Properties of Glass-Reinforced Plastic Under Compressive States of Stress," ASME Paper No. 63-WA-236, American Society of Mechanical Engineers, New York, Dec. 1963.

[4] Hom, K., Buhl, J. E., and Couch, W. P., "Hydrostatic Pressure Tests of Unstiffened and Ring-Stiffened Cylindrical Shells Fabricated of Glass-Filament Reinforced Plastics," David Taylor Model Basin Report 1745, David Taylor Research Center, Annapolis, MD, Sept. 1963.

[5] Camponeschi, E. T., Jr., "Compression of Composite Materials: A Review," Report DTRC-87-050, David Taylor Research Center, Annapolis, MD, Nov. 1987.

[6] Irion, M. N. and Adams, D. F., "Compression Creep Testing of Unidirectional Composite Materials," Composites, Vol. 12, No. 2, 1981, pp. 117–123.

[7] Timoshenko, S. P. and Gere, J. M., Theory of Elastic Stability, McGraw-Hill, New York, 1961.

[8] Camponeschi, E. T., Jr., "Through-Thickness Strain Response of Thick Composites in Compression," Report DTRC SME-89-67, David Taylor Research Center, Annapolis, MD, 1989.

[9] Pagano, N. J. and Pipes, R. B., "Influence of Stacking Sequence on Laminate Strength," *Journal of Composite Materials,* Vol. 5, No. 1, 1971, pp. 50–57.

[10] Shuart, M. J., "Failure of Compression-Loaded Multi-Directional Composite Laminates," AIAA Paper No. 88-2293, AIAA/ASME/ASCE/AHS Twenty-Ninth Structures, Structural Dynamics and Materials Conference, 1988.

[11] Berg, J. S. and Adams, D. F., "An Evaluation of Composite Material Compression Test Methods," Report UW-CMRG-R-88-106, University of Wyoming, Laramie, WY, 1988.

[12] Camponeschi, E. T., Jr., "Compression Response of Thick-Section Composite Materials," Ph.D. thesis, University of Delaware, Newark, DE, Aug. 1990.

[13] Argon, A. S., "Fracture of Composites," *Treatise on Materials Science and Technology,* Vol. 1, 1972, pp. 106–114.

[14] Budiansky, B., "Micromechanics," *Computers and Structures,* Vol. 16, Nos. 1–4, 1983, pp. 6–10.

Ram C. Madan[1]

Influence of Low-Velocity Impact on Composite Structures

REFERENCE: Madan, R. C., **"Influence of Low-Velocity Impact on Composite Structures,"**
Composite Materials: Fatigue and Fracture (Third Volume), ASTM STP 1110, T. K. O'Brien,
Ed., American Society for Testing and Materials, Philadelphia, 1991, pp. 457–475.

ABSTRACT: Composite materials are sensitive to out-of-plane loads because they are weaker
through the thickness than in the plane of lamination. Consequently, composite structure subjected to impact may suffer significant damage, which results in a loss in strength. The effect of
impact energy on different thicknesses of laminate with respect to ultimate compressive strength
was evaluated using the NASA ST-1 specimen test. Impact damage was measured using ultrasonic C-scans and quality was assessed by visual observations. A semiempirical model for impact
damage was proposed to determine residual strength in terms of laminate properties, static influence coefficients, and impact energy.

   To determine residual compression strength, single- and multiple-impact damage was
inflicted on bonded stiffened structures. The results were verified by the residual strength
method, which uses laminate properties and measured damage size to predict the residual
strength of composites. The correlation between predicted and actual strength was very encouraging. Although the damage was inflicted with different impactor masses, velocities, and impact
energy, the data indicated that residual strength is a function only of damage present and does
not reflect the manner in which the damage is inflicted. Finally, a semiempirical relationship was
developed for damaged areas in stitched laminates.

KEY WORDS: durability, stitched composites, dry forms, impact damage, damage tolerance,
residual strength, damage size, impact energy, stitch step, stitch spacing, composite materials,
fracture, fatigue (materials)

   The response of materials and structures to impulsive loading is complex. Practical impact
problems often involve impactors and targets whose behavior is influenced by their finite
boundaries. As the intensity of impact energy increases, the material is driven from the elastic
into the plastic stage. This process involves large deformations, exothermal processing, and
fractures, resulting in target failure through a variety of mechanisms. The impact response of
composites depends on various combinations of materials, layups, and fabrication processes,
including the properties of the impactor.

   It is essential to understand the effect of impact by foreign objects on structural strength
when using composites for heavily loaded primary structural components, such as wings and
fuselage. Aircraft structures damaged by large impact energy can also experience significant
changes in stiffness at the component level. Within a wing, for example, severe skin damage,
such as panel detachment or rupture, can reduce the torsional stiffness below the flutter
requirements of the operating envelope. Severance of a wing spar could have a similar effect.
This paper identifies some factors affecting low-velocity impact damage and describes possible
courses in controlling damage growth in composites.

---

[1] Principal engineer, Douglas Aircraft Company, McDonnell Douglas Corporation, Long Beach, CA
90846.

Impact Damage

The two major types of composite damage are matrix cracks and through-the-thickness cracks (fiber breakage) where delamination is the subset of matrix cracks. A matrix crack usually grows in a certain preferred direction and remains finite in size because of the laminate's internal reinforcement structure. The internal geometrical constraints of the laminate (the unidirectional fibers and the parallel interfaces of the laminating plies) regulate the way the matrix cracks propagate.

Although the thickness of most laminates is small compared to their width and length dimensions, significant three-dimensional stresses do occur near areas where there is an abrupt geometrical change. Intraply and interply matrix cracks are mutually interacting during the course of laminate loading. This interaction can precipitate disintegration of the lamination structure before significant fiber damage occurs.

A laminate's residual strength and modulus after delamination depend on the stacking sequence and the location of the delamination in the laminate. In certain instances, the layup of a delaminated sublaminate also influences the strength and modulus. For example, in PEEK/AS4 $(0,90,45,-45)_s$ laminate, the residual modulus value is higher for delamination between 0 and 90° plies than for delamination occurring between 90 and 45° plies (see Table 1). There is no further reduction in modulus value if there is a delamination between 45 and

TABLE 1—*AS4/PEEK laminate modulus after delamination.*

Layup No.	Degree, °	Rotated Stacking Sequence				Residual Modulus, GPa (Msi)		
		1^a	2	3		1	2	3
		LAYUP A: $[(90 - \theta)/-\theta/-(45 + \theta)/(45 - \theta)]_s$						
1	−90	0	90	45	−45	55.20 (8.01)	48.06 (6.97)	48.06 (6.97)
2	−75	−15	75	30	−60	37.20 (5.40)	33.99 (4.93)	52.20 (8.02)
3	−60	−30	60	15	−75	36.98 (5.36)	33.99 (4.93)	51.41 (7.03)
4	−45	−45	45	0	90	48.06 (6.97)	48.06 (6.97)	48.28 (7.46)
5	−30	−60	30	−15	75	55.20 (8.01)	33.99 (4.93)	51.41 (7.46)
6	−15	−75	15	−30	60	51.41 (7.46)	33.99 (4.93)	55.20 (8.01)
7	0	90	0	−45	45	48.29 (7.03)	48.06 (6.97)	48.06 (6.97)
8	15	75	−15	−60	30	51.41 (7.46)	33.99 (4.93)	36.98 (5.36)
9	30	60	−30	−75	15	55.20 (8.01)	33.99 (4.93)	37.20 (5.40)
10	45	45	−45	90	0	48.06 (6.97)	48.06 (6.97)	55.20 (8.01)
11	60	30	−60	75	−15	36.98 (5.36)	33.99 (4.93)	37.20 (5.40)
12	75	15	−75	60	−30	37.20 (5.40)	33.99 (4.93)	36.48 (5.36)
13	90	0	90	45	−45	55.20 (8.01)	48.06 (6.97)	48.06 (6.97)
		LAYUP B: $[-(45 + \theta)/-\theta/(45 - \theta)/(90 - \theta)]_s$						
14	−90	45	90	−45	0	48.06 (6.97)	47.77 (6.93)	55.20 (8.01)
15	−75	30	75	−60	−15	36.98 (5.36)	38.45 (5.77)	37.20 (5.40)
16	−60	15	60	−75	−30	36.98 (5.36)	38.45 (5.77)	36.98 (5.36)
17	−45	0	45	90	−45	55.20 (8.01)	47.77 (6.93)	48.06 (6.97)
18	−30	−15	30	75	−60	37.20 (5.40)	48.87 (7.09)	55.20 (8.01)
19	−15	−30	15	60	−75	36.98 (5.36)	48.87 (7.09)	51.41 (7.46)
20	0	−45	0	45	90	48.06 (6.97)	47.77 (6.93)	48.49 (7.03)
21	15	−60	−15	30	75	55.20 (8.01)	38.45 (5.77)	51.41 (7.46)
22	30	−75	−30	15	60	51.41 (7.46)	38.45 (5.77)	55.20 (8.01)
23	45	90	−45	0	45	48.49 (7.03)	47.77 (6.93)	48.06 (6.97)
24	60	75	−60	−15	30	51.41 (7.46)	48.87 (7.09)	36.98 (5.36)
25	75	60	−75	−30	15	55.20 (8.01)	48.87 (7.09)	37.20 (5.40)
26	90	45	90	−45	0	48.06 (6.97)	47.77 (6.93)	55.20 (8.01)

^{a} Denotes delamination location.

$-45°$; that is, the delaminated sublaminate consists of three (0,90,45) plies instead of two (0,90) plies. In a $(-30,15,60,-75)_s$ laminate of the same material, the residual modulus increases as the number of plies of delaminated sublaminate increases.

The impact damage is a function of stacking sequence and the amount of impact energy, which determines the depth of the delamination. When NASA ST-1 test coupons [1] were employed, experimental results for 3501/AS4, PEEK/AS4, 1808I/IM6, and 8551-7/IM7 showed that residual compressive strength and strain after impact are a function of impact energy, $(U)^{1/2}$. This can be seen in Fig. 1 for laminates fabricated with 1808I/IM6 composite materials. Figure 2 depicts the residual stress for laminates with different thicknesses and fabricated with different materials. These figures show that there is a finite level of impact energy where no damage occurs in the laminate. However, a slight increase in this level of impact energy will inflict damage on the laminate. This level of impact energy is referred to as threshold energy for that laminate, and it is determined by the material system, thickness, and stacking sequence of the laminate at which the laminate behaves like an undamaged laminate. The ultimate stress and strain decrease with impact energy $(U)^{1/2}$ until the laminate is no longer affected by single low-velocity impact damage.

Test Procedure

Carbon/epoxy (8551-7/IM7) laminates of different thicknesses (32, 36, 48, and 64 plies) were fabricated and tested. All the laminates were quasi-isotropic, representing NASA ST-1 test coupons with different thicknesses. Table 2 summarizes the laminates and the impact energy levels of each thickness tested. The impact tests were conducted on all the laminates using a Dynatup 8200 impact machine. Three specimens were tested for each thickness and impact energy level, and the average of the three results is reported here. The tup (impactor) head was hemispherical, 12.7 mm (½ in.) in diameter. Impact and rebound velocity were measured electronically. All specimens were C-scanned for quality assurance before the strain gages were attached.

The maximum load point depends on the severity of the impact. If the impact causes no damage, the maximum load just indicates the peak resistive force provided by the plate to stop the tup, which is an indication of the plate's flexural stiffness. If the impact is high enough to cause damage, the first drop in the load time trace indicates damage initiation (such as delamination), and the peak load may indicate where back surface fiber breakage occurs. Impact energy corresponding to the maximum load is the amount of energy the specimen can absorb before failing [2]. Total energy is the amount of energy the specimen absorbs during the complete test. In case of abrupt specimen failure, the total energy corresponds to the maximum load.

Impacts on composites usually do not cause visible damage, yet may cause internal damage, such as delamination or debonding. With these materials, the maximum load is less important than the load and energy at the onset of damage. The energy to reach maximum load and failure is independent of the rate at which the load is applied (velocity). The first sharp decrease in the load trace diagram represents the incipient damage (that is, decrease in modulus of the laminate) as shown in Fig. 3. At high impact energy that produces fiber breakage, a large drop in force is observed on the force-time plots at the peak force. This drop in force occurs at roughly the same load level for any given laminate of any given material and also has about the same drop in load.

A postimpact compression test was performed on a Baldwin Universal testing machine. The compression strength was calculated based on the average thickness and width, and the results are shown in Fig. 2. A typical instrumented impact test output is shown in Fig. 3 for a 32-ply laminate impacted by 27.2 J (20 ft·lb) of energy.

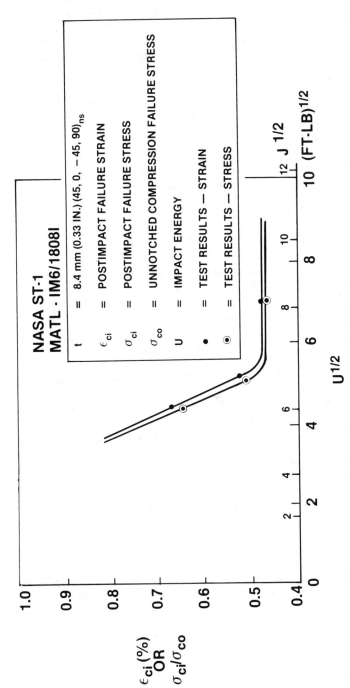

FIG. 1—*Effect of impact energy on stress ratio and strain.*

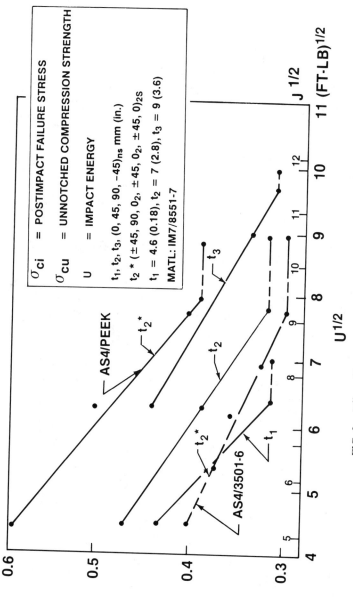

FIG. 2—*Effect of impact energy on failure stress.*

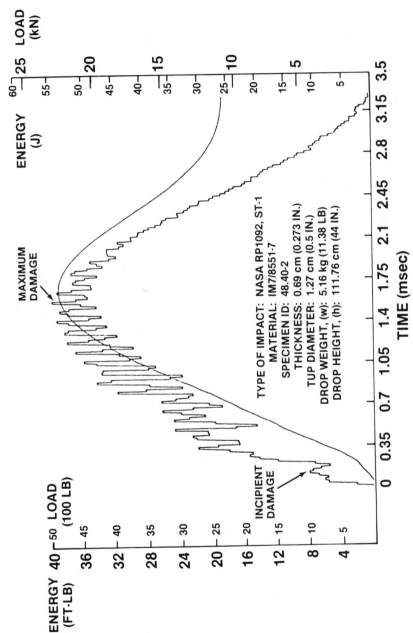

FIG. 3—*Impact force-time response and calculated energy curve.*

TABLE 2—*Impact energy levels.*

Specimen ID	Laminate Thickness, mm (in.)	Impact Energy, J (ft·lb)			
32	4.6 (0.180)	27.1 (20)	40.7 (30)	54.2 (40)	67.8 (50)
36	5.1 (0.202)	27.1 (20)	40.7 (30)	54.2 (40)	
48	7 (0.275)	27.1 (20)	54.2 (40)	81.3 (60)	108.5 (80)
64	9 (0.358)	54.2 (40)	81.3 (60)	108.5 (80)	135.6 (100)

Damage Tolerance Panels

The multistringer impact test panels were assembled as two- and three-stringer panels (Fig. 4). The panels and stringers were laid up with a 44/44/12 percentage of plies in 0, ±45, and 90° directions [3]. This layup was consistent with the requirements of a high-aspect-ratio wing, where stiffness is critical. The skin panels were all 54 plies thick, while the stringers were constructed with either 36- or 18-ply flanges and 72-ply blades.

The impact test requirements, currently proposed through combined study for the Air Force Wright Aeronautical Laboratories (AFWAL) by Northrop and Boeing [4] for the Mil-A-Prime handbook, call for ultimate strength to be maintained following an impact level, up to a maximum of 136 J (100 ft·lb), that just causes visual damage, not to exceed a 2.5-mm-deep (0.1 in.) dent. The laminates tested were thick enough to preclude visual damage below the maximum level; therefore, all panels were impacted at the 136 J (100 ft·lb) level. Locations for external skin impacts were midbay between stringers, over the blade, and over the stringer flange edge. As shown in Fig. 4, the internal impact tests were applied laterally on the stringer blade. The tests verified that the midbay impact causes maximum damage to the panel (as compared to other cases). While thinner flanged stiffeners remained intact with the skin, the thicker flanged stiffeners were disbonded from the skin with a couple of skin plies; that is, the flange did not peel off at the flange/skin bond line, but instead took a couple of the top skin plies with it. Thus, midbay impact and the thin-flanged stiffeners were selected for further study.

Two types of panels, each with a different flange thickness (blade stiffeners), were fabricated from IM6/1808I, IM7/8551-7, and other materials (AS4/3501-6). The panels fabricated with 36 plies had a 5.8 mm (0.23 in.) flange thickness, while flanges fabricated with 14 or 18 plies had a 2.8 mm (0.11 in.) thickness [3]. All the specimens were impacted with a 136 J (100 ft·lb) energy level at midbay, using a 25.4-mm (1.0-in.) hemispherical impactor. Visible delamination occurred on the far side of all of the skin panels. The panels were nondestructively inspected and then tested in compression to ultimate failure. Test results indicated that by increasing the skin thickness of thin-flanged panels fabricated with IM6/1808I (interleaf) material, the stress at failure can be increased significantly without altering stiffener dimensions. Table 3 summarizes the results of all three- and five-stringer panels. The blade-stiffened panels (ID No. 14 through 28 in Table 3) were fabricated and tested at Douglas, while the channel- and hat-stiffened specimens were fabricated and tested by other investigators [5–8]. The residual strength analysis in Ref 3 was used to determine the strength of damaged blade-stiffened panels that had been impact-tested. The analysis is strictly applicable for damages due to fiber breakage. Figure 5 illustrates a typical load-versus-displacement curve for a thin-

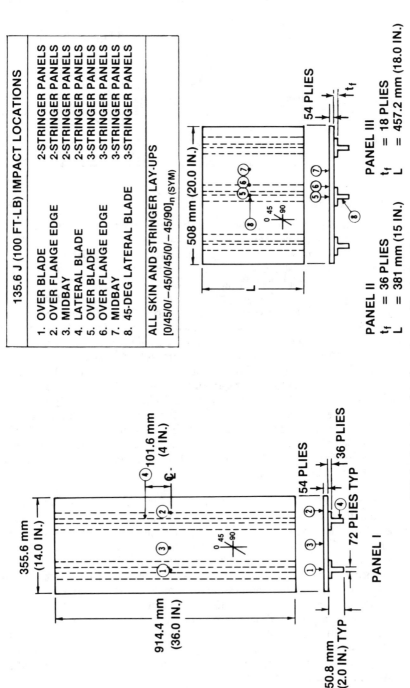

FIG. 4—*Impact panel configurations and impact locations.*

TABLE 3—*Residual compression strength of postimpact stiffened panels.*

Panel ID	Material	Length, Width, cm (in.)	Thickness Skin, Flange (No. of Plies)	Impact Location	Impact Energy, J (ft·lb)	Failure Stress, MPa (ksi)	Failure Strain, %	No. of Stiffeners
1	T700/3620	40 (15.75) 39 (15.4)	18, 9	mid-stringer	4.3 (3.21)	· · ·	0.385	3, hat
2	T700/3620	40 (15.75) 39 (15.4)	18, 8	stringer edge	4 (3.05)	· · ·	0.385	3, hat
3	T700/3620	40 (15.75) 39 (15.4)	18, 9	midbay	4.2 (3.15)	· · ·	0.494	3, hat
4	T700/3620	40 (15.75) 39 (15.4)	18, 9	stringer edge	12.26 (9.05)	· · ·	0.298	3, hat
5	carbon/ epoxy tape	51 (20) 34.3 (13.5)	41, 41	midbay	12.2 (9.0)	· · ·	0.52	3, blade adhesively bonded
6	carbon/ epoxy	48.1 (18.9) 36.6 (14.4)	41, 41	midbay	14.2 (10.51)	· · ·	0.37	3, blade adhesively bonded
7	carbon/ epoxy	48.8 (18.9) 57.2 (22.5)	48, 48	mid-stringer	13.4 (9.87)	· · ·	0.47	3, hat
8	AS4/3501-6	61 (24) 45.7 (18)	48, 24	midbay, mid-stringer, stringer edge	135.5 (100) EACH	· · ·	0.37	3, channel mechanically fastened
9	AS4/3501-6	61 (24) 45.7 (18)	48, 24	midbay	135.5 (100)	· · ·	0.38	3, channel mechanically fastened
10	AS4/3501-6	61 (24) 45.7 (18)	48, 24	midbay, mid-stringer, stringer edge	112.5 (83) EACH	· · ·	0.38	3, channel mechanically fastened
11	AS4/3501-6	61 (24) 45.7 (18)	48, 24	centers of two bays	81.3 (60) EACH	· · ·	0.32	3, channel mechanically fastened
12	AS4/3501-6	61 (24) 45.7 (18)	48, 24	centers of two bays	54.2 (40)	· · ·	0.35	3, channel mechanically fastened
13	AS6/5245-6	61 (24) 45.7 (18)	48, 24	midbay, mid-stringer, stringer edge	135.5 (100) EACH	· · ·	0.39	3, channel mechanically fastened
14	IM6/1808I	51 (20) 48 (19)	39, 14	midbay	135.5 (100)	292 (42.3)	· · ·	3, blade
15	IM6/1808I	51 (20) 48 (19)	54, 14	midbay	135.5 (100)	361 (52.3)	· · ·	3, blade adhesively bonded
16	IM6/1808I	51 (20) 48 (19)	66, 14	midbay	135.5 (100)	477 (69.1)	· · ·	3, blade adhesively bonded
17	IM6/1808I	51 (20) 48 (19)	40, 14	midbay	135.5 (100)	271 (39.3)	· · ·	3, blade adhesively bonded
18	IM6/1808I	51 (20) 48 (19)	54, 36	midbay	135.5 (100)	225 (32.6)	0.433	3, blade adhesively bonded
19	IM6/1808I	51 (20) 48 (19)	54, 36	mid-stringer	135.5 (100)	267 (39)	0.400	3, blade adhesively bonded
20	IM6/1808I	51 (20) 48 (19)	54, 18	mid-stringer	135.5 (100)	321 (46.5)	0.50	3, blade adhesively bonded
21	IM6/1808I	51 (20) 48 (19)	54, 18	midbay	135.5 (100)	267 (39)	· · ·	3, blade adhesively bonded

TABLE 3—*Continued*.

Panel ID	Material	Length, Width, cm (in.)	Thickness Skin, Flange (No. of Plies)	Impact Location	Impact Energy, J (ft·lb)	Failure Stress, MPa (ksi)	Failure Strain, %	No. of Stiffeners
22	IM6/1808I	51 (20) 48 (19)	54, 18	mid-stringer	135.5 (100)	323 (47)	0.444	3, blade adhesively bonded
23	IM6/1808I	51 (20) 48 (19)	54, 18	midbay	271 (100)	184 (26.7)	0.333	3, blade adhesively bonded
24	IM6/1808I	51 (20) 48 (19)	54, 18	side of blade	144 (106)	425 (617)	0.689	3, blade adhesively bonded
25	IM7/8551-7	51 (20) 48 (19)	63, 18	midbay	135.5 (100)	302 (43.8)	· · ·	3, blade adhesively bonded
26	IM7/8551-7	51 (20) 48 (19)	63, 18	midbay	135.5 (100)	326 (47.3)	· · ·	3, blade adhesively bonded
27	IM6/1808I	142 (56) 84 (33)	54, 18	midbay	135.5 (100)	· · ·	0.410	3, blade adhesively bonded
28	IM6/1808I	142 (56) 84 (33)	54, 18	midbay	135.5 (100)	· · ·	0.424	3, blade adhesively bonded

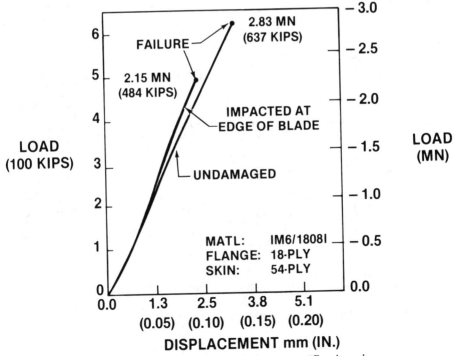

FIG. 5—*Load-displacement for three-stringer stiffened panels.*

FIG. 6—*Compression test on three-stringer, adhesively bonded composite panel.*

flanged panel. The three-stringer panel failed in the post-impact compression test, as shown in Fig. 6.

Another Douglas study showed that low-mass/high-velocity impact obtained greater strength reduction and included a larger damaged area than the high-mass/low-velocity impact. In the case of high-mass/low-velocity impact with the same energy level, global structural response of a smaller transient deformation gradient occurred. Therefore, it was deduced that the localized deformation was responsible for the larger damage size, and the greater reduction in strength was induced by the low-mass/high-velocity impact.

In the case of repeated impacts at the same impact energy, Wyrick et al. [9] showed that the peak force decreases linearly with each subsequent impact. The steady decline in peak force was apparently due to previously damaged material in the specimen, which cushioned the impactor during the next impact. An increase in peak force was observed for a few initial impacts. This increase may be attributed to the thin layer of unreinforced resin at the impacted surface, which had been compacted by the low-level impacts. Until fiber damage occurs, the compacted surface provides a harder surface for the next impact. For low-level impacts, no perforation occurred and the plates retained their overall rigidity, which was demonstrated by residual tensile strength tests.

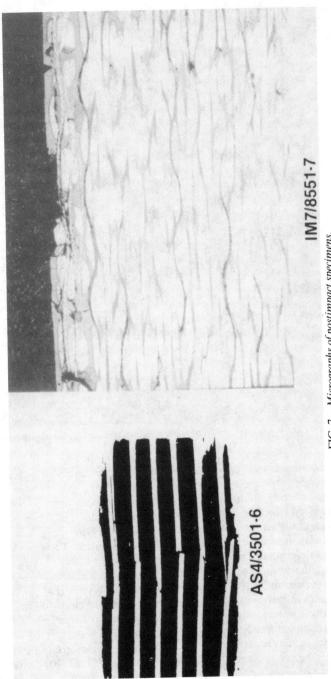

FIG. 7—Micrographs of postimpact specimens.

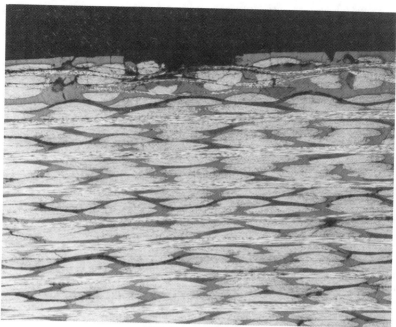

FIG. 7—*Continued.*

Liu [*10*] tested two composite plates made of two different materials and with the same layup. The same impact conditions were used for both tests. Both plates had peanut-shaped delaminations but different delamination areas. With the same impacting energy, Kevlar/epoxy had a larger delamination area than carbon/epoxy, which had a larger delamination area than glass/epoxy. This inconsistency was caused by the differences in interlaminar stress and strength among the various materials. Figure 7 shows the difference in damage for two material systems tested at Douglas under the same impact energy levels. The specimen fabricated with the toughened composite system (IM7/8551-7) incurred delamination only, while the specimen fabricated with the brittle composite system (AS4/3501-6) incurred fiber breakage.

Impact Mechanics

For completeness, a high-velocity impact dynamics derivation has been included, even though it does not relate to the subject of this paper. It is assumed that a single impact-generated impulse is a triangular (sawtooth) pulse. Delamination at a free surface far from the impact area can be treated as one-dimensional stress pulse propagation without a change in shape or intensity. An incident-compressive pulse of magnitude, f_m (dependent on wave velocity), with λ wavelength is reflected at the far surface with a tensile wave of magnitude, f_t. This will always occur at the leading edge of the reflected wave.

$$f_t = f_m - f_i \tag{1}$$

where f_i is the compressional-incident stress at the same point as the leading edge of the reflected wave. This equation is true at any given time during the reflection of the wave. If f_m

$> F_m$, there will be an instant when $f_i = F_m$ and the delamination will occur with fracture of the far surface ply. F_m is the fracture strength of the laminate determined from the test.

At this instant

$$F_m = f_m - f_i \tag{2}$$

or

$$f_i/(\lambda - 2t_1) = f_m/\lambda$$

$$f_i = f_m(\lambda - 2t_1)/\lambda \tag{4}$$

where t_1 is the delamination thickness of the laminate from the far surface (as delamination initiates at the far surface of the laminate).

Using Eqs 1, 3, and 4 gives t_1 in terms of known values of F_m, f_m, and λ.

$$t_1 = 0.5F_m\lambda/f_m \tag{5}$$

In the case of $F_m = f_m$, delamination thickness is $\lambda/2$, but if ply thickness is greater than $\lambda/2$, there will be a damaged ply without delamination. On the other hand, if the ply thickness is less than $\lambda/2$, the two-ply sublaminate will delaminate at that instant. If $F_m \gg f_m$, multiple delaminations can occur. The number of delaminations can be determined by the following relationship [11] with each thickness of $\lambda/2$

$$n = F_m/f_m \tag{6}$$

Padilla [5] conducted impact tests with 6.8 J (5 ft·lb) energy on carbon/epoxy $(45,0_2,45_2,45)_s$ laminates that were fatigued at 12 000 simulated flight hours. All 3.7-mm-thick (0.147-in.) laminates failed at the same compression-after-impact (post-fatigue) strength, even when they were fatigued at different peak compressive strain levels (0.4 through 0.6%). There was a 17% decrease in residual strength for 6.8 J (5 ft·lb) impact energy after the laminates went through the fatigue cycles. There was an increase of 20% in impact-damage area from the pre-fatigue to the post-fatigue tests for carbon/epoxy laminates.

A carbon/epoxy laminate of $\pm45°$ fibers has been found capable of carrying considerably higher strain loads with damage than laminates containing $0°$ plies [6]. An all $\pm45°$ laminate is an optimum configuration for carrying shear loads, but it has a low axial stiffness. Kevlar/epoxy has a relatively high structural efficiency in tension applications since the axial strength and specific tension modulus have high values. The material exhibits a nonlinear stress-strain response at strains in excess of 0.35% and behaves almost like a perfectly plastic material. Williams et al. [7,8] showed that a compression-loaded predominantly $\pm45°$ Kevlar/epoxy laminate (with 0 and $90°$ plies) is damage tolerant.

Linear damage size (d) (see Fig. 8) is determined here by the following equation, in terms of the static influence coefficient (a), which is defined as the deflection per unit of static load at the point of impact; panel width (w); and impact energy (U)

$$d/t = A_1\sqrt{Ua/t^2} + B_1 \tag{7}$$

or

$$d/w = A_1\sqrt{Ua/t^2} + B_1$$

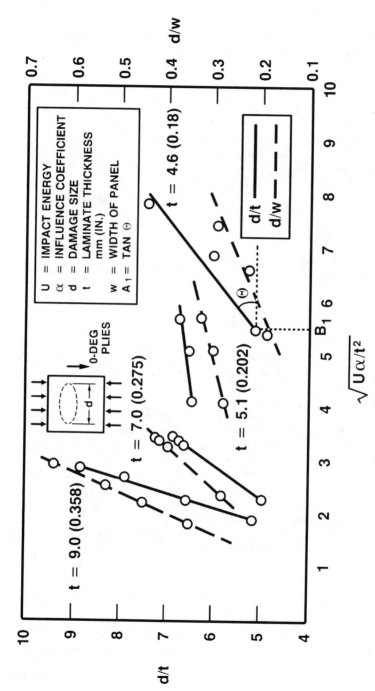

FIG. 8—*Effect of impact energy on damage size.*

where A_1 and B_1 (empirical constants to be determined experimentally) are functions of laminate material, thickness, layup, and impactor material, size, shape, and mass. Figure 8 shows the test data for 32-, 36-, 48-, and 64-ply laminates as per NASA ST-1 test specimen results. A_1 is the slope of the straight line for each laminate, while B_1 assists in determining the threshold impact energy just before the onset of damage.

Stitched Composites

The laminates were fabricated with dry uniwoven carbon fabric. They were stitched in 0 and 90° directions, with equal stitch spacing and the same stitch step and thread. After stitching, the dry forms were impregnated and cured under normal conditions. Compression-after-impact test results showed that stitching improved damage tolerance of the composites [12]. It was observed that, with an increase of stitch spacing to 6.3 mm (0.25 in.), the damage can be reduced to 30% of that for unstitched laminates under the same impact conditions. A semi-empirical formulation (Eq 8) was developed to determine impact damage in stitched composites.

$$A_{di}/A_{dio} = C_1 + C_2/s + C_3/s^2; \quad 0.125 \text{ in.} < s < 2.0 \text{ in.} \tag{8}$$

where

s = stitch spacing,
C_1 = 1.10,
C_2 = -0.20, and
C_3 = 0.01.

Figure 9 shows the relationship between Eq 8 and the test data. Stitch spacing was limited to the 0.318 to 5.08 cm range (0.125 to 2.0 in.) because laminate integrity is reduced significantly when stitch spacing is less than 0.318 cm (0.125 in.), and stitching does not reduce damage area when stitch spacing is greater than 5.08 cm (2.0 in.).

Results of a post-impact fatigue study conducted at Douglas are summarized in Table 4. Compression tests were conducted on unimpacted, impact-damaged, and fatigue-cycled post-impact specimens. AS4 biwoven and hybrid (uniwoven AS4 and S2-glass) fabrics with 3501-6 resin systems were used. The impacted laminates were fatigue-conditioned at 10 Hz to one million cycles before ultimate compression failure tests were conducted. The results, listed in Table 4, show that the post-impacted specimens regained strength through fatigue cycling. These laminates depict stress relaxation behavior under fatigue cycling. The stitched laminates showed a better post-impact strength than unstitched laminates. To determine the effect of fatigue frequency, one of the panels (Table 4, ID No. 4) was tested at 5 Hz. The results, compared with identical panel ID No. 3 at 10-Hz frequency, indicate a better residual compression strength (σ_{if}/σ_0 = 0.57 for 5 Hz and 0.48 for 10 Hz).

Concluding Remarks

The objective of the Douglas composite technology development program is to design, manufacture, and test composite specimens representing commercial aircraft components that are capable of satisfying strength, stiffness, damage tolerance, and durability requirements for the lowest cost.

Compression-after-impact tests were performed on two-, three-, and five-stringer panels, which were impacted with 136 J (100 ft·lb) of energy. When tested, all panels showed that midbay impact caused critical damage and produced the lowest residual panel strength. The

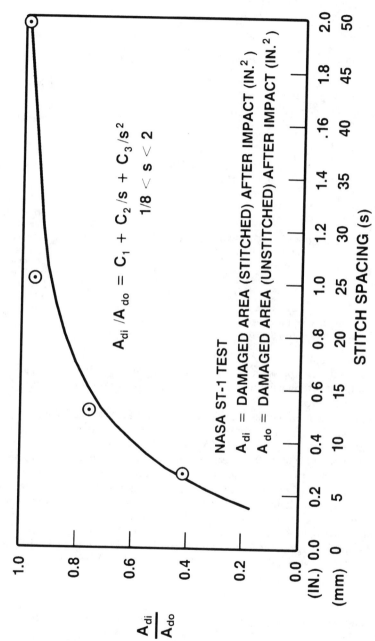

NASA ST-1 TEST

A_{di} = DAMAGED AREA (STITCHED) AFTER IMPACT (IN.^{2})

A_{do} = DAMAGED AREA (UNSTITCHED) AFTER IMPACT (IN.^{2})

$$A_{di}/A_{do} = C_1 + C_2/s + C_3/s^2$$

$$1/8 < s < 2$$

FIG. 9—*Influence of stitch spacing on damaged area after impact.*

TABLE 4—*Post-impact fatigue study.*

ID No.	Material, % AS4, % S2-Glass	Ply Layup	Thickness, mm (in.)	Fatigue Frequency, Hz	Post Impact, σ_i/σ_0	Post Fatigue σ_{if}/σ_0	Remarks
1	100, 0	$(0, 45, 90, -45)_s$	7.3 (0.286)	10	0.34	0.38	resin: 3501-6
2	90, 10	$(0/90)_{32}{}^a$	6.7 (0.264)	10	0.48	$\cdots$	impact: 53.3 J/cm (100 ft·lb/in.) thickness
3	90, 10	$[(0/90)_4, (0,45, 90, -45)_{2s}]_s$	7.8 (0.306)	10	0.46	0.48	σ_i = compression-after-impact strength
4^b	90, 10	$[(0/90)_4, (0,45, 90, -45)_{2s}]_s$	7.8 (0.306)	5	0.52	0.57	σ_0 = ultimate compression failure strength
5	80, 20	$(0/90)_{32}$	6.6 (0.259)	10	0.48	0.52	σ_{if} = postimpact and after-fatigue compression strength
6	80, 20	$[(0/90)_4, (0,45, 90, -45)_{2s}]_s$	77 (0.304)	10	0.45	0.48	1 million fatigue cycles except specimen ID No. 3 went through 1.42 million cycles

[a] Biwoven fabric (0/90); other uniwoven fabric.
[b] 1 to 4 and 37 to 40 plies were stitched.

failure sequence, under compression loading, was panel buckling, stiffener debonding, stiffener crippling, and catastrophic failure. The impact damage location did not appear to influence the final panel failure mode.

Test results showed that an increase of skin thickness, with the same stiffener dimensions, provides a significant increase in postimpact residual compressive strength (see Table 3, ID Nos. 14, 15, and 16). Post-impact testing (NASA ST-1) showed that the residual compressive strength for fatigued (10 Hz and 10^6 cycles) specimens was 10% higher than for unfatigued specimens. These specimens depicted stress relaxation behavior under fatigue cycling.

Furthermore, test results showed that the stitched laminates have a better toughness impact compressive strength than the unstitched laminates [12]. A semiempirical formulation was proposed to determine the damaged area ratio of the stitched and unstitched laminates. A linear relationship exists between post-impact compression residual strength, strain, and the square root of impact energy for each thickness. A similar linear relationship exists between nondimensional damage size and the square root of nondimensional impact energy.

References

[1] *Standard Tests for Toughened Resin Composites,* NASA Reference Publication 1092, National Aeronautics and Space Administration, Washington, DC, July 1983.
[2] Kessler, S. L. et al., *Instrumented Impact Testing of Plastics and Composite Materials, ASTM STP 936,* S. L. Kessler, Ed., American Society for Testing and Materials, Philadelphia, 1987.
[3] Madan, R. C., "Composite Transport Wing Technology Development," NASA Contract NAS1-17970, NASA Contractor Report CR-178409, Douglas Aircraft Company, NASA Langley Research Center, Hampton, VA, Feb. 1988.

[4] Horton, R. E. and Whitehead R., "Damage Tolerance of Composites," Interim Report No. 8, AFWAL Contract F33615-82-C-3212, Air Force Wright Aeronautical Laboratories, Dayton, OH, 1 March–31 Aug. 1986.

[5] Padilla, V. E., "Low-Velocity Damage Tolerance of Thin Graphite/Epoxy Laminates," Fifth Conference on Fibrous Composites in Structural Design, New Orleans, LA, MCAIR Report 81-009, 27–29 Jan. 1981.

[6] Rhodes, M. D. and Williams, J. G., "Concepts for Improving the Damage Tolerance of Composite Compression Panels," *Proceedings,* 5th DoD/NASA Conference on Fibrous Composites in Structural Design, Vol. 2, Jan. 1981, pp. 301–341.

[7] Williams, J. G., Waters, W. L., and Starnes, J. H., Jr., "Application of Kevlar/Epoxy in Damage-Tolerant Compression-Loaded Structures," 6th Conference on Fibrous Composites in Structural Design, AMMRC MS 83-2, Army Material and Mechanics Research Center, Watertown, MA, Nov. 1983.

[8] Williams, J. G., Anderson, M. S., Rhodes, M. D., Starnes, J. H., Jr., and Stroud, W. J., "Recent Developments in the Design, Testing, and Impact Damage Tolerance of Stiffened Composite Panels," NASA TM-80077, NASA Langley Research Center, Hampton, VA, April 1979.

[9] Wyrick, D. A. and Adams, D. F., "Residual Strength of Carbon/Epoxy Composite Material Subjected to Repeated Impact," *Journal of Composite Materials,* Vol. 22, Aug. 1988.

[10] Liu, D., Impact-Induced Delamination, "A View of Bending Stiffness Mismatching," *Journal of Composite Materials,* Vol. 22, July 1988.

[11] Rinehart, J. S., Rinehart and Associates, Golden, CO, Air Force Special Weapon Center, AFSWC-TR-60-7 CAD236719, 1960.

[12] Pelstring, R. M., and Madan, R. C., "Stitching to Improve Damage Tolerance of Composites," Douglas Paper 8155, Douglas Aircraft Company, 34th International SAMPE Symposium and Exhibition, Society for the Advancement of Material and Process Engineering, Reno, NV, May 1989.

Ernest F. Dost,[1] *Larry B. Ilcewicz,*[1] *William B. Avery,*[1] *and Brian R. Coxon*[2]

Effects of Stacking Sequence on Impact Damage Resistance and Residual Strength for Quasi-Isotropic Laminates

REFERENCE: Dost, E. F., Ilcewicz, L. B., Avery, W. B., and Coxon, B. R., "**Effects of Stacking Sequence on Impact Damage Resistance and Residual Strength for Quasi-Isotropic Laminates,**" *Composite Materials: Fatigue and Fracture (Third Volume), ASTM STP 1110,* T. K. O'Brien, Ed., American Society for Testing and Materials, Philadelphia, 1991, pp. 476–500.

ABSTRACT: Residual strength of an impacted composite laminate is dependent on details of the damage state. Stacking sequence was varied to judge its effect on damage caused by low-velocity impact. This was done for quasi-isotropic layups of a toughened composite material. Experimental observations on changes in the impact damage state and post-impact compressive performance were presented for seven different laminate stacking sequences. The applicability and limitations of analysis compared to experimental results were also discussed.

Post-impact compressive behavior was found to be a strong function of the laminate stacking sequence. This relationship was found to depend on thickness, stacking sequence, size, and location of sublaminates that comprise the impact damage state. The post-impact strength for specimens with a relatively symmetric distribution of damage through the laminate thickness was accurately predicted by models that accounted for sublaminate stability and in-plane stress redistribution. An asymmetric distribution of damage in some laminate stacking sequences tended to alter specimen stability. Geometrically nonlinear finite element analysis was used to predict this behavior.

KEY WORDS: composite materials, impact, graphite/epoxy, stacking sequence, residual strength prediction, compression after impact strength, toughened matrix, pulse-echo ultrasonics, fracture, fatigue (materials)

Design loads in structures made of laminated carbon fiber reinforced plastic (CFRP) are limited by degraded performance following low-velocity impact. This degradation is due to a combination of matrix and fiber failures caused by an impact event. The type and distribution of damage defines a characteristic damage state (CDS) that is a complex function of material (for example, delamination resistance and fiber strength), laminate (for example, stacking sequence), structural (for example, boundary conditions), and extrinsic (for example, impact energy) variables. A general analysis method to determine CDS currently does not exist; however, some progress has been made for specific cases [1–5].

Post-impact behavior of a laminate is related to the CDS and combined damage interactions. The most critical damage type may change as a function of load conditions. For example, matrix cracks couple with delaminations to form sublaminates that buckle under compressive loads and cause a drop in compression after impact (CAI) strength [6–8]. The

---

[1] Specialist engineer, senior specialist engineer, and senior specialist engineer, respectively, The Boeing Company, Seattle, WA 98124.

[2] Director of research, Integrated Technologies, Inc., Bothell, WA 98011.

behavior of a tensile loaded structure is relatively insensitive to this matrix damage; however, fiber failures will affect both tension [1,9] and compression strength.

The primary objective of the current work was to evaluate the effects of stacking sequence on post-impact performance for a toughened composite material. Tests were performed to determine impact CDS and CAI strength. This work also documents an evaluation of the applicability and limitations of analysis as applied to experimental results. Details of the sublaminate stability analysis method discussed throughout this study can be found elsewhere [6].

The quasi-isotropic stacking sequences used in experiments were chosen to obtain variations from the CDS found in earlier work [2,6]. One variation considered $\pi/6$-type laminates (that is, 0, 90, ± 30, and ± 60 plies). Changes in m and n for $[45_m, 90_m, -45_m, 0_m]_{ns}$ were also studied to judge the effects of increasing ply group thickness and heterogeneous stacking sequences, respectively. Finally, another class of heterogeneous stacking sequence for $\pi/4$-type laminates was chosen to create large delaminations at two through-thickness locations, effectively breaking the laminate into three sublaminates. Each sublaminate in this case had strong shear/extension and bending/twisting couplings.

Characteristic Damage State (CDS)

Many experimental studies have been performed to judge post-impact performance of composite laminates [see 10–14], yielding valuable data bases. Some of this work has resulted in semiempirical methods for predicting CAI [14]. A comprehensive review of past work was given in Ref 15. This review indicated a need to establish damage tolerance analysis capability for efficient composite structural design.

Some analysis methods to predict laminate post-impact response to an applied in-plane load have recently been developed and verified by experiments [1,6–9]. Each of these methods model impact damage as reduced stiffness inclusions that result in load redistribution. The nature of load redistribution is different for fiber and matrix damage. Fibers are major load carrying members that directly affect local stiffness when broken. Coupled transverse cracks and delaminations effectively reduce the stiffness of the damage zone as a function of sublaminate stability.

The most important steps in residual strength analysis are identification of the CDS and simulation of its post-impact response. Figure 1 shows a schematic diagram classifying CDS that have been observed in flat laminates following low-velocity impact by spherical objects. Planar and cross sectional views of CDS are given in the figure.

Symmetric Damage States

Three classes of CDS consisting of symmetric damage through the laminate cross section are shown in Fig. 1. Damage size and type (fiber, matrix, or combined) depend on variables such as delamination resistance and impact energy. The use of analysis methods from Refs 1, 6 as applied to predict CAI for laminates with symmetric damage types was discussed in Ref 7. The most common damage observed in experiments with a stacking sequence used for material screening tests (that is, $[45,0,-45,90,]_{ns}$) was matrix damage.

Plate boundary conditions, laminate thickness, and material form (for example, braided composites) are among the variables that suppress delamination, causing damage dominated by fiber failure. Fiber damage, when present, tends to concentrate at the impact site. Cairns et al. [1,9] demonstrated a method to predict the tensile behavior of a laminate with damage dominated by fiber failure. The upper limit to fiber damage diameter, which set a lower limit on residual strength, was roughly the size of a hole caused by impactor penetration.

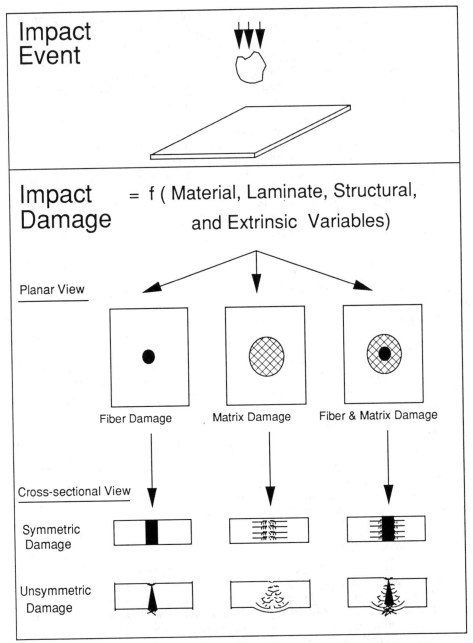

FIG. 1—*Schematic diagram classifying symmetric and asymmetric impact damage states.*

Matrix damage is also centered at the impact site but tends to radiate away from this point to a size dependent on delamination resistance. The current authors [6–8] presented a sublaminate stability approach for predicting the CAI of laminates with impact damage dominated by symmetric matrix failures. The laminates studied had a CDS consisting of sublaminates that were approximately four plies thick. Each sublaminate had similar size and ply stacking sequence. Analysis indicated that these sublaminates become unstable at the same load. Effective reduced stiffness of the sublaminates was estimated by assuming that they support only the buckling load. The failure criterion used for the sublaminate stability analysis method only considered the in-plane stress redistribution (that is, the potential for delamination growth was not modeled). Tests with tough and brittle materials yielded similar CAI versus damage diameter curves, indicating that toughness was only important during the impact event for cases studied.

The most general classification of symmetric damage involves both fiber and matrix failure. A combination of the two analysis methods [1,6] would account for both types of damage. Ilcewicz et al. [7] showed data for some composite materials that had both fiber and matrix damage. The CAI of specimens containing "small" damage sizes (that is, damage diameters less than 2.5 cm) appeared dominated by fiber failures that occur within the impact contact patch rather than the matrix damage radiating from this patch. Larger damage sizes became dominated by the matrix damage and sublaminate stability. These observations were deduced by comparing test data with predictions from the sublaminate stability analysis model [6] that assumes matrix damage dominance. The model compared very well with data for damage diameters greater than 2.5 cm, but over predicted results (by up to 20%) for smaller damage diameters.

Sublaminate stability and its influence on effective reduced stiffness depends on the thickness, stacking sequence and diameter of sublaminates. Smaller ply thickness yields thinner sublaminates, and hence, lower sublaminate stability for a given stacking sequence and damage diameter. Therefore, tests with varying ply thickness (while holding laminate thickness constant) can be used to help evaluate the competing effects of local fiber failure and sublaminate stability on reduced stiffness. Hypothetically, CAI test results should deviate from sublaminate stability predictions at different damage diameters depending on ply thickness. Tests performed with a toughened composite material were compared with analysis predictions to support this hypothesis (Figs. 2 and 3). Experimental methods used to obtain these data are the same as discussed in the following section.

Figures 2 and 3 compare sublaminate stability analysis with CAI test results for specimens with ply thicknesses equal to 0.193 and 0.150 mm, respectively. Local fiber failures, matrix cracks, and delamination were found during microscopic evaluation of cross sections taken from impacted specimens with both ply thicknesses. Fiber damage was localized to a zone roughly the size of the impactor indentation (Fig. 4), while matrix cracks and delamination spanned the full planar damage diameter (Fig. 5). This type of damage has also been observed in previous work and was most pronounced in laminates with toughened matrices [16].

Data in Fig. 2 suggest that CAI for small damage sizes may be dominated by fiber failures for laminates with ply thickness equal to 0.193 mm (that is, deviations from predictions occurred at damage diameters less than 3.2 cm). This trend is similar to that reported in Ref 7 for a toughened material with approximately the same ply thickness. Data for the thin ply thickness in Fig. 3 agreed with predictions for the entire data range (that is, damage diameters as small as 1.9 cm). Apparently, the decreased sublaminate stability associated with relatively thin ply thicknesses dominates the reduced stiffness of smaller impact damage diameters.

Asymmetric Damage States

In general, the planar view of damage size (that is, as quantified by through-transmission ultrasonics) does not give sufficient detail. Additional planar and through-thickness inspection are needed to distinguish fiber and matrix damage distributions. Pulse-echo ultrasonics and cross sectional microscopy are two methods of obtaining additional information on the CDS. Referring back to Fig. 1, asymmetric damage states are possible for each damage type discussed in the last subsection.

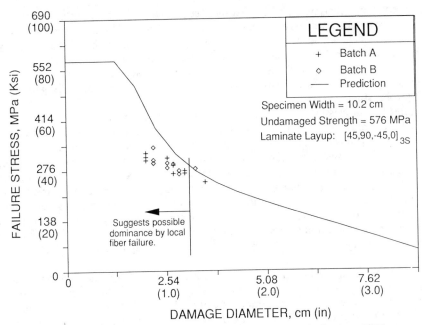

FIG. 2—*Experiments versus predictions for laminates with ply thickness = 0.193 mm.*

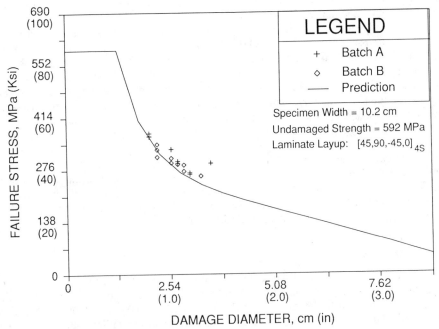

FIG. 3—*Experiments versus predictions for laminates with ply thickness = 0.150 mm.*

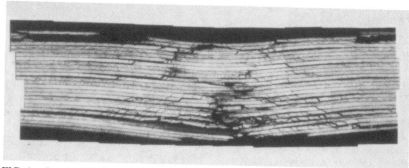

FIG. 4—*Cross section (cut through center) of impacted specimen with localized fiber failure.*

Many factors can affect the CDS symmetry. Test observations have indicated thin laminates and heterogeneous stacking sequences tend to have asymmetric CDS with damage concentrated opposite the impacted surface [13]. Very thick laminates are also expected to have asymmetric damage, but with damage concentrating closer to the impacted surface. Work by the current authors have indicated that delamination resistant materials have a stronger tendency for asymmetric CDS than brittle materials tested with the same impact variables.

Most post-impact CDS observed for laminates typically have some asymmetry, but a desire for simpler analysis methods has led to simulation of the CDS with symmetric damage whenever possible. Predictions of post-impact tensile behavior may be relatively insensitive to the symmetry of CDS providing the planar extent of fiber damage is accurately modeled. Simulation of asymmetric damage is expected to be most critical for accurate models of compression or shear performance where local and global stability can affect the apparent strength. For example, asymmetric delamination size leads to a need to model load redistribution as a function of the stability of each sublaminate. A panel imperfection caused by asymmetric CDS can also suppress local load redistribution by promoting global instability. This will be discussed in greater detail later.

Experimental Procedure

Specimen Fabrication and Machining

All specimens were fabricated from Hercules IM7/8551-7, grade 190 (fiber volume = 0.57 and ply thickness = 0.19 mm), carbon epoxy tape. Seven panel layups were fabricated including:

1. $[45,90,-45,0]_{3s}$
2. $[45_2,90_2,-45_2,0_2]_{2s}$

FIG. 5—*Cross section (cut slightly off center) of impacted specimen with delamination and transverse cracks.*

3. $[45_3,90_3,-45_3,0_3]_s$
4. $[30,60,90,-60,-30,0]_{2s}$
5. $[30,60,90,-30,-60,0]_{2s}$
6. $[45,(90,-45)_3,(0,45)_2,0]_s$
7. $[45,(0,-45)_3,(90,45)_2,90]_s$

Test specimens were obtained by laying up and curing flat panels of sufficient size to yield the required number of specimens. Laminate stacking was performed on a polyurethane-coated workbench. After stacking, the laminate was placed on a 0.64-cm-thick aluminum caul plate that had been sanded, cleaned, and coated with a 0.025-mm-thick layer of FEP film, providing a smooth, wrinkle-free surface. Bagging was performed by first placing lengths of fiberglass cord on the caul plate around the laminate circumference. Then three to four plies of 25.4-mm-wide (1-in.-w) fiberglass tape were stacked around the laminate circumference such that the cord protruded between tape and caul plate at each corner. Protrusion of the cord through the tape provides an outlet for vacuum yet minimizes resin bleed during cure. Next, another layer of FEP film followed by two to three plies of bleeder paper were placed on top of the laminate. A nylon film vacuum bag was then added, and sealed at the fiberglass tape with vacuum bag sealant. The vacuum connection was made to the bleeder paper.

Curing was performed using standard autoclave procedures. The bagged laminate was placed into an autoclave and a vacuum of 81.3 KPa was applied to the laminate. Then 586 KPa of autoclave pressure was applied, with the laminate being vented at approximately 138 to 207 KPa. The autoclave temperature was raised at a rate of 2.78°C/min to 179°C, held at temperature for 2 h, then cooled to room temperature at a rate of 2.78°C/min. The laminate was then removed from the autoclave and unbagged. Nondestructive inspection (NDI) was performed to assess laminate quality. This was done using through-transmission ultrasonics at a frequency of 5 MHz.

Compression-after-impact (CAI) specimens were machined to 10.2 by 15.2 cm. Initially, slightly oversized specimens were cut with a carbide-tipped band saw. These were then ground to the required dimensional tolerances using a water-cooled surface grinder. Coupon orientation was such that the 15.2 cm length corresponds to the 0° direction of laminate stacking sequence.

Impact Procedure

The impact procedure started by placing a coupon in the support fixture consisting of a sandwiched wood/aluminum block with a 7.62 by 12.7 cm cutout. Specimens were oriented such that the tool surface was impacted. Coupons were held with clamps on both sides along the 12.7-cm length to prevent coupon rebound during impact. The test fixture approximates simply-supported boundary conditions. A drop tower, consisting of a steel tube supported in a frame structure, was used to perform the impact event. A steel tup with a 1.59-cm-diameter spherical head was inserted in the tube such that it rested against the fixtured specimen center. Impact was performed by dropping a 5.44 kg (12 lb) weight down the tube, impacting the tup, and driving it into the laminate. The weight was caught on the rebound in order to prevent a second impact event. Different drop heights were used to obtain the desired range of impact energy. After impact, each coupon was labeled with the drop height.

Nondestructive Inspection of Impact-Damaged Coupons

Both sides of impact-damaged coupons were scanned using a time-of-flight, pulse-echo, ultrasonic technique. The NDI scans were performed using a Dynacon UDRPS ultrasonic system, operated at 10 MHz. The system graphics yielded color images of the damaged area, with each color representing a different depth through the thickness of the laminate. Color photographs of the NDI images were obtained by photographing the video screen with a 10.2 by 12.7 cm camera. Vericolor Type L and Polaroid Type 58 film were used for photographs.

The specimen impact face will be referred to as the front surface in all photographs. The other face of the specimen will be denoted the back surface. Since specimens were impacted on the laminate tool side (that is, bottom side of laminate layup), angle ply orientations appearing in front and back photographs have signs reversed from those denoted in the stacking sequence convention used for layup identification (that is, $[45,90,-45,0]_{ns}$ appears as $[-45,90,45,0]_{ns}$ in front side photographs). This point is important in interpreting results from pulse-echo scans.

Compression After Impact Testing

Post-impact tests were performed to determine CAI for a range of impact levels. Specimen dimensions were measured to the nearest 0.1 mm prior to testing. Note that different numbers of specimens were tested for each laminate layup. All tests were performed in machine stroke control using a head displacement rate of 0.13 cm/min and a 1 point/s data acquisition rate.

The CAI specimens were mounted in a side-supported fixture developed by The Boeing Company for compression residual strength tests [12]. A slight fixture modification was incorporated as described in SACMA Recommended Method SRM 2-88 [17]. The fixture was placed on a fixed compression platen in a servohydraulic universal testing frame with 22 680 kg (50 000 lb) capacity and centered using previously positioned alignment guides. A gimbaled upper platen was aligned in series with the load cell and load was zeroed. A ±1.27-cm linear variable differential transformer (LVDT) was used to measure out-of-plane deflection. The LVDT was mounted on a frame support and positioned 1 to 2 cm from the center of visible damage on the back of specimens. The LVDT position was chosen in an attempt to avoid areas where broken or buckled surface fibers would distort the measurement. Data were collected on a PDP 11/73 with an MTS 468 test processor using MTS BASIC.

A list covering drop height, impact energy, damage diameter, and failure load for all tests is given in Table 1.

Results and Discussion

This section is broken into three parts. The first discusses CAI versus drop weight impact energy curves obtained in experiments. These plots yield qualitative data on the overall effect of stacking sequence, but do not separate variables affecting impact damage resistance from those important to post-impact strength. The remaining two sections of the paper will concentrate on post-impact CDS characterization and CAI analysis predictions. By taking an approach of characterizing the CDS first, factors critical to post-impact performance were separated from those affecting damage created during the impact event. This was thought to be crucial because some factors affect both. For example, increased ply group thickness typically results in larger planar delamination areas for a given impact event, but also higher CAI for a given sublaminate area.

Although instrumented impact test methods were not used in the current study, they are recommended for future work to obtain quantitative data on energy absorbed during the impact event. This information, when coupled with microscopy and NDI, can lead to an understanding of how different variables combine to affect energy dissipation and damage accumulation during an impact event.

TABLE 1—*CAI test data for IM7/8551-7.*

Layup	Drop Height, cm	Impact Energy, J	Damage Diameter, cm	Failure Load, MN
$[45_2,90_2,-45_2,0_2]_{2s}$	30.5	16.3	3.00	213.8
$[45_2,90_2,-45_2,0_2]_{2s}$	45.7	24.4	4.42	179.6
$[45_2,90_2,-45_2,0_2]_{2s}$	61.0	32.5	5.72	156.0
$[45_2,90_2,-45_2,0_2]_{2s}$	76.2	40.7	6.70	142.2
$[45_3,90_3,-45_3,0_3]_s$	30.5	16.3	4.10	137.9
$[45_3,90_3,-45_3,0_3]_s$	45.7	24.4	5.76	100.7
$[45_3,90_3,-45_3,0_3]_s$	61.0	32.5	6.86	88.9
$[45_3,90_3,-45_3,0_3]_s$	76.2	40.7	8.11	89.4
$[45,(90,-45)_3,(0,45)_2,0]_s$	10.2	5.4	0.00	160.3
$[45,(90,-45)_3,(0,45)_2,0]_s$	20.3	10.8	2.07	160.6
$[45,(90,-45)_3,(0,45)_2,0]_s$	30.5	16.3	2.73	155.6
$[45,(90,-45)_3,(0,45)_2,0]_s$	45.7	24.4	4.31	126.3
$[45,(90,-45)_3,(0,45)_2,0]_s$	61.0	32.5	5.20	114.0
$[45,(90,-45)_3,(0,45)_2,0]_s$	76.2	40.7	6.00	100.2
$[45,(0,-45)_3,(90,45)_2,90]_s$	10.2	5.4	0.00	160.2
$[45,(0,-45)_3,(90,45)_2,90]_s$	20.3	10.8	1.95	175.1
$[45,(0,-45)_3,(90,45)_2,90]_s$	30.5	16.3	2.85	127.4
$[45,(0,-45)_3,(90,45)_2,90]_s$	45.7	24.4	3.08	115.3
$[45,(0,-45)_3,(90,45)_2,90]_s$	38.1	20.3	3.13	121.7
$[45,(0,-45)_3,(90,45)_2,90]_s$	50.8	27.1	3.85	110.7
$[45,(0,-45)_3,(90,45)_2,90]_s$	76.2	40.7	4.37	93.2
$[45,(0,-45)_3,(90,45)_2,90]_s$	63.5	33.9	4.37	101.7
$[45,(0,-45)_3,(90,45)_2,90]_s$	88.9	47.5	4.37	85.5
$[45,(0,-45)_3,(90,45)_2,90]_s$	61.0	32.5	4.42	107.9
$[45,(0,-45)_3,(90,45)_2,90]_s$	76.2	40.7	4.94	111.8
$[45,90,-45,0]_{3s}$	10.2	5.4	0.00	188.9
$[45,90,-45,0]_{3s}$	20.3	10.8	0.44	187.7
$[45,90,-45,0]_{3s}$	30.5	16.3	1.21	175.3
$[45,90,-45,0]_{3s}$	45.7	24.4	1.85	168.1
$[45,90,-45,0]_{3s}$	76.2	40.7	2.76	131.3
$[30,60,90,-60,-30,0]_{2s}$	17.8	9.5	1.24	195.4
$[30,60,90,-60,-30,0]_{2s}$	33.0	17.6	2.32	188.7
$[30,60,90,-60,-30,0]_{2s}$	33.0	17.6	2.23	175.2
$[30,60,90,-60,-30,0]_{2s}$	40.6	21.7	2.58	193.1
$[30,60,90,-60,-30,0]_{2s}$	48.3	25.8	3.18	168.2
$[30,60,90,-60,-30,0]_{2s}$	55.9	29.8	3.47	147.1
$[30,60,90,-60,-30,0]_{2s}$	66.0	35.3	3.39	144.7
$[30,60,90,-60,-30,0]_{2s}$	76.2	40.7	3.63	131.1
$[30,60,90,-60,-30,0]_{2s}$	91.4	48.8	3.77	123.3
$[30,60,90,-60,-30,0]_{2s}$	121.9	65.1	3.52	115.7
$[30,60,90,-30,-60,0]_{2s}$	15.2	8.1	0.00	176.6
$[30,60,90,-30,-60,0]_{2s}$	25.4	13.6	1.85	173.7
$[30,60,90,-30,-60,0]_{2s}$	35.6	19.0	2.85	148.5
$[30,60,90,-30,-60,0]_{2s}$	45.7	24.4	2.60	142.4
$[30,60,90,-30,-60,0]_{2s}$	61.0	32.5	3.10	123.5
$[30,60,90,-30,-60,0]_{2s}$	76.2	40.7	3.59	123.3

CAI Test Results

Figure 6 shows CAI data as a function of the drop weight impact energy (that is, drop height times drop weight). Data scatter for each type of laminate layup was small compared to the total range of results. This indicated that laminate stacking sequence is critical to CAI. Note that all other material, laminate, structural, and extrinsic variables were held constant for the tests. The one exception was for the $[45_2,90_2,-45_2,0_2]_{2s}$ laminate that had 32 plies instead of 24 plies. Despite the additional thickness, this laminate did not have the highest CAI for a given impact energy, again indicating the importance of stacking sequence.

The strongest effect of stacking sequence on CAI occurred for drop weight impact energies between 15 and 25 J in Fig. 6. The effects of stacking sequence diminished at lower impact energies because stability drives coupon behavior for small damage sizes. Any relationship between impact energy and the sensitivity of CAI to stacking sequence is expected to be unique to the material studied (for example, a brittle material would be sensitive in a different range of impact energies). The calculated undamaged specimen buckling stresses for each stacking sequence are also shown in Fig. 6. These stresses were determined by a finite element code that uses an eight-node, reduced Ahmad plate bending element that accounts for transverse shear [18]. Note that the highest values of CAI data are bounded by specimen stability predictions (that is, data trends level off for low-impact energies).

Laminates with the stacking sequence $[45,90,-45,0]_{3s}$ were found to yield relatively high CAI per given impact energy. The only variation between this and the stacking sequence $[45,0,-45,90]_{3s}$ used in past work with IM7/8551-7 material [6,7], was a switch in the location of 0 and 90° plies. Superposing past results with those in Fig. 6 indicated little difference between the two stacking sequences. Both laminates have plies stacked in sequence such that

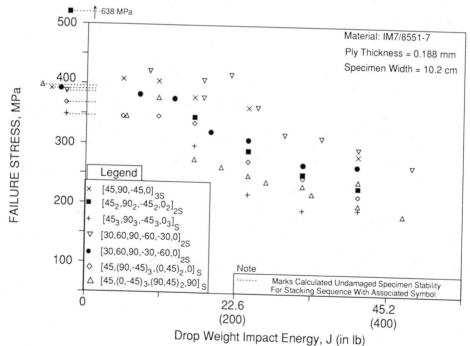

FIG. 6—*Post-impact compression performance as a function of stacking sequence.*

the orientation of each ply always differs by a constant angle from the previous ply (moving from top to bottom of the laminate). The ply stacking sequence of $[45,90,-45,0]_{3s}$ rotates in a $+45°$ sense, while $[45,0,-45,90]_{3s}$ rotates in a $-45°$ sense.

The stacking sequence $[30,60,90,-60,-30,0]_{2s}$ yielded the highest CAI results in Fig. 6 for the impact energies studied. A small variation to this stacking sequence was also studied with the layup $[30,60,90,-30,-60,0]_{2s}$ (that is, -60 and $-30°$ ply locations were switched). Results for this stacking sequence indicate a significant degradation in the relationship between CAI and impact energy. The plies in $[30,60,90,-60,-30,0]_{2s}$ are stacked in a sequence such that each ply is rotated $+30°$, relative to the previous ply. The -30 and $-60°$ plies in $[30,60,90,-30,-60,0]_{2s}$ disrupt this sequence, apparently degrading performance.

The stacking sequence $[45_2,90_2,-45_2,0_2]_{2s}$ yielded CAI results that fall in the middle of data shown in Fig. 6. The ply group thickness is doubled for this laminate, while holding the sequence of ply orientations the same as in $[45,90,-45,0]_{3s}$. Note that the $[45_2,90_2,-45_2,0_2]_{2s}$ laminate was thicker than all others studied. Despite the added thickness, which should help impact damage resistance, this laminate had CAI that fell below that of $[45,90,-45,0]_{3s}$ for a given impact energy. The increased ply group thickness appears to have an overall negative effect on CAI performance. Results for $[45_3,90_3,-45_3,0_3]_s$, which fall at the bottom of data in Fig. 6, substantiates this tendency. Similar results were reported in the literature for a brittle material [13].

The CAI tended to increase as ply group thickness decreased for $[45_3,90_3,-45_3,0_3]_s$, $[45_2,90_2,-45_2,0_2]_{2s}$, and $[45,90,-45,0]_{3s}$ laminates shown in Fig. 6. This trend may be misleading because both ply group thickness and the heterogeneity of stacking sequences were altered at the same time. This was done to keep laminate thickness in a range between 0.40 and 0.65 cm. Since delaminations do not form between adjacent plies with the same orientation, laminates with thicker ply groups have less planes for delamination. This resulted in large planar damage areas. This result, coupled with the observation that the first level of damage in toughened composites is smaller than remaining levels, promoted asymmetric CDS through-the-thickness of heterogeneous laminates. The asymmetric CDS may have promoted different failure modes in $[45_2,90_2,-45_2,0_2]_{2s}$ and $[45_3,90_3,-45_3,0_3]_s$ than in $[45,90,-45,0]_{3s}$ laminates. This will be discussed later in comparisons with analysis.

Two other heterogeneous laminate stacking sequences, $[45,(90,-45)_3,(0,45)_2,0]$ and $[45,(0,-45)_3,(90,45)_2,90]_s$, were also tested. Both performed below the average of those shown in Fig. 6. These stacking sequences were designed to resist delamination in all but two locations through the laminate thickness using a "strength of materials" approach to CDS formation as discussed in Ref 2. The two laminate stacking sequences are identical, with the exception of a switch between 0 and 90° plies. The layup with 0° plies stacked towards the center of the laminate did somewhat better than the other. This may be due to stacking the 0° plies such that they become part of a more stable sublaminate located at the CDS center.

Note that some past CAI tests performed with the layup $[45,(90,-45)_3,(0,45)_2,0]_s$ indicated an opposite trend in CAI performance for a brittle composite material (T300/934). Two explanations seem possible. First, the added interlaminar toughness of IM7/8551-7 coupled with the stacking sequence generated delamination resistance of the $[45,(90,-45)_3,(0,45)_2,0]$ layup may have led to fiber failure as an alternative mechanism of impact energy dissipation. A second explanation was discussed earlier when it was noted that the CDS for heterogeneous IM7/8551-7 layups were asymmetric, conceivably promoting a specimen stability failure mode.

NDI Test Results

Although much damage is seen when inspecting an impacted laminate, only damage thought to be of structural consequence is considered in our discussions of CDS. The CDS structure is generally independent of impact energy, within certain bounds. For example, it was observed that a parameter defined by normalizing the delamination area at each ply interface by the total delamination area (that is, sum of delamination areas at all interfaces) was independent of impact energy for a given set of impact variables (for example, stacking sequence and boundary conditions). Therefore, CDS that spans a total area that is close to circular can be quantified by an independent length variable such as damage diameter.

The CDS for a set of impact variables used in material screening tests was described in earlier work [6,7]. These tests use a $[45,0,-45,90]_{ns}$ laminate stacking sequence. Since impact occurs on the laminate tool side, the stacking sequence becomes $[-45,0,45,90]_n$, when defining CDS. As discussed earlier, any fiber damage caused by impact tends to concentrate at the core of the CDS. A network of matrix cracks and delaminations comprise the remainder of the CDS. Delaminations at each ply interface are connected to those at neighboring ply interfaces by transverse matrix cracks. In a planar view, double-lobed delaminations formed at each interface. These delaminations are wedge shaped due to the $\pi/4$ difference in orientation of neighboring plies, breaking the CDS into octants. The ply orientation angles increase in $\pi/4$ increments from the impacted surface to the center. The stacking sequence and CDS is reflected at the center. Ply orientations decrease by $\pi/4$ with each ply from the center to the back side. This pattern causes a CDS with interconnected delaminations spiraling toward the center, reversing direction, and proceeding out toward the back side.

The CDS just described splits the laminate into separate sublaminates. These sublaminates are connected in a fashion similar to a spiral staircase, but are conceptualized as circular disks to simplify the analysis. The sublaminates near the outer surfaces vary in thickness from 2 to 5 plies. The next set of sublaminates are 4 plies in thickness with stacking sequence varying stepwise around the damage. This type of sublaminate can repeat several times, depending on the number of plies in the stacking sequence. Damage that occurs approaching both sides of the laminate midplane results in two discontinuous sublaminates and a symmetric core sublaminate that varies in thickness from two to eight plies. The total number of sublaminates for a $[45,0,-45,90]_{ns}$ laminate stacking sequence is $(2n+1)$. This can be generalized for other repeating stacking sequences that increment by either decreasing or increasing ply angles if a sum of the difference between adjacent angles in the repeat element equals zero (that is, $[\alpha,\beta,\phi, \ldots, \theta]_{ns}$ where $\{\beta - \alpha\} + \{\phi - \beta\} + \cdots + \{\alpha - \theta\} = 0.0$). Absolute values of each difference should also not exceed 90°.

Figure 7 presents both front and back surface pulse-echo scans of impact damage in toughened (IM7/8551-7) and brittle (T300/934) laminates built with the stacking sequences $[45,0,-45,90]_{3s}$ and $[45,90,-45,0]_{3s}$, respectively. Higher impact energies were used with IM7/8551-7 to get damage sizes similar to T300/934. The varying shades of grey in the damage region indicate delaminations at different through-thickness locations, with darker shades indicating shallower delamination depth.

Delaminations appearing near the front surface in IM7/8551-7 are small relative to those seen through the rest of the laminate thickness. This is the main difference between T300/934 and IM7/8551-7 matrix damage CDS. The only other difference is eliminated by accounting for opposite directions of CDS rotation for the two stacking sequences. Referring back to Fig. 1, photos of CDS in Fig. 7 suggest that the matrix damage in T300/934 is symmetric while that for IM7/8551-7 is slightly asymmetric. Post-impact tests indicate that the small amount of asymmetry in the CDS of IM7/8551-7 makes little difference in post-impact performance for $[45,0,-45,90]_{ns}$, $n \geq 3$ [6,7].

To further demonstrate the concept of CDS in this study, a laminate with ply angles based on a $\pi/6$ division of a circle was generated. The stacking sequence $[30,60,90,-60,-30,0]_{2s}$ is similar to the stacking sequence previously described (that is, a constant change in ply orientation, $+30°$, is used for each lamina added to the stack). The NDI scans in Fig. 8 show that the delaminations created in this laminate also have a stair-stepped appearance. Each step

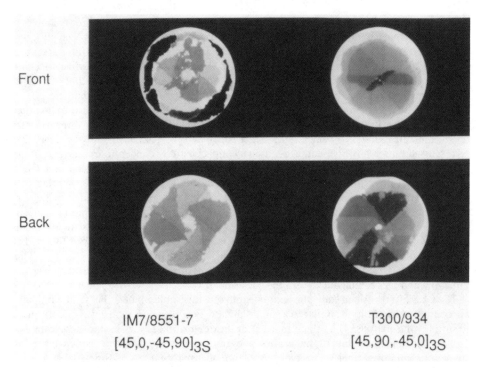

Front

Back

IM7/8551-7
[45,0,-45,90]₃S

T300/934
[45,90,-45,0]₃S

FIG. 7—*A comparison of front and back surface pulse-echo scans for two material types.*

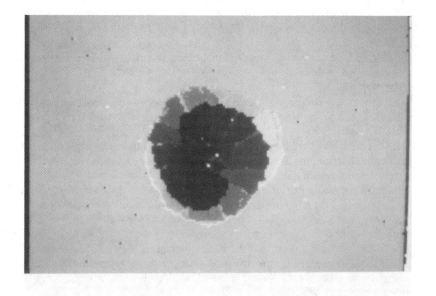

[30,60,90,-60,-30,0]₂S

FIG. 8—*A pulse-echo scan for [30,60,90, − 60, − 30,0]₂s.*

appears as a $\pi/6$ wedge. A six-ply-thick basic repeat sublaminate is created by matrix damage in a $[30,60,90,-60,-30,0]_{ns}$ laminate. Since $n = 2$ in this case, the total number of sublaminates through the thickness is five. Note that delamination size near the front surface increases with depth, reaching maximum size after a 180° rotation. The same trend was seen in Fig. 7.

Figure 9 presents NDI scans for two similar stacking sequences, $[45_2,90_2,-45_2,0_2]_{2s}$ and $[45_3,90_3,-45_3,0_3]_s$. Each of these is a variation of the $\pi/4$ stacking sequence previously described (that is, the location of 0 and 90° plies are switched and ply group thickness has been increased). The NDI scans are shown for both laminates subjected to four different impact energies. The damage size increases with impact energy, but the CDS remains constant.

The CDS of the $[45_2,90_2,-45_2,0_2]_{2s}$ laminate is similar to that of the $[45,0,-45,90]_{3s}$ laminate described earlier. The one difference is in the direction of damage rotation due to a switch in 0 and 90° ply group locations. The $\pi/4$ wedge-shaped delaminations occur at interfaces of dissimilar plies and not between those of the same orientation. As a result, sublaminate thickness doubles (that is, essentially the same result as increasing ply thickness). A total of five sublaminates occur for $[45_2,90_2,-45_2,0_2]_{2s}$.

A highly asymmetric damage state can be seen in Fig. 9 for the $[45_3,90_3,-45_3,0_3]_s$ laminate. When counting plies starting from the impact surface, large delaminations occur in three wedge locations; between plies 15 and 16, plies 18 and 19, and plies 21 and 22. As was the case in the preceding examples, $[45_3,90_3,-45_3,0_3]_{ns}$ has ply angles that change sequentially through the laminate thickness. However, a high degree of heterogeneity in stacking sequence is brought on by choosing $n = 1$. This leads to the asymmetrical CDS shown in Fig. 9. As with other IM7/8551-7 layups just discussed, the first group of delaminations that occur near the impact surface are small. In addition, a complete rotation of the CDS does not occur before reaching the laminate center. This apparently causes most impact energy dissipation to occur in creation of large delaminations near the back surface (that is, front surface delaminations are insignificant in size when compared to those in the back half of the laminate).

The large delaminations occurring in the back half of the $[45_3,90_3,-45_3,0_3]_s$ laminate effectively break it into two sublaminates. The sublaminate located toward the back side has octants containing 0, 3, 6, and 9 plies while the thicker, base sublaminate has octants containing 24, 21, 19, and 15 plies, respectively. The structure of these sublaminates were used in finite element analysis to be described later.

Figures 7 through 9 showed that the planar size of delaminations appearing closest to the impacted surface were relatively small compared to delaminations in the rest of the laminate. Similar laminate thicknesses were considered for all cases, resulting in the small delaminations occurring at different depths for laminates with thicker ply groups. Whether or not this is unique to a tough material, as was the case for laminates in Fig. 7, remains to be shown with additional studies.

The final stacking sequences studied were designed to create large delaminations at two through-thickness locations and suppress them at other interfaces. NDI scans for the two stacking sequences, $[45,(90,-45)_3,(0,45)_2,0]_s$ and $[45,(0,-45)_3,(90,45)_2,90]_s$, and three impact energy levels are presented in Fig. 10. Again, the total size of the damage increases with impact energy; however, the relative sizes of individual delaminations stay commensurate to overall damage size for a given impact energy. Delaminations occur between plies 7, 8, and 9 and plies 17, 18, and 19 for both stacking sequences. A change in the shape and orientation of delaminations between the two laminates is apparent in these photos.

Delaminations in Fig. 10 separate both laminates into three sublaminates. All of these sublaminates are highly unbalanced, causing strong shear/extensional and bending/twisting coupling. This is thought to have a strong influence on postimpact performance. The CDS for both of these laminates is again seen to be asymmetrical, with significantly larger delaminations occurring in the back half of the laminate.

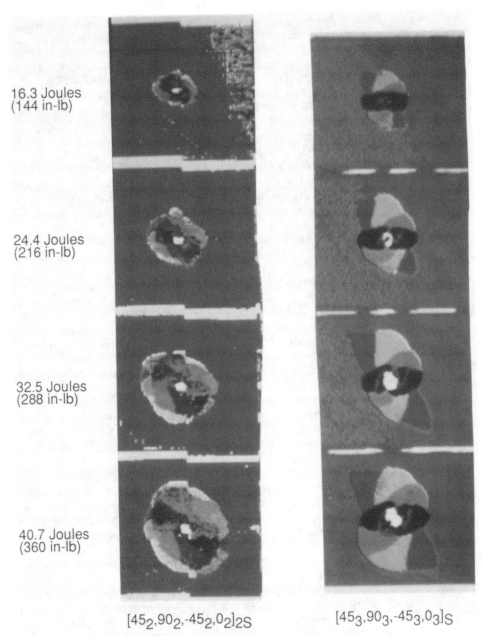

16.3 Joules
(144 in-lb)

24.4 Joules
(216 in-lb)

32.5 Joules
(288 in-lb)

40.7 Joules
(360 in-lb)

$[45_2,90_2,-45_2,0_2]_{2S}$ $[45_3,90_3,-45_3,0_3]_S$

FIG. 9—*Pulse-echo scans for* $[45_2,90_2,-45_2,0_2]_{2s}$ *and* $[45_3,90_3,-45_3,0_3]_s$.

One significant difference in the $[45,(90,-45)_3,(0,45)_2,0]_s$ and $[45,(0,-45)_3,(90,45)_2,90]_s$ laminates, is the location of the major load bearing plies. In $[45,(90,-45)_3,(0,45)_2,0]_s$, the $0°$ plies are located in the central core sublaminate; and in $[45,(0,-45)_3,(90,45)_2,90]_s$, they appear in surface sublaminates. Since the core is relatively thick and nearly symmetric, it would be expected to remain relatively stable under compressive loading. Buckling of surface sublami-

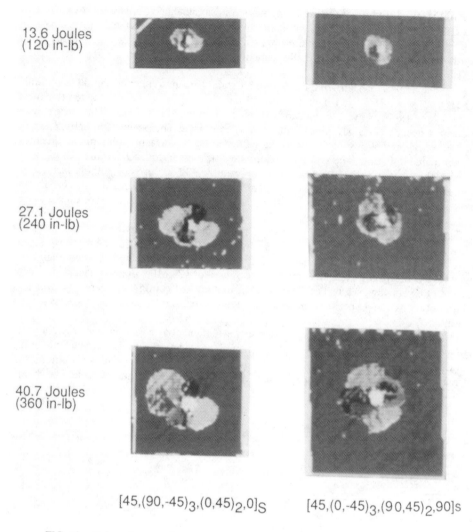

13.6 Joules
(120 in-lb)

27.1 Joules
(240 in-lb)

40.7 Joules
(360 in-lb)

[45,(90,-45)$_3$,(0,45)$_2$,0]$_S$ [45,(0,-45)$_3$,(90,45)$_2$,90]$_S$

FIG. 10—*Pulse-echo scans for* [45,(90,−45)$_3$,(0,45)$_2$,0] *and* [45,(0,−45)$_3$,(90,45)$_2$,90]$_s$.

nates is expected to be most detrimental to the CAI of [45,(0,−45)$_3$,(90,45)$_2$,90]$_s$ specimens. This appeared to be the case (see Fig. 6).

Comparisons with Analysis

The analysis method used for comparison with experiments is documented in Refs 6 and 7. In summary, five basic steps are followed in applying the method. First, the CDS is identified and simulated as a sublaminate with ply stacking sequence and thickness representing an average of those appearing in the real CDS. Second, a sublaminate stability analysis is performed using damage diameter as an independent variable characterizing the planar size of the CDS. This is done using a modification to the buckling analysis method described in Ref 19. The modification accounts for sublaminates with asymmetric ply stacking sequences [6,7]. Third, effective reduced stiffness of the impact damage zone is calculated using results from sublaminate stability analysis. Fourth, the in-plane stress concentration associated with the reduced stiffness is determined. Finite elements are used for this step in order to account for specimen width/damage size interactions. Finally, a maximum strain failure criteria is applied to predict CAI.

Note that Steps 2 through 4 of the sublaminate stability analysis method should be modified for the most general CDS in which sublaminate parameters (for example, diameter, thickness, and stacking sequence) vary significantly through the laminate thickness. The more general model is currently being developed under NASA contract at The Boeing Company.

Theoretical predictions for four stacking sequences with increasing sublaminate thickness are compared in Fig. 11. The average sublaminate thickness, τ, for each laminate in Fig. 11 is equal to the repeating ply group thickness (for example, $[30,60,90,-60,-30,0]_{ns}$ has τ equal to the total thickness of six plies). The residual strength is shown to increase with τ for a given damage size. These predictions assume that n is large enough such that the repeating sublaminates dominate the laminate behavior (that is, approximated as $n \geq 3$). Note that all curves in Fig. 11 show a strong interaction between specimen width and damage size.

The compression strength used in Fig. 11 was measured using a face-supported compression test fixture for IM7/8551-7 laminates with a $[45,90,-45,0]_{3s}$ stacking sequence. Figure 11 assumes a constant undamaged compressive strength of 478 MPa, due to a lack of test data for the other stacking sequences. The effect of undamaged compression strength, which is commonly considered a performance parameter rather than a material property, was reviewed in Ref 7.

As discussed in the previous section, ply stacking sequence of wedges that comprise the average sublaminate vary sequentially depending on planar location. This is characteristic of the spiral stair-stepped CDS of laminates in which the ply stacking sequence has orientations that increase or decrease sequentially [2,6]. Each layup in Fig. 11 has this characteristic. The chang-

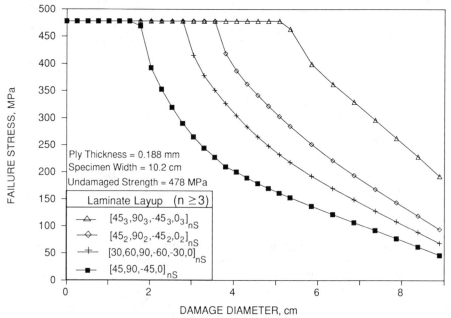

FIG. 11—*Sublaminate stability/reduced stiffness predictions of CAI for IM7/8551-7.*

ing stacking sequence is simulated for stability analysis by calculating a buckling load for each permutation of the stacking sequence that appear in different wedges of the sublaminate. The sublaminate buckling load is then taken as the average of these calculations.

A limit was placed on the applicability of curves in Fig. 11 by the requirement that $n \geq 3$. Each curve in Fig. 11 uses τ equal to that of the repeating ply group. As discussed earlier, the number of sublaminates for the class of laminates shown in Fig. 11 is $(2n + 1)$. Dividing the total number of plies in the laminate by $(2n + 1)$ yields τ. When $n \geq 3$, τ is greater than or equal to 0.86 times the repeating group thickness. This was felt to be a reasonable cutoff for assuming the full repeat group thickness in calculations. This assumption has been shown to yield good predictions of experimental data [6-8]. A close evaluation of these studies indicates that the best comparisons with data occurs with the largest values of n.

Values of $n < 3$ result in laminate CDS dominated by surface and core sublaminates with thickness significantly less than that of the repeating ply group. This effectively lowers the residual strength, relative to predictions in Fig. 11. Since most of the laminate stacking sequences studied in the current work had values of $n < 3$, the analysis method was "scaled" to account for smaller τ. All steps in the scaled analysis are the same as discussed in Ref 6 with the exception that the ply thickness used in calculations was reduced so that sublaminate thickness equals τ for a given laminate. For example, calculations for $[30,60,90, -60, -30,0]_{2s}$ would use τ equal to 0.8 times the thickness of six plies. As before, sublaminate buckling load is taken as the average of calculations for each permutation of the repeating six ply stacking sequence. The only difference is in the use of a scaled ply thickness in each calculation.

Scaled analysis results, assuming circular sublaminates, were compared to experimental data for the first four laminates listed in the legend of Fig. 6. The total planar damage area (that is, plan view C-scan damage area) from NDI experiments was converted to an equivalent damage diameter, assuming a circular planar damage shape. This was an accurate representation of the shape in all cases except the $[45_3,90_3, -45_3,0_3]_s$ laminate.

Figure 12 shows a comparison between $[45,90, -45,0]_{3s}$ experimental data and predictions. Data from past work [6] with $[45,0, -45,90]_{3s}$ is also shown in the figure. Analysis results are identical for both layups. Note that there is little difference in CAI versus damage diameter data for the two stacking sequences. The unscaled and scaled predictions shown in Fig. 12 indicate the difference between assuming a sublaminate thickness equal to 4 and 3.43 plies, respectively. Considering the data scatter, both predictions appear to compare equally well with experiments. As discussed, the difference in analysis curves would continue to decrease for $n > 3$ and increase for $n < 3$.

Figure 13 shows a comparison between experimental data and predictions for the $[30,60,90, -60, -30,0]_{2s}$ laminate. The analysis predictions compare very well with the data, indicating that the scaling procedure is also accurate for this laminate and the range of damage diameters tested. Since relatively small damage diameters were studied with this laminate, none of the results were strongly affected by specimen width.

As discussed earlier, the $[30,60,90, -60, -30,0]_{2s}$ laminate layup had the best overall post-impact strength versus impact energy. This may indicate an optimum sublaminate structure when considering the combined effects of post-impact strength and impact damage resistance. When the CAI failure mechanism is the same as assumed by a sublaminate stability approach, thick sublaminates yield the highest CAI per planar damage area as was seen in Fig. 11. Experimental trends from this study showed that the planar area of each sublaminate created for a given impact energy was inversely related to the total number of sublaminates for a given stacking sequence. This may suggest that the total amount of delamination area created per unit thickness (that is, determined by adding the delamination area between all sublaminates through the thickness) is relatively independent of stacking sequence. The thin sublaminate structures yield the most delamination sites per unit laminate thickness, and hence, the small-

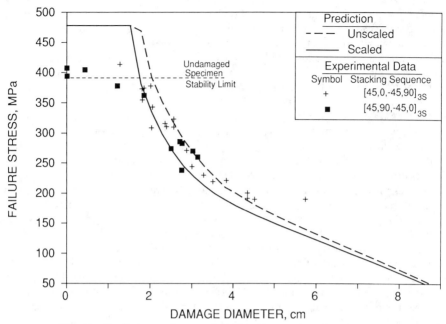

FIG. 12—*Predictions and test results for [45,90,−45,0]ₛ and [45,0,−45,90]ₛ.*

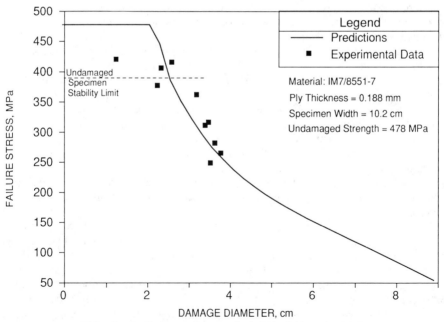

FIG. 13—*Predictions and test results for [30,60,90,−60,−30,0]₂ₛ.*

est planar areas for each sublaminate. As a result, sublaminate thickness indirectly affects another stability parameter (that is, sublaminate area) during the impact event. Additional work is recommended to confirm this hypothesis and to develop a method for determining the optimum stacking sequence for a given laminate thickness.

A limited amount of data was collected for $[45,90,-45,0]_{3s}$ and $[30,60,90,-60,-30,0]_{2s}$ specimens without damage and with small damage sizes (that is, those sizes that had a sublaminate buckling stress higher than undamaged strength). These data points were used to verify predictions of undamaged specimen stability that are also shown in Figs. 12 and 13.

Figures 14 and 15 show analysis and test results for $[45_2,90_2,-45_2,0_2]_{2s}$ and $[45_3,90_3,-45_3,0_3]_s$ laminates, respectively. Experimental damage areas measured for these laminates were much larger than those of $[45,90,-45,0]_{3s}$ and $[30,60,90,-60,-30,0]_{2s}$ specimens for a given impact energy. Despite a limited amount of test results, data trends are shown to vary significantly from sublaminate stability predictions for these two laminates. The most striking difference occurs for large damage areas where the finite width effect is not evident in data. Past work with brittle laminates and the $[45,0,-45,90]_{ns}$ ($n \geq 3$) verified that finite width effects are strong for CAI specimens [6,7]. The behavior in Figs. 14 and 15 suggest a different specimen failure mechanism.

In an attempt to better interpret experimental results shown in Fig. 15, a geometrically nonlinear finite element analysis was performed using STAGSC-1 [20]. The damage state shown in Fig. 9 for the $[45_3,90_3,-45_3,0_3]_s$ stacking sequence was modeled discretely, with stacking sequence varying by octant. The damage was modeled as circular to simplify the analysis, although as seen in Fig. 9, it would be more appropriately described as elliptical. Experimental data were quantified by the NDI planar area for comparisons to analysis.

The finite element grid used for models is shown in Fig. 16. This grid has 1827 nodes, 96 triangular (STAGS 322) elements, and 1344 quadrilateral (STAGS 411) elements, for a total

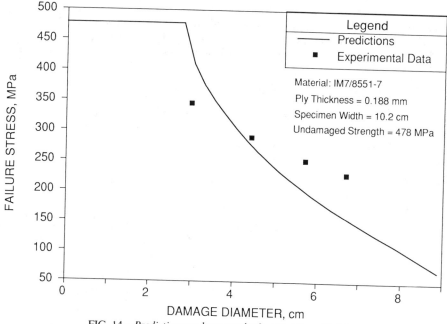

FIG. 14—*Predictions and test results for $[45_2,90_2,-45_2,0_2]_{2s}$.*

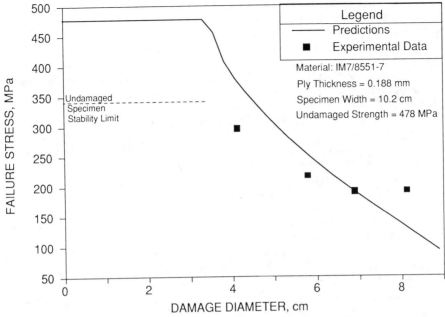

FIG. 15—*Predictions and test results for* $[45_3, 90_3, -45_3, 0_3]_s$.

of 12 699 active degrees of freedom. The 5.04-cm-diameter sublaminate was modeled with a second level of elements 0.127 mm above the base laminate. The sublaminate was connected to the laminate at its edge. Specimen boundary conditions were W and RV fixed on the sides; U, V, W, RU, and RW fixed on the bottom; and V, W, RU, and RW fixed on the top. The displacements U, V, W correspond to the X, Y, Z directions, respectively; and RU, RY, RW correspond to rotation about the X, Y, Z axes. A uniform U displacement, applied along the top of the model in the $-X$ direction, was used for loading.

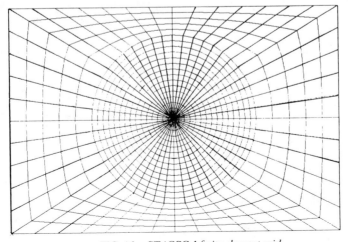

FIG. 16—*STAGSC-1 finite element grid.*

Nonlinear analysis of an undamaged specimen with two different imperfection amplitudes were also performed for the purpose of comparison. The Mode 1 eigenvector from a bifurcation buckling analysis was used with amplitudes of 1 and 10% of the laminate thickness (0.0432 and 0.432 mm, respectively) for the initial imperfection.

Geometrically nonlinear analysis performed on the discrete sublaminate model demonstrated that specimen stability was affected by sublaminate stability. Outward buckling in the +Z direction of the sublaminate forced overall coupon instability in this direction. Kardomateas [21] predicted and confirmed similar postbuckling behavior for composites with thin film delaminations.

Predicted postbuckling behavior for models with and without damage are shown in Fig. 17. The shape of the out-of-plane displacement versus load curve for the model with damage indicates that local sublaminate instability acts as a strong "imperfection." For example, results from the undamaged analysis with a 0.432-mm imperfection amplitude are well below those of the damage model.

Figure 18 shows LVDT measurements of out-of-plane specimen displacements. The amount of "imperfection" imparted to the specimen increases with damage size. The damaged "main laminate" stability curve in Fig. 17 falls between experimental data curves for 13.2 and 26.1 cm^2 damage areas in Fig. 18. This is expected since the damage area used in the model was 20.0 cm^2. Note that the out-of-plane displacement corresponding to a surface stress of 478 MPa (undamaged compressive strength) in the model was 0.95 mm. This also compares well with the experimental data for 13.2 and 26.1 cm^2.

The post-impact behavior of $[45_3,90_3,-45_3,0_3]_s$ laminates may best be termed a change in specimen stability rather than residual strength. This is thought to be due to the asymmetric CDS. An apparent limit to the effect of damage on specimen stability is reached for the larger damage sizes shown in Fig. 18. This pattern is also demonstrated by a leveling off of the appar-

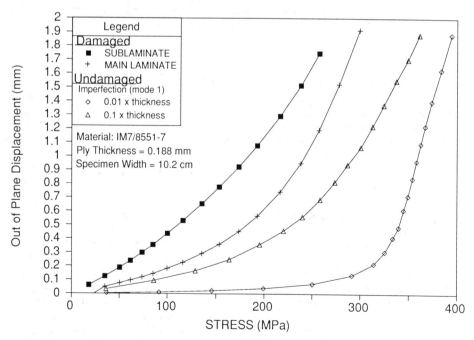

FIG. 17—*Damaged and undamaged specimen stability predictions for* $[45_3,90_3,-45_3,0_3]_s$.

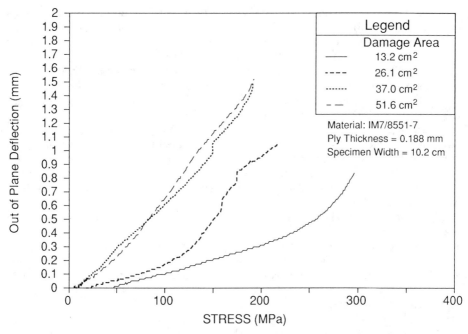

FIG. 18—*Experimental measurements of specimen stability for* $[45_3, 90_3, -45_3, 0_3]_s$.

ent CAI versus damage diameter data shown in Fig. 15. Specimen failure that relates to stress distribution around a soft inclusion would continue to decrease for larger damage areas due to the strong finite width effects (for example, predictions in Fig. 15).

Conclusions

Characteristic damage states (CDS) were classified for impacted laminates. A planar classification included local fiber failures, matrix damage, and a combination of both. Symmetric and asymmetric damage size distributions through the laminate thickness were considered in each case. Results from tests with a toughened material and small damage sizes showed that local fiber failures and matrix damage found in the CDS compete for dominance of compression after impact strength (CAI).

Specimens for seven different quasi-isotropic laminate stacking sequences were impacted with drop weights and tested for CAI. All but one of the laminates had the same thickness. Results from plots of CAI versus drop weight impact energy indicated a strong influence of stacking sequence. Predictions of undamaged specimen stability were shown to form an upper bound on measured CAI results.

The CDS for each laminate stacking sequence was determined with the help of ultrasonics. Pulse-echo scans for a given stacking sequence showed that the ratios of individual ply delamination areas to the total planar damage area was close to a constant for the range of impact energies studied. Laminates with homogeneous stacking sequence and ply orientation angles that increment from ply to ply in a repeatable pattern were found to yield a sublaminate size distribution that was close to symmetric through the laminate thickness. The total number of sublaminates for this class of stacking sequence was found to be $(2n + 1)$, where $2n$ is the number of repeating ply groups in the laminate (for example, $2n = 8$ for $[45, 90, -45, 0]_{4s}$).

The damage size distribution for heterogeneous stacking sequences (that is, laminate with plies of same orientation stacked together, such as $[45_3,90_3,-45_3,0_3]_s$) were found to be asymmetric through the thickness.

An analysis method described in earlier work [6,7] was compared with CAI experimental results for five stacking sequences. This method, which accounts for the in-plane stress redistribution following sublaminate buckling, was originally limited to stacking sequences with ($n \geq 3$). The method was modified to estimate behavior in laminates with ($n < 3$). Predictions were found to be accurate for $[45,90,-45,0]_{3s}$, $[45,0,-45,90]_{3s}$, and $[30,60,90,-60,-30,0]_{2s}$ laminates. The $[45_2,90_2,-45_2,0_2]_{2s}$ and $[45_3,90_3,-45_3,0_3]_s$ laminates did not compare as well with the analysis. A geometrically nonlinear finite element analysis that discretely modeled damage in $[45_3,90_3,-45_3,0_3]_s$ specimens indicated that the large asymmetric CDS distributions alter specimen stability. Test results on out-of-plane displacements and failure loads confirmed this prediction.

Acknowledgments

The authors wish to acknowledge J. Gosse, P. Mori, J. Quinlivan, P. Smith, and P. Whalley of The Boeing Company for technical support. Preparation of the manuscript, analysis, and $\pi/6$-type laminate tests were funded by NASA (Contract NAS1-18889, under the technical leadership of W. T. Freeman). The remaining experimental data bases were provided by The Boeing Company.

Use of commercial products or names of manufacturers in this report does not constitute official endorsement of such products or manufacturers, either expressed or implied, by The Boeing Company or The National Aeronautics and Space Administration.

References

[1] Cairns, D. S., "Impact and Post-Impact Response of Graphite/Epoxy and Kevlar/Epoxy Structures," PhD thesis, Telac Report #87-15, Dept. of Aeronautics and Astronautics, Massachusetts Institute of Technology, Cambridge, MA, Aug. 1987.

[2] Gosse, J. H. and Mori, P. B. Y. in *Proceedings,* Third Technical Conference of American Society for Composites, Technomic Publishing Co., Lancaster, PA, 1988, pp. 344–353.

[3] Shivakumar, K. N., Elber, W., and Illg, W., *Journal,* American Institute of Aeronautics and Astronautics, March 1985, pp. 442–449.

[4] Wu, H. T., "Impact Damage of Composites," PhD thesis, Dept. of Aeronautics and Astronautics, Stanford University, Stanford, CA, 1986.

[5] Zukas, J. A., Nicholas, T., Swift, H. F., Greszczuk, L. B., and Curran, D. R., *Impact Dynamics,* Wiley, New York, 1982.

[6] Dost, E. F., Ilcewicz, L. B., and Gosse, J. H. in *Proceedings,* Technical Conference of American Society for Composites, Technomic Publishing Co., Lancaster, PA, 1988, pp. 354–363.

[7] Ilcewicz, L. B., Dost, E. F., and Coggeshall, R. L. in *Proceedings,* Twenty-First International SAMPE Technical Conference, Society for Advancement of Material and Process Engineering, Covina, CA, 1989, pp. 130–140.

[8] Avery, W. B. in *Proceedings,* Twenty-First International SAMPE Technical Conference, Society for Advancement of Material and Process Engineering, Covina, CA, 1989, pp. 141–147.

[9] Cairns, D. S. and Lagace P. A., "Residual Tensile Strength of Graphite/Epoxy and Kevlar/Epoxy Laminates with Impact Damage," TELAC Report 88-3, Massachusetts Institute of Technology, Cambridge, MA, 1988.

[10] Walter, R. W., Johnson, R. W., June, R. R., and McCarty, J. E. in *Fatigue of Filamentary Composite Materials, ASTM STP 636,* R. Evans, Ed., American Society for Testing and Materials, Philadelphia, 1977, pp. 228–247.

[11] Starnes, J. H., Rhodes, M. D., and Williams, J. G., "Effect of Impact Damage and Circular Holes on Compression Strength of a Graphite/Epoxy Laminate," NASA-TM-78796, NASA Langley Research Center, Hampton, VA, Oct. 1978.

[*12*] Byers, B. A., "Behavior of Damaged Graphite/Epoxy Laminates Under Compression Loading," NASA CR-159293, NASA Langley Research Center, Hampton, VA, Aug. 1980.

[*13*] Guynn, E. G. and O'Brien, T. K. in *Proceedings,* Twenty-Sixth Structures, Structural Dynamics and Materials Conference, AIAA-85-0646, American Institute of Aeronautics and Astronautics, April 1985, pp. 187–196.

[*14*] Horton, R. et al., "Damage Tolerance of Composites, Final Report," AFWAL-TR-87-3030, Vol. 3, Air Force Wright Aeronautical Laboratories, Dayton, OH, May 1988.

[*15*] Baker, A. A., Jones, R., and Callinan, R. J., *Composite Structures,* Vol. 4, 1985, pp. 15–44.

[*16*] Boll, D. J., Bascom, W. D., Weidner, J. C., and Murri, W. J., *Journal of Material Science,* Vol. 21, 1986, pp. 2667–2678.

[*17*] SACMA Recommended Method SRM 3-88, "Open-Hole Compression Properties of Oriented Fiber-Resin Composites," Suppliers of Advanced Composite Materials Association, Arlington, VA, 1989.

[*18*] Weaver, W., Jr., and Johnston, P. R. in *Finite Elements for Structural Analysis,* Prentice-Hall Inc., Englewood Cliffs, NJ, 1984, pp. 214–220.

[*19*] Shivakumar, K. N. and Whitcomb, J. D., *Journal of Composite Materials,* Vol. 19, 1985, pp. 2–18.

[*20*] Almroth, B. O., Brogan, F. A., and Stanley, G. M., *Structural Analysis of General Shells, Vol. 2 User Instructions for STAGSC-1,* Lockheed Palo Alto Research Laboratory, Palo Alto, CA, Jan. 1983.

[*21*] Kardomateas, G. A. in *Proceedings,* Twenty-Ninth Structures, Structural Dynamics, and Materials Conference, AIAA-88-2260, American Institute of Aeronautics and Astronautics, 1985, pp. 382–390.

Clarence C. Poe, Jr.[1]

Relevance of Impacter Shape to Nonvisible Damage and Residual Tensile Strength of a Thick Graphite/Epoxy Laminate

REFERENCE: Poe, C. C., Jr., **"Relevance of Impacter Shape to Nonvisible Damage and Residual Tensile Strength of a Thick Graphite/Epoxy Laminate,"** *Composite Materials: Fatigue and Fracture (Third Volume), ASTM STP 1110*, T. K. O'Brien, Ed., American Society for Testing and Materials, Philadelphia, 1991, pp. 501–527.

ABSTRACT: An investigation was made to determine the relevance of impacter shape to nonvisible damage and tensile residual strength of graphite/epoxy cases for the solid rocket motors of the Space Shuttle. Impacters were dropped onto 30.5-cm (12-in.)-long rings (short cylinders) that were 76.2 cm (30 in.) in diameter and 36 mm (1.4 in.) thick. The kinetic energies ranged from 17.0 to 136 J (12.5 to 100 ft·lb). Some rings were filled with inert propellant and some were empty. A 5 kg (11 lb) impacter was used with a 12.7-mm (0.5-in.)-diameter hemisphere and a sharp corner attached. The rings were impacted numerous times around the circumference and cut into 51-mm (2-in.)-wide specimens. Because of a shortage of rings, impacts with a 6.3-mm (0.25-in.)-diameter bolt-like rod were simulated by quasi-statically pressing the rod against the face of individual specimens. All of the specimens were uniaxially loaded to failure in tension. Results from a previous impact study using a 25.4-mm (1.0-in.)-diameter hemisphere were also considered. For the range of impact energies investigated, the corner and rod always made visible damage on the surface, but the hemispheres did not. The damage on the surface consisted of a crater shaped like the indenter, and the damage below the surface consisted of broken fibers that appeared to result from shear failure of the matrix. The damage initiated when the contact pressure exceeded a critical level but did not become visible on the surface until an even higher pressure was exceeded. The contact pressures for the rod and corner exceeded the critical level to cause visible damage. The kinetic energy to initiate damage and to cause visible damage on the surface increased approximately with hemisphere diameter to the third power. For a given kinetic energy, the residual strengths did not vary significantly with indenter shape. However, the reduction in strength for nonvisible damage increased dramatically with increasing hemisphere diameter, 9 and 30% for the 12.7-mm (0.5-in.) and 25.4-mm (1.0-in.)-diameter hemispheres, respectively. Factors of safety for nonvisible damage increased with increasing kinetic energy. With the factor of safety for the filament-wound case (1.4), the maximum allowable kinetic energy was 123 J (91 ft·lb). The effects of hemisphere diameter on impact force, damage size, damage visibility, and residual tensile strength were predicted quite well assuming Hertzian contact and using maximum stress criteria and a surface crack analysis.

KEY WORDS: low-velocity impact, nonvisible impact damage, impacter shape, residual tensile strength, graphite/epoxy composite, motor case, composite materials, fracture, fatigue (materials)

Nomenclature

a	Depth of impact damage or equivalent surface crack, m (in.)
A_{11}, A_{22}, A_{12}	Constants in Hertz equation, Pa (psi)
c	Half the length of impact damage or equivalent surface crack, m (in.)

[1] Senior research engineer, NASA Langley Research Center, Hampton, VA 23665-5225.

E_1, E_2	Young's moduli of isotropic impacter and target, Pa (psi)
E_x, E_y	Young's moduli of filament-wound case (FWC) laminate, Pa (psi)
E_r, E_z	Young's moduli of transversely isotropic body, Pa (psi)
$f\left(\dfrac{a}{h}, \dfrac{a}{c}, \dfrac{c}{W}, \phi\right)$	Correction to stress intensity factor for finite width and thickness
F	Contact or impact force, N (lb)
F_{max}	Maximum contact force during impact, N (lb)
G_{xy}	Shear modulus of FWC laminate, Pa (psi)
G_{zr}	Shear modulus of transversely isotropic body, Pa (psi)
h	Thickness, m (in.)
k_b	Spring constant for beam deflection, N/m (lb/in.)
k_1, k_2	Factors in Hertz's equation, Pa^{-1} $(\text{psi})^{-1}$
K_Q	Fracture toughness, Pa $\text{mm}^{1/2}$, (psi $\text{in.}^{1/2}$)
KE_{eff}	Effective kinetic energy, $\frac{1}{2} M v_1^2$, J (ft·lb)
m_1, m_2	Mass of impacter and target, kg (lb)
M	Effective mass, kg (lb)
n_0	Factor in Hertz's equation, Pa (psi)
p	Contact pressure, Pa (psi)
p_c	Average contact pressure, Pa (psi)
Q	Shape factor for an elliptical crack
r_c	Contact radius, m (in.)
R_1	Radius of impacter or indenter, m (in.)
S_{xc}	Hoop stress at failure, Pa (ksi)
t	Time, s
u	Relative displacement between impacter and target, m (in.)
u_i	Indentation of target, m (in.)
u_b	Beam deflection of target, m (in.)
v_1, v_2	Velocity of impacter and target, m/s (in./s)
W	Width of specimen in test section, m (in.)
z_0	Depth from surface where damage initiates
ζ	Ratio z/r_c
ζ_0	Ratio z_0/r_c
ν_{xy}, ν_{yx}	Poisson's ratios of FWC laminate
ν_r, ν_{rz}	Poisson's ratios of transversely isotropic semi-infinite body
ν_1, ν_2	Poisson's ratio of isotropic impacter and target
τ_u	Shear strength, Pa (psi)
τ_{max}	Maximum principal shear stress, Pa (psi)
ϕ	Parametric angle of ellipse

Subscripts

x, y	Cartesian coordinates (The x-direction is the axial direction of the cylinder or hoop direction of the FWC laminate.)
r, θ, z	Cylindrical coordinates (The z-direction is normal to the laminate.)

Recently, the National Aeronautics and Space Administration (NASA) developed several sets of solid rocket motors with graphite/epoxy cases to use in lieu of existing motors with steel cases for the Space Shuttle. These light-weight motor's were to have been used for certain missions that required a lower mass at launch. (The program was canceled before the first flight.) The cases were made using a wet filament-winding process, hence the name filament-wound case (FWC). It was desired, but not required, that the FWCs be reusable like the steel cases. Each light-weight motor would have consisted of four FWCs, a forward case, two center cases, and an aft case. The FWCs were 3.66 m (12 ft) in diameter and were joined together with steel pins. The forward and center FWCs were 7.6 m (25 ft) in length, and the aft FWC was somewhat shorter. The thickness of the membrane region away from the ends was approximately 36 mm (1.4 in.). The ends were thicker to withstand the concentrated pin loads. The FWCs are designed primarily for internal pressure caused by the burning solid propellant. However, the motors are subjected to bending when the main engines of the orbiter ignite, causing relatively large compressive stresses in the aft FWCs.

For graphite/epoxy pressure vessels with thin walls, burst pressure is reduced significantly by low-velocity impacts [1]. However, because of the stoutness of the FWCs, it was not expected that low-velocity impacts by tools and equipment could seriously reduce strength. However, dropped tools and equipment are not the only threat; the potential for handling accidents is also significant. Consider that the FWCs were to have been manufactured at one location, shipped by truck to another and loaded with solid propellant, and then shipped by rail to the launch site and assembled. Moreover, one empty FWC was dropped in the factory. Each of the longer FWCs had a mass of about 4500 kg (10 000 lb) empty and about 140 000 kg (300 000 lb) when filled with solid propellant. The potential energy or kinetic energy of even an empty FWC when lifted 1 m (39 in.) would be over 44 000 J (33 000 ft·lb). Thus, handling accidents have the potential for causing more serious impact damage than dropped tools and equipment.

In previous investigations conducted at Langley Research Center [2–5], the effect of low-velocity impacts on the tensile strength of an FWC was determined for impacters of various masses with a 25.4-mm (1.0-in.)-diameter hemisphere attached to the end. The impacters were dropped from various heights onto 36-mm (1.4-in.)-thick graphite/epoxy rings (short cylinders) to simulate falling tools and equipment. Impact energies ranged from 38 to 447 J (28 to 329 ft·lb). One of the rings contained inert solid propellant; the other was empty. The rings were impacted numerous times around the circumference and cut into specimens. Then, the specimens were loaded uniaxially in tension to failure. The impacts damaged the laminate, and the residual tensile strength was reduced accordingly. The damaged region contained broken fibers, the loci of which resembled cracks. Up to about 100 J (74 ft·lb), the damage was not visible on the surface, but the tensile strength was reduced as much as 30%. Even for the largest energy 447 J (329 ft·lb), the damage was localized and only about 6 mm (0.24 in.) deep. The damage size (breadth and depth) and residual tensile strength were predicted assuming Hertzian contact and using maximum stress criteria and surface crack analysis. However, the size and nature of the damage could not be positively determined nondestructively to verify the predictions of damage [6–9]. Thus, a number of specimens were taken from an actual FWC and damaged by simulated impacts and deplied [10]. The maximum size of the damaged region was in good agreement with that predicted by maximum stress criteria. The broken fibers appeared to have been caused by shear failure of the epoxy.

The present investigation was made to determine the relevance of impacter shape to nonvisible damage and the associated reduction in strength. Accordingly, impact tests were conducted with the following additional indenters: a 12.7-mm (0.5-in.) diameter hemisphere, a sharp corner, and a 6.3-mm (0.25-in.) diameter bolt-like rod. The hemisphere and corner were attached to the end of a 5 kg (11.1 lb) impacter and dropped from various heights onto a ring containing inert solid propellant and an empty ring. These rings were taken from the same cylinder as those in Refs 2–9. The rings were impacted numerous times around the circumference and cut into 51-mm (2-in.)-wide specimens as before. Because no rings remained, impacts with the 6.3-mm (0.25-in.)-diameter bolt-like rod were simulated by quasi-statically pressing the rod against the face of individual specimens. The specimens were then loaded uniaxially in tension to failure. The effects of indenter shape on impact force, damage size, damage visibility, and residual tensile strength were predicted assuming Hertzian contact and using the maximum stress criteria and surface crack analysis reported in Refs 3–5. Factors of safety for strength reduction with nonvisible damage were calculated in terms of kinetic energy. With a factor of safety of 1.4, the maximum allowable kinetic energy was 123 J (91 ft·lb). This paper is a summary of Ref 11 in which the test results and material properties are tabulated.

Material

A 0.76-m (30-in.)-diameter, 2.13-m (7-ft)-long cylinder was made by Hercules Inc. to represent the region of an FWC away from the ends. The thickness of the cylinder was the same as that of an FWC, 36 mm (1.4 in.). The directions of the hoop and helical layers in the cylinder are shown in Fig. 1. Notice that the hoop and helical directions were rotated 90° to provide straight specimens for uniaxial loading in the hoop direction. For this reason, the hoop layers could not be wound using the set process but had to be hand laid using prepreg tape. (Reference to hoop layers in this paper is reference to fibers in the hoop direction of the FWC but in the longitudinal direction of the cylinder.) However, the helical layers were wound using a wet process like that used to make an FWC. The graphite fiber and winding resin were also the same used to make the FWCs. The fiber was AS4W-12K graphite, and the winding resin was HBRF-55A epoxy (Hercules Inc.'s designations). The prepreg tape was actually a unidirectional broadgoods. The epoxy in the prepreg was Hercules Inc.'s MX-16. Mechanical and physical properties of the constituents and laminae are given in Refs 2–5 and 11.

From outside to inside, the orientations of the layers were $\{(\pm 56.5)_2/0/[(\pm \underline{56.5})_2/0]_3/$ $[(\pm 56.5)_2/0]_7/(\pm 56.5/0_2)_4/(\pm 56.5)_2/\text{cloth}\}$, where the 0° layers are the hoops and the $\pm 56.5°$ layers are the helicals. (The layer angles are measured from the axis of the cylinder.) The underlined $\pm \underline{56.5}°$ helical layers are about 1.6 times as thick as the other helical layers. The cloth

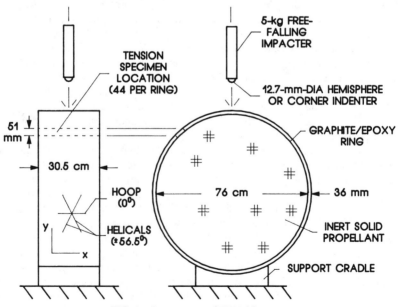

FIG. 1—*Impact tests of FWC-like ring.*

TABLE 1—*Laminate constants.*

Property	Value
E_x, GPa (Msi)	30.6 (4.44)
E_y, GPa (Msi)	39.0 (5.66)
G_{xy}, GPa (Msi)	19.7 (2.86)
ν_{xy}	0.351
ν_{yx}	0.447

layer at the inner surface is a plain weave. The layup is balanced (equal numbers of $+56.5$ and $-56.5°$ layers) but not symmetrical about the midplane. More hoop layers are near the inner surface than the outer surface.

The elastic constants of the cylinder are given in Table 1. They were predicted with lamination theory. In general, bending and stretching couple in an unsymmetric laminate like that of the FWC. However, for the axisymmetric case of a cylinder containing pressure, bending and stretching do not couple. Thus, the elastic constants were calculated assuming that the laminate is symmetric. The x- and y-directions in the subscripts of the elastic constants correspond to the axial and hoop directions of the cylinder (see Fig. 1).

Certain commercial materials are identified in this paper in order to specify adequately which materials were used. In no case does such identification imply that the materials are necessarily the only ones or the best ones available for the purpose.

Test Apparatus and Procedure

Impact Tests

The cylinder was cut into seven 30.5-cm (12-in.)-long rings. The rings were impacted by free-falling masses (see Fig. 1). Several rings were cut into specimens for other investigations. Inert solid propellant was cast in several of the rings. The masses of an empty and a filled ring were 40 kg (89.1 lb) and 288 kg (635 lb), respectively. During impacts, the rings lay on a thin rubber sheet in a shallow aluminum cradle. The bottom of the empty ring was secured to the concrete floor with a cross-bar to prevent the ring from "leaping" off the floor when impacted. Each ring was impacted every 59 mm (2.3 in.) of circumference, giving 44 impact sites. The damaged regions did not overlap.

The free-falling impacter was a 51-mm (2-in.)-diameter steel rod with indenters of several shapes attached to one end. The indenters were a 12.7-mm (0.5-in.)-diameter hemisphere and a sharp corner. The tip of the corner had a radius less than 0.25 mm (0.01 in.). The mass of the impacter was 5.0 kg (11.1 lb), including the indenter. The impacter was instrumented to determine the maximum impact force. Drop heights were varied from 36 to 274 cm (14 to 108 in.) to give kinetic energies from 17 to 136 J (13 to 100 ft·lb). After each impact, the ring was rotated to present a new site. After the impacts were completed, each ring was cut into 44 tension specimens that were oriented as shown in Fig. 1. The center of each specimen coincided with an impact site. The cut edges were ground flat and parallel so that the width and length of the specimens were 51 mm (2.0 in.) and 31 cm (12 in.), respectively.

Simulated Impact Tests

No rings remained for impact tests with the rod. Thus, nine of the 51-mm (2.0-in.) by 30.5-cm (12-in.) tension specimens that remained from other investigations were used instead for simulated impact tests with the rod. The impacts were not made by dropping weights because of a potential difference between the dynamic response of the ring and that of the smaller tension specimen. The rod had a diameter of 6.3 mm (0.25 in.). The corner of the rod (intersection of the side and end) had a radius of 1.3 mm (0.05 in.) to represent the end of a bolt. The impacts were simulated by mounting the rod in the upper grip of a hydraulic testing machine and slowly pressing the rod against the outside face of each specimen. Stroke was programmed to increase linearly with time. The maximum strokes were 3.4, 6.5, and 9.5 mm (0.135, 0.255, and 0.375 in.) for three specimens each. Load and stroke were recorded on an x-y recorder.

It is believed that simulated impacts for a thick laminate like that of the FWC are equivalent to actual impacts for the following reasons. Simulated and actual impacts on a filled and empty ring with the 25.4-mm (1.0-in.)-diameter hemisphere resulted in about the same residual tensile strengths [3,4]. Simulated impacts on pieces from an actual FWC [10] and actual impacts on the rings [3,4] resulted in damage of about the same depth for corresponding levels of impact force for the same hemisphere. Hertz's quasi-static equations for contact radius and indentation were in agreement with test data for impacts of the rings [3,4] as well as simulated impacts of the FWC pieces [10].

X-Ray Tests

The impacted face of each specimen (including those with simulated impacts) was soaked in a zinc iodide penetrant for 30 s, and radiographs were made from the impacted side and the edge. The penetrant was contained by a circular dam on the surface of the specimen. The depth of impact damage in the radiographs was measured. After the specimens were X-rayed, circular arcs were ground into the specimens' edges to reduce the width in the test section to 38 mm (1.5 in.) (Fig. 2). The width of the test section was reduced to assure that failure originated at the impact damage. Without the more narrow test section, the failure of specimens with small damage seemed to originate at the grips where the stresses were elevated by the large grip pressure [2–4].

Residual Strength Tests

The specimens were uniaxially loaded to failure in tension with a 445-kN-capacity (100 kips) hydraulic testing machine. Stroke was programmed to increase with time at the rate of 0.0076 mm/s (0.0003 in./s). Time to failure was several minutes. Hydraulically actuated grips that simulate fixed-end conditions were used. Otherwise, uniaxial loading would cause bending because the FWC laminate is not symmetrical. Strain gages on the inner and outer faces indicated negligible bending [2,3].

Analysis and Results

Impact Force

The impact force increases with time to a maximum value and then decreases to zero, much as a haversine. For impacts to the thick composite rings, the time for the impact force to reach a maximum value was usually less than one millisecond [3,4]. Assuming Hertzian contact and Newtonian mechanics, the maximum impact force, F_{max}, is given by

$$\frac{1}{2k_b}F_{max}^2 + \frac{2}{5}R_1^{-1/3}n_0^{-2/3}F_{max}^{5/3} - KE_{eff} = 0 \tag{1}$$

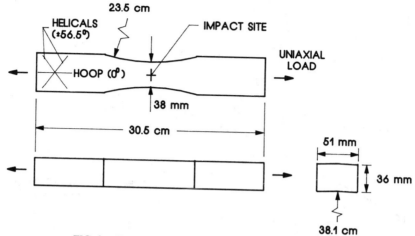

FIG. 2—*Tension specimen used to measure residual strength.*

where R_1 is the radius of the impacter, n_0 is a Hertzian constant defined by Eq 19, and k_b is a spring constant for the ring defined by Eq 17. The effective kinetic energy, KE_{eff} is defined by

$$KE_{\text{eff}} = \tfrac{1}{2}Mv_1^2 \tag{2}$$

where M is the effective mass defined by Eq 15 and v_1 is the velocity of the impacter. The derivation of Eq 1 is given in the Appendix. Values of n_0 and k_b were determined from tests. For either filled or empty rings, $n_0 = 4.52$ GPa (656 ksi). For filled rings and empty rings, respectively, $k_b = 6.34$ MN/m (36.2 kips/in.) and 5.08 MN/m (29.0 kips/in.). Details are given in the Appendix.

The results from the formulation of the impact problem here and in Ref *12* are significantly different. In Ref *12*, it was assumed that the velocities of the impacter and target are equal during contact. Conservation of momentum was used to obtain the velocity during contact. Using the same initial and final conditions as those in the Appendix, one arrives at an equation with a form similar to that of Eq 1 but with

$$KE_{\text{eff}} = \tfrac{1}{2}m_1v_1^2/(1 + \tfrac{1}{4}m_2/m_1) \tag{3}$$

rather than Eq 2, which also can also be written

$$KE_{\text{eff}} = \tfrac{1}{2}m_1v_1^2/[1 + m_1/(\tfrac{1}{4}m_2)] \tag{4}$$

where $m_2/4$ is the effective mass of the target.

The mass ratio $m_1/(m_2/4)$ in Eq 4 is inverted relative to that in Eq 3. Thus, the effect of target mass on impact force in Ref *12* is opposite to that in the present formulation.

Values of impact force are plotted against effective kinetic energy in Fig. 3 for a 25.4-mm (1.0-in.) diameter hemisphere [3,4]. Each symbol is an average of several tests. The impact forces for the empty and filled rings coalesce quite well for a given value of the effective kinetic energy. For all values of KE_{eff} less than that labeled nonvisible damage (NVD), the impacts did not cause visible damage on the surface. For values of KE_{eff} greater than this threshold, the

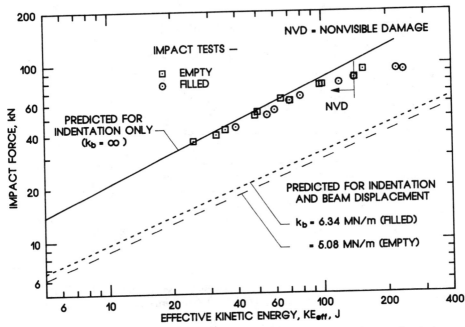

FIG. 3—*Impact force versus effective kinetic energy for 25.4-mm (1.0-in.)-diameter hemisphere.*

impacts caused a visible crater. Near the threshold, the craters were very shallow but perceptible. The depth of the craters increased with increasing KE_{eff}. Values of impact force calculated with Eq 1 are also plotted. Curves are shown for three values of beam stiffness: $k_b = 5.08$ MN/m (29.0 kips/in.), 6.34 MN/m (36.2 kips/in.), and ∞. For $k_b = \infty$, the ring does not deflect at all in a global sense. When damage is nonvisible, the predicted curve for $k_b = \infty$ agrees with the test results quite well. But when damage is great enough to be visible, the actual impact forces are somewhat less than the predicted values. Values of impact force predicted with quasi-static values of k_b are much too small. The quasi-static values of k_b greatly overestimate the deflection of the rings during impact. Thus, the natural vibration periods of the rings were probably the same order of magnitude as the duration of the impacts, which was only a few milliseconds [3,4].

Values of impact force are plotted against effective kinetic energy KE_{eff} in Fig. 4 for the 12.7-mm (0.5-in.)-diameter hemisphere, the corner, and the rod. For the simulated impacts with the rod, the area under the load-displacement curve was used for KE_{eff}. Results for the 25.4-mm (1.0-in.)-diameter hemisphere in Fig. 3 are also plotted for comparison. Each symbol is an average of several tests. Values of impact force are smallest for the rod and corner and largest for the 25.4-mm (1.0-in.)-diameter hemisphere. Filled and open symbols represent visible and nonvisible damage, respectively. The damage on the surface consisted of craters shaped like the indenters. The corner and rod made craters for all impacts. For the hemispheres, the thresholds for NVD are labeled as in Fig. 3. Values of impact force calculated with Eq 1 are also plotted for the hemispheres. The calculations were made using $k_b = 219$ MN/m (1250 kips/in.). The test values of impact force for nonvisible damage agree a little better with Eq 1 for $k_b = 219$ MN/m (1250 kips/in.) than for $k_b = \infty$. Still, the term containing k_b in Eq 1 contributes very little, which is evident by comparing the curve in Fig. 4 with that in Fig. 3 for $k_b = \infty$. The effect of hemisphere diameter in Fig. 4 is modeled quite well by Eq 1 when

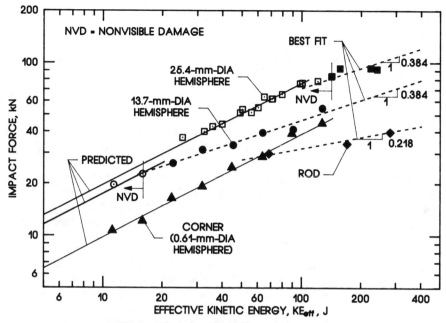

FIG. 4—*Effect of impacter shape on impact force.*

impacts did not make craters. However, for both hemispheres, impact forces are less than predicted when impacts did make craters. The discrepancy increases with increasing KE_{eff}. The dashed lines, which were fit to the data with visible damage, were assumed to have the same slope for both hemispheres. The slopes of the lines that represent test data with visible damage in Fig. 4 vary with indenter shape. Thus, impact force for one indenter shape is not always less or greater than that for another shape for all values of KE_{eff}.

Calculations made with Eq 1 are also plotted in Fig. 4 for the corner, which was represented as a hemisphere with a diameter of 0.61 mm (0.024 in.). This diameter was chosen to give the best agreement between predictions and tests. The actual radius of the tip was about half this value, approximately 0.25 mm (0.01 in.). The test data are modeled quite well by Eq 1 even though the corner made a deep crater on the surface. The agreement is certainly fortuitous because the limit on indentation in Hertz's equation (Eq 18 in the Appendix) is violated. Nevertheless, the corner was represented as a small hemisphere in subsequent calculations of damage size and residual strength to determine the limits of such a representation.

Damage

For Hertzian contact [13], the contact pressure between a hemispherical impacter of radius R_1 and a semi-infinite, transversely isotropic body is given by

$$p = \frac{3}{2} p_c \left(1 - \frac{r^2}{r_c^2} \right)^{1/2} \tag{5}$$

where r is the radius (polar coordinate) measured from the center of the contact site, r_c is the contact radius given by

$$r_c = \left(\frac{F_{max} R_1}{n_0} \right)^{1/3}$$ (6)

and p_c is the average pressure given by

$$p_c = \frac{F_{max}}{\pi r_c^2}$$ (7)

Using the theory of elasticity, Love [14] obtained a closed-form solution for the internal stresses in a semi-infinite, isotropic body produced by the "hemispherical" pressure given by Eq 5. The problem is axisymmetric. Even though the composite is made of orthotropic layers, the results for the isotropic half space should at least be applicable in a qualitative sense when the contact radius is large compared to layer thickness. Contours of maximum shear stress = 228 MPa (33.0 ksi) and maximum compressive stress = 587 MPa (85.1 ksi) were calculated using the equations in Ref 14 with $v_2 = 0.3$ and plotted in Fig. 5 for various values of average contact pressure. The maximum compressive stress is in the plane of the composite layers, but the shear stress is generally not. Along the center line, for example, the maximum shear stress lies in a plane that is inclined at 45° to the surface. The depth from the surface and distance from the center of contact are normalized by the contact radius r_c. For a given contact pressure, each contour represents the outermost extent of damage according to the maximum shear or maximum compressive stress criterion. Since the stresses from Love's solution do not account for damage, they are valid only for predicting the onset of damage. Thus, the contours only give the approximate size of the damage region. Notice that damage initiates on the axis of symmetry. For the maximum shear stress criterion, damage initiates below the surface at a

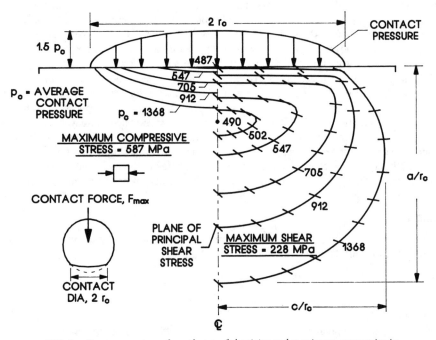

FIG. 5—*Damage contours from theory of elasticity and maximum stress criteria.*

normalized depth of 0.482 for $p_c = 490$ MPa (71.1 ksi); whereas, for the maximum compressive stress criterion, damage initiates at the surface for $p_c = 487$ MPa (70.6 ksi). Therefore, damage initiates at a critical contact pressure, independent of impacter radius.

The maximum depth of the damage contours for maximum shear stress in Fig. 5 is given by

$$\left(\frac{1}{2} - \nu_2\right)(1 + \zeta^2)^{-1} + (1 + \nu_2)\left[\zeta \tan^{-1}\left(\frac{1}{\zeta}\right) - \zeta^2(1 + \zeta^2)^{-1}\right] - \frac{4\tau_u}{3p_c} = 0 \qquad (8)$$

where $\zeta = a/r_c$. The depth $\zeta_0 = z_0/r_c$, which corresponds to the location of τ_{max} and damage initiation, is given by[2]

$$3\zeta_0 - (1 + \nu_2)(1 + \zeta_0^2)\left[(1 + \zeta_0^2)\tan^{-1}\left(\frac{1}{\zeta_0}\right) - \zeta_0\right] = 0 \qquad (9)$$

The size of the damage in Fig. 5 and Eqs 8 and 9 are normalized by the contact radius. The contact radius, r_c, can be written in terms of the average contact pressure, p_c, instead of the maximum impact force, F_{max}. Using Eq 7 to eliminate F_{max} in Eq 6 results in

$$r_c = \frac{\pi p_c R_1}{n_0} \qquad (10)$$

Thus, r_c increases in proportion to p_c and impacter radius R_1. Hence, the absolute size of the damage contours in Fig. 5 increases in proportion to p_c and R_1. Thus, for a given pressure above the critical level to initiate damage, the depth and width of damage increase in proportion to impacter radius.

Since damage is predicted to initiate on or near the surface for a constant value of average contact pressure, damage also should become visible on the surface for a constant value of average contact pressure, at least for the hemispheres. Accordingly, contact pressures are plotted against impact force in Fig. 6 for the various indenters. The contact pressures were calculated using Eqs 6 and 7 and the measured values of impact force in Fig. 4. The corner was again represented as a 0.61-mm (0.024-in.)-diameter hemisphere. For the rod, the contact area was assumed to be equal to the full area of the rod's cross section—not the reduced area at the end. Filled and open symbols represent visible and nonvisible damage, respectively. The lines were fit to the data. Indeed, the critical contact pressure to cause visible damage for the 25.4-mm (1.0-in.)-diameter hemisphere was only about 7% greater than that for the 12.7-mm (0.5-in.)-diameter hemisphere. The dashed horizontal line in Fig. 6 represents an average contact pressure of 705 MPa (102 ksi). Contact pressures for all tests with the corner and rod exceeded 705 MPa (102 ksi) and damage was likewise visible. The slope of the curve for the rod was greater than that for the other indenters because contact area was constant for the rod but not for the other indenters.

Fiber Damage

To verify radiographic and ultrasonic measurements of impact damage in the rings, specimens from an actual FWC were impacted and pyrolyzed. The pyrolysis removed most of the

---

[2] The equation in Refs 5 and 10 that corresponds to Eq 8 in this report contains an error. The author also discovered a discrepancy between Eqs 8 and 9 in this report and results reported in Ref 14. It is believed that the sign of the second term in Eq 20a of Ref 14 should be reversed.

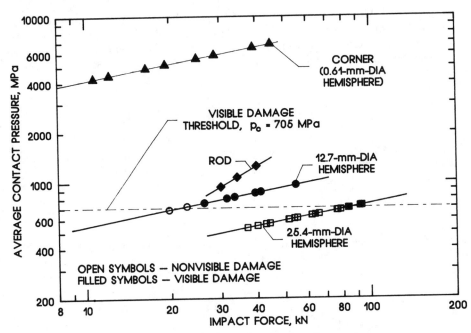

FIG. 6—*Effect of indenter shape on contact pressure.*

epoxy, facilitating the separating of layers or deplying of the composite. The results were reported in Ref *10*. The impacts were simulated by quasi-statically pressing hemispheres against the specimen's face, much like the simulated impacts with the rod in the present investigation. Photographs of the outermost nine deplied layers from one specimen are shown in Fig. 7. The layers contain broken fibers, the loci of which resemble cracks. The layer samples were 38 by 38 mm (1.5 by 1.5 in.). The "cracks" were visible in the 15 outermost layers, of which the outermost nine are shown in Fig. 7. These cracks were mostly parallel to the direction of fibers in the neighboring layers. By coincidence, the cracks appear to be nearly normal to the fibers in the layer in which they reside. When the fibers in the neighboring layers were not parallel to each other, the direction of the cracks wandered between the direction of the fibers in each neighboring layer. For this sample, a 50.8-mm (2.00-in.)-diameter hemisphere was used to produce a contact force of 267 kN (60.0 kips) and a contact pressure of 648 MPa (94.0 ksi). The simulated impact caused a visible crater on the surface.

Several specimens in Ref *10* were not deplied but were sectioned through the thickness and examined using a scanning electron microscope. Photomicrographs of two sections from the same depth are shown in Fig. 8. This impact was also simulated with the 50.8-mm (2.00-in.)-diameter hemisphere, producing a contact force of 200 kN (45.0 kips) and a contact pressure of 589 MPa (85.4 ksi). A crater was visible on the surface. The section on the left in Fig. 8 is directly below the contact area, and the section on the right is outside the contact area. The sections are normal to the plane of the hoop layers and parallel to the direction of the hoop fibers. An edge of the outermost hoop layer can be seen at the center of both sections. The section on the left reveals several matrix cracks in the helical layers on both sides of the hoop layer. The cracks are located in the plane of maximum shear stress. The matrix shear cracks continue across the hoop layer and thus break the fibers. This combination of matrix shear

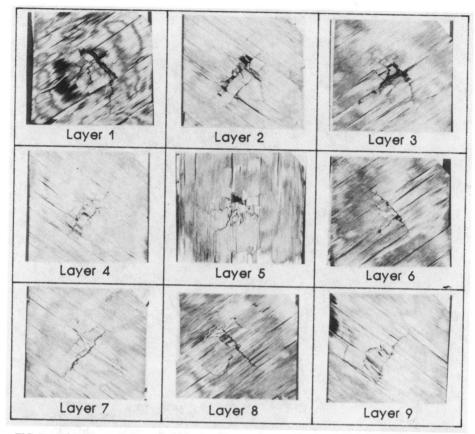

FIG. 7—*Fiber damage in outermost nine layers after deply (simulated impact with 50.8-mm-diameter hemisphere and 648 MPa contact pressure).*

cracks and broken fibers was typical in the other sections as well. Fiber kinking in the hoop layer, which is associated with in-plane compression failure, can be observed in the section on the right. A rather large void lies above the kinked fibers, and what appears to be a delamination lies below. This delamination is probably a lack of bond caused by the manufacturing process rather than by the simulated impact. This type of delamination was not unusual in the many FWCs that were made. Thus, the fiber kinking is probably a result of the in-plane compressive stress due to the simulated impact combined with the lack of matrix support to the fibers.

The maximum depths of fiber breaks for the deplied specimens from Ref *10* are plotted in Fig. 9. Each symbol is an average of several tests. The damage depth was normalized by the contact radius. The filled symbols indicate visible damage on the surface, and the open symbols indicate nonvisible damage. The maximum depths of the contours in Fig. 5 are also plotted in Fig. 9 for comparison. The compression allowable in Fig. 5 is based on the failing strain and Young's modulus of the composite. The shear allowable was chosen to give an upper bound to the damage depths in Fig. 9. Compression tests [*10*] gave an ultimate shear strength of 310 MPa (45.0 ksi), which would have only moved the threshold for damage initiation from

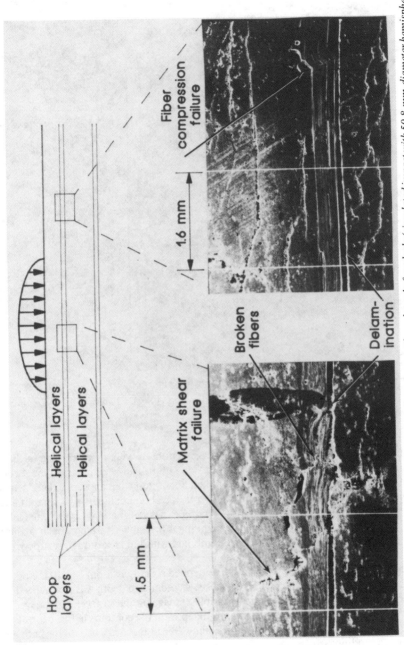

FIG. 8—Photographs of damage in cross section containing outermost hoop layer before deply (simulated impact with 50.8-mm-diameter hemisphere and 589 MPa contact pressure).

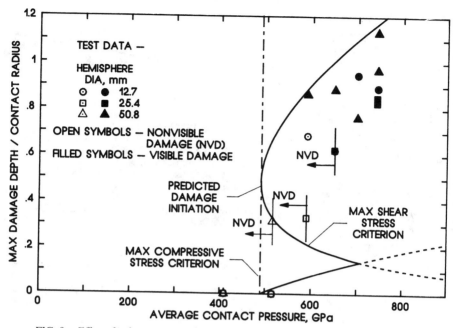

FIG. 9—*Effect of indenter shape on damage depth in deplied layers (simulated impact).*

490 MPa (71.1 ksi) to 668 MPa (96.9 ksi) in Fig. 9 and would have given a lower bound to the test data. The envelope of predicted damage depth is represented by the solid line. The damage far below the surface corresponds to the maximum shear stress criteria, and the damage near the surface corresponds to the maximum compressive stress criteria. The maximum stress criteria represent the data quite well considering that the composite is neither homogeneous nor isotropic. Widths of the contours in Fig. 5 are approximately 1.6 times the depth [*10*]. The widths (crack lengths) from the tests were two to three times the depths. Hence, the maximum stress criteria tended to underpredict the width of damage, probably an effect of the transversely isotropic nature of the composite.

The maximum depths of impact damage from radiographs of the impacted specimens are plotted against average contact pressure in Fig. 10. Recall that impacts with the rod were simulated. Each symbol is an average of several tests. The data for the 25.4-mm (1.0-in.)-diameter hemisphere are from Refs *3* and *4*. The damage depth was normalized by the contact radius calculated with Eq 6. The filled symbols indicate visible damage on the surface, and the open symbols indicate nonvisible damage. Logarithmic scales are used because of the wide range of contact pressures for the various indenters. The maximum depths of the contours in Fig. 5 are also plotted in Fig. 10 for comparison. As with the deply data for simulated impacts, the radiographic data for the hemispheres and corner follow the maximum stress criteria quite well. The agreement for the corner is better than anticipated considering that it was represented as a small hemisphere. The rod produces uniform displacements on the surface, which result in infinite stresses at the edge of the rod compared to finite stresses at the edge of the contact region of the hemisphere. (See Ref *15* for an excellent review of the contact problem.) Thus, the rod acts like a punch and slices through the composite. Consequently, the depth of damage is greater than that predicted for a hemisphere.

Values of average contact pressure are plotted against indentation (displacement of the actu-

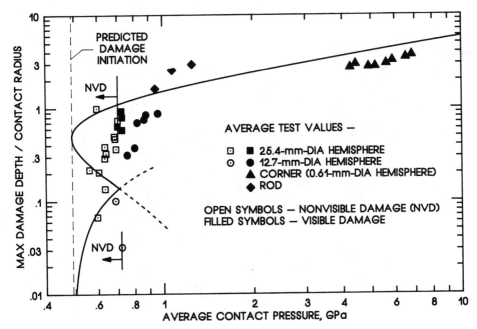

FIG. 10—*Effect of indenter shape on maximum damage depth in radiographs.*

ator) in Fig. 11 for three of the simulated impact tests with the rod. Again, the full area of the rod's cross section was used to calculate pressure. These curves are typical for the rod. Notice that each curve is smooth until reaching pressures between 600 and 650 MPa (87 and 94.3 ksi) where a jump in indentation occurs. These jumps probably correspond to damage initiation. These pressures are only slightly more than the lowest pressure that resulted in damage for the hemispheres in Fig. 10. Thus, the critical level of contact pressure to initiate damage is essentially the same for the rod and hemispheres. Since the pressures to initiate damage are about the same for the two indenter shapes, it is expected that the pressures to cause visible damage are also about the same. Notice that the indentation in Fig. 11 increases dramatically once damage develops because the rod penetrates like a punch.

In Fig. 12, values of indentation and damage depth in radiographs for the rod are plotted against average contact pressure. Except for the largest value of indentation, the damage extends ahead of the rod, much as it does for the hemispheres. Some specimens were delaminated completely at the bottom of the damage as a result of the simulated impact. On the other hand, large delaminations were not found in specimens impacted with the other indenters. Perhaps the large amount of material pushed ahead of the rod caused the delamination.

Impact damage from the 25.4-mm (1.0-in.)-diameter hemisphere was difficult to see in radiographs [3–9]. A comparison between damage depth determined by deply [10] and radiographs for the 12.7 mm (0.5 in.) and 25.4-mm (1.0-in.)-diameter hemispheres is made in Figs. 13 and 14. The radiographic data for the 25.4-mm (1.0-in.)-diameter hemisphere are from Refs 3 and 4. The impacts were simulated for the deply data but not for the radiographic data. The maximum depth is normalized by contact radius and plotted against average contact pressure. The contact radius was calculated with Eq 6. The predictions made with the maximum stress criteria are also plotted for a reference. The deply and radiographic results are in good agreement for the 25.4-mm (1.0-in.)-diameter hemisphere. However, for the 12.7-mm

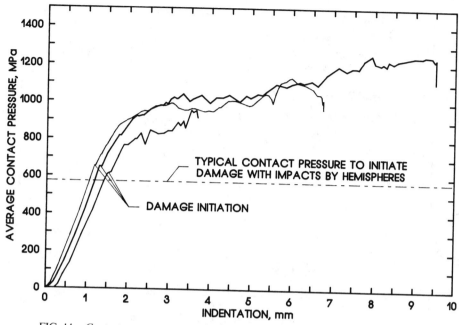

FIG. 11—*Contact pressure versus indentation for simulated impact tests using the rod.*

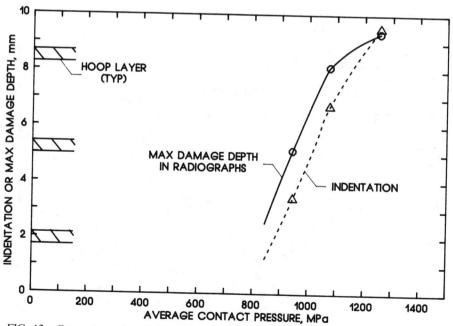

FIG. 12—*Comparison of maximum damage depth in radiographs and indentation for simulated impact tests using the rod.*

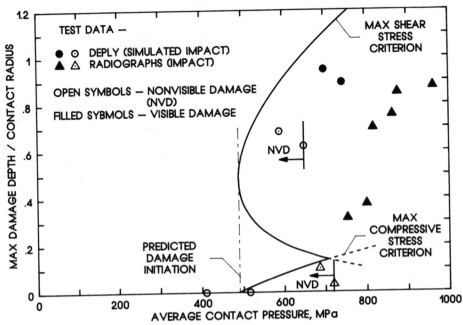

FIG. 13—*Comparison of maximum damage depth from deply and radiographs for 12.7-mm-diameter hemisphere.*

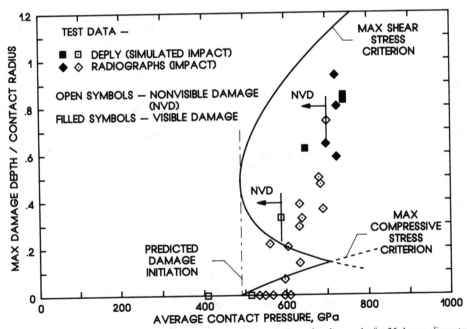

FIG. 14—*Comparison of maximum damage depth from deply and radiographs for 25.4-mm-diameter hemisphere.*

(0.5-in.)-diameter hemisphere, the contact pressure to cause a given value of damage depth is greater for the radiographs than for the deply tests. On the other hand, the thresholds for non-visible damage for radiographs and deply tests are in agreement for the two hemispheres. The difference between results for the radiographs and deply tests is not likely due to the difference between impacts and simulated impacts but rather due to the X-ray opaque dye not penetrating to the innermost damage [3,4,6]. In impact tests with the 12.7-mm (0.5-in.)-diameter hemisphere where the contact pressures were high, local heating was sufficient to pyrolyze the epoxy matrix. The higher temperatures may have made the damage less penetrable. Pressures were generally lower for the 25.4-mm (0.5-in.)-diameter hemisphere and no pyrolysis was noticed.

The critical contact pressure to cause visible damage did vary somewhat with hemisphere diameter. Average contact pressures from Figs. 9 and 10 are plotted against hemisphere diameter in Fig. 15. Different symbols are used to indicate visible and nonvisible damage. The average contact pressure to cause visible damage increased somewhat with decreasing hemisphere diameter, particularly for the deply tests where the variation of diameter is greatest. Because the contact radius increases in proportion to hemisphere radius for a given contact pressure (Eq 10), the surface area of damaged material increases in proportion to the square of the hemisphere diameter. Thus, the surface damage caused by a large hemisphere is more visible than that caused by a small hemisphere. Hence, there is no contradiction between damage initiation being independent of hemisphere radius and damage visibility increasing with increasing hemisphere radius.

The hemisphere diameter is plotted against effective kinetic energy for constant values of contact pressure in Fig. 16. The curves were calculated using Eqs 1, 6, and 7 for the pressures that correspond to thresholds for damage initiation 490 MPa (71.1 ksi) and damage visibility 705 MPa (102 ksi). A value of $k_b = 219$ MN/m (1250 kips/in.) was used. Equation 1 should

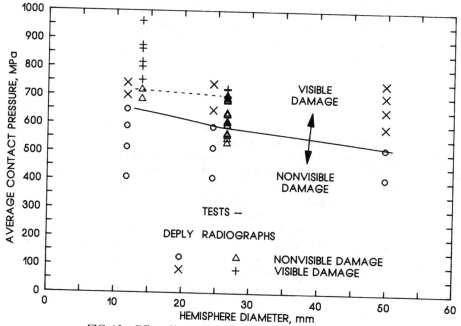

FIG. 15—*Effect of hemisphere diameter on visibility of damage.*

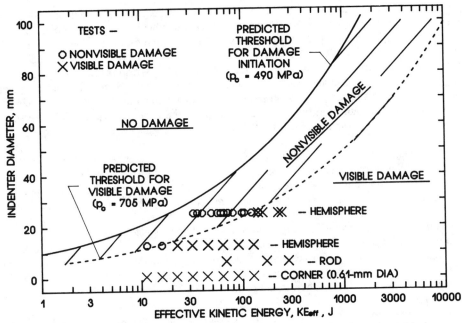

FIG. 16—*Indenter diameter and effective kinetic energy to initiate damage and to cause visible damage.*

be reasonably accurate for $p_c \le 705$ MPa (102 ksi), since damage is small. The curves divide the graph into regions of no damage, nonvisible damage, and visible damage. Along these curves, the effective kinetic energy increases approximately with hemisphere diameter to the third power. Thus, for large hemisphere diameters, very large energies are necessary to initiate damage. Test data for the two hemispheres and the corner and rod are plotted for comparison. The agreement between the tests and predictions verifies the use of simple mechanics and Hertzian contact to predict the effect of indenter size and shape.

Residual Strength

When the impacted specimens were loaded to failure in tension, those with very shallow damage (outermost helical layers or less) failed catastrophically in one stage [3–5]. But, those with deeper damage (one or more hoop layers) failed in two stages: first, the damaged layers failed and delaminated from the undamaged layers; and then, with additional load, the undamaged layers failed. The two stages of failure were referred to as first- and remaining-ligament failure. Only the first-ligament strengths will be considered here because they are the lowest. The first-ligament strengths were predicted by representing the impact damage as a semielliptical surface crack in a plane normal to the hoop direction. The depth of the equivalent crack was calculated with Eq 8 assuming $\nu_2 = 0.3$ and $\tau_u = 228$ MPa (33 ksi). The length of the equivalent crack was assumed to be two times the depth (semicircular). At failure, the critical stress intensity factor or fracture toughness, K_Q, is given by

$$K_Q = S_{xc} \left(\frac{\pi a}{Q} \right)^{1/2} f\left(\frac{a}{h}, \frac{a}{c}, \frac{c}{W}, \phi \right) \tag{11}$$

where

Q = shape function,
S_{xc} = hoop stress at failure,
a = equivalent crack depth,
$2c$ = equivalent crack length, and
$f(a/h, a/c, c/W, \phi)$ = correction factor for the finite thickness, h, and width, W, of the specimen.

The parametric angle, ϕ, was taken as 0 or 90°, whichever gave the largest value of the correction factor. The functionals, Q and $f(a/h, a/c, c/W, \phi)$ are given in Refs 3–5 and 16. The value of fracture toughness, K_Q, for the cylinder was 0.949 GPa mm$^{1/2}$ (27.3 ksi in.$^{1/2}$) [3–5]. A general fracture toughness parameter and the failing strain of the fibers was used to predict K_Q.

For shallow surface cracks or impact damage, Eq 11 gives strengths that are greater than those from tests. To correct this discrepancy, a line was drawn tangent to Eq 11 and passing through the undamaged strength at a depth corresponding to the outermost hoop layer [5]. (Strengths were not reduced unless the hoop layers were damaged.)

Values of strength associated with first-ligament failure were divided by the undamaged strength of 345 MPa (50.1 ksi) and plotted against impact force in Fig. 17. The test values of strength, which are plotted as symbols, were calculated by dividing the failure loads by the thickness h = 36 mm (1.4 in.) and width W = 38 mm (1.5 in.) of the test section (see Fig. 2). Each symbol is an average of several tests. A different symbol was used for each indenter with open symbols representing nonvisible damage and filled symbols representing visible damage. The test data for the 25.4-mm (1.0-in.)-diameter hemisphere was taken from Refs 3 and 4. Calculations with Eq 11 were plotted as lines, which are dashed to indicate nonvisible damage, p_c < 705 MPa (102 ksi), and solid to indicate visible damage, $p_c \geq$ 705 MPa (102 ksi). The

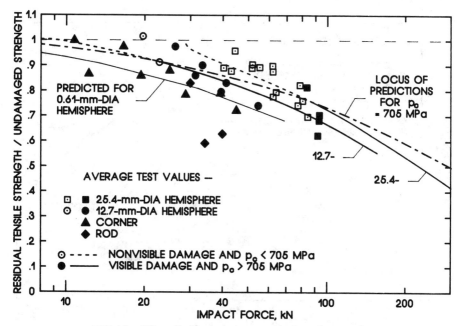

FIG. 17—*Effect of indenter shape on residual tensile strength.*

strengths for the corner were predicted by representing it as a 0.61-mm (0.024-in.)-diameter hemisphere. No strengths were predicted for the rod. The predictions and tests are in reasonably good agreement, even for the corner. The residual strengths for the rod are lowest because of the penetration and hence deep damage. Notice that the strength was reduced as much as 9 and 30% by the 12.7-mm (0.5-in.) and 25.4-mm (1.0-in.)-diameter hemispheres, respectively, without causing visible damage. The locus of predictions for p_c = 705 MPa (102 ksi), which represents the lowest strengths with nonvisible damage, is also plotted in Fig. 17 as a dash-dot line. The dash-dot curve is in good agreement with the data on the average.

The factor of safety for impact damage is given by the ratio of undamaged strength to damaged strength. Reciprocals of the strength ratios for the test data in Fig. 17 are plotted against KE_{eff} in Fig. 18. The curve for p_c = 705 MPa (102 ksi) in Fig. 17 is plotted as a solid line labeled W = 38 mm. The values of KE_{eff} were calculated with Eqs 1 and 2 using k_b = 219 MN/m (1250 kips/in.). The predicted curve for p_c = 705 MPa (102 ksi) corresponds to a factor of safety for nonvisible damage. For this factor of safety, all nonvisible damage would be acceptable, but all visible damage would have to be detected and repaired. The predicted curve and open symbols are in agreement on the average. The dashed curve was drawn through the locus of highest open symbols using the predicted curve as a guide. The factor of safety for the membrane of the FWC, which is 1.4, is shown as a horizontal line. It intersects the dashed curve at $KE_{eff} \leq$ 123 J (91 ft·lb). Thus, the FWC must be protected for $KE_{eff} \leq$ 123 J (91 ft·lb).

The specimens in this investigation were 38 mm wide; those from Refs 3 and 4 were 33 and 38 mm wide. The solid curve in Fig. 18 was calculated assuming that W = 38 mm. Differences between calculations for W = 33 and 38 mm were insignificant. The dash-dot curve in Fig. 18 was calculated for W = ∞, which is more representative of an FWC. The difference between calculations for W = 38 mm and W = ∞ is only significant for very large values of effective kinetic energy where the length of the equivalent surface crack in Eq 11 exceeds about one half the specimen width.

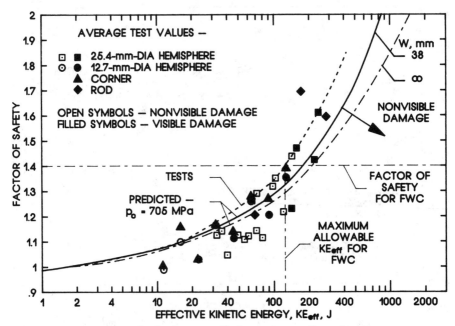

FIG. 18—*Factor of safety for nonvisible damage versus effective kinetic energy.*

Although the investigation here and in Refs *2–4* was conducted with impacts from small objects like tools and equipment in mind, designs should also account for the structure moving. The results here can be applied directly because the velocity, v_1, in Eq 2 can be taken as the relative velocity of the impacter and target. For an effective kinetic energy exceeding 2000 J (1500 ft·lb), the predicted minimum factor of safety in Fig. 18 would exceed two. The potential energy or kinetic energy of one of the FWCs can easily exceed 2000 J (1500 ft·lb), particularly for an FWC containing solid propellant. Thus, the factor of safety for nonvisible damage can be greater for handling accidents than for dropped tools and equipment.

The analytical methods used here to predict impact damage and residual strength could be used to perform sensitivity studies and to evaluate material improvements analytically, for example, variations in n_0, τ_u, and K_Q. Design curves, like the dashed curve in Fig. 18, can be also established experimentally for other composites and other thicknesses. Kinetic energies need only be large enough to exceed the threshold for visible damage.

Conclusions

The relevance of impacter shape to nonvisible damage and residual tensile strength was determined for thick graphite/epoxy rocket motor cases of the Space Shuttle. The cases were represented by rings (short cylinders) that were 30.5 cm (12 in.) long and 36 mm (1.4 in.) thick. The rings were cut from a 2.13-m (7-ft)-long cylinder that was wet-wound on a 76.2-cm (30-in.)-diameter mandrel using AS4 graphite fibers and an epoxy resin. A 5-kg (11.1-lb) instrumented impacter was dropped from various heights onto the rings, which were either empty or filled with inert solid propellant. A 12.7-mm (0.5-in.)-diameter hemisphere and a sharp corner were attached to the end of the impacter. Because no rings remained, impacts with a 6.3-mm (0.25-in.)-diameter bolt-like rod were simulated by quasi-statically pressing the rod against the face of individual specimens. After the rings were impacted numerous times around the circumference, they were cut into 51-mm (2-in.)-wide specimens that were loaded uniaxially in tension to failure. The damage was always local to the impact site and never extended into neighboring impact sites. Previous results with a 25.4-mm (1.0-in.)-diameter hemisphere were included in the investigation.

For the hemispheres, damage initiated and became visible on the surface when the peak contact pressure during the impact event exceeded a critical level. The damage on the surface consisted of a crater shaped like the indenter and the damage below the surface consisted of broken fibers that appeared to be caused by shear failure of the matrix. The damage initiated below the surface before it became visible on the surface. The pressure level to cause visible damage was about 1.4 times that to initiate damage. The extent of nonvisible damage and the resulting reduction in strength increased with increasing hemisphere diameter. For the 12.7-mm (0.5-in.) and 25.4-mm (1.0-in.)-diameter hemispheres, the reductions in strength were 9 and 30%, respectively. The corner and rod caused visible damage for even the smallest values of kinetic energy. The corner behaved somewhat like a hemisphere in that the resistance to indentation increased with increasing indentation. On the other hand, the rod acted like a punch and sliced through the composite with little resistance after a critical level of contact pressure was exceeded. The contact pressure to initiate damage was about equal for the rod and hemispheres.

The depth and width of impact damage was predicted assuming Hertzian contact and maximum stress criteria. Internal stresses were calculated using theory of elasticity. The impact damage was represented as a semicircular surface crack, and residual strengths were predicted using surface crack analysis. The corner was represented as a hemisphere with a diameter of 6.3 mm (0.25 in.). Strengths were not predicted for the rod. Factors of safety predicted for nonvisible damage increased with increasing kinetic energy of impact. The size of damage and residual strengths for the hemispheres and corner were predicted quite well. Ground handling accidents can involve much more energy than tool drops, requiring a much larger factor of safety for nonvisible damage.

APPENDIX

Derivation of Equation for Maximum Impact Force

The mass and velocity of the impacter and ring are denoted by m_1, v_1 and m_2, v_2, respectively. Because the rings are large, only a portion of the mass will act at the contact point. For example, when a simply supported beam is impacted transversely at the center [12], the "effective" mass is approximately one half. Based on rebound velocities for impacters of various masses, the effective masses of the empty and filled rings were found to be approximately one fourth the total mass [3,4].

In the following formulation of the impact problem, the mass of the ring, m_2, was replaced by the effective mass, $m_2/4$. The mass of the impacter is assumed to be fully effective, that is, concentrated at its end. During contact, Newton's law gives

$$m_1 \frac{dv_1}{dt} = -F, \quad \frac{1}{4} m_2 \frac{dv_2}{dt} = -F \qquad (12)$$

Denoting the relative displacement between the impacter and ring by u, the relative velocity can be written

$$\frac{du}{dt} = v_1 + v_2 \qquad (13)$$

Differentiating Eq 13 with respect to time and substituting Eq 12 into the result

$$\frac{d^2u}{dt^2} = -\frac{F}{M} \qquad (14)$$

where

$$M = \frac{1}{\dfrac{1}{m_1} + \dfrac{4}{m_2}} \qquad (15)$$

If the duration of the impact is long compared to natural periods of vibration, the relative displacement, u, in Eqs 13 and 14 can be written

$$u = u_b + u_i \qquad (16)$$

where u_b is the beam deflection of the ring and u_i is the local indentation for Hertzian contact. The steel impacter is assumed to be rigid compared to the composite ring. The beam component is represented by a linear spring

$$u_b = \frac{F}{k_b} \qquad (17)$$

where k_b is a spring constant.

For a spherical, isotropic body in contact with a semi-infinite body that is homogeneous and transversely isotropic, the local indentation from Ref 13 is

$$u_i = R_1^{-1/3} \left(\frac{F}{n_0} \right)^{2/3}$$

(18)

where R_1 is the radius of the sphere. The term, n_0, is defined as

$$n_0 = \frac{4}{3(k_1 + k_2)}$$

(19)

where

$$k_1 = \frac{1 - \nu_1^2}{E_1}$$

$$k_2 = \frac{A_{22}^{1/2}\{[(A_{11}A_{22})^{1/2} + G_{zr}]^2 - (A_{12} + G_{zr})^2\}^{1/2}}{2G_{zr}^{1/2}(A_{11}A_{22} - A_{12}^2)}$$

$$A_{11} = \frac{E_z}{1 - \dfrac{2\nu_{rz}^2 E_z}{(1 - \nu_r)E_r}}$$

$$A_{22} = \frac{\left(\dfrac{E_r}{E_z} - \nu_{rz}^2 \right) A_{11}}{1 - \nu_r^2}$$

$$A_{12} = \frac{\nu_{rz} A_{11}}{1 - \nu_r}$$

and ν_1 and E_1 are the elastic constants of the isotropic sphere and ν_r, ν_{rz}, E_r, E_z, and G_{zr} are the elastic constants of the transversely isotropic semi-infinite body. Equation 18 is accurate only when $u_i < R_1$.

Substituting Eqs 17 and 18 into Eq 16 and differentiating with respect to time, the relative velocity is

$$\frac{du}{dt} = \frac{1}{k_b} \frac{dF}{dt} + \frac{2}{3} R_1^{-1/3} n_0^{-2/3} F^{-1/3} \frac{dF}{dt}$$

(20)

Multiplying Eq 14 by Eq 20 and using the identity $d(x^m)/dt = mx^{m-1}dx/dt$

$$\frac{M}{2} \frac{d[(du/dt)^2]}{dt} = -\frac{1}{2k_b} \frac{d(F^2)}{dt} - \frac{2}{5} R_1^{-1/3} n_0^{-2/3} \frac{d(F^{5/3})}{dt}$$

(21)

Integrating Eq 21 with initial conditions $du/dt = v_1$ and $F = 0$ and final conditions $du/dt = 0$ and $F = F_{max}$ yields

$$\frac{1}{2k_b} F_{max}^2 + \frac{2}{5} R_1^{-1/3} n_0^{-2/3} F_{max}^{5/3} - KE_{eff} = 0$$

where

$$KE_{eff} = \tfrac{1}{2} M v_1^2$$

Calculations with Eq 19 give $n_0 = 4.69$ GPa (680 ksi) for the approximately transversely isotropic ring and steel impacter [10]. Values of n_0 were also determined experimentally by directly measuring contact radii and displacements. From the measurements of contact radii, the average value of n_0 was 3.98 GPa (577 ksi) [3–5]; and, from the displacement measurements, the average value of n_0 was 4.52 GPa (656 ksi)

[10]. These values of n_0 agree quite well. The value, $n_0 = 4.52$ GPa (656 ksi), was used in all subsequent calculations.

Values of the spring constant, k_b, were 5.08 MN/m (29.0 kips/in.) and 6.34 MN/m (36.2 kips/in.), respectively, for an empty ring and a filled ring [3,4]. They were determined from quasi-static load-displacement curves. The value of k_b for the filled ring is only 25% greater than that for the empty ring. Thus, the inert solid propellant, which has a Young's modulus of 1.2 to 34 MPa (0.18 to 5.0 ksi) depending on loading rate, did not contribute substantially to the static compliance of the ring.

References

[1] Loyd, B. A. and Knight, G. K., "Impact Damage Sensitivity of Filament-Wound Composite Pressure Vessels," *1986 JANNAF Propulsion Meeting,* New Orleans, LA, CPIA Publication 455, Vol. 1, Johns Hopkins University, Laurel, MD, Aug. 1986, pp. 7–15.

[2] Poe, C. C., Jr., Illg, W., and Garber, D. P., "A Program to Determine the Effect of Low-velocity Impacts on the Strength of the Filament-wound Rocket Motor Case for the Space Shuttle," NASA TM-87588, NASA Langley Research Center, Hampton, VA, Sept. 1985.

[3] Poe, C. C., Jr., Illg, W., and Garber, D. P., "Tension Strength of a Thick Graphite/epoxy Laminate after Impact by a ½-In.-Radius Impacter," NASA TM-87771, NASA Langley Research Center, Hampton, VA, July 1986.

[4] Poe, C. C., Jr., and Illg, W., "Strength of a Thick Graphite/Epoxy Rocket Motor Case After Impact by a Blunt Object," *Test Methods for Design Allowables for Fibrous Composites, ASTM STP 1003,* C. C. Chamis, Ed., American Society for Testing and Materials, Philadelphia, 1989, pp. 150–179; also in NASA TM-89099, NASA Langley Research Center, Hampton, VA, Feb. 1987, and in *1987 JANNAF Composite Motor Case Subcommittee Meeting,* CPIA Publication 460, Feb. 1987, pp. 179–202.

[5] Poe, C. C., Jr., C. E. Harris, and D. H. Morris, "Surface Crack Analysis Applied to Impact Damage in a Thick Graphite/Epoxy Composite," *Surface Crack Growth: Models, Experiments and Structures, ASTM STP 1060,* W. G. Reuter, J. H. Underwood, and J. A. Newman, Jr., Eds., American Society for Testing and Materials, Philadelphia, 1989, pp. 194–212; Also in NASA TM-100600, NASA Langley Research Center, Hampton, VA, April 1988.

[6] Poe, C. C., Jr., Illg, W., and Garber, D. P., "Hidden Impact Damage in Thick Composites," *Proceedings,* Review of Progress in Quantitative Nondestructive Evaluation, D. O. Thompson and D. E. Chimenti, Eds., Vol. 5B, Plenum Press, New York, 1986, pp. 1215–1225.

[7] Madaras, E. I., Poe, C. C., Jr., Illg, W., and Heyman, J. S., "Estimating Residual Strength in Filament Wound Casings from Non-Destructive Evaluation of Impact Damage," *Proceedings,* Review of Progress in Quantitative Nondestructive Evaluation, Vol. 6B, Plenum Press, New York, 1986, pp. 1221–1230.

[8] Madaras, E. I., Poe, C. C., Jr., and Heyman, J. S., "Combining Fracture Mechanics and Ultrasonics NDE to Predict the Strength Remaining in Thick Composites Subjected to Low-Level Impact," *Proceedings,* 1986 Ultrasonics Symposium, B. R. McAvoy, Ed., Institute of Electrical and Electronic Engineers, New York, Vol. 86CH2375-4, No. 2, pp. 1051–1059.

[9] Madaras, E. I., Poe, C. C., Jr., and Heyman, J. S., "A Nondestructive Technique for Predicting the Strength Remaining in Filament Wound Composites Subjected to Low-Level Impact," *Proceedings, 1987 JANNAF Composite Motor Case Subcommittee Meeting,* Hampton, VA, CPIA Publication 460, Feb. 1987, Johns Hopkins University, Laurel, MD, pp. 249–258.

[10] Poe, C. C., Jr., "Simulated Impact Damage in a Thick Graphite/Epoxy Laminate Using Spherical Indenters," American Society for Composite Materials, Technomics Publishing Co., Inc., Lancaster, PA, Nov. 1988; also in NASA TM-100539, NASA Langley Research Center, Hampton, VA, Jan. 1988.

[11] Poe, C. C., Jr., "Relevance of Impacter Shape to Nonvisible Damage and Residual Tensile Strength of a Thick Graphite/Epoxy Laminate," NASA TM-102599, NASA Langley Research Center, Hampton, VA, Jan. 1990.

[12] Timoshenko, S. P., *History of Strength of Materials, with a Brief Account of the History of Theory of Elasticity and Theory of Structure,* Dover Publications, Inc., New York, 1982, pp. 178–180.

[13] Greszczuk, L. B., "Damage in Composite Materials due to Low Velocity Impact," *Impact Dynamics,* Wiley, New York, 1982, pp. 55–94.

[14] Love, A. E. H., "The Stress Produced in a Semi-Infinite Solid by Pressure on Part of the Boundary," *Philosophical Transactions,* Royal Society, London, Series A, Vol. 228, 1929, pp. 377–420.

[15] Johnson, K. L., "One Hundred Years of Hertz Contact," *Proceedings,* The Institution of Mechanical Engineers, Vol. 196, No. 39, London, England, 1982, p. 366.
[16] Newman, J. C., Jr., and Raju, I. S., "Stress-Intensity Factor Equations for Cracks in Three-Dimensional Finite Bodies," *Fracture Mechanics: Fourteenth Symposium—Volume I: Theory and Analysis, ASTM STP 791,* J. C. Lewis and G. Sines, Eds., American Society for Testing and Materials, Philadelphia, 1983, pp. I-238—I-268.

Ronald B. Bucinell,[1] Ralph J. Nuismer,[2] and Jim L. Koury[3]

Response of Composite Plates to Quasi-Static Impact Events

REFERENCE: Bucinell, R. B., Nuismer, R. J., and Koury, J. L., **"Response of Composite Plates to Quasi-Static Impact Events,"** *Composite Materials: Fatigue and Fracture (Third Volume), ASTM STP 1110,* T. K. O'Brien, Ed., American Society for Testing and Materials, Philadelphia, 1991, pp. 528–549.

ABSTRACT: A two degree of freedom, spring-mass model is used to investigate the response of filament-wound composite plates subjected to transverse impacts. From this model, it is concluded that it is the ratio of impactor to target frequencies that determines whether the structural response to impact is quasi-static or dynamic in nature. In addition to the model development, a test matrix is designed to isolate the effects of impactor mass, velocity, and energy on quasi-static impact response and damage formation. These data compared well with the model predictions.

KEY WORDS: composite materials, quasi-static impact, dynamic impact, Rayleigh-Ritz energy method, two degree of freedom model, drop tower, response, damage, post-impact compression panel, fatigue (materials), fracture

Nomenclature

$A_{I,II}$	Two degree of freedom system amplitudes
E	Kinetic energy of impactor
E_{ii}	Young's modulus
F	Contact force
F_{max}	Maximum contact force
G_{ij}	Shear modulus
k_b	Bending contribution to the structural stiffness
k_{eff}	Effective impact stiffness
k_m	Membrane contribution to the structural stiffness
k_s	Shear contribution to the structural stiffness
k_1	Linear contact stiffness
k_2	Structural stiffness
m	Impactor mass for one degree of freedom system
m_1	Impactor mass for two degree of freedom system
m_2	Target mass for two degree of freedom system
n	Hertz contact stiffness
P	Transverse static load
r	Impactor radius

[1] Research engineer, Materials Sciences Corporation, Blue Bell, PA 19422.
[2] Staff scientist, Hercules Aerospace, Magna, UT 84044.
[3] Aerospace engineer, Composites Laboratory, Air Force Astronautics Laboratory, Edwards Air Force Base, CA 93523-5000

t	Time
t_{dur}	Duration of impact event
v	Impactor velocity
v_0	Initial impactor velocity
$X_{I,II}$	Two degree of freedom system mode shapes
x_1	Displacement of impactor mass
x_2	Displacement of target mass
$\dot{x}_1$	Velocity of impactor mass
$\dot{x}_2$	Velocity of target mass
$\ddot{x}_1$	Acceleration of impactor mass
$\ddot{x}_2$	Acceleration of target mass
α	Contact deformation
δ	Static structural displacement
ω	Natural frequency of the one degree of freedom impact event
$\omega_{I,II}$	Roots of the frequency equation for the two degree of freedom system
ω_1	Natural frequency of the two degree of freedom impact event
ω_2	Natural frequency of the target
ν_{ij}	Poisson's ratio

In typical aerospace environments, composite materials are subjected to a wide range of foreign body impact events. Accurate modeling of an impact event is necessary for a reliable assessment of structural degradation. The nature of the impact event, meaning whether the response can be classified as quasi-static, dynamic, or a combination of both, can influence the extent of the degradation. The identification of the physical parameters or properties that determine the nature of an impact response have been elusive.

When the duration of an impact event is many times longer than the time for generated stress waves to travel to the outer boundary of the plate and return, the effects of higher vibrational modes are small and can be neglected [1]. This type of impact response is considered to be quasi-static in nature. As the duration of the impact event becomes shorter and becomes less than the time for generated stress waves to rebound from the outer boundary, the higher vibrational modes have a significant effect on the response [1]. This type of impact response is considered dynamic in nature.

In the literature, there appears to be some confusion over what physical parameters or properties, of the target and impactor, influence the nature of the impact response. Many investigators [2–5] stipulate that the level of impactor velocity determines the nature of the impact response. Nuismer et al. [6], with the use of a Rayleigh-Ritz energy model [7], found that the ratio of the natural frequencies of the impact and target determined the nature of the impact response and that velocity only influences the magnitude of the response.

In this paper, the apparent discrepancy over which parameters or properties influence the nature of impact response is investigated. The experimental approach taken is to develop a test matrix designed to isolate the effects of impactor mass, velocity, and energy. This enables direct observations of the effects of these parameters on the nature of impact response and the formation of damage.

In addition to the experimental investigation, a two degree of freedom, spring-mass model is used to aid in the understanding of the nature of impact response. This model is used because it is a generalization of all other models. A number of other investigators have used spring-mass models to study impact, [4,5,8,9]. Some included one degree of freedom systems where the mass of the target is negligible compared to that of the impactor or the mass of the target and impactor move as one after impact [8]. Other models include two degree of freedom systems where both the mass of the target and impactor are considered; the contact spring can be linear or nonlinear Hertzian [4,5]; and the target stiffness is modeled as an effective bending stiffness or simply ignored if the mass of the target is very large compared to that of the impactor [9]. The model that is presented in this paper starts with a general two mass-two spring system. The structural frequency is expanded in terms of the ratio of the natural frequencies of the impact and target as suggested in Ref 6. This expansion facilitates the isolation of the parameters and properties that influence the nature of the response.

The discussion in this paper begins with explanations of the test specimens and apparatus used in the experimental investigation. This is followed by a detailed discussion of the test matrix design. The details of the two-degree of freedom model are then presented. This presentation included arguments for identifying the parameters and properties that influence the nature of impact response. Finally, the correlation between the two-degree of freedom model and the experimental data is discussed.

Experimental Investigation

The objective of this experimental investigation is to isolate the effects of impactor mass, velocity, and energy on the nature of impact response and damage formation. Flat, filament-wound, post-impact compression (PIC) coupons are used in the investigation because they are a recognized impact evaluation specimen.

Equipment

All impact testing was conducted at the Astronautics Laboratory on Edwards Air Force Base in California. A DynaTup drop weight impact tower equipped with an IBM/AT computer, 15.9-mm (⅝-in.)-diameter impactor tup, and a load cell, was used to perform all impacts. The fixture used to constrain the test specimens during testing is seen in Fig. 1. The post-impact detection of damage in the specimens was performed by the Advanced Methods Group at Hercules Aerospace in Utah using a gated-pulse echo ultrasonic inspection technique.

Specimen Preparation

In this investigation, materials typical of aerospace structure were used. Test specimens were made of Hercules IM7X/M fiber and HBRF-55A resin. Lamina properties for this material system are

$$E_{11} = 152. \text{ GPa} = 22.0 \text{ Msi}$$
$$E_{22} = 6.48 \text{ GPa} = 0.94 \text{ Msi}$$
$$G_{12} = 2.76 \text{ GPa} = 0.40 \text{ Msi} \tag{1}$$
$$\nu_{12} = 0.272$$
$$\nu_{23} = 0.311$$

The geometry of the panels was $[\pm 18/90_2/\pm 18]_s$. The thickness of each of the plates was approximately 3.56 mm (0.14 in.), and the fiber volume fraction of the plates was approximately 60%.

HBRF-55A is a wet winding resin. Manufacturing wet-wound flat plates is difficult. Each ply of the plates, in this investigation, was wound on a 762-mm (30-in.)-diameter mandrel, cut, laid flat, and placed into a freezer. The plies were placed in the freezer to stop the B-staging process. Once all the lamina were wound, they were taken out of the freezer and stacked in the appropriate sequence. Two such panels were manufactured. These panels were then cut into

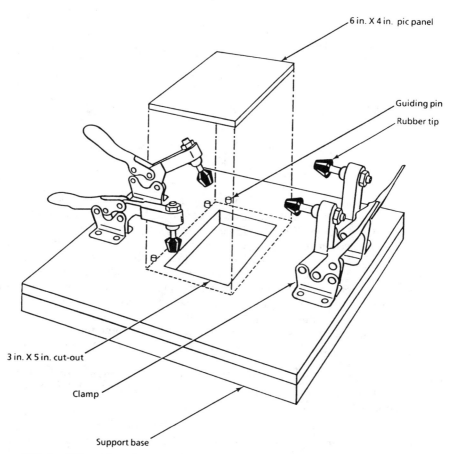

FIG. 1—*PIC panel test fixture used to constrain test specimens during drop tower impacts.*

508 by 508 mm (20 by 20 in.) plates. The plates were then placed between cual plates, vacuum bagged, and cured under pressure. The cured plates were then cut into 152 by 102 mm (6 by 4 in.) specimens.

Test Matrix Development

The objective of the test matrix is to isolate the effects of impactor mass, m, velocity, v, and kinetic energy, E, on response and damage formation. These parameters are fundamentally related by

$$E = \tfrac{1}{2}mv^2 \qquad (2)$$

Equation 2 implies that impactor energy is dependent on mass and velocity; this is true from an experimental viewpoint. Therefore, a two-dimensional table with mass being varied along the horizontal axis and velocity being varied along the vertical axis will result in the internal elements of the table representing various energy levels (see Table 1). If Table 1 is to be used

TABLE 1—*Test matrix that isolates the effects of impactor mass, velocity, and energy.*

	v_1	v_2	v_3	v_4	v_5	v_6
m_1	E_1	E_{10}	$\cdots$	E_4	E_2	E_6
	(1)	(2)		(3)	(4)	(5)
m_2	E_2	E_3	E_4	$\cdots$	E_7	E_8
	(6)	(7)	(8)		(9)	(10)
m_3	E_3	$\cdots$	E_5	E_7	$\cdots$	E_9
	(11)		(12)	(13)		(14)
m_4	E_4	E_5	E_6	E_8	E_9	$\cdots$
	(15)	(16)	(17)	(18)	(19)	

to study the independent effects of the parameters under consideration, the additional constraint of constant energy level along the diagonals of the table must be imposed. The derivation of expressions for elements in Table 1 becomes an exercise in solving a system of algebraic equations; however, there are more unknowns than equations. Thus, further constraints must be placed on this system of equations if they are to be solved.

In addition to the previously mentioned constraints, there are physical constraints that result from the equipment being used in this investigation. Drop towers use plates to adjust the impactor mass. This results in a discrete mass increment constraint on the test matrix. Drop towers also have maximum velocity contraints that result from the limited height of the impactor rails. Although there is no physical constraint on the minimum velocity of the impactor, there is a practical constraint on accurately predicting impactor heights for extremely low velocities.

With the addition of the physical constraints, the system of equations represented in Table 1 can now be solved. The valves for mass, velocity, and energy found in Eqs 3, 4, and 5, respectively, result from this solution.

$$m_1 = 4.99 \text{ kg} = 11 \text{ lb}_m = 0.342 \frac{\text{lb s}^2}{\text{ft}} \quad m_2 = 7.71 \text{ kg} = 17 \text{ lb}_m = 0.528 \frac{\text{lb s}^2}{\text{ft}} \quad (3)$$

$$m_3 = 15.1 \text{ kg} = 33 \text{ lb}_m = 1.034 \frac{\text{lb s}^2}{\text{ft}} \quad m_4 = 25.1 \text{ kg} = 55 \text{ lb}_m = 1.717 \frac{\text{lb s}^2}{\text{ft}}$$

$$v_1 = 1.13 \text{ m/s} = 3.71 \text{ ft/s} \quad v_2 = 1.58 \text{ m/s} = 5.19 \text{ ft/s} \quad v_3 = 2.04 \text{ m/s} = 6.69 \text{ ft/s} \quad (4)$$
$$v_4 = 2.53 \text{ m/s} = 8.31 \text{ ft/s} \quad v_5 = 3.81 \text{ m/s} = 11.6 \text{ ft/s} \quad v_6 = 4.57 \text{ m/s} = 15.0 \text{ ft/s}$$

$$E_1 = 3.19 \text{ J} = 2.35 \text{ ft·lb} \quad E_2 = 4.92 \text{ J} = 3.63 \text{ ft·lb} \quad E_3 = 9.64 \text{ J} = 7.11 \text{ ft·lb}$$
$$E_4 = 16.0 \text{ J} = 11.8 \text{ ft·lb} \quad E_5 = 31.4 \text{ J} = 23.1 \text{ ft·lb} \quad E_6 = 52.1 \text{ J} = 38.4 \text{ ft·lb} \quad (5)$$
$$E_7 = 48.4 \text{ J} = 35.7 \text{ ft·lb} \quad E_8 = 80.4 \text{ J} = 59.3 \text{ ft·lb} \quad E_9 = 157 \text{ J} = 116. \text{ ft·lb}$$
$$E_{10} = 6.24 \text{ J} = 4.60 \text{ ft·lb}$$

In Table 1, the numbers in parentheses designate test condition. At each test condition, six replicates were performed. Six replicates were used in order to minimize anomalous effects that result from the experimental procedures and to acquire significant estimates of the experimental variance.

Two Degree of Freedom Model

To investigate the physics of the impact event, the two degree of freedom spring-mass model illustrated in Fig. 2 is employed. This is the simplest system that includes both the contact and

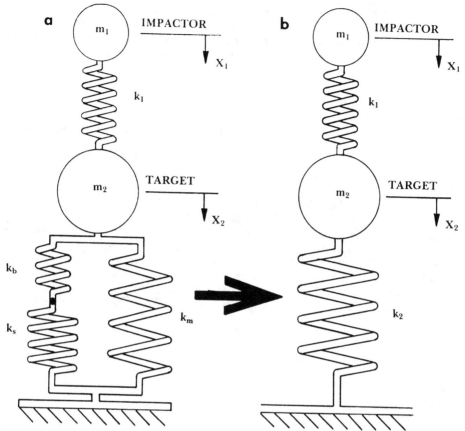

FIG. 2—*Two degree of freedom representation of a projectile in contact with a target during an impact event.*

structural response. This model does ignore the distribution of mass in a real structure; however, it serves a very important role in separating out the effects of the various impact parameters on the system response.

The spring-mass model seen in Fig. 2a represents an impactor of mass, m_1, striking a plate of mass, m_2, at a velocity, v_0. The force, F, generated during this impact event is a result of two types of deformations [4]. The first type of deformation represents the difference between the displacement of the target, x_2, and impactor, x_1, while the bodies are in contact, and is referred to as the contact deformation (α). Associated with the contact deformation is the contact stiffness k_1. The second type of deformation is the transverse deflection of the plate, x_2. The transverse deflection is the result of the bending, transverse shear, and membrane deformations [5]. Associated with these deformations are the bending, k_b, transverse shear, k_s, and membrane, k_m, stiffnesses. These stiffnesses combine to form the structural stiffness, k_2.

Structural Stiffness Simplification

For isotropic and specially isotropic plates, closed-form expressions for the bending, shear, and membrane stiffnesses exist [5]. Attempting closed-form solutions for these stiffnesses in an orthotropic rectangular plate, like those being used in this investigation, would be an arduous task. A simpler approach for obtaining the structural stiffness is to employ the finite element technique. Using shear deformable shell elements and the appropriate boundary conditions, the structural stiffness, k_2, can be calculated by applying a transverse load, P, at the center of the plate and measuring the transverse deflection, δ, directly under that load,

$$k_2 = \frac{P}{\delta} \tag{6}$$

This approach results in the simplification of the model seen in Fig. 2a to the model seen in Fig. 2b.

Linear Contact Stiffness

The contact load and deformation are related by the well-known, nonlinear, Hertz law [8]. The nonlinear nature of Hertzian contact makes a numerical solution necessary [5]. Numerical solutions impede the fundamental understanding of the impact event; therefore, the relationship between contact load and deformation must be linearized.

To start the contact stiffness linearization development, consider the one degree of freedom spring mass system seen in Fig. 3. This system represents a mass, m, with a velocity, v, striking an infinite half space. The spring represents the contact stiffness. For the case of a linear contact spring, k, the maximum force, F_{max}, and impact event duration, t_{dur}, are given by

$$F_{max} = \sqrt{2kE} \tag{7}$$

where

$$\omega = \sqrt{k/m}$$
$$E = \tfrac{1}{2}mv^2$$

and

$$t_{dur} = \frac{\pi}{\omega} = \pi\sqrt{m/k} \tag{8}$$

For the case of a nonlinear contact spring, n, the Hertz contact law is given by [8]

$$F(x) = nx^{3/2} \tag{9}$$

For this case, the maximum force, F_{max}, and event duration, t_{dur}, are [8]

$$F_{max} = 1.733n^{2/5}E^{3/5} \tag{10}$$

and

$$t_{dur} = 1.609(m/n)^{2/5}v^{-1/5} \tag{11}$$

The following observations are made when the linear (Eqs 7 and 8) and nonlinear (Eqs 9 and 10) solutions to the system in Fig. 3 are compared. The magnitude of the force applied to the target depends on the contact stiffness and initial kinetic energy of the impactor for both cases; however, the force magnitude for the nonlinear case is no longer linearly dependent on the velocity of the impactor as it is for the linear case. In both cases, the duration of the applied

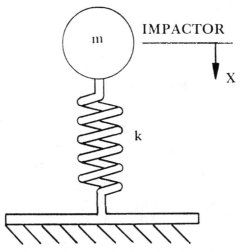

FIG. 3—*One degree of freedom representation of a rigid sphere impacting a specially isotropic target.*

force is dependent on the mass and contact stiffness; however, in the nonlinear case there is an additional dependency on the initial velocity.

The dependency of the maximum contact force and duration on the initial velocity of the impactor is very weak. By equating the maximum forces found in Eqs 7 and 10, an "equivalent" linear contact stiffness can be formed

$$k_1 = 1.502E^{1/5}n^{4/5} \qquad (12)$$

Substituting this result into Eq 8 results in a contact duration for the linearized system

$$t_{dur} = 1.374 \left(\frac{m}{n}\right)^{2/5} v^{-1/5} \qquad (13)$$

Comparing Eq 11 to Eq 13, it is seen the the contact duration given by the "equivalent" linear analysis results in 85% of the contact duration given by the "exact" nonlinear analysis. This leads to the conclusion that the nonlinearity is not that significant in influencing the impact response and that the response can be reasonably estimated using an "equivalent" linear analysis. This conclusion is further supported in Ref 6 where a point by point comparison was made between responses with and without nonlinear contact springs. The effects of linearization on the spring-mass model illustrated in Fig. 2 are discussed further in subsequent sections of this paper.

Two Degree of Freedom System Solution

The two degree of freedom problem illustrated in Fig. 2b has not been simplified to include an "equivalent" linear contact spring and a single structural spring. The equations of motion for this system are easily derived from Newton's second law of motion and are given in Eq 14

$$k_1(x_2 - x_1) = m_1\ddot{x}_1 \qquad (14)$$
$$-k_1(x_2 - x_1) - k_2x_2 = m_2\ddot{x}_2$$

The initial conditions appropriate for the problem are found in Eq 15

$$x_1(0) = x_2(0) = 0$$
$$\dot{x}_1(0) = v_0, \quad \dot{x}_2(0) = 0$$

(15)

where v_0 is the initial velocity of the impactor as it makes contact with the target. The solution to this system of equations is given in Eq 16

$$x_1(t) = v_0(A_I \sin \omega_I t + A_{II} \sin \omega_{II} t)$$
$$x_2(t) = v_0(X_I A_I \sin \omega_I t + X_{II} A_{II} \sin \omega_{II} t)$$

(16)

In Eq 16, the amplitudes are given by

$$A_I = 1/(1 - X_I/X_{II})\omega_I$$
$$A_{II} = 1/(1 - X_{II}/X_I)\omega_{II}$$

(17)

the mode shapes by

$$X_I = 1 - (m_1/k_1)\omega_I^2$$
$$X_{II} = 1 - (m_1/k_1)\omega_{II}^2$$

(18)

the roots of the frequency equation by

$$\omega_{I,II}^2 = \frac{(k_1 + k_2)m_1 + k_1 m_2 \pm \sqrt{[(k_1 + k_2)m_1 + k_1 m_2]^2 - 4k_1 k_2 m_1 m_2}}{2m_1 m_2}$$

(19)

and the contact force by

$$F(t) = k_1(x_2 - x_1)$$
$$= -m_1 v_1(\omega_I^2 A_I \sin \omega_I t + \omega_{II}^2 A_{II} \sin \omega_{II} t)$$

(20)

From this solution, it is seen that the mode shapes (Eq 18) and frequencies of motion (Eq 19) are totally independent of the impactor's initial velocity, v_0. The impactor's initial velocity only influences the magnitude of the resulting displacements (Eq 16) and contact force (Eq 20) but not the "type" of response.

Further insight into the parameters that influence the type of impact response can be gained through manipulation of the two degree of freedom solution. As suggested in Ref 6, the roots of the frequency equation (Eq 19) are expanded in terms of the ratio of the individual natural frequencies of the impactor mass contact spring (ω_1) and the target mass-spring (ω_2)

$$\omega_1 = \sqrt{k_1/m_1}$$

(21)

$$\omega_2 = \sqrt{k_2/m_2}$$

(22)

Asymptotic expansions for the cases of $(\omega_1/\omega_2)^2 \ll 1$ and $(\omega_1/\omega_2)^2 \gg 1$ are then performed.

Quasi-Static Impact ($(\omega_1/\omega_2)^2 \ll 1$)

This case could be referred to as large mass impact, because the inequality, $(\omega_1/\omega_2)^2 \ll 1$, also implies

$$m_1 \gg \left(\frac{k_1}{k_2}\right) m_2 \tag{23}$$

where the contact stiffness is typically several orders of magnitude larger than the structural stiffness, k_2. As discussed previously, this type of impact response is sometimes referred to as low-velocity impact. This may be a result of the observation that large mass impactors in general need lower velocities to cause substantial damage levels in composites. However, as previously shown, the velocity does not directly determine the "type" of response obtained in an impact event.

Expanding the roots of the frequency equation (Eq 19) in terms of ω_1/ω_2 results in

$$\omega_{I,II}^2 = \frac{1}{2} \omega_2^2 \left\{ (K+1) + \left(\frac{\omega_1}{\omega_2}\right)^2 \pm \sqrt{\left[(K+1) + \left(\frac{\omega_1}{\omega_2}\right)^2\right]^2 - 4\left(\frac{\omega_1}{\omega_2}\right)^2} \right\} \tag{24}$$

where

$$K = \frac{k_1}{k_2}.$$

Substitution of Eq 24 into the solution of the two degree of freedom system and keeping only the highest order terms, the solution for the contact load is found to be

$$F(t) = -\sqrt{k_{eff} m_1 v_1^2} \sin \omega_1 t - \frac{K}{1+K} \sqrt{k_{eff} m_1 v_1^2} \left(\frac{\omega_1}{\omega_2}\right) \sin \omega_{II} t \tag{25}$$

where

$$k_{eff} = \frac{k_1 k_2}{k_1 + k_2}.$$

The first term on the right-hand side of Eq 25 is exactly the response of a mass, m_1, hitting an "effective" spring (k_1 attached directly to k_2 in series) as shown in Fig. 4a. The second term represents the oscillation of mass, m_2, in the manner illustrated in Fig. 4b. Careful examination of Eq 25 shows that the first term (or mode) dominates the solution. The second term has an ω_1/ω_2 term that is assumed to be much less than one in this expansion. This diminishes the effect of the second term. Therefore, the duration of the event is determined by the period of the first term in Eq 24 and is given by

$$t_{dur} = \pi \sqrt{\frac{m_1}{k_{eff}}} \tag{26}$$

Comparing the event duration to the structural wave reflection time, which is on the order of

$$t_{ref} = \sqrt{\frac{m_2}{k_2}} \tag{27}$$

it is seen that $t_{dur} \gg t_{ref}$. This implies that multiple wave reflections occur during the impact event that leads to the conclusion that the impact event is quasi-static in nature [1].

Also of interest is an expression for the maximum load for the quasi-static case. From Eq

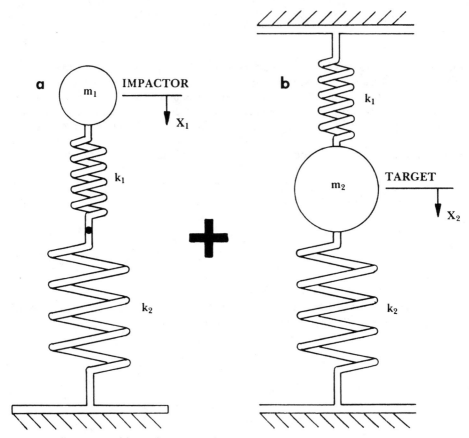

FIG. 4—*Illustrations of the modes associated with the response of the two degree of freedom system in the quasi-static domain of impact.*

25, again assuming the first term to be dominant, the expression for the maximum load is found to be

$$F_{\text{max}} = \sqrt{2k_{\text{eff}}E} \qquad (28)$$

where

$E = \frac{1}{2}m_1 v_0.$

Equations 26 and 28 will be useful when comparing the model to the experimental results.

Dynamic Impact $((\omega_1/\omega_2)^2 \gg 1)$

This case could be referred to as small mass impact because the inequality, $(\omega_1/\omega_2)^2 \gg 1$, also implies

$$m_1 \ll \left(\frac{k_1}{k_2}\right) m_2 \qquad (29)$$

This type of impact response is sometimes referred to as high-velocity impact. This may be a result of the observation that small mass impactors in general need higher velocities to cause substantial damage levels in composites. However, as previously discussed, the velocity does not directly determine the "type" of response obtained in an impact event.

Expanding the roots of the frequency equation (Eq 19) in terms of ω_2/ω_1 results in

$$\omega_{I,II}^2 = \frac{1}{2}\,\omega_1^2 \left\{ (M + 1) + \left(\frac{\omega_2}{\omega_1}\right)^2 \pm \sqrt{\left[(M + 1) + \left(\frac{\omega_2}{\omega_1}\right)^2\right]^2 - 4\left(\frac{\omega_2}{\omega_1}\right)^2} \right\} \quad (30)$$

where

$$M = \frac{m_1}{m_2}.$$

Substitution of Eq 30 into the solution of the two degree of freedom system and keeping only the highest order terms results in a solution for the dynamic response of the two degree of freedom system seen in Fig. 2b. The response for the dynamic case, $(\omega_1/\omega_2)^2 \gg 1$, is exactly the response of the two mass-two spring system in the limit as $k_2 \to 0$, superimposed on the larger plastic impact response. The first portion of the response, the limit as $k_2 \to 0$, is the vibrational response of m_1 hitting a mass, m_2, that has no structural stiffness, k_2. This is illustrated in Fig. 5a. The second portion of the response, the plastic impact, is the response of the lumped mass, $m_1 + m_2$, on the structural spring, k_2, after m_1 plastically impacts m_2. This is illustrated in Fig. 5b.

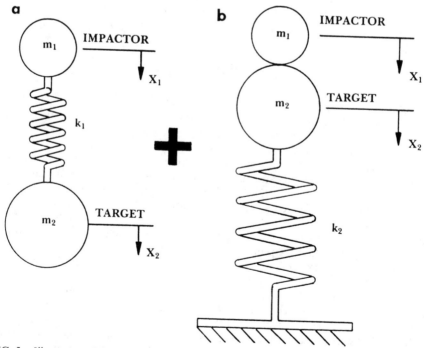

FIG. 5—*Illustration of the modes associated with the response of the two degree of freedom system in the dynamic domain of impact.*

The asymptotic solution for the contact load is given in Eq 31.

$$F(t) = -\frac{M}{1 + M} \sqrt{\frac{k_1 m_1 v_1^2}{1 + M}} \left(\frac{\omega_2}{\omega_1}\right) \sin \omega_1 t - \sqrt{\frac{k_1 m_1 v_1^2}{1 + M}} \sin \omega_{II} t \qquad (31)$$

The first term on the right-hand side of Eq 31 contains an ω_1/ω_2 term that approaches zero in the asymptotic expansion; therefore, the magnitude of the contact force response is dominated by the second term. Considering only this term, the expansion for the maximum contact force is given by

$$F_{max} = \sqrt{\frac{2k_1 E_1}{1 + M}} \qquad (32)$$

If it is assumed that the impact event terminates when the sign of the contact force changes, the expression for the impact event duration is given by

$$t_{dur} = \pi \sqrt{\frac{m_1 m_2}{k_1(m_1 + m_2)}} \qquad (33)$$

A comparison of Eq 33 with the structural wave reflection time (Eq 27) indicates that $t_{dur} \ll t_{ref}$. This verifies that this type of impact event is truly dynamic in nature [1].

The assumption that the impact event terminates when the sign of the contact force changes requires further consideration for the case of dynamic impact events. The asymptotic solution for the impactor velocity during a dynamic impact event is given by

$$\dot{x}_1(t) = \frac{M}{1 + M} v_0 \cos \omega_1 t + \frac{1}{1 + M} v_0 \cos \omega_{II} t \qquad (34)$$

Note that when the impactor to target mass is small ($M = m_1/m_2 \ll 1$), the second term on the right-hand side of Eq 34 dominates the velocity response, and when this ratio is large ($M = m_1/m_2 \gg 1$), the first term dominates the velocity response. For the case of the small impactor-to-target mass ratio, the frequency of the force response and the impactor velocity response are the same, but are out of phase. At the same time the contact force changes sign, the impactor velocity is negative. This indicates that the impactor has separated from the target and is traveling away from the target. This is the scenario used to develop Eq 33. For the case of the large impactor mass ratio, the frequency of the velocity response is much greater than the frequency of the contact force response. For this case, at the time the contact force changes sign the impactor velocity is still positive. This indicates that the impactor has separated from the target, but the impactor is still traveling towards the target. For this case multiple impacts will occur until the momentum of the impactor is changed. For this scenario, Eq 33 will only predict the duration of the first contact between the target and impactor.

There are recognized shortcomings to the two degree of freedom model presented here. The lumped-mass model does not account for the actual mass and stiffness distribution of the structure that gives rise to higher vibrational modes. The model ignores the nonlinearity inherent in the contact stiffness. Also, the model does not explicitly account for the impactor-target loss of contact. The alternative to making these simplifications is to construct finite element or Reiligh-Ritz type models. These models will accurately model the impact event; however, these methodologies result in numerical solutions that do not lend themselves to gaining insight into the effects of parameters on impact response. It is the opinion of the authors that these shortcomings will not affect the general conclusions derived from the presented model. Previous studies support this opinion [6,10].

Discussion of Results

The data generated using the test matrix found in Table 1 and the predictions of the two degree of freedom model can now be compared. The proper application of the two degree of freedom model requires the determination of the domain of impact for each of the test conditions under consideration. Once the domain of impact has been established, a qualitative and quantitative assessment of the models correlation with the experimental data can be performed. A strong correlation between the model and the experimental data will support the hypothesis that the domain of impact response is determined by the ratio of the impact to target frequencies.

Calculation of the Domain of Impact

It is hypothesized here that the domain of impact is determined by the ratio of the impact to target frequencies. The impact and target frequencies are defined in Eqs 21 and 22, respectively. This ratio is used subsequently to determine which of the previously discussed asymptotic models is appropriate for comparison with the experimental data.

The effective target mass, m_2, and the target stiffness, k_2, used in the calculation of the target frequency, ω_2, are constant for all test conditions in this investigation. The mass of a typical plate, m_p, in this study is

$$m_p = 46.83(10^{-3})\text{kg} = 103.2(10^{-3})\text{lb}_m \tag{35}$$

The effective mass of a plate, m_2, is the portion of the mass that contributes to inertial effects and is approximately one fourth the total mass [11], therefore

$$m_2 = \tfrac{1}{4}m_p = 11.71(10^{-3})\text{kg} = 25.80(10^{-3})\text{lb}_m \tag{36}$$

The structural stiffness, k_2, of the plate used in this investigation, calculated using the previously discussed finite element approach, is

$$k_2 = 1.128(10^6)\text{N/m} = 77.28(10^3)\text{lb/ft} \tag{37}$$

Equations 36 and 37 can now be substituted into Eq 22 to calculate the target frequency, ω_2

$$\omega_2 = \sqrt{k_2/m_2} = 9819 \text{ Hz} \tag{38}$$

The impactor mass, m_1, linearized contact stiffness, k_1, and the impactor frequency, ω_1, vary between test conditions. The impactor mass used for each test condition is summarized in Table 2. The value of the linear contact spring is calculated from Eq 12 using the Hertz contact stiffness that is constant for the plates being used in this investigation. The value of the Hertz contact stiffness is calculated using the development by Greszczuk [4], for contact between a specially isotropic target and an infinitely stiff spherical impactor

$$n = \tfrac{4}{3}\sqrt{r}\, E_{zz} = 1.539(10^9)\text{N/m}^{3/2} = 58.23(10^6)\text{lb/ft}^{3/2} \tag{39}$$

where

r = radius of the impactor and
E_{zz} = transverse modulus of the plates.

Using this value for the Hertz contact stiffness and the appropriate energy level, the linearized contact stiffness can be calculated from Eq 12. This calculation is summarized in Table 2. Now the respective impactor masses and linear contact stiffnesses can be substituted into Eq 21 to determine the impact frequencies for each of the test conditions. This calculation, along with the calculation of the ratio of the impact to target frequencies squared, $(\omega_1/\omega_2)^2$, are summarized in Table 2.

The ratios of the impact to target frequencies summarized in Table 2 clearly indicate that all the test conditions under consideration are quasi-static in nature. Therefore, the quasi-static asymptotic expansion of the two degree of freedom model will be compared to the experimental data.

Qualitative Comparison

The hypothesis that the domain of an impact event is primarily influenced by the ratio of the impact-to-target frequencies can best be explored through a qualitative comparison of the asymptotic expansion of the model and the experimental data. A qualitative comparison enables the accuracy of the trends in the model to be examined without concern for the precision of the predictions. This type of comparison is facilitated by the structure of the test matrix seen in Table 1.

The impact-to-target frequency ratios listed in Table 2 indicate that all the impact events under consideration should respond in a quasi-static manner. Equation 25 suggests that the

TABLE 2—Summary of the parameters used in the calculations of impact domain and response behavior.

Test Condition	Energy Level		Linear Contact Stiffness, k_1		Effective Contact Stiffness, k_{eff}		Impactor Mass, m_1		ω_1,	$\omega_1/$
	J	ft·lb	n/m	lb/ft	n/m	lb/ft	kg	lb$_m$	Hz	ω_2
1	3.12	2.35	42.38 (10⁶)	2.90 (10⁶)	1.099 (10⁶)	75.3 (10³)	4.99	11.0	2914	0.30
2	6.24	4.60	74.09 (10⁶)	5.077 (10⁶)	1.111 (10⁶)	76.1 (10³)	4.99	11.0	3853	0.39
3	16.0	11.8	58.52 (10⁶)	4.010 (10⁶)	1.107 (10⁶)	75.8 (10³)	4.99	11.0	3425	0.35
4	31.4	23.1	66.95 (10⁶)	4.588 (10⁶)	1.109 (10⁶)	76.0 (10³)	4.99	11.0	3663	0.37
5	52.1	38.4	77.04 (10⁶)	5.279 (10⁶)	1.111 (10⁶)	76.2 (10³)	4.99	11.0	3929	0.40
6	4.92	3.63	48.47 (10⁶)	3.321 (10⁶)	1.102 (10⁶)	75.5 (10³)	7.71	17.0	2507	0.26
7	9.64	7.11	52.84 (10⁶)	3.621 (10⁶)	1.104 (10⁶)	75.7 (10³)	7.71	17.0	2618	0.27
8	16.0	11.8	58.52 (10⁶)	4.010 (10⁶)	1.107 (10⁶)	75.8 (10³)	7.71	17.0	2755	0.28
9	48.4	35.7	80.83 (10⁶)	5.539 (10⁶)	1.112 (10⁶)	76.2 (10³)	7.71	17.0	3238	0.33
10	80.4	59.3	92.47 (10⁶)	6.337 (10⁶)	1.114 (10⁶)	76.4 (10³)	7.71	17.0	3463	0.35
11	9.64	7.11	52.84 (10⁶)	3.621 (10⁶)	1.104 (10⁶)	75.7 (10³)	15.1	33.3	1871	0.19
12	31.4	23.1	66.94 (10⁶)	4.588 (10⁶)	1.109 (10⁶)	76.0 (10³)	15.1	33.3	2106	0.21
13	48.4	35.7	80.83 (10⁶)	5.539 (10⁶)	1.112 (10⁶)	76.2 (10³)	15.1	33.3	2314	0.24
14	157.	116.	102.3 (10⁶)	7.010 (10⁶)	1.116 (10⁶)	76.4 (10³)	15.1	33.3	2603	0.27
15	16.0	11.8	58.52 (10⁶)	4.010 (10⁶)	1.107 (10⁶)	75.8 (10³)	25.1	55.3	1527	0.16
16	31.4	23.1	66.94 (10⁶)	4.588 (10⁶)	1.109 (10⁶)	76.0 (10³)	25.1	55.3	1633	0.17
17	52.1	38.4	77.04 (10⁶)	5.279 (10⁶)	1.112 (10⁶)	76.1 (10³)	25.1	55.3	1752	0.18
18	80.4	59.3	92.47 (10⁶)	6.337 (10⁶)	1.114 (10⁶)	76.3 (10³)	25.1	55.3	1919	0.19
19	157.	116.	102.3 (10⁶)	7.010 (10⁶)	1.116 (10⁶)	76.4 (10³)	25.1	55.3	2019	0.21

CONSTANT MASS M₁

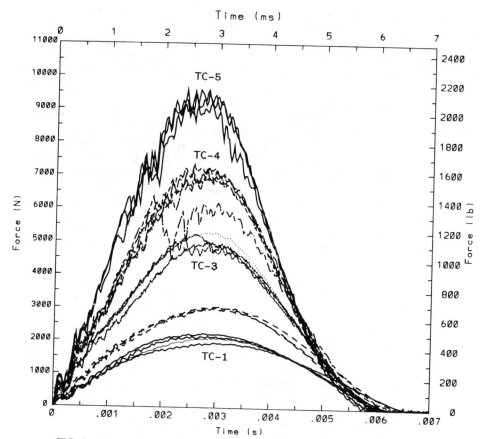

FIG. 6—*Contact force versus time for test conditions with common impactor mass.*

quasi-static contact force versus time response should appear as a single half sine wave. This differs from the dynamic contact force versus time response suggested by Eq 33 for the mass ratios involved. As previously discussed, the dynamic response would appear as a multi-peaked, possibly discontinuous, wave. The quasi-static, single half sine wave, behavior is observed in the experimental response curves illustrated in Fig. 6. The curves in Fig. 6 represent the five test condition in the Table 1 test matrix where the impactor mass was held constant at 4.99 kg (11 lb$_m$) (TC-1, TC-2, TC-3, TC-4, and TC-5).

The coincident intersection of the curves illustrated in Fig. 6 with the abscissa, to the right of the origin, implies that the impact event durations for these test conditions are equal. The quasi-static expression for the impact event duration, Eq 26, also suggests that the impact event durations for impact events with equal masses should be equal. The variables on the right-hand side of this equation (impactor mass, m_1) and effective stiffness, k_{eff}, are constant for the test conditions illustrated in Fig. 6. The constant nature of k_{eff} may not at first be obvious. Examination of the expression for k_{eff}, in Eq 25, indicates that when the linear contact stiffness, k_1, is much greater than the structural stiffness, k_2 (as is the case in this investigation),

k_{eff} will be a constant approximately equal to k_2. The constant nature of k_{eff} is reflected in Table 2.

In the dynamic domain of impact, the impact event duration, Eq 33, is dependent on the variables k_1, m_1, and m_2. The values of m_1 and m_2 are constant for the test conditions illustrated in Fig. 6. The values of k_1 in Table 2 vary significantly between these test conditions. This infers that if the test conditions illustrated in Fig. 6 responded in a dynamic manner, the corresponding impact event durations would vary significantly. This leads to the conclusion that the impact events in this investigation are responding in the quasi-static manner predicted by the impact to target frequency ratio.

It is suggested by Eq 28 that the maximum contact force generated during a quasi-static impact event is a constant for impacts with equal impactor kinetic energies. Equations 12 and 32 suggest that this trend also applies in the dynamic domain of impact. Illustrated in Fig. 7 are the experimental contact force versus time response curves for the three test conditions (TC-3, TC-8, and TC-15) along the 16.0 J (11.8 ft·lb) diagonal of the test matrix. The magnitudes of the maximum contact forces appear to follow the trend suggested by the model; however, a slight deviation from this trend is noticeable. As the mass of the impactor in these events increases, the magnitude of the maximum contact force also increases slightly. This slight increase in the maximum contact force correlates directly to the deviation observed between the experimentally measured impactor velocity and the target velocity values

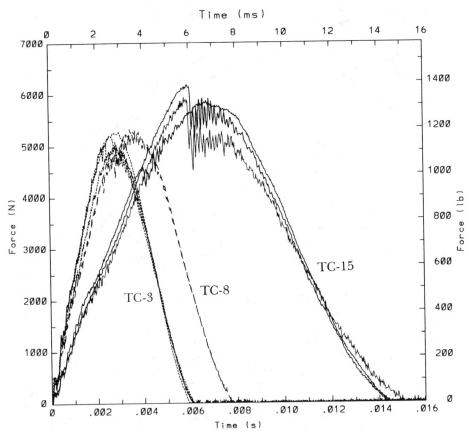

FIG. 7—*Contact force versus time for test conditions with common impactor energy.*

TABLE 3—*Comparison of target and experimentally measured impactor parameters for the test conditions along the (16.O-U) (11.8-ft·lb) energy diagonal of the test matrix.*

Test Conditions	Kinetic Energy of Impactor				Impactor Velocity				Impactor Mass,	
	Target,		Measured,		Target,		Measured,			
	J	(ft·lb)	J	(ft·lb)	J	(ft·lb)	J	(ft·lb)	kg	lb$_m$
TC-3	16.0	11.8	15.60	11.51	2.53	8.31	2.50	8.20	4.99	11.0
TC-8	16.0	11.8	16.05	11.89	2.04	6.69	2.04	6.70	7.71	17.0
TC-15	16.0	11.8	16.61	12.25	1.13	3.71	1.15	3.77	25.1	55.3

(Table 3). The impactor velocity error causes the experimental impactor kinetic energy to deviate from the target value (see Table 3) which, in turn, causes the magnitudes of the contact forces to vary from the predicted trend. The root cause of these errors has been identified as friction between the guide rails of the impact tower and the impactor sled. The effect of the guide rail friction is more significant as the mass decreased.

The impact-to-target frequency ratio predicts a quasi-static response in all the impact events under investigation in this study. The trends predicted by the quasi-static model correlate well with the behavior observed in the experimental data. This correlation supports the hypothesis that the domain of impact is dependent on the impact-to-target frequency ratio and not on the velocity of the impactor.

Quantitative Comparison

The precision of the two degree of freedom model in predicting the experimental response of the impact conditions under consideration can now be examined. Based on the previous discussion, the quasi-static model will be used to predict the experimental response. The discussion of the precision of the two degree of freedom model will be limited to the impact event duration and maximum contact force parameters.

Means of the experimentally determined impact event durations are plotted versus impactor mass in Fig. 8. At each of the impactor mass levels, multiple symbols appear that represent the various velocities associated with the test conditions under investigation in this study. Statistical analyses conducted on the data sets with common impactor mass concluded that no significant difference could be detected between these impact events. This supports the discussion in the previous section that suggested, for the quasi-static domain of impact, that the impact event duration was only a function of the impactor mass.

The dashed line in Fig. 8 represents the relationship between impactor mass and impact event duration as predicted by Eq 26. The impact event duration at all mass levels is predicted accurately by this equation. Only two test conditions deviate significantly form the prediction curve.

The 4.57 m/s (15 ft/s) data point in the 7.71 kg (17.0 lb$_m$) impactor mass data set falls well above the prediction and the other data points in the data set. The experimental contact force versus time curves for this test condition all exhibited a protracted decreasing slope; in addition, all specimens showed signs of extensive impactor penetration. These observations lead to the conclusion that the deviation from the prediction is a result of friction between the impactor and the specimen.

The other data point to deviate significantly from the prediction is the 3.46-m/s (11.63-ft/s), 25.1-kg (55.3-lb$_m$), test condition. This data point falls well below the prediction. Review of the test records indicate that during these impact events, impactor penetration occurred and the impactor lodged itself in the specimen during the rebound. The lodging of the impactor in

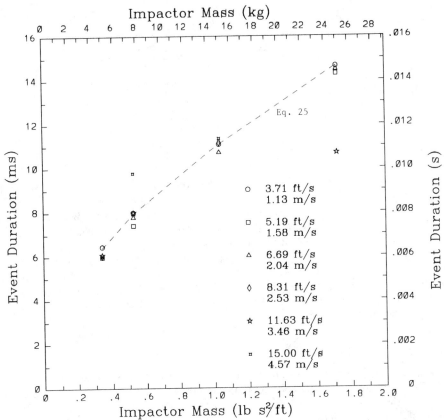

FIG. 8—*Comparison of the experimentally measured impact event duration with the analytical prediction.*

the specimen caused the impactor tup to prematurely sense a reversal of load. The point of load reversal was used in this investigation to determine impact event duration.

The maximum contact force generated during an impact event is plotted versus the experimentally measured impactor kinetic energy in Fig. 9. The dashed line in Fig. 9 represents the predicted maximum contact force versus kinetic energy relationship suggested by Eq 28. The prediction correlates well with the experimental data up to the 48-J (35-ft·lb) energy level. Above this energy level, the experimental data levels off. A post-impact examination of the specimens revealed that permanent indentations caused by the impactor also started to occur in the specimens at about the 48-J (35-ft·lb) impactor energy level. The simultaneous occurrence of these phenomena suggests that the leveling off load could be associated with the transverse compressive strength of the specimens. If it is, this would suggest that a transition from an elastic impact response to a plastic impact response is occurring at the impactor energy level of about 48 J (35 ft·lb).

Damage Formation

All the specimens tested in this program were C-scanned after being impacted. In Fig. 10, the mean delamination area detected by C-scan is plotted versus the impactor energy level for

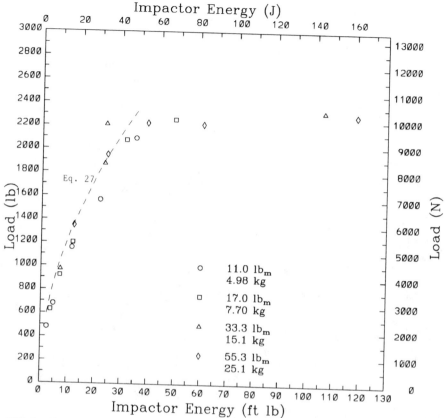

FIG. 9—*Comparison of the experimentally measured contact force with the analytical prediction.*

the specimens in this investigation. At about the 48-J (35-ft·lb) energy level, a leveling off of delamination area occurs. This suggests that delaminations are only formed during the elastic response phase of an impact event. Once the response transitions to the plastic phase, all energy goes into creating damage directly under the impactor.

The test conditions in this investigation all responded in the predicted quasi-static manner. This supports the hypothesis that the domain of an impact event is determined by the impact to target frequency ratio. Although the hypothesis was not tested in the dynamic domain of impact, the work presented here illustrates that the impactor mass, velocity, and energy level do not directly affect the type of plate response.

Conclusions

The objective of the investigation discussed in this paper was to determine the properties or parameters that influence the nature of the impact event. The approach taken was to develop a test matrix that would isolate the impact parameters and then compare the resulting experimental responses to the asymptotic solutions of a two degree of freedom model. The following conclusions resulted from this investigation:

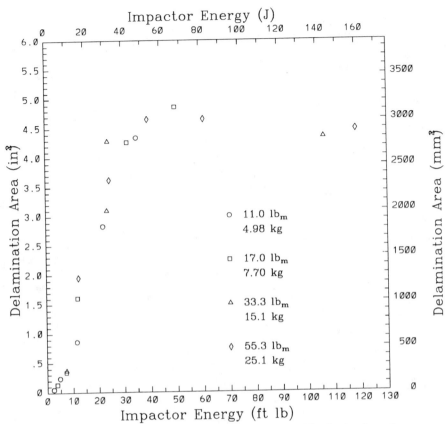

FIG. 10—*Illustration of the effect of impactor energy on delamination formation.*

1. In the quasi-static domain of impact events, the effects of higher vibrational modes can be neglected [*1*], therefore the two degree of freedom, spring-mass model presented here is an accurate representation of quasi-static impact events. The experimental and analytical results show that quasi-static impact events all have low impact-to-target frequency ratios. This leads to the conclusion that dynamic effects start to become more significant as the impact-to-target frequency ratio increases. This conclusion is supported by the two degree of freedom model presented; however, accurate representation of dynamic effects during impact requires the inclusion of higher vibrational modes that the two degree of freedom model presented here does not model.

2. The two degree of freedom model accurately predicts the trends and response of quasi-static impact events up to the point of impactor penetration.

3. At the point of impactor penetration, the impact deformation transitions from elastic to plastic. This transition point appears to be related to the transverse compressive strength of the specimens. Once the transition occurs, delamination formation ceases.

Acknowledgments

The authors would like to acknowledge Dr. David Cohen of Hercules Aerospace for his assistance in modeling the structural stiffness of the plates using a finite element technique.

The authors would also like to acknowledge Dr. William J. Murri and Mr. Bradley W. Shermon of Hercules Aerospace for their assistance in ultrasonically inspecting the specimens. The authors are appreciative of Mr. John O. Watson of Hercules Aerospace assistance in performing the impact testing. Finally, the authors would like to thank the U.S. Air Force for allowing us to use the facilities at the Astronautics Laboratory (AL) to conduct this study.

References

[1] Lord Rayleigh, "On the Prediction of Vibrations by Forces of Relatively Long Duration, with Application to the Theory of Collisions," *Philosophical Magazine,* Series 6, No. 11, 1906, pp. 283–292.

[2] Sjöblom, P. O., Hartness, J. T., and Cordell, T. M., "On Low-Velocity Impact Testing of Composite Materials," *Journal of Composite Materials,* Vol. 22, 1988, pp. 30–52.

[3] Schonberg, W. P., Keer, L. M., and Woo, T. K., "Low Velocity Impact of Transversely Isotropic Beams and Plates," *International Journal of Solids and Structures,* 1988.

[4] Greszczuk, L. B., "Damage in Composite Materials due to Low Velocity Impact," *Impact Dynamics,* Wiley, New York, 1982.

[5] Shivakumar, K. N., Elber, W., and Illg, W., "Prediction of Impact Force and Duration due to Low-Velocity Impact on Circular Composite Laminates," *Journal of Applied Mechanics,* Vol. 52, 1985, pp. 674–680.

[6] Nuismer, R. J., Bucinell, R. B., Morgan, M. E., and Cairns, D. S., "Scaling Impact Response and Damage in Composite Rocket Motor Cases," *Proceedings,* JANNAF Composite Motor Case Meeting, Precidio of San Francisco, San Francisco, CA, 17–19 May 1988.

[7] Cairns, D. S. and Lagace, P. A. "Transient Response of Graphite-Epoxy and Kevlar/Epoxy Laminates Subjected to Impact," *Proceedings,* AIAA/ASME/ASCE/AHS 29th Structural Dynamics and Materials Conference, Paper No. 88-2328, 18–20 April 1988, Williamsburg, VA.

[8] Goldsmith, W., *Impact, the Theory and Physical Behavior of Colliding Solids,* Edward Arnold Ltd., London, 1960.

[9] Poe, C. C., Illg, W., and Garber, D. P., "Tension Strength of a Thick Graphite/Epoxy Laminate after Impact by a ½ in. Radius Impactor," NASA Technical Memorandum 87771, NASA Langley Research Center, Hampton, VA, July 1986.

[10] Bucinell, R. B., Madsen, C. B., Nuismer, R. J., Benzinger, S. T., and Morgan, M. E., "Experimental Investigation of Scaling Impact Response and Damage in Composite Rocket Motor Cases," *Proceedings,* JANNAF Composite Motor Case and Structures and Mechanical Behavior Meeting, Jet Propulsion Laboratory, Anaheim, CA, 7–9 Nov. 1989.

[11] Leissa, A. W., "Vibrations of Plates," NASA-SP-160, NASA Langley Research Center, Hampton, VA, 1969.

Eugene T. Camponeschi, Jr.[1]

Compression of Composite Materials: A Review

REFERENCE: Camponeschi, E. T., Jr., **"Compression of Composite Materials: A Review,"** *Composite Materials: Fatigue and Fracture (Third Volume), ASTM STP 1110,* T. K. O'Brien, Ed., American Society for Testing and Materials, Philadelphia, 1991, pp. 550–578.

ABSTRACT: This paper is a literature review of research on the compressive response of fiber-reinforced composite materials. The review is organized by subject and the papers are included in one of three main sections; (1) Compression Test Methods, (2) Failure Theories and Failure Mechanisms, and (3) Experimental Investigations. The compression test method section includes a discussion of 15 compression test methods. The section on failure theories and failure mechanisms discusses failure theories and observations including those dealing with fiber microbuckling, transverse tension failure, and fiber kink-band formation. The section on experimental investigations summarizes data that is available in the literature. Tables are presented that list information on the evaluation of test methods and summarize experimentally developed data. These tables provide an index for quick reference to the research discussed in this paper.

KEY WORDS: composite materials, compression testing, compression failure, compression response, fracture, fatigue (materials)

The compressive response of fiber-reinforced composite materials has been the subject of investigation since the development of these materials. Even with this long-term interest, this area is still one of the least understood in the field of composites today. Many factors influence the compressive response of composite materials, and considered together or separately they can trigger a number of failure modes. These factors occur at the structural level (coupon geometry), the macrostructural level (lamina level), and the microstructural level (fiber-to-matrix level). On the microstructural level, the presence of local inhomogeneities and defects, which are difficult to characterize and model, influence the failure mechanisms that will dominate the response of a composite in compression more than in any other state of stress. Therefore, manufacturing plays a significant role in determining the compressive response of composite materials, and will play an even more significant role as the section thickness of composites increases. Constituent properties, laminate orientation, specimen geometry, method of load introduction, fiber waviness, voids, and stress concentrations all have been shown to play a role in determining the predominant failure mode governing compression failure. These failure modes include global Euler buckling, microbuckling, transverse tension, fiber kinking, fiber compression failures, matrix compression failures, or delamination.

The objective of this review paper is to present a summary of the recent research efforts that have concentrated on the compressive response of composite materials. The literature available in this area is voluminous, so the most recent work in the area (1980 to the present) was researched in the preparation of this document. This paper is still a thorough review of the compressive response of composites since important results of studies prior to 1980 are

[1] Project engineer, David Taylor Research Center, Code 2802, Annapolis, MD 21402.

included in recent compression research efforts, and the early papers that have significantly influenced our current understanding are included here for completeness.

This paper is not organized chronologically, but rather by subject to provide an understanding of four often asked questions concerning compression response:

1. How should composite materials be tested in compression?
2. How do composite materials fail under a compressive load?
3. What failure theories describe these failures?
4. What compression data is available for composite materials?

Tables as well as text have been prepared to present this information in the most accessible manner.

Although the papers reviewed here help to answer the preceding four questions, a more important question that is often overlooked in papers dealing with compression is: What is compression failure strength? Unlike isotropic materials, for composites this becomes more of a structural question than a fundamental material one at the coupon level. Implicit in many of the papers reviewed here is the assumption that an externally applied compressive load results in a state of uniaxial compressive stress in the gage-section of the composite specimen. Compression failure is then a result of this state of stress. Efforts to achieve a state of uniaxial compressive stress have directed the development of test methods towards designs with the shortest possible gage-length, but not all test methods achieve this objective. This is a worthy goal and the right approach for determining intrinsic material properties, but is this the information that is most useful in the final application of the data? The answer may lie in observing the extreme efforts that are necessary to induce "material compression failure" even in the most controlled environment, the test laboratory. Are these controlled conditions to be expected when a composite component is subjected to its service environment?

These observations tell us that there is no single definition of compression failure strength. The two extremes are global elastic Euler buckling and material compression failure. Between these two extremes lies a transition region in which the state of stress that exists is a combination of bending induced stresses and compressive stresses. To illustrate this issue consider Fig. 1. This figure shows the spectrum of response of a fiber-reinforced laminated composite to externally applied compressive loads. The term "structural" is used here in the sense that a laminate is considered as a structural element and not in the sense of an end-use structure. The response of a laminate to an externally applied load is on the "structural" level, the resulting state of stress on an element in the strength-of-materials or laminated plate sense is on the

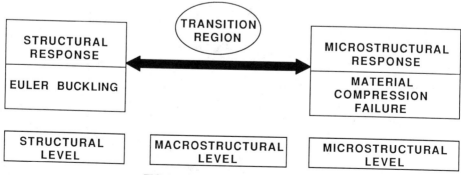

FIG. 1—*General compression response.*

"macrostructural" level, and the response on the fiber-matrix interaction level is called the "microstructural" level.

The results of an experimental program [1] that show the extreme sensitivity of specimen gage length/thickness ratio and its influence of the compressive response of 0° graphite/epoxy specimens are shown in Fig. 2. This figure shows a line best-fit through compression data that has been generated in an Illinois Institute of Technology Research Institute (IITRI) test fixture that has been shown to produce the most consistent and reliable data at one specific gage length. This figure clearly indicates that what is reported as compressive strength from the IITRI fixture is simply the strength from a narrow region of a curve that is very sensitive to specimen geometry, suggesting there is no intrinsic material compression strength.

The perspective of the current literature with respect to Fig. 1 is that the theoretical papers in the open literature fall within the microstructural response level, while the papers dealing with test methods and experimental data fall within the microstructural response level and the transition region. That is, the theoretical papers assume a state of uniaxial compression, but the experimental studies cover test methods and specimens with widely varying gage lengths, gage-length support methods, and with differing load-transfer schemes. The goals of research programs warrant the use of many test methods, but the implications of a test fixture and specimen configuration should be considered.

Even though composite material compression tests are difficult to conduct and subject to the many factors just discussed, methods to compare and rank composite materials based on their compression strength are necessary. Using this information to design composite components for compressive loading is not a straightforward task. The preceding discussion is intended to alert the reader to the issues to be considered when conducting compression tests,

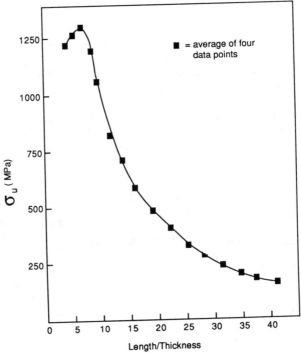

FIG. 2—*Ultimate strength versus length-to-thickness ratio.*

reviewing compression related research, or using compression data for component design. As a conclusion, the last section of this paper provides a synopsis of our current compression testing practices and the most recent philosophy of compression failure processes. Guidelines for designing compression tests that require a deviation from accepted practices are suggested.

Compression Test Methods

General

Many investigators have been interested in the effect of test configuration and specimen configuration on compression response of advanced composites. Because compression test results are sensitive to many experimental parameters, some have attempted to determine the most suitable test method for specific material systems or laminate stacking sequences. Others have investigated specimen design to determine the effect of load introduction and gage length on the state of stress in the test section and the buckling stability of the specimen. This section will discuss papers that have dealt with these concerns. The discussion is divided into a section on the evaluation of test fixtures and a section on the evaluation of compression specimens. A summary of the papers concerning the evaluation of compression test methods is included as Table 1.

Test methods found in the discussion to follow are too numerous to describe here; therefore, Table 2 has been provided for the interested reader. This table is a list of the test methods included in the papers reviewed along with the reference that provides a drawing (or schematic) of the method. Table 2 also groups the test methods by classification. In the most general sense, the test methods can be grouped by gage-length restraint (supported or free) then by means of load introduction (shear-loaded or end-loaded), or by load introduction then by gage-length restraint as done here. Other authors have divided the test methods into more specific groups.

Two recent publications [2,3] have listed and discussed the numerous test methods proposed for compression tests of fiber-reinforced composites. Whitney, Daniel, and Pipes [2] include a section on compression testing in their text on experimental mechanics. They classify compression test methods into three types by specimen description; Type I being those with a short and unsupported test section length; Type II being those with a long and fully supported test section length; and Type III are those with straight-sided coupons bonded to a honeycomb core. In Type I they describe the Celanese, Illinois Institute of Technology Research Institute (IITRI), Northrop, and National Bureau of Standards (NBS) [4] methods, in Type II the Southwest Research Institute (SWRI) and the Lockheed [5] methods, and in Type III the edgewise honeycomb and four-point bend honeycomb methods.

The discussion by Whitney et al. points up that Type I fixtures all yield acceptable data. However, the Celanese test method requires extreme precision in mounting the specimens in the fixture, the mass of the IITRI test method requires prolonged soak periods for elevated temperature testing, and all Type I fixtures require specimen edges or tab surfaces or both be close to perfect in parallelism. They report that the Type II test methods yield data compatible with Type I data for laminates, but yield consistently lower strength of unidirectional specimens than the Type I fixtures. For Type III test methods, the authors point out values for ν_{xy} are often higher than values from tension tests, and that these methods usually yield higher strength than any of the other test methods.

Abdallah [3] conducted a literature review that also discussed available compression test methods. In addition to the methods discussed in Ref 1, Abdallah reviews the Wright-Patterson, Rockwell, Narmco, Boeing, Texaco Experiment, Inc. (TEI), Royal Aircraft Establishment (RAE), ASTM Test Method for Compressive Properties of Rigid Plastics (D 695-89),

TABLE 1—*Compression test method evaluations.*

Ref	Authors	Fixtures/Specimens Evaluated	Comments
2	Whitney et al.	Celanese, IITRI, Northrop, NBS, SWRI, Lockheed, edgewise sandwich beam, four-point bend sandwich beam	discussion of compression test methods
3	Abdallah	Celanese, IITRI, NBS, Northrop, SWRI, Lockheed, Wright-Patterson, Rockwell, TEI, Narmco, ASTM D 695-89, Boeing, Fed. Std. 406, RAE, sandwich beam	literature review
6	Adsit	ASTM D 695-89, Celanese, IITRI, sandwich beam	experimental; ASTM round-robin results
7	Clark et al.	IITRI, laterally restrained, capped end-loaded	experimental; 3 thicknesses, 3 widths and 3 laminate orientations tested. No one fixture acceptable for all configurations
8	Hofer et al.	IITRI	experimental; original description and evaluation of IITRI fixture
9	Lamothe et al.	Celanese, IITRI, end-loaded cylinders	experimental; Celanese and IITRI unsatisfactory for metal matrix specimens tested
10	Irion et al.	IITRI style and capped end-loaded fixtures	experimental; fixtures evaluated under static and creep loading
11	Daniel et al.	101.6-mm-dia, 25.4-mm-wide, 6 to 8-ply rings	experimental; fixture for high strain rate (20 to 500 1/s) testing evaluated
12	Bogetti et al.	IITRI	theoretical; reports admissible length-to-thickness ratios considering uniform compressive stress in gage section and buckling stability
13	Reiss et al.	specimens subjected to combined compression and bending	theoretical, reports admissable length-to-width ratios for combined stress to decay to a state of uniaxial compressive stress
14	Rehfield et al.	tabbed shear-loaded specimens	theoretical; closed form approximate solution for stresses at tab/specimen interface, tab terminations, and in gage section
15	Woolstencroft et al.	Celanese, RAE, ASTM D 695-89, modified Celanese, BAe	theoretical/experimental; finite element analysis used to determine S_{xx}, S_{yy}, S_{zz}, at specimen midlength
16	Shuart	sandwich beam	theoretical/experimental; 3-D finite element analysis of sandwich beam to determine the effect of the honeycomb core
1	Smoot	IITRI	experimental; compressive strength versus gage length reported

TABLE 2—*Compression test method summary.*

| Fixtures | Shear-Loaded Gage Length | | End-Loaded Gage Length | | Ref |
	Supported	Free	Supported	Free	
Celanese	. . .	X	. . .	. . .	2
IITRI	. . .	X	. . .	. . .	8,2
Sand Flex	½	½	. . .	. . .	2
NASA	. . .	. . .	. . .	X	7
NBS	. . .	. . .	. . .	X	4,2
Rockwell	. . .	. . .	. . .	X	3
Wright Patt	. . .	. . .	. . .	X	3
RAE	. . .	. . .	. . .	X	3
TEI	. . .	. . .	. . .	X	8,3
Narmco	. . .	. . .	. . .	X	8,3
ASTM D 695	. . .	. . .	X	. . .	. . .
Northrop	. . .	. . .	X	. . .	2
Boeing	. . .	. . .	X	. . .	3
SWRI	. . .	. . .	X	. . .	2
Lockheed	½	. . .	½	. . .	5

and Federal Standard Test Method 406. Abdallah classifies compression test methods as Group I, specimens loaded through friction (shear-loaded); Group II, end-loaded specimens; Group III, end-loaded with lateral restraint; and Group IV, other test methods such as sandwich beams, rings, and tubes. He then discusses the specimen and fixture configuration for each test method and summarizes the results of published test method evaluations.

After his review of independent test method evaluations, Abdallah concluded that no method is universally accepted throughout the testing community, but that the two best compression test fixtures are the IITRI and the sandwich beam in four-point bending. Based on the performance of the IITRI fixture, Abdallah recommends a new design that is claimed to be an improvement on the current IITRI design. His design is smaller than the current one and incorporates linear bearings between the wedge grips and the grip housing to eliminate friction that may occur there. The new fixture has not been manufactured so fabrication or experimental problems are not resolved. The less massive wedge housing would need to be evaluated for adequate stiffness.

Fixture Evaluation

Adsit [6] reports on the results of an ASTM Committee D-30 (High-Modulus Fibers and Their Composites) round-robin test of various compression test methods currently in use. At the time of testing, the only compression test method recommended by ASTM for aligned fiber composites was ASTM Test Method for Compressive Properties of Unidirectional or Crossply Fiber-Resin Composites (D 3410-87), the Celanese test method. Results are presented for four unidirectional material systems (E-glass/1002, AS/3501, T300/5208, T300/934) tested by nine laboratories. The four test methods evaluated were ASTM D 695-89, ASTM D 3410-87 (Celanese method), IITRI, and the Sandwich Beam four-point bend test. Material was prepared by one source and distributed to each investigator to tab, cut, instrument, and test. Consequently, material inconsistencies were eliminated in this study. However, specimen preparation sensitivity is not.

The results from this investigation indicate that modulus of elasticity is not a function of loading method. Strength data for the end-loaded ASTM D 695-89 test method were lower than for the other three test methods and are attributed to premature failure from stress buildup on fibers that are not the same length on the specimen ends. Consequently, ASTM D 695-89 is not recommended for advanced composites. The strength results from this investigation were the same for the Celanese, IITRI, and sandwich beam test methods. The paper does point out that the Celanese (conical) fixture can be misused by not controlling the total specimen thickness. Poisson's ratio was not determined during these tests.

The final recommendation from this round-robin evaluation is that ASTM D 3410-87 be modified to include the IITRI fixture and the sandwich beam fixture in addition to the already recommended Celanese fixture. This modification is to be included in the forthcoming revision of ASTM D 3410-87.

Clark and Lisagor [7] evaluated the effects of specimen thickness, laminate orientation, support arrangement, and method of load transfer on graphite/epoxy composites. The three test methods they investigated were a shear-loaded method (IITRI), an end-loaded method (NASA), and a face-supported method. Specimens of 12.5-, 25-, and 50-mm widths, 8-, 16-, and 24-ply thickness, and [0], [±45], and [0/±45,90] orientation were evaluated in the IITRI and face-supported fixtures. The [0/±45/90], 16-ply specimens were tested in the NASA fixture.

Clark and Lisagor conducted a thorough experimental investigation and included in their results a discussion of Uniformity of Load Transfer, Compressive Strength and Stiffness, Comparison of Strength by Fixtures, Specimen Thickness Effects, and Failure Modes. Their major findings included the results of a study on the sensitivity of IITRI generated stress-strain response to tab flatness. Three plots show stress-strain response for as-fabricated specimens, specimens with tabs ground flat and parallel within ±100 μm, and specimens with tabs ground flat within ±25 μm. Strain response from four gages was acceptable only after the final grinding.

A wide range of strength of [±45] style specimens resulted from compression tests in the IITRI fixture. This is attributed to the effect of the high Poisson's ratio in specimens of the tested widths. The IITRI fixture yielded the highest strength of the three for 16-ply, 25-mm-wide quasi-isotropic specimens. The average value of modulus was approximately the same for all fixtures.

The general conclusion from this paper is that no single test fixture is adequate for all of the specimens tested, the IITRI fixture provided the most consistent data for unidirectional specimens, and the face-supported fixture produced the most consistent data for the [±45] specimens. Finally, the data produced by the NASA end-loaded fixture was not substantially different from the IITRI and face-supported fixtures, so the simplicity of the NASA specimen and fixture has led to its use at NASA Langley for additional compression testing of composite materials.[2]

The development of the IITRI fixture is discussed in a paper by Hofer and Rao [8]. The design was an attempt to eliminate shortcomings in test methods used at the time. The Celanese fixture was the most promising fixture at that time, and the IITRI fixture eliminated the cone-to-cone seating problems often associated with it. As part of their investigation, Hofer and Rao compared the compressive moduli of aluminum, brass, and steel to the respective tensile moduli to determine the frictional performance of the system. Their tests results indicated frictionless performance of the IITRI test fixture.

Lamothe and Nunes [9] investigated the effect of test method on the compressive response of laminates of five material systems, each with a different fiber orientation. They tested T300/

<hr/>

[2] Private communication with M. J. Shuart, NASA Langley Research Center, Hampton, VA, June 1987.

5208 graphite/epoxy $[0]_{16}$ specimens, S2/SP-250 glass/epoxy $[0/\pm45/90]_{2s}$ specimens, Kevlar 285 weave/Cycom 4143 Aramid/epoxy specimens, unidirectional FP alumina/aluminum specimens, and unidirectional FP alumina/magnesium specimens in Celanese and IITRI fixtures. For the organic matrix composites, they found the IITRI fixture easier to use and more flexible in testing specimens of differing thicknesses. The two test methods yielded the same modulus results and the IITRI yielded somewhat higher strength results although a small specimen population was tested.

Neither the Celanese nor IITRI fixture was suitable for testing the metal matrix composites (MMCs). Tab failures occurred in specimens of suggested design. Thinner specimens were then manufactured to induce compressive failure, but the higher length-to-thickness ratio caused global buckling failure. Reduction of the gage section length to reduce length/thickness made strain measurement difficult, so a new specimen was designed for the MMCs. The specimen was a short, cylindrical end-loaded type with capped ends.

Finally in the area of test fixture evaluation, Irion and Adams [10] designed and evaluated two fixtures for creep testing of composites while Daniel and LaBedz [11] evaluated a new dynamic compression test fixture. The results of these studies are discussed in the section on Experimental Investigations.

Specimen Evaluation

References *14, 15, 16,* and *17* are theoretical studies that deal with the effect of tabbing, load introduction, and specimen geometry on the state of stress in the gage section of compression specimens. Reference *18* describes the reusable sandwich beam test method and Ref *19* is an experimental study determining the effect of gage length on the compressive response of graphite/epoxy tested in the IITRI fixture.

Bogetti, Gillespie, and Pipes [12] were interested in the effect of specimen design on the compressive response of IITRI specimens. While most investigators design to an upper limit on gage-length to prevent Euler buckling, their study considered both the upper and lower bounds of specimen gage length. Gage length must be short enough to preclude buckling failure, but long enough for the state of stress to decay to a state of pure compression in the test section. An elasticity solution based on Saint-Venant's principle and a finite element method are used to establish the gage-length lower bound. The gage-length upper bound was determined based on a conservative estimate of critical buckling strain assuming pinned end conditions and including the effects of transverse shear deformation. The analyses allow for variations in specimen geometry (length/thickness) and varying material anisotropy, E_x/G_{xz} (longitudinal stiffness/interlaminar shear stiffness).

The results from this study demonstrate that ASTM D 3410-87-recommended specimen geometries for shear-loaded test methods (IITRI and Celanese) are appropriate for boron/epoxy, graphite/epoxy (E_x/G_{xz} = 30.5), and glass/epoxy. However, as E_x/G_{xz} ratios increase, the gage-length upper limit decreases and the gage-length lower limit increases so caution must be exercised when generating compressive test data for highly anisotropic materials. A graph showing allowable length/thickness ratios as a function of E_x/G_{xz} is included and provides a convenient means for selecting specimen geometry for the IITRI and Celanese type test methods.

Reiss, Yao, and Clark [13] used the principle of minimum complementary energy to determine the effect of constraining action by grips on the state-of-stress in the gage section of compressively loaded composite coupons. Their analysis assumed clamped ends that were allowed to undergo pure compression and pure bending. They present results for the analysis of [0/ $\pm45/90]_s$, $[\pm45]_s$, and [0] graphite/epoxy laminates. Their results include values of σ_x, σ_y, and τ_{xy} within the gage section of the specimen and a recommended length-to-width ratio that allows for a state of uniaxial compressive stress at the midlength of the gage section. A length-to-width ratio of 0.75 is recommended for the $[0/\pm45/90]_s$ laminates, 1.5 for the $[\pm45]_s$ laminates, and it was found that the constrained edge influence is limited to a small region near the corners of the gage section for the [0] laminates.

Rehfield, Armanios, and Changli [14] were also interested in the design of compression specimens. However, they analyzed end-loaded specimens with tabs. The tabs on these specimens are claimed to eliminate the need for test fixtures such as the IITRI or Celanese, prevent end brooming, and stabilize against buckling. The analysis developed in this program is in closed form and based on the assumption that classical bending theory can be used to estimate transverse and normal strains. The analysis is used to estimate τ_{xz} and σ_{xz} at the tab specimen interface, axial strain at the tab termination, and strain uniformity in the gage section. The results in the paper are presented for end-loading a standard size IITRI specimen, although a specimen of this geometry is likely to buckle under simple end loading. The results from the closed form solution were compared to a finite element model and the results compared graphically. The results of this paper would be more useful if the analysis had been used to suggest a tabbed, end-loaded specimen design as proposed by the authors, since the standard IITRI specimen will not perform satisfactorily under these loading conditions. No specimen design is suggested.

Woolstencroft, Curtis, and Haresceugh [15] used a three-dimensional finite element analysis to determine σ_x, σ_y, and σ_z for the shear-loaded Celanese, end-loaded RAE, and end-loaded ASTM D 695-89 specimens. All three specimen types yielded negligible σ_z stresses. Both the ASTM D 695-89 and Celanese experienced transverse compressive stress, with only the RAE specimen having uniform σ_x and negligible σ_y.

The sandwich beam in four-point bending has been a very popular compression test method for advanced composites. One concern with this method is the effect the honeycomb core has on the state of stress in the composite facing. Shuart [16] theoretically determined the biaxial state of stress in the composite skin and the percentage of load carried by the core. A three-dimensional linear finite element analysis was used. The analysis indicated the transverse stress in the composite face was negligible for the honeycomb core (aluminum or titanium) and graphite/polytimide materials if ribbon direction modulus of the honeycomb is no greater than 1765 kPa. The analysis also showed that less than 1% of the load is carried by the honeycomb core, even for 90° specimens.

Although compression test results generated by the sandwich beam method are highly acceptable, the main disadvantage of the method is the large amount of material required for the specimen and the cost of the sandwich beam that can only be used for one test. This weak point motivated Gruber, Overbeeke, and Chou [17] to design and evaluate a "reusable sandwich beam" compression test method. Their design uses a 254-mm-long by 12.7-mm-wide specimen with tabs on one surface that is clamped to a sandwich beam constructed of aluminum honeycomb, a graphite/epoxy tension face, and a plexiglass core under the specimen gage section. A Kevlar/glass phenolic hybrid composite system was evaluated in the reusable sandwich beam and the results were consistent with tests run in an IITRI fixture.

Smoot [1] tested AS1/3501-6 graphite/epoxy in the IITRI fixture using unidirectional specimens of standard width (6.4 mm) and standard tab length while varying the gage length-to-thickness ratio from (4.4 to 47.1). Sixteen length/thickness ratios were tested, four specimens at each ratio. The major result of this investigation is evident from a plot of ultimate compression strength versus length/thickness ratio, Fig. 2. For length/thickness ratios greater than 8.7, the strength drops rapidly with small increases in length/thickness. There was also a decrease in strength for decreasing length/thickness ratios between 8.7 and 4.4, but the decrease was not as rapid. Since the compressive strength does not approach an upper bound and remain constant, this datum points up the sensitivity of compression response to specimen geometry and suggests that there may not be an ultimate compressive strength for unidirectional composites. If ultimate compression strength is defined as the highest compressive strength than can be measured in a given test fixture, then the range of gage lengths for specimens tested in the IITRI fixture is very limited.

Smoot also reports changes in the failure modes and end condition factor (pinned versus clamped) for increasing length/thickness ratios. Calculated from the Euler buckling equation, the end condition factor changes from clamped for large length/thickness ratios to pinned for small length/thickness ratios.

Delamination and Damage Tolerance

An area in which compression testing has played an important role recently is in determining the delamination resistance and damage tolerance of composite materials. It has been recognized that the presence of delaminations and impact damage have a greater effect on the compressive response than the tensile response of composite laminates. Consequently, compression tests are conducted to quantify these effects. The substantial amount of work that has been done in this area warrants a critical review of its own, so this topic will not be addressed here. However, the information discussed in this paper can and should be applied to research programs in the area of delamination growth and damage tolerance. In the case of delamination growth papers such as those by Gillespie and Pipes [18]; Rothschilds, Gillespie, and Carlsson [19]; and Whitcomb [20], the experiments and models they report on concentrate on the delamination growth of thin sublaminates caused by Euler buckling. Investigators aim to avoid Euler buckling in efforts to determine basic compression strength and stiffness. While providing a clearer understanding of the delamination growth process and the influence of constituent properties and laminate toughness, this failure mechanism surfaces more in the area of structural compression response than in coupon response.

Compression after impact studies such as those presented by Williams and Rhodes [21] and Manders and Harris [22] are certainly influenced by material inhomogeneities, test methods, and constituent properties as discussed in this paper. However, a review of work in this area would concentrate on compressive strength after impact, projectile mass and velocity, and area of damage and open up a whole new set of discussions not included in the current paper.

The omission of papers covering work in the area of delamination growth and damage tolerance will not affect the review of papers that concentrate on understanding compression response in its simplest form as intended here.

Failure Theories and Failure Mechanisms

General

Failure theories to predict the compressive strength of fiber-reinforced composites began to appear in the literature in the early to mid 1960s. Since the emphasis of this paper is on recent literature, a comprehensive review of the literature from that time to the present will not be included. However, the relevance and motivation for current theories cannot be understood independent of past developments, so as an introduction, a summary of early papers that have had a significant impact on our understanding of compression response will follow.

Although many failure modes have been observed and suggested to explain compression failure, the analytical models that have been the foundation of our current understanding include fiber buckling models, transverse tension models, and fiber kinking models.

Fiber Buckling—A large percentage of the proposed theories are based on the stability of fibers in a flexible matrix. One of the earliest and the most referenced publication on stability-related compression failure is the paper by Rosen [23]. In this paper, Rosen presents a two-

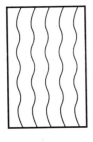

SHEAR MODE

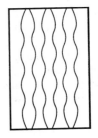

EXTENSIONAL MODE

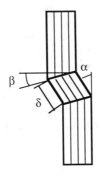

KINK BAND GEOMETRY

FIG. 3—*Microbuckling failure modes and kink-band geometry.*

dimensional model of columns supported by an elastic foundation as the basis for compression failure of unidirectional composites. Basing his solution on energy principles, Rosen proposed that failure constituted the short-wavelength buckling of the fibers in two modes, an extensional mode and a shear mode, Fig. 3. For fiber volume fractions less than 30%, the extensional mode dominates and the fibers buckle out of phase, and, for fiber volume fractions greater than 30%, the shear mode dominates and the fibers buckle in phase. Two expressions for the ultimate compressive strength result, and the lower value is used for the strength. The equation for the shear failure mode results in the lower strength for most composite materials.

The strength predictions provided by Rosen's analysis yield theoretical strengths that have been shown to be higher than observed experimental results for fiber-reinforced composites. Experimental efforts to investigate the microbuckling theory followed, and the most comprehensive study was conducted by Greszczuk [24–28]. Greszczuk manufactured highly controlled composite specimens of laminae (plate)-reinforced, circular-fiber (rod)-reinforced, and actual graphite fiber-reinforced composites. The study investigated the effects of different constituent properties, fiber array, bowed, unbonded, and misaligned fibers. Greszczuk concluded microbuckling models can predict the compression failure of ideal laminae-reinforced composites, but not the failure of circular-fiber composites or actual composites. He concluded further that laminae-reinforced composites consisting of a low-modulus resin fail by microbuckling, but, as resin modulus increases, the failure mode changes and failure is governed by the compressive strength of the reinforcement. Refinements of the fiber buckling theory have been proposed that include nonlinear effects [29], fiber curvature [29,30], and partial bonding

[24,31]. A historical review of the proposed compression failure theories that deal with the short-wavelength buckling phenomenon can be found in a publication by Shuart [32].

Transverse Tension—Another phenomenon considered to explain the discrepancy between theoretical and experimental compression failure results is the presence of transverse tensile stress in unidirectional composites subjected to compression loading. Even though the resulting transverse tensile stress is small, it can be significant enough to cause failure in unidirectional composites due to their low transverse strength. Greszczuk [24] used an interactive failure theory relating transverse tensile failure to uniaxial compressive stress to predict ultimate compression strength.

Fiber Kinking—Kink band formation in composites subjected to compressive loads is also a failure mechanism that has been proposed as contributing to the low compressive strength of composites. Hahn et al. [33] point up that anisotropic materials are susceptible to kink banding and reference the appearance of kink bands in oriented rods of polyethylene, oriented rods of nylon, anisotropic rocks, paperboard, and wood. Since elastic buckling analyses alone are not adequate for predicting the strength of fiber reinforced composites, and the experimentally observed formation of kink bands in composites subjected to compression involves plastic deformation of the resin, kink band theories have become popular for strength predictions.

Argon [34] suggests that the regions in a composite in which fibers are not aligned with the compression axis will form a failure nucleus that undergoes kinking and occurs at a stress lower than the ideal buckling strength derived by Rosen. Berg and Salama [35] observed kinking in graphite/epoxy composites subjected to fatigue loading.

Weaver and Williams [36] proposed a kinking model that involves buckling of fibers on an elastic foundation, fracture of the fibers at a critical strain, movement of the kink across the width of the specimen, and finally the rotation of the kink towards the axis of principal stress.

Evans and Alder [37] conducted a thermodynamic analysis of kinking in composites in which minimization of plastic work determines kink inclination and minimization of elastic strain energy determines kink boundary orientation. General kink band geometry is shown in Fig. 3.

Recent Theories and Observations

In a series of papers, Hahn, Williams, Sohi, and Moon [33,38,39] have attempted to explain the failure processes in unidirectional and quasi-isotropic compressively loaded composite laminates. The main thrust of their research was to design and carefully conduct compression experiments that would explain the characteristics of compression failure as a function of the constituent properties in reinforced composites. Their experimental observations led them to propose a nonlinear microbuckling theory [38] and an analysis that predicts kink band geometry and compression failure based on kink band formation [33].

In Ref 33, Hahn, Sohi, and Williams evaluated the compressive response of fiber bundles embedded in two epoxy resin systems. The fiber bundles tested were E-glass, T300 graphite, T700 graphite, P75 graphite, Kevlar 49, and FP alumina. They observed that bundle failure characteristics were the same as those reported for the same single fibers embedded in a matrix. The highly anisotropic Kevlar and P75 fibers failed by kinking of microfibrils, and the E-glass, T300, T700, and FP alumina fibers failed by microbuckling. They also found that fiber buckling was uniformly distributed in specimens with the lower resin modulus (2.13 GPa), and quite localized in the specimens with higher resin modulus (3.45 GPa).

Unidirectional composite laminates were tested in compression by Hahn et al. in Refs 33 and 38. These laminates had the following constituents: T300 graphite fibers in epoxy resins of varying modulus, PPS and PEEK, T700 graphite fibers in epoxy resins of varying modulus, and S2-glass and Kevlar 49 fibers were tested in one epoxy.

From the results of their experiments on fiber bundles and unidirectional composites, Hahn and coworkers concluded the following about compression failure in unidirectional composites. If the fibers are weak in compression, they fail in compression before buckling and cause laminate failure. If the fibers are not weak in compression, failure initiates as microbuckling from the instability of fibers that have the least lateral support due to free-boundaries, the formation of longitudinal splitting, the presence of voids, or at stress concentrations due to test hardware. If the matrix is soft, the fibers will fail in bending due to continued microbuckling. If the resin is stiff, failure takes the form of kinking. Regardless of the mode of incipient failure, Hahn et al. further conclude that all compression failures of composite materials of the class that they studied exhibit final failure in the form of kink bands. That is, although microbuckling of compression overstressing may initiate failure, each lead to the formation of kink bands in final failure when the resin system is of adequate stiffness. They also state that microbuckling is the cause of kinking in most composite systems.

Sohi, Hahn, and Williams [39] also investigated the response of 24-ply $[45/0/-45/90]_{3s}$ quasi-isotropic laminates reinforced with T300 and T700 graphite fibers. The resins in this study were the same four epoxy systems used in the unidirectional laminates evaluated in Refs 33 and 38. This choice of constituent properties allowed a comparison of failure sequences for laminates with matrices ranging in toughness from 5208 (failure strain 1.4%) to BP907 (failure strain 4.8%). Compression tests were conducted in an IITRI test fixture. Failure was catastrophic for all of the laminates although in some specimens failure was arrested by loading to a certain level and then unloading. For the partially failed specimens, failure was observed to be initiating as kinking of the 0° ply. Failure then proceeded as delamination and sublaminate buckling. The effect of resin toughness on the failure progression was that failure was quite sudden and arrest of fiber kinking was difficult for the T300/5208 (brittle) laminates, while when the T300/BP907 laminate was loaded to 81% of the ultimate compressive strength, failure was limited to kinking of the 0° and no delamination was present. Although the tougher resin resists the propagation of delamination, the BP907 resin allowed fiber kinking at lower strains than the other resins. This observation again points up the dependence of microbuckling initiation on resin modulus, and signals the need for awareness that the lower modulus usually associated with tougher resins means a tradeoff between delamination resistance and microbuckling initiation. Compressive strength increased with resin tensile modulus for the quasi-isotropic laminates and their strains-at-failure were higher than for the unidirectional laminates.

Sohi et al. also investigated quasi-isotropic laminates with holes in their study and found failure initiation to be fiber kinking in 0° plies. As in the case of the unnotched laminates, the failure progression following fiber kinking was dependent on the resin toughness.

Budiansky [40] presented a review and discussion on the application of kink band theories to the compression failure of composites. He proposes a solution for the maximum composite compression strength based on plastic kink band formation. He concludes that the factors affecting kink strength of composites are understood, that high shear stiffness and strength are desirable, and the sensitivity to fiber misalignment is high. Budiansky also proposes solutions for kink band angle and width, but adds while our understanding of kink band geometry is qualitatively sound, quantitively we need to include the influence of random imperfections in our analyses.

While most analyses to determine the compression response of composites are for unidirectional systems, Shuart [32] developed an analysis for symmetric off-axis and angle-ply laminates. He uses a linear analysis to determine laminate stress and strain and short-wavelength (microbuckling) mode shape. A nonlinear analysis including the effects of initial imperfections (fiber-waviness) is used to determine in-plane and interlaminar shear strains. Both analyses are based on the principle of minimum potential energy and are two-dimensional models of plates on elastic foundations. Results presented include the short-wavelength mode shapes for off-axis and angle-ply laminates from the linear analysis. Laminate stresses, out-of-plane displacements, and interlaminar shear strains are generated by the nonlinear analysis and reported for [0], [0/90], [±45], and [+45/0/−45/90] graphite/epoxy laminates. Significant interlaminar shearing strains occur at stresses lower than those causing short-wavelength buckling, and a compression failure criterion is described that includes in-plane shear, interlaminar shear, and short-wavelength buckling failure mechanisms.

Piggott and Harris [41] conducted an experimental study to determine the effect of resin properties on compression strength of composites. Short pultruded solid cylinders were tested with high-strength graphite fibers, high-modulus graphite fibers, E-glass fibers, and Kevlar 49 fibers. The cylinders were manufactured such that the polyester resin was in various stages of cure resulting in varying degrees of modulus and strength. The fiber volume fraction of the composites in this study was 30%. The graphical results included in this paper are for the E-glass composites and show the compressive strength is a strong increasing function of matrix yield strength up to 60 MPa. Results also show that the composite modulus follows a rule of mixtures relationship up to a fiber volume fraction of 45%, and the composite strength follows a rule of mixtures relationship up to 30% fiber volume fraction.

Piggott and Wilde [42] then conducted a study to determine the effect of fiber strength on the compressive strength of aligned fiber-reinforced epoxy. The composites were reinforced with steel rods hardened to different degrees to control their strength. From their results, they developed a rule of mixtures relationship for ultimate compressive strength in terms of fiber and matrix volume fractions, matrix compressive strength, and the flexural strength of steel. The theoretical and experimental results were in good agreement for the steel/epoxy (V_f = 15%, 34%) composites evaluated.

As a result of the experimental results Piggott developed, he found the need to formulate a new theory for compressive strength. He recognized the different failure mechanisms that contribute to failure, and developed his theory accordingly in Ref 43. Six governing equations are proposed with each accounting for a different failure mechanism, and the one that yields the lowest strength is considered the active one. All of the models are rule-of-mixtures type that represent ultimate strength (or modulus) as a function of fiber and matrix strengths, stiffnesses, and volume fractions. The first model assumes compression failure of the fibers, is used for composites reinforced with ductile fibers, and is the rule-of-mixtures formula discussed in Ref 42. Four more failure models are proposed that stem from the presence of fiber waviness. The first determines the reduction of composite modulus as a function of fiber curvature. Three others predict composite strength based on the response of the matrix and fiber-matrix interface due to increasing fiber bending under the influence of compressive loads. These models are for matrix yielding (matrices with low yield strength), transverse splitting (low adhesive strength), and transverse compression. The sixth model accounts for fiber-to-fiber interactions present in hybridized composites.

Piggott compares his models with experimental data for glass/polyester and Kevlar/polyester and finds good agreement. This work is an interesting effort to explain the compressive response of composites with models that are simple yet account for a number of different failure mechanisms.

Sinclair and Chamis [44] proposed a series of failure theories in a manner similar to Piggott. Their theory included four governing equations to account for different failure mechanisms. The objective of their work was to describe the compressive strength of a composite from its tensile and flexural strengths. The first mechanism modeled was global Euler buckling using the classical fixed-end column equation not including the effects of transverse shear. Next, they determined the compressive strength from the flexural strength assuming a rectangular stress distribution in a beam in three-point bending. The third model predicts delamination-controlled compressive strength from apparent interlaminar shear strength and is an empirical curve fit from NASA Lewis data. The final model assumed fiber compressive strength failure and is a rule-of-mixtures type micromechanics model. Sinclair and Chamis conducted IITRI compression tests on unidirectional AS/PR288 and T300/5208 graphite/epoxy laminates for comparison with their failure theories. The results for the AS/PR288 laminates indicate that the strength is predicted by the interlaminar shear model, therefore, strength is controlled by delamination. The compressive strength of the T300/5208 laminates fall into three groups, a high, medium and low-strength group. The failure of the high-strength group was predicted by fiber compressive failure, the medium strength group by flexure or delamination, and the low strength group by Euler buckling (unsupported gage length assumed to extend into tabbed region).

Batdorf and Ko [45] developed an expression to determine the failure of unidirectional composites under combined compression and shear. Their analysis is based on fiber kinking and subsequent plastic yielding of the matrix as in the work by Budiansky [40] and Hahn et al. [33]. This analysis differs in that it includes the effect of combined loads. When shear stress is neglected, this solution reduces to Budiansky's solution exactly (assuming elastic-perfectly plastic matrix response), and to Hahn's solution (assuming matrix strain-hardening) within a factor of fiber volume fraction. Limited experimental comparison shows the theory may work well for materials that can survive large shear strains.

Chou and Kelly [46] indicate the importance of including transverse shear effects when analyzing the compression response of composite laminates. The effect of transverse shear in a composite must be considered because of the high longitudinal-stiffness-to-shear-stiffness ratio. Including transverse shear reduces the load or strain required to cause global Euler buckling and can be significant for unidirectional composites.

Oriented polymer composites are not conventional fiber-reinforced composite materials, however, they exhibit similar mechanical characteristics because of their anisotropy. For this reason, DeTeresa, Porter, and Farris [47] choose to apply the classical elastic stability analysis discussed in the fiber buckling section to the buckling of extended chain polymers. Their model predicts the compressive loads required for buckling of a long single polymer chain. In this analysis, the polymer chain was modeled as series of rigid links of equal length connected by elastic hinges of equal stiffness. Using a differential equation for the static equilibrium of the chain, DeTeresa showed that the curvature of the buckled chain is proportional to the angular change between hinges (or bond angle deformations). The final result is that the differential equation describing the buckling of the link-hinge chain and a continuous column are completely analogous. In addition, the bending stiffness of the link-hinge chain is the product of the link length and the hinge stiffness and can be substituted into the classical buckling load equation for elastic modulus times moment of inertia (EI), the bending stiffness of a continuous column. An oriented polymer fiber is then analyzed by considering single chains that interact through lateral bonding. As in the case of fiber-reinforced composites, compressive strength of an oriented polymer based on the microbuckling of the chains is higher than experimentally observed strengths. Suggested explanations for the discrepancy include buckling of chains at the surface of the fiber at one-half of the predicted load, the presence of voids, areas of lateral interaction where elastic support is lower than modeled, presence of residual stresses, and the misalignment of chains. DeTeresa also points up that the presence of these local inhomogeneities are sites for the initiation of kink bands.

In a short paper on the compressive response of aramid reinforced composites, Van Dreumel [48] suggests that the use of aramid composites for compressive loading should not be ruled out just because they have low compressive strength. He points up that in a structural buckling analysis, the Euler buckling strength of a graphite composite is only 1.3 times that of an aramid composite. Since aramid has a lower density than graphite, he states that an aramid panel will be thicker providing greater handleability and greater impact resistance.

The work by Kim [49] compared the tensile and compressive response of off-axis and angle-ply laminates to results predicted by the Tsai-Wu tensor polynomial failure theory. The experimental compression results compared favorably with the failure theory when the value of the interaction term (F_{12}) was based on the von Mises yield criteria for isotropic materials ($F_{12} = -0.5$). This paper is the only one reviewed that discusses compression failure on the macrostructural level. A review of the literature on macromechanical failure theories may provide compression failure data, but the current search of compression-related literature did not reveal any work other than that of Kim. A comprehensive review of macromechanical failure theories is presented by Nahas [50], and this work would be an excellent source of information for analyzing compression failure on a macromechanical level using interactive failure theories. Nahas does not refer to any experimental compression work that evaluates the applicability of the described failure theories.

The last five papers to be discussed in this section deal with experimental investigations in which the authors made careful observations of the failure characteristics of composite laminates loaded compressively. In all of these papers, the laminates tested contained holes.

Potter and Purslow published two papers [51,52] dealing with the response of graphite/epoxy laminates subjected to exposure to moisture and temperature. Their laminates were 30-mm wide by 160-mm-long, 24-ply Fibredux 914/XAS panels with 50% 0° plies, 50% ±45° plies, and 4.83-mm holes. The specimens were supported around the perimeter of the test section with a 20-mm-wide by 44-mm-long unsupported test section. The first paper reported the results for tests run under room temperature dry (RT-D) and hot-wet (H-W) conditions. Some specimens tested in the hot-wet conditions were preloaded to 66.7% of their hot-wet ultimate strength to determine the effect of preload on subsequent compression response. The second paper reported results from tests on specimens tested room temperature wet (RT-W) and high-temperature dry (H-D) to compliment the results from the first program. One unique aspect of the program by Potter and Purslow was their attempt to arrest the compression failure process in its early stages for inspection purposes. This was accomplished by monitoring changes in strain in the vicinity of the hole using metal foil strain gages. Their observation was that at specimen failure strain close to the hole suddenly changes about 20 to 50 ms prior to final failure. By monitoring changes in strain response, initiation of final failure was detected and the test machine actuator was reset by a subroutine in the program used to generate the displacement control ramp.

Both papers by Potter and Purslow contain very complete discussions of the failure characteristics including scanning electron microscopy (SEM) and low magnification photographs. Their conclusions state that under RT-D, RT-W, and H-D there is little or no notch sensitivity in compression. The ply cracking and delamination around the hole serve to relieve the stress concentration caused by it. In the H-W condition, notched-strength was lower than unnotched-strength and the absence of axial splitting suggested no stress relief occurred around the hole. From the specimens subjected to a preload prior to compression testing, it was concluded that the level of preload applied had no effect on the environmental test that followed.

Shuart and Williams [53] investigated the compressive response of ±45-dominated laminates with holes and impact damage. Six ±45°-dominated laminates were evaluated with various percentages (0–33%) of 90° plies. All specimens were 127-mm-wide and 254-mm-long, and tested either unflawed, with holes (6.4 to 50.8-mm-diameter), or after impact loading. The primary compression failure mechanism for the all ±45° laminates with holes was in-plane matrix-shearing. For the ±45°-90° style laminates with a hole, failure mechanisms were a combination of matrix-shearing and delamination. The percentage of delamination failure increased with increasing percentage of 90° plies. Matrix-shearing is defined as in-plane shear failure and originated at the hole boundaries and extended outward parallel to the fiber directions. The failure mechanisms for the impact-damaged specimens was predominantly delamination, but matrix shearing was observed.

Glass/epoxy all ±45 birefringent specimens with holes were also tested, then illuminated with polarized light to observe stress concentration patterns. Regions of maximum stress were indicated at the hole boundary on the horizontal axis of the specimen that correspond to the areas where failure initiated. Visual inspection revealed that fiber kinking occurred along the boundaries of the matrix-shear bands.

Knauss and Henneke [54] studied the compressive failure of graphite/epoxy plates with holes. Unflawed plates and plates with holes from 1.6 to 38-mm in increments of 3 mm were tested to failure in compression. The 24- and 48-ply (±45/90/0) style laminates were tested in both quasi-isotropic ($E_x/E_y = 1.0$) and orthotropic ($E_x/E_y = 2.015$) configurations. The panels in this investigation were large, 127-mm-wide by 254-mm-long, so buckling failure modes were experienced for the panels with lower stiffness (transverse direction for the 48-ply orthotropic panel) and the panels with high l/t ratios (both 24-ply laminates). The sequence of events leading to failure for the 48-ply orthotropic plates consisted of local delamination, local buckling, and panel collapse. A threshold existed between $0.504 \leq$ diameter/thickness ≤ 1.01 at which failure changed from buckling to compression. The 48-ply quasi-isotropic laminates failed suddenly and catastrophically. Buckling failure dominated for all of the 24-ply laminates, with hole dimension having little effect on buckling load. From the experience the authors gained in testing the 48-ply laminates, the dimensions of the 24-ply quasi-isotropic laminate were changed so that the failure mode would be strength-dominated. When tested, panel failure was typical of that observed in the 48-ply quasi-isotropic panels.

Experimental Investigations

Papers discussed in this section are concerned with experimental investigations in which the principal objective was to investigate the compressive response of a specific composite material system. A summary of the papers reviewed in this section is contained in Tables 3 and 4. Table 3 also contains a summary of all papers reviewed in this literature survey that contain experimental results.

Static Compression

The investigations by Guess et al. [55] and Leach et al. [56] deal with the through-thickness properties of Kevlar and graphite reinforced epoxy systems. Guess' program was designed to generate data to be used in the design of thick-walled spherical internal pressure vessels in which a composite laminate will be in compression in the radial direction. The laminates in this investigation were cured such that void content varied from 1 to 20% by volume. Guess' 25.4-mm-long by 11.4-mm-diameter cylinders were cut from a thick quasi-isotropic Kevlar fabric laminate. Neat 5209 resin, Kevlar 49 (181 fabric)/5208, Kevlar 49 (181 fabric)/5209, Kevlar 49 (328 fabric)/5208, and Kevlar 49 (3281 fabric)/5209 systems were tested.

Guess found increasing porosity in Kevlar reinforced laminates caused increased yielding at moderate stress levels, a decrease in initial modulus, a reduction in axial compressive strength, and an increase in axial failure strain. Guess also found that most specimens failed along inclined planes breaking through layers of fabric.

Leach et al. reported on the through-thickness response of S2-glass, Kevlar 49, and T300 epoxy composites. The matrix was a Brunswick Corp. LRF (Lincoln Resin Formulation)-215

TABLE 3—*Static compression.*

Ref	Authors	Material	Orientation	Test Configuration	Comments
6	Adsit	E-glass/1002, AS/3501, T300/5208, T300/934	[0]	IITRI, Celanese, ASTM D 695-89, sandwich beam	all materials tested in all fixtures for ASTM Committee D-30 round-robin; 3 widths and 3 thicknesses tested
7	Clark et al.	T300/5208	[0] [±45] [0/±45/90]	IITRI, end-loaded (NASA), face supported	
8	Hofer et al.	MOD II/5206, T300/5208, Hercules 3002M, boron/epoxy	[0], [90], [0/±45/0/90]	IITRI	tests at 21 and 177°C dry
9	Lamothe et al.	T300/5208, S2/SP-250, Kev/CYC 5143, FP/Al, FP/Mg.	$[0]_{16}$ $[0/\pm 45/90]_{2s}$ $[0/90]_{8s}$	Celanese and IITRI for all materials and end-loaded cylinders for metal matrix specimens	Celanese and IITRI unsatisfactory for metal matrix specimens tested static and creep data reported
10	Irion et al.	AS/3501-6 S2/3501-6	[0] $[0]_{25}$ $[90]_{25}$ $[0]_{22}$ $[90]_{22}$	6.4 to 9.5-mm-wide, 83 to 90-mm-long, capped, end-loaded; 12.7-mm-wide, 114-mm-long shear loaded	
15	Woolstencroft et al.	XAS/914C (graphite/epoxy)	[0]	Celanese, RAE, D 695	
16	Shuart	HTS1/PMR-15 (graphite/polyamide)	[0], [90], [±45], [0/±45/90] $[0]_{10}$	sandwich beam	tests conducted from −157 to 316°C
1	Smoot	AS1/3501-6		IITRI, 6.4 mm wide, 1.4 mm thick, gage length from 6.4 to 64 mm	compressive strength versus gage length reported
33	Hahn et al.	E-glass, T300, T700, P75, Kevlar 49, and FP Alumina fiber bundles in two epoxy resins. Laminates shown is Ref *40*	single fiber bundles and [0]	IITRI, and single fiber bundles in epoxy resin blocks	Refs *35, 40,* and *41* report complimentary data

TABLE 3—*Continued.*

Ref	Authors	Material	Orientation	Test Configuration	Comments
38	Hahn et al.	T300 in 5208, BP907, 4901/MDA, 4901/mPDA, PEEK and PPS. T700 in BP907, 4901/MDA, 4901/mPDA. Kev/epoxy and S2-glass/epoxy	[0]	IITRI	Refs 35, 40, and 41 report complimentary data
39	Sohi et al.	T300 in 5208, BP907, 4901/MDA, 4901/mPDA. T700 in BP907, 4901/MDA, 4901/mPDA	$[(45/0/-45/90)]_{3s}$	IITRI	Refs 35, 40, and 41 report complimentary data
41	Piggott et al.	E-glass, HMS and HTS graphite, Kevlar, all in polyester	[0]	End-loaded solid cylinders	degree of resin cure varied
42	Piggott et al.	steel/epoxy	[0]	end-loaded solid cylinders	strengths of steel rods varied
44 49	Sinclair et al. Kim	AS/PR288 T300/5208 AS/3501-5A	[0], [+15] to [90], [±15] to [90] in 15° increments	IITRI dogbone, 13 by 75 mm gage length, lateral restraint, thickness not given	results compared to Tsai-Wu failure theory
51	Potter et al.	Fibredux 914/XAS (graphite/epoxy)	24-ply laminate, 50% 0° and 50% ±45° plies	30-mm-wide, 100-mm-long, end-loaded, perimeter of gag-section clamped, 4.83-mm holes	some specimens preloaded, RTD and HW tests
52	Purslow et al.	Fibredux 914/XAS (graphite/epoxy)	24-ply laminate, 50% 0° and 50% ±45° plies	30-mm-wide, 100-mm-long, end-loaded, perimeter of gag-section clamped, 4.83-mm holes	RTW and HD tests
53	Shuart et al.	AS4/3502	48-ply [±45] style laminates with 0, 8, 17, 25, and 33% 90° plies	127-mm-wide, 254-mm-long plates, clamped ends simply supported along length, end-loaded	specimens unnotched and with varying hole sizes tested

54	Knauss et al.	T300/5208	48 and 24-ply (±45/0/90) style laminates	127-mm-wide, 254-mm-long plates, capped, simply supported along length, end-loaded	through-thickness compression, void content 1 to 20% through-thickness tests
55	Guess et al.	Kev 49 328 Fabric/5209 Kev 49 181 Fabric/5208	[0/90/±45]s [0,8,16,24 . . .]	end-loaded solid cylinders, 25.4-mm-long, 11.4-mm-diameter	
56	Leach et al.	S2/LRF-215 (Brunswick Ep.), Kev 49/LRF-215, T300/LRF-215	[0], [0/90]	38-mm-square, 25.4-mm-thick "near cubes" from filament-wound rings, curved loading blocks	
57	Ditcher et al.	Ciba-Giegy Fibredux 914C-HTS-5	[0/90] [90/0]	5-mm-wide, 60-mm-long, capped, end-loaded, lateral restraint, 1 and 2-mm-thick	
58	Hayashi	Epikote 828	. . .	end-loaded solid cylinders, 15-mm-long, 15-mm-diameter	0.5 mm by 60 mm, Al end blocks bonded to specimen, lateral restraint, 1 and 2-mm-thick shear modulus versus compressive stress reported
59	Kar et al.	AS1/3501-6	[(±45/90/0)₃ (90/0/ ±45) (+45/0)s0₂]s [0], [90], [±45], [0/ ±45/90]	specimen size and fixture description not given	
16	Shuart	HTS1/PMR-15 (graphite/ polyamide)		sandwich beam	tests conducted from −157 to 316°C

TABLE 4—Dynamic compression.

Ref	Authors	Material	Orientation	Test Configuration	Comments
11	Daniel et al.	SP288/AS SP288/T300	[0] [90] [0] [90]	rings, 101.6-mm-diameter, 25.4-mm-wide, 6 to 8-plies-thick, load applied explosively through a liquid	dynamic compression tests, 20 to 500 s⁻¹
60	Highsmith et al.	AS/3502	$([0/\pm45/90]_s)_6$ $([0/90/\pm45]_s)_6$ $([0/+45/90/-45]_s)_6$	25.4-mm-wide, 102-mm unsupported gage length, shear loaded	R = 10
61	Matondang et al.	T300/914C	$[\pm45]_{8s}$ $[0_2/\pm45/0_2/\pm45/\overline{90}]_s$	16-mm-wide, 210-mm-long shear loaded, lateral restraint	R = −1, influence of anti-buckling guides studied
62	Ramkumar	T300/5208	$[0/45/90/-45]_{8s}$ $[45/90/-45/0]_{8s}$ $[90/45/0/-45]_{8s}$	38-mm-wide, 152-mm-long 38-mm unsupported gage length, shear loaded	R = 10, specimens unflawed and with imbedded delaminations
63	Mohlin et al.	T300/1034E, neat epoxy	$[\pm45/0_2/\pm45/90/0_3/\pm45/0_2]_s$	24-mm-wide, 104-mm-long 24-mm unsupported gage length, shear loaded, 6-mm holes	R = ∞, room temperature dry; room temperature wet
64	Grimes	AS/3501-6	$[0]_{16}$ $[90]_{16}$ $[\pm45]_{4s}$ $[(\pm45)_5/0_{16}/90_4]_c$	Atmur fixture, 50-mm-wide, 200-mm-long shear-loaded, lateral restraint	room temperature dry; room temperature wet; high temperature dry; high temperature wet;
65	Grimes et al.	AS1/3501-6	$[(\pm45)_5/0_{16}/90_4]_c$ $[(\pm45)_5/0_{14}/90_4]_c$ $[(\pm45)_3/0_{16}/90_4]_c$	Atmur fixture, 50-mm-wide, 200-mm-long, shear-loaded, lateral restraint	low temperature wet; room temperature dry/wet; high temperature dry/ wet
66	Walsh et al.	T300/5208 T300/5209	$[0_2/\pm45]_{5s}$ $[0/45/0/-45]_{5s}$ $[0/\pm45/90]_{5s}$	25.4-mm-wide, 152-mm-long, 51-mm unsupported gage length, 6.4-mm holes, IITRI fixture	R = 0.1,10 Hz
67	Han	fiberglass/polyester	Combination random mat and unidirectional	10.2-mm-wide, 10.2-mm-thick, 30.5-mm-long, capped, end-loaded	low temperature fatigue

epoxy. Specimens were cut from filament-wound unidirectional and cross-ply tubes with a 145-mm inner diameter and 25.4-mm wall thickness. The simple curvature on the upper and lower surfaces was left intact on the compression specimens and compensated for with matching machined loading blocks. End load was applied directly to the specimens with no end caps to prevent brooming. Transverse modulus and strength are reported along with plots of the resulting stress strain curves.

Ditcher and Webber [57] examined the effects of specimen thickness, orientation, and edge effects on the compressive response of cross-ply carbon epoxy laminates. Their study included [0/90]$_s$ and [90/0]$_s$ laminates of two thicknesses. Experimental compressive response was compared to theoretical predictions from a linear and nonlinear analysis. Experimental longitudinal stress-strain response and ultimate compressive strength results were found to compare favorably to nonlinear laminated plate analysis and a linear analysis that included edge effects. Transverse strain was significantly affected by the interlaminar free-edge effects in the 2-mm [0/90]$_s$ specimens and varied substantially from the nonlinear laminated plate theory, which did not include free-edge effects. The free-edge analysis does account for an increase in transverse strain within a laminate thickness of the edge, and free-edge effects would influence 80% of the 2-mm-thick, 5-mm-wide specimens. The transverse strains for the [90/0]$_s$ laminates show only small increases due to free-edge effects.

Hayashi [58] was interested in the effect of compressive stress on the shear response of epoxy resin since the compressive response of laminated composites is dependent upon the shear modulus of the matrix. To determine the effect, he designed a 60-mm-long specimen with two 15-mm-long, 15-mm-diameter cylindrical gage sections. Using this specimen, Hayashi was able to measure shear modulus as a function of applied compressive stress. Epikote 828 epoxy was evaluated in this study and the results indicate shear modulus decreases with increasing compressive stress. For compressive stress just beyond the proportional limit, a 24% reduction is reported.

Little work has been conducted to determine the compressive creep response of fiber-reinforced composites, so Irion and Adams [10] studied this problem. Two fixtures were designed for creep testing based on the advantages and disadvantages of current static compression test fixtures. The first is a Celanese/IITRI type shear loading fixture that is claimed to be less bulky than the IITRI while more stable than the Celanese fixture. The other fixture is an end-loaded design with four steel blocks to clamp and support the specimen ends, and bearing supported guide rods for alignment of the steel blocks. Longitudinal and transverse data are reported for graphite/epoxy and glass/epoxy systems. Results indicate that longitudinal graphite and glass reinforced composites creep little, if at all. Transverse specimens exhibited higher strains than longitudinal ones with long periods of no creep followed by abrupt jumps. Expanded strain versus time plots were recorded in this study and the irregularities in the curves suggest local failures occur almost continually, but are quickly arrested. The tests in this study were run for less than 200 h and most specimens tested did not attain a steady-state creep rate. Both of the fixtures designed for this program worked satisfactorily for creep testing.

The investigation by Kar, Herfert, and Kessler [59] microscopically compared the failure surfaces of AS1/3501-6 specimens tested under room temperature dry and elevated temperature wet conditions. Comparison with other work is difficult since no specimen geometry or test fixture description is included in the paper.

Dynamic Compression

The previous discussions on compression test methods and failure characteristics in this paper are for the static response of composite materials, and the majority of the research on compressive response has been for static loading. The number of proposed test methods and the complexity of the failure process for static testing hints at the difficulties in trying to understand the dynamic compressive response of fiber-reinforced composite materials. For instance, in the fatigue work to be discussed later, no two investigations were performed with test specimens of the same geometry or in the same fixtures. A complete discussion of the recent papers on compression fatigue will not be included here, but the papers will be briefly reviewed and summarized in Table 4.

Highsmith, Stinchcomb, and Reifsnider [60] present an overview of the damage development processes in composite laminates subjected to fatigue loading. The paper deals mainly with tension-tension fatigue since the knowledge base is more developed for that regime; however, tension-compression and compression-compression fatigue response is discussed in light of tension-tension response and limited experimental work.

The experimental configuration for compression fatigue studies is very inconsistent as seen in the summary of the studies discussed here (Table 4). Many investigators use the aid of antibuckling guides to prevent global Euler buckling of specimens with high length-to-thickness ratios. Matondang and Schütz [61] studied the effect of antibuckling guides on compression fatigue results. The effect of two guides were evaluated on $[\pm45]_{8s}$ and $[0_2/\pm45/0_2/\pm45/90]_s$ graphite/epoxy specimens.

The effect of delamination growth in graphite-epoxy laminates was investigated by Ramkumar [62] and Mohlin et al. [63]. Ramkumar nondestructively tracked the growth of imbedded Teflon circular and through-width delamination in 64-ply T300/5208 laminates. Mohlin et al. nondestructively monitored the growth of delaminations around holes in graphite/epoxy composites. Mohlin et al. complimented their experimental program with an analysis of delamination growth using a strain-energy release rate based finite element program.

Grimes [64] and Grimes and Dusablon [65] report a very comprehensive study on the compression response of graphite/epoxy composites. They develop a shear-loaded laterally restrained fixture called the Atmur fixture. Reference 64 includes static and compression-compression fatigue results for 0°, 90°, $[\pm45]$, and $[(\pm45)_5/0_{16}/90_4]_c$ laminates tested room temperature dry, room temperature wet, hot dry, and hot wet. Their second study [65] reports RTD, RTE, HD, HW, and cold wet compression-compression fatigue data for the $[(\pm45)_5/0_{16}/90_4]_c$ laminates with ply dropoffs.

Walsh and Pipes [66] studied the influence of compression-compression fatigue on graphite/epoxy laminates with holes. T300/5208 and T300/5209 materials systems were evaluated in $[0_2/\pm45]_{5s}$, $[0/45/0/-45]_{5s}$, $[0/\pm45/90]_{5s}$, and $[90/0/\pm45]_{5s}$ stacking sequences. They modified the IITRI test fixture to accept 25.4-mm-wide specimens for their investigation.

Han [67] evaluated fiberglass/polyester in a random-mat/unidirectional combination at room temperatures and cryogenic temperatures. He studied the cryogenic response of fiberglass/polyester for the application of composites as the supporting structure for a superconductive energy storage magnet.

Daniel and LaBedz [11] report results on the only study in which the compressive response of composites at high strain rates (20 to 500 s^{-1}) was investigated. They conduct tests on rings 101.6-mm in diameter, 25.4-mm-wide, and 6 to 8-plies-thick, subjected to a load developed by an explosive charge and transferred to the specimen through water.

Summary and Conclusions

Compression Test Methods and Experimental Investigations

Compression testing of composite materials is a process that is inconsistent from program to program and investigator to investigator. However, recommendations and guidelines can be deduced from a review of the published work in the area. Although no single test method will adequately satisfy the many objectives of compression testing programs, some programs will benefit from the selection of tried and tested methods.

For testing unidirectional coupons, shear-loading (Celanese, IITRI, sandwich beam) methods are by far the most popular and critically evaluated. The choice of one of these methods when specimen size and environmental conditions allow will help assure avoiding problems that have been experienced in the past, and will result in data that can be compared to those of other programs. The choice of another of the twelve test methods in Table 2 may produce favorable results, but the use of these will be guided by less technology-wide experience.

To summarize, the three shear-loaded test methods discussed in this paper, the IITRI and Celanese methods are fundamentally the same test methods, with the IITRI being the more popular and an improved version of the Celanese. Both methods require the investment in the test fixture, tab surfaces that are flat and parallel to a high tolerance, and specimen thickness variations in the Celanese can cause line versus surface loading on the conical wedges. The IITRI fixture is inconvenient for environmental testing because the mass of the fixture makes long soak times necessary. Both of these methods also require a specimen gage length long enough to allow the transition of the stress state from shear to compression, and all methods require a gage length short enough to preclude global Euler buckling (including the effects of transverse shear). The sandwich beam method requires an expensive specimen and does not yield acceptable values of Poisson's ratio.

The remaining twelve test methods listed in Table 2 are end-loaded type methods. Although simple in concept, they require an awareness that specimen ends must be flat and perpendicular to the load axis, and provisions to prevent end-brooming must be made.

When planning a compression testing program, specimen design is the next most important consideration. The recommended specimen dimensions for the fixtures just discussed are usually for unidirectional specimens and may need adjustments for testing other than 0° coupons. The effect of specimen width becomes important when laminate testing is performed. The high Poisson's ratio of ±45° type specimens has been shown to influence the state of stress in the gage section when specimen widths recommended for 0° specimens are chosen. Free-edge effects must also be considered for laminates since recommended specimen width is 6.4 to 12.7 mm for 0° specimens.

Finally, consider program objectives that require the design of unique fixtures of specimens due to limitations in existing designs. For example, in considering a compression test program for thick-section laminates, it may not be possible to utilize shear-loaded test methods since a specimen of this type becomes limited by the tab-specimen shear strength. No matter what the requirements of a test program require, adhering to a few guidelines will help assure accurate and reliable results. For shear-loaded compression test methods, consider

1. the flatness and parallelism of tabs,
2. the upper and lower limits on gage length,
3. Poisson and free-edge effects in laminates, and
4. no stress concentrations or bending induced by the fixture.

For end-loaded compression test methods, consider

1. the flatness and perpendicularity of specimen ends,
2. provisions to prevent end-brooming,
3. the upper limit on gage length,
4. Poisson and free-edge effects in laminates, and
5. no stress concentrations or bending induced by the fixture.

Failure Theories and Failure Mechanisms

Efforts to develop a theoretical understanding of the compressive response of composite materials have primarily focused on the microstructural response of unidirectional materials. A summary of the current understanding of this response is shown in Fig. 4. This figure indicates the influence of constituent properties and local inhomogeneities on compression response, and shows a number of proposed failure modes. The shaded boxes show the failure sequence that has been suggested for unidirectional graphite and glass fiber-reinforced composites. The failure sequence can be summarized as follows: for composites with strong fibers and fiber volume fractions greater than 30%, microbuckling in a shear mode (in-phase buckling) will occur, initiated by regions of local inhomogeneities (voids, stress concentrations, areas of weak matrix, free boundaries), and lead to final failure in the form of fiber kinking or fiber overstressing in bending.

The analysis of compression failure as shown in Fig. 4 is based on the assumption that externally applied loads result in a state of uniaxial compression on the macro scale, yet allows for local inhomogeneities. However, as discussed in the beginning of this paper, a state of uniaxial compression stress may not be encountered as a result of an externally applied compressive load. This observation suggests that although the understanding of compressive response has improved over the last 25 years, this understanding is in a very limited portion of the entire range of possible compression responses. In terms of Fig. 1, all of the reviewed work on compression failure theories fall in the microstructural response block.

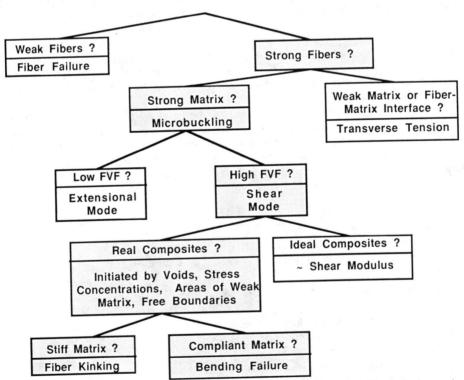

FIG. 4—*Microstructural compression response (FVF = fiber volume fraction).*

References

[1] Smoot, M. A., "Compressive Response of Hercules AS1/3501-6 Graphite/Epoxy Composites," Report No. CCM-82-16, University of Delaware, Newark, DE, June 1982.

[2] Whitney, J. M., Daniel, I. M., and Pipes, R. B., *Experimental Mechanics of Fiber Reinforced Composite Materials*, SEM Monograph No. 4, Society for Experimental Mechanics, 1982.

[3] Abdallah, M. G., "State of the Art Advanced Composite Materials: Compression Test Methods," *Proceedings*, JANAF, CMCS, SM&BS meeting, Nov. 1984, California Institute of Technology, Pasedena, CA.

[4] Kasen, M. B., Schramm, R. E., and Read, D. T., "Fatigue of Composites at Cryogenic Temperatures," *Fatigue of Filamentary Composite Materials, ASTM STP 636*, R. Evans, Ed., American Society for Testing and Materials, Philadelphia, 1977, pp. 141–151.

[5] Ryder, J. T. and Black E. D., "Compression Testing of Large Gage Length Composite Coupons," *Composite Materials: Testing and Design (Fourth Conference), ASTM STP 617*, American Society for Testing and Materials, Philadelphia, 1977, pp. 170–189.

[6] Adsit, N. R., "Compression Testing of Graphite/Epoxy," *Compression Testing of Homogeneous Materials and Composites, ASTM STP 808*, R. Chait and R. Papirno, Eds., American Society for Testing and Materials, Philadelphia, 1983, pp. 175–186.

[7] Clark, R. K. and Lisagor, W. B., "Compression Testing of Graphite/Epoxy Composite Materials," *Test Methods and Design Allowables for Fibrous Composites, ASTM STP 734*, C. C. Chamis, Ed., American Society for Testing and Materials, Philadelphia, 1981, pp. 34–53.

[8] Hofer, K. E. and Rao, P. N., "A New Static Compression Fixture for Advanced Composite Materials," *Journal of Testing and Evaluation*, Vol. 5, No. 4, July 1977, pp. 278–283.

[9] Lamothe, R. M. and Nunes, J., "Evaluation of Fixturing for Compression Testing of Metal Matrix and Polymer/Epoxy Composites," *Compression Testing of Homogeneous Materials and Composites, ASTM STP 808*, R. Chait and R. Papirno, Eds., American Society for Testing and Materials, Philadelphia, 1983, pp. 241–253.

[10] Irion, M. N. and Adams, D. F., "Compression Creep Testing of Unidirectional Composite Materials," *Composites*, Vol. 12, No. 2, April 1981, pp. 117–123.

[11] Daniel, I. M. and LaBedz, R. H., "Method for Compression Testing of Composite Materials at High Strain Rates," *Compression Testing of Homogeneous Materials and Composites, ASTM STP 808*, R. Chait and R. Papirno, Eds., American Society for Testing and Materials, Philadelphia, 1983, pp. 121–139.

[12] Bogetti, T. A., Gillespie, J. W., Jr., and Pipes, R. B., "Evaluation of the IITRI Compression Test Method for Stiffness and Strength Determination," *Composites Science and Technology*, Vol. 32, No. 1, 1989, pp. 57–76.

[13] Reiss, R., Yao, T. M., and Clark, R. K., "Effect of Load Introduction in Compression Testing of Composite Laminates," *Compression Testing of Homogeneous Materials and Composites, ASTM STP 808*, R. Chait and R. Papirno, Eds., American Society for Testing and Materials, Philadelphia, 1983, pp. 200–220.

[14] Rehfield, L. W., Armanios, E. A., and Changli, Q., "Analysis of Behavior of Fibrous Composite Compression Specimens," *Recent Advances in Composites in the United States and Japan, ASTM STP 864*, J. R. Vinson and M. Taya, Eds., American Society for Testing and Materials, Philadelphia, 1985, pp. 236–280.

[15] Woolstencroft, D. H., Curtis, A. R., and Haresceugh, R. I., "A Comparison of Test Techniques Used for the Evaluation of the Unidirectional Compressive Strength of Carbon Fibre-Reinforced Plastic," *Composites*, Vol. 12, No. 4, Oct. 1981, pp. 275–280.

[16] Shuart, M. J., "An Evaluation of the Sandwich Beam Compression Test Method for Composites," *Test Methods and Design Allowables for Fibrous Composites, ASTM STP 734*, C. C. Chamis, Ed., American Society for Testing and Materials, Philadelphia, 1981, pp. 152–165.

[17] Gruber, M. B., Overbeeke, J. L., and Chou, T. W., "A Reusable Concept for Composite Compression Test," *Journal of Composite Materials*, Vol. 16, 1982, pp. 162–171.

[18] Gillespie, J. W. and Pipes, R. B., "Compressive Strength of Composite Laminates with Interlaminar Defects," *Composite Structures*, Vol. 2, No. 1, 1984, pp. 49–69.

[19] Rothschilds, R. J., Gillespie, J. W., Jr., and Carlsson, L. A., "Instability-Related Delamination Growth in Thermoset and Thermoplastic Composites," *Composite Materials: Testing and Design (Eighth Conference), ASTM STP 972*, J. D. Whitcomb, Ed., American Society for Testing and Materials, Philadelphia, 1988, pp. 161–179.

[20] Whitcomb, J. D., "Strain-Energy Release Rate Analysis of Cyclic Delamination Growth in Compressively Loaded Laminates," *Effects of Defects in Composite Materials, ASTM STP 836*, American Society for Testing and Materials, Philadelphia, 1984, pp. 175–193.

[21] Williams, J. G. and Rhodes, M. D., "Effect of Resins on Impact Damage Tolerance of Graphite/ Epoxy Laminates," *Composite Materials: Testing and Design (Sixth Conference), ASTM STP 787,* I. M. Daniel, Ed., American Society for Testing and Materials, Philadelphia, 1982, pp. 450–482.

[22] Manders, P. W. and Harris, W. C., "A Parametric Study of Composite Performance in Compression-After-Impact Testing," *SAMPE Journal,* Society for the Advancement of Material and Process Engineering, Vol. 22, No. 6, Nov. 1986, pp. 47–52.

[23] Rosen, B. W., "Mechanics of Composite Strengthening," *Fiber Composite Materials,* American Society for Metals, Metals Park, OH, 1964.

[24] Greszczuk, L. B., "Failure Mechanisms of Composites Subjected to Compression Loading," Report No. AFML-TR-72-107, Air Force Materials Laboratory, Wright Patterson Air Force Base, OH, Aug. 1972.

[25] Greszczuk, L. B., "Compression Strength and Failure Modes of Unidirectional Composites," *Analysis of the Test Methods for High Modulus Fibers and Composites, ASTM STP 521,* American Society for Testing and Materials, Philadelphia, 1973, pp. 192–217.

[26] Greszczuk, L. B., "Microbuckling of Lamina Reinforced Composites," *Composite Materials: Testing and Design (Third Conference), ASTM STP 546,* American Society for Testing and Materials, Philadelphia, 1974, pp. 5–29.

[27] Greszczuk, L. B, "Microbuckling of Circular Fiber-Reinforced Composites," *AIAA Journal,* American Institute of Aeronautics and Astronautics, Vol. 13, No. 10, Oct. 1975, pp. 1311–1318.

[28] Greszczuk, L. B., "On Failure Modes of Unidirectional Composites Under Compressive Loading," *Fracture of Composite Materials, Proceedings,* 2nd USA-USSR Symposium, Bethlehem, PA, March 1981.

[29] Wang, A. S. D., "Non-Linear Microbuckling Model Predicting the Compressive Strength of Unidirectional Composites," ASME Paper 78-WA/Aero-1, American Society of Mechanical Engineers, 1978.

[30] Davis, J. G., Jr., "Compressive Strength of Fiber-Reinforced Composite Materials," *Composite Reliability, ASTM STP 580,* American Society for Testing and Materials, Philadelphia, 1975, pp. 364–377.

[31] Kulkarni, S. V., Rice, J. R., and Rosen, B. W., "An Investigation of the Compressive Strength of Kevlar 49/Epoxy Composites," *Composites,* Vol. 6, No. 5, 1975, pp. 217–225.

[32] Shuart, M. J., "Short-Wavelength Buckling and Shear Failures for Compression-Loaded Composite Laminates," NASA TM-87640, NASA Langley Research Center, Hampton, VA, Nov. 1985.

[33] Hahn, H. T., Sohi, M., and Moon, S., "Compression Failure Mechanisms of Composite Structures," NASA Contractor Report 3988, NASA Langley Research Center, Hampton, VA, June 1986.

[34] Argon, A. S., "Fracture of Composites," *Treatise on Materials Science and Technology,* Vol. 1, 1972, pp. 106–114.

[35] Berg, C. A. and Salama, M., "Fatigue of Graphite Fibre-Reinforced Epoxy in Compression," *Fiber Science and Technology,* Vol. 6, 1973, pp. 79–117.

[36] Weaver, C. W. and Williams J. G., "Deformation of a Carbon-Epoxy Composite Under Hydrostatic Pressure," *Journal of Materials Science,* Vol. 10, 1975, pp. 1323–1333.

[37] Evans, A. G. and Alder, W. F., "Kinking as a Mode of Structural Degradation in Carbon Fiber Composites" *Acta Metallurgica,* Vol. 26, 1978, pp. 725–738.

[38] Hahn, H. T. and Williams, J. G., "Compression Failure Mechanisms in Unidirectional Composites," *Composite Materials: Testing and Design (Seventh Conference), ASTM STP 893,* J. M. Whitney, Ed., American Society for Testing and Materials, Philadelphia, 1981, pp. 115–139.

[39] Sohi, M. M., Hahn, H. T., and Williams, J. G., "The Effect of Resin Toughness and Modulus on Compression Failure Modes of Quasi-Isotropic Graphite/Epoxy Laminates," *Toughened Composites, ASTM STP 937,* N. J. Johnston, Ed., American Society for Testing and Materials, Philadelphia, 1987, pp. 37–60.

[40] Budiansky, B., "Micromechanics," *Computers and Structures,* Vol. 16, No. 1–4, 1983, pp. 6–10.

[41] Piggot, M. R. and Harris, B. "Factors Affecting the Compression Strength of Aligned Fiber Composites," *Advances in Composite Materials, Proceedings,* Third International Conference on Composite Materials, Paris, France, Vol. 1, 1980, pp. 305–313.

[42] Piggott, M. R and Wilde, P., "Compressive Strength of Aligned Steel Reinforced Epoxy Resin," *Journal of Materials Science,* Vol. 15, 1980, pp. 2811–2815.

[43] Piggott, M. R., "A Theoretical Framework for the Compressive Properties of Aligned Fibre Composites," *Journal of Materials Science,* Vol. 16, 1981, pp. 2837–2845.

[44] Sinclair, J. H. and Chamis, C. C., "Compressive Behavior of Unidirectional Fibrous Composites," *Compression Testing of Homogeneous Materials and Composites, ASTM STP 808,* R. Chait and R. Papirno, Eds., American Society for Testing and Materials, Philadelphia, 1983, pp. 155–174.

[45] Batdorf, S. B. and Ko, R. W. C., "Stress Strain Behavior and Failure of Uniaxial Composites in Combined Compression and Shear," Report No. UCLA-ENG-85-25, University of California, Los Angeles, July 1985.

[46] Chou, T. W. and Kelly, A., "The Effect of Transverse Shear on the Longitudinal Compressive Strength of Fibre Composites," Journal of Materials Science, Vol. 15, 1980, pp. 327–331.

[47] DeTeresa, S. J., Porter, R. S., and Farris, R. J., "A Model for the Compressive Buckling of Extended Chain Polymers," Journal of Materials Science, Vol. 20, 1985, pp. 1645–1659.

[48] Van Dreumel, W. H. M., "A Short Note on the Compressive Behavior of Aramid Fibre Reinforced Plastics," Report No. VTH-LR-341, Technische Hogeschool, Delft Department of Aerospace Engineering, Jan. 1982.

[49] Kim, R. Y., "On the Off-Axis and Angle-Ply Strength of Composites," Test Methods and Design Allowables for Fibrous Composites, ASTM STP 734, C. C. Chamis, Ed., American Society for Testing and Materials, Philadelphia, 1981, pp. 91–107.

[50] Nahas, M. N., "Survey of Failure and Post-Failure Theories of Laminated Fiber-Reinforced Composites," Journal of Composites Technology and Research, Vol. 8, No. 4, Winter 1986, pp. 138–153.

[51] Potter, R. T. and Purslow, D., "The Environmental Degradation of Notched CFRP in Compression," Composites, Vol. 14, No. 3, July 1983, pp. 206–225.

[52] Purslow, D. and Potter, R. T., "The Effect of Environment on the Compression Strength of Notched CFRP—A Fractographic Investigation," Composites, Vol. 15, No. 2, April 1984, pp. 112–120.

[53] Shuart, M. J. and Williams, J. G., "Compressive Behavior of ±45-Dominated Laminates with a Circular Hole or Impact Damage," AIAA Journal, American Institute of Aeronautics and Astronautics, Vol. 24, No. 1, 1986, pp. 115–122.

[54] Knauss, J. F. and Henneke, E. G., "The Compressive Failure of Graphite/Epoxy Plates with Circular Holes," Composites Technology Review, Vol. 3, Summer 1981, pp. 64–75.

[55] Guess, T. R. and Erickson, R. H., "Transverse Compressive Stress-Strain Behavior of Thick Kevlar/Epoxy Laminates," Compression Testing of Homogeneous Materials and Composites, ASTM STP 808, R. Chait and R. Papirno, Eds., American Society for Testing and Materials, Philadelphia, 1983, pp. 221–240.

[56] Leach, J. A. and Grover, R. L., "Mechanical Behavior of Thick Filament Wound Composites Tested in Transverse Compression," Proceedings, 14th National SAMPE Conference, Society for the Advancement of Materials and Process Engineering, Oct. 1982.

[57] Ditcher, A. K. and Webber, J. P. H., "Edge Effects in Uniaxial Compression Testing of Cross-Ply Carbon-Fiber Laminates," Journal of Composites, Vol. 16, May 1982, pp. 228–243.

[58] Hayashi, T., "Shear Modulus of Epoxy Resin Under Compression," Recent Advances in Composites in the United States and Japan, ASTM STP 864, J. R. Vinson and M. Taya, Eds., American Society for Testing and Materials, Philadelphia, 1985, pp. 676–684.

[59] Kar, R. J., Herfert, R. E., and Kessler, R. T., "Fractographc and Microstructural Examination of Compression Failures in Wet Compression Graphite/Epoxy Coupons," Composite Materials: Testing and Design (Seventh Conference), ASTM STP 893, J. M. Whitney, Ed., American Society for Testing and Materials, Philadelphia, 1986, pp. 140–157.

[60] Highsmith, A. L., Stinchcomb, W. W., and Reifsnider, K. L., "Effect of Fatigue-Induced Defects on the Residual Response of Composite Laminates," Effects of Defects in Composite Materials, ASTM STP 836, American Society for Testing and Materials, Philadelphia, 1984, pp. 194–215.

[61] Matondang, T. H. and Schütz, D., "The Influence of Anti-Buckling Guides on the Compression-Fatigue Behaviour of Carbon Fibre-Reinforced Plastic Laminates," Composites, Vol. 15, No. 3, July 1984, pp. 217–221.

[62] Ramkumar, R. L., "Compression Fatigue Behavior of Composites in the Presence of Delaminations," Damage in Composite Materials, ASTM STP 775, K. L. Reifsnider, Ed., American Society for Testing and Materials, Philadelphia, 1982, pp. 184–210.

[63] Mohlin, T., Blom, A. F., Carlsson, L. A., and Gustavsson, A. I., "Delamination Growth in a Notched Graphite/Epoxy Laminate Under Compression Fatigue Loading," Delamination and Debonding of Materials, ASTM STP 876, W. S. Johnson, Ed., American Society for Testing and Materials, Philadelphia, 1985, pp. 168–188.

[64] Grimes, G. C., "Experimental Study of Compression-Compression Fatigue of Graphite/Epoxy Composites," Test Methods and Design Allowables for Fibrous Composites, ASTM STP 734, C. C. Chamis, Ed., American Society for Testing and Materials, Philadelphia, 1981, pp. 281–340.

[65] Grimes, G. C. and Dusablon, E. G., "Study of Compression Properties of Graphite/Epoxy Composites with Discontinuities," Composite Materials: Testing and Design (Sixth Conference), ASTM

STP 787, I. M. Daniel, Ed., American Society for Testing and Materials, Philadelphia, 1982, pp. 513–540.

[66] Walsh, R. M. and Pipes, R. B., "Compression Fatigue Behavior of Notched Composite Laminates," Report No. CCM-81-10, University of Delaware, Newark, DE, June 1981.

[67] Han, K. S., "Compressive Fatigue Behavior of a Glass Fibre-Reinforced Polyester Composite at 300 K and 77 K," *Composites*, Vol. 14, No. 2, April 1983, pp. 145–150.

Fatigue and Fracture

Don C. Curtis,[1] *Mark Davies,*[1] *D. Roy Moore,*[1] *and Barbara Slater*[1]

Fatigue Behavior of Continuous Carbon Fiber-Reinforced PEEK

REFERENCE: Curtis, D. C., Davies, M., Moore, D. R., and Slater, B., **"Fatigue Behavior of Continuous Carbon Fiber-Reinforced PEEK,"** *Composite Materials: Fatigue and Fracture (Third Volume), ASTM STP 1110,* T. K. O'Brien, Ed., American Society for Testing and Materials, Philadelphia, 1991, pp. 581–595.

ABSTRACT: This paper examines fatigue behavior for AS4/PEEK and IM6/PEEK laminates. Tensile fatigue data in the form of stress-log n curves are shown for multiangle ply laminates at 23 and $-55°C$. Fatigue behavior is also examined for ± 45 laminates at different cooling rates from the melt during consolidation. Compressive fatigue data are reported for 40-ply unidirectional laminates at test temperatures of 23 and 120°C. Again, the influence of cooling rate from the melt is explored. The paper also attempts to discuss an interpretation of the results in terms of the expected failure mechanisms that occur for these laminates and the chosen stress fields.

KEY WORDS: composite materials, fatigue (materials), tension, compression, thermoplastic composites, CF/PEEK, failure mechanisms, fracture

Continuous carbon fiber reinforced thermoplastics have received considerable attention in the scientific literature over the last decade [1]. The number of publications on their fatigue performance is however a minor fraction of the overall total. In part, this is to be expected because the early accounts of mechanical performance inevitably focus on simpler and short-term characteristics. The pace for picking up on fatigue properties is still relatively slow, however, because the complexities of approach and measurement introduce aspects that hitherto might not have been recognized.

Some of these complexities can be illustrated in the efforts at simply making a comparison between fatigue strength of carbon fiber (CF) composites based on thermoplastic and thermosetting matrix materials. In terms of a classical description of fatigue strength in the form of $\sigma v \log n$, some authors have apparently shown that thermoplastic composites such as carbon fiber reinforced poly ether ether ketone (CF/PEEK) [2,3] exhibit better fatigue strength than composites based on thermosetting materials. On the other hand, the reverse conclusion has been also observed [4]. In addition, if the description of fatigue is articulated in terms of a "Paris law" crack growth plot, then Gustafson [5], and Trethewey et al. [6] show CF/PEEK to be superior for both Mode I and II tests, respectively, while Russell et al. [7] conclude an opposite view for the Mode II case, again for comparisons with CF/thermosetting systems. Attempts to rationalize some of these data have been superficially made by one of the authors [8], but it remains likely that an inadequate understanding of fatigue behavior lies at the root of the problem.

This barrier to a better understanding of fatigue behavior lies in both material science issues and testing techniques. For example, an optimum testing technique might involve selecting

[1] Senior experimental physicist, research physicist, research associate, and senior experimental physicist, respectively, I.C.I., Advanced Materials, Wilton Materials Research Centre, Wilton, England, United Kingdom.

appropriate test conditions such as frequency of load waveform or specimen geometry, both of which have already been shown to be important [9]. Alternatively, the combination of fiber type, interface type, matrix type, and layup might be the overriding items. Yet understanding fatigue performance must relate to the purpose for which fatigue results are required. In this sense, a clear division can be seen between the end-user requiring properties on some specific geometry or layup; to the material supplier wishing to know the relationship between fatigue performance and choice of fiber type, matrix type, and so on. Naturally, considerable common ground must exist between the two groups of people because whatever the purpose in obtaining fatigue properties, they have to be obtained on a specific sample in a specific form. To this end, it would therefore follow that an ability to interpret fatigue results from a view of both end-user and material developer could be profitable. This is the overall aim of this study.

This paper contains fatigue properties for a particular composite material (CF/PEEK). These results are presented in a way that is relevant to an end-user and also in a manner that enables the various mechanisms that occur during fatigue to be discussed, since such features are important in material design and improvement. The concept of this approach will be portrayed in this paper through the presentation of a wide range of fatigue properties for CF/PEEK.

The failure mechanisms that occur in fatigue of continuous carbon fiber composite components depend on both the applied stress field and the layup, but include interlaminar fracture (delamination between the ply regions), intralaminar fracture (matrix cracking that can occur within the structure of the laminate not just within an inter-ply region), debonding, and translaminar fracture (fiber fracture). It is possible for these mechanisms to occur independently as well as in an interconnected manner but no generalities exist. Three specific combinations of stress field and layup geometry will be used in this particular study. Limiting the subject to this choice is somewhat arbitrary and other combinations require additional work. Tensile fatigue of $(\pm 45)_{ns}$ laminates can be shown to include interlaminar and intralaminar fracture [10]. Tensile fatigue of $(-45/0/+45/90)_{ns}$ laminates starts with intralaminar fracture in the 90° plies followed by matrix cracking in the 45° plies and then delamination crack growth between the ply layers [11]. Compressive fatigue of unidirectional laminates involves matrix cracking (in the form of shear-induced cracks), debonding between fiber and matrix, and fiber failure (probably in the form of buckling of fibers) [12–16]. Therefore, fatigue experiments for these three types of laminate can furnish information on a combination of the basic failure mechanisms mentioned earlier. In addition, by varying the test temperature or the cooling rate from the melt in the preparation of the laminate or even the type of carbon fiber, then considerably more understanding on the role of these mechanisms can ensue.

The description of the various mechanisms cited in the literature do not always relate to CF/PEEK. Nevertheless, it is assumed that the individual mechanisms are dominated by the layup and stress field, albeit that the balance and mix will be related to the fiber type, matrix type, or interface. Consequently, it will be possible to approach failure mechanisms by adopting this knowledge. Such might be the view from a material supplier or designer in contemplating a fatigue study.

These fatigue experiments viewed by the end-user could provide quite a different aspect. The potential use of angle ply laminates for aerospace structures brings with it an interest in the fatigue strength of these laminates, that is, $(\pm 45)_{ns}$, $(-45/0/+45/90)_{ns}$. The option to have data at ambient temperatures, particularly in contemplating the flight of an aircraft, would require fatigue properties in a temperature range of -55 to 23°C. In addition, fatigue strength plots for laminates prepared at different cooling rates from the melt will add practical significance. Compressive fatigue of a thermoplastic composite might focus interest at both 23°C and at some elevated temperature where stiffness of the resin will be less.

In a practical sense, one set of experiments can satisfy both views of the fatigue performance of a composite material. This concept will be approached in this fatigue study for CF/PEEK.

Materials and Experimental Procedures

Materials

The materials of this study are based on Imperial Chemical Industries' Aromatic Polymer Composite range of composites. Two specific materials are involved, namely AS4/PEEK and IM6/PEEK. Both composites are based on a thermoplastic matrix material, poly ether ether ketone (PEEK) with 61% by volume continuous carbon fiber reinforcement. One composite contains Hercules AS4 fiber, while the other contains Hercules IM6 fiber. Prepregs of these materials are consolidated into laminates using a compression molding process where the manufacturer's recommended processing conditions are used. This particularly relates to the rate of cooling from the melt during the consolidation process. Any changes to these conditions are described as "slow" or "fast" cooled materials, since in these cases, the cooling from the melt has been purposefully prolonged or shortened, respectively, in order to change the balance of residual molding stresses (in an undefined manner) as well as the level of matrix crystallinity (the slower the cooling from the melt, the higher the level of crystallinity).

In summary, the following materials have been used in this work:

AS4/PEEK = $(\pm 45)_{ns}$, where $n = 2$ and 4, cooling rate about 40°C/min (standard).
AS4/PEEK = $(-45/0/+45/90)_{2s}$, cooling rate about 40°C/min (standard).
AS4/PEEK = $(0)_{40}$, cooling rates 80°C/min (fast) and 0.5°C/min (slow).
IM6/PEEK = $(\pm 45)_{2s}$, cooling rates 50°C/min (standard) and 1°C/min (slow).

All laminates were ultrasonically C-scanned in order to confirm satisfactory consolidation.

Experimental Procedures

Fatigue tests were conducted in a "zero"-tension load controlled mode or a "zero"-compression load controlled mode. Two types of fatigue machine were used; either an Instron servohydraulic 8033 machine or one of four pneumatic fatigue machines built in our laboratory [9]. The fatigue tests involved stress-log n measurements on unnotched specimens, in the temperature range -55 to 120°C. Square load waveforms were generally employed in order to enable tests to be conducted on any of the fatigue sites. Inter-test site variability was negligible as discussed in a previous publication [9]. In the majority of cases, temperature rise during fatigue was measured in order to establish whether test conditions gave rise to any autogenous heating. This was achieved through contact probes attached to the test specimens. No forced cooling of specimens during testing was employed.

Specimen preparation and design proved to be a critical aspect of the fatigue measurements. The tensile fatigue tests employed both end-tabbed coupon specimens and width-waisted specimens; these are shown in Fig. 1. Discussion of the specimen design and performance is presented elsewhere [9]. However, either specimen could be used to generate fatigue data provided that data stemming from failures in the vicinity of the clamped regions were ignored.

Specimen and test design for compressive fatigue had to be developed during the course of this work. A number of specimen designs were tried where the aim was to achieve a buckling failure in the central region without damage elsewhere, that is, no end crushing and no delamination away from the central region. For example, attempts to use the specimen geometry according to ASTM Test Method for Compressive Properties of Rigid Plastics (D 695-89) with

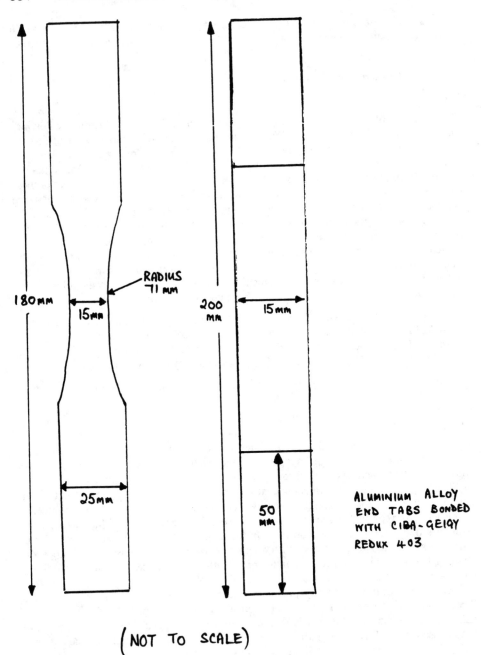

(NOT TO SCALE)

FIG. 1—*Tensile fatigue specimens, width-waisted and coupon.*

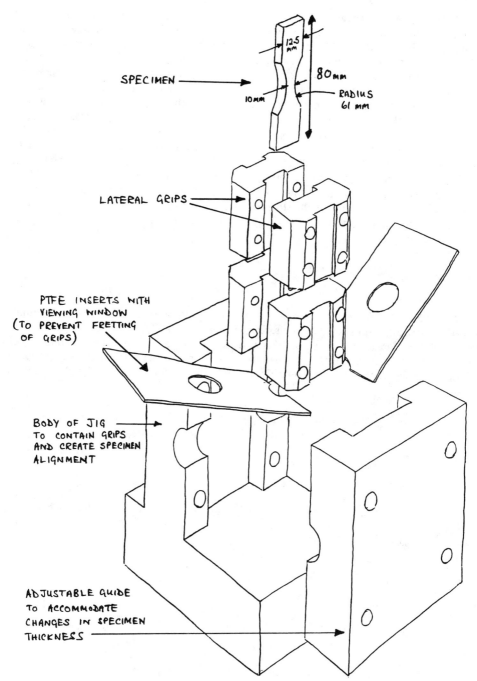

FIG. 2—*Specimen and jig configuration for compressive fatigue.*

its thicker regions at the end of the specimen proved successful for short-term fatigue, but at long times under load, it merely gave rise to a delamination fracture at the interface of the change in specimen thickness. This rendered it ineffective for obtaining compressive fatigue data. A successful specimen design emerged in the form of a short width-waisted specimen that was mounted into a special jig for gripping and supporting the specimen. The specimen and the jig are shown in Fig. 2. The role of the jig was to support the specimen during the fatigue test, ensuring that the center region of the specimen was as unrestrained as possible. The success of the specimen is noted by the type of failure obtained in compression, namely, failure at the waisted region with collapse of the fiber reinforcement angled across the width of the waist. Static tests on this waisted specimen in comparison with the compressive strength obtained with the ASTM D 695-89 approach at 23°C gave agreement to within about 10%, albeit that our waisted specimen regularly gave lower strength values.

In the preparation of all specimens, it was necessary to use a diamond cutting disk for the preparation of parallel strips and then to use a silicon carbide grinding wheel in the preparation of the waisted regions of the specimen. These techniques ensured minimal damage to the edges of the specimens.

Presentation of Results

Tensile Fatigue

The Influence of Some Test Factors—In a previous publication [9], we discussed the influence of a range of test parameters on the fatigue behavior of AS4/PEEK, angle-ply laminates. Some of these data are a reference for this work, and therefore we will review the influence of waveform frequency in order to present and discuss the results that relate to this paper. Figure 3 shows fatigue strength versus log number of cycles to failure for $(-45/0/+45/90)_{2s}$ laminates

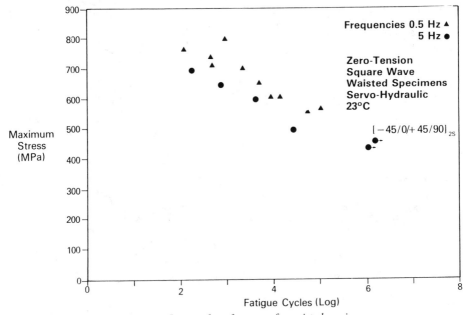

FIG. 3—*The influence of test frequency for waisted specimens.*

of AS4/PEEK at two frequencies (0.5 and 5 Hz). The applied stress being aligned with the 0° fibers. Autogenous heating was observed at 5 Hz with a temperature rise of up to 25°C [9], but at 0.5 Hz the temperature rise was always less than 3°C.

It is known that the failure mechanisms that occur in a quasi-isotropic laminate (as discussed earlier) include matrix cracking in the 90° plies, matrix cracking in the 45° plies, and delamination between the ply layers. The sequence of these events is not known in terms of the fatigue tests, although the data included in the fatigue curves of Fig. 3 are assumed to reflect all of these mechanisms. However, the fact that frequency of load waveform influences the fatigue function and that a temperature rise occurs in the tests conducted at the higher frequency implies that at least one of these mechanisms is temperature dependent.

A similar analysis can be conducted for the (± 45) laminates with 16 plies. Figure 4 shows the fatigue curves at the same two waveform frequencies, where the applied stress is at 45° to the fiber direction. On this occasion, the temperature rise at 0.5 Hz is typically 30°C and at 5 Hz is 140°C. As can be seen from Fig. 4 at 10 000 cycles, the change in fatigue strength is around 80%. There are known to be two failure mechanisms that occur in the tensile fatigue of this laminate, namely, delamination (interlaminar) crack growth between the plies and matrix cracking between the 45° fibers (intralaminar). Reference to the fatigue curves in Fig. 4 indicates that at least one of these mechanisms is temperature dependent. In addition, there is an indication that at long times under load that the fatigue strength will be similar for the two test frequencies. It has been also observed and reported that the temperature rise during fatigue is stress dependent [9], so it is possible that at the lower stress levels associated with these long-term data that the temperature dependence is not so large if at all. This might be an explanation as to the way the fatigue curves at these two waveform frequencies converge at the long times. It can be noted that for both types of laminate system discussed so far that there are two common failure mechanisms. These are delamination and matrix cracking in the 45° fibers.

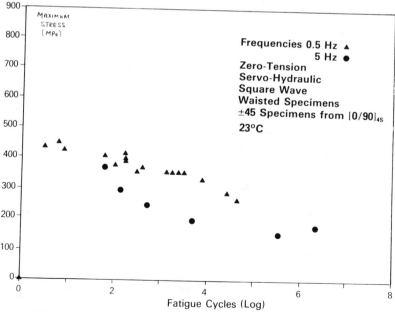

FIG. 4—*The influence of frequency on fatigue strength of ± 45 specimens.*

The Influence of Temperature—Zero-tension fatigue has been conducted on $(-45/0/+45/90)_{2s}$ laminates at $-55°C$ in order to study the influence of temperature on practically relevant laminates but also to investigate any change in mechanisms. Figure 5 shows the results from these tests by plotting normalized fatigue strength against log n. The normalized strength is the ratio of fatigue strength to static strength, thus enabling interlaminate variability to be removed from the data. It is now completely clear from the data in Fig. 5 that the trends with time under load at the two temperatures are the same. (This is not to say that the fatigue strengths are the same!) Consequently, the mechanisms that dictate fatigue behavior at 23°C (as previously discussed) are likely to be the same as those mechanisms that dictate fatigue performance at $-55°C$. However, the contribution of at least one of these mechanisms will have changed in order to produce absolute differences in the fatigue curves at the two temperatures. At present, it is not apparent as to which one or more of these mechanisms is exhibiting this temperature dependence out of the three processes of matrix cracking in the 90° plies, matrix cracking in the 45° plies, or delamination between the plies.

The Influence of Fiber Type and Cooling Rate—The influence of fiber type is examined through a comparison of fatigue strength for AS4/PEEK and IM6/PEEK in the form of $(\pm 45)_{2s}$ laminates (that is, eight plies). Both failure mechanisms that occur for this layup can be influenced by a change in fiber type simply due to the contribution from fibers of different strengths. In addition, the interface between matrix and fiber could affect the intralaminar fracture for the 45° fibers while the delamination toughness could be altered by a change in fiber type and, therefore, the interlaminar mechanism may also change. Standard cooling rates were used in the preparation of the laminates. Figure 6 shows the data for 0.5 Hz square waveform at 23°C, and although the data are obtained on different geometry specimens, it is clear from previous work that this does not affect a comparison of fatigue strength for these materials [9]. It is clear that within experimental scatter that the IM6/PEEK material exhibits the higher strength and that both materials show a similar dependence of fatigue strength with log

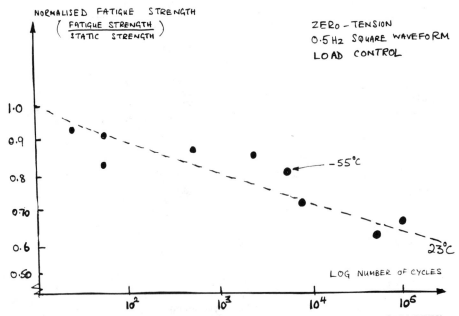

FIG. 5—*Influence of temperature on the fatigue of* $[-45/0/+45/90]_{2s}$ *laminates of AS4/PEEK.*

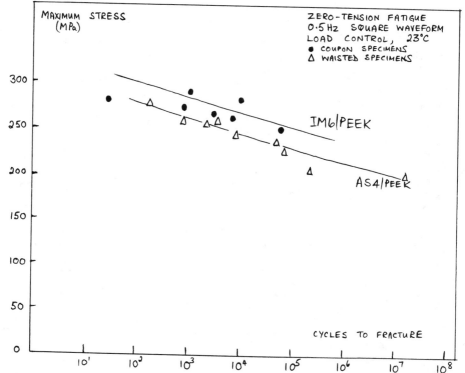

FIG. 6—*Comparison of fatigue curves for* [±45]₂ₛ *laminates of standard cooled IM6/PEEK and AS4/ PEEK.*

number of cycles. Microscopy of fractured surfaces shows good interfacial bonding between both types of fiber and PEEK. Consequently, the higher strength fiber (IM6) is conferring a higher strength on the composite, but not necessarily influencing the balance of the expected delamination and matrix cracking mechanisms.

The influence of cooling rate is examined for IM6/PEEK (± 45)₂ₛ laminates (eight ply). Standard and slow-cooled panels resulted in matrix crystallinities of 26 and 40%, respectively [17]. Figure 7 shows the fatigue data obtained at 0.5 Hz square waveforms on parallel-sided specimens at 23°C. The different cooling rates can be seen to have no influence on either the fatigue strength or the fatigue failure mechanisms for these ± 45 laminates.

Compressive Fatigue

The Influence of Temperature—Compressive fatigue was conducted on unidirectional laminates containing 40 plies. There are primarily three failure mechanisms that are expected in zero-compression tests that include matrix cracking by a shear process between fibers, fiber failure through some buckling mechanism, and a matrix cracking mechanism between fiber and matrix.

The fast-cooled sheets of AS4/PEEK were used in order to compile some compressive fatigue data and to examine the influence of temperature. Figure 8 shows the fatigue data at 23 and 120°C obtained in load control at 0.5 Hz square waveforms. The data show a similar dependence of fatigue strength with log number of cycles at the two temperatures, although as

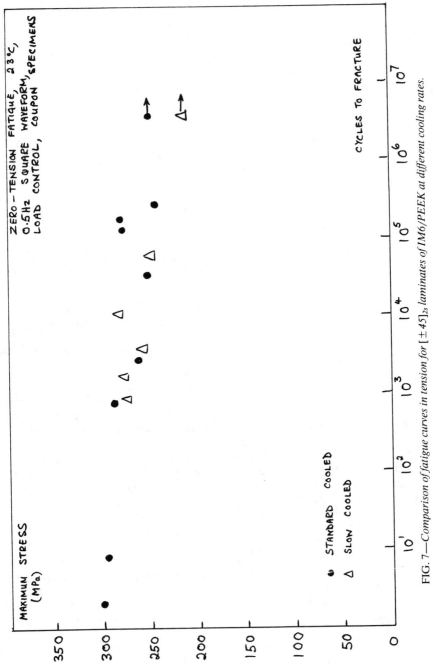

FIG. 7—*Comparison of fatigue curves in tension for* [±45]$_{2s}$ *laminates of IM6/PEEK at different cooling rates.*

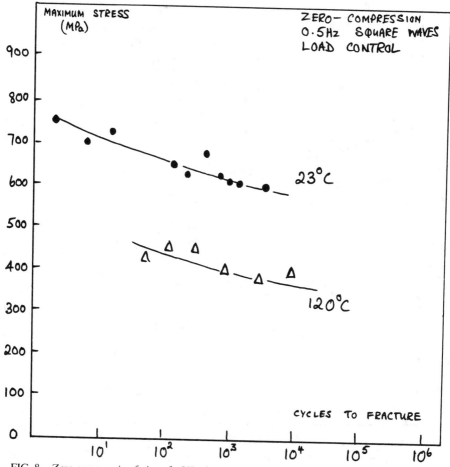

FIG. 8—*Zero-compression fatigue for* [0]$_{40}$ *laminates of fast-cooled AS4/PEEK at 23 and 120°C.*

expected, the curve at 120°C indicates a smaller fatigue strength. It is suggested that the lower stiffness of the matrix at 120°C encourages a higher contribution in absolute terms from the fiber buckling mechanism. In which case, it is implicit that the relative contributions from the matrix cracking mechanisms is unchanged in this temperature regime. This is quite an encouraging view of the compressive fatigue performance of AS4/PEEK. Such a view would be further validated if longer term fatigue data were available at the higher test temperature, but these are at present unavailable due mainly to the experimental difficulties in conducting longer term compressive fatigue tests at these high temperatures.

The Influence of Cooling Rate—Zero-compressive fatigue obtained under the same test conditions were also conducted on the fast and slow-cooled unidirectional laminates at both 23 and 120°C. Figure 9 shows the compressive fatigue curves at 23°C for slow and fast-cooled laminates, while Fig. 10 shows similar data but at 120°C. Curves have been fitted by eye to these data, and within the scatter of the data the curves could be considered to be parallel. At both temperatures, the slow-cooled material exhibits the higher fatigue strength. One possible explanation of this lies in the higher matrix stiffness resulting from the slower cooling from the melt leading to greater resistance to fiber buckling during compression.

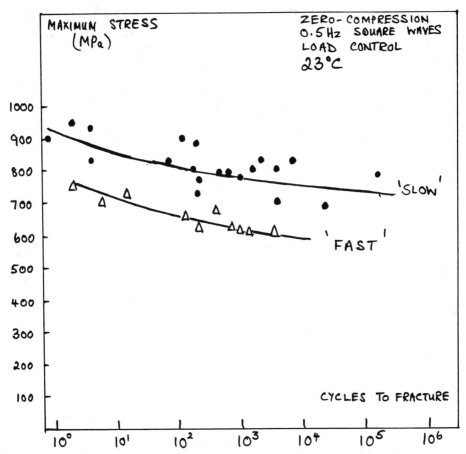

FIG. 9—*Comparison of compressive fatigue for* [0]$_{40}$ *laminates of AS4/PEEK at 23°C for different cooling rates.*

The ratio of strength (slow to fast cooled) can be calculated on the basis of the curves being parallel; this ratio then being the same at all cycles to failure. This strength ratio is 1.2 at both temperatures, implying that the mechanistic process that produces the reduced compressive fatigue strength for the fast-cooled laminate is merely changing the contribution of one of the failure mechanisms. As just mentioned, this is likely to be due to fiber buckling.

Concluding Comments

One of the purposes of this paper was to present and discuss a range of fatigue functions for CF/PEEK laminates. In this context, fatigue functions are presented for both AS4/PEEK and IM6/PEEK in zero-tension fatigue on angle ply laminates at temperatures of 23 and −55°C. In addition, zero-compressive fatigue data are presented for unidirectional laminates at 23 and 120°C. Overall, the influence of cooling rate from the melt in the consolidation process is also examined.

It is hoped that the expression of these fatigue functions will add confidence to any potential end-use for these materials. In particular, there is little discouraging news relating to the fatigue

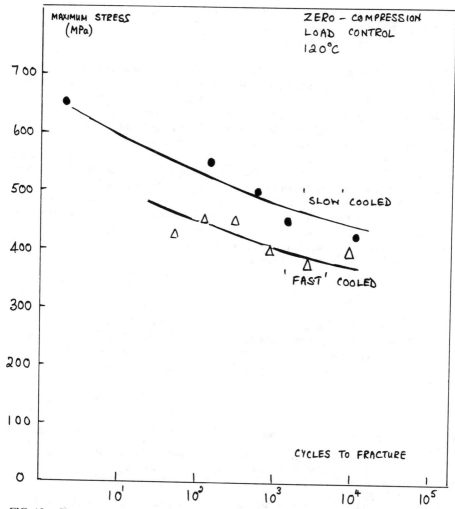

FIG. 10—*Comparison of compressive fatigue for* [0]$_{40}$ *laminates of AS4/PEEK at 120°C for different cooling rates.*

behavior as expressed by the stress-log n functions for these composites. Perhaps from an end-users point of view, a number of comments can be summarized.

1. Load waveform frequency can influence the fatigue strength of laboratory coupons, particularly for matrix dominated layups. It should not be assumed that components made from the same layups will be similarly influenced particularly as it is difficult to imagine large structures being dynamically stressed in the same way and at the same frequencies as laboratory specimens. Nevertheless, fatigue results from laboratory work should be challenged in terms of their relevance to downstream application.

2. Tensile fatigue results are presented for CF/PEEK at 23 and −55°C. These results for quasi-isotropic laminates show that the fatigue behavior at low temperatures need be no more of a concern than the performance at 23°C. This is particularly encouraging in contemplating the performance of CF/PEEK at high altitudes.

3. Compressive fatigue curves are presented for unidirectional CF/PEEK at 23 and 120°C. These data provide a foundation for a designer exploring the feasibility of high temperature applications for these composites where a buckling failure criterion might be a critical design aspect.

In addition, we have started a process of trying to understand the fatigue behavior through identifying the fundamental failure mechanisms that occur during the fracture of various laminates. It is apparent that this understanding is incomplete. For example, it has often been possible to identify a combination of mechanisms, but seldom possible to isolate a single process. Nevertheless, by examining the influence of different material parameters and test parameters, the start of some useful observations can then be made.

Our initial choice of laminates for this work was motivated by a desire to balance end-user interests with a mechanistic study. In order to expand the mechanistic elements, it is apparent that the choice of laminate will be critical in future work. For example, tensile fatigue of 0/90 laminates will involve primarily one failure mechanism, that of matrix cracking in the 90° plies. In addition, a double cantilever beam geometry will be dominated by delamination crack growth. Therefore, adoption of these approaches together with further variation in the test and material parameters should provide a clearer understanding of the mechanisms. These are the focal points of our ongoing studies.

Acknowledgments

The work conducted on the IM6/PEEK materials is part of a wider creep and fatigue study being coordinated in the Department des Materiaux in the Ecole Polytechnique Federale de Lausanne, Lausanne, Switzerland. The authors acknowledge the supply of these laminates from this group and some associated data on the levels of matrix crystallinity.

References

[1] Cogswell, F. N., *Thermoplastic Composites,* L. A. Carlsson, Ed., Elsevier, New York, Chapter 1, in press.
[2] Hartness, J. T. and Kim, R. Y., in *Proceedings,* Twenty-eighth National SAMPE Symposium, Society for the Advancement of Material and Process Engineering, April 1983.
[3] Dickson, R. F., Jones, C. J., Harris, B., Leach, D. C., and Moore, D. R., *Journal of Materials Science,* Vol. 20, 1985, pp. 60–70.
[4] Curtis, P. T., "An Investigation of the Mechanical Properties of Improved Carbon Fibre Composite Materials," RAE Technical Report 86021, Royal Aircraft Establishment, April 1986.
[5] Gustafson, C.-G., "Initiation and Growth of Fatigue Damage in Graphite/Epoxy and Graphite/PEEK Laminates," PhD thesis, Report No. 88-9, ISSN 0280-4646, The Royal Institute of Technology, Stockholm, Sweden, 1988.
[6] Trethewey, B. R., Gillespie, J. W., and Carlsson, L. A., *Journal of Composite Materials,* Vol. 22, 1988, p. 459.
[7] Russell, A. J. and Street, K. N., in *Toughened Composites, ASTM STP 937,* N. J. Johnston, Ed., American Society for Testing and Materials, Philadelphia, 1987, pp. 275–294.
[8] Moore, D. R., *Thermoplastic Composites,* L. A. Carlsson, Ed., Elsevier, New York, Chapter 10, in press.
[9] Curtis, D. C., Moore, D. R., Slater, B., and Zahlan, N., *Composites,* Vol. 19, No. 6, 1989, p. 448.
[10] Ogin, S. L., Smith, P. A., and Beaumont, P. W. R., *Composite Science and Technology,* Vol. 22, 1985, pp. 23–31.
[11] Beaumont, P. W. R., "The Fatigue Damage Mechanics of Composite Laminates," CUED/C/MATS/TR, Vol. 139, Cambridge University Engineering Department, Cambridge, U.K., 1987.
[12] Reifsnider, K. L. and Jamison, R. D., *International Journal of Fatigue,* Oct. 1982, p. 187.
[13] Jamison, R. D., Schulze, K., Reifsnider, K. L., and Stinchomb, W. W., *A85-30501,* 1983, pp. 13–24.

[*14*] Curtis, P. T. and Moore, B. B., RAE Technical Report 82031, Royal Aircraft Establishment, 1982.

[*15*] Sturgeon, J. B., Rhodes, F. S., and Moore, B. B., RAE Technical Report 78031, Royal Aircraft Establishment, 1978.

[*16*] Barnard, P. M., Butler, R. J., and Curtis, P. T., *Proceedings,* Third International Conference on Composite Structures, Paisley, Scotland, 1985.

[*17*] Davies, P., private communication, Department des Materiaux, Ecole Polytechnique Federale de Lausanne, Lausanne, Switzerland, 1989.

Mark T. Kortschot[1] and Peter W. R. Beaumont[2]

Damage-Based Notched Strength Modeling: A Summary

REFERENCE: Kortschot, M. T. and Beaumont, P. W. R., "**Damage-Based Notched Strength Modeling: A Summary,**" *Composite Materials: Fatigue and Fracture (Third Volume), ASTM STP 1110,* T. K. O'Brien, Ed., American Society for Testing and Materials, Philadelphia, 1991, pp. 596–616.

ABSTRACT: The effect of subcritical damage on the notched strength of cross-ply graphite-epoxy specimens has been studied using 914C/T300 laminates of the family $(90_j/0_j)_{ns}$ ($j = 1,2$, $n = 1,2,4$). Damage was monitored using quasi-real-time radiography and dynamic scanning electron microscopy. In addition, the strain field in the vicinity of the notch tip was mapped using a high-resolution optical technique. A strong relationship between subcritical damage and notched strength is evident in all of the data. A theoretical model that is based on observed damage is briefly summarized. Finite element analysis was used to determine the effect of splitting and delamination on the stress distribution in the 0° ply. All specimens were found to fail when the maximum stress in the 0° ply equaled the strength of this ply. The model predicts the effect of both notch size and layup on strength without the need for empirical parameters.

KEY WORDS: notched strength, hole size effect, subcritical damage, radiography, scanning electron microscopy, strain measurement, finite element analysis, linear elastic fracture mechanics, splitting, delamination, damage mechanics, composite materials, fracture, fatigue (materials)

The strength of graphite-epoxy laminates containing notches or holes has proven very difficult to predict because of the complex way in which these materials fail. There have been numerous models attempting to relate notched strength to notch size [1–3], layup [4,5], laminate thickness [6,7] and other parameters. Although these models are often successful in explaining the behavior of a limited class of specimens, no truly general model has been developed.

It has long been recognized that subcritical damage plays a role in determining notched strength [8,9]. A variety of techniques have been used to study the accumulation of subcritical damage: radiography [10,11], strain gaging [12], ultrasonics [13], and scanning electron microscope (SEM) fractography [14]. In virtually all graphite-epoxy laminates, matrix cracks form in the notch tip region prior to catastrophic failure. Although the effects of such cracking on the notch tip stress distribution are well documented, there have been few attempts to incorporate observed notch tip damage into a strength model.

In fact, most notched strength models are based on notch tip stress distributions calculated using assumptions of homogeneous, anisotropic behavior, without reference to subcritical damage. While this may be acceptable in the case of certain laminates where subcritical damage is suppressed [7], it not a valid assumption for most laminates. In order to achieve a truly general notched strength model, the physical basis of failure must be considered, and an extensive study of subcritical damage is an essential element of this process.

[1] Assistant professor, Department of Chemical Engineering and Applied Chemistry, University of Toronto, Toronto, Ontario, Canada.
[2] Lecturer, Cambridge University Engineering Department, Cambridge, United Kingdom.

In this paper, some experimental results are presented that illustrate the dependence of notched strength on subcritical damage. A comprehensive notched strength model based on these results [15–19] is summarized briefly here. Finite element analysis based on observed damage was used to determine modified notch tip stress distributions. A simple failure criterion incorporating the modified stress distributions and an independently measured 0°-ply strength was found to be applicable for all specimens tested in the study. The failure criterion is complemented by theoretical predictions of damage growth, resulting in a model capable of predicting the effect of both notch size and layup on notched strength without the need for empirical parameters.

Experimental Procedure

The laminates used in the study were manufactured by DFVLR, Braunschweig, West Germany, from 914C/T300 graphite-epoxy prepreg. The $(90_j/0_j)_{ns}$ ($j = 1,2$; $n = 1,2,4$) laminates were either guillotined or milled to the correct width depending on the thickness of the laminate. Notches (double-edge-notched (DEN) configuration) were machined using a 60° wheel cutter. Specimens were between 5 and 40 mm wide, and had a notch-to-specimen width ratio, $2a/w$, of 0.5.

Dynamic Scanning Electron Microscopy and Optical Strain Measurement

Previous fractographic studies of fibrous composites using the SEM have been primarily confined to "post mortems." Dynamic microscopy, however, can provide an alternative source of information about the accumulation of crack tip damage in these materials. In this study, DEN specimens were fractured during SEM observation using a miniature, fully instrumented tensile rig. The rig was interfaced with a BBC microcomputer to provide displacement control and data-logging and analysis facilities. The apparatus and procedures have been described fully elsewhere [20].

The tests were conducted in Cambridge Instruments Model 100 and 200 scanning electron microscopes. Specimens were positioned edge-on to the electron beam (Fig. 1), so that the entire length of the crack tip was visible. The specimens were limited to widths of either 5 or 10 mm, and the gage length (between grips) was fixed by the size of the apparatus at 35 mm.

A high-resolution optical strain measurement system was developed for the purposes of

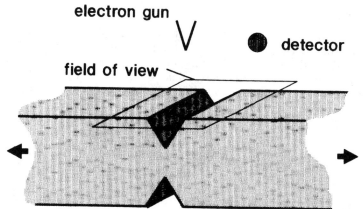

FIG. 1—*Edge-on view of a DEN specimen obtained with the dynamic SEM rig.*

examining the strains near the notch tip during the dynamic SEM work [21]. A metal mesh with a 65-μm hole spacing was rested on the specimen surface but not attached. The mesh does not deform during the test and by observing the motion of the specimen surface with respect to the mesh at very high magnification, various components of strain can be measured over gage lengths of 65 μm to 2 mm with resolution of up to 1 $\mu\epsilon$ (depending on the gage length). The term "gage length" refers to the distance between reference points on the specimen surface, as for an extensometer. Small gage length measurements are extremely valuable in evaluating rapidly changing strain fields near the notch tip.

This technique was used to measure the strain at the base of a square notch in DEN (0)$_4$ and (90/0)$_s$ specimens. The notches were 2 mm wide by 2.5 mm deep in a 10-mm-wide specimen. The strain on the face of (0/90)$_s$ V-notched DEN specimens both with and without subcritical notch tip damage was also measured to determine the effect of splitting on the notch tip stress concentration factor.

Radiography

Quasi-realtime radiography was used to monitor damage during tension tests. Tension was applied using a small hand-cranked tensile rig that fitted into the chamber of a Scanray Corporation Torrex 120D X-ray machine. Specimens were held horizontally in a set of bolt tightened serrated plates and the notches were surrounded by a rubber reservoir filled with zinc-iodide penetrant. Loads of up to 10 kN were measured with an accuracy of $\pm$20 N by a full bridge circuit mounted on the drive shaft.

Stress was applied in 100 MPa increments at a rate of 100 MPa/min. After each increment of stress was applied, the reservoir of penetrant was removed and the notches were rinsed and dried. The entire rig, with the specimen still under full tension, was transferred to the X-ray machine, and a contact print of the specimen was taken on Kodak Industrex MX film. Typical exposure times were about 1 to 2 min at 20 kV and 4 mA. The notches were then re-immersed in penetrant, and precisely 5 min after the previous increment of stress was applied, the stress was again increased. Sample strengths ranged between 200 and 600 MPa, and the test duration ranged from 10 to 30 min. Because the matrix cracks formed while the specimen was immersed in penetrant, penetration was found to be equal or superior to the penetration obtained in a 30-min post-tension soak.

Experimental Results and Discussion

Notched Strength

The effect of notch size on specimen strength was measured for (90/0)$_s$ and (90/0)$_{2s}$ laminates. The ratio of notch length to specimen width, $2a/w$, was fixed at 0.5, and the whole specimen was scaled in order to vary the notch length.

Fracture toughness, K_{Ic}, values for these specimens are presented in Fig. 2. K_{Ic} was calculated using the standard formula for finite width isotropic DEN specimens given by Paris and Sih [22]

$$K_{Ic} = \sigma_f (\pi a)^{1/2} \left[\frac{2w}{\pi a} \left[\tan \frac{\pi a}{2w} + 0.1 \sin \frac{\pi a}{w} \right] \right]^{1/2} \tag{1}$$

where

σ_f = the remote failure stress,
a = crack length, and
w = the specimen width.

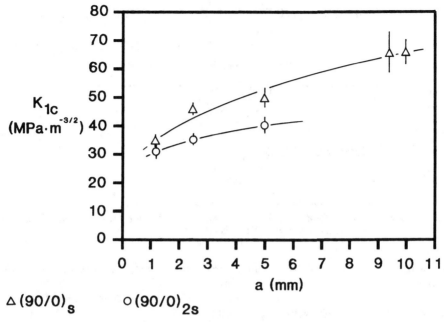

$\triangle\,(90/0)_s$ $\circ\,(90/0)_{2s}$

FIG. 2—*Fracture toughness values for cross-ply DEN specimens with 2a/w = 0.5.*

Isotropic K_{1c} calculations such as Eq 1 are frequently used for composite materials, although the value is often referred to as a "nominal" fracture toughness [11,23,24]. The results of Fig. 2 illustrate that calculated fracture toughness increased with increasing crack size, a phenomenon reported in other studies of graphite-epoxy laminates [11,24]. This indicates that linear elastic fracture mechanics may be inapplicable for through-thickness notches in thin, cross-ply, graphite-epoxy specimens. It should be noted that Harris and Morris have obtained reasonably constant K_{1c} values for cross-ply laminates of more than 64 plies [7]. For laminates with fewer plies, they reported increased fracture toughness values, and this was attributed to more pronounced 0°-ply splitting in the thinner laminates.

A schematic of splitting in a unidirectional specimen is given in Fig. 3. In this specimen, subcritical damage would clearly alter the notch tip stress distribution substantially. In fact, unidirectional graphite-epoxy specimens normally split into the grips, and the failure stress is simply determined by the unnotched strength and the cross section of the remaining ligament.

In cross-ply laminates, the pattern of subcritical damage is somewhat more complex. In addition to 0°-ply splits, transverse ply cracks (TPCs) and 0/90 delaminations also form (Fig. 4). The 90°-ply spans the splits and transfers tensile stress into the section of the 0° ply over the notch that would otherwise by isolated by the splits. It seems likely that the subcritical damage in cross-ply specimens also has a significant effect on the notch tip stress distribution and hence the strength of the specimen.

Dynamic Scanning Electron Microscopy

Figure 5 shows a view of the notch tip (see Fig. 1) of $(90/0)_{2s}$ and $(90_2/0_2)_s$ specimens at 95% of their failure load. The fibers of the 0° ply are exposed by the splitting/retraction mechanism illustrated in Fig. 3, and it is this retraction that comprises most of any measured crack-tip

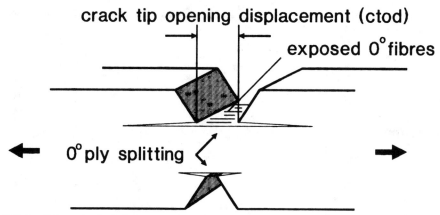

FIG. 3—*Schematic of split formation in a unidirectional specimen. The crack-tip opening displacement is caused by retraction of the unstressed material isolated from the ligament by the split.*

opening displacement (CTOD). No self-similar crack growth (that is, in the plane of the starter notch) was observed during the course of this study.

There is a difference between the notched strengths of the two specimens in Fig. 5, and this appears to be due to the difference in the amount of subcritical damage. Previous studies have shown that 0°-ply splitting leads to a notch blunting effect [12,25,26], and this is the cause of the strength difference.

To monitor the notch blunting effect directly, the strain in the exposed fibers must be measured and this can be done using the optical technique. Unfortunately, the mesh must be

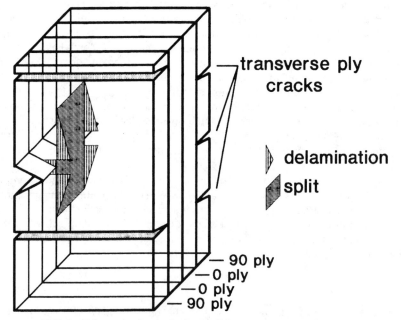

FIG. 4—*Schematic of typical subcritical damage in a (90/0)ₛ laminate.*

$\left[(90/0)_{2s}, \ 2a/w = .5, \ w = 10 \ mm, \ mag. = 54x, \right.$
$\left. \sigma = 300 \ MPa, \ 96\% \ of \ failure \ stress \right]$

$\left[(90_2/0_2)_s, \ 2a/w = .5, \ w = 10 \ mm, \ mag. = 54x, \right.$
$\left. \sigma = 443 \ MPa, \ 95\% \ of \ failure \ stress \right]$

FIG. 5—Comparison of the (near) TDS of two different layups.

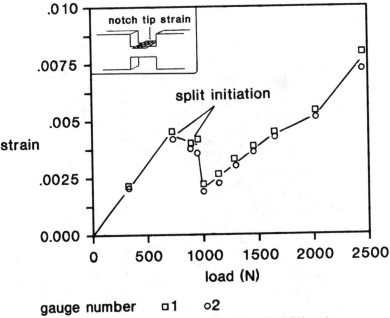

FIG. 6—*Strain at the root of a square notch in a* (0)₄ *DEN specimen.*

placed initially on a flat section of unstrained material, so specimens with square notches were used in the study of notch tip strain. Figure 6 illustrates the effect of splitting in the $(0)_4$ laminate on the strain at the tip of the notch. Although the load was increased monotonically, the strain in the notch tip fibers dropped substantially after the initial pop-in of the splits. This is a direct measurement of notch blunting resulting from splitting. Note that the precision of the optical technique is indicated by the similarity of two independent strain readings.

The procedure was repeated for three 10-mm-wide $(90/0)_s$ square-notched DEN specimens. The 0°-ply splits formed in Specimen 1 at 1530 N but did not appear in the other two specimens. The onset of splitting was clearly visible in the edge-on view of the specimen. In the specimens that did not split, the strain-load curve was basically linear as would be expected (Fig. 7). The deviation from linearity for the curve representing the other specimen indicates that the stress concentration factor was reduced by the formation and propagation of splits. This specimen also proved to be stronger, suggesting that splitting and the resultant crack blunting led to improved notched strength.

The optical technique was also used to map strain as a function of distance from the notch tip on the face of two $(0/90)_s$ specimens. Specimens with exterior 0° plies were used since transverse ply cracks interfere with the strain readings over very small ($\simeq 300\ \mu$m) gage lengths. The specimens were 10 mm wide with 2.5 mm V-notches. One of the specimens was prestressed to 315 MPa, and subsequent radiography was used to determine that the average split length was 5.0 mm, giving a value for ℓ/a of 2.0. Strain readings were taken with 75 and 250 MPa remote stress on the undamaged and damaged specimens, respectively; these stresses gave maximum sensitivity but did not cause any further damage growth. The strain readings were then normalized by the applied stress and are presented in Fig. 8. A set of finite element results

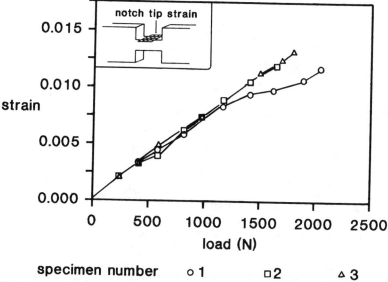

FIG. 7—*Strain at the root of the notch for three identical $(90/0)_s$ specimens. Specimen 1 developed $0°$ ply splits at 1530 N, but no splits developed in the other two specimens.*

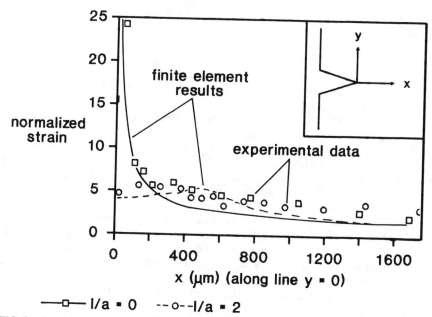

FIG. 8—*Strain on the face of two $(90/0)_s$ V-notched specimens. One specimen contained no subcritical damage and the other specimen was prestressed to 315 MPa resulting in 5 mm splits $(\ell/a) = 2.0$).*

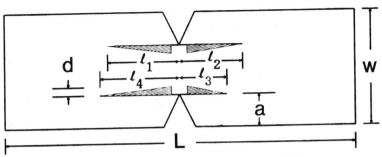

FIG. 9—Measurements used to quantify subcritical damage.

has been superimposed on the experimental data. The model used to generate these results will be described in the next section of this paper.

In the undamaged specimen (with no splits), the stress rose sharply in the vicinity of the notch tip as predicted by the finite element model (or by anisotropic elasticity). Unfortunately, polishing reduced the thickness of the surface 0° ply by 22%, and this elevated the experimental results by a similar amount.

The notch tip 0°-ply stress was finite in the specimen containing splits, and the maximum stress occurred 200 μm away from notch. The location of the maximum stress corresponds with the tip of the triangular delamination zone (see Figs. 4, 9, and 10). The finite element results match the observed strains very well in this case—surface polishing was more controlled resulting in a thickness reduction of only a few percent.

Both the experimental and finite element results indicate that the notch tip stress intensity is completely removed by splitting, even in the cross-ply laminates. It is not surprising, therefore, that attempts to calculate a fracture toughness (for example, Fig. 1) for thin laminates have met with little success.

Radiography

The accumulation of subcritical damage was also studied using quasi-real-time radiography. For each specimen, damage growth was monitored as a function of applied stress. Figure 9 illustrates the specimen geometry and the damage measurements obtained from each specimen. All specimens contained the damage features illustrated schematically in Fig. 4.

The SEM work showed that the stress distribution in the critical 0°-ply fibers at the notch tip was modified by subcritical damage. Catastrophic failure must be precipitated by the notch tip stress immediately prior to failure, and hence it must depend on the terminal damage state (TDS). In view of this, any trends in notched strength data, such as the notch size effect, must be reflected in trends in the TDS measurements.

By studying several series of damage growth radiographs it was determined that the TDS, characterized by the value of ℓ/a, could be measured reliably on a radiograph of the fractured specimen. In Fig. 10, radiographs of specimens of four different sizes (notch lengths) have been scaled to the same size to illustrate the differences in TDS. Notched strength was found to be proportional to the terminal value of ℓ/a.

The data for all $(90/0)_s$ specimens have been plotted in Fig. 11. The "notch size effect" is clearly related to the accumulation of subcritical damage and the TDS. In order to generate a physically realistic model for the notch size effect, the effect of damage must be quantitatively accounted for in some way.

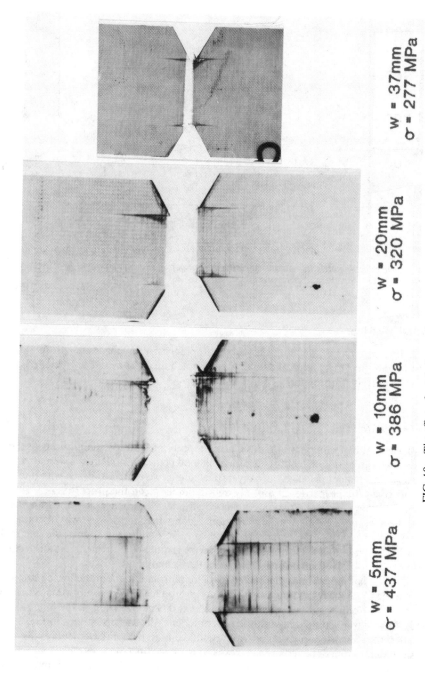

w = 5mm
σ = 437 MPa

w = 10mm
σ = 386 MPa

w = 20mm
σ = 320 MPa

w = 37mm
σ = 277 MPa

FIG. 10—*The effect of notch size on damage and strength is illustrated.*

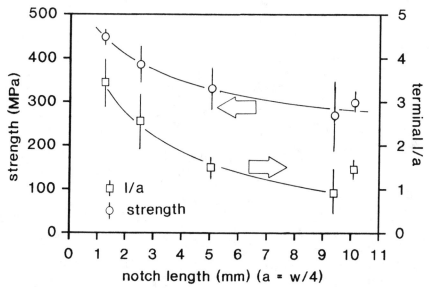

FIG. 11—*The effect of notch size on damage and strength for all (90/0)$_s$ specimens with 2a/w = 0.5. Each data point represents three to six specimens.*

Theoretical Modeling

This section contains a brief summary of a comprehensive notched strength model that incorporates the effect of subcritical damage on the notch tip stress field. For a detailed development refer to Refs *15* through *19*.

Finite Element Modeling

One quarter of the specimen was modeled using appropriate boundary conditions, as shown in Fig. 12. Two-dimensional, eight-noded quadrilateral plane stress elements were used with the MARC finite element package. Two layers of elements were superimposed with one layer of elements given the properties of a 0° ply and the other given the properties of a 90° ply. Corresponding elements in the two layers shared all nodes and hence deformed identically except in the area representing the delamination region where the 90° ply was disconnected from the 0° ply. In addition, the elements of the 90° ply were continuous across the split in the 0° ply, as illustrated in Fig. 4. Although this geometry is quite complicated, it is a direct representation of the observed subcritical damage in (90/0)$_s$ laminates.

The finite element model was constructed to yield information about the tensile stress distribution in the 0° ply near the notch tip. The model produced stress contour maps that displayed finite maximum stresses for all values of $\ell/a \neq 0$. The strains from the maps for $\ell/a = 0$ and $\ell/a = 2$ were compared with the experimentally determined strains in Fig. 8. The strain fields matched reasonably well providing confidence in the finite element results.

A mesh similar to that depicted in Fig. 12 was constructed for ℓ/a ratios of 0, ½, 1, 2, and 4. Each configuration provided a different stress contour map, with a unique maximum stress (very large for $\ell/a = 0$). The stress concentration factor (K_t) was determined as the **maximum**

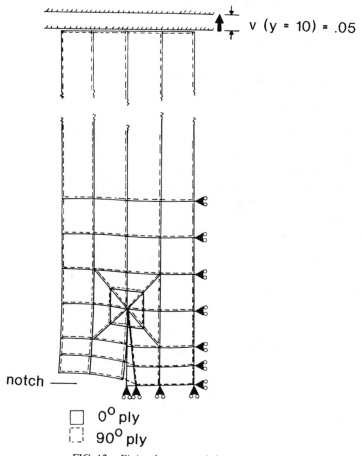

$v \ (y = 10) = .05$

notch —

☐ $0°$ ply

▢ $90°$ ply

FIG. 12—*Finite element mesh for* $\ell/a = 1$.

stress in the 0° ply divided by the remote stress on the laminate for each ℓ/a ratio. The relationship between K_t and ℓ/a was approximated very closely by the empirical function

$$K_t = 8.16(\ell/a)^{-0.284} \tag{2}$$

In order to derive a failure criterion, that is, to discover what aspect of the 0°-ply stress field is constant at failure, the terminal stress distributions for a variety of specimens must be compared. To do this in a simplified way, the peak stress in the 0° ply immediately prior to failure (σ_{0p}) was calculated for all specimens. A value for terminal ℓ/a was obtained from a radiograph of the failed specimen and substituted into Eq 2. The remote stress at failure ($\sigma_{\infty f}$) was then multiplied by K_t to obtain σ_{0p}.

The peak stress was between 2 and 2.5 GPa for all specimens even though the TDS and notched strength varied widely. This indicates that a suitable failure criterion might be obtained by assuming laminate failure when the maximum tensile stress in the load bearing 0° ply equals the strength of this ply. The predicted failure stress can be expressed as

$$\sigma_{\infty f}(\text{predicted}) = \frac{\sigma_{0f}}{K_t} \qquad (3)$$

where

$K_t = 8.16(\ell/a)^{-0.284}$, and ℓ/a determined from the terminal radiograph;

$\sigma_{\infty f}$ = remote failure stress of the laminate; and

σ_{0f} = failure stress of $0°$ ply.

The strength of the $0°$ ply could be determined as an empirical parameter to provide the best fit of Eq 3, and this value would be in the range 2 to 2.5 GPa. Alternatively, it could be measured using an unnotched tensile coupon of unidirectional material, but the volume of material tested in this way would be much larger than the volume of $0°$ ply exposed to the peak stress in the DEN specimens. There have been many studies indicating a relationship between specimen size and strength in graphite-epoxy laminates, and this relationship is usually modeled using Weibull statistics [27,28].

A Weibull strength analysis was performed for the $0°$ ply using both $(0)_8$ and $(90/0)_s$ laminates. The details of this analysis are beyond the scope of this discussion, but may be found in Ref 17. Essentially, by measuring the variability of strength for a large number of unnotched specimens, the dependence of $0°$-ply strength on specimen volume can be determined. The general form of this relationship is

$$\sigma = \sigma_{\text{ref}} \left[\frac{V_{\text{ref}}}{V} \right]^{1/m}$$

where

σ = the strength of a volume (V) of material,

σ_{ref} = the mean strength of a set of reference specimens of volume V_{ref}, and

m = the Weibull modulus ($15 < m < 30$ for most composites).

In the present analysis, an equivalent volume, KV, is used to account for the varying stress field near the notch tip.

Thirty-six specimens of $(90/0)_s$ laminate were tested to establish the values for the Weibull parameters for the central $0°$ ply. The $(90/0)_s$ material was chosen so that the effect of transverse ply cracks on $0°$-ply strength would be included in the Wiebull parameters. The result of this study may be summarized by the equation

$$\sigma_{0f} = 1.88 \text{ GPa} \left[\frac{7.4 \text{ mm}^3}{KV} \right]^{1/20} \qquad (4)$$

where KV is the effective volume of $0°$ ply exposed to the peak tensile stress in the DEN specimens. The value of KV must be obtained from a numerical integration of the tensile stress contours produced by the finite element analysis for each value of ℓ/a.

It is important to note that this prediction of $0°$-ply strength, σ_{0f}, is completely independent of the notched strength data, and yet it provides a result in the required range. For example, for 20-mm-wide $(90/0)_s$ DEN specimens with $\ell/a = 1.0$, KV is equal to 0.045 mm^3, and the strength of this volume of $0°$ ply is predicted to be 2.43 MPa (Eq 4). Combining Eqs 2, 3, and 4 provides a prediction of notched strength based solely on the damage observed in the terminal radiograph, the TDS.

In Fig. 13, the predicted strength is plotted against the actual strength for 43 of the specimens tested in this study. The data represent specimens with widths ranging from 5 to 40 mm and

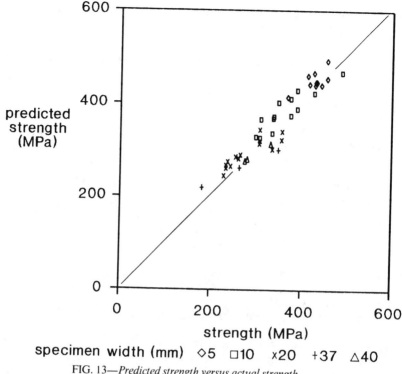

specimen width (mm) ◇5 □10 ×20 +37 △40

FIG. 13—*Predicted strength versus actual strength.*

three different layups of the (90/0$_{ns}$ family, but for every specimen, notched strength was accurately predicted solely from a knowledge of the TDS. The results indicate the maximum tensile stress criterion is sufficient to predict failure provided the true stress distribution at failure is known. It is, however, possible that a more sophisticated failure criterion, involving transverse and shear stresses, might provide an even better correlation.

Damage Growth Model

In order to predict the strength of a notched specimen using Eqs 3 and 4, a radiograph of the TDS is required. If subcritical damage propagation is modeled theoretically, then the strength of a particular specimen can be predicted directly using the subcritical damage as an intermediate step in the calculation.

It has been established that crack growth parallel to the fibers of a fiber composite (that is, splits and delaminations) can be described by linear elastic fracture mechanics [29,30]. Since the path of the subcritical damage in cross-ply DEN specimens is known in advance, it is possible to relate the global energy released to the energy absorbed by the split and delamination faces. Energy release may be calculated using the compliance of the finite element mesh (suitably scaled for specimen dimensions) for each value of ℓ/a. In fact, the finite element analysis gave a constant value for $\partial C/\partial \ell$ as the split length increased. The finite element model does not compute the energy released by transverse ply cracks, so the energy absorbed by these cracks will also be omitted from the calculation. The effect of thermal residual stresses on energy release has been also ignored in this model. Refer to Ref *18* for a description of the effect of this simplification.

Consider one fourth of a (90/0) DEN specimen. The energy released during split propagation is

$$E_r = \tfrac{1}{2}\delta Pu \qquad (5)$$
$$E_r = \tfrac{1}{2}P^2\,\delta C$$

where

P = remote load on specimen,
u = remote displacement, and
C = compliance of specimen.

The energy absorbed during an increase in split length from ℓ to $\ell + \delta\ell$

$$E_{ab} = G_s t\,\delta\ell + G_d k\ell\,\delta\ell \qquad (6)$$

where

G_s = the energy absorbed per unit area of split,
G_d = the energy absorbed per unit area of delamination,
t = thickness of 0° ply, and
k = constant determining delamination zone shape ($= \tan \phi$, see Fig. 9).

Both the delamination and split propagate in mixed-mode that is assumed to remain constant throughout the test and thus G_s and G_d are treated as material constants. Examination of the deformed finite element meshes and radiographs of deformed specimens indicates that the splits propagate in Mode I/Mode II, and the delamination propagation is primarily Mode III. Recent finite element work by Reddy et al. [31] supports these observations.

Equating $E_r = E_{ab}$

$$\tfrac{1}{2}P^2\,\delta C = G_s t\,\delta\ell + G_d k\ell\,\delta\ell \qquad (7)$$

$$\ell = \frac{P^2(\partial C/\partial\ell)}{2G_d k} - \frac{G_s}{G_d}\left[\frac{t}{k}\right] \qquad (8)$$

where $(\partial C/\partial\ell)$ (suitably scaled) is obtained from the finite element model.

Since the delamination propagates in Mode III and the split propagates in Mode I/Mode II, it is expected that the critical strain energy release rate for delamination, G_d, will be as much as ten times the value for the split, G_s [32]. In addition, the area of the delamination is greater than the area of the split for splits more than 1 mm long. Considering both of these observations, it is apparent that the energy required to create the split surface is insignificant compared to the energy required to create the delamination, and thus the second term in Eq 8 may be ignored.

The load is proportional to the stress multiplied by the specimen width, and for specimens of constant thickness and $2a/w$, the load is proportional to $\sigma \times w$ or $\sigma \times a$. Thus, $P^2 \propto \sigma^2 a^2$. Furthermore, scaling arguments can be used to demonstrate that $\partial C/\partial\ell \propto 1/w$ for specimens of fixed thickness, and for specimens of constant $2a/w$, $\partial C/\partial\ell \propto 1/a$. This relationship has been verified through finite element analysis. Equation 8 reduces to

$$\ell/a \propto \frac{\sigma^2}{G_d} \tag{9}$$

Figure 14 shows the experimental data for the split growth of $(90/0)_s$ specimens of four different widths. Each data point represents three specimens. Equation 9 (with appropriate constants evaluated) represents the data extremely well, yielding a value for G_d of 400 J/m². This value is quite reasonable for a delamination propagating primarily in Mode III.

Given that ℓ/a depends only on remote stress (Eq 9), and that K_t depends only on ℓ/a for geometrically similar $(90/0)_s$ specimens (Eq 2), the peak stress in the 0° ply, σ_{0p}, must depend only on remote stress. Figure 15 gives a plot of σ_{0p} versus remote stress. The data points represent a mean value for three specimens and were obtained by multiplying the remote stress by the stress concentration factor determined from a radiograph of split length. The theoretical line is plotted using the theoretical split length from Eq 9 to determine the stress concentration factor.

The specimens should fail when the concentrated stress reaches the breaking stress of the 0° ply and the remote failure stress can then be read from the abscissa of Fig. 15.

It is apparent from the slope of the line in Fig. 15 that for $(90/0)_s$ specimens, **small variations in the strength of the 0° ply are magnified by the progressive crack blunting mechanism.** If the 0° ply is marginally stronger, the specimen lasts longer, and the stress concentration is further blunted by damage growth. In this case, the relationship between 0°-ply strength and specimen strength is no longer linear. An application of Weibull analysis to the 0° ply can result in a predicted specimen strength variation of 40 to 50%, which is sufficient to describe the variation of strength with notch size typically reported in the literature.

To model the notch size effect, the strength of the 0° ply must be calculated for each specimen size, and the strength of the specimen is then obtained from Fig. 15. Numerically, this is accomplished by substituting expressions for σ_{0f} and K_t in Eq 3.

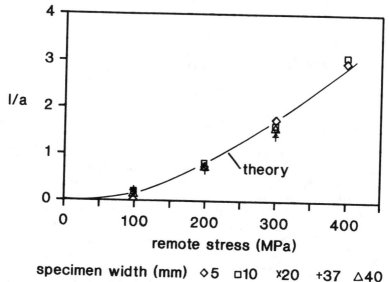

FIG. 14—Split growth as a function of applied stress.

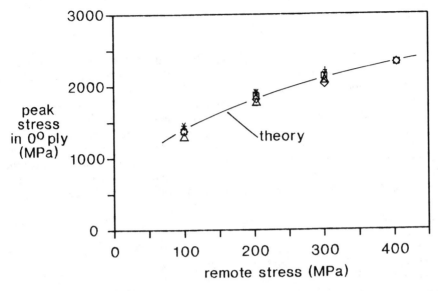

specimen width (mm) ◇5 □10 ×20 +37 △40

FIG. 15—*Peak stress in 0° ply versus applied stress.*

(Eq 3) $\sigma_{\infty f}(\text{predicted}) = \dfrac{\sigma_{0f}}{K_t}$

where

$\sigma_{0f} = f(w, \ell/a)$ (Eq 4),
$K_t = f(\ell/a)$ (Eq 2),
$\ell/a = f(\sigma_\infty^2)$ (Eq 9), and
$\sigma_\infty = \sigma_{\infty f}$ at failure.

Using the full form of the relevant equations, the result is

$$\sigma_{\infty f}(\text{predicted}) = Qa^{-0.162} \qquad (10)$$

where Q is a dimensional constant and can be evaluated by using the full form of Eqs 4, 2, and 9.

The result of this analysis (with Q evaluated) is given in Fig. 16, together with the averaged experimental data for (90/0)$_s$ specimens of various widths. The dashed lines represent informal error limits on the theory, which arise from uncertainty in determining the effective volume of 0° ply exposed to the peak stress, KV, and hence uncertainty in the strength of the 0° ply, σ_{0f}. The model slightly overestimates notched strength, but the trend of the data has been accurately predicted. It is important to note that the model contains no parameters adjusted to the notched strength data, and thus represents a considerable departure from conventional notched strength models.

The model could be applied to (90/0)$_s$ specimens of a different material system using a single radiograph to determine the delamination zone shape and the delamination energy of the system. Unfortunately, the mixed-mode nature of the delamination propagation prevents the

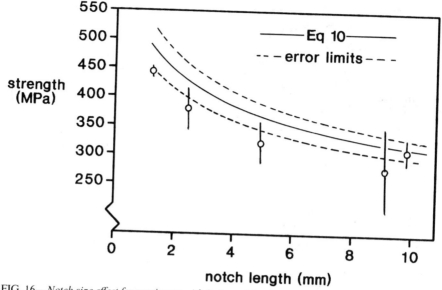

FIG. 16—*Notch size effect for specimens with* 2a/w = 0.5. *The theoretical curve is not based on empirical parameters.*

measurement of G_d independently, and G_d must be measured for every material using at least one split growth measurement.

A fundamental weakness of many phenomenological models of notched strength is that they are incapable of predicting the effect of layup on strength. In this work, the notched strength of specimens with a variety of layups, that is, $(90/0)_s$, $(90/0)_{2s}$, and $(90/0)_{4s}$, were included in the data of Fig. 13. For every layup, strength was successfully predicted using a radiograph of the TDS, and this demonstrates that the dependence of strength on layup is entirely due to the effect of layup on damage growth. This effect has been noted in other studies, but has not yet been modeled quantitatively.

The notched strength model developed here can be applied to specimens with any layup in the $(90_j/0_j)_{ns}$ family by predicting the effect of layup on the damage growth rate. This procedure can be illustrated by comparing $(90/0)_{2s}$ and $(90_2/0_2)_s$ specimens (see Fig. 5). The in-plane elastic properties of these two laminates are identical and thus the energy released during an increment of split propagation is the same for specimens of both laminates, provided the split length is identical in every $0°$ ply (probably not the case for thick laminates [6]) and the delamination shape remains constant. The amount of energy absorbed, however, depends on the total amount of split and delamination surface created. Since the $(90/0)_{2s}$ specimen has three times the number of 0/90 interfaces, three times the area of delamination is associated with a given amount of split growth. Clearly, the growth rate for a particular layup must depend both on the thickness of the specimen and the number of 0/90 interfaces.

By substituting modified growth rates into Fig. 15, the dependence of notched strength on layup can be modeled. This model contains an intermediate calculation of TDS, as before, and for a detailed analysis, refer to Ref *19*. In Fig. 17, the model predictions are compared for five different layups. The average strengths for three 20-mm-wide specimens of each layup have been normalized by the strength of the $(90/0)_s$ specimens. The model correctly ranks the effect of layup on notched strength but the accuracy of the results is not entirely satisfactory,

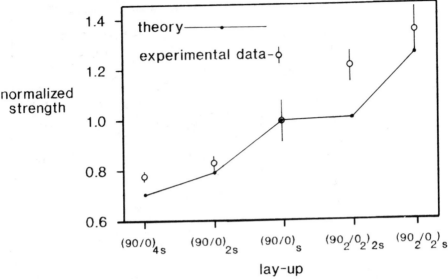

FIG. 17—*The effect of layup on notched strength.*

which indicates that the mechanisms of growth are not completely accounted for in the simple model. Nevertheless, the effect of layup on notched strength is clearly caused by the dependence of subcritical damage growth on layup. It is doubtful that a truly general notched strength model can be constructed without accounting for split and delamination growth quantitatively.

Conclusions

A variety of techniques have been used to illustrate the effect of subcritical damage on the notched strength of thin, cross-ply, graphite-epoxy laminates. Measurements of strain near the notch tip for several $(90/0)_s$ specimens showed that 0° ply splitting reduces the notch tip stress concentration factor. Since specimen failure is precipitated by failure of the 0° ply at the notch tip, the modification of the notch tip stress distribution is significant.

A comprehensive notched strength model for thin cross-ply laminates was reviewed. Simple two-dimensional finite element analysis was sufficient to determine the effect of subcritical damage on the notch tip stress field. With access to the true terminal stress distributions for a variety of specimens, a very simple tensile stress failure criterion was found to be universally applicable.

A simple model describing split growth in cross-ply laminates was also described. The model, based on a global energy release rate, successfully accounted for growth rates for cross-ply laminates with a variety of layups. The combination of the failure criterion and the damage growth model allowed reasonably accurate predictions of notched strength from first principles. Notched strength was modeled successfully without the need for empirical parameters adjusted using the strength data.

Acknowledgments

The authors would like to thank the British Council for financial support and the European Space Agency for partial financial support. We would like to acknowledge many useful discussions with Dr. P. A. Smith, Dr. S. M. Spearing, and Prof. M. F. Ashby. We would also like to thank Mr. B. C. Breton and Dr. W. C. Nixon for their help with the scanning electron microscopy.

References

[1] Whitney, J. M. and Nuismer, R. J., "Stress Fracture Criteria for Laminated Composites Containing Stress Concentrations," *Journal of Composite Materials,* Vol. 8, 1974, pp. 253–265.

[2] Waddoups, M. E., Eisenmann, J. R., and Kaminski, B. E., "Macroscopic Fracture Mechanics of Advanced Composite Materials," *Journal of Composite Materials,* Vol. 5, 1971, pp. 446–454.

[3] Awerbuch, J. and Madhukar, M. S., "Notched Strength of Composite Laminates: Predictions and Experiments-A Review," *Journal of Reinforced Plastics and Composites,* Vol. 4, 1985, pp. 3–159.

[4] Harel, H., Marom, G., Fischer, S., and Roman, I., "Effect of Reinforcement Geometry on Stress Intensity Factor Calibrations in Composites," *Composites,* Vol. 11, 1980, pp. 69–72.

[5] Lagace, P. A., "Notch Sensitivity and Stacking Sequence of Laminated Composites," *Composite Materials: Testing and Design (Seventh Conference), ASTM STP 893,* J. M. Whitney, Ed., American Society for Testing and Materials, Philadelphia, 1986, pp. 161–176.

[6] Harris, C. E. and Morris, D. H., "Fracture Behavior of Thick, Laminated Graphite/Epoxy Composites," NASA Contractor Report 3784, National Aeronautics and Space Administration, Washington, DC, 1984.

[7] Harris, C. E. and Morris, D. H., "Effect of Laminate Thickness and Specimen Configuration on the Fracture of Laminated Composites," *Composite Materials: Testing and Design (Seventh Conference), ASTM STP 893,* J. M. Whitney, Ed., American Society for Testing and Materials, Philadelphia, 1986, pp. 177–195.

[8] Harris, C. E. and Morris, D. H., "Role of Delamination and Damage Development on the Strength of Thick Notched Laminates," *Delamination and Debonding of Materials, ASTM STP 876,* W. S. Johnson, Ed., American Society for Testing and Materials, Philadelphia, 1985, pp. 424–447.

[9] Harris, B., Dorey, S. E., and Cooke, R. G., "Strength and Toughness of Fibre Composites," *Composite Science and Technology,* Vol. 31, 1988, pp. 121–141.

[10] Peters, P. W. M., "On the Increasing Fracture Toughness at Increasing Notch Length of 0/90 and 0/±45/0 Graphite/Epoxy Laminates," *Composites,* Vol. 14, 1983, pp. 365–369.

[11] Garg, A. C., "Fracture Behavior of Cross-Ply Graphite/Epoxy Laminates," *Engineering Fracture Mechanics,* Vol. 22, 1985, pp. 1035–1048.

[12] Daniel, I. M., "Strain and Failure Analysis of Graphite/Epoxy Plates with Cracks," *Experimental Mechanics,* 1978, pp. 246–253.

[13] Yeung, P. C., Stinchcomb, W. W., and Reifsnider, K. L., "Characterization of Constraint Effects on Flaw Growth," *Nondestructive Evaluation and Flaw Criticality for Composite Materials, ASTM STP 696,* R. Pipes, Ed., American Society for Testing and Materials, Philadelphia, 1979, pp. 316–338.

[14] Garg, A. C., "Interlaminar and Intralaminar Fracture Surface Morphology in Graphite/Epoxy Laminates," *Engineering Fracture Mechanics,* Vol. 23, 1986, pp. 1031–1050.

[15] Kortschot, M. T., "Damage Mechanics of Carbon Fibre Composites," Ph.D. thesis, University of Cambridge, Cambridge, U.K., 1988.

[16] Kortschot, M. T. and Beaumont, P. W. R., "Damage Mechanics of Composite Materials I: Measurements of Damage and Strength," *Composite Science and Technology,* Vol. 39, 1990, pp. 289–301.

[17] Kortschot, M. T. and Beaumont, P. W. R., "Damage Mechanics of Composite Materials II: A Damage Based Notched Strength Model," *Composite Science and Technology,* Vol. 39, 1990 pp. 303–326.

[18] Kortschot, M. T., Ashby, M. F., and Beaumont, P. W. R., "Damage Mechanics of Composite Materials III: Prediction of Damage Growth and Notched Strength," *Composite Science and Technology,* Vol. 40, 1991, pp. 147–165.

[19] Kortschot, M. T., Ashby, M. F., and Beaumont, P. W. R., "Damage Mechanics of Composite Materials IV: The Effect of Lay-up on Damage Growth and Notched Strength," *Composite Science and Technology,* Vol. 40, 1991, pp. 167–179.

[20] Kortschot, M. T., Beaumont, P. W. R., Nixon, W. C., and Breton, B. C., "Crack Advancement in a Carbon Fibre-Epoxy Composite Observed by Dynamic Scanning Electron Microscopy," *Proceedings,* International Conference on Composite Materials VI/ECCM II, London, 21–26 July 1987.

[21] Kortschot, M. T., "High Resolution Strain Measurement by Direct Observation in the Scanning Electron Microscope," *Journal of Materials Science,* Vol. 23, 1988, pp. 3970–3972.

[22] Paris, P. C. and Sih, G. C., "Stress Analysis of Cracks," *Fracture Toughness Testing and Its Applications, ASTM STP 381,* American Society for Testing and Materials, Philadelphia, 1964, pp. 30–81.

[23] Owen, M. J. and Bishop, P. T., "Critical Stress Intensity Factors Applied to Glass Reinforced Polyester Resin," *Journal of Composite Materials,* Vol. 7, 1973, pp. 146–159.

[24] Zimmer, J. E., "Fracture Mechanics of a Fiber Composite," *Journal of Composite Materials,* Vol. 6, 1972, pp. 312–315.

[25] Tirosh, J., "The Effect of Plasticity and Crack Blunting on the Stress Distribution in Orthotropic Composite Materials," *Journal of Applied Mechanics,* 1973, pp. 785–790.

[26] Bishop, S. M., "Deformation of Notched Carbon Fibre Composites," Technical Report 73124, Royal Aircraft Establishment, 1973.

[27] Harlow, D. G. and Phoenix, S. L., "The Chain-of-Bundles Probability Model for the Strength of Fibrous Materials I: Analysis and Conjectures," *Journal of Composite Materials,* Vol. 12, 1978, pp. 195–214.

[28] Hitchon, J. W. and Phillips, D. C., "The Effect of Specimen Size on the Strength of CFRP," *Composites,* Vol. 9, 1978, pp. 119–124.

[29] Wu, E. M., "Application of Fracture Mechanics to Anisotropic Plates," *Journal of Applied Mechanics,* 1967, pp. 967–974.

[30] Wang, S. S., "Fracture Mechanics for Delamination Problems in Composite Materials," *Journal of Composite Materials,* Vol. 17, 1983, p. 210.

[31] Reddy, E. S., Wang, A. S. D., and Zhong, Y., "Simulation of Matrix Cracks in Composite Laminates Containing a Small Hole," *Proceedings,* Damage Mechanics in Composites, Winter Meeting, American Society of Mechanical Engineers, Boston, 13–18 Dec. 1987, pp. 83–92.

[32] Donaldson, S. L., "Interlaminar Fracture Due to Tearing (Mode III)," *Proceedings,* International Conference on Composite Materials VI/ECCM II, London, 21–26 July 1987, pp. 3.274–3.283.

Mark Spearing,[1] Peter W. R. Beaumont,[1] and Michael F. Ashby[1]

Fatigue Damage Mechanics of Notched Graphite-Epoxy Laminates

REFERENCE: Spearing, M., Beaumont, P. W. R., and Ashby, M. F., **"Fatigue Damage Mechanics of Notched Graphite-Epoxy Laminates,"** *Composite Materials: Fatigue and Fracture (Third Volume), ASTM STP 1110,* T. K. O'Brien, Ed., American Society for Testing and Materials, Philadelphia, 1991, pp. 617–637.

ABSTRACT: A modeling approach is presented that recognizes that the residual properties of composite laminates after any form of loading depend on the damage state. Therefore, in the case of cyclic loading, it is necessary to first derive a damage growth law and then relate the residual properties to the accumulated damage.

The propagation of fatigue damage in notched laminates is investigated. A power law relationship between damage growth and the strain energy release rate is developed. The material constants used in the model have been determined in independent experiments and are invariant for all the layups investigated. The strain energy release rates are calculated using a simple finite element representation of the damaged specimen. The model is used to predict the effect of tension-tension cyclic loading on laminates of the T300/914C carbon-fiber epoxy system. The extent of damage propagation is successfully predicted in a number of cross-ply laminates of the form $(90_i/0_j)_{ns}$ and the quasi-isotropic laminate $(90/+45/-45/0)_s$. The dependence of damage on load amplitude and specimen size is also well described.

Residual strength is calculated as a function of damage dimensions for $(90/0)_s$ specimens using a stress-based failure criterion in conjunction with a Weibull dependence of the 0° ply strength on the volume under stress.

KEY WORDS: fiber reinforced composite laminates, damage, fatigue (materials) growth law, strain energy release rate, toughness, finite element method, residual strength, Weibull statistics, composite materials, fracture

Nomenclature

a	Notch length, for central notches tip-to-tip notch size $= 2a$
C	Compliance
$\dfrac{\partial C}{\partial \ell}$	Compliance change with split length
FE	Finite element, used as a subscript to denote data from FE mesh
δE_r	Increment of energy released due to damage growth
δE_{ab}	Energy absorbed by increment of damage extension
$G, \Delta G$	Strain energy release rate, range of strain energy release rate
G_c	Critical strain energy release rate, apparent toughness
G_d, G_s	Energy per unit area absorbed in delamination, splitting
k	Damage dimension in $(90/+45/-45/0)_s$ laminates
ℓ	Split length
ℓ_0	Initial split length, after first cycle of a fatigue test

[1] Graduate student, lecturer, and Royal Society research professor, respectively, Cambridge University Engineering Department, Cambridge, United Kingdom.

$d\ell/dN$	Rate of crack growth with load cycles
$i,j,n,$	Subscripts in laminate notation, for example, $(90_i/0_j)_{ns}$
m	Exponent in fatigue damage growth model
N	Number of load cycles
$P, \Delta P$	Applied load, load range
t	Ply thickness (0.125 mm)
t_{lam}	Laminate thickness
v, V	Volume
V_0	Reference volume
W	Specimen width
α	Angle of delamination at split tip
β	Weibull modulus
λ	Constant in fatigue damage growth model
ρ	Notch root radius
σ	Stress
σ_0	Stress in 0° ply
σ_{0f}	Failure strength of 0° ply
σ_∞	Remote applied stress (P/wt_{lam})
σ_{ref}	Reference stress

Many authors have investigated the response of fiber-composite laminates to cyclic loading, and the general characteristics of the damage associated with fatigue is well documented. Attempts have been made to model damage growth using methods similar to fatigue crack growth in homogeneous media. Success has been achieved for delamination growth [1,2] and the propagation of transverse ply (matrix) cracks [3]. Difficulties have been encountered when trying to apply data obtained for one laminate to others of different ply orientations, although made of the same fiber/resin system.

It is not so much the damage but the deterioration of properties that is of concern to the design engineer. Success has been achieved in relating stiffness reduction to the current damage state [4,5], but the effect of damage on residual strength and fatigue lifetime has proved less easy to analyze. Often, approaches have been used that ignore the damage and treat fatigue data in the form of S-N curves and similar diagrams, from which purely empirical models result.

This paper embodies the philosophy of modeling damage suggested by Kortschot and Beaumont [6] for the case of quasi-static loading. They contest that in order to predict strength (or other properties) it is necessary to take account of the damage and how it evolves. Reifsnider [7] proposes similar ideas in his critical element philosophy. Modeling the effect of fatigue loading is consequently a two-part process. First, the damage growth is modeled and then the mechanical behavior is related to the damage state. The work described here is largely concerned with damage growth originating from drilled notches in relatively thin laminates. A brief description of the relationship between damage and residual strength is also included. Despite the particular nature of the damage in these cases, evidence is produced that suggests that the approach can be extended to other cases of fatigue damage growth.

Experimental

The laminates were made of the Ciba Geigy Fibredux T300/914C carbon fiber-epoxy system, fabricated at DFVLR, Braunschweig, West Germany. Several cross-ply laminates of the general configuration, $(90_i/0_j)_{ns}$, and the quasi-isotropic laminate, $(90/+45/-45/0)_s$, were

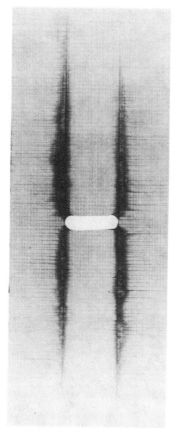

FIG. 1—*A ZnI$_2$ enhanced X-ray radiograph of a (90/0)$_s$ specimen, tested at* R = 0.1, σ_{max} = 324 MPa, N = 1 × 10^6 *cycles. Specimen width is 24 mm in all radiographs.*

produced. Individual plies had a thickness of 0.125 mm. The specimens were center-notched rectangular coupons. A range of specimen and notch sizes ($2a$) were used. Specimen widths (W) were in the range 12 to 72 mm with coupon width-to-notch-size ratios ($W/2a$) of either three or nine. The notches were round-ended, formed by drilling two holes and joining them to form an ellipsoidal notch. A gage-length (L) of 150 mm was used throughout. Fatigue tests were performed at a frequency of 10 Hz using an Instron 1271 servohydraulic machine.

The appearance of fatigue damage in the cross-ply laminates was similar to that observed in quasi-static tensile loading by Kortschot and Beaumont. Figure 1 shows an X-ray of damage in a (90/0)$_s$ specimen, subjected to a maximum cycling stress of 324 MPa and an R-ratio (minimum cycle stress/maximum cycle stress) of 0.1. The damage consisted of splits in the 0° ply growing tangentially to the notch tips, delaminations between the 90 and 0° plies, and transverse ply cracking in the 90° plies over the regions bounded by the splits and the specimen edges. Figure 2 shows fatigue damage in (90/0)$_{2s}$ and (90$_2$/0$_2$)$_s$ laminates where the damage patterns are similar in form, but very different in extent. In general, the fatigue damage grew to greater dimensions than observed in a single quasi-static load application. In all cases, damage was observed to grow in a self-similar manner and hence the damage could be characterized by a single dimension, such as the split length (ℓ). Figure 3 shows an idealization of the

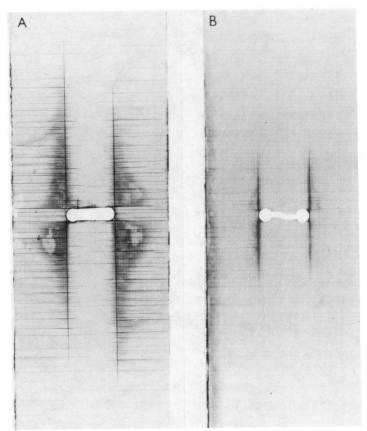

FIG. 2—(a) *Radiograph of a* $(90_2/0_2)_s$ *specimen, tested at* R = 0.1, σ_{max} = 208 MPa, N = 1 × 10^6 *cycles. (b) Radiograph of a* $(90/0)_{2s}$ *specimen, tested at* R = 0.1, σ_{max} = 208 MPa, and N = 1 × 10^6 *cycles.*

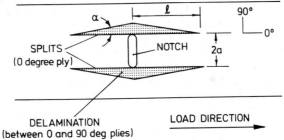

FIG. 3—*Idealization of the damage pattern in a notched cross-ply laminate, showing damage dimensions. Transverse ply cracks are not modeled.*

notch-tip damage pattern for cross-ply laminates of this material system. In most cases, the angle (α) made by the delamination at the tip of the split was observed to lie in the range 3 to 7°.

Fatigue Damage Growth Modeling

From the observations of fatigue damage, it seems that matrix cracking is the dominant mode of failure. There appears to be little fatigue fracture of the fibers. The high fiber volume fraction of the laminates makes it difficult to distinguish between cracking of the matrix and de-cohesion of the fiber-matrix interface. The high fiber density also implies that the matrix experiences a significant degree of constraint. Despite these considerations, it seems reasonable to suppose that fatigue damage consists of several interacting, planar cracks whose formation and growth depend on the fatigue behavior of the matrix.

Experimental data for fatigue crack growth in a wide range of materials, including epoxies (Hertzberg and Manson [8]) has been accurately described by Paris's law

$$\frac{d\ell}{dN} \propto (\Delta K)^m \tag{1}$$

where

$d\ell/dN$ = the crack growth rate with cycling,
ΔK = the range of stress intensity factor in each cycle, and
m = an empirical material constant.

For many cases of damage propagation, a stress intensity factor is an unattractive parameter to use for modeling and predictive purposes. Cracking in composites tends to result from local mixed-mode loading of the crack fronts, for which it is difficult to derive stress intensity factors. Consequently, investigators have employed the strain energy release rate (G) as a parameter in crack growth models for composites. This can be more easily calculated by considering the changes in the global energy of the system, which result from a compliance increase due to crack extension. For the mixed mode of loading and multiple cracking that constitutes damage growing from a notch in a composite laminate, the strain energy release rate offers a simple method of analyzing the driving forces for crack propagation. The crack growth equation would therefore be of the form

$$\frac{d\ell}{dN} \propto (\Delta G)^{m/2} \tag{2}$$

(since $\Delta G \propto \Delta K^2$). However, this is insufficient for describing the present case of combined split and delamination growth. As the split length increases, the associated delamination area grows with a dependence on ℓ^2, which implies an increasing resistance to further crack advance. Therefore, it is more appropriate to use an equation of the form

$$\frac{d\ell}{dN} = \lambda \left[\frac{\Delta G}{G_c} \right]^{m/2} \tag{3}$$

where G_c is the current critical strain energy release rate or effective toughness for the damage growing under a static applied load, and λ is an empirical constant.

Damage Growth in (90/0)ₛ in Quasi-Static Loading

Kortschot and Beaumont [6] used an energy balance method, equating the strain energy released (as the compliance increases) to the energy absorbed by the split and delamination cracks, but omitting the transverse ply cracks. No attempt is made to predict the path of the damage; the damage patterns used for analysis are those observed experimentally. By symmetry, it is only necessary to consider one quadrant of the specimen; width $= W/2$, length $= L/2$, and consisting of one 90° and one 0° ply.

The total energy absorbed during an increase in split length from ℓ to $\ell + \partial\ell$ is given by

$$\Delta E_{ab} = G_s t \partial\ell + G_d \tan \alpha \ell \partial\ell \qquad (4a)$$

where

$G_s =$ energy absorbed per unit area of split,
$G_d =$ energy absorbed per unit area of delamination,
$t =$ 0° ply thickness, and
$\alpha =$ the delamination angle at the split tip.

The energy released during split propagation is

$$\Delta E_r = \tfrac{1}{2}P^2 \partial C \qquad (4b)$$

where

$P =$ the applied load and
$\partial C =$ the change in specimen compliance

Equating ΔE_r with ΔE_{ab}, and rearranging gives

$$\frac{1}{2}\frac{P^2}{t}\frac{\partial C}{\partial\ell} = G_s + G_d \frac{\ell \tan \alpha}{t} \qquad (5)$$

Here, the left-hand side is the energy release rate, and the right-hand side is the effective toughness.

This model is incorporated into the following work on fatigue crack growth in order to predict the extent of damage growth (ℓ_0) in the first cycle of the fatigue loading.

Fatigue Damage Growth in (90/0)ₛ

Equation 5 gives the terms of the fatigue damage propagation model [3]. The relationship between ΔG and the change in compliance is well known

$$\Delta G = \frac{1}{2}\frac{(\Delta P)^2}{t}\frac{\partial C}{\partial\ell} \qquad (6a)$$

where $\Delta P = (2t)(W/2)(\Delta\sigma_\infty)$ for a one half thickness (two-ply) quadrant of the specimen ($\Delta\sigma_\infty$ is the remote applied stress range). Inserting the effective toughness given by

$$G_c = G_s + G_d \frac{\ell \tan \alpha}{t} \qquad (6b)$$

into Eq 3 gives

$$\frac{d\ell}{dN} = \lambda \left[\frac{1/2(\Delta P)^2 \left(\frac{\partial C}{\partial \ell} \right)}{G_s t + G_d \ell \tan \alpha} \right]^{m/2} \tag{7}$$

If the initial split length is ℓ_0 (during the first load cycle), the split length after N cycles of constant load amplitude is given by the integrated form of Eq 7

$$\ell = \frac{1}{G_d \tan \alpha} \left[\lambda (\Delta G t)^{m/2} \left[\frac{m + 2}{2} \right] (G_d \tan \alpha) N + (G_s t + G_d \ell_0 \tan \alpha)^{(m+2)/2} \right]^{2/(m+2)} \tag{8}$$

The simplicity, the model is kept in a differential form.

Damage Growth in $(90_i/0_j)_{ns}$ Laminates

For one quadrant of a general $(90_i/0_j)_{ns}$ laminate where half the laminate thickness $= (i + j)nt$, there are $(2n - 1)$ 90/0 interfaces that can delaminate and a total thickness of 0° plies $= njt$. Therefore, the general form of Eq 7 becomes

$$\frac{d\ell}{dN} = \lambda \left[\frac{\frac{1}{2} (\Delta \sigma_\infty)^2 (W/2)^2 (n(i + j)t)^2 \frac{\partial C}{\partial \ell}}{G_s njt + G_d(2n - 1)\ell \tan \alpha} \right]^{m/2} \tag{9}$$

Values of G_s and G_d have already been determined by Kortschot and Beaumont [6] from monotonic tensile tests on this material ($G_s = 158$ and $G_d = 400$ Jm^{-2}, respectively). The delamination angle, α, has been measured at about 3.5° and ℓ_0 can be determined from Kortschot and Beaumont's analysis of quasi-static damage growth [4]. The constants, λ and m, can be obtained by experimental calibration. Finally, it is necessary to evaluate $\partial C/\partial \ell$, for which a finite element model was used. Details are given in the Appendix in which the computations show $\partial C/\partial \ell$ is independent of ℓ.

Results and Discussion

Damage Growth in $(90/0)_s$ Laminates

Figure 4 shows split growth data for several $(90/0)_s$ specimens subjected to three different peak cycling stresses, with $R = 0.1$. The split lengths are normalized with respect to the notch size, a. Also shown is the integrated form of the crack growth equation for these stress levels. Gustafson and Hojo [9] have found that $m = 14$ approximately, for Mode II delamination growth in this material system. A good fit to the data was obtained using $\lambda = 8.0 \times 10^{-4}$ and $m = 14$. The curves do not intersect the origin because there is quasi-static split growth (ℓ_0) in the first cycle.

The effect of maintaining a constant peak load but varying the load amplitude is illustrated in Fig. 5. Data for split lengths after 10^6 load cycles is shown for specimens cycled with a peak applied stress of 324 MPa and R ratios between 0.1 and 0.7. The model provides a reasonable description of the observed decrease in split length with increasing R ratio. There is some evidence of an additional mean stress effect that is not predicted by the model.

The growth law can be normalized with respect to the notch size. A weak dependence on

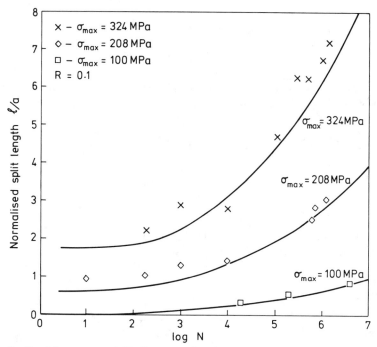

FIG. 4—*Predicted damage growth (Eq 8) compared with experimental data for (90/0)$_s$ specimens with* W = 24 *mm and* a = 4 *mm. Each data point refers to a different specimen.*

the notch size is predicted. For the range of specimen sizes examined, it can be shown that ℓ/a depends only upon the applied stress, number of cycles, and the $W/2a$ ratio. Figure 6 illustrates this, displaying data for specimens with three different widths, but keeping $W/2a = 3$, with the predicted curves for the three specimen sizes. The experimental data and the predicted growth curves are practically coincident.

For laminates with different values of $W/2a$, a new finite element computation of $\partial C/\partial \ell$ is required. Figure 7 shows data for specimens with $W/2a = 3$ and 9. The model predicts that there will only be a small variation in the propagation of damage for values of $W/2a$ greater than three, which is supported by the experimental evidence.

For the range of tensile loading conditions employed in the experimental study, the model appears to accurately describe the propagation of fatigue damage in (90/0)$_s$ laminates without the need to recalibrate the model for different specimen configurations.

Damage Propagation in $(90_i/0_j)_{ns}$ Cross-Ply Laminates

Using the values of $\partial C/\partial \ell$ obtained from the finite element model and the general form of the damage growth Eq 7, damage growth for a variety of cross-ply laminates was predicted and compared with experimental data. The empirically determined parameters, G_s, G_d, m, and λ, were unchanged for all the layups examined (α is assumed to remain constant).

Figure 8 shows the effect of increasing the ply thickness. For a $(90_i/0_i)_s$ laminate, ΔG scales with the ply thickness (it), but G_c is largely dependent on the number of delaminating interfaces ($2n - 1$), which remains as 1 irrespective of the value of i. Therefore, i controls the rate

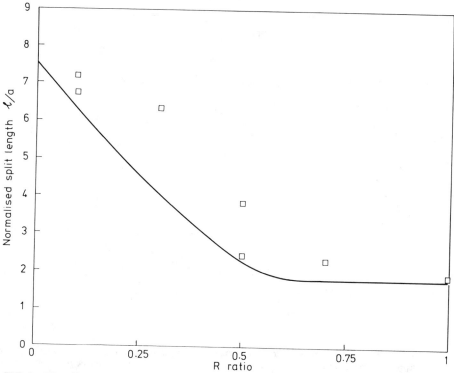

FIG. 5—*The effect of changing the amplitude of the cycling stress. Predicted damage growth (Eq 8) and experimental data for (90/0)ₛ specimens with* W = 24 mm, a = 4 mm, σ_{max} = 324 MPa, *and* N = 1 × 10⁶ *cycles.*

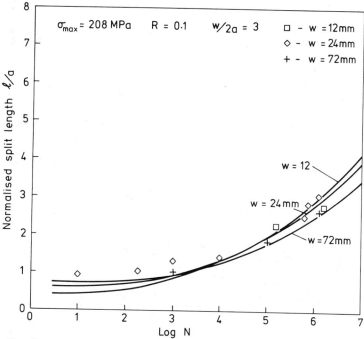

FIG. 6—*The effect of changing the specimen dimensions but maintaining a constant value of* W/2a = 3. *Predicted damage extent (Eq 8) and data for (90/0)ₛ.*

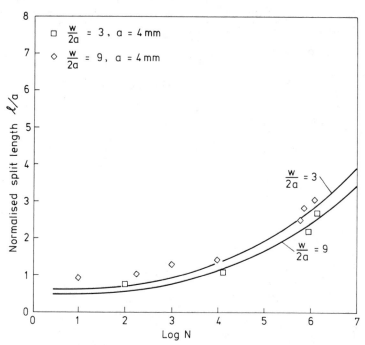

FIG. 7—*The effect of changing the specimen aspect ratio (W/2a), maintaining a constant value of* a = *4 mm, varying the specimen width. Predicted damage extent (Eq 8, with appropriate* $\partial C/\partial \ell$ *from Table 1) and data for* (90/0)$_s$.

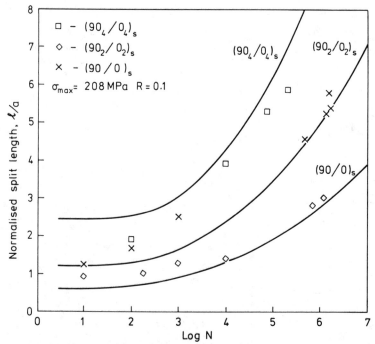

FIG. 8—*Predicted damage growth (Eq 9) and data for notched cross-ply laminates of the form* (90$_i$/0$_i$)$_s$, W = 24 mm, and a = 4 mm.

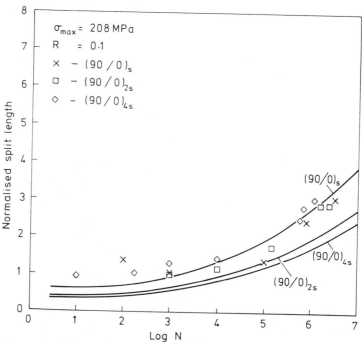

FIG. 9—*Predicted damage growth (Eq 9) and data for notched cross-ply laminates of the form* $(90/0)_{ns}$, *W* = 24 *mm, and* a = 4 *mm.*

of damage propagation. Figure 9 shows the effect of increasing the number of plies but maintaining a constant ply thickness in $(90/0)_n$ laminates. G_c increases with the number of delaminations $(2n - 1)$, but ΔG scales with the number of plies, $2n$. This results in a predicted gradual decrease in damage growth rate as the number of plies increases. The experimental data does not support this prediction; instead, the damage growth appears independent of the value of n. This is probably because the assumption that the extent of damage is equal in all $0°$ plies and $(90/0)$ interfaces is incorrect [10]. For a few specimens, the outer 90/0 plies were ground off, and it was apparent that the extent of damage in these plies was greater than in the interior layers. The X-ray technique superimposes the damage in the different layers, making it difficult to distinguish damage between the different plies without resorting to other techniques. Nonetheless, an acceptable agreement between theory and experiment was obtained.

Figure 10 shows the predicted damage growth curves and experimental data for $(90_i/0_j)_s$ laminates. A good agreement between experiment and theory is obtained.

Residual Strength

The modeling described by Kortschot and Beaumont [6] was used to predict residual strength of $(90/0)_s$ specimens ($W/2a = 3$, $W = 24$ mm). They used a two-part Weibull model to account for the statistical variability of laminate strength. The Weibull parameters were found by performing strength tests on unnotched $(90/0)_s$ specimens; the model has, therefore, been calibrated independently of the tests on notched laminates. The Weibull model reduces to

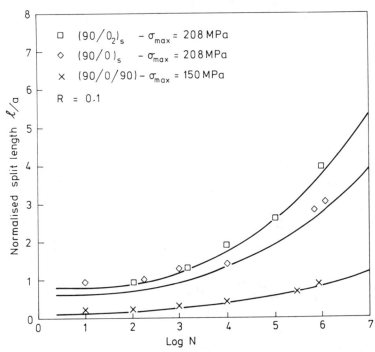

FIG. 10—*Predicted damage growth (Eq 9) and experimental data for notched cross-ply laminates of the form $(90_i/0_j)_s$, W = 24 mm, and a = 4 mm.*

$$\sigma_{0f} = \frac{\sigma_0 V_0^{1/\beta}}{\sigma_{ref}} \left[\int_{v=0}^{v=V} \sigma(v)^\beta \, dv \right]^{-1} \qquad (10)$$

where

$\sigma(v)$ = stress acting over a volume, dv;

σ_{ref} = a reference stress in the distribution, here taken to be the maximum stress;

σ_{0f} = failure strength of a volume (V) of the 0° ply;

σ_0, V_0 = reference stress and volume from experiments on unnotched $(90/0)_s$ specimens where $\sigma_0 = 1.88$ GPa and $V_0 = 7.4$ mm^3; and

β = Weibull modulus where $\beta = 20$ from unnotched $(90/0)_s$ strength tests.

The stress-volume integrals were evaluated numerically, by measuring the areas enclosed by the notch tip stress contours generated by the finite element model. The stress-volume integral was expressed as a continuous function of split length by interpolation between the values of ℓ/a for which FE meshes were constructed. Residual strength could therefore be calculated as a function of split length. The model assumes that it is the longitudinal tensile stress that governs failure. This is not a valid assumption in general, but it seems reasonable for the specimens and loading conditions presented here. In Fig. 11, measured and predicted residual strength is plotted as a function of split length. The predicted residual strength dependence on damage is superimposed. It can be seen that the model can accurately predict the residual strength. The notch tip blunting effect is the dominant factor governing the residual strength, and the residual strength increased throughout the duration of each fatigue test. As a consequence, no fatigue failures were observed.

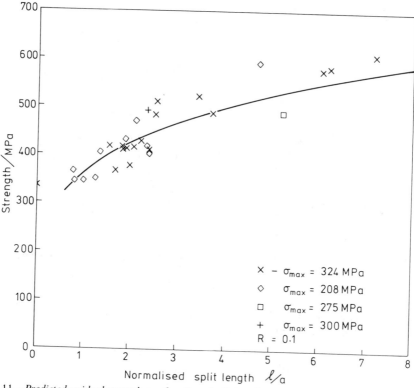

FIG. 11—*Predicted residual strength as a function of split length and experimental data for* (90/0)ₛ *laminates, W = 24 mm, a = 4 mm, and R = 0.1.*

Damage Propagation in (90/+45/−45/0)ₛ Laminates

For $(90/+45/-45/0)_s$ laminates, it is assumed that the damage evolution is controlled by the splits in the 0° plies and delamination at the interface of the 0° ply with the innermost −45° plies. The overall damage pattern is shown in Fig. 12, after 2.5×10^6 cycles, $\sigma_\infty = 200$ MPa and $R = 0.1$. The delamination at the $0/-45$ interfaces can be identified by grinding off the outer $(90/+45)$ plies. Figure 13 shows the central four plies of the specimen in Fig. 12. The damage was observed to grow in a self-similar manner throughout the test duration. A reasonable geometrical approximation to the damage pattern is the double triangular pattern superimposed on the X-ray (Fig. 13), and this pattern will be used for subsequent analysis.

Considering one eighth of the specimen ($W/2$, $L/2$, $90/+45/-45/0$), the energy absorbed by the splits and the delamination at the $-45/0$ interface for an increment of split growth is given by

$$\delta E_{ab} = G_s l \delta \ell + G_d \frac{(\ell + k)}{2} \delta \ell \tag{11}$$

where k is the distance from the tip of the notch tip edge of the delamination with the center line of the notch (see also Fig. 20). Since it was observed that the delamination was bounded by off-axis ply cracks tangential to the notch tip for a notch tip radius, ρ

$$k = (\sqrt{2} - 1)\rho \tag{12}$$

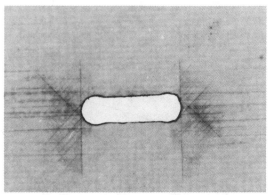

FIG. 12—*X-ray radiograph of a (90/+45/−45/0)$_s$ specimen, tested at* R = 0.1, σ_{max} = 200 MPa, N = 2.5 × 10^6 *cycles, and* W = 24 mm.

The energy released due to the increase in specimen compliance, δC, is still given by

$$\delta E_r = \frac{1}{2}P^2\delta C \qquad (13)$$

For split growth under quasi-static loading, the energy balance: $\delta E_r = \delta E_{ab}$ is satisfied. Therefore, the equivalent to Eq 4 for this laminate is

$$G_s + G_d \frac{(\ell + k)}{2t} = \frac{1}{2}\frac{P^2}{t}\frac{\partial C}{\partial \ell} \qquad (14)$$

from which

$$\ell = \frac{P^2}{G_d}\frac{\partial C}{\partial \ell} - \frac{2tG_s}{G_d} - k \qquad (15)$$

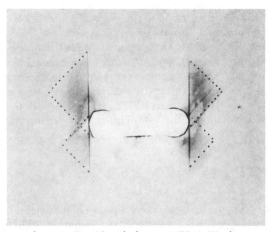

FIG. 13—*The specimen shown in Fig. 13 with the outer (90/+45) plies ground away to reveal the delamination pattern at the (−45/0) interface.*

Fatigue Damage Growth

Equation 3 still describes the fatigue damage growth, the strain energy release rate is given by Eq 5, and from Eq 11

$$G_c = G_s + G_d \frac{(\ell + k)}{2t} \qquad (16)$$

From this, Eq 3 becomes

$$\frac{d\ell}{dN} = \lambda \left[\frac{\frac{1}{2}(\Delta P)^2 \left(\frac{\partial C}{\partial \ell}\right)}{G_s t + G_d \frac{(\ell + k)}{2}} \right]^{m/2} \qquad (17)$$

where $\partial C/\partial \ell$ is unknown for this laminate and damage pattern and a finite element mesh was therefore constructed to determine $\partial C/\partial \ell$. Details are given in the Appendix. As before, $\partial C/\partial \ell$ is independent of ℓ, as shown in Fig. 14.

Figure 15 shows data for split growth in the $(90/+45/-45/0)_s$ laminate under a quasi-static monotonically increasing load. The damage appears to "pop in" at a stress above 100 MPa. This is not predicted by the model, but the subsequent damage growth can be modeled by setting: $k = 0.414$ mm ($\rho = 1$ mm), $G_d = 300$ Jm^{-2}, $G_s = 158$ Jm^{-2}, and $\partial C/\partial \ell = 1.57 \times 10^{-7}$ N^{-1}. The value for G_d differs from that found for cross-ply materials.

If λ and m are material constants for the matrix, they should retain the values from the cross-ply model ($\lambda = 8.0 \times 10^{-4}$, $m = 14$). G_d, G_s, and k have been established from the quasi-static tests; therefore, it should be possible to apply the model without any further calibration. The predicted split growths and experimental data are plotted for two different load levels ($\sigma_{max} = 100$ and 200 MPa) in Fig. 16. A good agreement is achieved.

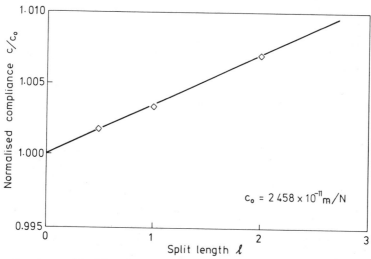

FIG. 14—*Compliance of the $(90/+45-45/0)_s$ meshes as a function of split length ($(W/2)_{FE} = 3$ m, $a_{FE} = 1$ m, $t_{FE} = 1$ m).*

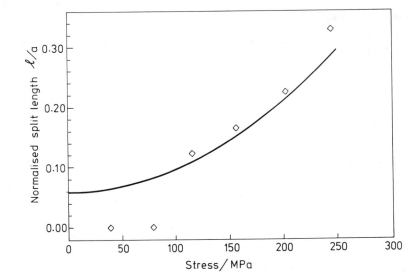

FIG. 15—*Predicted quasi-static split growth (Eq 15) as a function of applied stress, and experimental data for $(90/+45/-45/0)_s$, W = 24 mm and* a *= 4 mm.*

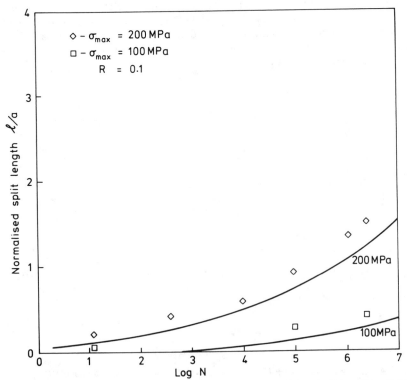

FIG. 16—*Predicted fatigue damage growth (Eq 17) and experimental data for $(90/+45/-45/0)_s$, W = 24 mm,* a *= 4 mm, and* R *= 0.1.*

Conclusions

An equation of the form $d\ell/dN = \lambda[\Delta G/G_c]^{m/2}$ describes fatigue damage growth from notches in a variety of laminates, without recalibration between layups. The effective toughness, G_c, which increases with split length, has been modeled in detail.

In conjunction with a model for the dependence of strength on damage, it is possible to predict residual strength as a function of loading conditions.

The method is generally applicable, although it has only been described for one particular material system. Similar analyses may be performed for other laminated materials under different loading regimes.

Acknowledgments

The authors would like to acknowledge the contribution of Professor M. T. Kortschot, who participated in many helpful discussions at the start of this work. The project was partially funded by the European Space Agency and Engineering Systems International. Mr. Spearing holds a Science and Engineering Research Council studentship.

APPENDIX

Finite Element Model

The form of the finite element (FE) model is similar to that employed by Kortschot and Beaumont [6], consisting of two identical layers of two-dimensional plane-stress elements superimposed. The layers have the in-plane properties of the 90 and 0° plies; the elastic properties of a single ply are

$E_{11} = 135\,\text{GPa}$
$E_{22} = 9.6\,\text{GPa}$
$G_{12} = 5.8\,\text{GPa}$
$v_{12} = 0.31$
$v_{21} = 0.022$

where E_{11} and E_{22} are the moduli measured parallel and perpendicular to the fiber directions, respectively, G_{12} is the in-plane shear modulus, and v_{12} and v_{21} are the principal Poisson ratios.

The split in the 0° layer is modeled using nodal pairs. The two layers share the same nodes everywhere except in the delaminated region, where separate nodes are generated in the 90° ply. Only one quadrant of the specimen, consisting of one 0° and one 90° ply is modeled and appropriate symmetry conditions are applied. Since the model is essentially two-dimensional, there is no means of incorporating through-thickness stresses. In reality, such stresses may affect delaminations and may cause some variation in their shape and growth rate with distance from the midplane. Because this investigation concerns center-notched specimens, some modifications were required to Kortschot and Beaumont's model for edge-notched specimens. In particular, the symmetry conditions were rearranged. The mesh uses a square-ended notch, which is a reasonable approximation to a blunted round tip. A typical mesh ($\ell/a = 2$) is shown in Fig. 17. For all meshes, the overall length was $15.167a$. Meshes with different values of $W/2a$ were also constructed.

The mesh was loaded by applying a displacement of $0.05a$ to the top edge of the specimen. Figure 18 shows the resulting stress contours in the 0° ply in the region of the split. The model predicts a decreasing stress concentration at the notch tip with increasing split length. In Fig. 19, the compliance of the meshes is plotted against the split length for delamination angles of 3.5 and 7°. In both cases, the compliance, C, is directly proportional to the split length, ℓ, so that $\partial C/\partial \ell$ is independent of split length. The relationship between the value of $\partial C/\partial \ell$ obtained from the FE model ($\partial C/\partial \ell | \text{FE}$) and those for the actual $(90/0)_s$ specimen is given by the scaling equation

$$\frac{\partial C}{\partial \ell} = \frac{(W/2)_{\text{FE}}(2t)_{\text{FE}}}{(W/2)(2t)}\frac{\partial C}{\partial \ell}\bigg|_{\text{FE}} \tag{18}$$

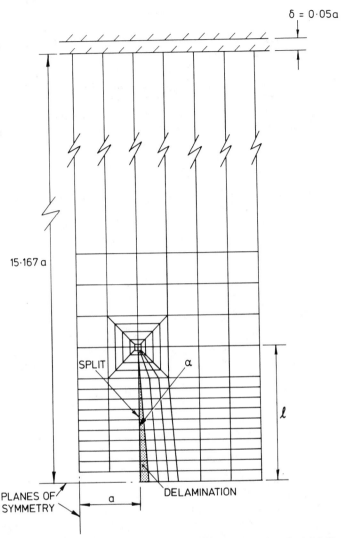

FIG. 17—*Finite element mesh representing one quadrant of a center-notched (90/0)$_s$ specimen. The mesh consists of two superimposed layers of elements in the same configuration, representing the 90° and 0° plies separately.*

The same FE model can be used to derive data for all laminates of the form $(90_i/0_i)_{ns}$ in which there is an equal proportion of 90 and 0° plies. The scaling equation becomes

$$\frac{\partial C}{\partial \ell} = \frac{(W/2)_{FE}(2int)_{FE}}{(W/2)(2int)} \frac{\partial C}{\partial \ell}\bigg|_{FE} \tag{19}$$

The method can be extended to laminates with unequal thicknesses of 0 and 90° plies by altering the relative ply thicknesses of the layers of the finite element mesh. Table 1 shows values of $\partial C/\partial \ell$ for a range of laminates and different values of $W/2a$.

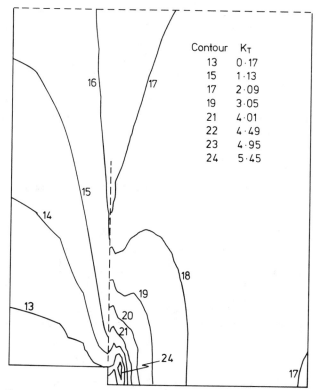

FIG. 18—*Tensile stress contours in the 0° ply, the stresses are represented as stress concentrations (K_T) with respect to the remote stress applied to the laminate. The peak stress is finite and is found near the notch tip (Contour Number 24).*

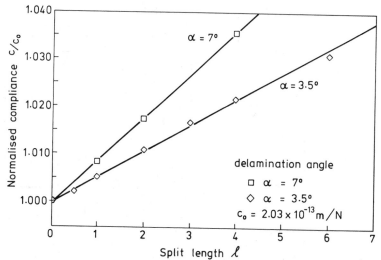

FIG. 19—*Compliance of the (90/0)$_s$ meshes as a function of split length ((W/2)$_{FE}$ = 3 m, a$_{FE}$ = 1 m, and t$_{FE}$ = 1 m).*

TABLE 1—Values of $\dfrac{\partial C}{\partial \ell}\bigg|_{FE}$ for various laminates and specimen configurations.

Layup	$W/2a$	W_{FE}, m	ℓ_{FE}, m	$\dfrac{\partial C}{\partial \ell}\bigg|_{FE}$, N^{-1}
$(90/0)_s$	3	6	1	1.45×10^{-13}
$(90/0)_s$	6	12	1	3.30×10^{-14}
$(90/0)_s$	9	18	1	1.52×10^{-14}
$(90_2/0)_s$	3	6	1	1.20×10^{-13}
$(90/0_2)_s$	3	6	1	$8.0 \ \times 10^{-14}$
$(90/+45/-45/0)_s$	3	6	1	7.62×10^{-14}

NOTE—For these meshes $a = 1$ m, and hence $\partial C/\partial \ell$ must be scaled accordingly.

Finite Element Representation of $(90/+45/-45/0)_s$

Figure 20 shows the finite element mesh used for this damage pattern. The method employed was the same as for the cross-ply laminates. Two identical layers were used, one representing the 0° ply with the appropriate properties, the other represents the $(90/+45/-45)$ plies with equivalent laminate properties derived from the unidirectional ply properties using laminated plate theory. The equivalent $(90/+45/-45)$ ply was three times as thick as the 0° ply. The two layers were separated in the delaminated region

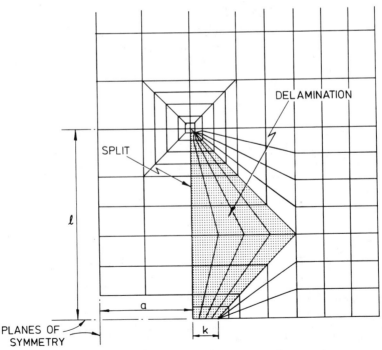

FIG. 20—Part of the finite element mesh representing the damaged region of a center-notched $(90/+45/-45/0)_s$ specimen. The mesh consists of two superimposed layers of elements of the same configuration, representing the homogenized $(90/+45/-45)$ plies and 0° plies, separately.

(shown shaded in Fig. 20); elsewhere, the elements of the two layers shared the same nodes. Figure 14 shows a graph of compliance versus split length for the series of meshes used in this study. The compliance varies linearly with increasing split length, implying $\partial C/\partial \ell$ is independent of split length. The method of scaling was the same as for the cross-ply specimens using Eq 11. A value of $\partial C/\partial \ell = 1.57 \times 10^{-7}\,N^{-1}$ was calculated.

References

[1] O'Brien, T. K. in *Damage in Composite Materials, ASTM STP 775,* K. Reifsnider, Ed., American Society for Testing and Materials, Philadelphia, 1982, pp. 140–167.

[2] Trethewey, B. R., Gillespie, J. R., and Carlsson, L. A., *Journal of Composite Materials,* Vol. 22, 1988, pp. 459–483.

[3] Ogin, S. L., Smith, P. A., and Beaumont, P. W. R., *Composite Science and Technology,* Vol. 24, 1985, pp. 23–31.

[4] Poursartip, A., Ashby, M. F., and Beaumont, P. W. R., *Composite Science and Technology,* Vol. 25, 1986, pp. 193–218.

[5] Talreja, R., *Engineering Fracture Mechanics,* Vol. 25, 1986, pp. 751–762.

[6] Kortschot, M. T. and Beaumont, P. W. R., "Damage Mechanics of Carbon Fibre Composite Materials," *Composite Science and Technology,* to be published.

[7] Reifsnider, K. L., *Engineering Fracture Mechanics,* Vol. 25, 1986, pp. 739–749.

[8] Hertzberg, R. W. and Manson, J. A., *Fatigue of Engineering Plastics,* Academic Press, New York, 1980.

[9] Gustafson, G-C. and Hojo, M., *Journal of Reinforced Plastics and Composites,* Vol. 6, 1987, pp. 6–52.

[10] Harris, C. E. and Morris, D. H., *Fracture of Thick Laminated Graphite Epoxy Composites,* NASA Contractor Report 3784, NASA Langley Research Center, Hampton, VA, 1984.

Dawei Lai[1] and Claude Bathias[2]

Hole Effect and Compression Fatigue of T300/N5208 Composite Materials

REFERENCE: Lai, D. and Bathias, G., "**Hole Effect and Compression Fatigue of T300/N5208 Composite Materials,**" *Composite Materials: Fatigue and Fracture (Third Volume), ASTM STP 1110,* T. K. O'Brien, Ed., American Society for Testing and Materials, Philadelphia, 1991, pp. 638–658.

ABSTRACT: Mechanical testing of notched high-performance composites were performed in compression. The compact tension specimen is modified to be used in compression loading. It is shown that the strain energy released rate is an appropriated criteria in compression. A model of the fatigue damage zone is proposed.

KEY WORDS: composite materials, compression loading, fatigue (materials), hole effect, damage analysis, damage modeling, fracture

Carbon fiber-reinforced composite materials are known for their good mechanical behavior, but it has been discovered [1,2] that they have two patent weaknesses: poor compression and stress concentration resistance, which have hindered, in spite of all efforts, the development of their application in the aeronautical industry. The compression damage behavior is therefore one of the most important factors in the composite materials design, especially in the presence of stress concentrations. In order to predict the failure of composite parts containing a notch, under cyclic compression loading, a research program involving these two parameters was proposed. One of the particular difficulties in compression studies on composite materials results from parasite loading: bending and buckling. Some authors have proposed the use of anti-buckling apparatus [3,4]. However, Matondang and Schütz [5] have shown that these systems can lead to modifications of damage conditions or make observations very difficult. To avoid these problems, a technique, not used previously for composites, inspired by linear fracture mechanics was developed. This involved adapting a compact tension specimen to compression tests. This technique facilitates not only the mechanical tests, but above all allows us to choose the fracture mechanics parameters for the damage characterization, since it was shown that, for the case of tension tests, fracture mechanics can be applied correctly to carbon fibers composite materials [6,7]. The static and fatigue compression-compression (C-C) and fatigue compression-tension (C-T) tests are performed on this specimen. The development of damage and mechanical behavior is followed by stopping the test at different loading levels or at different numbers of cycles. The micrographic sections are taken at different damage stages to examine more closely the damage development and then to relate it to the variation of the mechanical behavior. As the damage zone does not show a two-dimensional form and does not show a regular form in *z*-direction, and the energy of damage progression is a function of

[1] Assistant professor, Division Mecanique, Université de Technologie de Compiègne, 60206 Compiègne, France.
[2] Professor, Conservatoire National des Arts et Metiers (CNAM), Paris, France.

TABLE 1—*Mechanical properties of the material.*

Ply Angle	E_1, MPa	E_2, MPa	σ_{1c}, MPa	σ_{2c}, MPa	ν_{12}	ν_{21}	σ_{12}, MPa	τ_c, MPa
0°	140 000	1 000	1500	60	0.3	0.02	5 500	100
90°	1 000	140 000	60	1500	0.02	0.3	5 500	100
±45°	18 000	18 000	180	180	0.75	0.75	36 000	. . .
laminate	74 000	37 000	750	480	0.36	0.18	17 000	205

damage volume, three-dimensional analysis is developed. The rate of dissipated energy per unit damage volume is then calculated.

The material used in this study is a carbon/epoxy T300/N5208 laminate that has macroscopically orthotropical properties. The stacking sequence, which is widely used in the aeronautical industry, is

$$((0/45/0/-45/90/45/0/-45/0/90/-45/0/45/0)_2)_s$$

The mechanical properties of each ply and of the laminate are shown in Table 1.

Test Method

The compact tension specimen was transformed to adapt to compression testing. A hole of radius, ρ, varying between 2.5 and 5.2 mm was introduced at the notch tip. Such a specimen allows us to easily combine compression and stress concentration effects.

Figure 1 gives the dimensions of the specimen and test setup.

In order to avoid contacts, and to mount the compliance transducer, a V-notch is introduced into the back of the specimen. The mechanical tests are performed on a MAYES hydraulic machine of capacity ±50 000 N. The technique of acoustic emission is used to follow the damage progression and to particularly spot the damage threshold. To isolate the specimen from the machine, in order to detect correctly the acoustic emission when the damage occurs, the loading axis are covered by two sleeves of acetal. These, assuring an improved stress distribution near the loading holes, help to avoid undesirable damage at these zones due to the load introduction. Mechanical parameters are measured by an opening displacement transducer, a lateral deformation extensometer surmounted by a magnetic induction transducer placed near the notch, and a vertical strain gage at 8.5 mm from the notch tip.

Mechanical tests were conducted in static compression loading and under C-C and C-T fatigue loading with a frequency of 5 Hz. During the test, the damage development was observed by stopping at different loading levels or at different numbers of cycles. For this purpose, the X-ray technique with opacifier liquid made of zinc iodine (ZnI) was used. The damage development was related to the variation of macroscopic mechanical parameters measured by the devices described earlier. The micrographic sections were taken before and after the final failure for closer observation of damage progression and to help understand the mechanisms that occurred.

To characterize the damage by fracture mechanics, in order to establish criterion for the compression failure, the specimens with different notch lengths and different hole diameters were used. To verify the failure criterion established and to justify the validity of the use of compact specimens for the compression studies, some mechanical tests were performed with rectangular specimens that are more classical.

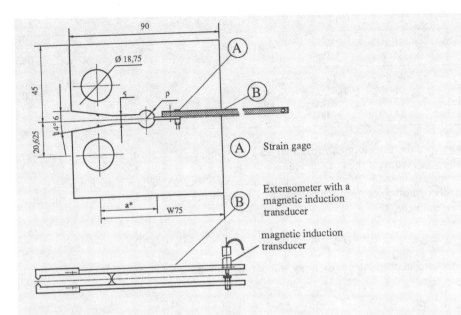

FIG. 1—*Compact specimen and test setup.*

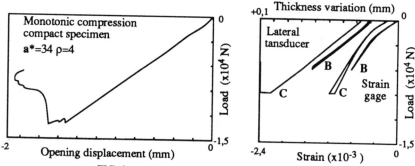

FIG. 2—*A typical monotonic compression curve.*

Mechanical Properties in Compression

Monotonic Loading

The cracking of the specimen under monotonic compression loading can be considered as macroscopically elastic. Figure 2 shows that during the test the slope of compression load-opening displacement curve is linear, although the first damages occur very early, at about 60 to 70% of ultimate load (P_u). This signifies that the stiffness of the specimen, or the response of the opening displacement transducer, is not sensitive to the first damage.

Table 2 summarizes the results of mechanical tests carried out on different specimens for several notch lengths, and Fig. 3 shows the evolution of the ultimate stress at the notch tip as a function of a/W (stress calculation is discussed later).

It can be seen clearly that the failure stress is not a simple function of notch length. The radius of curvature of the notch tip, ρ, has an important influence. This conclusion is logical because, in the case where $\rho \neq 0$, the stress field at the vicinity of the notch tip is not only governed by the notch length but also by ρ.

Fatigue Loading

In general, the damage evolution of the compression cyclic specimen does not end by unstable propagation. As for the monotonic compression specimen, the compliance transducer is

TABLE 2—*Failure load of compression specimens.*

Compact Specimen					Rectangular Specimen				
a, mm	a/W	ρ, mm	$P_f \times 10^3$ N/mm	$\sigma_{ymaxc} \times 10^3$ MPa	a, mm	a/W	ρ, mm	$\sigma_f \times 100$ MPa	$\sigma_{ymaxc} \times 10^3$ MPa
25.2	0.336	4	−2.483	−1.019	8.0	0.16	4	−3.06	−0.925
29.2	0.389	4	−2.274	−1.056	4.0	0.08	4	−4.31	−0.877
34.0	0.453	4	−1.917	−1.050	18.0	0.18	4	−2.47	−1.140
40.1	0.535	4	−1.513	−1.056	23.0	0.23	4	−1.98	−1.096
44.2	0.589	4	−1.257	−1.064	10.2	0.102	5.2	−3.40	−0.978
34.2	0.456	2.5	−1.713	−1.196	9.9	0.099	2.5	−2.72	−1.109
34.5	0.460	5.2	−2.102	−1.029					

NOTE—σ_{ymaxc} is the normal stress in the vertical direction and at the notch tip calculated by the method discussed later.

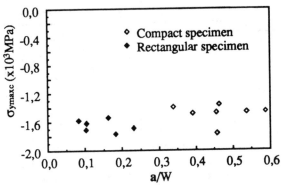

FIG. 3—*Evolution of the failure stress at the notch tip versus* a/W.

not sensitive to the first damage. Figure 4 gives typical curves for compression fatigue. We can see that the significant stiffness degradation of the specimen occurs at about three fourths of the life time in the case of C-T loading, and, four fifths of the life time in the case of C-C loading. The growth of the compliance due to damage is more pronounced for the second case than for the first one. This leads us to think that the damage mode depends on the fatigue ratio, R, in such fashion that when $R = 1$, the damage mode becomes that of monotonic compression.

For compact specimens, $W = 75$, $a^* = 34$ mm, and $\rho = 4$ mm, some observations have to be noted. Under C-C fatigue loading, the "no damage" threshold is found at 60 to 70% of P_u. In the case of C-T loading, the lifetime approaches infinity when the maximum applied load is below 42% of P_u. The endurance limit is lower in the second situation than in the first one. So we can conclude that the C-T fatigue is the most dangerous loading mode for the material.

Evolution of Damage Around the Hole

As discussed earlier, the compliance transducer is not sensitive to the early damage, in both monotonic compression loading and cyclic compression loading cases. By contrast, the lateral transducer and the strain gage show an excellent sensitivity to the early damage. In Fig. 4, we can distinguish three steps on the curve of lateral transducer. From the beginning, the curve increases with an prominent slope, and reveals the initiation and the progression of damage. The X-ray photo (above the curves) taken in this step gives the confirmation. The increase of this curve continues until a plateau that corresponds to the second step. In this step, damages stably propagate. Only the strain gage can reveal this, because it is placed further from the notch tip than the lateral transducer. We can visualize this stable propagation with the two other X-ray photos. In the last step, the damage progresses quickly, until final failure. All three curves recorded by the compliance transducer, the lateral transducer, and the strain gage have a very strong slope. It should be pointed up that the slope of the strain gage curve changes sign, and the curve can decrease until it is below 0. This signifies that the concerned region is in a tension state. The last X-ray photo shows the final form of the damage zone. This one has a quasi-elliptical form.

Microscopic Aspect of Damage Zone in Compression

The X-ray technique is very efficient to visualize the damage formation and its progression, but this technique gives only projected images through the whole specimen thickness. To

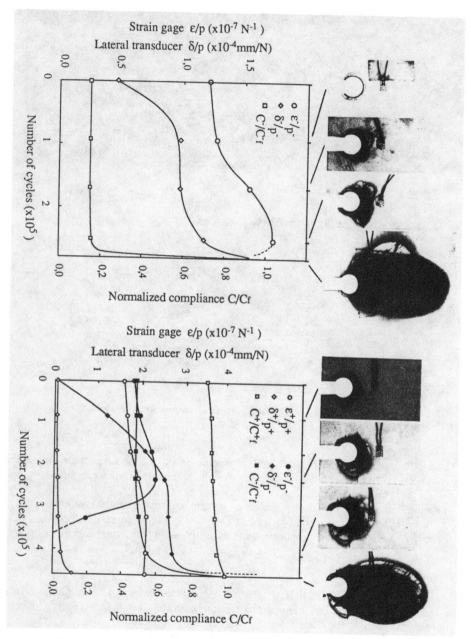

FIG. 4—*Evolution of fatigue damage.*

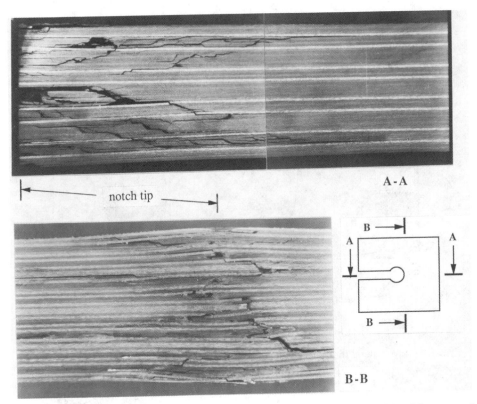

FIG. 5—*Micrographic aspect of final failure of monotonic compression specimen (a^* = 34 mm, ρ = 4 mm, P_u = -1917 N/mm).*

observe the damage progression closely and to help understand the mechanisms, micrographic sections were taken before and after final failure. The schematic of micrographic sections is given in Fig. 5.

From different microscopical observations, a general conclusion can be given: the damage mechanisms change from one loading mode to another, but in all cases, the delamination is the principal mode of damage propagation in the compression laminate specimens.

Monotonic Loading

The delamination is initiated first in the most external interfaces. But this initiation is very limited before final failure. The micrographic section taken at 95% of P_u shows that the delaminations are confined to the notch tip and in the external interfaces.

When the applied load reaches P_u, the closing of the notch becomes very hard and the energy accumulated at the notch tip attains the critical value, the breakdown of the specimen occurs unstably. Figure 5 shows a photo of the micrographic section A-A of a failed specimen. We can differentiate three zones. The first, near the notch tip, corresponds to a generalized crushing damage zone. In the second zone, the crushing damages tend to be transformed to delaminations and then to propagate in the external plies, because in these locations the necessary propagation energy is the lowest. The change of propagation planes happens by breaking adjacent laminas, even for the 90° plies that are very difficult to break since the fibers are in the delamination direction. The change of delamination planes by breaking 90° plies is possible because the energy liberation due to the crushing damage of the first zone is very sudden and important. The delaminations, limited now in the external interfaces, progress in the third zone that is in tension.

From these observations, it is reasonable to think that final failure of the monotonic compression specimen occurs suddenly at the last moment, under the crushing form at the notch tip, and this crushing damage begins at same time in the all specimen thicknesses. This consideration is very important for the application of fracture mechanics to characterize the monotonic compression failure, which will be discussed later.

Cyclic Loading

Since the applied load is lower than the critical value for the monotonic generalized damage, the first damages of cyclic loading, confined at the notch tip and limited in the external interfaces, can not progress quickly. But because the resin is sensitive to cyclic loading, these delaminations can propagate slowly. The exterior plies grow progressively weak due to the degradation of the resin between the fibers, and begin to bend with an increased amplitude as a function of the number of cycles. This induces the delamination of the second adjacent plies, and then the third, the fourth . . . , until final failure. For this reason, the sudden breaking is not observed in the compression fatigue specimen.

In the A-A section of a C-C fatigue specimen (Fig. 6), we can observe that in the second zone of damage progression there is no fracture of the 90° plies. That is, the accumulated energy at the damage tip is not sufficient to break these plies for changing the propagation direction.

In the A-A section of a C-T fatigue specimen (Fig. 7), there is very good regularity of the damage distribution. The first damage state is practically absent, and the delamination tips can be included inside a quasi-parabolic curve.

In the B-B micrographic sections, we can easily compare the regularity of the damage distribution of C-C and C-T specimens that is quasi-parabolic, with the monotonic compression specimen. It should also be pointed up that in these sections, we can distinguish more clearly the difference between C-C failure and C-T failure. In the first case, the microcracks do not exist, and the damages are essentially delamination and fractures due to bending and buckling in these plies. In the second case, by influence of the positive loading, microcracks are formed in the 90 and ±45° plies. The microcracks are aided by the delamination of the adjacent plies and aid mutually the progression of these delaminations. This is why the C-T fatigue is the most dangerous loading mode for laminated materials.

Characterization of Compression Damage

The damage mechanisms discussed earlier lead us to think that it is reasonable to try to use linear fracture mechanics to characterize the compression failure of this material. For this purpose, it is convenient to assume generalized plane stress in the determination of stress field.

Calculation of Stress and Displacement Fields

Since the considered medium is a macroscopically orthotropic plate containing a notch, to calculate the stress field in the specimen, especially near the notch tip, a theoretical analysis was necessary. To simplify the calculation, we have replaced the notch by an elliptical hole (Fig. 8) of which the curvature radius of the tip is same as that of the hole. The validity of this

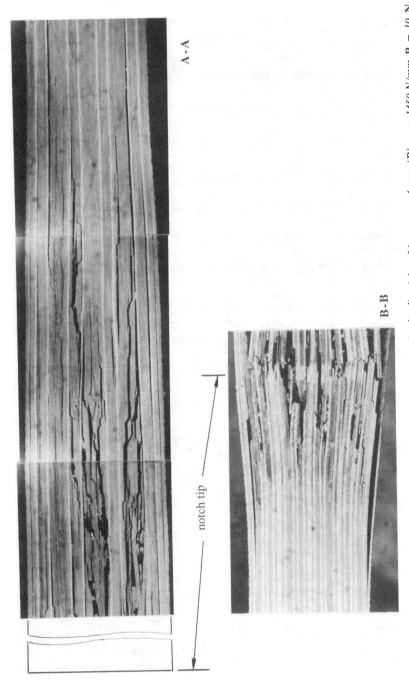

FIG. 6—*Final failure of compact specimen under compression-compression loading (*$a*$ *= 34 mm,* ρ *= 4 mm,* $|P|_{max}$ *= 1450 N/mm,* R *= 10.* N_f *= 2.75 · 10⁵ cycles).*

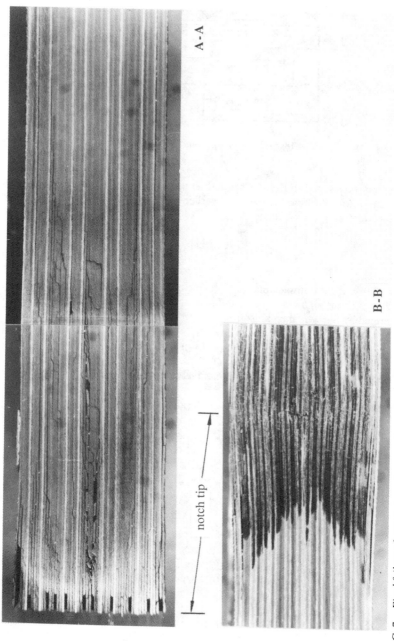

FIG. 7—*Final failure of compact specimen under tension-compression loading* ($a^* = 34\ mm$, $\rho = 4\ mm$, $|P|_{max} = 930\ N/mm$, $R = -1$, $N_f = 4.59 \cdot 10^5\ cycles$).

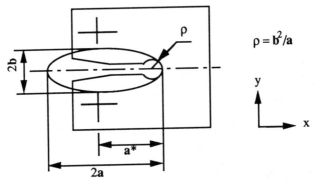

$\rho = b^2/a$

FIG. 8—*Approximation of the notch by an elliptical hole.*

approximation is discussed in Ref 8. The method used was based on the so called "modified mapping-collocation" technique developed earlier by Bowie and Freese [9]. The following mapping formulation [8] was proposed

$$z_j = \omega(\zeta_j) = \frac{L_j}{2}\left(\zeta_j + \frac{m_j}{\zeta_j}\right) \tag{1}$$

where $z_j = x + \mu_j y$, the complex planes representing the physical region of the specimen; $L_j = a + i\mu_j b$, $m_j = (a + i\mu_j b)/(a - i\mu_j b)$; and $\mu_j = \alpha + i\beta$, the roots of the characteristic equation determined by the specimen anisotropy

$$a_{11}\mu^4 - 2a_{16}\mu^3 + (a_{66} + 2a_{12})\mu^2 - a_{26}\mu + a_{22} = 0 \tag{2}$$

This mapping carries the unit circle $\zeta = \sigma$ and its exterior into an ellipse along the modified notch and its exterior, respectively.

The representation of the stress function can be given as follows

$$2Re\left\{\eta_{1i}\left[S \cdot \Phi_2(\zeta_1) + T \cdot \overline{\Phi}_2\left(\frac{1}{\zeta_1}\right)\right] + \eta_{2i}\Phi_2(\zeta_2)\right\} = \overline{X}_i \tag{3}$$

or

$$2Re\left\{\eta_{1i}\left[S \sum_{n=\infty}^{\infty} a_{n2}^*\zeta_1^n + T \sum_{n=\infty}^{\infty} \overline{a}_{n2}^*\zeta_1^{-n}\right] + \eta_{2i} \sum_{n=\infty}^{\infty} a_{n2}^*\zeta_2^n\right\} = \overline{X}_i \tag{4}$$

where $S = (\overline{\mu}_1 - \mu_2)/(\mu_1 - \overline{\mu}_1)$; $T = (\overline{\mu}_1 - \overline{\mu}_2)/(\mu_1 - \overline{\mu}_1)$; $\overline{X}_i$ represents the boundary conditions in the direction i ($i = 1,2$); η_{1i}, η_{2i} are constants depending the nature of boundary conditions $\overline{X}_i$

1. for force resultant conditions, $\eta_{11} = \mu_1$, $\eta_{21} = \mu_2$, $\eta_{12} = \eta_{22} = 1$ and
2. for displacement conditions, $\eta_{j1} = p_j = a_{11}\mu_j^2 + a_{12} - a_{16}\mu_j$, $\eta_{j2} = q_j = a_{11}\mu_j + a_{22}/\mu_j - a_{26}$ ($j = 1,2$); and $a_{n2}^* = c_{n2}^* + id_{n2}^*$, complex coefficients to identify Eq 4.

This representation of stress function allows us to satisfy the automatically load-free condition along the notch boundary.

The following procedure has been carried out and is considered efficient in the resolution of Eq 4:

1. Truncation of the infinite expansion in Eq 4 to a suitable finite number of positive and negative powers, n_1, n_2, respectively.
2. Selection of points around the external boundary in a fairly uniform manner and writing force boundary conditions at each of the direct locations. The number of points was so arranged that the number of boundary equations thus obtained was about twice or triple the number of unknowns, a_{n2}^*.
3. Selection of the unknowns in such a manner that the equations in Step 2 are satisfied in a least-square sense.

A computer code was developed in order to determine the coefficients of the stress function, a_{n2}^* and to calculate the stress fields by the following relationship

$$\sigma_{ij} = 2 \sum_{n=-n1}^{n2} nc_{n2}^* Re \left\{ \eta_{1i} \frac{S\zeta_1^{n-1} - T\zeta_1^{-n-1}}{\omega_1'(\zeta_1)} + \eta_{2i} \frac{\zeta_2^{n-1}}{\omega_2'(\zeta_2)} \right\}$$
$$- 2 \sum_{n=-n1}^{n2} nd_{n2}^* Im \left\{ \eta_{1i} \frac{S\zeta_1^{n-1} + T\zeta_1^{-n-1}}{\omega_1'(\zeta_1)} + \eta_{2i} \frac{\zeta_2^{n-1}}{\omega_2'(\zeta_2)} \right\} \quad (5)$$

where $\eta_{ji} = \mu_j^2$ when $\sigma_{ij} = \sigma_{xx}$, $\eta_{ji} = \mu_j$ when $\sigma_{ij} = \sigma_{xy}$, and $\eta_{ji} = 1$ when $\sigma_{ij} = \sigma_{yy}$.

Calculation of the Rate of Dissipated Energy for Monotonic Failure Initiation

Since the monotonic compression failure occurs immediately after its initiation, the characterization or the establishment of the criterion for this damage state is essential. With the knowledge of the stress and displacement fields, the rate of dissipated energy can be determined by integrating the expression $\sigma_{2i} du_i$ from B to A [10]

$$G_1 = \frac{\partial U}{\partial a^*} = \int_B^A \sigma_{2i} du_i$$

The significance of a^*, B, and A are indicated in Fig. 9a.

In practice, the damage does not begin on all heights of the notch; the integration must be effected from B to A^*. Point A^* or the height of damage zone is determined by the point criterion of Nuismer-Whitney [11] applied in the vertical direction and in the reverse manner. That is, knowing the compression resistance of the material, σ_{yc}, and applied stress state, σ_{ij}, we determine the interval d_1 in which the stress $\sigma_{yy} \geq \sigma_{yc}$ (Fig. 9b).

Since the damage height, d_1, is not zero when damage occurs, the energy of failure initiation must be carried on an energy by unit damage volume. To avoid confusion, we named this energy L_1

$$L_1 = \frac{1}{d_1} \int_B^{A^*} \sigma_{2i} du_i \quad (6)$$

The index 1 signifies that, macroscopically, the loading is in Mode I. The compression failure occurs when $L_1 \geq L_{1c}$. In the calculation of L_1, if the applied stress is much less, the height of the damage zone d_1 can be inferior to 0. In this case, it is clear that the failure can not be initiated.

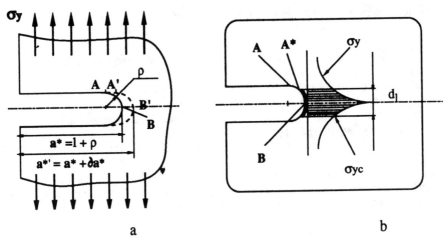

FIG. 9—*Calculation of the dissipated energy per unit damage volume.*

Calculation of Cyclic Damage Progression

As discussed earlier, the damage under a compression cyclic loading has an elliptical form in the specimen plane, and the delamination tips can be included inside a family of parabolic curves. A three-dimensional physical model of damage zone is then proposed in Fig. 10a and the damage progression is shown in Fig. 10b. The sections parallel to the middle plane (xoy) of the laminate give an elliptical form of which one of two foci always coincides with the center of the hole at the notch tip. The corresponding equation is

$$\frac{y^2}{B^2(z)} + \frac{[x - C(z)]^2}{A^2(z)} = 1 \tag{7}$$

where z is the distance between slice plane and middle plane, and

$$A^2(z) = B^2(z) + C^2(z) \tag{8}$$

The section in the specimen symmetrical plane (xoz) has a parabolic form whose the equation is given by

$$\alpha z^2 + \beta = A(z) + C(z) \tag{9}$$

It is clear that β describes the distance between the center of the hole and the top of the parabola (Fig. 10b). Since one of the two foci of the ellipse always coincides with the center of the hole, we can assume that, in addition to three equations, there is a fourth relationship between $A(z)$ and $C(z)$, that is

$$A(z) = C(z) + \rho \tag{10}$$

Thus, knowing β and α, we can determine fully the form of the damage zone. For this fact, we adopt a hypothesis that the parabola moves globally in the x-direction without change in form, that is, α is a constant depending only on the loading mode. Hence, we can determine α from

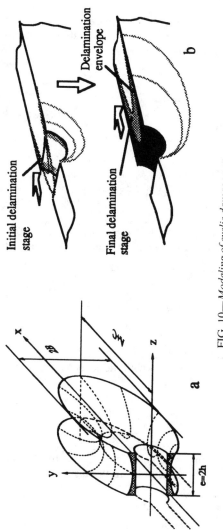

FIG. 10—*Modeling of cyclic damage zone.*

the micrographic section in the xoz plane after the final failure of the specimen and β by using the relationships

$$
\begin{aligned}
\alpha h^2 + \beta &= A(h) + C(h), \\
&= 2A(h) - \rho, \\
&= 2A - \rho, \text{ and} \\
h &= e/2.
\end{aligned}
$$

(11)

where $2A$ is the length of the largest ellipsis that can be observed on the X-ray photographs. An analytical expression for A is obtained by interpolating the measured values. The volume of the damage zone is calculated by multiple integration of Eqs 7 and 9. Since the problem is doubly symmetrical, we need only calculate a quarter part of the volume and then multiply by 4. That is

$$
V = 4 \int_{z_0}^{h} \int_{C(z)-A(z)}^{C(z)+A(z)} \int_{0}^{B(z)/A(z) \cdot \sqrt{A^2(z)-[x-C(z)]^2}} dx\,dy\,dz - 2\pi\rho^2(h-z_0)
$$

$$
= \pi \left\{ 4\,(hBA - z_0 b'a') + (3\beta + 2\rho)(hB - z_0 b') \right.
$$

(12)

$$
\left. + \frac{\beta\sqrt{\rho}}{2\sqrt{\alpha}}(3\beta + 4\rho)\,\mathrm{Ln}\left[\frac{h\sqrt{\alpha} + \sqrt{\alpha h^2 + \beta}}{z_0\sqrt{\alpha} + \sqrt{\alpha z_0^2 + \beta}}\right]\right\}\frac{1}{8} - 2\pi\rho^2(h-z_0)
$$

where a', b', and z_0 correspond to the half-long axis, half-short axis of the smallest ellipse, and the distance between this one and the middle plane, respectively. They are given by

$$
\begin{aligned}
z_0 &= \begin{cases} \sqrt{\rho - \beta} & \beta < \rho \\ 0 & \beta \geq \rho \end{cases} \\
a' &= (\alpha z_0^2 + \beta + \rho)/2 \\
b' &= \sqrt{\rho}\,\sqrt{\alpha z^2 + \beta}
\end{aligned}
$$

(13)

The rate of the dissipated energy due to a infinitesimal increase of damage volume, dV, can be calculated by the following formula with the consideration that the stiffness in the damage volume is null

$$
L_1 = \frac{1}{2}\,P^2\,\frac{d\mathrm{Comp}}{dV}
$$

(14)

where Comp is the specimen compliance. Since Comp and V evolve with the time, t, or the number of cycles, N, L_1 can be determined in this fashion

$$
\frac{d\mathrm{Comp}}{dV} = \frac{d\mathrm{Comp}/dN}{dV/dN}
$$

$$
\frac{dV}{dN} = \pi\,\frac{d\beta}{dN}\left\{\frac{3\sqrt{\rho}}{4}\,[h\sqrt{\alpha h^2 + \beta} - z_0\sqrt{\alpha z_0^2 + \beta}]\right.
$$

(15)

$$
\left. + \frac{\sqrt{\rho}}{4\sqrt{\alpha}}(3\beta + 2\rho)\,\mathrm{Ln}\left[\frac{h\sqrt{\alpha} + \sqrt{\alpha h^2 + \beta}}{z_0\sqrt{\alpha} + \sqrt{\alpha z_0^2 + \beta}}\right]\right\}
$$

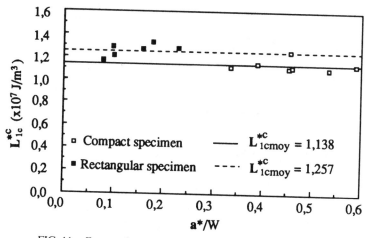

FIG. 11—*Energy of monotonic compression damage initiation.*

Application and Discussions

1. Monotonic Damage—The results of calculation are shown in Fig. 11 for both the compact specimens and the rectangular specimens. The failure initiation energy can be considered as constant for all specimens, except there is a small difference between compact specimens and rectangular specimens. We think that this difference comes from the use of the antibuckling devices for the rectangular specimens, since some authors have found that the antibuckling devices can cause some small changes in the material behavior. So it is convenient to take $1.138 \cdot 10^7$ J/m³ as the energy of the monotonic failure initiation. Of course, this energy can also be used to establish the catastrophic failure criterion because of the severity of the monotonic damage progression.

2. Fatigue Damage—With the definition of the rate of dissipated energy per unit volume, Eq 6, we can summarize all results of endurance tests in the same curve $L_{1max} - N_f$, where L_{1max} corresponds to the maximum load in the cyclic loading. Figure 12a shows the endurance curve of the notched material under the C-C loading, and Fig. 12b gives results for C-T specimens. It should be noted that the rate of dissipated energy used here has only symbolic significance, by the reason that the damage mode in the monotonic specimen and in the cyclic specimen is quite different. We have chosen this parameter for the fatigue damage characterization because it should show a relationship between the monotonic failure energy and cyclic failure energy. Thus, with parameter L_{1max}, we can give an idea of the compression fatigue resistance of this notched laminate material.

As an example of the application of the fatigue damage model, we have interpolated the specimen compliance, Comp, and the damage length, A, as functions of number of cycles for a specimen ($a^* = 34$ mm, $\rho = 4$ mm), which are given

$$\text{Comp} = \frac{0.053}{(46.03 - N/10^4)^2} + \left(\frac{N/10^4 + 7}{65}\right)^{5.5} + 1.185 \quad (\times 10^{-7} \text{ m/N})$$

$$A = \frac{0.9}{45.99 - N/10^4} + 0.0815(N/10^4)^{1.08} + 3.982 \quad (\times 10^{-3} \text{ m})$$

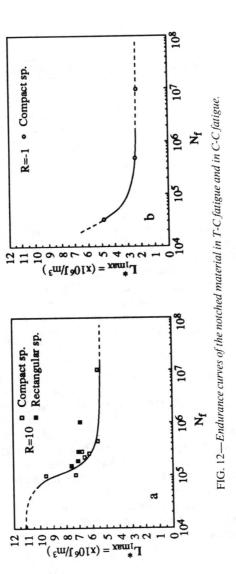

FIG. 12—*Endurance curves of the notched material in T-C fatigue and in C-C fatigue.*

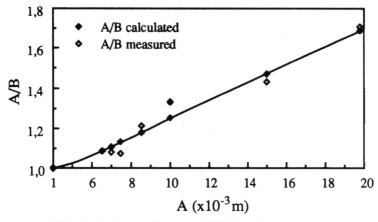

FIG. 13—*Evolution of the ratio* A(N)/B(N) *as function of* A(N).

To verify the validity of this consideration on the form of damage zone, we calculated the ratio between $A(z_0,N)$ and $B(z_0,N)$

$$\frac{A(z_0,N)}{B(z_0,N)} = \frac{1}{\sqrt{1 - [1 - \rho/A(z_0,N)]^2}}$$

Figure 13 gives the comparison between the measured values and calculated values. Since the results are quite satisfactory, the consideration is valid.

The measurement made on the micrographic section gives

$$\alpha = 0.74 \text{ mm}^{-1}$$

The final results of the calculation for the damage progression are illustrated in the Fig. 14.

We can discern three different stages on this curve. The first, which has a very high slope, corresponds to the beginning of damage. The second, less vertical, represents the state of stable propagation, and this leads into the last stage of damage until final failure. The second stage is the most interesting for the characterization of fatigue damage. Since this part is a straight line, we can choose Paris's law to describe this stable progression of damage, that is

$$\frac{dV}{dN} = 7.34 \cdot 10^{-18} \, (L_{1\text{max}})^{0.862}$$

The units for V, N, and $L_{1\text{max}}$ are m^3, cycle, and J/m^3, respectively.

From the calculation, the parabola that describes the delamination tops enters fully into the zone beyond the notch tip at about $4.2 \cdot 10^5$ cycles. When the damage begins its third stage of propagation, the parabola top is still 2.07 mm behind the notch tip. At $3.2 \cdot 10^5$ loading cycles, the damage has penetrated into the specimen at a depth of 1.34 mm in the thickness direction. That is in excellent agreement with the micrographic section made on a compact specimen loaded in the same conditions. The critical value of $L_{1\text{max}}$ is about $1.35 \cdot 10^6$ J/m^3, which is ten times smaller than the monotonic value. This is the consequence of the difference between the damage modes. In fact, the necessary energy of failure initiation for the monotonic specimen

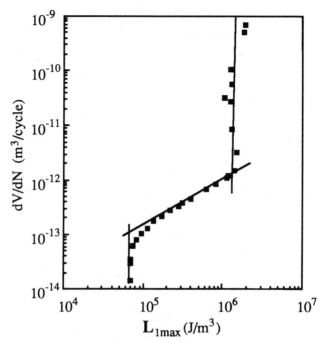

FIG. 14—*Tension-compression fatigue curve.*

is devoted to create a zone of crushing damage in an intact region. By contrast, the critical damage energy for a cyclic specimen is used to propagate the delaminations.

Conclusions

The mechanical properties, especially the damage behavior of the notched laminate material under compression loading, are very specific. Specifically, we must stress the following points:

1. The notched laminate can fail under compression loading in both monotonic and cyclic loading. In the first case, breaking is quasi-elastic. In the second case, damage progression does not end by a sudden "fracture."

2. In the monotonic compression specimen, the first damage by delamination is formed relatively early, at about 65% of ultimate load. But these delaminations are confined to the notch tip, limited to external interfaces, and have no influence on the specimen stiffness. Only the lateral transducer and strain gage placed at 1 and 8.5 mm, respectively, from the notch tip can detect their presence.

3. The critical damage in the monotonic compression specimen is crushing. The crushing damage occurs very suddenly at the notch tip and throughout the specimen thickness. This damage is then transformed immediately to delamination, which corresponds to final failure. The characterization of the critical damage is then essential in the study of monotonic compression failure. Because critical damage has a homogeneous form throughout the specimen thickness, we conclude that the critical damage mechanism is independent of specimen geometry and stacking sequence. Therefore, we have considered that it is sufficient to assume generalized plane stress in the determination of the stress field.

4. Since the resin is sensitive to the cyclic loading, the exterior delaminated plies of the fatigue compression specimen are weakened by repeated bending, and tend to buckle easily. This leads to the delamination of the next interior plies. There is therefore a penetration processes of delamination into the center of the specimen. The cyclic damage has an elliptical form in the specimen plane, and the tops of these can be described by a parabola in the notch symmetrical plane, perpendicular to the loading.

5. Linear fracture mechanics (LFM) are applied successfully, because the material has an elastic behavior in compression. But application of LFM requires particular attention, because the studied laminate has anisotropic properties, and here there is not a true crack under tension loading but a notch under compression loading. The curvature radius of the notch tip is not null. Therefore, it was necessary to proceed with a pseudo three-dimensional analysis for the monotonic compression specimens (that is, length times width times height of damage zone), and a full three-dimensional calculation for the cyclic specimens.

6. The rate of dissipated energy of monotonic failure initiation is about $1.14 \cdot 10^7$ J/m^3. The result is reproducible on both compact and rectangular specimens for the particular laminate tested in this study. This verifies the fact that the mechanism of critical damage (or final failure initiation) is independent of specimen geometry. However, the independence of L_{1c} with stacking sequence has to be verified experimentally later.

7. The endurance limit of the C-C specimens is about $5.2 \cdot 10^6$ J/m^3, and of the C-T specimen is $2.2 \cdot 10^6$ J/m^3. The second is two times lower than the first. So that the C-T fatigue is the most dangerous loading mode for the material.

8. The cyclic damage propagation can be described by Paris's law; for example, C-T specimen: $dV/dN = 7.34 \cdot 10^{-18} (L_{1max})^{0.862}$. The critical value of L_{1max} is about $1.35 \cdot 10^6$ J/m^3, which is ten times smaller than the monotonic value. This is the consequence of the difference between the damage modes.

The use of the compact specimen for the compression study on carbon/epoxy laminate materials is a very interesting technique. This specimen does not present any problem for mechanical tests and for results since it has been verified that the results are reproducible on other specimens. The advantages will be still more marked when the mechanical tests are conducted with an alternate cyclic loading.

References

[1] Schütz, D. and Gerhartz, J., "Fatigue Properties of Unnotched, Notched and Jointed Specimens of a Graphite/Epoxy Composite," *Fatigue of Fibrous Composite Materials, ASTM STP 723,* American Society for Testing and Materials, Philadelphia, 1981, pp. 31–47.
[2] Tsangarakis, N., "Fatigue Failure of an Orthotropic Plate with a Circular Hole," *Journal of Composite Materials,* Vol. 18, Jan. 1984, pp. 47–57.
[3] Lamothe, T. H. and Nunes, J., "Evaluation of Fixturing for Compression of Metal Matrix and Polymer/Epoxy Composite," *Compression Testing of Homogeneous Materials and Composites, ASTM STP 808,* R. Chait and R. Papirno, Eds., American Society for Testing and Materials, Philadelphia, 1983, pp. 241–253.
[4] Adsit, N. R., "Compression Testing of Graphite/Epoxy," *Compression Testing of Homogeneous Materials and Composites, ASTM STP 808,* R. Chait and R. Papirno, Eds., American Society for Testing and Materials, Philadelphia, 1983, pp. 175–186.
[5] Matondang, T. H. and Schütz, D., "The Influence of Anti-buckling Guides on the Compression-Fatigue Behaviour of Carbon Fiber-Reinforced Plastic Laminates," *Composite,* Vol. 15, No. 3, July 1984, pp. 217–221.
[6] Bathias, C., Esnault, R., and Pellas, T., "Application of Fracture Mechanics to Graphite Fiber-reinforced Composites," *Composites,* July 1981, pp. 195–200.
[7] Wang, A. S. D., "Fracture Mechanics of Sublaminate Cracks in Composite Materials," *Composite Review,* Vol. 6, No. 2, Summer 1984, pp. 45–62.

[8] Lai, D., "Contribution a la Modélisation de la Rupture des Plaques Stratifiées Entaillées sous un Chargement Monotone ou Cyclique Comprenant une Sollicitation de Compression," Ph.D. thesis, Université de Technologie de Compiègne, France, Dec. 1988.

[9] Bowie, O. L. and Freese, C. E., "Central Crack in Plane Orthotropic Rectangular Sheet," *International Journal of Fracture Mechanics,* Vol. 8, 1972, pp. 49–58.

[10] Bowie, O. L. and Neal, D. M., "The Effective Crack Length of an Edge Slot in a Semi-Infinite Sheet under Tension," *International Journal of Fracture Mechanics,* 1967, Vol. 3, pp. 111–119.

[11] Nuismer, R. J. and Whitney, J. M., "Stress Fracture Criteria for Laminate Composites Containing Stress Concentrations," *Journal of Composite Materials,* July 1974, pp. 253–265.

Erhard Krempl[1] and Deukman An[1]

Effect of Interlaminar Normal Stresses on the Uniaxial Zero-to-Tension Fatigue Behavior of Graphite/Epoxy Tubes

REFERENCE: Krempl, E. and Deukman, A., **"Effect of Interlaminar Normal Stresses on the Uniaxial Zero-to-Tension Fatigue Behavior of Graphite/ Epoxy Tubes,"** *Composite Materials: Fatigue and Fracture (Third Volume), ASTM STP 1110,* T. K. O'Brien, Ed., American Society for Testing and Materials, Philadelphia, 1991, pp. 659–666.

ABSTRACT: During the past several years, the Mechanics of Materials Laboratory of Rensselaer Polytechnic Institute (RPI) has developed a method of obtaining biaxial fatigue data under axial/torsion loading. A thin-walled tubular specimen can be made from prepregs by a layup procedure and tested in an MTS servohydraulic axial/torsion testing machine with computer control. We have provided completely reversed load-controlled fatigue data on graphite/epoxy materials under uniaxial and combined loadings using $[\pm 45]_s$ and $[0, \pm 45]_s$ layups. The edgeless specimen eliminates suspected end effects and can be used for tests involving significant compressive loading.

Interlaminar normal stresses were thought to influence fatigue performance by enhancing delamination. To check on this hypothesis, zero-to-tension fatigue tests were run on graphite/ epoxy $[\pm 45]_s$ tubes with and without internal pressurization. The pressure levels were chosen so as to compensate the suspected interlaminar tensile stresses. Fatigue test results in the range from 10^4 to 10^6 cycles with and without pressurization were within the same reasonable scatter band. In the course of testing, it was discovered that performance could be considerably improved by providing a restraint in the hoop direction by either inserting a mandrel or by including a 90° layer on the inside and outside. Fatigue tests with $[90, \pm 45]_s$ specimens under zero-to-tension loading showed a significantly improved fatigue performance.

KEY WORDS: biaxial fatigue, tubular specimens, composite materials, interlaminar stresses, layups, fatigue performance, fracture, fatigue (materials)

The evaluation of the fatigue properties of epoxy matrix composites is traditionally performed using flat coupon test specimens. They are easy to make, and valuable fatigue data have been generated with them. However, substantial loading in compression will cause buckling of this specimen and very little fatigue data are available for negative R-ratios. Moreover, elasticity theory predicts high local stresses at the edges whose influence on the fatigue performance is very difficult to quantify. Also, the use of coupon specimens restricts the loading to uniaxial conditions, and no possibility exists to ascertain the biaxial fatigue performance of the composite materials.

In the future, composites are expected to replace conventional materials in high performance and other applications. Because of this trend, there will be an increased need for additional fatigue data involving negative R-ratios or biaxial loading or both. Consequently, con-

[1] Professor and head, and graduate student, respectively, Mechanics of Materials Laboratory, Rensselaer Polytechnic Institute, Troy, NY 12180-3590. Mr. An is now at Pusan National University, Pusan, Korea.

siderable efforts have been expended to design and construct a tubular specimen and associated fixtures [1]. With this specimen design and with the help of an MTS axial-torsion servohydraulic, computer-controlled mechanical testing machine, the fatigue performance of [±45]$_s$ graphite/epoxy tubes has been evaluated [1,2]. All tests were performed under completely reversed, load-controlled conditions and involved combined axial/torsion loading for in-phase and out-of-phase loading [2]. In contrast to metals, no significant effect of out-of-phase loading was found on the fatigue life of graphite/epoxy. Other tests included the determination of the fatigue performance of near unidirectional [±5]$_s$ Kevlar/epoxy specimens [3] for negative R-ratios and of [0,±45]$_s$ graphite/epoxy material for completely reversed load-controlled loading [3,4].

Interlaminar Stresses

In a discussion of the preceding work, the importance of interlaminar normal stresses [5] was emphasized. It was felt that the curvature in the tubular specimens promotes interlaminar normal stresses that would contribute to the delamination. Therefore, material properties obtained with the tubular specimen would be those of the material/tube configuration rather than the desired material properties.

One of the great difficulties with composite materials is the establishment of "true material properties." Even the widely accepted strip specimen is prone to edge effects and material data obtained with this specimen are tainted by such possible effects whose influence on fatigue performance is very hard to quantify.

However, it was felt that the criticism brought forward in Ref 5 was valid and deserved the separate investigation described later.

Specimen Design and Magnitude of Interlaminar Normal Stresses

The specimen used in the investigation is shown in Fig. 1. For the present investigation, graphite/epoxy with a [±45]$_s$ layup was used. The specimens were manufactured from prepregs at RPI using the techniques described in detail in Refs 1 and 2. The prepreg material was Fiberite hy-E1048 A1E with Union Carbide T-300 graphite fibers and low-temperature cure 948 A1 resin.

A strength of materials analysis of the interlaminar normal stress, σ_3, at the interface between the first and second plies of the tube was suggested in Ref 5 and is reproduced in Fig. 2. The result is

$$\sigma_3 = \sigma_1(\cos^2 \theta)t/r \tag{1}$$

where σ_1 is the stress in the fiber direction, $\theta = \pi/4$ is the angle between the fiber and the cylinder axis, $t = 0.15$ mm is the thickness of the ply, and $r = 12.7$ mm is the nominal radius of the tube used in the calculation.

In previous tests on [±45]$_s$ specimens, the axial strength was on the order of 150 MPa [2]. From this value, the stress in the fiber direction, σ_1, and the interlaminar normal stress, σ_3, can be calculated. It is below 0.69 MPa. Because of this magnitude of the interlaminar normal stress, shop air pressure was sufficient to compensate the normal stress and it was decided to build an internal pressurization device with this pressure source. (In the design of the fixture in Ref 1, the possibility of internal pressurization had been already anticipated and pressure ports had been introduced.)

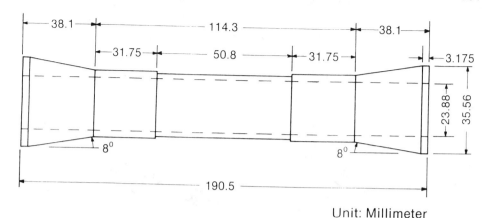

Unit: Millimeter

FIG. 1—*Specimen for a* [±45]ₛ *layup. The thickness dimensions of the gage section increase for the* [90, ±45]ₛ *specimen.*

Pressurization Equipment

At the outlets of the shop air, a check valve, an automatic pressure shutoff, a pressure gage, and a regulator were installed. A switch was connected to the actuator piston of the MTS servohydraulic axial-torsion testing system. It activated the automatic shutoff when the piston dropped below a certain level. With this arrangement, fatigue testing could be performed around the clock. To prevent leakage of air during the test and after failure occurred, a rubber bladder was inserted into the tube. An O-ring and a remachined plug provided sealing at the upper end of the fixture.

Test Results

A static test with an internal pressure of 0.55 MPa was performed and resulted in an ultimate tensile strength of 135 MPa. This value was below the average of previously tested unpressurized specimens. As part of this test, the stiffness of the specimen with and without internal pressure was measured and was found to be unchanged.

All fatigue tests were performed at a frequency of 5 Hz under zero-to-tension axial load-controlled conditions ($R = 0$) using the MTS axial-torsion servohydraulic testing machine of the Mechanics of Materials Laboratory. (Within the accuracy of the testing machine, zero load was obtained.) A total of 16 specimens were tested with and without constant internal pressure. The purpose of the tests with internal pressure was to compensate for the maximum interlaminar tensile stresses predicted by Eq 1. Therefore, tests with internal pressure can be thought of as having no interlaminar tensile stresses. In all tests, failure was defined as separation of the specimen into two pieces.

The test results showing the nondimensional stress amplitude, the internal pressure, and the fatigue life are presented in Table 1 and Fig. 3. In reporting stresses, actual cross-section dimensions are used.

It is evident that the internal pressurization has no discernible effect on the fatigue life.

Discussion

From the results of Table 1 and Fig. 3, no direct influence of internal pressure on the fatigue life is evident. It appears, therefore, that the interlaminar normal stresses that develop during

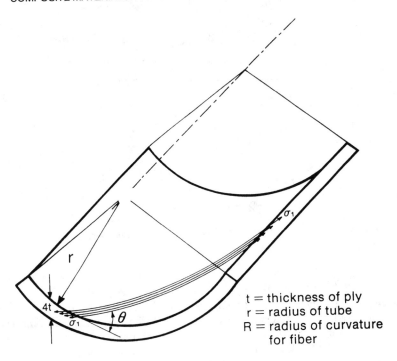

t = thickness of ply
r = radius of tube
R = radius of curvature
    for fiber

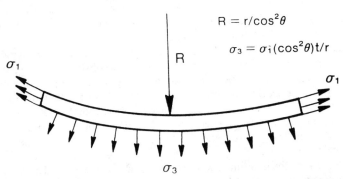

$$R = r/\cos^2\theta$$

$$\sigma_3 = \sigma_i(\cos^2\theta)t/r$$

FIG. 2—*Estimate of interlaminar normal stresses, after W. Elber* [5].

tensile loading between the plies do not affect the fatigue performance for $R = 0$. Although no direct experimental evidence is available, it is not unreasonable to expect that similar conclusions would be true for other fatigue loading conditions. For $R = -1$, for example, interlaminar stresses of equal magnitude are expected to develop between the outermost ply and the ply next to it when the maximum compressive stress is reached. From the results of the

TABLE 1—*Uniaxial fatigue test results for* [± 45]$_s$ *graphite/epoxy* R = 0 *at 5 Hz.*

Specimen Number	Amplitude, S (%)[a]	Fatigue Life, $\log N_f$	Internal Pressure, P_i, MPa	Failure Location
11	21.43	>6	0.36	. . .
11	21.43	>6	0	. . .
10	24.72	4.59	0	U[c]
3	24.72	5.03	0	U
8	24.72	5.51	0	U
2	24.72	5.6	0	D[c]
12	24.72	5.75	0	U
6	24.72	4.75	0.41	U
4	24.72	5.41	0.41	U
9	24.72	5.69	0.41	D
5	24.72	5.55	0.41	U
13	26.79	4.73	0	U
14	26.79	3.59	0.45	U
7	26.79	5.02	0.45	D
M3[b]	26.79	5.62	0	C[c]
M2[b]	26.79	5.75	0.45	C
1	3.0	4.29	0.50	C

[a] Stress amplitude/ultimate tensile strength (145 MPa) $\times$ 100%.
[b] Specimens M3 and M2 are from a different batch.
[c] U, D, and C denote upper, lower, and center part of the gage section, respectively.

present tests, there is reason to believe that this interlaminar tensile stress does not have a significant influence on fatigue performance either.

In a completely reversed torsional test, the fatigue life was found to be highest [2] of all the test conditions. The stresses in the fiber and, consequently, the interlaminar normal stress are very high in this case. Despite this fact, fatigue performance is very good owing to the fiber domination of the deformation in torsional loading. So, if interlaminar normal stresses have a significant effect, it should manifest itself under torsional loading. No such effect is evident.

An elementary analysis and the test results show that the addition of internal pressure does not change the stiffness of the tube. However, the stiffness can be appreciably changed if the lateral movement of the tubes is restrained by leaving the metal mandrel needed in the manufacture inside the tube during testing. As a first approximation, it can be assumed that in this case the hoop strain is nearly zero. An elementary analysis shows that (e is the axial strain)

$$\frac{e_{\text{without}}}{e_{\text{with}}} = 1/(1 - a^2) \tag{2}$$

where $a = (A_{12}^2/A_{11}A_{22})^{1/2}$ with the A_{ij} denoting the stiffness matrix of the composite laminate [6]. Substitution of nominal values for a graphite/epoxy material shows the ratio to be approximately 2.3. Only an axial stress was assumed to act in calculating e_{without}; as a first approximation zero hoop strain is used to compute e_{with}. A slightly smaller ratio would be obtained if the compliance of the mandrel is considered. This detailed analysis is not needed in the present context.

Some preliminary static tension tests and some fatigue tests were performed with the mandrel inside the tube, and the results are shown in Table 2. (Tests with the mandrel inside are not recommended as routine tests.) It is evident that the addition of the mandrel has caused a significant improvement in the static and fatigue strengths. This result shows that the restraint of the lateral motion causes beneficial effects that are not attributed to the elimination of the

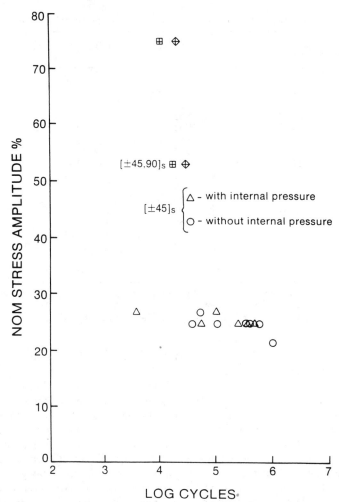

FIG. 3—*Fatigue test results for* R = 0 *with and without internal pressure for* [± 45]*s specimens. Results for* [90, ± 45]*s specimens without internal pressure are also shown.*

interlaminar tensile stresses. (The compressive stresses developed between the mandrel and the innermost ply of the tube during tensile loading will help to cancel the interlaminar tensile stress.) Rather, the significant effects are the reduction of lateral motion and the increase in effective axial stiffness that reduces the axial strain by a factor of up to 2.3 for load control. If it is assumed that the strain is the controlling factor for fatigue, then the stiffness increase is sufficient to explain the increase in fatigue life observed at the same stress amplitude.

In Ref 2, the role of microbuckling in initiating fatigue failure was discussed. Based on the observation of unequal static torsional strength upon reversal of the torque, and some limited observations of fatigue tests in progress, it was suggested that microbuckling initiates delamination. In the absence of torque, compressive loading is required to facilitate microbuckling. No compression was, however, used in this study. If microbuckling plays a significant role, then there should be a significant influence of R-ratio on fatigue life. Such an influence was not observed for axial loading for $R = 0$ and $R = -1$ in Ref 7. Almost no influence of R-ratio on the fatigue life up to 10^6 cycles was observed.

TABLE 2—*Test results with and without lateral restraint.*

Static Tension Condition	Ultimate Strength, MPa
Aluminum mandrel	196
Aluminum mandrel	200
Steel mandrel	235
No mandrel	117
No mandrel	125
No mandrel	145

Fatigue, R = 0.5 Hz, Condition	Stress Amplitude, S %[a]	Fatigue Life Cycles
Steel mandrel	0.65	4×10^5
No mandrel	0.60	4×10^3

[a] Stress amplitude/ultimate tensile strength (145 MPa) $\times$ 100%.

The addition of transverse plies on the inside and on the outside of the tube would also increase the axial stiffness and would also restrain the lateral motion. Accordingly, four specimens were made using a $[90, \pm 45]_s$ layup. They were used to run two static tests and two fatigue tests. The results are reported in Tables 3 and 4. Although the tensile strength value is invalid due to a grip failure, it can be asserted that the static strengths were considerably improved when compared to the $[\pm 45]_s$ layup (In Ref 2, tensile and compressive strength were found to be equal and the average tensile strength was determined to be 145 MPa, see Refs 2 and 7). It can be seen from Table 3 and Fig. 3 that the fatigue strength improved dramatically.

The test results with the $[90, \pm 45]_s$ layup confirmed the hypothesis that the restraint of the lateral motion improves stiffness, strength, and fatigue performance. Although manufacture of the specimens is difficult, the transverse layer is extremely beneficial for performance and should be always included when optimum properties in axial loading are sought. It remains to be seen what properties will result when biaxial loadings are performed with such a specimen.

TABLE 3—*Test results for $[90, \pm 45]_s$ specimens for two static tests.*

Specimen No.	σ_{max}, MPa	Remark
C1	204	grip failure; tension
C2	−289	upper part of gage section; compression

TABLE 4—*Test results for $[90, \pm 45]_s$ specimens for two fatigue tests, R = 0 at 5 Hz.*

Specimen No.	Amplitude, S %[a]	Cycles	Remark
C4	37	$2 \cdot 10^6$	run out
	45	$4 \cdot 10^5$	run out
	53	$4.6 \cdot 10^5$	run out
	59	$5 \cdot 10^5$	run out
	75	$2.6 \cdot 10^4$	failed lower part of gage section
C5	75	$1.1 \cdot 10^4$	failed upper part of gage section

[a] Stress amplitude/ultimate strength (145 MPa) $\times$ 100%.

Conclusion

Fatigue tests with and without internal pressure yielded approximately the same fatigue life of $[\pm 45]_s$ graphite/epoxy tubes for $R = 0$. These tests did not therefore confirm the suspected detrimental effect of interlaminar tensile stresses on the fatigue performance of thin-walled tubes. Limited tests showed that the addition of 90° plies on the inside and the outside significantly improved the static and fatigue strengths of the tubes. This finding should be investigated by further experiments.

Acknowledgment

This research was supported by NASA and AFOSR under NASA Grant NGL 33-018-003. Dr. M. Greenfield and Dr. A. Amos were technical monitors in the respective agencies. Eric C. Coffin and Volker Paedelt manufactured the $[90, \pm 45]_s$ specimens.

References

[1] Krempl, E. and Niu, T. M., "Graphite/Epoxy $[\pm 45]_s$ Tubes. Their Axial and Shear Properties and their Fatigue Behavior under Completely Reversed Load Controlled Loading," *Journal of Composite Materials,* Vol. 16, 1982, pp. 172–187.
[2] Niu, T. M., "Biaxial Fatigue of Graphite/Epoxy $[\pm 45]_s$ Tubes," D. Eng. thesis, Rensselaer Polytechnic Institute, Troy, NY, May 1983.
[3] Krempl, E., Elzey, D. M., Hong, B. Z., Ayar, T., and Loewy, R. G., "Uniaxial and Biaxial Fatigue Properties of Thin-Walled Composite Tubes," *Journal of the American Helicopter Society,* Vol. 33, 1988, pp. 3–10.
[4] Ayar, T., "Biaxial Fatigue Study of Near-Unidirectional Composite Tubes for Negative R-Ratios," Masters thesis, Rensselaer Polytechnic Institute, Troy, NY, May 1984.
[5] Elber, W., private communication with E. Krempl, 7 Jan. 1985.
[6] Tsai, S. W. and Hahn, H. T., *Introduction to Composite Materials,* Technomic Publishing Company, Inc., Westport, CT, 1980.
[7] An, D., "Cumulative Damage Theory in Multiaxial Fatigue of Graphite Epoxy $[\pm 45]_s$ Composites and Weight Function Theory for a Rectilinear Anisotropic Body," PhD thesis, Rensselaer Polytechnic Institute, Troy, NY, Dec. 1986.

John H. Underwood,[1] Ian A. Burch,[2] and Sri Bandyopadhyay[2]

Effects of Notch Geometry and Moisture on Fracture Strength of Carbon/Epoxy and Carbon/Bismaleimide Laminates

REFERENCE: Underwood, J. H., Burch, I. A., and Bandyopadhyay, S., "**Effects of Notch Geometry and Moisture on Fracture Strength of Carbon/Epoxy and Carbon/Bismaleimide Laminates,**" *Composite Materials: Fatigue and Fracture (Third Volume), ASTM STP 1110,* T. K. O'Brien, Ed., American Society for Testing and Materials, Philadelphia, 1991, pp. 667–685.

ABSTRACT: Composite panels of various ply orientations and 1 to 2 mm thickness were tested in tension with a center notch to determine the translaminar fracture toughness and in bending with no notch to determine the bulk flexural strength and modulus. Carbon/epoxy panels with a variety of notch lengths and widths were subjected to laboratory air conditions before testing. Carbon/bismaleimide panels with one basic notch configuration were tested in three pretest exposure conditions: laboratory air, a 400-h exposure in a controlled moisture chamber, and a 4000 h exposure in a natural tropical environment.

The tests were performed to investigate the following effects on fracture and mechanical properties and on micro and macrofailure mechanisms of the laminates.

(*a*) Effects of notch width and length and ply orientation on translaminar fracture toughness were studied in 0/90 and 0/±60 carbon/epoxy laminates, including effects of notches produced by ballistic penetration.

(*b*) Effects of extended exposure to moisture and sunlight on fracture and mechanical properties were studied in various 0/90 layups of carbon/bismaleimide laminates. A direct comparison of controlled chamber and natural tropical moisture effects was made for two materials.

(*c*) The load-deflection macrobehavior of notched panels was noted and analyzed. Linear-elastic and *J*-integral analyses were used to determine critical values of stress intensity factor, *K*, at the onset of self-similar translaminar fracture. A simple splitting model was used to describe the interlaminar failure that competed with and often prevented the translaminar fracture.

(*d*) Scanning electron microscope fractography was used to investigate and contrast the microfailure mechanisms of the laminates. The differences in micromechanisms between carbon/epoxy and carbon/bismaleimide were related to differences in the macrofracture behavior of the materials.

Analysis was performed of the test procedures for determination of fracture toughness and flexural strength and modulus. The method used for fracture toughness and associated *K*, *J*, and displacement expressions were proposed for more general use in measuring translaminar fracture toughness of composites. Use of the four-point bending test for flexural strength and modulus revealed an error that can occur in this type of test that is not fully addressed in ASTM Test Methods for Flexural Properties of Unreinforced and Reinforced Plastics and Electrical Insulating Materials (D 790–86).

KEY WORDS: composite materials, fatigue (materials), fracture, laminar composites, carbon/bismaleimide panels, fracture toughness, moisture effects, tropical exposure, test methods, failure mechanisms, carbon/epoxy panels

[1] Visiting scientist, Materials Research Laboratory, Defence Science and Technology Organisation, Melbourne, 3032, Australia; currently at Army Armament Research, Development and Engineering Center, Watervliet, NY 12189-4050.

[2] Experimental officer and senior research scientist, respectively, Materials Research Laboratory, Defence Science and Technology Organisation, Melbourne, 3032, Australia.

The primary objective of this work is to investigate the translaminar fracture behavior of carbon/epoxy and carbon/bismaleimide laminates, as affected by certain configurational and environmental effects. Cross-ply laminates of the thickness and orientation shown in Table 1 were subjected to three types of environment and tested for fracture toughness, flexural strength, and flexural modulus. Flexural tests of this type of composite material are commonly used to obtain design information for applications of all sorts. In contrast, fracture mechanics tests and analyses are not commonly used to describe translaminar fracture of composites; their use can be subject to criticism [1]. Because of this question of the use of fracture mechanics with composites, an additional objective of the work is to identify some material and test conditions for which fracture mechanics can be used to describe the translaminar fracture behavior of these composite materials. This additional objective is closely related to the armament applications of the work here. Since preexisting damage must be assumed for many armament components, particularly battlefield and aircraft components that are subjected to ballistic penetration, translaminar fracture toughness is clearly a critical material property. This would seem to be true for many military and aerospace applications.

As the investigation proceeded, some significant differences in fracture behavior became apparent, as described in upcoming sections. Therefore, a third objective was established, that is, to relate the differences in macroscopic fracture behavior based on load-displacement traces to the differences in microscopic fracture mechanisms observed from scanning electron micrographs of the fracture surfaces. If definitive relationships of this sort could be made, they would help with decisions of practical import, such as material selection and design of components, and also in identifying conditions for which fracture mechanics can be well used with composite materials.

It is emphasized that the objectives and related applications of this work deal with fracture of composite materials that is translaminar in basic nature, that is, with damage and continuing fracture proceeding across fibers and laminae and in line with the preexisting notch, see Fig. 1. This is a quite different phenomenon from interlaminar fracture, in which crack growth occurs by delamination between laminae. A summary of some of the considerable body of work on interlaminar fracture toughness testing was recently given by O'Brien and co-workers [2].

Materials and Specimens

Two types of laminate were tested, carbon/epoxy and carbon/bismaleimide, see Table 1. The carbon/epoxy was purchased from the 3M Company as a cured 0/90 sheet and a cured $0/\pm60$ sheet, both of ply type SP-286/T2 using Hercules high-strength HT-S fibers. Tensile

TABLE 1—*Material and test conditions.*

Matrix	Thickness B, mm	Orientation	Environment Prior to Test		
			Lab Air	Moisture Chamber	Tropical Exposure
Epoxy	1.1	$[(0/90)_3/0]_T$	X	...	...
Epoxy	1.1	$[(90/0)_3/90]_T$	X	...	...
Epoxy	1.1	$[0/\pm60/0/\pm60/0]_T$	X	...	...
Bismaleimide	1.8	$[(0/90)_5/0]_T$	X	X	X
Bismaleimide	1.8	$[(90/0)_5/90]_T$	X	...	...
Bismaleimide	1.8	$[(0_2/90)_3/0_2]_T$	X	X	X
Bismaleimide	1.8	$[(90_2/0)_3/90_2]_T$	X	X	...

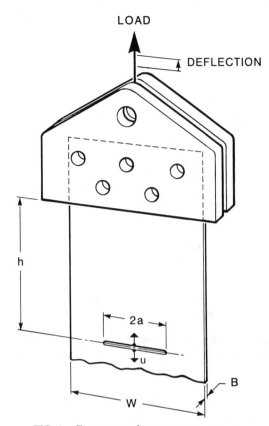

FIG. 1—*Fracture toughness test arrangement.*

panels, nominally 100 mm by 300 mm, were cut from the sheets for fracture toughness specimens, as shown in Fig. 1. The specimens for flexure tests were the broken halves from the fracture tests, using care to avoid damaged areas (Fig. 2). Fracture toughness and flexure tests of $[0/\pm60/0/\pm60/0]_T$, $[(0/90)_3/0]_T$, and $[(90/0)_3/90]_T$ laminates were performed.

The carbon/bismaleimide was Fiberrite X-86 prepreg tape, cured at 180°C and 170 kPa pressure between 300 by 400 mm platens for 4 h, followed by postcuring at 240°C for 6 h. The 0° and 90° fibers were Celanese G-40 high strength and G-50 high modulus fibers, respectively. Three fracture toughness specimens, nominally 90 by 300 mm, were made from each cured blank. The flexure test specimens were typically a 50 by 100 mm undamaged section of the fracture test specimen. Fracture toughness and flexure tests of $[(0/90)_5/0]_T$, $[(90/0)_5/90]_T$, $[(0_2/90)_3/0_2]_T$, and $[(90_2/0)_3/90_2]_T$ laminates were performed.

Test Methods

Environmental Exposure

The baseline exposure to environment before the tests was the so-called laboratory air exposure. This was simply typical laboratory conditions, that is, 20°C and 50% relative humidity, for about three months in the case of carbon/bismaleimide and several years for the carbon/epoxy.

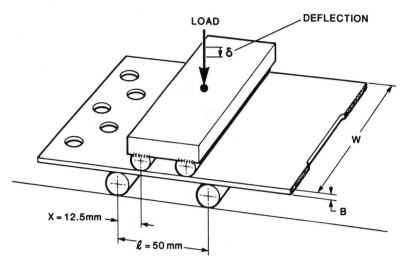

FIG. 2—*Flexural strength and modulus test arrangement.*

The moisture chamber exposure of the carbon/bismaleimide was accomplished at Aeronautical Research Laboratories, Melbourne, Australia. The specimens were exposed to 85°C and 97% relative humidity for 400 h and then placed in a plastic bag. Careful weighing before exposure and just before testing showed the following increase in weight due to moisture absorption.

0/90 specimens = 0.79%, 0.10% standard deviation
0_2/90 specimens = 0.67%, 0.15% standard deviation

Natural tropical environment exposure of the carbon/bismaleimide was accomplished at Materials Research Laboratory—Innisfail, Queensland, Australia. The specimens were exposed in a horizontal orientation in an open setting a few miles inland from the ocean at a latitude of 17 S. The exposure was from 28 May through 10 November 1987 for a total of 4000 h. Mean weather conditions for this period are typically 22°C and 79% relative humidity. Weighing before exposure and just before testing showed the following change in weight.

0/90 and 0_2/90 specimens = −0.02%, 0.03 standard deviation

Since the standard deviation is larger in magnitude than the indicated decrease in weight, a negligible decrease in weight is indicated. A fine loose powder was observed on the top surface of the specimen following exposure. This powder is believed to be a product of degradation of the outer surface of bismaleimide and could explain the weight-change results.

Fracture Toughness Tests

The center-notch tensile panel shown in Fig. 1 was chosen for fracture toughness tests of the thin sheet material because compression stresses are not present in this configuration in the areas ahead of the notch. In testing of thin carbon/polymer laminates, compression can lead

TABLE 2—*Summary of fracture toughness tests.*

Material Type	Notch Length $2a/W$	Notch Radius, r, mm	Environment	Behavior	Fracture Toughness,[a] MPa m$^{1/2}$ K_{max}	K_{Jmax}
0/±60 Ep	0.51	0.02–0.3	laboratory	linear	22.6(1.4)	. . .
	0.50	3.0	laboratory	linear	29.1(2.0)	. . .
0/90 Ep	0.3–0.7	0.15–0.3	laboratory	linear	19.2(3.3)	. . .
	0.37	bullet	laboratory	linear	21.4(1.7)	. . .
90/0 Ep	0.49	0.3	laboratory	linear	16.6(1.6)	. . .
0/90 Bs	0.67	0.3	laboratory	splits	84.1(0.4)	349(30)
	0.68	0.3	chamber	splits	85.8(2.8)	364(32)
	0.68	0.3	tropical	splits	84.1(8.2)	281(99)
90/0 Bs	0.68	0.3	laboratory	nonlinear	45.3(2.3)	. . .
0$_2$/90 Bs	0.50	0.3	laboratory	splits	57.7(15.8)	283(80)
	0.67	0.3	laboratory	splits	92.2(6.4)	425(55)
	0.69	0.3	chamber	splits	74.4(6.7)	344(46)
	0.65	0.3	tropical	splits	89.5(1.2)	328(78)
90$_2$/0 Bs	0.68	0.3	laboratory	linear	26.4(1.3)	. . .
	0.67	0.3	chamber	linear	29.7(2.0)	. . .

[a] Mean value of three replicates; standard deviation indicated ( ).

to local fiber and overall structural buckling, either of which can ruin a test. A test configuration similar to the center-notch panel that would be also suitable with regard to compression stress ahead of the notch is the double-edge-notch specimen, but this configuration does not relate as directly to a panel with an internal notch due to a ballistic penetration. Recent work [3] described the use of a pressurized cylinder as a fracture toughness test configuration. This can meet the compressive stress requirement and is, of course, well suited to pressure vessel applications.

Fracture toughness tests were performed for 15 cases of material, notch configuration, and environment, as summarized in Fig. 1 and Table 2. Three or more replicates were performed for each case, using a displacement rate of 1 mm/min. A comparison of results from two cases, 90/0 carbon/epoxy and 90/0 carbon/bismaleimide, was included in a recent summary of fracture mechanisms of composite laminates [4]. All of the results are described in Table 2 and subsequent figures and discussion. The initial choice of a measuring point for fracture toughness in the tests is the maximum value of applied K during the test, designated K_{max}. The notch used in most of the tests was cut with a jeweler's saw that produced a 0.3 mm wide kerf. The notch type was varied in the first four cases listed in Table 2 to determine some effects of notch geometry. Notch radii, r, from 0.02 and 3.0 mm were tested for 0/±60 carbon/epoxy by using sharpened notches (with a razor blade) and blunted notches (by intersecting a drilled hole). Notch lengths with $2a/W$ from 0.3 to 0.7 for 0/90 carbon/epoxy were tested and the results were compared with those from a notch produced by an oblique penetration of a rifle projectile. A 5.5-mm-diameter projectile with about 1000 m/s velocity was used at an angle of 77° from normal incidence.

The expression used here for stress intensity factor, K, is based on the early Feddersen expression [5] with a modification [6] to account for eccentricity, e, of the crack centerline relative to the specimen center line. The expression is as follows, with other nomenclature defined in Fig. 1.

$$K = [P/BW][\pi a \sec(\pi a/\{W - 2e\})]^{1/2} \qquad (1)$$
$$0 \le a/W \le 1; e/W \le 0.15; h/W \ge 1.0$$

The effects of crack eccentricity, e/W, and height-to-width ratio, h/W, on K are quite small [6] and are well represented by the preceding expression. All specimen dimensions were within the preceding ranges, and the calculation of K for the specimens is believed to be accurate within about 1%.

An expression for the crack surface displacement, u, shown in Fig. 1 is available from Tada, Paris, and Irwin [7] that fits the collocation results of Newman [8] within 2.5%. A similar expression was developed here that fits the collocation results within 0.5% and is somewhat less complex, as follows

$$uEB/P = -0.14(2a/W) - 1.04(2a/W)^2 + 0.21(2a/W)^3 - 2.14(\ln[1 - 2a/W]) \qquad (2)$$
$$0 \le 2a/W \le 1; h/W \le 1.0$$

where E is elastic modulus and the other parameters are as shown in Fig. 1. This expression is believed to be accurate within about 1%.

The K and displacement expressions of Eqs 1 and 2 were developed for isotropic materials. For composite laminates, these expressions may be useful only for layups with approximately equal numbers of plies in two or more principal directions, such as 0/90, 0/±45, and 0/±60 laminates.

Flexure Tests

Flexural strength and modulus tests were performed by four-point-bend loading, as indicated in Fig. 2. This test was chosen to avoid the more severe stress concentration effects of the single central load point in the three-point test. Note in Fig. 2 that the support pins were free to rotate; the load pins were fixed, to help maintain a repeatable test configuration. Flexural tests were performed for 15 cases of material and environment, as summarized in Table 3. Three or more replicates were tested for each case, using a displacement rate of 6 mm/min.

The expression used for flexural strength, S_B, follows directly from the usual definition of outer fiber stress in bending, using the nomenclature of Fig. 2

$$S_B = 3PX/WB^2 \qquad (3)$$

where P is the total applied load. The expression for flexural modulus, E_B, is available from Roark and Young [9]

$$E_B = (1 - \mu)Pl^3[3(X/l) - 4(X/l)]/[4\delta WB^3] \qquad (4)$$

where μ is Poisson's ratio. For the tests here $X/l = 0.25$, and Eqs 3 and 4 reduce to the expressions in ASTM Test Methods for Flexural Properties of Unreinforced and Reinforced Plastics and Electrical Insulating Materials (D 790-86). A potential problem was noted while analyzing the flexural modulus tests here, which is not fully addressed in ASTM D 790-86. Some ratios of loading pin diameter to loading span can cause a significant nonlinearity in the load versus deflection curve. This can affect the results of flexure tests, as discussed in the upcoming section.

TABLE 3—*Summary of flexural strength and flexural modulus tests.*

Material Type	Environment	Flexural Strength,[a] S_B, MPa	Flexural Modulus,[a] E_B, GPa
0/±60 Ep	laboratory	678(22)	82.4(3.8)
0/90 Ep	laboratory	691(11)	85.0(1.8)
90/0 Ep	laboratory	414(12)	53.1(1.0)
0/90 Bs	laboratory	482(43)	89.0(8.7)
	chamber	491(96)	87.0(14.0)
	tropical	484(30)	82.5(11.0)
90/0 Bs	laboratory	473(33)	63.0(1.7)
0_2/90 Bs	laboratory	486(39)	101.3(17.0)
	chamber	397(78)	88.0(24.1)
	tropical	473(71)	90.3(9.0)
90_2/0 Bs	laboratory	279(23)	32.7(4.0)
	chamber	300(14)	33.3(3.2)

[a] Mean value of three replicates; standard deviation indicated ( ).

Discussion of Results

Fracture Toughness

Material and Orientation Effects—The general fracture behavior of the two types of laminates is summarized in Table 2 and Figs. 3 and 4. A greatly simplified description is that the carbon/epoxy gave linear load-deflection curves, low values of K_{max}, and fracture confined to the plane of the notch (Figs. 3a and b), whereas the carbon/bismaleimide gave nonlinear load-deflection curves, higher values of K_{max}, and significant splitting out of the notch plane (Figs. 3d, e, and 4). The exceptions to this description were the carbon/bismaleimide 90/0 (Fig. 3c) and 90_2/0 specimens, which gave more linear curves, lower K_{max} values, and much less splitting than the other carbon/bismaleimide specimens.

The most prevalent orientation effect was the correspondence between fracture toughness and the number of 0° plies in the laminate. Observations of this effect can be made from the results listed in Table 4. This table shows the ratio of K_{max} from a 90/0 type layup to K_{max} from the corresponding 0/90 layup (the same material rotated 90°). Also shown is the ratio of the number of 0° plies in the 90/0 layup to the number in the corresponding 0/90 layup. Note that

TABLE 4—*Ply orientation effect on fracture toughness.*

Orientation/ Matrix	Notch Length, $2a/W$	Fracture Toughness, K_{max}, MPa m$^{1/2}$	Ratio of K_{max}	Ratio of 0° Plies
$[(0/90)_3/0]_T$ Ep	0.50	20.0		
$[(90/0)_3/90]_T$ Ep	0.49	16.6	0.83	0.75
$[(0/90)_5/0]_T$ Bs	0.67	84.1		
$[(90/0)_5/90]_T$ Bs	0.68	45.3	0.54	0.83
$[(0_2/90)_3/0_2]_T$ Bs	0.67	92.2		
$[(90_2/0)_3/90_2]_T$ Bs	0.68	26.4	0.29	0.27

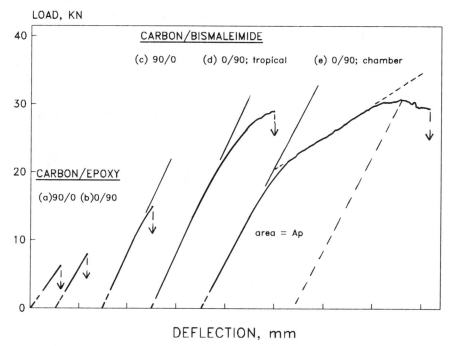

FIG. 3—*Load versus deflection behavior of notched tensile panels.*

the ratio of K_{max} is little different from the ratio of 0° plies for the three sets of laminates. The largest difference in these ratios was for 0/90 and 90/0 carbon/bismaleimide. This may be due to the fact that the 0/90 samples showed splitting and the 90/0 samples did not; this may have increased K_{max} for the 0/90 samples and affected the ratio of K_{max} for this comparison.

The splitting behavior, shown schematically in Fig. 4, was associated with high values of K_{max} and also with extensive nonlinear strain energy dissipation by the specimen, see Fig. 3e. The use of K_{max} as a measure of fracture toughness does not include the nonlinear effects, so calculations of applied J-integral were made for this purpose. The procedure followed that of Rice, Paris, and Merkle [10], who gave the following expression for applied J for a center cracked panel, using the nomenclature here

$$J = K^2/E + (A_P - P\delta_P/2)/B(W/2 - a) \tag{5}$$

where K is applied stress intensity factor and A_p is the area under the curve following a given plastic displacement, δ_p, see Fig. 3e. When used here, the nonlinear displacement is referred to as permanent displacement, since it is clearly different from the plastic deformation that occurs in metals. The applied J at maximum load was calculated using Eq 5, converted to K using $J = K^2/E$, and reported as shown in Table 2. Note that, as would be expected from the load-displacement plots of Fig. 3, and applied K values calculated from J at maximum load, designated K_{Jmax}, are significantly higher than the K_{max} values, which include only elastic strain energy. Applied J provides a measurement of fracture toughness for materials and orientations that display the important splitting behavior that is characterized by extensive permanent strain energy before failure. Even though the J integral analysis of Eq 5 was not intended for

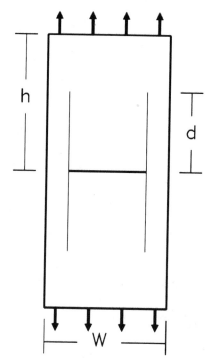

FIG. 4—*Typical splitting of 0/90 and 0_2/90 carbon/bismaleimide panels.*

anisotropic materials [*10*], it may still be useful in characterizing the permanent strain energy that appears to control the fracture behavior in some cases.

The splitting behavior can be described in another manner. The approximately bilinear load versus displacement behavior of Fig. 3e and the end result of the splitting as sketched in Fig. 4 can be related. If, after considerable splitting, the load on the specimen were carried by only the material in the two unnotched ligaments, then a reduced apparent elastic modulus, E_A, would be produced that depended upon the notch length-to-specimen width ratio, $2a/W$, as follows

$$E_A/E = 1 - 2a/W \qquad (6)$$

where E is the modulus of the unnotched and unsplit panel. The ratio of the reduced apparent modulus to the initial modulus in Fig. 3e is 0.34, and the value of $(1 - 2a/W)$ for that specimen was 0.32. This relatively close agreement supports the preceding model, which involves an unloading of the material above and below the notch and a bilinear load-displacement behavior. This bilinear elastic slope behavior may be a useful way of characterizing splitting behavior during fracture testing of cross-ply laminates.

Notch Configuration Effects—Figures 5, 6, and 7 highlight some of the effects of notch length, root radius, and means of production on the measured fracture toughness of carbon/epoxy panels. The effect of notch length over the range $0.25 \geq 2a/W \geq 0.70$ is shown in Fig. 5 for 0/90 carbon/epoxy. Results are shown for two notch radii, $r = 0.30$ and 0.15 mm, produced using different sizes of saw blade. Regression lines for each set of results show less change

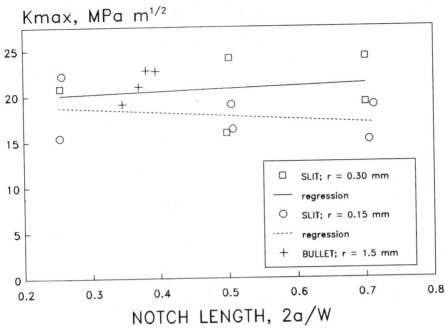

FIG. 5—*Effects of notch configuration on fracture toughness for 0/90 carbon/epoxy.*

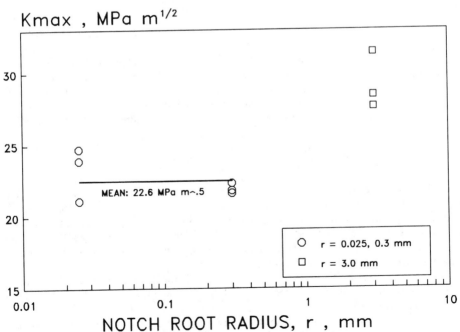

FIG. 6—*Effects of notch radius on fracture toughness for 0/±60 carbon/bismaleimide; nominal 2a/W = 0.5.*

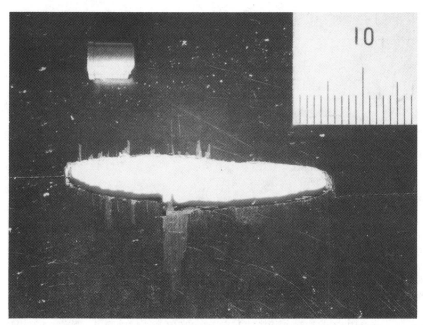

FIG. 7—*Photo of 0/90 carbon/epoxy panel with ballistic penetration; pieced together after tensile loading to failure; penetrator also shown.*

of K_{max} with $2a/W$ than there is scatter in the data. If there is any significance to the placement of the regression lines, the placement is as would be expected, that is, the K_{max} values are lower for smaller root radius. Values of fracture toughness for 0/90 carbon/epoxy laminates are available from the literature that are in general agreement with the results in Fig. 5. Mao [*11*] reports center-notch fracture toughness of 20 to 26 MPa m$^{1/2}$ for $2a/W$ of 0.28 and 0.50, respectively.

Results from four panels with a bullet hole for a notch are compared in Fig. 5 with the saw-cut notch results. Figure 7 shows one of the panels, pieced together after the test, with the projectile also shown. Even though the root radius of the notch produced by ballistic penetration was considerably larger than that of the saw cuts, the K_{max} values were within the scatter of the saw-cut results. This indicates that, for this material, damage from ballistic penetration can be essentially equivalent to that from relatively sharp notches.

The effect of notch root radius over the range $0.025 \geq r \geq 3.0$ mm on measured K_{max} is shown in Fig. 6 for $0/\pm60$ carbon/epoxy. The mean of the six results for $r = 0.025$ and 0.3 mm is shown as the horizontal line. The K_{max} values for the $r = 3.0$ mm tests are significantly higher, which could simply be an indication that higher toughness results from radii of this size. It may be also an indication of the relatively clean radius produced by a sharp 6.0 mm drill compared with the rough radius produced by a saw cut.

Environmental Effects—A comparison of fracture toughness results showing the effects of environment is given in Table 5. Included in the table is a simple and direct test of statistical significance, that is, comparing the difference in the mean value of K_{max} with the larger of the standard deviations for the two sets of results. For the 0/90 material, no significant difference in fracture toughness is indicated for either chamber or tropical exposure. Note, however, in Fig. 3 that the permanent displacement for one of the 0/90 tropical exposure tests was considerably less than that for the 0/90 chamber test. This difference is reflected in the K_{Jmax} values

TABLE 5—*Environmental effects on fracture toughness for carbon/bismaleimide; nominal 2a/W =
0.67, r = 0.3 mm.*

Orientation	Environment	Fracture Toughness, K_{max}, MPa m$^{1/2}$	Difference in Fracture Toughness, MPa m$^{1/2}$	Largest Standard Deviation, MPa m$^{1/2}$
$[(0/90)_5/0]_T$	laboratory	84.1	. . .	. . .
	chamber	85.8	1.7	2.8
	tropical	84.1	0.0	8.2
$[(0_2/90)_3/0_2]_T$	laboratory	92.2	. . .	. . .
	chamber	74.4	17.8	6.7
	tropical	89.5	2.7	6.4
$[(90_2/0)_3/90_2]_T$	laboratory	26.4	. . .	. . .
	chamber	29.7	3.3	2.0

in Table 2, where the mean value for tropical exposure is 281 MPa m$^{1/2}$, lower than the values
for laboratory and chamber exposure. The large standard deviation for this value for tropical
exposure is caused by the fact that the other two tropical results were much higher, in the same
range as the laboratory and chamber results.

The only significant difference in results that is clearly attributed to environment is the lower
K_{max} for $0_2/90$ material due to moisture chamber exposure, see Table 5. The difference in mean
K_{max} is about three times the amount of the larger standard deviation. In addition, referring to
Table 2, K_{Jmax} is significantly lower for the $0_2/90$ chamber results than for the laboratory results.
This indicates that the amount of permanent displacement that can be sustained before failure
is reduced as a result of moisture chamber exposure. Note (in Table 2) that the same was true
following tropical exposure, but the K_{Jmax} values that showed this decrease following tropical
exposure included considerable scatter.

Failure Mechanisms—The drastic differences in macrofracture behavior of the two types of
laminate, as summarized in Table 2 and Fig. 3, led us to look for concomitant differences in
the micromechanisms of fracture. One of the authors has described experiments and models
of failure mechanisms in glass/polyester materials [12]. The approach here was to study the
fracture surfaces using scanning electron microscopy (SEM). Figures 8 and 9 show some key
results. The low magnification fractographs of Fig. 8 show the fracture surface ahead of the
notch (the notch tip is at the left for each photo) for 90/0 laminates of each type. The fracture
surfaces are as drastically different as the fracture toughness results. The carbon/epoxy is relatively flat and featureless with no evidence of individual fiber pullout, whereas the carbon/
bismaleimide is a severely contorted surface due primarily to extensive fiber pullout.

Higher magnification fractographs in Fig. 9 show that the only evidence of pullout for the
carbon/epoxy is a limited amount of pullout of fiber bundles before failure of the intact bundles. In contrast, extensive individual fiber pullout is observed for carbon/bismaleimide. The
relatively smooth surface of the fibers in the carbon/bismaleimide, which can be seen in Fig.
9, may have contributed to the extensive pullout. Whatever the cause, extensive fiber pullout
involves significant interfacial area and therefore would be expected to result in extensive dissipation of energy during fracture and a high fracture toughness. It is interesting to note that
Evans and Marshall [13] presented fractographs nearly identical in appearance to Fig. 9b but
from a ceramic/ceramic composite that showed greatly improved toughness over that of the
ceramic matrix. They attributed the increase in toughness to the same mechanism under discussion here, that is, extensive individual fiber pullout and related dissipation of energy.

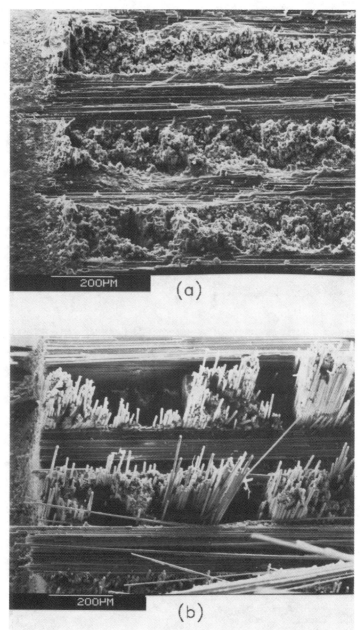

FIG. 8—*Low magnification SEM fractographs at notch tip: (a) 90/0 carbon/epoxy; little fiber pull-out, and (b) 90/0 carbon/bismaleimide; extensive fiber pull-out.*

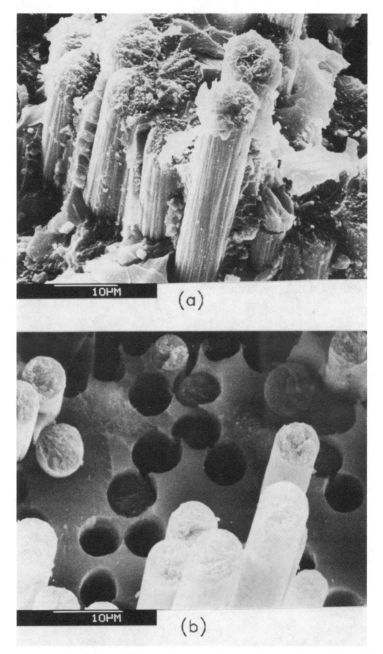

FIG. 9—*High magnification SEM fractographs, 3 mm from notch tip: (a) 90/0 carbon/epoxy; predominantly failure of fiber bundles, and (b) 90/0 carbon/bismaleimide; predominantly pull-out of individual fibers.*

Flexure Tests

General Behavior—Typical load-versus-deflection plots obtained from the two types of laminate are shown in Fig. 10. A clear difference in the two materials was seen in the final failure behavior. The carbon/epoxy specimen shown failed abruptly (typical of all tests) at the point of contact of one or both of the load pins. The carbon/bismaleimide specimens often showed load variations that corresponded to progressive failure through the thickness. This distinction between abrupt and progressive failure was generally the same as observed in fracture testing, see again Fig. 3.

Correction for Rotation—The plots of all flexure tests had one disturbing feature in common, that is, a continuously increasing slope. This was not the usual toe region nonlinearity, as described in ASTM D 790-86, because it continued until failure. The cause of the nonlinearity was found to be the rotation of specimen relative to the loading and support pins, as sketched in Fig. 11. Using plane geometry, it can be shown that the initial moment span, X, is changed due to rotation of pins relative to specimen, to a smaller value, X_r,

$$X_r = X - D \sin \alpha \tag{7}$$

where D is the pin diameter and α is the angle of rotation. For small angles $\alpha = \delta/X$ and an expression for X_r/X is

$$X_r/X = 1 - \delta D/X^2 \tag{8}$$

where δ is the specimen deflection, see Fig. 2. For the carbon/epoxy test in Fig. 10a, for example, with $\delta = 3.3$ mm, $D = 12.5$ mm, and $X = 12.5$ mm, the ratio of actual-to-calculated moment span, X_r/X, is 0.74. This 26% decrease in the moment span would be directly reflected

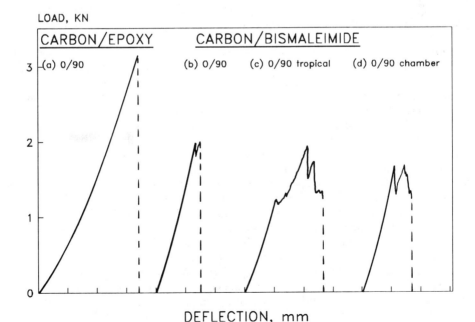

FIG. 10—*Load versus deflection behavior of flexure specimens.*

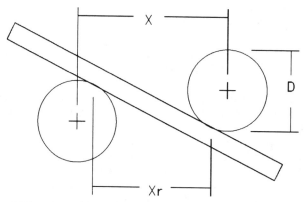

FIG. 11—*Rotation of loading pins relative to flexure specimen.*

in flexural strength, S_B, determined from the test, if Eq 3 were used with no modification for rotation. In the results here, Eqs 8 and 3 were combined to include rotation effects in the calculation of S_B. The flexural modulus was calculated from the slope at a load of 0.3 kN, at which point there is little effect of rotation on the slope, so no modification was used.

A review of ASTM D 790-86 showed that, although the previously mentioned rotation error is reduced by the requirement of a maximum pin size, some significant errors are still possible. For example, for the test of Fig 10a using the D = 6.0 mm maximum pin size specified in Method D 790-86, the value of X_r/X is 0.87, indicating that a 13% error in calculation of flexural strength would still occur.

A further demonstration of the need for correction of rotation errors in flexural tests is given in Fig. 12. The rotation correction given in Eq 8 was applied to the results of Fig. 10a, and plots of uncorrected and corrected flexural stress are compared. Note that the corrected values are very nearly a straight line, as would be expected for a linear elastic test, and that the corrected flexural strength is much reduced.

Corrected Results—The corrected flexural strength results and the flexural modulus results are listed in Table 3. A general comparison with these results can be made from published data for various quasi-isotropic carbon/epoxy laminates; with no effect of environment, flexural strengths of about 500 MPa and moduli of about 50 GPa are typical [*14*], in reasonable agreement with results here. Ply orientation effects on strength and modulus are as expected in the results of Table 3, that is, strength and modulus are closely related to the number of 0° plies. Stacking sequence of the plies may have also affected the strength and modulus results, but the tests provided little basis for determination of this effect.

One result of flexural modulus can be compared with a modulus value determined from a tensile panel. One of the carbon/epoxy specimens whose mean flexural modulus was 85.0 GPa (see Table 3) was also instrumented so that modulus could be measured as part of the fracture toughness test. Small aluminum blocks were bonded to the specimen surface about 5 mm above and below the notch centerline to accommodate a clip gage of the type used in ASTM Test Method for Plane-Strain Fracture Toughness of Metallic Materials (E 399-83). The slope of the plot of this displacement, u, (see Fig. 1) versus load was used with Eq 2 and the specimen dimensions to calculate a modulus of 34.3 GPa. The significantly higher value in bending (85.0 GPa) is a reflection of the dominance of the outer layers of a laminate in determining flexural modulus. This is particularly true for the relatively thin, seven-ply laminate under discussion, in which the outer layers make up a larger share of the thickness than in a thicker laminate.

The effect of environment on flexural properties of carbon/bismaleimide was investigated

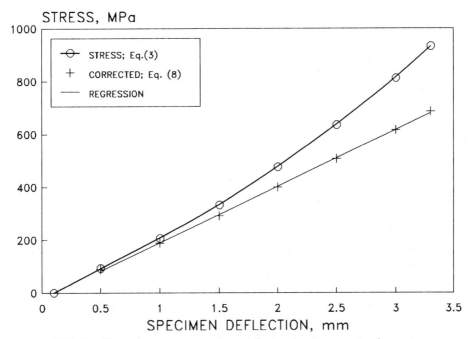

FIG. 12—*Flexural stress versus specimen deflection showing correction for rotation.*

for three ply orientations, 0/90, $0_2/90$, and $90_2/0$, see Table 3. No significant effect on flexural modulus was found; some lower values were noted following environmental exposure, but the differences were smaller than the standard deviations. A statistically significant decrease in flexural strength was found for the $0_2/90$ material following moisture chamber exposure (0.67% weight increase, due to 85°C, 97% relative humidity exposure for 400 h). The mean strength decreased from 486 to 397 MPa, an 18% drop. These results can be compared to typical data from the literature [14] for carbon/epoxy, which can absorb up to 1.5% water due to 82°C, 95% relative humidity, and suffer a 10 to 15% decrease in flexural strength. Specific data are available for comparison with the carbon/bismaleimide results here. For a 0° laminate of Hercules AS-4 fibers in Narmco 5230-3 matrix, the flexural strength and modulus both increased about 21% to 1696 MPa and 135 GPa following 864 h at 71°C and 95% relative humidity and a 0.75% weight gain [15].

Considering the results here and data from the literature, no significant trend of moisture effects on flexural properties is apparent. The one case of a significant change in flexural strength in these tests was offset by an opposite effect in the literature data.

Summary

Characterization of the fracture behavior and flexural properties of carbon fiber laminates has shown the following.

(a) Center-notched panels of carbon/epoxy and $90_2/0$ orientations of carbon/bismaleimide show linear elastic translaminar fracture at K_{max} values up to about 30 MPa $m^{1/2}$. Carbon/bismaleimide panels with a larger share of 0° fibers show mixed translaminar and interlaminar splitting fracture and considerable permanent deformation, at K_{max} values of 45 to 90 MPa $m^{1/2}$ and K_{Jmax} values up to 400 MPa $m^{1/2}$.

(b) Significant environmental effects on fracture were noted for $0_2/90$ carbon/bismaleimide following 85°C, 97% relative humidity chamber exposure, causing a 20% decrease in K_{max}, from 92 to 74 MPa m$^{1/2}$.

(c) Scanning electron microscope fractography showed clear distinctions between the low toughness carbon/epoxy, which had a relatively featureless fracture surface with no fiber pull-out, and the high toughness carbon/bismaleimide, which had a rough surface and extensive fiber pull-out.

(d) Flexural strength and modulus for both types of laminate varied between 300 and 700 MPa and 30 and 100 GPa, respectively, with higher values for higher shares of 0° fibers. A significant decrease in flexural strength was noted for $0_2/90$ carbon/bismaleimide following moisture chamber exposure, causing an 18% decrease, from 486 to 397 MPa.

Evaluation and development of test procedures for determining fracture and flexural properties of the laminates have shown the following.

(*a*) The center-notched tensile panel was found to be entirely suitable for fracture toughness tests of carbon fiber laminates. Expressions for K, J, and displacement were identified and modified. Notch configuration effects on test results were found to be small.

(*b*) Interlaminar splitting at the notch tips was found to be critical, being both the cause and a convenient indicator of the high toughness mode of deformation and fracture for these carbon fiber laminates.

(*c*) The source of a significant error in flexure testing was identified as rotation of the loading pins relative to the specimen. An expression was given that can be used to include effects of rotation in calculation of strength and modulus.

Acknowledgments

The authors are pleased to acknowledge the work of B. Yaiser of U.S. Army ARDEC in preparing the composite samples for test, J. Hill of Materials Research Laboratory-Innisfail in conducting the tropical exposure tests, D. Saunders of the Aeronautical Research Laboratory, Melbourne, in conducting the moisture-chamber tests, B. Baxter of the Materials Research Laboratory, Melbourne, in performing the ballistic penetration tests, and M. Butler of the Materials Research Laboratory in preparing the manuscript.

References

[1] Underwood, J. H., written discussion for "Damage Mechanics Analysis of Matrix Effects in Notched Laminates," by C.-G. Aronsson and J. Backlund, *Composite Materials: Fatigue and Fracture, ASTM STP 907*, H. T. Hahn, Ed., American Society for Testing and Materials, Philadelphia, 1986, pp. 156–157.
[2] O'Brien, T. K., Johnson, N. J., Raju, R. S., Morris, D. H., and Simonds, R. A., "Comparisons of Various Configurations of the Edge Delamination Test for Interlaminar Fracture Toughness," *Toughened Composites, ASTM STP 937*, N. J. Johnston, Ed., American Society for Testing and Materials, Philadelphia, 1988, pp. 199–221.
[3] Anderson, C. E., Cardinal, J. W., and Kanninen, M. F., "A Fracture Mechanics Analysis for the Failure of Filament-Wound Pressurized Containers Subjected to Intense Energy Deposition," *Lightening the Force—The Role of Mechanics, Proceedings*, Army Symposium on Solid Mechanics, West Point, NY, 1986.
[4] Bandyopadhyay, S., Gellert, E. P., Silva, V. M., and Underwood, J. H., "Microscopic Aspects of Failure and Fracture in Cross-Ply Fiber-Reinforced Composite Laminates," *Journal of Composite Materials*, Vol. 23, 1989, pp 1216–1231.
[5] Feddersen, C. E., "Evaluation and Prediction of the Residual Strength of Center Cracked Tension Panels," *Damage Tolerance in Aircraft Structures, ASTM STP 486*, American Society for Testing and Materials, Philadelphia, 1971, pp. 50–78.

[6] Tada, H., Paris, P. C., and Irwin, G. R., *The Stress Analysis of Cracks Handbook,* Paris Productions, Inc., St. Louis, MO, 1985, pp. 11.1–11.2.

[7] Tada, H., Paris, P. C., and Irwin, G. R., *The Stress Analysis of Cracks Handbook,* Paris Productions, Inc., St. Louis, MO, 1985, pp. 2.4–2.5.

[8] Newman, J. C., Jr., "Crack-Opening Displacements in Center-Crack, Compact, and Crack-Line Wedge-Loaded Specimens," TN D-8268, NASA Langley Research Center, Hampton, VA, July 1976.

[9] Roark, R. J. and Young, W. C., *Formulas for Stress and Strain,* McGraw-Hill, New York, 1975, pp. 96–108.

[10] Rice, J. R., Paris, P. C., and Merkle, J. G., "Some Further Results of J-Integral Analysis and Estimates," *Progress in Flaw Growth and Fracture Toughness Testing, ASTM STP 536,* American Society for Testing and Materials, Philadelphia, 1973, pp. 231–245.

[11] Mao, T. H., "Tensile Fracture of Graphite/Epoxy Laminates with Central Crack," *Fifth International Conference on Composite Materials,* The Metallurgical Society, 1985, pp. 383–390.

[12] Bandyopadhyay, S. and Murthy, P. H., "Experimental Studies on Interfacial Shear Strength in Glass Fibre Reinforced Plastics Systems," *Materials Science and Engineering,* Vol. 19, 1975, pp. 139–145.

[13] Evans, A. G. and Marshall, D. B., "Mechanical Behavior of Ceramic Matrix Composites," plenary lecture, Seventh International Conference on Fracture, Houston, TX, 1989.

[14] Springer, G. S., "Moisture and Temperature Induced Degradation of Graphite Epoxy Composites," *Environmental Effects on Composite Materials,* G. S. Springer, Ed., Technomic Publishing Co., Lancaster, PA, 1984, pp. 6–19.

[15] Rondeau, R. A., Askins, D. R., and Sjoblom, P., "Development of Engineering Data on New Aerospace Materials," Report UDR-TR-88-88, University of Dayton Research Institute, Dayton, OH, Dec. 1988.

Michael R. Piggott[1] and Patrick W. K. Lam[2]

Fatigue Failure Processes in Aligned Carbon-Epoxy Laminates

REFERENCE: Piggott, M. R. and Lam, P. W. K., **"Fatigue Failure Processes in Aligned Carbon-Epoxy Laminates,"** *Composite Materials: Fatigue and Fracture (Third Volume), ASTM STP 1110,* T. K. O'Brien, Ed., American Society for Testing and Materials, Philadelphia, 1991, pp. 686–695.

ABSTRACT: Experiments carried out on aligned carbon fiber-reinforced epoxies published elsewhere show that during tensile-tensile fatigue the slope of the S-N curve, β, is increased by reducing the adhesion between fibers and polymers, and decreased by reducing the internal microstresses in the composite. Also, β is increased by plasticizing the polymer. During fatigue, the flexural modulus decreases at a greater rate than the tensile (Young's) modulus. The Poisson's ratio and the hysteresis loop energy both increase during fatiguing. Circular holes drilled in the specimen become gradually more elliptical, with the major axis at right angles to the stressing (and fiber) direction. Final failure involves extensive splitting, and scanning electron microscopic examination of the fractures reveals powdered polymer present on the fracture surfaces. These results suggest that initial fiber waviness may be important. At the antinodes of the waves, there are transverse stresses which could lead to fiber debonding and splitting of the matrix. Once debonded, the fibers could slide within the debonded regions, giving rise to matrix attrition. The cyclic stressing could then cause the powdered matrix thus formed to vibrate and gradually transfer itself to the inside of the curves of the fiber profiles, increasing the fiber curvature (and hence increasing Poisson's ratio, and the width of the hole in the specimen and accounting for the decline in Young's modulus). Eventually the fiber curvature could increase to such an extent that the resulting flexural stresses, combined with the tensile stresses, are enough to cause fiber failure in the regions where this process has developed farthest. Then final failure could occur by connection of these regions through splits in the matrix. (These splits must also develop continuously during fatigue, to account for the gradual loss in flexural modulus.) The process is accelerated by poor fiber-matrix adhesion, and slowed down by reducing the internal stresses which are present due to the cure shrinkage of the resin.

KEY WORDS: composite materials, fatigue (materials), failure, fracture, carbon-epoxies, fiber composites

Fatigue failure in carbon-epoxy laminates has been extensively studied, most notably by Reifsnider and co-workers [1]. Damage is initiated in plies in which the fibers are oriented at an angle to the applied stress axis. Cracks grow and multiply in these layers until the load is effectively carried almost entirely by the on-axis fibers. It is thought that these cracks may play a role in the eventual failure of the on-axis fibers, and hence final composite fracture. However, the evidence for this is not conclusive.

For unidirectional composites, Talreja [2] has constructed a theory based on an observation by Dharan [3] that the fatigue failure strain of the polymer matrix may be correlated with the fatigue life of the composite. (Unfortunately, there is an error in the Dharan's reference, which

[1] Professor, Advanced Composites Physics & Chemistry Group, Department of Chemical Engineering, University of Toronto, Toronto, Ontario M5S 1A4 Canada.
[2] Research scientist, Nova Husky Research Corporation, Calgary, Alberta, T2E 7K7, Canada.

was not picked up in his more recent paper [4].) In any case, such factors as fiber-matrix adhesion, residual internal stresses, and probably most important, imperfections in the composite, are likely to have an important role in initiating damage in aligned fiber composites. Hence, these factors should be taken into account also.

It has been already demonstrated that imperfections, particularly waviness of the fibers, are very important in controlling the compressive properties of composites [5]. This is because they have a disturbing effect on the internal stresses, by introducing new stresses which are perpendicular to the applied stress. Then off-axis stresses can cause splitting of the composite [6].

In the case of fatigue in uniaxial composites, the off-axis stresses induced by curved fibers are a probable cause of progressive damage. A series of experiments has been recently described in which a number of techniques were used to show that damage was occurring during fatigue and to monitor this damage. The observations for the case of tensile-tensile fatigue ($R = 0.1$) will be discussed, and a sequence of damage processes will be deduced therefrom. For experimental details, etc., the reader is referred to the original papers [7–9].

Evidence of Polymer Matrix Damage

Visual examination of the composites that had failed due to fatiguing showed that extensive splitting had taken place as well as fiber fracture. This was true of the composites tested in tension-tension, as well as those that had been subject to compression stresses during fatigue. In addition, scanning electron microscopic (SEM) examination of the failed composites revealed a great deal of powdered polymer on the fractured surfaces.

This polymer damage was probably progressive. During the tension-tension fatigue tests ($R = 0.1$) some of the specimens were released temporarily, and their flexural and tensile (Young's) moduli were measured. Both moduli decreased continuously, but the flexural modulus decreased at a faster rate than the tensile modulus, Figs. 1 and 2 [9]. This loss in tensile

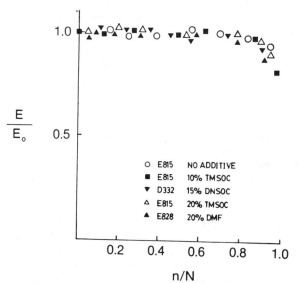

FIG. 1—*Relative change in Young's modulus during fatigue. (n/N indicates fraction of fatigue life; results from five composites.)*

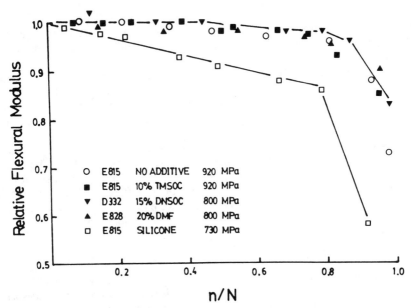

FIG. 2—*Relative change in flexural modulus during fatigue for five composites* [9].

modulus has been also observed by Rotem [10] in reversed fatigue loading ($R = -1$) of laminates.

(In Fig. 1, tests on five composites are included. The resins were DGEBA type: Shell EPON 815 (E815 on legend) and 828 (E828), and also DOW DER 332 (D332). In three cases, they had expanding monomers (DNSOC and TMSOC) added to them and copolymerized with the epoxy to decrease the shrinkage stresses. In one case, EPON 828 had a plasticizer, dimethyl formamide (DMF), added. In Fig. 2, in addition to the preceding, in one case the fibers were silicone coated before making the composite. The fourth column (920 MPa, etc.) in the legend indicates the maximum stresses.)

The tensile modulus, E_1, was initially given approximately by the Rule of Mixtures

$$E_1 = V_f E_f + V_m E_m \tag{1}$$

where V_f and V_m are the fiber and polymer volume fractions, and E_f and E_m the corresponding Young's moduli. Loss in E_1 can result from fiber fragmentation, or destressing of the fibers due to complete loss of adhesion along the whole fiber length, or loss of orientation. It is unlikely that matrix damage plays a major role here.

In the case of the flexural modulus, E_{fl}, on the other hand, matrix damage such as splitting of the composite parallel to the fibers will have a large effect. Thus for a bar having a thickness, t, which splits into two equal pieces of thickness, $t/2$, the flexural rigidity, initially $E_{fl}t^3/12$ per unit width, will be reduced to $2E_{fl}(t/2)^3/12 = E_{fl}t^3/48$, or one quarter of the original value. A split which does not extend the whole length of the specimen will obviously have a smaller effect, but flexural modulus is clearly very sensitive to internal splitting.

It will be noted that flexural modulus decreases sharply at the end of the fatigue life, and this effect is much greater than with the tensile modulus. Since splitting was observed in the final failure, the loss in E_{fl} is probably largely due to the development of matrix splits. Thus, it seems likely that this type of matrix damage is progressive during the fatigue process, since E_{fl} decreases continuously.

Fiber debonds will contribute to this process. In some experiments, debonding was promoted by coating the fibers with silicone oil. Composites made with these fibers showed the most rapid loss of E_{fl}, see Fig. 2. Any debonding will transfer extra stress onto the web of polymer between the fibers, and hence promote early failure there.

Evidence of Fiber Damage

Although the failure of the composites was accompanied by a great deal of splitting, at final failure the fibers themselves were broken when the two halves of the composite separated.

Loss in Young's modulus and loss in tensile strength are indicators of loss in fiber strength, or fiber breakage. Unfortunately, it is hardly feasible to measure tensile strength at all stages of the fatigue process, since strengths can be highly variable, and the specimen is destroyed by the test anyway, so cannot be used for further fatiguing. Some tensile strength tests were carried out at about half the fatigue life, but these showed no significant loss in strength.

The Young's modulus can be continuously monitored without interrupting the fatigue test, by measuring the orientation of the stress-strain hysteresis loop during the fatigue process. This gives the average modulus during the fatigue cycle, and gave similar results to those obtained when the fatigue process was interrupted (compare Fig. 3 with Fig. 1)

If we assume that the loss in modulus is due to the fibers breaking into short lengths, we can very approximately estimate the average fiber length at any stage of the fatigue process. Thus, the Young's modulus of an aligned short fiber composite is given approximately by [11]

$$E_1 = V_f E_f \left(1 - \frac{\tanh(ns)}{ns} \right) + V_m E_m \tag{2}$$

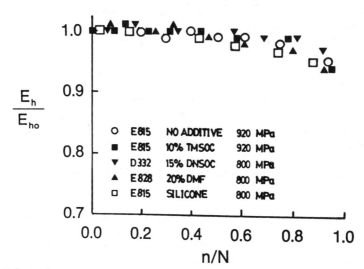

FIG. 3—*Relative change in Young's modulus during fatigue for five composites, as indicated by hysteresis loop orientation* [9].

Here

$$n^2 = 2E_m/E_f(1 + \nu_m) \ln(2\pi/\sqrt{3} \ V_f) \tag{3}$$

and ν_m is the Poisson's ratio of the polymer matrix and s is the fiber aspect ratio. For V_f in the range about 0.3 to 0.6 and $s > 30$, corresponding to a fiber length $L > 240 \ \mu m$

$$E_1 \cong V_f E_f(1 - \sqrt{E_f/E_m} \ /s) + V_m E_m \tag{4}$$

A deviation of as little as 5% in Young's modulus requires very short fibers indeed. Thus for carbon fibers with $E_f = 228$ GPa and $V_f = 0.5$ in a polymer with $E_m = 2.3$ GPa

$$\Delta E_1/E_1 \cong 9.90/s \tag{5}$$

where ΔE_1 is the difference between E_1, for continuous fibers and E_1 for short fibers. $\Delta E_1/E_1 = 0.05$ requires an aspect ratio of about 200, corresponding to a length of about 1.6 mm.

The microscopic evidence does not support such severe fiber attrition: to attain a loss of 5% in Young's modulus requires that the average fiber length is about 1.6 mm. Consequently, we have to look elsewhere for an explanation of the loss in modulus.

Evidence of Geometrical Imperfection

The loss of Young's modulus could be due, in part at least, to loss of fiber alignment. If a composite is tested at an angle, ϕ, to the fiber direction, the Young's modulus, E_ϕ, is given by Ref 12

$$1/E_\phi = \frac{\cos^4 \phi}{E_1} + \left(\frac{1}{G_1} - \frac{2\nu_{12}}{E_1} \right) \sin^2 \phi \cos^2 \phi + \frac{\cos^4 \phi}{E_2} \tag{6}$$

where G_1 is the shear modulus, E_2 is the Young's modulus normal to the fiber direction, and ν_{12} is the in-plane Poisson's ratio of the composite. For the carbon-epoxy just described (which had $E_1 \cong 115$ GPa), ϕ needs only to be about 1.6° for $E_\phi = 0.95 \ E_1$. Such a small deviation could develop during fatigue.

Evidence that slight fiber curvature does develop during fatigue comes from measurements of the sizes of pre-cut holes during the fatigue process. These holes, 4.55 ± 0.05 mm diameter, were carefully drilled through the centers of some of the pultrusions (the pultrusions were 20.0 mm wide). During the fatigue process, the holes became approximately elliptical, with the major axis at right angles to the fiber (and stressing) direction. The minor axis (x) hardly changed during fatigue, but the major axis (y) increased continuously, Fig. 4.

During fatigue, the Poisson's ratio also increased continuously, Fig. 5. This again could be due to the development of fiber waviness. (It was not possible to monitor ν_{12} for as long as the other properties because the strain gages used to estimate the transverse strains became detached).

The process that could take place is shown in Fig. 6. For clarity, the fiber waviness is greatly exaggerated. The "initial state" in the drawing corresponds to an average angular deviation of 1.6°, that is, that required to give, very roughly a 5% loss in Young's modulus, which does not actually develop until nearly the end of the fatigue process.

The increase in waviness will increase the specimen width, as well as reducing the Young's modulus: this would explain how the hole becomes more elliptical. The fibers will tend to

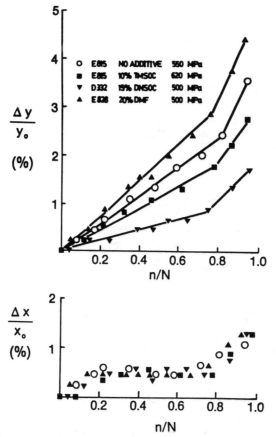

FIG. 4—*Increase in hole size during fatigue for four composites* [9].

become straighter as the stress applied to the composite is increased, and relax and become more curved as the stress is decreased during each fatigue cycle: this will account for the increase in the Poisson's ratio of the composite that was observed.

Overall Damage and Failure Mechanism

The evidence presented suggests that there are two important damage processes that develop during fatigue: (1) splitting of the polymer, accompanied by fiber debonding and (2) development of fiber waviness. Fiber damage and failure is probably not important, except at the end of the fatigue life.

It seems highly probable that the sequence of events is as follows. First debonding occurs where the off-axis stresses are highest. This will be at the antinodes of the fiber profiles, as shown in Fig. 6, will occur initially on the tensile side, but will then probably extend to the compression side. The fibers thus become completely debonded in these regions and can slide relative to the polymer, causing some attrition of the polymer. This results in increased energy absorption. Evidence that this occurs comes from the hysteresis loss shown in Fig. 7. This increases continuously during fatigue. This process continues with the fibers developing

0.25
0.20
0.15
ν_{12}
E815 NO ADDITIVE 920 MPa
E815 10% TMSOC 920 MPa
D332 15% DNSOC 800 MPa
E828 20% DMF 800 MPa
0.2
0.4
0.6
n/N

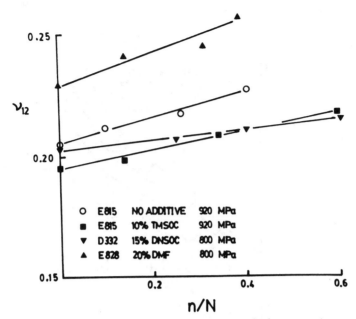

FIG. 5—*Increase in Poisson's ratio during fatigue for four composites.*

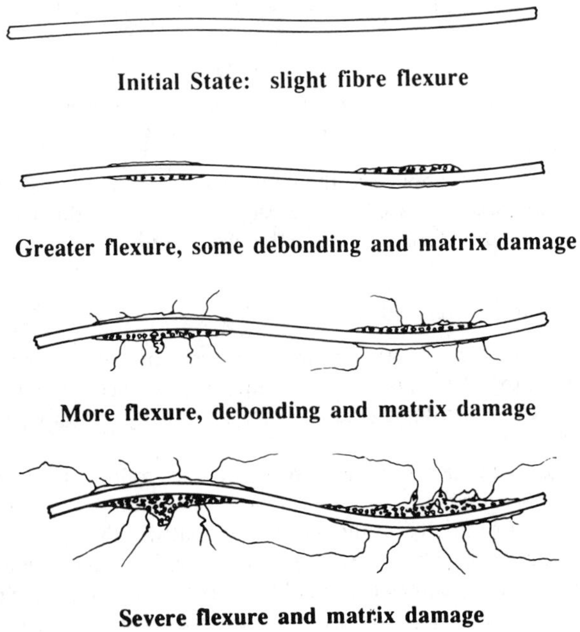

Initial State: slight fibre flexure
Greater flexure, some debonding and matrix damage
More flexure, debonding and matrix damage
Severe flexure and matrix damage

FIG. 6—*Development of damage during fatigue. Fiber curvature increased for clarity. (In the final state, when the composite is about to fracture, the curvature may be about the same as shown at the top.)*

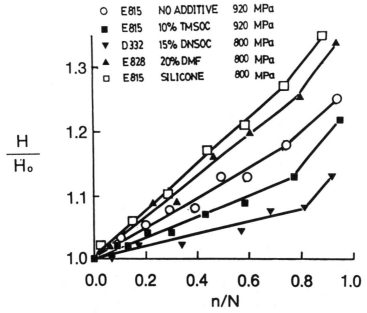

FIG. 7—*Increase in hysteresis loop energy during fatigue for five composites.*

increasingly wavy profiles, as shown, and the polymer becoming more damaged. At the same time, the matrix starts to split, probably initiated at the fiber debonds, where the web of matrix between the fibers now has extra stress on it normal to the fibers. This splitting develops and becomes extensive, as indicated by the gradual loss in flexural modulus.

A great deal of fiber flexure is not necessary to cause composite failure. The slope of the S-N curve is small, Fig. 8, with failure at 10^6 cycles occurring at about 71% of the static strength. The smallest amplitude sine wave shown in Fig. 6 (marked initial state) has an amplitude (a) which is equivalent to about 0.33 fiber diameters (d) and a wavelength (λ) which is about 47 diameters. If we neglect all other stresses, apart from those due to the slight curvature shown in the drawing, the fiber strain at the antinode is $2\pi^2 ad/\lambda^2$, which in this case comes to 0.29%, or nearly one quarter of the breaking strain for Hercules AS1 fibers. The curvature shown in the second drawing from the top is more than enough to reduce the fiber breaking strain to 71% of the static value, as needed for failure at 10^6 cycles. If all the fibers were this wavy at the final stage of the fatigue process then the composite would have a strength of only about 25% of its initial value. It is more likely that regions of fiber waviness develop in widely separated places. These fibers then fail, and the failed regions are connected by splits already present in the matrix.

The effect of silicone coating the fibers, and thus reducing the adhesion, is to promote early initiation of the debonding, and hence accelerates the whole process. The effect of reducing the polymer shrinkage is to reduce the local internal stresses, and hence delays polymer splitting and attrition.

Conclusion

The mechanism of failure in tensile-tensile fatigue of carbon fiber-reinforced polymers is probably initiated by fiber curvature, which is initially present in the composite. The varying

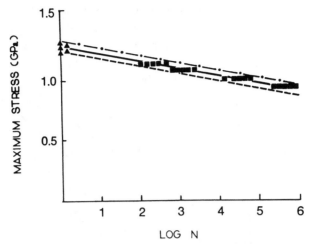

FIG. 8—*S-N curve for aligned fiber composite with* R = 0.1 *(tension-tension).*

stresses during fatigue start to damage the composite at the antinodes of the fiber profiles. This damage is probably in the form of fiber debonds, which generate matrix cracks and lead to comminution of the polymer. As the process proceeds, the fibers become more wavy as powdered polymer transfers itself to the insides of the antinodes of the fiber waves. Eventually, the flexural stresses in the fiber are great enough to cause early fiber failure in the highly wavy regions thus produced. These breaks are connected by splits in the polymer, so that final failure is facilitated. Reducing internal microstresses in the polymer delays the polymer failure process, and enhancing the fiber-matrix adhesion inhibits the initial fiber debonds that start the whole process in motion.

Acknowledgments

We are grateful to NSERC (Canada); OCMR Ontario, Canada; Du Pont; and the U.S. Air Force for financial support for our research. The original work on which this analysis was developed was supported by the Canadian Defence Research Establishment.

References

[1] Razvan, A., Bakis, C. E., Wagnecz, L., and Reifsnider, K. L., "Influence of Cyclic Load Amplitude on Damage Accumulation and Fracture of Composite Laminates," *Journal of Composites Technology and Research,* Vol. 10, 1988, pp. 3–10.

[2] Talreja, R., "Fatigue of Composite Materials: Damage Mechanisms and Fatigue Life Diagrams," *Proceedings,* Royal Society (London), Vol. A378, 1981, pp. 461–475.

[3] Dharan, C. K. H., "Fatigue Failure in Graphite Fibre and Glass Fibre Composites," *Journal of Materials Science,* Vol. 10, 1975, pp. 1665–1670.

[4] Talreja, R., "A Conceptual Framework for the Interpretation of Fatigue Damage Mechanisms in Composite Materials," *Journal of Composites Technology and Research,* Vol. 7, 1985, pp. 25–29.

[5] Martinez, G. M., Bainbridge, D. M. R., Piggott, M. R., and Harris, B., "The Compression Strength of Composites with Kinked, Misaligned and Poorly Adhering Fibres," *Journal of Materials Science,* Vol. 16, 1981, pp. 2831–2836.

[6] Piggott, M. R., "A Theoretical Framework for the Compressive Properties of Aligned Fibre Composites," *Journal of Materials Science,* Vol. 16, 1981, pp. 2837–2845.

[7] Lam, P. W. K. and Piggott, M. R., "The Durability of Controlled Matrix Shrinkage Composites I:

Mechanical Properties of Resin Matrices & Their Composites," *Journal of Materials Science,* Vol. 24, 1989, pp. 4068–4075.

[8] Lam, P. W. K. and Piggott, M. R., "The Durability of Controlled Matrix Shrinkage Composites II: Fatigue Properties of Carbon Fibre-Epoxy Copolymer Pultrusions," *Journal of Materials Science,* Vol. 24, 1989, in press.

[9] Lam, P. W. K. and Piggott, M. R., "The Durability of Controlled Matrix Shrinkage Composites III: Measurement of Damage During Fatigue," *Journal of Materials Science,* Vol. 24, 1989, in press.

[10] Rotem, A., "Stiffness Change of a Graphite-Epoxy Laminate Under Reverse Fatigue Loading," *Journal of Composites Technology and Research,* Vol. 11, 1989, pp. 53–58.

[11] Piggott, M. R., *Load Bearing Fibre Composites,* Pergamon, Oxford, 1980, p. 86.

[12] Piggott, M. R., *Load Bearing Fibre Composites,* Pergamon, Oxford, 1980, p. 74.

Yehia A. Bahei-El-Din[1]

Fracture of Fibrous Metal Matrix Composites Containing Discontinuities

REFERENCE: Bahei-El-Din, Y. A., **"Fracture of Fibrous Metal Matrix Composites Containing Discontinuities,"** *Composite Materials: Fatigue and Fracture (Third Volume), ASTM STP 1110*, T. K. O'Brien, Ed., American Society for Testing and Materials, Philadelphia, 1991, pp. 696–710.

ABSTRACT: The work of Dvorak et al. and Bahei-El-Din et al. on fracture of unidirectionally reinforced boron/aluminum specimens with a center notch is extended here for specimens with other types of discontinuities. The experimental results indicate that long, discrete plastic shear zones similar to those found at the notch tip are present in all specimens. The plastic zones grow from the discontinuity in the fiber direction. The measured fracture strength was not affected by geometry of the notch, it is only a function of the notch width/specimen width ratio. Finite element analysis of the specimens with center square hole provided the plastic zone length and the local stresses. The results verified that the criterion controlling fracture in center-notched specimens is applicable to other notch geometries. Namely, failure is controlled by a critical ratio of the largest principal stress to off-axis unnotched strength in the principal direction, in a small representative volume in the vicinity of the notch.

KEY WORDS: composite materials, fracture, notches, holes, plastic zone, fracture criterion, experiments, finite element analysis, fatigue (materials)

In their 1989 papers, Dvorak et al. [1] and Bahei-El-Din et al. [2] investigated the fracture behavior of center-notched unidirectionally reinforced boron/aluminum (B/Al) specimens and proposed a fracture initiation criterion. It was found that long, discrete plastic zones are present at the notch tip and extend in the fiber direction. The length of the plastic zones was from 3 to 17 times half the notch length. Similar plastic zones were found in a notched B/Al laminate by Post et al. [3]. These zones blunt the notch and cause a significant stress redistribution in the notched specimen.

Earlier studies of this subject focused on experimental measurement of the fracture strength of notched B/Al specimens of different geometries and evaluation of fracture toughness using linear elastic fracture mechanics [4–9]. In the presence of the large-scale yielding observed by Dvorak et al. [1] in notched B/Al composites, standard or modified linear elastic fracture mechanics approaches can not be used to predict the onset of fracture. The failure of these approaches to define a material property characterizing fracture in unidirectionally reinforced B/Al composites was clearly indicated by Reedy [7]. As suggested by Dvorak et al. [1] and Bahei-El-Din et al. [2], the onset of fracture must be inferred from analysis of the deformation field for each specimen geometry and the geometry of the plastic zone using continuum mechanics methods that admit plastic deformation of the composite.

The present work examines the fracture behavior of unidirectional B/Al composite speci-

[1] Associate professor, Structural Engineering Department, Cairo University, Giza, Egypt. Presently on leave at the Department of Civil Engineering, Rensselaer Polytechnic Institute, Troy, NY 12180.

mens containing notches of various geometries. We begin by describing the experimental procedure and present results pertaining to the plastic zones and the effect of the notch geometry on the fracture strength. This is followed by analysis of the local stress field in the vicinity of the notch and confirmation of the fracture criterion proposed by Dvorak et al. [1] and Bahei-El-Din et al. [2] for the notch geometries considered in the experimental part.

Experiments

Materials

The material used is a unidirectional boron/aluminum composite identical with that tested by Dvorak et al. [1]. It consists of six-ply 6061 aluminum panels reinforced with 0.142-mm continuous boron fibers. The panels were supplied by Amercom, Inc., Chatsworth, California, and received in the as-fabricated condition. The fiber volume fraction was 50 ± 3%. The thickness of the panels was 1.067 mm. All fracture specimens cut from the boron/aluminum panels were tested in the as-fabricated condition.

Specimen Geometry

Specimens 25.4 mm in width and 305 mm in length were cut from the composite panels with a diamond saw such that the fiber direction was along the length of the specimens. The length of the specimens was equal to the length of the panels and no saw cuts were made perpendicular to the fiber. A center hole or two edge notches were made in the specimens using the electrostatic discharge machining (EDM) technique. The edge notches were either perpendicular to the fiber or inclined by 45°. These notches were 0.2 mm wide. The holes were circular, square, or rectangular in shape and were cut symmetric with respect to the center of the specimen. Figure 1 shows the geometry of the specimens tested. The dimensions of the fracture specimens containing edge notches, circular, or square hole are given in Table 1. The dimensions of the specimens with rectangular holes are given in Table 2. Note that the width of the rectangular hole in all the specimens in Table 2 was constant whereas the length in direction of the fiber varied.

The ends of all specimens were bonded to 50.8-mm-long and 4-mm-thick annealed aluminum tabs. The free length of each specimen between grips was 203 mm. Prior to testing, four specimens, E6, D6, C5, and S6, Table 1, were selected for application of a fine bar code on one face of the specimen as described by Dvorak et al. [1]. The bar code consisted of 0.051 and 0.102 mm wide, parallel dark and bright lines arranged in a certain periodic sequence and photodeposited on the surface of the specimen perpendicular to the fiber direction. The deposited lines deformed with the specimen and revealed the geometry of the plastic shear zones [1].

All specimens were loaded in axial tension up to failure. Possible load eccentricities were limited to 0.05 mm by using an alignment table that permitted centering of the specimen and the grips with two micrometers [1]. Specimens with the bar code were photographed at several load levels.

Fracture Strength

Tables 1 and 2 list the strength magnitudes found for the fracture specimens. The overall ultimate stress at failure, σ_{ult}, and net ligament stress at failure, σ_{lig}, are given for each specimen. For comparison between the different specimen groups tested, the average ultimate stress,

TABLE 1—*Dimensions and results for 6061-F Al/B fracture specimens.*

Specimen	W^a, mm	$2c^b$, mm	$2c/W$	$\sigma_{ult}{}^c$, MPa	$\sigma_{lig}{}^d$, MPa
		TWO 90° EDGE NOTCHES			
E1	25.273	2.743	0.109	1007	1164
E2	25.273	2.819	0.112	1120	1260
E3	25.273	2.794	0.111	1020	1147
E4	25.451	7.518	0.295	779	1106
E5	25.222	7.722	0.306	790	1138
E6	25.349	12.725	0.502	569	1143
E7	25.375	12.624	0.498	594	1181
E8	25.375	17.958	0.708	385	1317
E9	25.324	17.805	0.703	359	1208
		TWO 45° EDGE NOTCHES			
D1	25.375	2.540	0.100	867	964
D2	25.375	2.540	0.100	915	1017
D3	25.197	4.724	0.187	829	1020
D4	25.603	4.851	0.189	796	981
D5	25.425	10.160	0.400	584	973
D6	25.298	10.135	0.401	631	1052
D7	25.425	15.342	0.603	408	1028
D8	24.790	15.494	0.625	390	1040
D9	25.044	19.787	0.790	251	1196
		CENTER CIRCULAR HOLE			
C1	25.222	1.295	0.051	1216	1282
C2	25.096	1.295	0.052	1154	1217
C3	25.298	5.088	0.201	829	1037
C4	25.502	5.088	0.200	795	993
C5	25.375	10.175	0.401	574	958
C6	25.197	10.175	0.404	568	953
C7	25.603	15.250	0.596	345	853
C8	25.298	15.250	0.603	374	941
		CENTER SQUARE HOLE			
S1	25.222	2.032	0.081	1038	1129
S2	25.857	2.032	0.079	1034	1122
S3	25.349	4.925	0.195	826	1026
S4	25.654	4.960	0.192	841	1040
S5	25.349	10.174	0.401	592	988
S6	25.502	10.176	0.399	605	1007
S7	25.400	10.166	0.401	592	987
S8	25.476	15.245	0.598	417	1038
S9	25.375	15.251	0.601	412	1033

[a] Specimen width.
[b] Notched width.
[c] Measured ultimate strength.
[d] Ligament stress $= \sigma_{ult}/(1 - 2c/W)$.

$\bar{\sigma}_{ult}$, and average ligament stress, $\bar{\sigma}_{lig}$, at failure are listed in Table 3. The specimen group indicated in the table refers to the specimens shown in Tables 1 and 2. The results are listed with respect to the $2c/W$ ratio for all specimen groups.

The results in Table 3 indicate that, in principle, the measured fracture strength is unaffected by the shape of the discontinuity. It is only a function of the $2c/W$ ratio. This is clearly indicated in Fig. 2 where the measured ultimate strength for all specimens listed in Tables 1 and 2 is normalized by the unnotched tensile strength of the B/Al composite and plotted versus

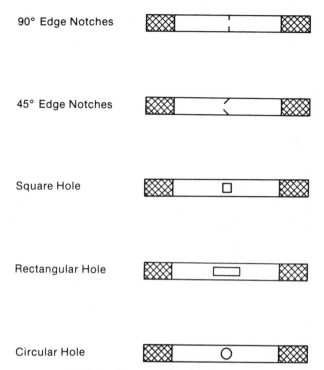

90° Edge Notches

45° Edge Notches

Square Hole

Rectangular Hole

Circular Hole

FIG. 1—*Geometry of specimens tested.*

the $2c/W$ ratio. Also included in Fig. 2 are the measured ultimate strength ratios reported by Dvorak et al. [1] for as-fabricated center-notched specimens of an identical B/Al composite. The fitting curve shown by the solid line was found using all the data points given in Fig. 2. The points fall into a narrow range and indicate that the relationship between fracture strength and the notched width/specimen width ratio ($2c/W$) can be represented by a single curve. This behavior was also observed in B/Al laminates [10,11].

TABLE 2—*Dimensions and results for 6061-F Al/B specimens containing center rectangular hole.*

Specimen	W^a, mm	$2c^b$, mm	$2d^c$, mm	$2c/W$	$\sigma_{ult}{}^d$, MPa	$\sigma_{lig}{}^e$, MPa
R1	25.273	10.211	5.131	0.404	579	972
R2	25.222	10.211	5.080	0.405	597	1003
R3	25.248	10.211	20.345	0.404	669	1123
R4	25.349	10.185	20.345	0.402	695	1162
R5	25.222	10.211	30.302	0.405	608	1021
R6	25.451	10.211	30.429	0.401	626	1045
R7	25.705	10.211	40.589	0.397	604	1002
R8	25.832	10.211	40.640	0.395	607	1004

[a] Specimen width.
[b] Hole width.
[c] Hole length.
[d] Measured ultimate strength.
[e] Ligament stress $= \sigma_{ult}/(1 - 2c/W)$.

TABLE 3—*Results of fracture tests on B/Al specimens.*

$2c/W$	Specimen Group	$\bar{\sigma}_{ult}$, MPa	$\bar{\sigma}_{lig}$, MPa
0.05	C	1185	1250
0.08	S	1036	1126
0.10	E	1049	1190
0.10	D	891	991
0.20	D	813	1001
0.20	C	812	1015
0.20	S	834	1033
0.30	E	785	1122
0.40	D	608	1013
0.40	C	571	956
0.40	S	596	994
0.40	R($d/c = 0.5$)	588	988
0.40	R($d/c = 2.0$)	682	1143
0.40	R($d/c = 3.0$)	617	1033
0.40	R($d/c = 4.0$)	606	1003
0.50	E	582	1162
0.60	D	399	1034
0.60	C	360	897
0.60	S	415	1036
0.70	E	372	1263
0.80	D	251	1196

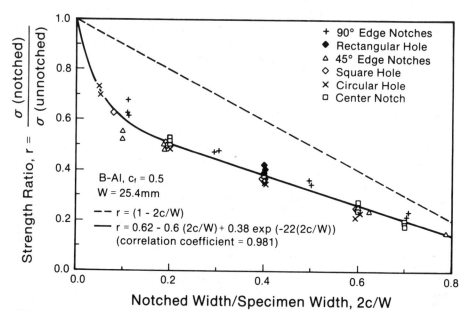

FIG. 2—*Experimentally measured relative fracture strength reduction caused by various shapes of notches in 25.4 mm wide B/Al specimens.*

The dashed line shown in Fig. 2 indicates the variation of the net ligament strength ratio with $2c/W$. The reduction from the net ligament strength is due to stress concentrations caused by the discontinuity. Dvorak et al. [1] attributed the reduction in net ligament strength in center-notched B/Al composites to local stresses induced by plastic zones emanating from the notch tips in the direction of the fiber longitudinal axis. The development of plastic zones in the B/Al specimens tested here is examined next.

Plastic Zones

The four specimens, E6, D6, C5, and S6, with the bar code were photographed at several load levels up to failure. Figure 3 shows the bar code distortions found in Specimen S6 at 0.968 of ultimate load. Similar bar code distortions were present in the photographs of Specimens E6, D6, and C5. As was found by Dvorak et al. [1] for center-notched specimens, plastic shear zones were present in all specimens. The zones are discrete and emanate from the notch tips, square hole corners, or circular hole circumference near the unnotched ligament and extend in the fiber direction. The actual length of the plastic zones cannot be measured from distortions of the bar code due to its limited resolution. However as shown by Bahei-El-Din et al. [2] and the subsequent analytical results, the plastic zone length is much larger than the notched width.

The photographed bar code distortions permitted measurement of the plastic zone width and the plastic shear strain magnitudes in the zones. The method used begins by magnifying the deformed bar code photographs and locating two points on each bar code line where a sharp bend from the originally straight configuration occurred [1]. In this way, the boundary of the plastic zone was determined, and the relative displacement in the fiber direction between each pair of boundary points was measured. Finally, the longitudinal shear strains in the plastic zone were computed at each bar code line from the measured relative displacement and the distance perpendicular to the fibers between the plastic zone boundary points.

A sample of these measurements is shown in Fig. 4 for Specimen S6 at different load levels. The curves in Fig. 4a indicate the left and right boundaries of the plastic zone. They terminate in direction of the x_2-axis when the shear strains are smaller than the 1% shear strain resolution of the bar code [1]. Since the elastic limit shear strain of the B/Al composite is smaller than 1%, the actual length of the plastic zones is larger than that implied by Fig. 4a. For increasing overall load, the visible plastic zone length increases, whereas the zone width decreases. In any case, the width of the plastic zones found in all notched specimens before fracture did not exceed 0.4 mm near the notch and 1.0 mm at the observed end of the plastic zone.

The decrease in the plastic zone width with increasing overall load seen in Fig. 4a indicates a change in the plastic deformation mechanism from the matrix-dominated mode (MDM), in which plastic deformation occurs by slip on matrix planes parallel to the fiber, to the fiber-dominated mode (FDM) with plastic slip occurring on many planes that may intersect the fibers [12]. The experiments of Dvorak et al. [13] on a unidirectional B/Al composite verified the existence of these two modes. Since the bar code lines are perpendicular to the fiber, they deform only where MDM plastic deformation of the composite takes place. Consequently, FDM plastic deformation may exist but not be detected by the bar code technique. However, the plastic strains developing under the fiber-dominated mode are much smaller than the elastic strains [13], and as such the FDM response of the composite can be approximated by the elastic response.

The measured relative displacements across the plastic zones are shown in Fig. 4b. The relative displacements computed with the subsequent analysis are also shown. The longitudinal shear strains computed from Figs. 4a and b appear in Fig. 4c. As expected, the shear strains increase for increasing applied load. The largest shear strains evaluated for all specimen geometries tested here were in the order of 0.15 to 0.5.

⇧ load direction

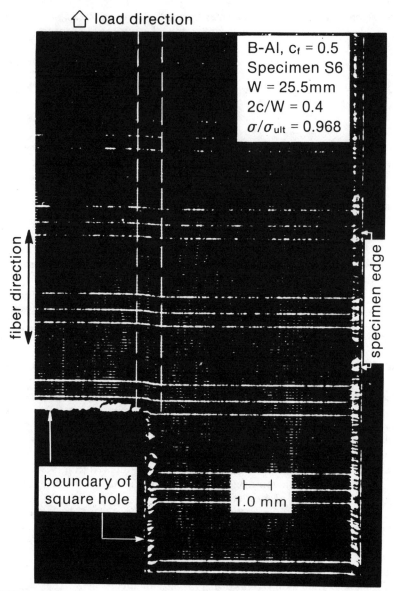

FIG. 3—*Distorted parts of the bar code found on Specimen S6 at 0.975 of ultimate load.*

Analysis

Modified Finite Element Analysis

To evaluate the local stresses in tension specimens containing discontinuities, we utilized the modified finite element analysis employed by Bahei-El-Din et al. [2]. The solution of a specific domain under uniaxial tension was obtained by the combination of two solutions. A finite element solution was used to find the stresses in specific specimens for their actual geometry, and an analytical solution was used to evaluate the local fields where large stress gradients

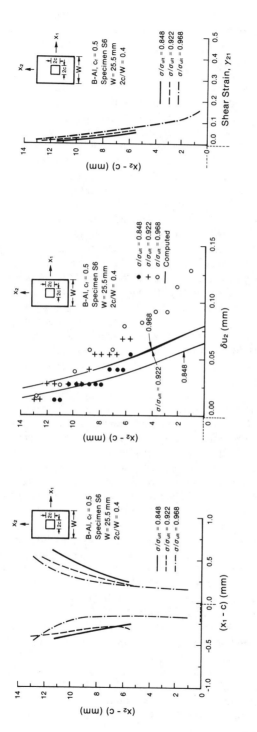

FIG. 4—*Plastic zone measurements found from bar code distortions on loaded Specimen S6. (a) plastic zone shapes, (b) relative displacements across the width of the plastic zones, and (c) shear strains.*

exist, for example at the notch tip or corner of a square hole. This decomposition is illustrated in Fig. 5 where the original geometry of a typical specimen with a center rectangular hole is modified to reflect the existence of the plastic shear zones observed experimentally. Inasmuch as large shear strains exist in the plastic zones, the shear stress in the zones has a constant magnitude equal to the flow shear stress, τ^*, in most of the plastic zone length except near the zone end away from the hole. The magnitude of τ^* used in the analysis was 96 MPa [1]. In Fig. 5b, the shear stress, τ^*, is removed within a small length $\pm \bar{l}$ measured from the hole corners leaving short vertical slots at each corner of the hole. The overall stress, $\bar{\sigma}$, was applied in small increments and the solution was obtained using the finite element method.

The finite element domain is shown in Fig. 6a. It represents one quadrant of the geometry shown in Fig. 5b. Displacement boundary conditions derived from symmetry of the specimen geometry and the applied load with respect to the planes $x_1 = 0$ and $x_2 = 0$ were applied together with a uniform displacement condition at the boundary $x_2 = L$, where L = half the specimen length. The finite element solution was simplified by limiting plastic deformation in the domain to one row of elements that were designated to represent the plastic shear zones. All other elements in the domain were assumed to remain elastic. This was suggested by the deformation of the bar code indicated earlier and the existence of the MDM and FDM deformation modes in B/Al composites [12,13]. The stress field computed for the specimens with a square hole using the finite element domain in Fig. 6a revealed no violations of the MDM yield condition except in the notched region at the vicinity of the hole. As indicated in Ref 2, the stresses calculated in the unnotched ligament ahead of the notch tip in center-notched specimens were not affected when the width of the discrete plastic zone in the finite element domain was increased by a factor of two. In the unnotched ligament ($c \leq x_1 \leq W/2, 0 \leq x_2 \leq L$), the stresses satisfied the FDM yield condition. In this mode, the composite response is very stiff and can be regarded as elastic. An example of the finite element mesh at the corner of a rectangular hole is shown in Fig. 6b.

The finite element solution was found using the ABAQUS program. In the elastic range, the composite was regarded as a homogeneous, transversely isotropic medium. Elastic-plastic response of the plastic zone elements was specified using Hill's anisotropic yield condition after adjusting the size of the anisotropic yield surface of the composite such that it is in contact with the bimodal yield surface [12] under axial tension, transverse tension, and longitudinal shear. Strain hardening in the plastic zone elements was described by the isotropic hardening option of ABAQUS. The large strain option of ABAQUS was also specified in the analysis.

The second part of the solution, Fig. 5c, was found by approximating the local stresses in the unnotched ligaments by the stresses found from the solution of the boundary value problem indicated in Fig. 7; a solitary crack of length $2\bar{R}$, equal to the width of the hole in the x_2 direction plus the total length of the plastic zone, R, on either side of the hole, in an infinite transversely isotropic elastic medium. The crack is subjected to a shear load within the distance $\bar{l}$. It consists of a linearly increasing part from $\tau = 0$ at $x_2 = \pm u$ to $\tau = \tau^*$ at $x_2 = \pm v$ and a constant part at $\tau = \tau^*$ within the length $\pm (\bar{l} - v)$, where $\bar{l} = \bar{l} + u$. The length $\pm u$ represents the width of the machined imperfection and is empty of the shear load. For example, $u = 0.1$ mm in a center notch, $u = d$ in a center rectangular hole, and $u = c$ in a center square hole, where c is half the side length of the hole. The distance, $\pm (v - u)$ is the opening displacement in the x_2 direction found in the finite element solution at the intersection of the plastic zone and the perimeter of the imperfection. The solution of the problem shown in Fig. 7 can be found in Ref 2. The magnitude of the length, $\bar{l}$, was selected so as to make the local stresses independent of $\bar{l}$. Actual values of $\bar{l}$ were selected much smaller than specimen dimensions, plastic zone length, and notch dimensions, but larger than the width of the representative volume element (see the next section) over which the stresses in the unnotched ligament were averaged.

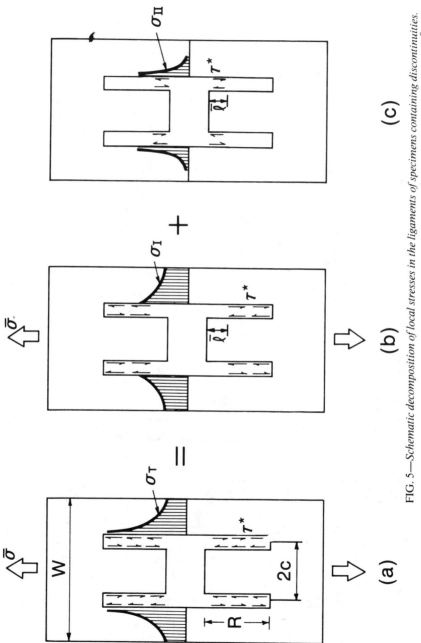

FIG. 5—*Schematic decomposition of local stresses in the ligaments of specimens containing discontinuities.*

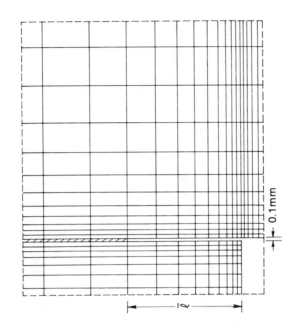

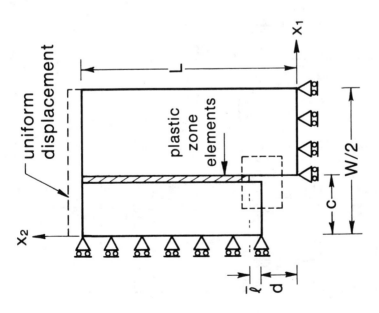

FIG. 6—*Finite element domain, mesh, and boundary conditions of a specimen containing a center rectangular hole.*

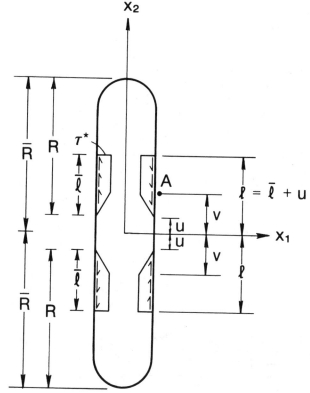

FIG. 7—*The geometry and load distribution in evaluation of the local fields.*

Results

The modified finite element analysis was performed by Bahei-El-Din et al. [2] on center-notched specimens. Here, the analysis was performed on Specimens S3 through S9 that contain a center square hole. The finite element part of the solution, Fig. 5b, provided the total length of the plastic zones that could not be measured from the bar code deformation. The end of the plastic zone away from the hole was determined by comparing the computed longitudinal shear strain in the plastic zone elements to the elastic limit longitudinal shear strain of the B/Al composite. The plastic shear zone terminates at the location where the shear strain equals the elastic limit value. As was found for center-notched specimens, the plastic shear zones computed for the specimens with center square holes were much longer than the notched width. Specifically, the length of the plastic zone measured from the corner of the square hole was 24 mm for Specimens S3 and S4, and 34 mm for Specimens S5 through S9.

The local stresses were evaluated in Specimens S3 through S9 to examine the fracture criterion suggested in Refs 1 and 2. They were found by superposition of the stress components given by the finite element solution and the corresponding stresses computed from the local problem, Figs. 5c and 7. Following Bahei-El-Din et al. [2], the total stresses were averaged within the length $c \leq x_1 \leq w_{RV} + c$, where w_{RV} is the width of a representative volume element of the composite. The representative volume element was found from a periodic hexagonal array model of the composite [14]. This provided $w_{RV} = D(2\pi/3\sqrt{3}c_f)^{1/2}$, where D is the fiber diameter and c_f is the fiber volume fraction. For the B/Al system under consideration, $w_{RV} = 220$ μm. The local normal stress, σ_{22}, in the axial x_2 direction and the longitudinal shear stress, σ_{21}, averaged over the width, w_{RV}, are denoted, respectively, by σ_{RV} and τ_{RV}.

For each specimen, the stress averages, σ_{RV} and τ_{RV}, corresponding to the measured overall fracture loads were used to evaluate the magnitude, σ_{max}, and direction, θ, of the maximum principal stress. Next, the unnotched off-axis tensile strength, σ_{ult}, of the composite corresponding to loading in the direction given by θ measured from the fiber direction was interpolated from the experimental data given in Ref 2 for $\theta = 0$, 45, and 90°. The largest magnitude of the normal stress ratio $\sigma_{max}/\sigma_{ult}$ among all points in the vicinity of the square hole corner was found and plotted versus $2c/W$. Invariably, the largest $\sigma_{max}/\sigma_{ult}$ ratio was found at the corner of the opened hole (Point A in Fig. 7). The result together with the data given by Bahei-El-Din et al. [2] for the center-notched as-fabricated specimens tested by Dvorak et al. [1] and Poe and Sova [6] are shown in Fig. 8 by the upper shaded band. It is seen that the normal stress ratio for all specimens with either discontinuity type fall into a narrow band with scatter sim-

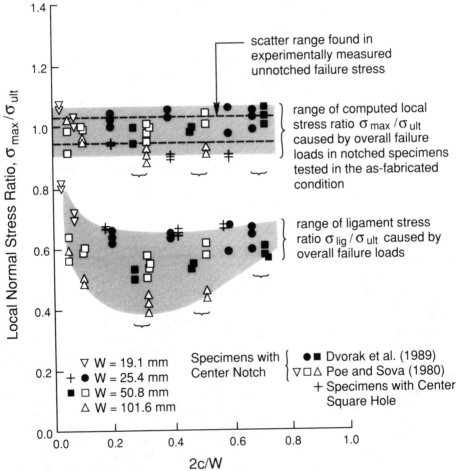

FIG. 8—*The critical stress/strength ratios computed in the unnotched ligament, at experimentally measured fracture loads.*

ilar to that found in unnotched strength magnitudes. This suggests that the onset of fracture in those specimens is controlled by a critical ratio of the local principal stress and the ultimate strength in the principal direction as proposed in Refs *1* and *2*. Since the overall fracture strength was independent of the notch geometry, Fig. 2, the fracture criterion applies to all as-fabricated B/Al fracture specimens. As suggested in Fig. 8, the critical stress ratio is approximately 1.0 in this case.

The lower shaded band in Fig. 8 shows the data points for the fracture specimens based on the ligament stress, σ_{lig}, corresponding to the experimental fracture strength, Table 1. The difference between the points in the upper band and the corresponding points in the lower band in Fig. 8 is the contribution of the plastic shear zone to the local stress. This contribution to the total local stress average is responsible for the strength reduction from the net ligament strength in notched specimens found in Fig. 2.

Conclusions

Long, plastic shear zones similar to those found in B/Al specimens containing a center notch were found in specimens containing other types of discontinuities such as edge notches and center holes. The plastic shear strain in the zone was in the order of 0.15 to 0.50 at the notch prior to fracture.

The measured fracture strength was found to be independent of the geometry of the discontinuity. It is only a function of the ratio between the notched width and the specimen width. Failure in the fracture B/Al specimens is controlled by a stress ratio between the largest principal stress and the ultimate strength in the principal direction. This was verified by calculation of the normal stress ratio for the specimens with square holes using a modified finite element analysis that reflects the presence of the plastic shear zones. Within an experimental error, the ratio computed for the specimens with square holes as well as that computed for center-notched specimens is constant for all notch width/specimen width ratios.

Acknowledgments

This work was supported by a grant from the Mechanics Division of the Office of Naval Research to Rensselaer Polytechnic Institute. Dr. Yapa Rajapakse served as program monitor. The author is indebted to Dr. George Dvorak for useful suggestions and discussions. Dr. Himanshu Nigam tested part of the specimens with two edge notches and Mr. Jer-Fang Wu assisted in the finite element calculations.

References

[1] Dvorak, G. J., Bahei-El-Din, Y. A., and Bank, L. C., *Engineering Fracture Mechanics,* Vol. 34, 1989, pp. 87–104.
[2] Bahei-El-Din, Y. A., Dvorak, G. J., and Wu, J. F., *Engineering Fracture Mechanics,* Vol. 34, 1989, pp. 105–123.
[3] Post, D., Czarnek, R., Joh, D., Jo, J., and Guo, Y., *Journal of Composite Technology & Research,* Vol. 9, 1987, pp. 3–9.
[4] Awerbuch, J. and Hahn, H. T., *Journal of Composite Materials,* Vol. 13, 1979, pp. 82–107.
[5] Jones, W. F. and Goree, J. G. in *Mechanics of Composite Materials 1983,* G. J. Dvorak, Ed., American Society of Mechanical Engineers, AMD 58, 1983, pp. 171–177.
[6] Poe, C. C. and Sova, J. A., "Fracture Toughness of Boron/Aluminum Laminates with Various Proportions of 0° and ±45° Plies," NASA Technical Paper 1707, National Aeronautics and Space Administration, Washington, DC, 1980.
[7] Reedy, E. D., *Journal of Composite Materials Supplement,* Vol. 14, 1980, pp. 118–131.
[8] Reedy, E. D., *Journal of Composite Materials,* Vol. 16, 1982, pp. 495–509.

[9] Wright, M. A. and Welch, D., *Fibre Science and Technology,* Vol. 11, 1978, pp. 447–461.
[10] Johnson, W. S., Bigelow, C. A., and Bahei-El-Din, Y. A., "Experimental and Analytical Investigation of the Fracture Process of Boron/Aluminum Laminates Containing Notches," NASA Technical Paper 2187, National Aeronautics and Space Administration, Washington, DC, 1983.
[11] Mar, J. W. and Lin, K. Y., *Journal of Composite Materials,* Vol. 11, 1977, pp. 405–421.
[12] Dvorak, G. J. and Bahei-El-Din, Y. A., *Acta Mechanica,* Vol. 69, 1987, pp. 219–241.
[13] Dvorak, G. J., Bahei-El-Din, Y. A., Macheret, Y., and Liu, C. H., *Journal of the Mechanics and Physics of Solids,* Vol. 36, 1988, pp. 655–688.
[14] Dvorak, G. J. and Teply, J. L. in *Plasticity Today: Modelling, Methods, and Applications, W. Olszak Memorial Volume,* A. Sawczuk and V. Blanchi, Eds., Elsevier, Amsterdam, 1985, pp. 623–642.

Peter Kantzos,[1] *Jack Telesman,*[1] *and Louis Ghosn*[2]

Fatigue Crack Growth in a Unidirectional SCS-6/Ti-15-3 Composite

REFERENCE: Kantzos, P., Telesman, J., and Ghosn, L., **"Fatigue Crack Growth in a Unidirectional SCS-6/Ti-15-3 Composite,"** *Composite Materials: Fatigue and Fracture (Third Volume), ASTM STP 1110,* T. K. O'Brien, Ed., American Society for Testing and Materials, Philadelphia, 1991, pp. 711–731.

ABSTRACT: An investigation was conducted to characterize and model the fatigue crack growth (FCG) behavior of a SCS-6/Ti-15-3 metal matrix composite. Part of the study was conducted using a fatigue loading stage mounted inside a scanning electron microscope (SEM). This unique facility allowed high magnification viewing of the composite fatigue processes and measurement of the near crack tip displacements. The unidirectional composite was tested in the $[0]_8$ (that is, longitudinal) and $[90]_8$ (that is, transverse) orientations. For comparison purposes, unreinforced matrix material produced by the identical process as the reinforced material was also tested.

The results of the study reveal that the fatigue crack growth behavior of the composite is a function of specimen geometry, fiber orientation, and the interaction of local stress fields with the highly anisotropic composite. In the case of $[0]_8$ oriented single edge notch (SEN) specimens and $[90]_8$ oriented compact tension (CT) specimens, the crack growth was normal to the loading direction. However, for the $[0]_8$ CT specimens, the crack grew mostly parallel to the loading and the fiber direction.

The unusual fatigue behavior of the $[0]_8$ CT specimens was attributed to the specimen geometry and the associated high tensile bending stresses perpendicular to the fiber direction. These stresses resulted in preferential cracking in the weak interface region perpendicular to the fiber direction. The interface region, and in particular the carbon coating surrounding the fiber, proved to be the composite's weakest link.

In the $[0]_8$ SEN, the crack growth was confined to the matrix, leaving behind unbroken fibers that bridged the cracked surfaces. As the crack grew longer, more fibers bridged the crack resulting in a progressive decrease in the crack growth rates and eventual crack arrest. The actual near-crack-tip displacement measurements were used in a proposed formulation for a bridging-corrected effective crack driving force, ΔK_{eff}. This parameter was able to account for most of the experienced bridging and correlated the $[0]_8$ SEN fatigue crack growth data reasonably well.

KEY WORDS: metal matrix composite, fatigue crack growth, crack bridging, fatigue mechanisms, composite materials, fracture, fatigue (materials)

The new generation of aerospace vehicles will require materials capable of withstanding high temperatures while retaining a high stiffness under relatively high loads. Continuous fiber, metal matrix composites (MMC) are candidate materials for such applications. One of these candidate materials is a Ti-15V-3Cr-3Al-3Sn matrix reinforced by continuous silicon-carbon (SiC) fibers. Before these materials become widely utilized in aerospace applications, their fatigue behavior has to be understood and reliable life prediction methodology has to be developed.

---

[1] Materials engineers, respectively, NASA Lewis Research Center, Cleveland, OH 44135.

[2] Structures engineer, Sverdrup Technology-NASA Lewis Research Center Group, Cleveland, OH 44135.

Relatively little information is available concerning the fatigue behavior of continuous fiber-reinforced composites. The objective of this study is to broaden this knowledge by investigating in detail the fatigue behavior of the previously mentioned composite system. The emphasis of the study was twofold: (1) identification of the fatigue crack initiation and propagation damage processes with special attention focused on identifying the composite constituent most susceptible to fatigue damage; and (2) identification of the appropriate crack driving force for this composite. To achieve these goals, part of the study was conducted using a fatigue loading stage mounted inside a scanning electron microscope (SEM). This unique facility, developed at NASA Lewis [1], allows real-time dynamic and static viewing of the fatigue crack initiation and crack propagation processes.

Material

The composite used in this study is a Ti-15V-3Cr-3Al-3Sn (Ti-15-3) alloy matrix reinforced by 145 μm average diameter, continuous SiC (SCS-6) fibers. All the composite specimens used in this study were obtained from a single, eight-ply, unidirectionally reinforced panel manufactured by Textron Specialty Materials Division. The panel size was 30 by 30 by 0.21 cm. Specimens were also machined out of a 30 by 30 by 1 cm panel of unreinforced matrix, using the same Ti-15-3 matrix foil and same consolidation techniques employed in composite production.

The matrix is a metastable beta phase alloy, chosen because it can be cold rolled to very thin sheets. This enables a more cost-effective processing of the composite. Sample micrographs of the reinforced composite are shown in Fig. 1. The SiC fiber is surrounded by a complex multilayer structure (Fig. 1b) that consists mainly of a multilayer carbon coating approximately 3 μm thick and an approximately 0.5 to 1-μm-thick reaction zone consisting mainly of brittle intermetallic phases [2]. The combination of the carbon layers and the reaction zone will henceforth be referred to as the interface region.

Heat Treatment

Specimens were tested both in the as-received and heat-treated condition. The heat treatment consisted of 24 h at 700°C in vacuum. The purpose of the heat treatment was to precipitate out the alpha phase. The heat treatment did not produce any additional noticeable fiber/matrix interactions and the reaction zone remained unchanged [3].

Mechanical Testing

Fatigue crack growth (FCG) testing was performed using single edge notch (SEN) and compact tension (CT) specimens. The SEN specimens were tested in the [0]$_8$ orientation only. The CT composite specimens were tested in two orientations: (1) longitudinal, [0]$_8$, with fibers oriented parallel to the loading direction; and (2) transverse, [90]$_8$, with fibers perpendicular to the loading direction. Unreinforced matrix specimens were also tested using CT specimens. The geometry of the specimens was as follows: the width (w) of the SEN specimens was 5 mm, the thickness was 2 mm, and the initial crack length was 1 mm. The width (w) of the CT specimens was 22.5 mm, the thickness was 2 mm, and initial crack length was 4.5 mm.

All machining was done using diamond tip tools. Prior to testing, all specimens were polished to enhance crack detection.

The test matrix is shown in Table 1. The SEN specimens were tested in the fatigue loading stage mounted inside an SEM. The testing was conducted at room temperature at a load ratio (R) of 0.1 with the maximum gross stress being 215 MPa. Test frequency was 5 Hz. Tests were

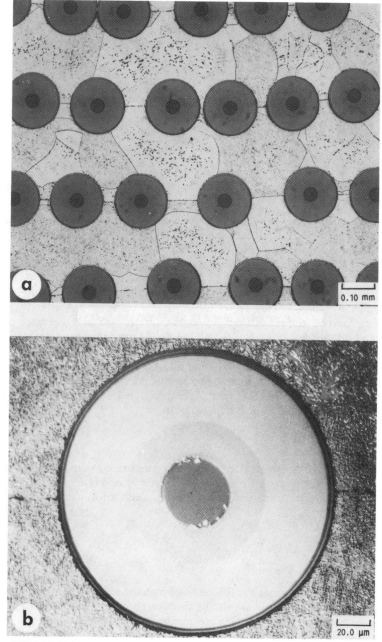

FIG. 1—*SCS-6/Ti-15-3 composite: (a) as-received etched to reveal microstructure, and (b) detailed view of heat-treated SCS-6 fiber.*

TABLE 1—*Test matrix.*

| Material Tested | R Ratio | Specimens Tested | | Maximum Load, KN |
		As-Received	Heat-Treated	
CT, [0]₈	0.1	2	1	1.23
SEN, [0]₈	0.5	2	. . .	2.24
CT, [90]₈	0.1	1	1	1.17
CT, unreinforced matrix	0.1	1	1	1.17
CT, unreinforced matrix	0.5	2	. . .	2.34

put on frequent holds at varied loads to obtain high magnification photomicrographs. From the micrographs, measurements of the near-crack-tip displacements were obtained as a function of the applied load and crack length. The resolution of the measurements is a function of the magnification of the individual micrograph with the typical resolution being approximately $\pm 0.5\ \mu$m. An extensometer was mounted in the crack mouth of the specimen to determine the changes in the compliance. The vacuum inside the SEM was approximately 1×10^{-6} torr.

Testing of the CT specimens was performed at room temperature and ambient environment using a computer controlled, closed loop, servohydraulic machine. The tests were performed using either a constant load range to determine the intermediate and higher crack growth rate regimes or a load shedding procedure per ASTM Test Method for Measurements of Fatigue Crack Growth Rates (E 647-88a) to determine the near-threshold regime. A test frequency of 5 Hz was used with load ratios of $R = 0.5$ and 0.1. A computerized data acquisition system was used to determine the crack length by measuring the specimen compliance through an extensometer mounted in the crack mouth of the specimen. Data were also gathered through periodical optical readings to ensure accurate results.

Post-Failure Analysis

Following each test, extensive metallography and fractography were performed in order to evaluate fatigue failure processes. Each sample fracture surface was examined in the scanning electron microscope, with some samples also being sectioned to determine the extent of damage.

Results

SEN Specimens, [0]₈ Orientation

The crack growth behavior of the [0]₈ SEN specimens tested in the in situ loading stage resulted in the crack growing primarily in the direction normal to the applied load. The crack length versus cycles plot (Fig. 2) reveals a progressive decrease in the fatigue crack growth rate with an increase in the crack length followed by an eventual crack arrest.

The high magnification observations of the fatigue processes revealed that crack initiation occurred in the first few cycles. The initiation occurred by a failure of the partly machined fibers at the notch tip (Fig. 3). The crack front subsequently propagated into the matrix ligament. Crack propagation in the matrix was always associated with formation of slip bands in the cyclic plastic zone ahead of the crack tip. The stable fatigue crack growth was mainly con-

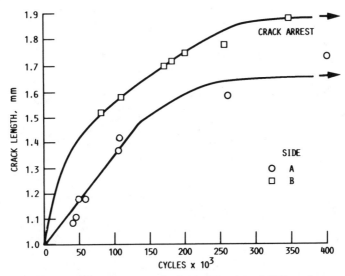

FIG. 2—*Crack length versus cycles for the* [0]$_8$ *oriented SEN specimen.*

fined to the matrix leaving in its wake unbroken SCS-6 fibers to bridge the two cracked surfaces.

A sequence of high magnification micrographs revealing the crack growth process observed for the SEN specimens is shown in Fig. 4. As shown in the first micrograph in the sequence (Fig. 4a), the main crack front has already approached a fiber and has started to grow around it. The reappearance of the crack front from behind the fiber to the outer specimen surface was preceded by the formation of slip bands (Fig. 4a). As the crack front approached the outer surface, the slip bands became more pronounced (Fig. 4b). A few thousand cycles later, the leading segments of the crack front broke through the outer surface (Fig. 4c). These crack segments were visible at maximum load (Fig. 4c); however, as the load was released the crack segments closed making it impossible to detect the crack tip region, even at high magnifications (Fig. 4d). After further crack growth in the matrix ligament (Fig. 4e), the entire ligament failed leaving in its wake unbroken fibers to bridge the crack. Once the entire matrix ligament between the two fibers was broken, a significant increase in the crack opening displacements

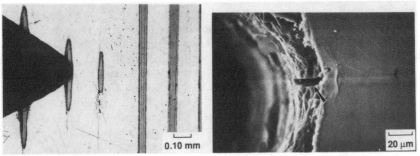

FIG. 3—*Crack initiation in the* [0]$_8$ *oriented SEN specimen:* (left) *prior to testing, and* (right) *after 150 cycles.*

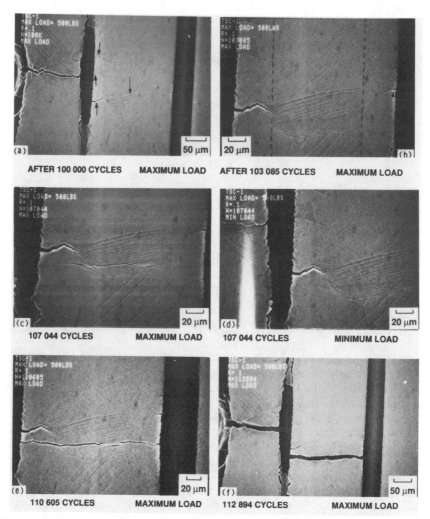

FIG. 4—*Crack growth sequence in the* [0]$_8$ *SEN specimen resulting in fiber bridging of the crack wake.*

(CODs) in the matrix ligament occurred due to the residual axial tensile stresses present in the matrix (Fig. 4f). At that point, the cracked matrix ligament remained open during the entire loading cycle. The behavior just described continued throughout the test until crack arrest was reached, resulting in four unbroken fiber rows bridging the two cracked surfaces (Fig. 5).

In cases where the fibers were partly exposed during the polishing process, some secondary cracking initiated at the fiber/matrix interface ahead of the crack tip. This secondary cracking was confined predominantly to the near surface region in the immediate vicinity of the exposed fiber. There was very little interaction observed between these small surface cracks and the main crack. Overall, their effect on the crack growth behavior of the main crack was negligible.

Using high magnification micrographs of the near-crack-tip region, CODs were measured as a function of the applied load and the distance behind the crack tip, t. One of the many data sets obtained is shown in Fig. 6. Shown in the figure is the measuerd COD range for a bridged

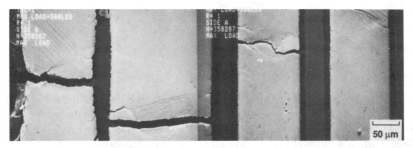

FIG. 5—*General appearance of the fatigued* [0]$_8$ *SEN specimens as observed in loading stage inside an SEM (358 287 cycles, maximum load).*

crack as a function of distance behind the crack tip, $\Delta u(t)_B$, and analytically derived COD range for an unbridged SEN specimen, Δu_{NB}, at the same applied stress intensity range, ΔK_{app}. The COD range, $\Delta u(t)_{NB}$, for an unbridged SEN specimen was calculated by the use of weight functions [4] as a function of distance behind the crack tip using the following integral equation

$$\Delta u(t)_{NB} = \frac{1}{E} \int h(t, \pm 0, c)\, \Delta K_{app}\, dc \qquad (1)$$

where E is the modulus of the composite, h is the Bueckner weight function, c is the crack length, and dc is the incremental crack length. The difference between the measured CODs and the predicted CODs for an unbridged SEN specimen is significant (Fig. 6) and can be used to quantify fiber bridging effect as will be shown later in this paper.

The extensometer mounted in the crack mouth of the specimen showed no apparent change in the compliance throughout the entire duration of the test. In comparison, according to the

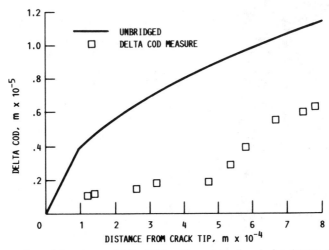

FIG. 6—*Comparison of the measured delta CODs with analytically calculated CODs for an unbridged SEN specimen.*

calculations performed, an unbridged SEN specimen would have been expected to experience an almost three-fold increase in the compliance for the amount of crack growth that has taken place.

CT Specimens, [0]₈ Orientation

In contrast to the SEN specimens, the overall direction of fatigue failure for $[0]_8$ oriented CT specimens was parallel to the loading direction and perpendicular to the machined notch, as shown in Fig. 7. The FCG data for the $[0]_8$ specimens are shown in Fig. 8 in the form of crack length (a) versus number of cycles (N) (Fig. 8a) and crack growth rate (da/dN) versus a (Fig. 8b). The data were not analyzed in terms of the applied ΔK due to the lack of K_I and K_{II} solutions for a crack in a CT specimen growing parallel to the loading direction. The FCG data, shown in Fig. 8, do reveal an increase of the FCG rate with an increase in the crack length indicating a probable increase of the crack tip stress field with crack growth.

A fractographic evaluation was performed to reveal the fatigue failure processes that resulted in this rather unusual behavior. The first row of fibers at the machined notch that were partly damaged by the machining process (Fig. 9) failed preferentially. The crack front continued for a short distance to grow in the plane of the first fiber row parallel to the loading direction (Fig. 7) before turning and growing perpendicular to the loading direction through the matrix ligament. As further seen in Figs. 7 and 9, after the crack front reached the undamaged second row of fibers, it again started to grow along the loading direction. After the crack reached a critical length, the specimens failed through debonding of the fibers and tearing of the matrix.

The stress field of the main crack resulted in the initiation of number of secondary cracks ahead of the crack tip (Fig. 10). As shown in the figure, the cross section of the specimen revealed a multitude of small secondary cracks in various stages of coalescence. Careful exam-

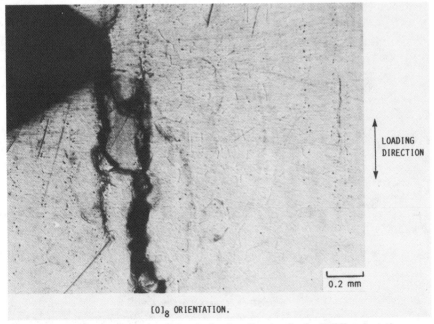

LOADING DIRECTION

0.2 mm

[0]₈ ORIENTATION.

FIG. 7—*Crack growth parallel to the loading direction for the CT* [0]₈ *orientation.*

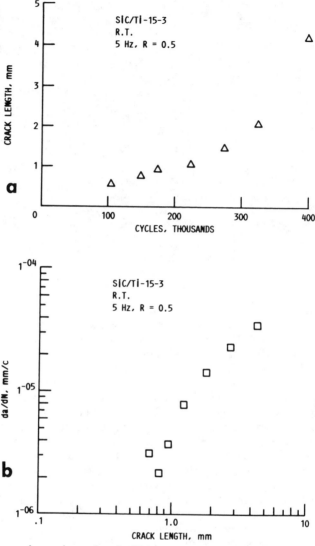

FIG. 8—*Fatigue crack growth rate data for CT* [0]₈ *specimens: (a) crack length versus cycles and (b) crack growth rate versus crack length.*

ination of the figure showed that the as yet uncoalesced cracks have all initiated in the interface region. Upon further fatigue loading, these small interface cracks propagated into the matrix ligaments and joined.

The microcracking appears to have initiated typically in the fiber/matrix interface region. A closer examination revealed a somewhat greater tendency for the cracks to initiate in the carbon coating of the fibers (Fig. 11) rather than the reaction zone adjacent to the coating. As shown in Fig. 11a, a crack presumably initiated between the carbon coating layers, cracked through the reaction zone and entered the matrix (crack width in the matrix is exaggerated due to the preferential etching of the matrix material adjacent to the crack). Irregardless of the

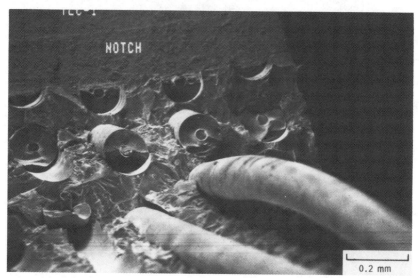

FIG. 9—*Crack growth in the* [0]₈ *CT specimen. Direction of loading parallel to fiber direction.*

precise crack initiation location, the microcracks usually tended to encompass the carbon coatings as well as the brittle intermetallic reaction zone (that is, the entire interface region). Post-failure analysis revealed carbon coating regions adhering to the matrix after the fiber has broken away, again indicating that cracking must have occurred between the fiber and the coating (Fig. 11b). These results support recent findings of Gabb et al. [5] who also identified the carbon coating as a site for preferential crack initiation. Johnson et al. [6], while not differentiating between the carbon layer and the reaction zone, also suggested that the weak interface region is responsible for degradation of fatigue life of this composite.

Post-failure fractographic analysis of the CT [0]₈ composite revealed that the interface region was not only the site for preferential crack initiation but was also the controlling factor that influenced crack propagation behavior. Throughout the test, the crack growth occurred preferentially in the interface region. The crack growth process consisted of the continuous cracking of the interface region at or ahead of the crack tip, followed by crack growth into the matrix ligaments and subsequent crack coalescence. This phenomenon is illustrated in Fig. 12. In particular, take note in the fractograph (Fig. 12b) of a high da/dN region just prior to final failure, and that stable crack growth has occurred in the areas adjacent to the interface. However, the remaining part of the matrix ligament failed through void coalescence during the catastrophic failure. This fractograph clearly shows the direction of localized crack growth to be from the interface region to the matrix ligament.

The crack propagation within the matrix was transgranular in nature (Fig. 12). The active fatigue failure mechanisms in the matrix included evidence of striation formation, slip band deformation, and some cleavage. Some secondary cracking was present in the matrix foil laminate boundaries perpendicular to the main crack front.

CT Specimens, [90]₈ Composite and Unreinforced Matrix Material

Contrary to the behavior of the CT [0]₈ specimens, for both the [90]₈ composite and the unreinforced matrix material CT specimens, the crack growth direction was perpendicular to the applied load. The comparison of the FCG behavior for the two materials is shown in Fig.

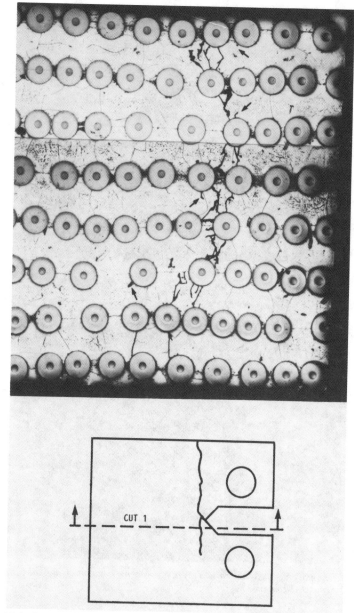

FIG. 10—*Multiple crack initiation and coalescence in the early stages of fatigue life in the CT* [0]$_8$ *composite. Arrows mark cracks initiating in the interface region.*

13. As shown for the ΔK range tested, the crack growth rates for the [90]$_8$ oriented composite are always considerably higher than those of the unreinforced material. Possible reasons for this type of behavior will be discussed later.

The FCG data in Fig. 13 are shown for both the heat-treated and as-received conditions. For the limited number of tests conducted, there appears to be no appreciable effect of heat treat-

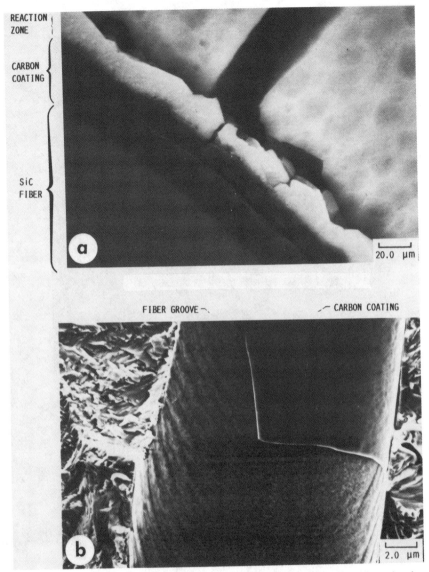

FIG. 11—*Preferential cracking of the carbon layer: (a) cracking emanating from the carbon layer and (b) carbon layer separated from the fiber during fatigue.*

ment on the FCG behavior of the unreinforced material. However, in the [90]₈ composite the heat-treated material exhibits higher FCG rates in at least the high ΔK region.

There were many similarities between the FCG processes observed for the [90]₈ CT specimens and those already described for the [0]₈ CT specimens. For the [90]₈ oriented composite, as in the [0]₈, the crack initiation occurred mostly in the fiber/matrix interface regions of fibers damaged by the machining process. Figure 14 shows a notch area where the fibers have pulled away from the matrix, leaving in its wake portions of the carbon coating. This indicates that crack growth was accompanied by the cracking of the carbon coating. In a manner very similar

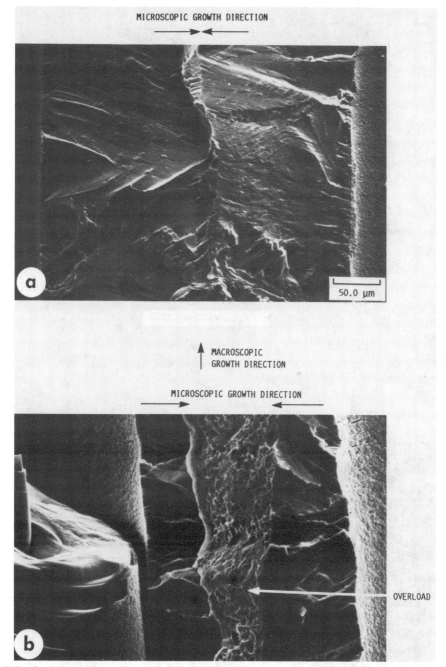

FIG. 12—*Microscopic crack growth direction from the interface into the matrix ligament in the* $[0]_8$ *CT specimens: (a) low* da/dN $[0]_8$ *and (b) high* da/dN $[0]_8$.

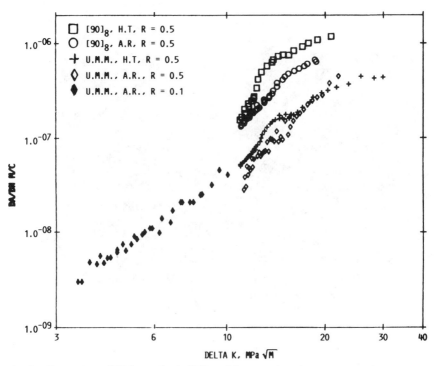

FIG. 13—*Comparison of FCG rates for the* [90]$_8$ *and the unreinforced matrix material in the as-received (AR) and heat-treated (HT) condition.*

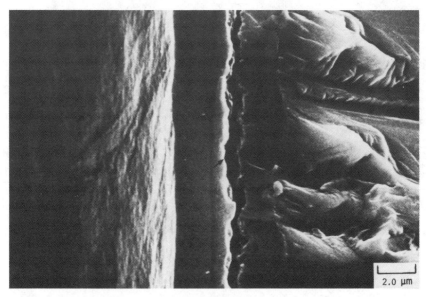

FIG. 14—*Separation of carbon coating from the fiber in* [90]$_8$ *CT specimen. In the center of the fractograph, note a crack between carbon coating and the interface.*

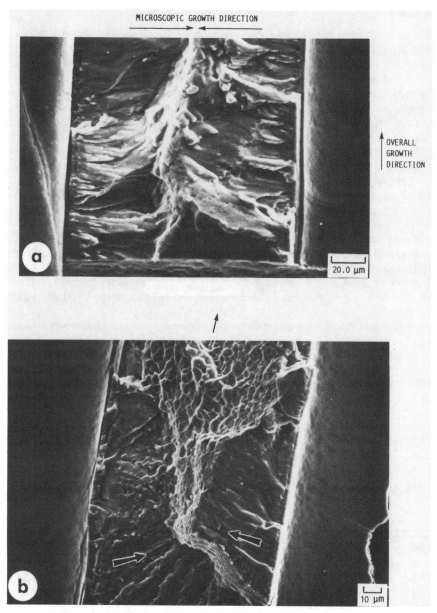

FIG. 15—*Microscopic crack growth from the interface into the matrix ligaments in the* [90]$_8$ *specimens:* (a) *low* da/dN *and* (b) *high* da/dN.

to the [0]$_8$ oriented material, crack growth occurred preferentially in the interface region, followed by growth into the matrix ligaments (Fig. 15). Again, the matrix deformed mainly by striation forming mechanisms and slippage. In the final stages of cracking, the same mechanisms of debonding and matrix tearing were observed as in the [0]$_8$ CT specimens. There were no major differences in the failure processes between the heat-treated and as-received composite.

Discussion

The results obtained in this study pose a number of important questions. (1) Why is the behavior of the $[0]_8$ oriented SEN and CT specimens so radically different? (2) What is the effective driving force for crack growth in the SEN specimens? (3) Why are the fatigue mechanisms of the $[0]_8$ and $[90]_8$ oriented CT specimens so similar?

As mentioned earlier, the crack growth direction was parallel to the loading direction for the $[0]_8$ CT specimens and normal to the loading direction in the $[0]_8$ SEN specimens. A comparison of the simple bending and shear stresses at the notch root for the two specimen geometries may explain the difference in the observed behavior. Both the bending and shear stresses were calculated at the notch root for each specimen geometry. The calculations were obtained using basic strength of materials equations for a loaded beam. As shown in Fig. 16, the bending and shear stresses were 13 and 2 times greater for the CT specimens in comparison to the SEN specimens, respectively. These higher bending stresses were able to preferentially crack the interface region and drive the crack front in the direction parallel to the loading direction. In the case of the SEN specimens, the considerably lower bending stresses and somewhat lower shear stresses were inadequate to crack the fiber/matrix interface region. The result was that the crack growth in the SEN specimens was confined to the matrix and bypassed the fibers leaving them unbroken in the crack wake with little evidence of failure in the fiber/matrix interface. As the crack grew longer in the SEN specimens, fiber bridging acted to reduce the bending stresses at the crack tip and thus the crack growth direction remained normal to the loading crack.

Fiber bridging in the SEN specimens reduced the near-tip crack displacements that are the driving force for crack extension. As the crack grew longer, more fibers bridged the crack wake. This resulted in a progressive decrease in the fatigue crack growth rates and finally led to full crack arrest (Fig. 2). The very efficient transfer of the load from the cracked matrix to the unbroken fibers resulted in no apparent change in compliance as measured by the extensometer mounted in the crack mouth. This points to the inadequacy of the far field measuring

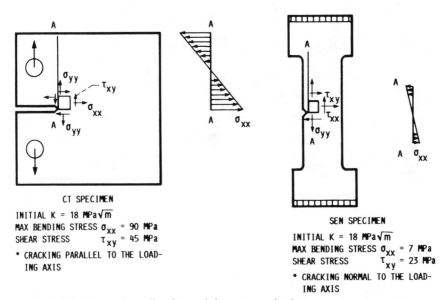

FIG. 16—*Comparison of bending and shear stresses for the two specimen geometries.*

devices to characterize the fatigue crack growth behavior of a complex material such as this composite.

The use of the fatigue loading stage inside the SEM allowed precise measurements of the near-tip CODs to be obtained. These measurements were used in an experimental formulation of the effect of fiber bridging on the crack driving force and the CODs in the $[0]_8$ SEN specimens as described later. Only the experimental formulation of crack bridging is described in this manuscript since the primary focus of the paper is the experimental study. The analytical modeling of the crack bridging, which was partly based on the shear lag model of Marshall, Cox, and Evans [7], was published elsewhere [8].

For a given crack length (c) and distance behind the crack tip (t), Eq 1 can be shown to result in the following

$$\Delta u_{NB} \; \alpha \; \Delta K_{app} \tag{2}$$

of for a more general case

$$\Delta u \; \alpha \; \Delta K \tag{3}$$

Since according to Eq 3 crack tip displacements are proportional to ΔK, and both the bridged and unbridged displacements are known, the following equation for an effective crack driving force can be applied

$$\Delta K_{eff} = \frac{\Delta u_B}{\Delta u_{NB}} \cdot \Delta K_{app} \tag{4}$$

In Fig. 17, the fatigue crack growth rate data for an SEN specimen are plotted as a function of both the global ΔK_{app} parameter and the calculated ΔK_{eff} parameter. For comparison purposes, also shown in Fig. 17 are the FCG rate data obtained from unreinforced matrix CT specimens. The FCG rates for the $[0]_8$ SEN specimens decreased as a function of the ΔK_{app}. For the same ΔK_{app}, the FCG rates were three orders of magnitude slower for the $[0]_8$ composite than the unreinforced material. This large difference in the measured FCG rates is due almost entirely to the crack bridging experienced by the SEN specimens. Some of the SEN fatigue crack growth data were recalculated in terms of the ΔK_{eff} parameter to account for crack bridging. As shown in Fig. 17, the corrected $[0]_8$ data are significantly closer to the data trends exhibited by the unreinforced matrix material. For the data points for which the ΔK_{eff} parameter was calculated, the stress intensity shifted from approximately 30 MPa$\sqrt{\text{m}}$ to 6 to 8 MPa$\sqrt{\text{m}}$. A stress intensity range of 3 to 4 MPa$\sqrt{\text{m}}$ produced comparable crack growth rates in the unreinforced matrix material. The remaining difference between the actual FCG rates in the unreinforced matrix material and the FCG rates of the $[0]_8$ SEN specimens, measured in terms of ΔK_{eff}, may in part be due to environmental effects. The unreinforced matrix material was tested in the ambient environment while the $[0]_8$ composite was tested in vacuum. In the low crack growth region, ambient environment has been shown to increase the FCG rates of titanium-based alloys [9]. Also, the COD measurements were obtained on the specimen surface and they may not fully represent crack tip opening displacements through the specimen thickness.

The Δu_b measurements used to calculate the ΔK_{eff} parameter were obtained at an approximate distance of 100 μm behind the crack tip. In order to check the sensitivity of the calculated ΔK_{eff} parameter to the location of the Δu_b measurements, the ΔK_{eff} parameter was calculated using displacement measurements obtained at various distances behind the crack tip. It was

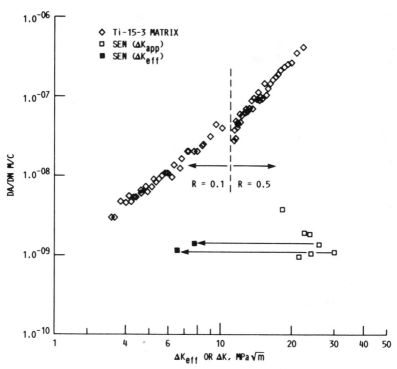

FIG. 17—*SEN crack growth data compared to unreinforced matrix material in terms of* ΔK_{app} *and the proposed* ΔK_{eff} *parameter.*

determined that as long as the measurements were within a relatively small distance from the crack tip, the ΔK_{eff} parameter remained essentially the same.

The difference between the ΔK_{app} and ΔK_{eff} is equal to the reduction in the applied stress intensity due to fiber bridging. For these particular SEN tests, the bridging resulted in a decrease of approximately 22 to 24 MPa $\sqrt{m}$ in the stress intensity range.

There have been many discussions in the past regarding the applicability of linear elastic fracture mechanics to characterize the FCG behavior of composites. However, in the case of the $[0]_8$ oriented SEN specimens, crack propagation was confined to the isotropic matrix material leaving in its wake unbroken fibers. The presence of the unbroken fibers in the crack wake can be thought of as a highly rigid mechanical constraint acting to reduce the crack tip opening displacement range in the isotropic matrix. Linear elastic fracture mechanics is well suited to describe this type of behavior.

One of the major findings of the study was the close similarity of the fatigue failure processes for the two different composite orientations tested using CT specimen geometry (compare Figs. 12 and 15). The most important event that led to this similarity in the fatigue behavior was the preferential cracking of the interface region, in particular, the carbon coatings surrounding the fibers. The cracking of the interface region occurred for both CT orientations tested, and promoted confinement of the crack growth to the fiber direction. While crack growth in the fiber direction is not surprising in the $[90]_8$ oriented CT specimens, it is somewhat surprising in the $[0]_8$ CT specimens. For the $[90]_8$ specimens, cracking of the weak interface region normal to the applied tensile load apparently occurred first. These microcracks later propagated into the matrix and linked together resulting in a dominant crack growing in

TABLE 2—*A comparison of bending and shear stresses for the CT* $[0]_8$ *specimens at initiation and failure.*

	Maximum Bending Stress, MPa	Maximum Shear Stress, MPa
Initiation	90	45
Failure	220	65

the fiber direction. For the $[0]_8$ CT specimens, with the main crack growing in the direction of the applied load, one would expect the shear stresses to control the FCG behavior. However, calculations of the shear and bending stresses shown in Table 2, reveal that the bending stress, σ_{xx}, perpendicular to the fiber direction, is initially twice that of the shear stress, τ_{xy}, parallel to the fibers, with the ratio increasing as the crack grows further in the fiber direction. These relatively high bending stresses must have been high enough to initiate cracking of the interface region perpendicular to the σ_{xx} direction. The cracking in the interface region was followed by propagation and linkage of the cracks in the matrix ligaments. The relatively high shear stresses in the $[0]_8$ CT specimens might have further contributed to keeping the crack front in the fiber direction by weakening the fiber/matrix interface. However, from the fractographic evidence as well as the previously mentioned analysis, it is evident that the tensile loads perpendicular to the fiber direction played the major role in controlling the FCG behavior of the CT $[0]_8$ specimens by preferentially cracking the interface region.

A schematic illustration of the crack initiation and propagation processes just discussed, for both composite orientations, is proposed and shown in Fig. 18. The process can be summarized as follows: (1) cracking of the interface region perpendicular to the resulting σ_{xx} stresses; (2) local debonding occurring in the failed interface region and growth of microcracks into the matrix; and (3) linking of the cracks in the matrix ligaments. This is a continuing process requiring continuous cracking of the interface region layer at or ahead of the crack front.

The FCG rate of the $[90]_8$ CT specimens was considerably greater than that of the unreinforced matrix for a given stress intensity tested (Fig. 13). The fatigue mechanisms within the matrix region (that is, formation of striations, slip deformation, etc.) were very similar for both the $[90]_8$ and the unreinforced material and thus probably did not contribute to the observed differences in the FCG rates. However, the ease of cracking of the interface region, in particular the carbon coating, is probably responsible for the increase in the crack growth rates. This is consistent with the hypothesis that the fiber/matrix interface has lower fatigue resistance than the matrix. This weak composite constituent therefore limited the fatigue cracking resistance of the $[90]_8$ composite.

Conclusions

1. Fatigue crack growth behavior of the SCS-6/Ti-15-3 composite is a function of specimen geometry, fiber orientation, and the interaction of local stress fields with the highly anisotropic composite.

2. For the $[0]_8$ SEN specimens, the crack growth was normal to the loading direction, while for the $[0]_8$ CT specimens, crack growth was along the fiber direction parallel to the loading direction. The difference in the fatigue behavior for the two specimen geometries was attributed to the significantly higher bending stresses in the interface region for the CT specimens.

3. Near crack tip opening displacements were measured for the SEN specimens. A formulation that uses both the actual bridged displacements and analytically derived unbridged displacements was proposed to determine the effective stress intensity parameter, ΔK_{eff}, to

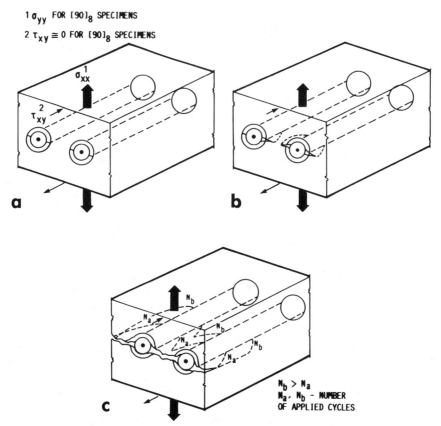

FIG. 18—*Proposed crack initiation and crack propagation processes for the CT specimen geometry:* (a) *crack initiation in the interface,* (b) *crack growth and local debonding, and* (c) *crack coalescence in the matrix.*

account for the effect of bridging on crack driving force. The new parameter accounted for most of the bridging.

4. The fatigue crack growth processes for the CT specimen geometry were controlled by the preferential failure of the fiber/matrix interface for both composite orientations tested. The interface proved to be the composite's weakest link.

References

[*1*] Telesman, J., Kantzos, P., and Brewer, D., "Insitu Fatigue Loading Stage Inside Scanning Electron Microscope," *Lewis Structures Technology 1988,* NASA CP 3003, NASA Lewis Research Center, Cleveland, OH, Vol. 3, 1988, pp. 161–172.
[2] Rhodes, C. G. and Spurling, R. A., "Fiber-Matrix Reaction Zone Growth Kinetics in SiC-Reinforced Ti-6Al-4V as Studied by Transmission Electron Microscopy," *Recent Advances in Composites in the United States and Japan, ASTM STP 864,* J. R. Vinson and M. Taya, Eds., American Society for Testing and Materials, Philadelphia, 1985, pp. 585–599.
[*3*] Lerch, B. A., Gabb, T. P., and MacKay, R. A., "A Heat Treatment Study of the SiC/Ti-15-3 Composite System," NASA TP 2970, Cleveland, OH, 1990.
[*4*] Bueckner, H. F., "Weight Functions for the Notched Bar," *Zeitschrift Angewandte Math. Mechematik und Mechanik,* Vol. 51, 1971, pp. 97–109.

[5] Gabb, T. P., Gayda, J., and MacKay, R. A., "Isothermal and Nonisothermal Fatigue Behavior of a Metal Matrix Composite," *Journal of Composite Materials,* Vol. 24, No. 6, June 1990, to be published.

[6] Johnson, W. S., Lubowinski, S. J., and Highsmith, A. L., "Mechanical Characterization of Unnotched SCS_6/Ti-15-3 Metal Matrix Composite at Room Temperature," *Thermal and Mechanical Behavior of Ceramic and Metal Matrix Composites, ASTM STP 1080,* J. M. Kennedy, H. H. Moeller, and W. S. Johnson, Eds., American Society for Testing and Materials, Philadelphia, 1990, pp. 193–218.

[7] Marshall, D. B., Cox, B. N., and Evans, A. G., "The Mechanics of Matrix Cracking in Brittle-Matrix Fiber Composites," *Acta Metallurgica,* Vol. 33, Nov. 1985, pp. 2013–2021.

[8] Ghosn, L. J., Telesman, J., and Kantzos, P., "Fatigue Crack Growth in Unidirectional Metal Matrix Composite," *Fatigue 90,* Vol. II, Kitagawa and Tanaka, Eds., July 1990, pp. 893–898.

[9] Shih, T. T. and Wei, R. P., "Load and Environment Interactions in Fatigue Crack Growth," *Prospects of Fracture Mechanics,* G. C. Sih, H. C. van Elst, and D. Broek, Eds., Noordhoff, Leyden, The Netherlands, 1974, pp. 231–248.

Bhaskar S. Majumdar[1] and Golam M. Newaz[1]

Thermomechanical Fatigue of a Quasi-Isotropic Metal Matrix Composite

REFERENCE: Majumdar, B. S. and Newaz, G. M., "**Thermomechanical Fatigue of a Quasi-Isotropic Metal Matrix Composite,**" *Composite Materials: Fatigue and Fracture (Third Volume), ASTM STP 1110,* T. K. O'Brien, Ed., American Society for Testing and Materials, Philadelphia, 1991, pp. 732–752.

ABSTRACT: In-phase thermomechanical fatigue (TMF) and elevated temperature isothermal fatigue (IF) experiments were conducted on a $[0, \pm 45, 90]_s$ Ti 15-3/SCS6 composite under load control. A correlation was obtained between the stabilized mechanical strain range and TMF life. TMF life was found to be significantly shorter than IF life when comparisons were made either on a stress-range or a gross mechanical strain-range basis. Damage modes were investigated using optical and scanning electron microscopy. The primary damage mode in IF specimens was transverse microcracking oriented perpendicular to the loading axis. In TMF specimens, although similar damage was observed very close to the fracture surface, the primary damage modes were inter-ply delamination and fiber-matrix debonding. Limited experiments were performed where damage was monitored using a replication technique on the specimen edges. These experiments confirmed delamination and debonding in TMF as well as thermally cycled specimens. Wavelength dispersive spectroscopic (WDS) analysis showed carbon-rich zones along many ply-to-ply interfaces, and these zones may have accelerated delamination cracking. Finite element analysis indicated that the maximum delamination stress occurred at the minimum temperature in the TMF experiments. The fatigue data of Ti 15-3/SCS6 composite, from this and other investigations, were plotted in the form of 0° fiber stress range versus logarithm of fatigue life. This phenomenological plot indicated a similar trend for various data sets, but the exact mechanisms remain to be determined.

KEY WORDS: thermomechanical fatigue, isothermal fatigue, fatigue life, titanium alloy, silicon carbide fiber, stress range, strain range, damage modes, delamination, fiber-matrix debonding, transverse microcracking, ply-to-ply interface, reaction zone, 0° fiber stress, residual stress, composite materials, fracture, fatigue (materials)

Advanced aerospace components require light-weight materials that can withstand extreme combinations of temperature, load, and environment. These demanding requirements have led to the development of advanced metal matrix composites (MMC).

The Ti 15-3/SCS6 composite is one such composite system that has potential application at temperatures up to 704°C (1300°F). The reinforcing fiber, SCS6, is a silicon carbon (SiC) fiber that is fabricated by Textron. It has an inner carbon core, surrounded by SiC, then by a complex outer shell consisting of alternate layers of carbon and silicon with nonstoichiometric composition. The matrix has the nominal composition Ti-15V-3Al-3Cr-3Sn, all in percent by weight, and it is a beta (bcc stabilized) titanium alloy. The alloy has very good forming properties, and foils of the alloy can be easily obtained by rolling at elevated temperatures. Thus, fabrication of long-fiber reinforced composites by foil consolidation is relatively straightfor-

---

[1] Principal research scientist and senior research scientist, respectively, Battelle Memorial Institute, Columbus, OH 43201.

ward with Ti 15-3 matrix material. Additionally, because the stabilized phase is a beta phase, the alloy does not suffer from problems associated with major phase transformations.[2]

The objective of the current work was to obtain some understanding of the thermomechanical fatigue (TMF) behavior of a quasi-isotropic Ti 15-3/SCS6 composite material. What differentiates this and other titanium-based composites from the better established aluminum-based MMC (such as aluminum-boron) is their higher matrix and interface strengths, and the capability to withstand high-temperature environments. However, because of large thermal expansion (α) mismatch between the matrix alloy ($\alpha = 9.72 \times 10^{-6}/°C$) and SiC ($\alpha = 4.86 \times 10^{-6}/°C$), high stresses and strains are anticipated at and around fiber-matrix interfaces when the composite in service is subjected to thermomechanical cycles over significant temperature intervals (typically much larger than for aluminum-alloy based MMC). Coincidentally, the fiber-matrix interface zone also is the region that is most significantly degraded by reaction between the SiC fibers and the highly reactive titanium in the matrix material. Thus, thermomechanical cycling constitutes a particularly severe type of loading for the composite material, and from a reliability standpoint it is important to evaluate fatigue performance under TMF conditions. The work presented herein, therefore, should have broad appeal in the rapidly developing field of high-temperature MMCs, including the new generation of aluminide based composites.

Among some of the recent work on fatigue of titanium-alloy based MMC, Johnson et al. [1,2] studied the tensile and fatigue behavior of uniaxial, cross-ply, and quasi-isotropic Ti 15-3/SCS6 composites at room temperature. The authors suggested that even in the nonuniaxial systems, the 0° plies played a significant role in controlling fatigue life. In particular, the authors suggested a model based on stress in the 0° fibers. Gayda et al. [3] studied the fatigue behavior of unidirectional (0°) Ti 15-3/SCS6 composites at elevated temperatures. They observed that, while on a stress range basis the isothermal fatigue life at 300°C (572°F) and 550°C (1022°F) was higher for the MMC than for the matrix alloy, the reverse was true when fatigue life was compared on a strain range basis. It was suggested that fatigue life probably was not matrix dominated, but more likely it was governed by fiber-matrix interface characteristics. In Ref 4, isothermal and in-phase and out-of-phase load-controlled bithermal TMF experiments were performed on the same unidirectional composite between 300 and 550°C. Both in-phase and out-of-phase fatigue lives were found to be shorter than IF life when data were compared on a stress range basis. On a strain range basis, in-phase and out-of-phase fatigue lives fell on the same curve, indicating no differences. It was suggested that this behavior was because TMF life was interface controlled.

The current work on in-phase TMF of an eight-ply [0, ±45,90]$_s$ Ti 15-3/SCS6 composite represents new understanding of the complex processes that govern TMF life of quasi-isotropic titanium-based composites. Both TMF and IF experiments were performed. The quasi-isotropic system is much more complex than a 0° uniaxial system, where the applied stresses of interest are significantly higher than for the matrix material. In this latter 0° uniaxial system, fiber-dominated failure can occur under in-phase TMF loading because thermal-mismatch induced stresses and applied stresses, both contribute to elevating the fiber stress. Hence, at short fatigue life, a 0° fiber-dominated fatigue mechanism might operate for a 0° system under TMF loading, provided of course that matrix cracking, interface damage, and oxidation attack of fibers do not significantly accelerate fatigue failure. On the other hand, as will be shown in

---

[2] In reality, the material is metastable, and long-term exposure at temperatures above 482°C (900°F) can lead to precipitation of α(hcp) phase [6]. Such precipitation slightly increases the room-temperature modulus and strengths, but effects on high-temperature properties, or the effect of thermal cycling in α-phase precipitation, are not adequately characterized.

this work on a quasi-isotropic system, extremely short fatigue lives were observed at applied stresses that were small compared with stresses required to cause 0° fiber failure. An important aspect of the study was to critically examine the damage mechanisms, so that a basis for a life-prediction model could be developed that accounted for actual damage processes.

Experiments

The material chosen for the investigation was a quasi-isotropic eight-ply $[0, \pm 45, 90]_s$ titanium-based composite; some initial experiments also were performed with a seven-ply $[0, \pm 45, \overline{90}]_s$ system,[3] where one 90° ply is missing. The reinforcing phase (40% by volume) was continuous 140-μm diameter SiC fibers, commercially designated as SCS6. The matrix material was a beta-phase Ti-15V-3Al-3Sn-3Cr (percent by weight) alloy. The composite was fabricated at Textron, Avco Specialty Metals, by hot isostatically pressing (HIP) alternate layers of fibers and thin foils of Ti 15-3, with ribbons of titanium used to hold the fibers in alignment. The width of each ply was approximately 0.192 mm.

Fatigue tests were performed on the composite using straight rectangular specimens of approximate size 1.6 mm thickness by 19.1 mm width by 152.5 mm length. In order to avoid stress concentration effects associated with exposed fibers, the gage length region was not reduced in cross section. This did not pose any problems, since all failures occurred near the center of the specimens. The specimens were gripped using aluminum end-tabs and water-cooled grips. The exposed length of the specimen between the grips was approximately 101.6 mm. Induction heating was used to heat the samples, and in a preliminary experiment five thermocouples were used to assure a uniform (within 4°C) heated length of approximately 40 mm. In later experiments, only two thermocouples were employed for monitoring and controlling temperature over the heated zone. The thermocouples were spot welded to the specimen; such spot welds did not cause any problems, since failure generally occurred away from the welds.

The specimen and fixtures were enclosed by a plexiglass chamber with O-ring seals. Argon flowed through the chamber at a positive pressure, and the gas flow was monitored by passing the exit gas stream through a water bath. The purpose of using argon was to reduce effects due to oxidation, although microstructural examination following testing indicated that oxidation was not prevented altogether.

The fatigue experiments were performed on a servohydraulic test system under load control at minimum to maximum load ratio (R-ratio) of 0.1. Strain was monitored over a 25 mm gage length using a strain-gage type of extensometer, with ceramic probes that were spring-loaded laterally against the edge of the specimen. Isothermal fatigue tests were performed at 649°C (1200°F)[4] at a loading frequency of 2 Hz. The temperature was equilibrated at 649°C for 30 min prior to start of mechanical loading. TMF experiments were performed with a load profile that was "in-phase" with the imposed saw-tooth temperature profile. Temperature cycling was performed between 315.5°C (600°F) and 649°C (1200°F) using linear heating and cooling rates; the cycle time was approximately 150 s. This long cycle time was necessitated by the cooling time for the specimen. No efforts were made to use forced air cooling, because of concerns regarding temperature gradients within the specimen. The load and temperature cycles were programmed on a personal computer, and data on load, temperature, and strain were stored at fixed time intervals.

The specimens were visually inspected before they were tested, and, except for one specimen that had some surface and delamination damage, none of the other specimens appeared to

---

[3] We thank McDonnell Douglas Corporation for kindly supplying us with the material.
[4] Original measurements were performed in English units.

have any initial significant damage. However, detailed microscopic or nondestructive examination was not performed prior to testing. Experiments were performed on the specimens in the as-received condition.

Following the fatigue experiments, the specimen fracture surfaces and edges were examined optically and also by using a scanning electron microscope (SEM). In some cases, the specimens were metallographically mounted and polished using diamond paste. Micrographs were taken to record the primary damage mechanisms. Few samples were metallographically polished prior to testing, and replicas were taken of the edges after a fixed number of fatigue cycles. These experiments were performed to evaluate whether some of the damage that was observed after testing (either before or after failure) was already present in the specimen before testing, or whether it occurred as a result of isothermal or TMF cycling. Selected microprobe analysis, utilizing wavelength dispersive spectroscopy, was performed on the composite material; these helped in evaluating the chemical or microstructural features that may have also contributed to microcracking under TMF and thermal cycling.

Results

Mechanical Test Results

Figure 1 is a plot of load, temperature, and strain versus time for a TMF specimen cycled between 30 and 300 MPa. The figure illustrates excellent in-phase control of load and temperature versus time. Good in-phase control also could be judged from the hysteresis loop, where the stress versus strain (thermal plus mechanical) was plotted for a particular fatigue cycle. When phase control was not perfect, the hysteresis loop had a dumbbell shape, with the strain remaining fairly constant over a limited stress interval; the reason was that thermal-expansion induced strain negated the applied-stress induced strain. On the other hand, when phase control was good, the hysteresis loop had sharp tips. An example of a TMF hysteresis

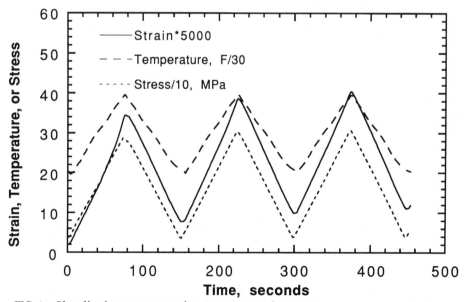

FIG. 1—*Plot of load, temperature and strain versus time for a TMF specimen, showing good in-phase control of local and temperature cycles.*

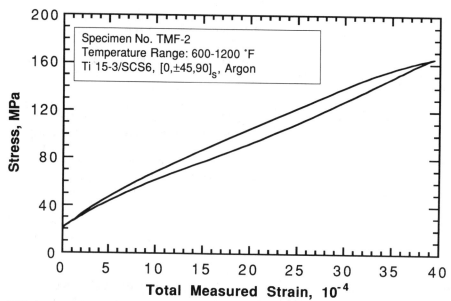

FIG. 2—*Typical stress-strain (total measured strain) hysteresis loop for a TMF specimen cycled between 315.5°C (600°F) and 649°C (1200°F).*

loop is illustrated in Fig. 2, and it shows the sharp tips at the stress reversal points. The hysteresis loops for an isothermally fatigued (649°C) specimen are shown in Fig. 3, and it indicates that the magnitude of inelastic strain was fairly small.

The TMF specimens generally had very low fatigue lives, at least for the stress levels that were chosen for performing the experiments. It was observed that the strain range was maximum in the first cycle, and that it decreased rapidly in a few cycles (approximately 15 cycles) to a stabilized constant value. At the same time, the mean strain increased by the maximum amount in the first cycle, and it too settled down to a constant value within a few fatigue cycles. This type of ratcheting behavior is typical of load controlled fatigue experiments, primarily where positive R-ratio is used. The larger strain in the first cycle, compared with subsequent cycles, was most likely because of plastic deformation coupled with some damage in the off-axis plies. With continued cycling, one would expect such inelastic deformation to decrease because of transfer of load from the more compliant and weaker off-axis plies (primarily 90° plies) to the stiffer and stronger 0° plies. This type of overall hardening behavior may have been responsible for reduction of ratcheting with number of cycles and subsequent attainment of steady-state strain range conditions. The stabilized strain range per cycle was determined for all the fatigue experiments.

Figure 4 is an S-N plot, where the imposed stress range is plotted versus the number of cycles to failure. Since the R-ratio was held constant at 0.1, the maximum stress corresponding to any data point can be obtained by dividing the ordinate by 0.9. The data corresponding to one-half cycle belong to specimens that failed under monotonic loading in the very first cycle. Data points with arrows refer to tests where cycling was discontinued before failure. There is some scatter in the data, and this may be due to inherent variability of the material, which is still in the developmental stage. The solid line in Fig. 4 corresponds to TMF specimens, and is a linear least squares fit with stress range as the dependent variable. The data corresponding

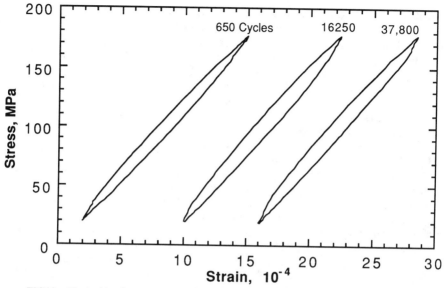

FIG. 3—*Typical load-strain hysteresis loop for an IF specimen cycled at 649°C (1200°F).*

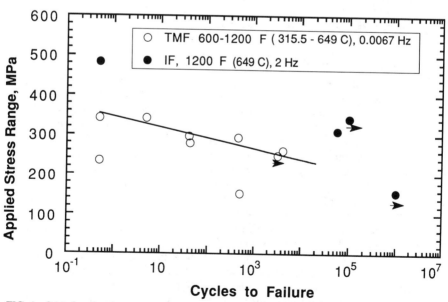

FIG. 4—S-N *plot, showing stress range plotted versus fatigue life. Both TMF and IF data are presented. The data corresponding to one-half cycle correspond to specimens that failed on monotonic loading. Arrows represent specimens where testing was discontinued before failure.*

to runout was not included in the regression analysis; the two data points that lie well below the curve also were excluded in the regression procedure since they appeared to be outliers.

Figure 4 illustrates the usual trend of increasing fatigue life with lower applied stress range levels. The data for isothermally fatigued specimens also are included in Fig. 4, and it is clear that, on a stress basis, the IF specimens had significantly longer life than TMF specimens. The reason for the stress-based large reduction in life of TMF specimens, compared with IF specimens, is not very clear at present. It may be related to new damage modes that were observed in TMF specimens or to the fact that local thermally induced stresses and strains were not accounted for in the TMF specimens. Another possible factor might be a strain-rate effect, since IF experiments were performed at 2 Hz while TMF experiments were performed at 0.0067 Hz. Work in progress is aimed at trying to resolve some of these issues. Although oxidation could not be prevented altogether, it did not appear to be a controlling factor since the IF experiments lasted longer than TMF experiments.

Two specimens were thermally cycled over the temperature range 315.5 to 649°C, without any applied load. Strain was monitored over a 25.4 mm gage length using a strain gage type of extensometer. The purpose of conducting these experiments was to determine the overall thermal expansion of the composite, and also to evaluate the type of damage that occurred under thermal cycling. It was observed that the thermal expansion strain was essentially linear and reversible over the temperature range of interest, and that there was only a slight change (less than 5%) in the thermal expansion coefficient with cycling. An average (over cycles) thermal expansion coefficient, α_c, of 8.01×10^{-6}/°C was obtained for the eight-ply composite. It also was observed that there was a slight irreversible contraction (approximately 0.5%) of the composite following approximately 2300 thermal cycles. The reason for the irreversible contraction is not clear at present, although it may be related to relaxation of residual stresses from processing, or to damages that occurred in the composite due to thermal cycling.

From the total measured strain ranges ($\Delta\epsilon_{tot,measured}$), which included strains both due to thermal expansion of the composite ($\Delta\epsilon_{th}$) and due to applied stresses ($\Delta\epsilon_m$), the strain range ($\Delta\epsilon_m$) due to applied mechanical stresses only were estimated using the relationship

$$\Delta\epsilon_m = \Delta\epsilon_{tot,measured} - \Delta\epsilon_{th} = \Delta\epsilon_{tot,measured} - \alpha_c \, \Delta T \qquad (1)$$

where ΔT is the temperature range, and α_c is the overall thermal expansion coefficient of the composite. The ΔT is zero for isothermal fatigue, and it is 333.5°C for TMF specimens.

The reason that an overall mechanical strain range ($\Delta\epsilon_m$) was calculated using Eq 1 is because it allowed approximate comparison between isothermal and thermomechanical fatigue data, if the composite was treated as a homogeneous body by itself. It must, of course, be realized that for the same location in the composite, and for the same overall $\Delta\epsilon_m$, the local mechanical strain range ($\Delta\epsilon_{m,local}$) will not be identical under IF and TMF conditions. Thus, Eq 1 cannot provide a rational basis for life prediction under general isothermal/non-isothermal loading conditions. It only provides comparison between IF and TMF data if the composite is treated as a material continuum in itself.

Figure 5 is a plot of the overall mechanical strain range ($\Delta\epsilon_m$) versus the number of cycles to failure for the in-phase TMF specimens. Figure 5 is the usual way in which low-cycle fatigue (LCF) data are represented in the literature of monolithic materials, because LCF lives have been found to be primarily inelastic strain-range controlled. Representing fatigue life on a strain range basis also is useful for composite materials, because deformation at any location is likely to be locally strain-controlled due to constraints by surrounding material with different constitutive properties. It must be recognized, however, that the strain ranges were obtained in this work by an indirect approach, in that experiments actually were performed under stress control. If experiments indeed were conducted under strain control, then greater

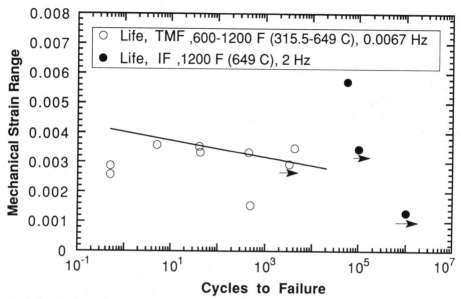

FIG. 5—*Mechanical strain range plotted versus fatigue life of TMF and IF specimens. Mechanical strain range was calculated from the total measured strain using Eq 1.*

fatigue life might have resulted, because strain control would accommodate greater crack-propagation life.

The plotting of strain range versus fatigue life has an added advantage when specimens are loaded isothermally. Provided there is proportional loading, the local strain at any location scales linearly with the far-field measured strain, even when there is nonlinear inelastic deformation. Thus, for the isothermal data in Fig. 5, a plot of local strain (at any particular location) versus fatigue life would be identical to Fig. 5, except for a different scale for the ordinate. Unfortunately, because of coupling of mechanical strains due to load and due to temperature, linear proportionality between local and far-field measured strains will not exist for TMF specimens under inelastic deformation conditions. Hence, a plot of local strain versus fatigue life for TMF specimens would in principle be different from Fig. 5. We have performed preliminary calculations of deformation under TMF conditions, using the nonlinear micromechanical METCAN code developed by Chamis et al. [5]. These calculations showed that for the inelastic strain ranges that were measured, the local strain ranges in the matrix adjacent to the 90° fibers had approximate linear relationships with the far-field 0° ply strains. Based on these calculations, it appears that even for the TMF specimens, a plot of local mechanical strain range ($\Delta\epsilon_{m,\text{local}}$ at any location) versus fatigue life would probably look similar to Fig. 5, except for a change in the scale for the ordinate.

Figure 5 shows fairly good correlation between the measured total strain range and fatigue life of TMF specimens. If we assume proportional loading, based on preliminary calculations using METCAN code, then Fig. 5 also indicates that TMF life may be controlled by local strain ranges. Figure 5 still shows a large disparity between TMF and IF data. This may partly result from the fact that the proportionality factors between local mechanical strain ranges and far-field (measured) mechanical strain ranges are different under TMF and IF, for the same failure location. However, preliminary calculations using METCAN indicate that such effects cannot explain the observed differences in fatigue life. Rather, it appears that differences in failure locations and mechanisms may have played a more dominant role in providing different fatigue lives for the two loading conditions, aided by large differences in strain rates. Results of microscopic analysis and investigation of failure mechanisms are described in the next section.

Fractography, Microstructure, and Damage

The fracture surfaces of the composite under TMF and IF loading were similar. In all cases, the fracture surfaces indicated overload failure. Fiber pullout was observed for the 0 and 45° plies. However, the main crack propagated perpendicular to the loading axis without significant crack-plane deviation, and in some cases an entire 90° fiber could be found spanning the fracture surface. For specimens that had extremely short fatigue lives (<100 cycles), the inner 90° fibers near the fracture surface were found projecting horizontally out of the matrix (see Fig. 7), indicating significant viscoplastic deformation of the matrix and fiber-matrix debonding.

The edges of the specimens were metallographically polished to determine the primary damage modes. Figure 6a is a photomicrograph of the edge of an IF specimen (stress range 341 MPa) that did not fail in 100 000 cycles. The loading direction is vertical. The figure shows transverse microcracks, oriented perpendicular to the loading direction, that were located primarily in the off-axis plies. This is the usual type of cracking that has been observed previously in isothermal fatigue of other composites, such as the aluminum-boron system.

Figure 6b is a photomicrograph of the edge of a seven-ply TMF specimen; in this figure, too, the load direction is vertical. In contrast to Fig. 6a, Fig. 6b illustrates that there was significant inter-ply delamination damage in the TMF specimen. The cracking type of damage was primarily in the 90 and 45° plies. Although not evident in the low magnification micrograph, significant fiber-matrix debonding was observed in the TMF specimens. The extent of debonding was more severe than in IF specimens, and it will be shown later that thermal cycling alone caused debonding and delamination-type cracking. It should be noted that not all TMF specimens indicated as severe delamination damage as in Fig. 6b. Micrograph 6c is taken from a different TMF specimen, and it shows that the extent of delamination damage was much less than in Fig. 6b. Additionally, near the fracture surface, transverse microcracks also were observed in TMF specimens. Their density was high near the fracture surface, but because of their localization it appears that they occurred just before or during fracture of the composite. In general, all the TMF specimens examined showed significant delamination and debonding damage compared with IF or untested specimens. Such damage tended to occur more frequently in off-axis plies with closely spaced fibers, indicating near-neighbor effects.

Figure 7 is a photomicrograph of the edge surface near the fracture plane of a TMF specimen, and it illustrates debonding and transverse microcracking in the matrix; the fiber belongs to a 90° ply. The major microcracks seem to initiate at the interface between the off-axis fibers and matrix. Also, as mentioned earlier, Fig. 7 shows a 90° fiber protruding from the edge of the specimen. We believe that such a protrusion was a result of lateral contraction of the matrix (in the 90° plies) because of time-dependent plastic elongation of the matrix in the loading direction. The protrusion of the fiber, by as much as 0.51 mm, also indicates that fiber-matrix compatibility of deformation was lost, most likely because of debonding between matrix and fibers—allowing each constituent to deform independently.

In order to confirm whether fiber-matrix debonding and inter-ply delamination could result from thermal cycling alone, the edge of a specimen was metallographically polished, and a replica was taken using acetate tape. The specimen was then subjected to approximately 2300 thermal cycles (315.5 to 649°C) without imposing any applied load, using linear heating and cooling cycles with a cycle time of approximately 150 s (same as that used in TMF tests). The thermally cycled specimen and replica of the original surface were then observed under SEM.

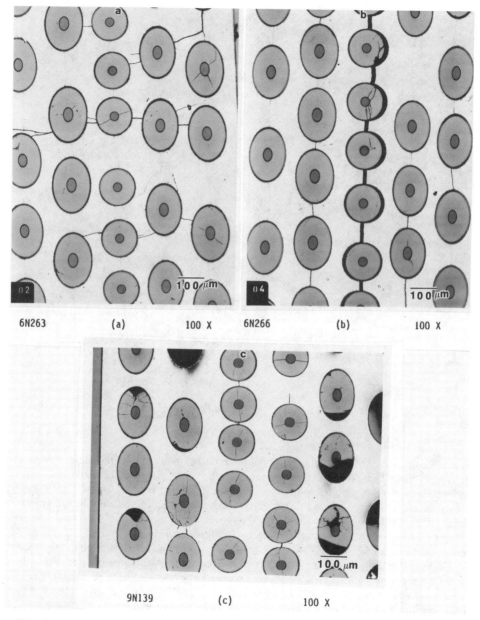

FIG. 6—(a) Micrograph of the edge surface of an IF specimen showing transverse microcracks; (b) micrograph of the edge surface of a TMF specimen showing primarily inter-ply delamination cracks; and (c) micrograph of the edge surface of another TMF specimen, showing variation in the magnitude of inter-ply delamination. The micrographs have been taken some distance away from the fracture surface. In all the figures, the loading axis is vertical.

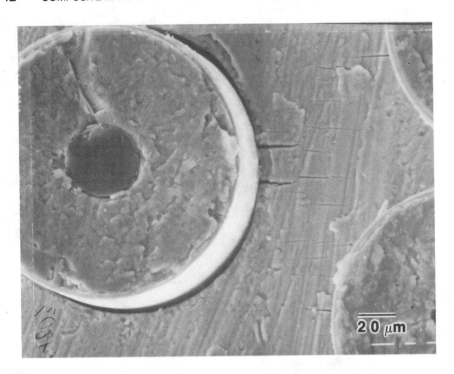

84803   500 X

FIG. 7—*Transverse microcracks around the 90° ply of a TMF specimen, taken very close to the fracture surface.*

Figure 8 compares the replica of a 45° ply of the untested material with the same region after thermal cycling. The loading axis is horizontal in the figure. Because Fig. 8a corresponds to the replica, while Fig. 8b corresponds to the actual specimen surface, the two micrographs should be mirror images of one another in the absence of damage. The identification mark is the crack in the fiber.

Figure 8 confirms that thermal cycling by itself was sufficient to initiate delamination; observe the two cracks in Fig. 8b while none are present in Fig. 8a. Also, note the significant fiber-matrix debonding as a result of thermal cycling alone. An intermediate replica, taken after the specimen had experienced approximately 530 cycles also showed similar damage, although it was less than after 2300 cycles. Similar to TMF specimens, it was observed that delamination cracking was more frequent between closely spaced fibers, indicating that local fiber concentration played an important role in raising the stress/strain amplitude in the reaction zone and matrix between adjacent fibers.

Limited wavelength dispersive spectroscopic (WDS) analysis was performed on metallographically polished specimens, using a JEOL 733 microprobe, to evaluate whether chemical features were responsible for delamination damage. Figure 9a is a back scattered electron (BSE) micrograph of the as-received composite. A 90° fiber can be observed at the bottom of the micrograph, while the Ti 15-3 matrix can be seen above it. The specimen axis is vertical. A ribbon-like slightly dark band may be observed rising at the 11:30 position from the fiber-

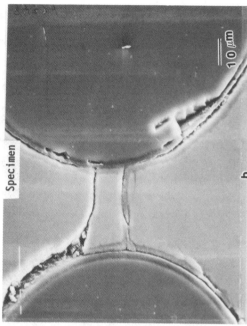

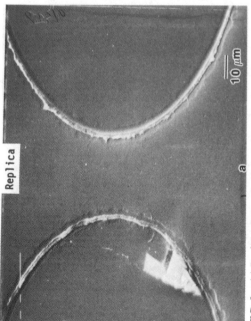

FIG. 8—(a) Replica of the polished edge of a sample before thermal cycling (no load); (b) micrograph of the sample in the same area as (a), after the specimen had experienced thermal cycling (without load). Both (a) and (b) are mirror images, and the crack in the fiber was used to identify the same location before and after thermal cycling. The loading direction is horizontal.

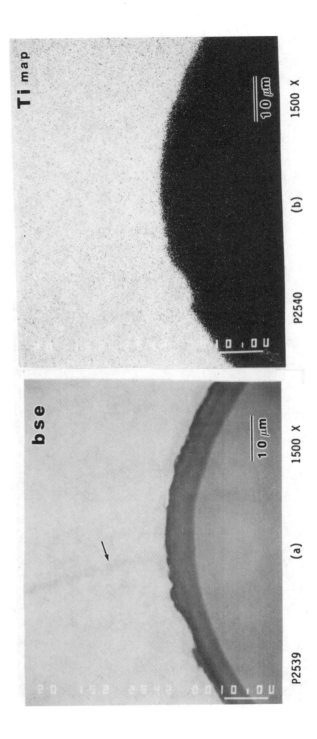

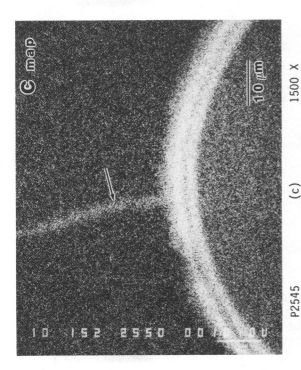

FIG. 9—(a) Back-scattered electron (BSE) micrograph of the as-received composite, with the fiber at the bottom of the micrograph. A ribbon-like slightly darker band may be observed rising at the 11:30 position from the fiber-matrix interface. (b) Titanium map showing titanium confined to the matrix material. (c) Carbon map, showing regions that are rich in carbon; note the outer carbon layers of the fiber, and the carbon-rich zone corresponding to the darkish band in Fig. 9a.

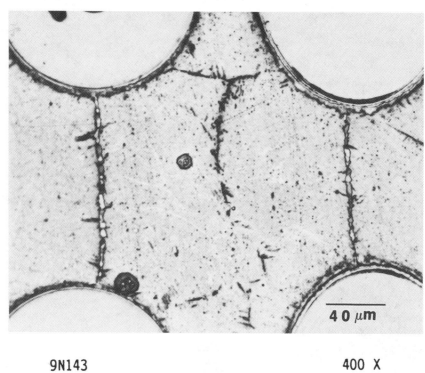

9N143   400 X

FIG. 10—*Optical micrograph of a polished and thermally etched specimen, showing a fine granular microstructure along ply-to-play interfaces.*

matrix interface, indicating that it is richer in lighter atomic weight elements. The thin band corresponds to a ply-to-ply interface;[5] it may be recalled that the composite is fabricated by hot isostatically pressing (HIP) layers of fibers between thin Ti 15-3 foils.

Figure 9b is a titanium map obtained using WDS analysis, and it confirms that the matrix contains titanium. Figure 9c is a carbon map, and it has a number of interesting features. First, it illustrates the alternate layers of carbon that were deposited on the fiber during its fabrication. A separate silicon map showed that the region between the circular carbon layers contains primarily silicon. More importantly, Fig. 9c shows that the thin band, corresponding to the ply-to-ply interface, is rich is carbon. Figure 10 is an optical micrograph of a polished and thermally etched specimen. The figure shows a fine-grained microstructure along the ply-to-ply interface[6] that is different than the large-grained matrix. It is likely that the presence of carbon can lead to formation of titanium carbides, causing local embrittling of the inter-ply region. Even high atomic concentration of carbon by itself would cause significant weakening of the titanium-alloy matrix. Thus, the segregation of carbon may partly explain acceleration of delamination cracking along ply-to-ply interfaces. Inter-ply delamination cracking also was observed away from inter-ply inhomogeneity in thermally cycled specimens, so that significant thermal stresses also played an important role in the cracking process.

[5] Optical observations also indicated microstructural variations along many ply-to-ply interfaces.
[6] We have not investigated the phase structure of the inter-ply layer.

Although inter-ply microstructural inhomogeneity was observed at many places, there were locations where inter-ply inhomogeneity could not be detected. The carbon-rich ply-to-ply interlayer is probably a type of contamination that occurs by diffusion of carbon from the fibers along the surface of matrix foils during consolidation. The current observations imply that further steps need to be taken to prevent carbon contamination along ply-to-ply interfaces during the consolidation of titanium-based composites.

Discussion

In this investigation, the in-phase TMF response of a quasi-isotropic Ti 15-3/SCS6 composite was evaluated, over the temperature range 315.5 to 649°C. Results were compared with the isothermal fatigue response at 649°C. It was observed that in-phase TMF life was significantly shorter than IF life, independent of whether comparisons were performed on applied stress range or gross mechanical strain range basis. The difference between IF life and TMF life in the current investigation also was significantly larger than that observed [3,4] previously for a unidirectional 0° Ti 15-3/SCS6 composite, where experiments were conducted over the temperature range 300 to 550°C.

Figures 4 and 5 indicate that there was some scatter in the data. As already indicated, such scatter may be a result of inherent variability of the material that is still in the developmental stage. In spite of the scatter, Fig. 5 appears to indicate a fairly good correlation between the gross mechanical strain range and TMF life. Although the "local" mechanical strain range would be different from the gross mechanical strain range at different locations of the composite, the good correlation suggests that TMF life is probably local strain range controlled. Thus, a plot of local strain range versus fatigue life would likely be similar to Fig. 5, except for a change in the scale for the ordinate. Further nonlinear FEM analysis is needed to evaluate more accurately the magnitude of local strains, and the degree of their proportionality to far-field mechanical strains.

The difference in damage modes between TMF and IF specimens is important, and partially explains the different fatigue lives under IF and TMF conditions. In the case of IF specimens, the primary damage mode was "transverse microcracking" in the 90° plies. On the other hand, the primary damage modes for TMF specimens were "fiber-matrix debonding" and "inter-ply delamination" in the off-axis plies, notably the 90° plies. Although transverse microcracking also was observed near the fracture surface of TMF specimens, they were too localized, and seemed to have originated slightly before or during fracture of the specimens. The occurrence of debonding and inter-ply delamination damage in 90° plies would increase their compliance, transferring load to the 0° plies. In turn, because of the higher loads, failure of 0° plies would be accelerated.

Thus, the failure scenario under TMF conditions appears to be early damage and loss of stiffness of 90° plies, followed by load transfer and cracking of 0° plies. Under IF conditions, delamination cracking and debonding were delayed, and the composite had to await formation of transverse microcracks before failure of 0° plies. An added complexity is that significant stress relaxation and accumulation of inelastic strains may have occurred in the matrix of TMF specimens because of extremely low strain rates ($\sim 5 \times 10^{-5}$/s). On the other hand, the strain rates in IF specimens were high (~ 0.2/s), so that matrix relaxation most likely was negligible. This strain rate effect also may have contributed to earlier damage in TMF specimens compared with IF specimens.

Calculations were performed using the finite element method (FEM) on a 90° composite to determine the local tangential and radial stresses in the matrix immediately adjacent to a 90° fiber. The calculations were limited to linear elastic analysis under generalized plane strain conditions. The stress-free state was assumed to be 815.5°C (1500°F)[7] (approximate processing temperature) and the residual thermal stresses were determined at 315.5 and 649°C on cooling from that temperature. The following parameters were used for the constituent properties: E_{fiber} = 399.9 GPa [3], α_{fiber} = 4.86 $\times$ 10^{-6}/°C [1,3], α_{matrix} = 9.9 $\times$ 10^{-6}/°C [6], E_{matrix} = 99.3 GPa at room temperature [6], E_{matrix} = 97.2 and 72.4 GPa at 300 and 550°C, respectively [3], and E_{matrix} = 67.6 GPa at 649°C (result of current test on the matrix material), volume fraction of fibers, V_f = 0.4. The Poisson's ratio was assumed to be independent of temperature, and equal to 0.25 and 0.35 for the fiber and matrix, respectively. The constitutive properties at intermediate temperatures were interpolated from the preceding data. Only preliminary results of FEM analysis are presented here.

Based on a stress-free state at 815.5°C, the residual delamination stress (tangential stress) at 315.5°C was calculated as 261 MPa. This stress compares favorably with a tangential residual stress of 276 MPa calculated in Ref 1 for a temperature interval of 537.8°C. The residual delamination stress at 649°C was 78 MPa. Thus, thermal cycling over the range 315.5 to 649°C can lead to a delamination stress range of 183 MPa. This stress range may appear low compared with the tensile strength of the composite (approximately 684 MPa at 315.5°C). However, many of the delamination cracks were observed between closely spaced fibers, where the local volume fraction (V_f) of fibers was high (as much as 0.70). This local increase in V_f could significantly enhance residual stresses. Additionally, the fiber-matrix reaction zone that was observed in this and previous work [7], as well as the carbon-rich inter-ply zones, are likely to possess lower fracture strength than the matrix. Thus, local elevation of stress due to higher local V_f, and existence of weaker embrittled zones, both may have contributed to delamination cracking.

The effect of mechanical load was evaluated using FEM calculations, by imposing a stress of 266 MPa to the 90° composite at 649°C. This stress corresponds approximately to the stress in the 90° plies of [0,$\pm$45,90]$_s$ composite, when the latter is subjected to an external stress of 341 MPa (the maximum stress range in the current set of experiments). The delamination stress at 649°C, resulting from an applied stress of 266 MPa to the 90° composite, was 134 MPa. Using these stress values, it is possible to obtain rough estimates of delamination stresses in the 90° plies of the [0,$\pm$45,90]$_s$ composite under IF and TMF conditions. As an example, maximum and minimum applied stresses of 379 and 38 MPa (the maximum values used in the current experiments) were selected. Under IF conditions, the delamination stresses were estimated to cycle between 95 and 242 MPa at 649°C. In contrast, for TMF conditions, the delamination stresses were estimated to cycle approximately between 278 MPa at 315.5°C and 242 MPa at 649°C. It is noteworthy that the maximum delamination stress was found to occur under TMF conditions at 315.5°C, at which temperature the reaction zone (both around the fiber and along the carbon-segregated ply-to-ply interface, see Fig. 9c) is expected to be more brittle than at 649°C; note that the delamination stress is located along the ply-to-ply interface. The delamination stress also is expected to be aided by stresses arising from possible modulus mismatch between the reaction zone (which is likely to contain carbides because of observed carbon segregation) and rest of the matrix. Such additional stresses from reaction-zone/matrix modulus mismatch are expected to be higher at 315.5°C than at 649°C, because of the higher modulus of each phase at the lower temperature. Additionally, while a constant hold at 649°C (under IF conditions) would relax the delamination stress, every excursion to 315.5°C (under thermal or TMF cycling) would rejuvenate the delamination stress at 315.5°C. Such stress

---

[7] The temperature corresponding to the stress-free state might in reality be slightly less than 815.5°C, because of stress relaxation during slow cooldown from the processing temperature. Since deformation is assumed to be linear elastic, linear interpolation can be used to obtain approximate values of residual stresses for a slightly different stress-free temperature.

cycling may explain why delamination damage was so dominant in TMF and thermally cycled specimens, whereas it was rare in IF specimens.

With regard to debonding, the maximum debonding stress (radial tensile stress) was found to occur at the maximum load, and it was identical under IF and TMF conditions. On the other hand, at minimum load, the radial debonding stress under TMF conditions was -103 MPa whereas it was -32 MPa under IF conditions. Thus, a partial reason for differences in debonding characteristics may be that in TMF specimens, the top and bottom of 90° fibers were locally subjected to greater debonding stress amplitudes and lower R-ratios, compared with IF specimens. Accurate fracture mechanics based assessment of debonding (essentially interface cracking) and delamination phenomena are necessary, before a complete understanding of damage in a quasi-isotropic composite can be obtained. In this regard, the approach of Tirosh et al. [8] appears to be a good beginning, although their calculations were restricted to rigid fibers. These issues are being currently addressed by the authors.

It was indicated earlier that an important consequence of damage in off-axis plies (notably 90° plies) of the $[0, \pm 45, 90]_s$ composite is that load is transferred to the 0° plies, making them more vulnerable to failure. As an extreme case, let us assume that the 90° plies cease to carry any load early in the life of TMF specimens, and that TMF life is dominated by failure of the two 0° plies. The 45° plies are assumed to retain their original as-fabricated strength and compliance. Under this scenario, we have calculated the 0° fiber stress range using the nonlinear micromechanics-based METCAN code [5], and determined how this stress range correlates with fatigue life [2]. For the IF specimens, it is assumed that the 90° plies retain their original strength and compliance. The results are compared with room-temperature data in Ref 2, and also with high-temperature IF and TMF data obtained from Refs 3 and 4. Only uniaxial 0° composites were tested in Refs 3 and 4, and we have used the constitutive properties of the Ti 15-3/SCS6 system to derive the 0° fiber stresses from the listed applied stress ranges. The 0° fiber stress ranges at room temperature, for various ply layups, were obtained directly from Fig. 17 in Ref 2.

Figure 11 is a plot of 0° fiber stress range versus fatigue life for the Ti 15-3/SCS6 system. In this plot, the 0° fiber stress range corresponding to the current TMF specimens was obtained by assuming a ply-discount scheme, whereby the two 90° plies were assumed not to carry any load while all the other plies carried load based on their stiffness, according to the rule of mixtures. The ply-discount scheme was also used for the monotonic test at 649°C on the quasi-isotropic system. The reason was that tensile failure of the composite would be preceded by failure of the inner 90° plies. Ply-discount scheme was not used for any other data. The terminology "Uni" in Fig. 11 corresponds to a uniaxial 0° composite; RT refers to room temperature.

Figure 11 indicates reasonably good correlation between the 0° fiber stress range and fatigue life for various experimental conditions. If data corresponding to fatigue life at half-cycle (monotonic loading) are considered, then the figure indicates appreciable reductions in static strength of SCS6 fibers with increasing temperature. The fiber strength reductions appear to be much larger than expected, since 649°C may be considered to be a low temperature for SiC.

The curve corresponding to room-temperature data is interesting, in that its shape is drastically different from the plot in Ref 2. In Ref 2, the 0° fiber stress range was plotted versus fatigue life on a linear scale, whereas Fig. 11 is a semilogarithmic plot; this latter representation is the usual way fatigue data are reported in the literature for metal fatigue. The room-temperature data in Fig. 11 shows that beyond 3000 cycles, large reductions in 0° fiber stress had negligible effect on fatigue life. It is difficult to ascribe this effect to fiber fatigue, since bulk SiC (which does not cross-slip easily) is not known to fatigue at low temperatures. One would expect a much flatter curve if fiber fatigue was involved.

The uniaxial Ti 15-3/SCS6 results at elevated temperature (from Refs 3 and 4) appear to

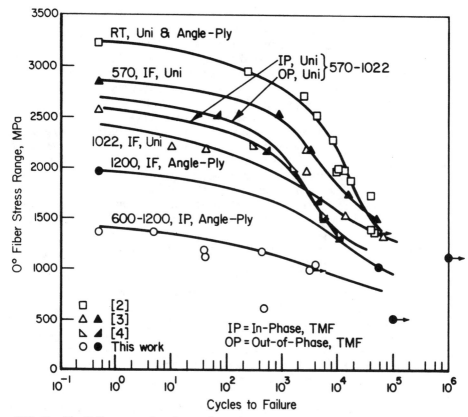

FIG. 11—*The 0° fiber stress plotted versus logarithm of fatigue life for Ti 15-3/SCS6 composite. The numbers 570, 600, 1022, and 1200 refer to the test temperatures in °F; the corresponding values are 300, 315.5, 550 and 649 in °C. Uni refers to a 0° uniaxial composite. Data from this and previous investigations are included in the plot.*

fall within a band. The band has a flat shape for fatigue life (N_f) less than 1000 cycles. In this domain, fatigue life may be dominated by failure of fibers. Beyond 1000 cycles, the band has a strong negative slope, possibly implying a different fatigue mechanism. In this domain, fatigue life may be matrix dominated or interface dominated [4].

For the current quasi-isotropic system, the IF data are somewhat consistent with the other results at RT, 300°C (572°F), and 550°C (1022°F), although additional data are necessary to establish the IF curve at 649°C (1200°F). More importantly, the 0° fiber stresses corresponding to the TMF specimens are very low; they are as much as 690 MPa less than for uniaxial IF specimens tested [3] at 550°C (which is only slightly lower than 649°C). The current TMF data are all the more significant because a 90° ply-discount scheme was used for the TMF specimens, whereas it was not used for the IF specimens; the only exception was the monotonic test result at 649°C on the quasi-isotropic system. If it is assumed that fatigue life is fiber dominated for N_f less than 1000 cycles, then the current TMF results imply drastic reductions in fiber strength due to TMF loading. It is difficult to rationalize such a possibility; hence, a straightforward explanation based on 0° fiber failure has to await further analysis. One possible reason for the observed TMF behavior may be that early cracking in off-axis plies, under slow strain-rate TMF conditions, may require a fracture mechanics analysis for determining the true 0°

fiber stress in the cracked composite. Pending those calculations, as well as evaluation of exact fatigue mechanisms for the other test conditions, the curves in Fig. 11 can currently only be considered phenomenological. In this regard, the similar shapes of curves for various test conditions appear to indicate a common trend, for which a theoretical explanation has yet to be developed.

There is one final note that may be important, and that is with regard to the test technique that was employed. The current experiments were performed under load control, and the first few cycles were observed to have higher strain range than the stabilized strain range for the test. On the other hand, most low-cycle fatigue experiments of metals are performed under strain control. This type of control system assures a constant strain range throughout a test. Although the small cross sections of composite specimens make it difficult to perform strain-controlled tests, such experiments may be needed to establish that fatigue lives are independent of the test technique employed.

Conclusions

The following conclusions can be made, based on the work presented here:

1. Under stress controlled loading, the measured strain amplitude of TMF and IF specimens stabilized to a constant value within a relatively short number of cycles. The stabilized value of mechanical strain range provided good correlation with thermomechanical fatigue life.

2. On an applied stress range basis or a gross mechanical strain range basis, TMF specimens had significantly shorter life than IF specimens.

3. Specimens that were thermally and thermomechanically cycled showed significant fiber-matrix debonding and delamination in 90 and 45° plies. The delamination cracks were aligned parallel to the loading direction, and appeared to be concentrated between closely spaced fibers.

4. The primary damage mode in isothermally fatigued specimens was transverse micro-cracks in 90° plies oriented perpendicular to the loading direction.

5. Transverse microcracks also were observed in TMF specimens, but these were localized very close to the fracture surfaces, and seemed to originate slightly before or during failure of the composite.

6. Microprobe analysis indicated carbon-rich zones along many of the ply-to-ply interfaces. These zones are expected to locally weaken the interply region, and may have accelerated delamination cracking.

7. FEM calculations indicated that delamination stress was highest for the TMF specimens, and it occurred at 315.5°C. This may explain why delamination cracks were dominant in thermally cycled and TMF specimens, and not in IF specimens cycled at 649°C.

8. Various fatigue data for the Ti 15-3/SCS6 composite were plotted in the form of 0° fiber stress range versus logarithm of fatigue life. Although the plot showed reasonably good correlation between the 0° fiber stress range and fatigue life for various sets of data, a satisfactory explanation for the trend of results is not yet available. In particular, the current TMF data appeared to indicate extremely low fiber strengths. A more rigorous analysis is needed to establish the phenomenological curves on a firm theoretical basis.

Acknowledgments

We thank McDonnell Douglas Corporation, in particular John Fogarty and Jim Sorenson, for making available the Ti 15-3/SCS6 composite specimens, and their active interest in this research area. We thank Dr. Frederick Brust of Battelle for the FEM calculations. We thank Norman Frey of Battelle for the experimental phase of the program.

References

[1] Johnson, W. S., Lubowinski, S. J., Highsmith, A. L., Brewer, W. D., and Hoogstraten, C. A., "Mechanical Characterization of SCS6/Ti 15-3 Metal Matrix Composites at Room Temperature," NASP Technical Memorandum 1014, NASA Langley Research Center, Hampton, VA, April 1988.

[2] Johnson, W. S., "Fatigue Testing and Damage Development in Continuous Fiber Reinforced Metal Matrix Composites," NASA TM-100628, NASA Langley Research Center, Hampton, VA, 1988.

[3] Gayda, J. G., Gabb, T. P., and Freed, A. D., "The Isothermal Fatigue Behavior of a Unidirectional SiC/Ti Composite and the Ti Alloy Matrix," NASA Technical Memorandum No. 101984, NASA Lewis Research Center, Cleveland, OH, April 1989.

[4] Gabb, T. P., Gayda, J., and Mackay, R. A., "Isothermal and Non-Isothermal Fatigue Behavior of a Metal Matrix Composite," submitted to *Journal of Composite Materials,* 1989.

[5] Chamis, C. C. and Hopkins, D., "Thermoviscoplastic Nonlinear Constitutive Relationships for Structural Analysis of High-Temperature Metal-Matrix Composites," *Testing Technology of Metal Matrix Composites, ASTM STP 964,* P. R. DiGiovanni and N. R. Adsit, Eds., American Society for Testing and Materials, Philadelphia, 1988, pp. 177–196.

[6] Rosenberg, H. W., "Ti 15-3 Property Data," *Beta Titanium Alloys in the 80's,* R. R. Boyer and H. W. Rosenberg, Eds., *Proceedings,* American Institute of Mechanical Engineers' Symposium, Atlanta, GA, TMS/AIME Publications, March 1983.

[7] Lerch, B. A., Hull, D. R., and Leonhardt, T. A., "As Received Microstructure of SiC/Ti-15-3 Composite," NASA Technical Memorandum No. 100938, NASA Lewis Research Center, Cleveland, OH, Aug. 1988.

[8] Tirosh, J., Katz, E., Lifschuetz, G., and Tetelman, A. S., "The Role of Fibrous Reinforcements Well Bonded or Partially Bonded on the Transverse Strength of Composite Materials," *Engineering Fracture Mechanics,* Vol. 12, 1979, pp. 267–277.

Rajiv A. Naik[1] and W. S. Johnson[2]

Observations of Fatigue Crack Initiation and Damage Growth in Notched Titanium Matrix Composites

REFERENCE: Naik, R. A. and Johnson, W. S., **"Observations of Fatigue Crack Initiation and Damage Growth in Notched Titanium Matrix Composites,"** *Composite Materials: Fatigue and Fracture (Third Volume), ASTM STP 1110,* T. K. O'Brien, Ed., American Society for Testing and Materials, Philadelphia, 1991, 753–771.

ABSTRACT: The purpose of this study was to characterize crack initiation and growth in notched titanium matrix composites at room temperature. Double-edge-notched or center-open-hole SCS-6/Ti-15-3 specimens containing 0° or both 0° and 90° plies were fatigued. The specimens were tested in the as-fabricated (ASF) and in heat-treated conditions. A local strain criterion using unnotched specimen fatigue data was used to estimate fatigue crack initiation with varying success. The initiation stress level was accurately predicted for both a double edge notched unidirectional specimen and a cross-plied center-hole specimen. The fatigue produced long multiple cracks growing from the notches. These fatigue cracks were only in the matrix material and did not break the fibers in their path. The combination of matrix cracking and fiber/matrix debonding appears to greatly reduce the stress concentration around the notches. The laminates that were heat treated showed a different crack growth pattern. In the ASF specimens, matrix cracks had a more tortuous path and showed considerably more crack branching. For the same specimen geometry and cyclic stress, the [0/90/0] laminate with a hole had far superior fatigue resistance than the matrix-only specimen with a hole.

KEY WORDS: metal matrix composite, silicon carbide fibers, fiber/matrix interface, composite materials, fatigue (materials), fracture

Titanium metal matrix composites (MMC) are being considered for high-temperature structural applications on man-rated aircraft. Prior to application, however, the fatigue and damage tolerance behavior of these materials must be well understood and be predictable. The purpose of this paper is to present an analytical and experimental study that describes some details of the fatigue crack initiation and growth in silicon carbide/titanium specimens containing either a crack-like slit or a circular hole. This study follows previous work on the unnotched, room-temperature behavior of silicon carbide/titanium composites [1] and on the effect of a high-temperature cycle on the mechanical behavior of this composite [2]. A brief review of the pertinent unnotched fatigue data from the previous reports will be given in the following section. This will be followed by a description of the material and specimens and the testing procedures used. Then the predicted crack initiation stress levels are compared to those observed experimentally. In addition, micrographs will show the crack growth progression as a function of applied load cycles. In some cases, the surface crack growth is compared to etched or polished specimens that reveal the subsurface damage in the matrix and fiber. Some rationale for the observed crack growth will be discussed.

[1] Research scientist, Analytical Services and Materials, Inc., Hampton, VA 23666.
[2] Senior research engineer, NASA Langley Research Center, Hampton, VA 23665-5225.

Background

Data for the applied maximum cyclic stress versus the number of load cycles to failure was presented for four layups of SCS-6/Ti-15V-3Cr-3Al-3Sn in Ref *1*. The room-temperature data for the as-fabricated (ASF) specimens are shown in Fig. 1. All of the layups tested, $[0]_8$, $[0_2/\pm45]_2$, $[0/\pm45/90]_2$, and $[0/90]_{2s}$, contained 0° plies. Since the fiber/matrix interfaces in the off-axis plies failed after a very few cycles [*1*], the 0° plies carried most of the load during the fatigue life of these specimens. Because of these interface failures in the off-axis plies, those laminates containing off-axis plies experienced a significant reduction in overall stiffness in the first few cycles. After the first few cycles, however, the stiffness remained constant for the rest of the fatigue life until just prior to specimen failure. The strain range was, therefore, also constant with the applied constant loading after the first few cycles. This strain range was multiplied by the fiber modulus of 400 GPa to yield the cyclic stress in the 0° fiber during essentially the entire fatigue life [*1*]. When the cyclic stress in the 0° fibers was calculated (using a micromechanics analysis [*3*]) for each data point in Fig. 1 and plotted against the number of cycles to fatigue failure, the data from the four different laminates collapsed into a narrow band, as shown in Fig. 2. This suggested that the fatigue life of a given laminate is a function of the stress in the 0° fiber. This is reasonable since none of the laminates tested will fail until the 0° plies fail. The cycles to failure as a function of the overall laminate strain range is also shown in Fig. 2. This data, in terms of the strain range, will be used later to predict local cracking using a strain range criterion.

Naik, Johnson, and Pollock [*2*] have shown that the mechanical behavior of an SCS-6/Ti-15-3 [0/90/0] laminate can be significantly effected by exposure to the high-temperature cycle associated with a superplastic forming/diffusion bonding (SPF/DB) process. The specific temperature cycle is described in the next section. Figure 3 shows S-N curves for the ASF and the SPF/DB materials. The data shows that the SPF/DB material experienced a 25% drop in static strength and a significant reduction in fatigue resistance. This degradation in mechanical behavior was attributed to a change in failure mode [*2*] caused by an increase in the fiber/matrix interface strength and an increase in the thermal residual stresses in the matrix surrounding the fiber. Thermal residual stresses were estimated by observing differences in the

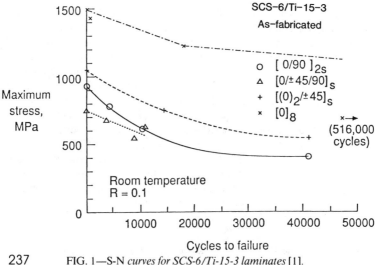

FIG. 1—*S-N curves for SCS-6/Ti-15-3 laminates* [1].

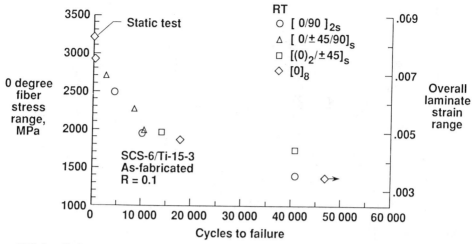

FIG. 2—*Cyclic stress range in 0° fiber and laminate strain range versus number of cycles to laminate failure* [1].

stress-strain behavior of the two materials [2]. Figure 4 shows a schematic of the failure processes described in Ref 2. In the ASF case, the interface was weak and the residual stresses were lower, thus allowing the matrix crack to debond along the length of the fiber and then continue on the other side of the fiber without breaking the fiber. In the SPF/DB case, the interface strength was strong enough, coupled with the higher thermal residual stresses [2], to cause the matrix crack to propagate through the fibers. As described in Ref 2, it can be shown using shear lag analysis [4] that the stress in the first unbroken fiber in the crack path is 35% higher when the crack grows without fiber/matrix debonding (SPF/DB case) compared to modeling fiber/

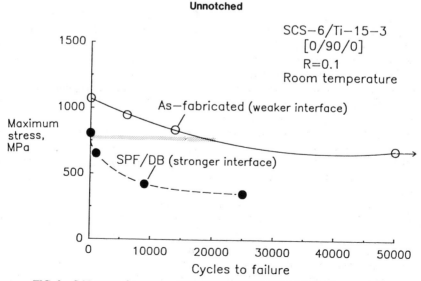

FIG. 3—*S-N curves for ASF and SPF/DB [0/90/0] SCS-6/Ti-15-3 laminates* [2].

Fiber failure behind crack tip

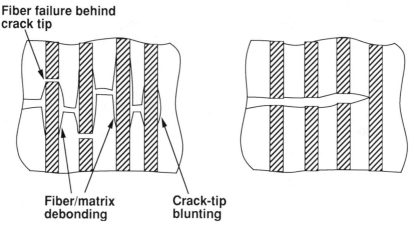

Fiber/matrix debonding   **Crack-tip blunting**

FIG. 4—*Failure mechanisms in 0° ply for ASF* (left) *and SPF/DB* [0/90/0] (right) *laminates* [2].

matrix debonding (ASF case). This explains, in part, the observed differences in the mechanical behavior of the ASF and SPF/DB specimens.

A similar change in the interface properties was found in Ref *1* for SCS-6/Ti-15-3 laminates that were aged at 482°C for 16 h. It was found that the aged specimens had a stronger fiber/matrix interface strength. This aging also increased the modulus of the Ti-15-3 material.

The present work will examine how fatigue cracks initiate and grow in the presence of a stress concentration and the applicability of the failure process observed in an unnotched coupon to the local area at a notch tip.

Experimental Procedures

Materials and Specimens

All specimens were made of SCS-6/Ti-15V-3Cr-3Al-3Sn (referred to as SCS-6/Ti-15-3). Ti-15-3 is a metastable beta titanium alloy [5]. The SCS-6 fibers are silicon carbide fibers that have a carbon core and a thin carbon-rich surface layer [6]. The typical fiber diameter is 0.142 mm. The composite laminates are made by hot-pressing Ti-15-3 foils (0.15 mm thick) between unidirectional tapes of SCS-6 fibers. The layups used in the present study were $[0]_8$, $[0/90]_{2s}$, and $[0/90/0]$. The thicknesses of the eight-ply and three-ply specimens were approximately 2 and 0.68 mm, respectively. The fiber volume fractions for the eight-ply and three-ply specimens were approximately 0.325 and 0.375, respectively.

All specimens were 19 mm wide and about 140 mm long and were cut using a diamond wheel saw. The $[0]_8$ specimens were machined using electro-discharge machining with edge notches that were 0.45 mm thick and 3 mm long (Fig. 5a). The $[0/90]_{2s}$ and $[0/90/0]$ specimens had center holes that were machined using a diamond core drill and were 6.35 mm in diameter (Fig. 5b).

All $[0]_8$ and $[0/90]_{2s}$ specimens were tested in the ASF condition, except one unidirectional edge notched specimen, which was tested in the aged condition. This aging was conducted at 482°C for 16 h.

The $[0/90/0]$ specimens were cut from two panels. One of the panels was in the ASF condition. The second panel was subjected to a thermal processing cycle that simulated a superplastic forming/diffusion bonding (SPF/DB) operation. This simulated SPF/DB cycle was per-

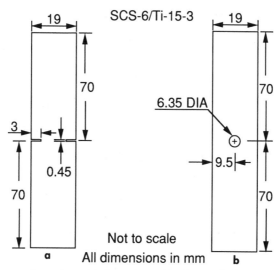

FIG. 5—*Specimen configuration and dimensions: (*a*) edge-notched specimen and (*b*) center-hole specimen.*

formed in a vacuum furnace and consisted of raising the temperature from ambient to 700°C at a rate of 10°C/min. After stabilizing at 700° C the temperature was further increased to 1000°C at a rate of 4°C/min. The panel was held at 1000°C for 1 h. It was then furnace cooled to 594°C at a rate of 8°C/min and held at that temperature for 8 h. Finally, it was furnace cooled to 150°C and held for about 10 h before cooling down to ambient temperature.

One rectangular unreinforced titanium specimen (19 by 140 mm) with a center hole (6.35 mm in diameter) was also tested. This unreinforced titanium specimen was a "fiberless composite" made by consolidating Ti-15-3 foils with the same temperature-time-pressure cycle used for the composite laminates.

Test Procedure and Equipment

The specimens were tested in a hydraulically-actuated, closed-loop, servocontrolled testing machine. The load, measured by a conventional load cell, was used for the feedback signal. All specimens were tested under constant amplitude fatigue at a frequency of 10 Hz and a stress ratio of R of 0.1. Two different loading approaches were used in the present study. To study crack initiation, some specimens were tested using an incremental loading approach. The $[0]_8$ and $[0/90]_{s2}$ ASF specimens, edge-notched and center-hole, were tested at a series of stress range values; at each stress range they were fatigued for 50 000 cycles before the stress range was increased (Table 1). To study crack growth, other specimens were tested at a constant stress range. The aged $[0]_8$ specimen and the $[0/90/0]$ ASF and SPF/DB specimens were tested at a constant stress range.

The fatigue tests were stopped periodically and the specimens were radiographed at 75% of the maximum cyclic stress with an industrial-type "soft" X-ray machine with a 0.25-mm-thick beryllium window and a tungsten target. The voltage was set at 50 kV for the $[0]_8$ and $[0/90]_{2s}$ specimens and at 40 kV for the $[0/90/0]$ specimens. A Kodak high-resolution X-ray film (Type M-II) was mounted to the specimen on the opposite side. The X-ray target-to-film distance was 610 mm. The $[0/90/0]$ specimens were exposed at 5 mA for 75 s, while the other

TABLE 1—*Test matrix and loading history (at* R = 0.1) *for the specimens tested at room temperature.*

Specimen Type and Condition	Figure	Cycles	Maximum Cyclic Stress, MPa
EDGE-NOTCHED SPECIMENS			
ASF, [0]₈	. . .	50 000[a]	115,122,160,168
ASF, [0]₈	8	50 000[a]	122,133,148,163,180,198,215, 239,262,290,315,350,383
ASF, [0]₈	7	50 000[a]	250,280,310,345,380,422
Aged, [0]₈	9 and 10	850 000	250
CENTER-HOLE SPECIMENS			
ASF, [0/90]₂ₛ	11 and 12	50 000 70 000	137 followed by 150
ASF, [0/90/0]	13	4 000 000	215
SPF/DB, [0/90/0]	14	2 680 000	215
ASF Matrix, Ti-15-3	. . .	31 000[b]	215

[a] Specimens were tested at each of the maximum cyclic stresses (at R = 0.1) for 50 000 cycles in the order in which they are listed.

[b] Only case where fatigue failure was observed; all other specimens did not fail after the completion of the preceding loading history.

specimens were exposed at 20 mA for 60 s. This procedure resulted in good contrast between the fibers and the matrix (see Fig. 6). A thin aluminum plate, 0.7 mm thick, was used as a filter and placed between the specimen and the X-ray tube. This resulted in good definition along the edges of the hole and the notches. Along with the X-rays, surface replicas and clip-gage readings were also taken periodically.

Before testing the specimens, the region around the hole and the notches was polished using fine-grain sandpaper to aid visual observations of fatigue crack initiation and growth. The polished surfaces also helped in getting good quality replicas of the cracks around the hole and the notches. Cellulose acetate film was used to make surface replicas at various stages of crack initiation and growth. The replicas were then studied under an optical microscope. For the open-hole specimens, a clip gage, located diametrically inside the hole, was used to record load-versus-hole elongation (along the loading direction) at various stages of the specimen history. Such a local load-elongation plot would show a marked change in the local stiffness if there were any fiber failures associated with the visible surface cracks.

After testing, some specimens were sectioned and polished to study the internal cracking in the matrix and the fibers. Other specimens were exposed to hydrofluoric acid to dissolve the surface layer of titanium to expose the fibers beneath.

Results and Discussion

Test results and discussions for the double edge-notched specimens are first presented for both the ASF and the aged specimens. These are followed by a description of the results for the [0/90]₂ₛ ASF and the [0/90/0] ASF and SPF/DB center-hole specimens. Finally, results for the center-open-hole "fiberless composite" specimen are presented.

Double Edge-Notched Specimens

Four double edge-notched specimens were tested: three ASF [0]₈ and one aged [0]₈ specimens. As mentioned earlier, the [0]₈ ASF specimens were tested using an incremental stress

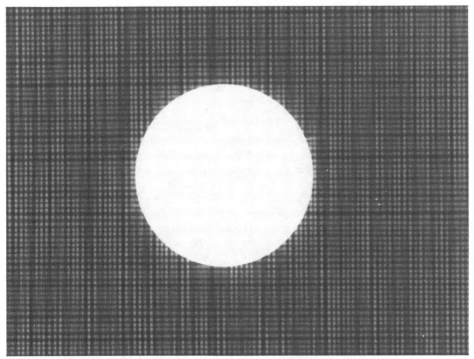

FIG. 6—*Radiograph of SCS-6/Ti-15-3 [0/90/0] center-hole specimen.*

range approach in which the specimen was fatigued at a series of stress ranges for 50 000 cycles (see Table 1). This incremental approach was used to determine the stress range at which cracks initiated at the notch tip. Thus, it was important to make a reasonable prediction of the crack initiation stress before starting the test. The data in Fig. 2 for unnotched SCS-6/Ti-15-3 laminates were used together with the computed notch stress concentration factor to make such a prediction. Extrapolating from the data in Fig. 2, the unnotched SCS-6/Ti-15-3 has an overall laminate strain range of 0.0033 at 50 000 cycles. Using a stress concentration factor of 5.61 for the edge-notched laminate, it can be shown that for an applied maximum cyclic stress of 122 MPa cracks would initiate at the notch tip after 50 000 cycles. The Appendix describes the calculations of the stress concentration factors and the maximum cyclic stress levels. To be certain that the crack initiation stress was seen, one of the $[0]_8$ ASF specimens was fatigued starting at a maximum cyclic stress of 115 MPa (Table 1). After 50 000 cycles at this stress level, there was no visible cracking observed at the notch tip. The stress level was then raised to 122 MPa and after 1000 cycles cracks initiated at the notch tip. This correlated with the predicted crack initiation stress level. A second ASF $[0]_8$ specimen was fatigued starting at a maximum cyclic stress of 122 MPa (Table 1), and crack initiation was observed after 25 000 cycles. This also agrees with the predicted crack initiation stress level. To study crack growth, a third ASF $[0]_8$ specimen was tested using the incremental stress range approach starting at a much higher value of maximum cyclic stress. Extrapolating from the data in Fig. 2, the unnotched $[0]_8$ laminate has an overall laminate strain range of 0.006 after 5000 cycles. Using the rule of mixtures and the computed stress concentration of 5.61, this corresponds to a maximum cyclic stress of 250 MPa. Cracking initiated in this specimen after 7500 cycles at a maximum cyclic stress of 250 MPa. In these cases, predictions made by the local strain range approach using unnotched fatigue data correlated fairly well with notched test data. The aged $[0]_8$ specimen was tested at a constant stress range for the entire test in order to study fatigue crack growth.

Figure 7 shows micrographs of surface replicas of the regions near the notch tips for an $[0]_8$ ASF specimen (see Table 1 for load history). All replicas were taken after 50 000 cycles at a particular maximum cyclic stress. Figure 7a shows crack growth in its early stages. Cracks start to grow from the corners of the notches at angles between 30 and 45° to the fiber direction. At higher loads, secondary cracks initiate from the notches and grow at angles approaching the fiber direction (Fig. 7b and c). There is also evidence of crack branching. The first 30 to 45° cracks change direction and tend to grow perpendicular to the fiber direction. At still higher loads (Fig. 7d), the secondary cracks become the dominant cracks and start growing almost parallel to the fiber direction. Again, there is more evidence of crack branching.

To determine if fiber failure was associated with the surface cracking seen in Fig. 7, some specimens were treated with hydroflouric acid after testing. This treatment dissolved the titanium from the surface. The acid first seeped into the surface cracks and, thus, etched deeper in regions where there were surface matrix cracks. This led to the fibers being more clearly visible in these regions. Such preferential etching of the surface layer preserved the original crack pattern, to some extent, while exposing the fibers directly beneath the surface cracks. Figure 8 shows micrographs of the acid-etched ASF $[0]_8$ specimen along with photographs of replicas showing corresponding surface cracking. All of the fibers underneath the surface cracks appear to be intact. The absence of fiber failure was also confirmed by the radiographs.

The crack growth in the aged $[0]_8$ specimen is shown by the photographs of surface replicas in Fig. 9. The cracks at the left notch tip initiate at about 30 to 45° to the fiber direction. The crack at the right notch tip started in a direction almost perpendicular to the fiber direction and then branched into two 30 to 45° cracks. The initial growth perpendicular to the fiber direction may have been due to a broken fiber that was found ahead of the notch, which could have forced the matrix to crack at that point. This fiber may have been broken by the machining process. After about 100 000 cycles these 30 to 45° cracks change direction and continue growing perpendicular to the fiber direction. Since the aged specimen was tested at a constant maximum cyclic stress of 250 MPa while the ASF specimens were tested at incremental levels of stress, the results from the two specimen types cannot be compared. However, as seen in Fig. 7a, the crack growth in the ASF specimens showed crack branching after 50 000 cycles at a maximum cyclic stress level of 280 MPa. After 75 000 cycles at 250 MPa, there were no signs of crack branching (Fig. 9b) in the aged specimen.

Figure 10 shows micrographs of the aged $[0]_8$ specimen before and after it was treated with hydrofluoric acid. All of the fibers underneath the surface cracks appear to be intact. There is only one broken fiber visible next to the right notch tip and, as just mentioned, this could have been the result of the notch machining process.

For both the ASF and aged $[0]_8$ notched specimens, the cracks grew only in the matrix leaving the fibers intact. The test stress levels were chosen by considering the stress concentration of the notch and the strain range at which fatigue failure had been observed for the unnotched laminate (Fig. 2). It was, therefore, expected that the fibers would break shortly after the matrix cracked. The lack of fiber failure suggests that the multiple matrix cracks with or without fiber/matrix debonding reduced the notch tip stress concentration to a level below the fatigue life of the SCS-6 fibers of about 1320 MPa at 50 000 cycles (see Fig. 2).

The initiation of matrix cracks at 45° to the fiber direction in both the ASF and aged $[0]_8$ specimens was also unexpected. According to a finite element analysis (see Appendix) of the notched specimen, the most highly stressed region lies along a line joining the two edge notches; however, matrix cracks grew along a 45° line from the notch tips. One explanation for the 45° cracks could be fiber/matrix debonding of the fiber next to the notch. According to

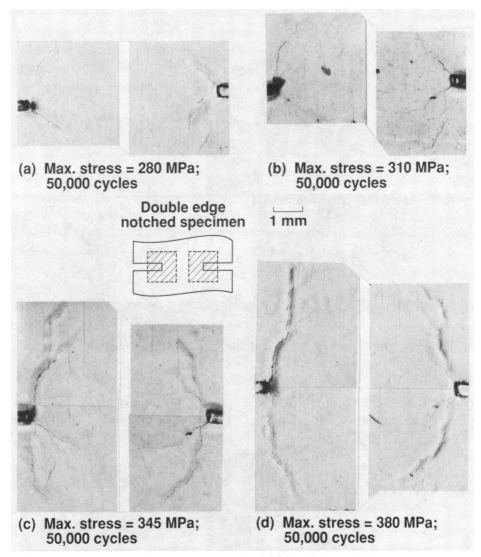

(a) Max. stress = 280 MPa;
    50,000 cycles

(b) Max. stress = 310 MPa;
    50,000 cycles

**Double edge
notched specimen**   1 mm

(c) Max. stress = 345 MPa;
    50,000 cycles

(d) Max. stress = 380 MPa;
    50,000 cycles

FIG. 7—*Surface replicas of damage growth in SCS-6/Ti-15-3 ASF [0]₈ specimen, stress ratio = 0.1.*

the micromechanics analysis in Ref 7, there are high tensile stresses acting normal to fiber just ahead of the notch. These stresses could lead to fiber/matrix debonding before the cyclic load cracks the matrix. Such a debond just ahead of the notch tip would redistribute the stresses locally causing a shift in the critical crack initiation region away from the net section.

Center-Hole Specimens

Three specimens containing center holes were tested: one ASF [0/90]$_{2s}$, one ASF [0/90/0], and one SPF/DB [0/90/0]. The [0/90]$_{2s}$ ASF specimen was tested using the incremental stress range approach described earlier. Both the [0/90/0] specimens were tested at a constant stress

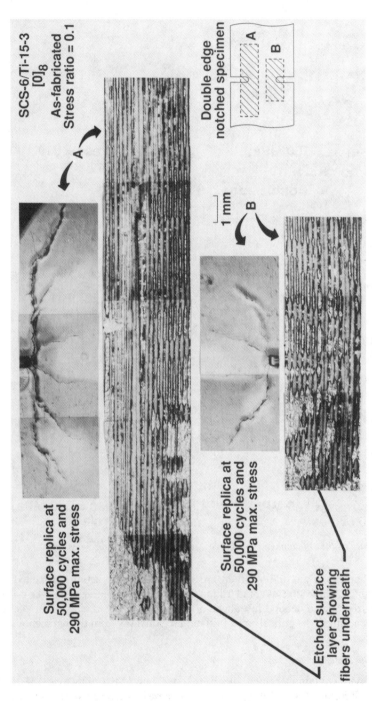

SCS-6/Ti-15-3
[0]₈
As-fabricated
Stress ratio = 0.1

Double edge notched specimen

A

B

1 mm

Surface replica at 50,000 cycles and 290 MPa max. stress

Surface replica at 50,000 cycles and 290 MPa max. stress

Etched surface layer showing fibers underneath

FIG. 8—Surface replicas and etched surface layer for ASF [0]₈ specimen.

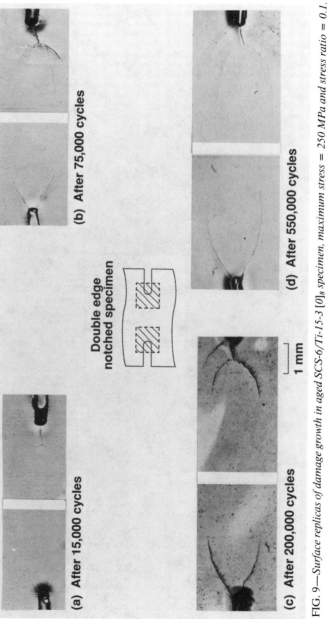

(a) After 15,000 cycles

(b) After 75,000 cycles

Double edge
notched specimen

(c) After 200,000 cycles

(d) After 550,000 cycles

1 mm

FIG. 9—Surface replicas of damage growth in aged SCS-6/Ti-15-3 [0]$_8$ specimen, maximum stress = 250 MPa and stress ratio = 0.1.

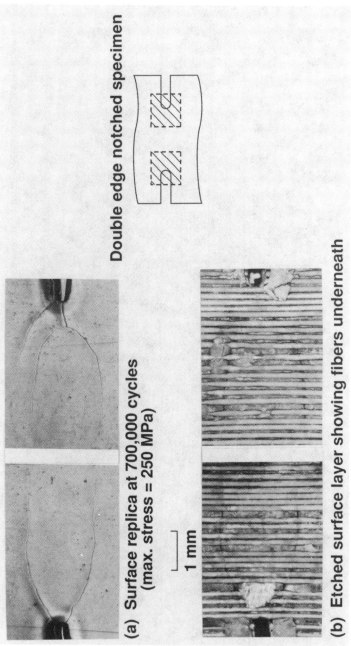

Double edge notched specimen

(a) Surface replica at 700,000 cycles (max. stress = 250 MPa)

1 mm

(b) Etched surface layer showing fibers underneath

FIG. 10—Surface replicas and etched surface layer for aged SCS-6/Ti-15-3 $[0]_8$ specimen, stress ratio = 0.1.

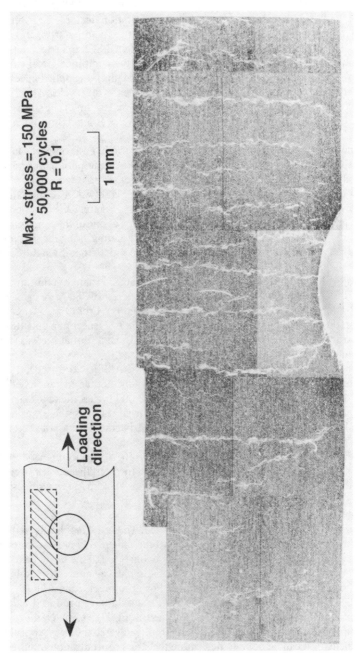

FIG. 11—*Surface damage for center-hole ASF SCS-6/Ti-15-3 [0/90]₂ₛ specimen.*

range in order to study fatigue crack growth and to directly compare the behavior of the ASF and the SPF/DB specimens.

For the $[0/90]_{2s}$ specimen, the data in Fig. 2 for unnotched SCS-6/Ti-15-3 laminates were used together with the computed hole stress concentration factor (see Appendix) and AGLPLY [3] to make a prediction of the crack initiation stress before starting the incremental stress testing. Extrapolating the data in Fig. 2, the unnotched SCS-6/Ti-15-3 has an overall laminate strain range of 0.0033 at 50 000 cycles. Using the computed stress concentration factor of 3.60 for the center hole $[0/90]_{2s}$ laminate (see Appendix), an applied maximum cyclic stress of 160 MPa would result in local crack initiation after 50 000 cycles. To be certain that the crack initiation stress was seen, testing was started at a maximum cyclic stress of 137 MPa (see Table 1). Crack initiation was observed after 1000 cycles at a maximum cyclic stress of 150 MPa.

Figure 11 shows a photograph of the multiple crack pattern that was observed in the center-hole $[0/90]_{2s}$ specimen. The cracks appear as light streaks in the photograph. This type of cracking was also observed by Harmon and Saff [8] for center-hole specimens of SCS-6/Ti-15-3. To study the internal damage, the specimen shown in Fig. 11 was sectioned and polished. Figure 12 shows micrographs of the damage in the 0° ply. Figure 12a shows cracks in the matrix but no fiber failure. Figure 12b shows a close-up view of one of these matrix cracks and the region around it. There is evidence of fiber/matrix debonding along the first fiber next to the hole and at both ends of the matrix cracks. This corresponds to the ASF failure mechanism described in Ref 2 (see Fig. 4).

The [0/90/0] ASF and SPF/DB specimens were tested at a constant maximum cyclic stress of 215 MPa. This corresponds to a local maximum cyclic stress of 780 MPa using a laminate stress concentration factor of 3.62 at the hole (see Appendix). From Fig. 3, this stress level should have produced local cracks in 17 000 cycles for the ASF specimen and 100 cycles for the SPF/DB. Surprisingly, for the ASF specimen tested, crack initiation was observed at 50 000 cycles and at 40 000 cycles for the SPF/DB specimen.

Figure 13 shows micrographs of surface replicas for the ASF [0/90/0] specimen. There is evidence of multiple matrix cracks similar to that seen in the double edge-notched ASF specimens. There is also crack branching. After about 1 000 000 cycles, the cracks had grown to the edges of the specimen. However, based on the clip gage readings, there was only a small change in the local stiffness. This indicated that the matrix cracks were not accompanied by fiber failure. This was also confirmed by the radiographs.

Figure 14 shows fatigue crack growth for the SPF/DB [0/90/0] specimen. Multiple cracks around the hole were observed as in the ASF specimen. However, these cracks were straighter than those in the ASF specimen and showed no branching. The clip gage readings showed only a small change in the local stiffness, indicating a lack of fiber failure. This was confirmed by the radiographs.

The lack of fiber failure was again most unexpected. The test stress levels were chosen by considering the stress concentration of the hole and the fatigue S-N curve of the unnotched laminate. The applied stress should have been high enough to fail the fibers next to the hole. Apparently, the stress concentration at the hole was also reduced after a few cycles by matrix cracking with or without fiber/matrix debonding. For example, for the ASF [0/90/0] specimen, an applied maximum cyclic stress of 215 MPa should have produced local cracking, including fiber failure, at approximately 17 000 cycles (see Fig. 3). However, the matrix cracks and the fiber/matrix debonding near the hole apparently lowered the stress concentration.

A simple calculation can be done to determine the stress concentration required to be just below the fatigue run-out life of the 0° fibers at 50 000 cycles. Extrapolating from the data in Fig. 2, the 0° fiber stress range at 50 000 cycles is approximately 1320 MPa. This corresponds to a strain range in the fiber, and also in the composite, of 0.0033. Thus, the strain range in the [0/90/0] composite must be at least 0.0033 in order to break fibers after 50 000 cycles.

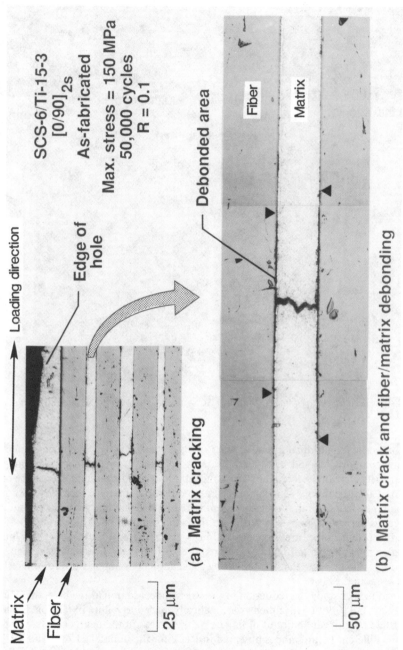

SCS-6/Ti-15-3
[0/90]₂s
As-fabricated
Max. stress = 150 MPa
50,000 cycles
R = 0.1

Loading direction

Edge of hole

Matrix

Fiber

25 μm

(a) Matrix cracking

Debonded area

Fiber

Matrix

50 μm

(b) Matrix crack and fiber/matrix debonding

FIG. 12—*Damage in 0° ply of [0/90]₂s ASF specimen with center hole.*

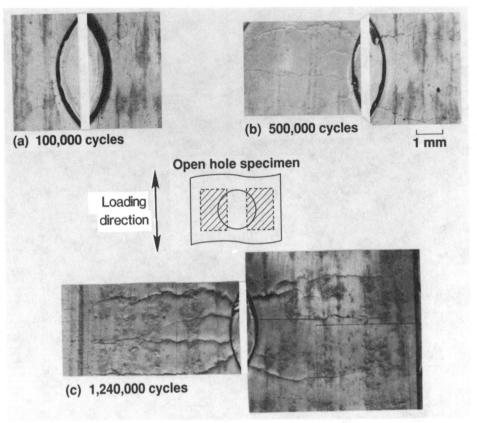

(a) 100,000 cycles

(b) 500,000 cycles

1 mm

Open hole specimen

Loading direction

(c) 1,240,000 cycles

FIG. 13—*Surface replicas of damage growth in SCS-6/Ti-15-3 [0/90/0] ASF specimen, maximum stress = 215 MPa.*

Using a composite longitudinal modulus of 156 GPa [2] and $R = 0.1$, the maximum cyclic stress in the composite should be at least 572 MPa in order to break fibers. Assuming the matrix carries its share of the load, the stress concentration would, therefore, have to be below 2.66 (computed as the ratio, 572 MPa/215 MPa) in order to not fatigue the fibers next to the hole beyond their fatigue limit. It does not seem unreasonable that fiber/matrix debonding at the hole could lower the stress concentration from the original 3.62 to below 2.66. These results indicate that if the stress concentration is reduced (for example, by fiber/matrix debonding), a much higher laminate stress is required before the 0° fibers and, subsequently, the composite will fail. Note that the preceding calculation does not account for the residual thermal stresses in the fiber.

In order to further study the reduction in the stress concentration at the hole, the ASF [0/90/0] and SPF/DB [0/90/0] specimens were statically loaded to failure in tension. Since the specimens had a significant amount of fatigue matrix cracks in the net section (see Figs. 13 and 14), the different failure modes observed in Ref 2 for the unnotched ASF and SPF/DB specimens (see Fig. 4) were not seen in these tests. Furthermore, since the 0° fibers carried most of the load, the two specimen types were expected to fail at about the same stress level at which the 0° fibers failed. The ASF and SPF/DB specimens failed, as expected, at a gross stress of 477 and 453 MPa, respectively. Since these failure stress levels were within 5% of each other, an

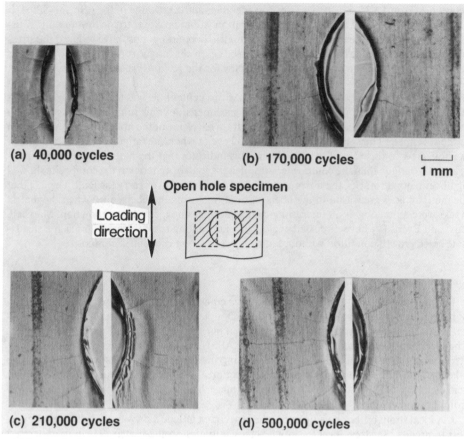

(a) 40,000 cycles

(b) 170,000 cycles

1 mm

Open hole specimen

Loading
direction

(c) 210,000 cycles

(d) 500,000 cycles

FIG. 14—*Surface replicas of damage growth in SCS-6/Ti-15-3 [0/90/0] SPF/DB specimen, maximum stress = 215 MPa and stress ratio = 0.1.*

average failure stress of 465 MPa was assumed for both the ASF and SPF/DB specimens. This failure stress together with the fiber strength can be used to compute two limiting cases for the stress concentration at the hole. Based on the results in Ref 2, the average fiber strength of the SCS-6 fibers used in the [0/90/0] laminates in this study is 3675 MPa. The two limiting cases are: (1) fiber failure occurs next to the hole with no matrix cracking so that the matrix continues to carry its share of the load or, (2) fiber failure occurs after severe matrix cracking, thereby shedding all the load to the fibers. In the first limiting case, consider the matrix around the hole to be intact. When the stress in the 0° fibers next to the hole reaches the fiber failure stress of 3675 MPa, the stress in the composite in that region can be calculated, using AGLPLY [3], as 1746 MPa. For an applied stress of 465 MPa (the average failure stress), this requires the stress concentration to be 3.75, computed as the ratio (1746 MPa/465 MPa). In the second limiting case, consider that, due to severe matrix cracking, all of the load is carried by the fibers in the two 0° plies. In this case, when the stress in the 0° fibers next to the hole reaches 3675 MPa, the stress in the composite next to the hole is 931 MPa (computed using a fiber volume fraction of 0.38 and the fact that two thirds of the cross section is occupied by the 0° plies). For an applied stress of 465 MPa, this corresponds to a stress concentration of 2.00. Therefore, in test specimens with matrix cracks in the net section, the stress concentration at the hole would be between 2.00 and 3.75. Since there was considerable matrix cracking in the ASF and SPF/DB [0/90/0] specimens, the actual stress concentration would be closer to the lower bound. Furthermore, as discussed earlier, the local fiber/matrix debonding must have reduced the stress concentration to a value below 2.66 in order to be below the fatigue run-out life of the 0° fibers at 50 000 cycles. The stress concentration factor for the [0/90/0] specimens was, therefore, between 2.0 and 2.66.

Finally, an unreinforced titanium specimen with a center hole was tested at the same stress range used for the [0/90/0] ASF and SPF/DB specimens. A crack initiated for this specimen at about 25 000 cycles and a single crack grew from each side of the hole along the net-section of the specimen. The specimen failed at 31 000 cycles when the fatigue crack on one side of the hole grew to the edge of the specimen. This indicates that the matrix alone is much less resistant to fatigue than the composite. Although the matrix material in the composite showed significant fatigue cracks, the fibers acted to retain stiffness and to carry the load. The fact that the matrix cracks grew around the fibers without fracturing the fibers, even when the matrix cracks were quite long, is an interesting, and somewhat unexpected, phenomenon. Marshall, Cox, and Evans [9] have discussed analytically how a long matrix crack can reach a steady-state crack growth condition without failing fibers in a ceramic matrix composite.

Conclusions

The purpose of this study was to characterize crack initiation and growth in notched titanium matrix composites at room temperature. Double edge-notched or center-hole SCS-6/Ti-15-3 specimens containing either unidirectional plies or both 0 and 90° plies were fatigued. The specimens were tested in the as-fabricated (ASF) or in one of two heat-treated conditions. Replicas of the surface cracks were taken during the fatigue testing. Radiographs were also taken periodically during testing to monitor fiber breaks. Several of the specimens were either acid etched or polished to reveal subsurface damage. The following conclusions were derived from this investigation.

1. A local strain criterion using unnotched specimen fatigue data was used with varying success to predict the stress level for fatigue crack initiation in the matrix. The initiation stress level was accurately predicted for both a unidirectional double edge-notched specimen and a cross-plied center-hole specimen.

2. The fatigue loading produced long multiple fatigue cracks growing from the notches. These fatigue cracks were only in the matrix material and did not break the fibers in their path.

3. The combination of matrix cracking and fiber/matrix debonding appears to greatly reduce the stress concentration around the notches.

4. The laminates that were heat treated (either aged or SPF/DB) showed fatigue behavior similar to the ASF specimens. However, the matrix cracks in the ASF specimen were more tortuous and showed considerably more crack branching.

5. For the same notch geometry and cyclic stress, the [0/90/0] laminate had far superior fatigue resistance than the matrix material alone.

APPENDIX

The stress concentration factors for the edge-notched specimens and the center-hole specimens were determined by two different techniques. In both cases, the specimen was assumed to be a homogeneous, orthotropic laminate and the laminate properties were obtained using AGLPLY [3] and the fiber and matrix properties from Ref 1.

The stress concentration factor for the edge-notched specimens was calculated from a two-dimensional finite element analysis. Isoparametric, quadrilateral elements were used with a very fine mesh refinement next to the notch. A convergence study for the mesh refinement was conducted by comparing the computed stress concentration factor for an isotropic specimen with handbook values [10]

The stress concentration factor for the center-hole specimens was calculated using the equations in Ref 11 for orthotropic laminates. The effects of finite width on the stress concentration were accounted for by the equations in Ref 11.

In order to calculate the maximum cyclic stress levels for crack initiation for both the edge-notch and the center-hole configurations, the AGLPLY [3] program was used to calculate the overall laminate stiffness for an unnotched laminate starting from the constituent fiber and matrix properties [1]. This overall stiffness was used together with the overall strain range for crack initiation and the stress concentration factor to compute the cyclic stress range and also the maximum cyclic stress level for crack initiation.

References

[1] Johnson, W. S., Lubowinski, S. J., and Highsmith, A. L., "Mechanical Characterization of Unnotched SCS-6/Ti-15-3 Metal Matrix Composites at Room Temperature," *Thermal and Mechanical Behavior of Ceramic and Metal Matrix Composites, ASTM STP 1080,* J. M. Kennedy, H. H. Moeller, and W. S. Johnson, Eds., American Society for Testing and Materials, Philadelphia, 1990, pp. 193–218.

[2] Naik, R. A., Johnson, W. S., and Pollock, W. D., "Effect of a High Temperature Cycle on the Mechanical Properties of Silicon Carbide/Titanium Metal Matrix Composites," *Proceedings,* American Society for Composites, Dayton, OH, 1989, pp. 94–103.

[3] Bahei-El-Din, Y. A. and Dvorak, G. J., "Plasticity Analysis of Laminated Composite Plates," *Journal of Applied Mechanics,* Vol. 49, 1982, pp. 740–746.

[4] Goree, J. G. and Gross, R. S., "Stresses in a Three-Dimensional Unidirectional Composite Containing Broken Fibers," *Engineering Fracture Mechanics,* Vol. 13, No. 2, 1980, pp. 395–405.

[5] Rosenberg, H. W., "Ti-15-3: A New Cold-Formable Sheet Titanium Alloy," *Journal of Metals,* Vol. 35, No. 11, Nov. 1986, pp. 30–34.

[6] Wawner, F. E., Jr., "Boron and Silicon Carbide/Carbon Fibers," *Fibre Reinforcements for Composite Materials,* A. R. Bunsell, Ed., Elsevier Science Publishers B. V., Amsterdam, The Netherlands, 1988.

[7] Chai, L. and Dharani, L. R., "A Micromechanics Model for Prediction of Failure Modes in Ceramic Matrix Composites," Final Report, NASA Contract No. NA53-25333, NASA Lewis Research Center, Cleveland, OH, Aug. 1988.

[8] Harmon, D. M. and Saff, C. R., "Damage Initiation and Growth in Fiber Reinforced Metal Matrix Composites," *Metal Matrix Composites: Testing, Analysis, and Failure Modes, ASTM STP 1032,* W. S. Johnson, Ed., American Society for Testing and Materials, Philadelphia, 1989, pp. 237–250.

[9] Marshall, D. B., Cox, B. N., and Evans, A. G., "The Mechanics of Matrix Cracking in Brittle Matrix Composites," *Acta Metallurgica,* Vol. 33, No. 11, 1985, pp. 2013–2021.

[10] Peterson, R. E., *Stress Concentration Design Factors,* Wiley, New York, 1953.

[11] Harmon, D. M., Saff, C. R., and Sun, C. T., "Durability of Continuous Fiber Reinforced Metal Matrix Composites," AFWAL-TR-87-3060, Air Force Wright Aeronautical Laboratories, Wright Paterson Air Force Base, OH, Oct. 1987.

Ricardo Osiroff,[1] *Wayne W. Stinchcomb,*[1] *and Kenneth L. Reifsnider*[1]

Damage and Performance Characterization of ARALL Laminates Subjected to Tensile Cyclic Loading

REFERENCE: Osiroff, R., Stinchcomb, W. W., and Reifsnider, K. L., **"Damage and Performance Characterization of ARALL Laminates Subjected to Tensile Cyclic Loading,"** *Composite Materials: Fatigue and Fracture (Third Volume), ASTM STP 1110,* T. K. O'Brien, Ed., American Society for Testing and Materials, Philadelphia, 1991, pp. 772–790.

ABSTRACT: The behavior of ARALL (ARamid ALuminum Laminates) subjected to tension-tension cyclic loading was experimentally investigated as a first step towards the understanding of the long-term behavior of ARALL laminates. Specifically, this work addresses fatigue damage mechanisms and relationships between damage and stiffness change, remaining strength, and life. The quasi-static and dynamic material response of unnotched ARALL-2 coupons were measured at normalized maximum stress levels (S) ranging from 0.4 to 0.9 of the ultimate tensile strength. The damage mechanisms and failure modes changed over this range of cyclic stresses. While at low cyclic stress levels, the fatigue properties of the fiber-reinforced plies are key factors; at high cyclic stress levels, the laminate's response is governed by the aluminum plies. Five distinct stages were recognized in the damage sequence. A shear lag analysis is presented to model the regular spacing of cracks called the characteristic damage state. Appropriate modifications were made to accommodate for the hybrid nature of the laminate.

KEY WORDS: composite materials, hybrid laminates, fatigue (materials), damage mechanisms, stiffness reduction, fracture

This phase of an investigation of the long-term behavior of ARALL (ARamid ALuminum Laminates) focuses on developing an understanding of the damage modes occurring in ARALL laminates subjected to cyclic mechanical loads. Destructive and nondestructive techniques were used to evaluate the state of the material. The different damage mechanisms were identified and correlated with the effects of mechanical loading on stiffness, remaining strength, and life. A shear lag model was modified to predict the characteristic damage state, a stable array of cracks that develops during cyclic loading, at certain stress levels, in the aluminum plies.

Long-Term Behavior of ARALL Laminates

ARALL Laminates

ARALL laminates consist of alternating thin aluminum sheets bonded by aramid fiber-reinforced adhesive film, Fig. 1. Primarily developed for civil aircraft fatigue and fracture critical structures, ARALL laminates offer many advantages over monolithic aluminum and conven-

[1] Graduate project assistant, professor, and professor, respectively, Materials Response Group, Engineering Science and Mechanics Department, Virginia Polytechnic Institute and State University, Blacksburg, VA 24061-0219.

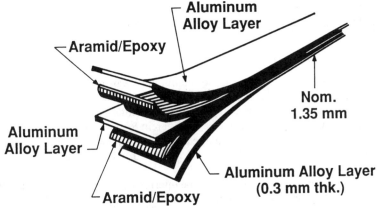

FIG. 1—*Schematic illustration of the 3/2 ARALL layup.*

tional fiber-reinforced composites. Particularly, it retains the plasticity, formability, ease of manufacture and supportability of metals. ARALL also offers a substantial improvement of the tensile fatigue response. These attributes contribute to a significant potential for structural weight reduction.

Vogelesang et al. [1] presented and discussed the concept of ARALL hybrid laminates; Bucci and Mueller [2] reviewed the extensive mechanical characterization of ARALL laminates. Significant analytical efforts complement the experimental endeavors. Teply [3] and Yeh [4] presented analytical models at the micromechanical, laminate, and structural levels that can be used to predict lamina engineering properties, process dependence, and residual stresses.

Fatigue Response of ARALL Laminates

The most common approach to the study of fatigue of ARALL laminates has been to treat the fatigue response as a fatigue crack growth, fracture mechanics problem. The emphasis has been on the growth of cracks in the aluminum layers originating from different notched configurations. The remarkable improvement in response ARALL laminates have shown when compared to monolithic aluminum is attributed to crack bridging; that is, to the closing forces exerted by the fibers at the flanks of cracks in the aluminum. Teply [3] and Marissen [5] have studied this mechanism extensively and concluded that the crack bridging effect due to unbroken aramid fibers in the wake of the crack in the aluminum is intensified by favorable residual stresses. Marissen's model is based on the hypothesis that crack growth is controlled by the stress intensity factors in the aluminum. Delamination of the aluminum/epoxy adhesive interface during fatigue occurs simultaneously with crack growth. Delamination growth rate is dependent on the load carried by the fiber plies, which in turn is related to the crack length.

Ritchie et al. [6] investigated the effect of crack-tip shielding from crack bridging on fatigue crack propagation using some novel experimental techniques. Crack bridging, which is promoted by controlled delaminations, was found to be the major contributor to the excellent crack growth resistance. Fatigue crack growth rates were found to be crack-size and history dependent and showed no unique correlation with the applied stress-intensity range.

It is evident from the experience of previous investigations that the fatigue response of ARALL laminates is a process greatly affected by the integrated effects of all damage modes, including cracking of the aluminum plies, delaminations, matrix cracking and fiber fracture.

The purpose of this paper is to study the fatigue response of ARALL as a hybrid composite material and to address the general forms of fatigue damage.

Stiffness as an Experimental Observable

Fatigue damage in composites consists of combinations of matrix cracks, broken fibers, interfacial cracks, and debonds that form a very complex damage state. The material system, stacking sequence, geometry, stress state, and environmental factors interact and affect the engineering properties in many intricate ways. Different loading conditions result in distinctive fatigue processes that cause changes in the local geometry and changes of local stress. The mechanics of fatigue in composite materials are reviewed in a recent volume edited by Reifsnider [7].

Available techniques of nondestructive testing interrogate the material and assess the damage state by measuring variations in material uniformity or material properties or both caused by imperfections. A variety of nondestructive methods (such as surface replication, X-ray radiography, vibrothermography, ultrasonic C-scans, acoustic emission, etc.) have been developed or modified specifically for application to composite materials [8].

Special attention has been given to the measurement of stiffness, since its reduction has been found to correlate well with damage in many cases and depends only on the material's condition [9]. O'Brien [10] presented applications of stiffness measurements for indirect assessment of damage growth for different configurations and loading conditions. Reifsnider and Stinchcomb [11] reviewed, described, and investigated the concept of stiffness change as a nondestructive fatigue damage parameter. In general, they found that stiffness change can be quantitatively related to the fatigue life and residual strength of composite laminates through various models of observed microdamage events and the damage patterns formed by those events. During cyclic loading, several damage modes may occur, resulting in a corresponding degradation of stiffness. Every mode is associated with a specific and distinct stage in a typical damage-versus-life fraction curve. As damage develops, the strength is changed. Fracture occurs when strength is reduced to the applied stress level. Osiroff and Stinchcomb [12] summarize some of the many ongoing investigations to understand the exact relationship between damage state, stiffness, residual strength, and fatigue life and to develop appropriate models.

Reifsnider and co-workers [13] take a hybrid approach; a stiffness-based cumulative damage model is used to predict residual strength and life. Subcritical and critical elements within a representative volume are identified. The former are involved in fatigue damage development, while the latter are responsible for the eventual failure. Stiffness changes and models of the damage events are used to estimate stress redistribution. Residual strength is evaluated at every point in time; and, with the aid of an appropriate failure theory, life is predicted.

Experimental Procedures and Results

Materials and Methods

The specimens tested in this study were cut from a 1.2 by 2.4-m panel of unstretched ARALL 2 laminate material. ARALL 2 laminates consist of alternating layers of 2024-T3 aluminum sheet and uniaxial aramid epoxy prepreg. The aluminum sheet is 0.3 mm thick and the prepreg is approximately 0.2 mm thick. Consolidation is obtained by processing in an autoclave at 121°C (250°F) and applying 69 000 N/m^2 (100 psi) pressure. Laminates typically range from two to five layers of aluminum; in this study, three layers of aluminum and two layers of reinforced epoxy were used. The two layers of aramid reinforced epoxy (denoted K/E)—prepreg SP366—are manufactured by 3M specially for ARALL laminates. A conceptual view of ARALL 2 was shown in Fig. 1.

The panel was C-scanned to determine its quality and consistency. Areas showing porosity, delaminations, or prepreg seams were discarded. It should be noted that 20 to 30% of the panel was discarded in this manner even though the panel was accepted by the manufacturer in its entirety. Preliminary tests of the discarded material show a decrease in ultimate tensile strength and elastic modulus of approximately 3%. The presence of defects may be much more damaging during fatigue and off-axis tests.

Two specimen sizes were accommodated in the test frames: 254 by 38 mm (10 by 1.5 in.) and 203 by 38 mm (8 by 1.5 in.), where the longer dimension is in the fiber direction. Monotonic and fatigue test results of both geometries do not indicate measurable differences.

Monotonic and fatigue tests were performed on an 89 000 N (20 kip) MTS servohydraulic, closed-loop testing machine, equipped with hydraulic, wedge action grips. Strains were measured with an MTS extensometer having a 25 mm (1 in.) gage length. The extensometer's knife edges are mounted on V-notched aluminum tabs bonded to the specimen with a compliant silicone adhesive. The use of strain gages was discontinued after several adhesive failures occurred due to the large strains. The development of cracks across the face of the specimen during fatigue testing makes the use of strain gages impractical and gives erroneous data. The extensometer provides a much better description of the average global strains compared to the local nature of the strain gage measurement.

All fatigue tests were conducted at room temperature in load control, at 10 Hz and $R = 0.1$ ($R = S_{min}/S_{max}$) using sinusoidal loading, where S_{max} and S_{min} are the maximum and minimum nominal applied stresses, respectively. Quasi-static tests were conducted to determine initial material properties and residual tensile strength in load control at a rate of 89 N/s (20 lb/s).

The coupons were gripped with MTS 641.35 hydraulic grips and the contact area was 50 by 38 mm (2 by 1.5 in.). Fatigue specimens were tested with 5.5×10^6 N/m² (800 psi) grip pressure with good results; that is, no significant damage or failures were induced by the grips. Monotonically loaded specimens were gripped at 2.07×10^6 N/m² (300 psi) and a layer of emory cloth was placed between the grip surface and the specimen to avoid grip-induced damage and premature failure.

Static and Fatigue Characterization

The average tensile strength of ARALL 2 laminates tested in the fiber direction is 717×10^6 N/m² (104 ksi). The stress-strain response is bilinear with a well-defined knee point. The yield stress is 372×10^6 N/m² (54 ksi) (0.2% offset method) and the initial elastic axial modulus is 64.1×10^9 N/m² (9.3 Msi).

The results of the cyclic loading testing program are presented as S (normalized stress level $= S_{max}/S_{ult}$) versus log N_f (cycles to failure). The uniaxial fatigue response of K/E [14] and that of 2024-T3 aluminum [15] are plotted in Fig. 2, along with S/N data for ARALL 2 ($R = 0.1$). To do so, the strains at maximum and minimum load were used to evaluate the initial axial stresses and stress ratios in the aramid/epoxy and aluminum plies. On the basis of the S/N curve, three stress levels of engineering interest resulting in lives of 10^5 to $>10^6$ cycles were chosen: $S = 0.4$, 0.45, and 0.5 whereas the knee in the stress-strain response is centered at $S = 0.44$.

Stiffness Monitoring

Secant and unloading moduli were monitored for all fatigue and residual strength specimens. Secant modulus is evaluated using the strains at minimum and maximum loads on the loading portion of the stress-strain curve for a particular cycle. Unloading modulus [16] is evaluated using the initial slope of the unloading portion of the stress-strain curve. Unless oth-

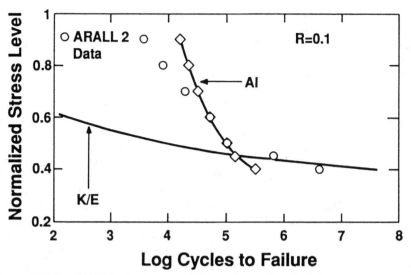

FIG. 2—*ARALL 2, aluminum and K/E superimposed tensile fatigue response.*

erwise specified, the ratio of the cyclic and virgin unloading moduli is reported (and referred to as normalized or reduced stiffness), since it seems to represent stiffness changes due to damage and is not affected by strain hardening effects [16].

The authors have found that life fraction (n)—the number of cycles divided by the number of cycles to failure—rather than log cycles, provides a better comparison of trends between specimens with different lives. Reduced stiffness-versus-life fraction data are an excellent indicator of the changing condition of the material throughout life. Excellent repeatability is achieved using this normalization process [17]. Reduced stiffness-versus-life fraction curves, such as shown in Fig. 3, were fitted using the least squares method with a modified power law of the form

$$E(n) = E_0 + \frac{E_s - E_0}{(1 + \tau/\lambda)^\mu} \tag{1}$$

where E_0 is the initial stiffness and E_s is the secondary stiffness attained after complete delamination and cracking of the aluminum plies. Two parameters define the curve: τ shifts the curve on the time ordinate and μ defines the slope. The dependence of the parameters on stress level, S, was fitted by

$$\mu(S) = 0.3687 + 9.1279E - 5 \cdot (S)^{-11.56} \tag{2}$$

and

$$\log \tau(S) = \left[\frac{10}{(1 + 10^{(4.75-10^5)})^{1.25}} \right] \tag{3}$$

λ is the transformed life fraction given by,

$$\log \lambda = (10 \cdot n) \tag{4}$$

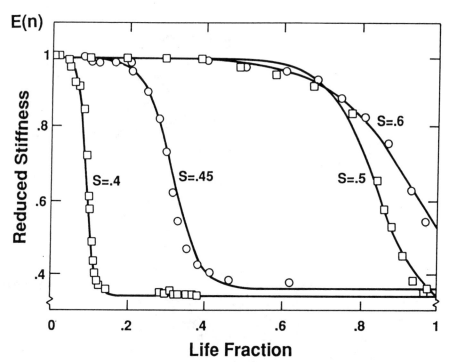

FIG. 3—*Generalized power law fit and data for reduced stiffness-versus-life fraction, S = 0.4 to 0.6.*

Using Eqs 2 and 3 in the reduced stiffness Eq 1, a three-dimensional surface was obtained, as shown schematically in Fig. 4. The surface shows the effect of cyclic stress level on the rate and amount of stiffness change due to damage during fatigue life. At high stress levels, stiffness change (damage) occurs late in life. At lower stress levels, stiffness change begins earlier in life and its magnitude at the end of life is greater than at higher stress levels.

Surface Analysis

Plies that delaminated during cyclic loading were analyzed using electron spectroscopy for chemical analysis (ESCA). The aluminum and K/E surfaces were scanned in narrow and broad frequency bands. The elements and relative quantities present on both sides were nominally identical and corresponded to an epoxy surface, indicating cohesive failure. Table 1 summarizes the elements and concentrations found on both delaminated surfaces.

TABLE 1—*ESCA surface analysis of delaminated ARALL-2.*

Element	Sensitivity Factor	Aluminum Side Concentration, %	Aramid/Epoxy Side Concentration, %
C	10.250	77.17	76.42
O 1	0.660	16.42	14.01
Al 1	0.185	0.00	0.00
Br 1	0.830	1.02	0.59
N 1	0.420	5.39	8.98

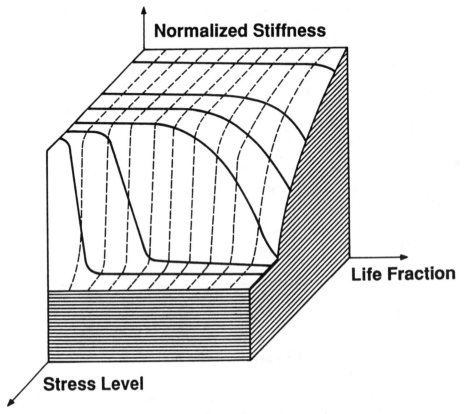

FIG. 4—*Normalized stiffness three-dimensional surface as a function of life fraction and stress level.*

Photography

Figure 5 shows magnified and enhanced pictures of the crack patterns as they developed in specimens cycled to a prescribed life fraction, as determined from the reduced stiffness. Cracks in the surface aluminum plies initiated at the edges of the specimen and grew perpendicular to the load direction. With additional cycles, more cracks developed creating a distinct crack pattern. Eventually, two cracks growing from opposite edges may join or interact in such a way to change their initial direction.

Table 2 summarizes the number of cracks per unit length initiating at the edges and indicates the mean number of cracks and crack spacing for different stress levels and stiffness ratios. The overall mean spacing is 8.76 mm (0.345 in.) and the coefficient of variance is 0.093.

Ultrasonic C-Scan

Figures 6, 7, and 8 show ultrasonic C-scan pictures of specimens cycled to a prescribed reduced stiffness at different stress levels. The dark tones indicate defects and the gray scale is proportional to the attenuation. Note, the extensometer tabs appear as dark spots.

The C-scans are similar for specimens tested at the same stress level and life fraction, and reasonably similar for different stress levels but similar life fraction. The C-scans give an overall indication of cracks in the aluminum plies and delaminations between the aluminum and fiber-reinforced layers. It should be noted that at $S = 0.5$, the long life-fraction specimens are less delaminated than at $S = 0.45$ or $S = 0.4$, as shown in Fig. 9.

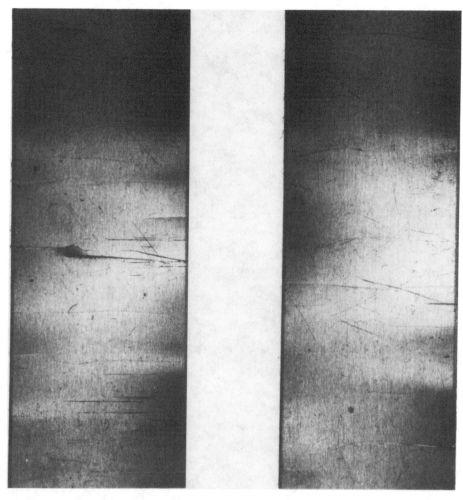

FIG. 5—*Cracked residual strength specimens (S = 0.4), after 10⁶ cycles.*

TABLE 2—*Cracks per 101.6 mm (4 in.) length in residual strength specimens of ARALL 2.*

$S = \dfrac{S_{max}}{S_{ult}}$	$E(n)$, reduced stiffness	Average Number[a]	Average Spacing, in.[b]
0.4	0.7	10.67 ± 1.77	0.375
0.4	0.34	11.08 ± 1.56	0.361
0.45	0.7	11.50 ± 1.00	0.348
0.45	0.34	12.83 ± 1.19	0.312
0.5	0.7	13.33 ± 3.42	0.300
0.5	0.58	10.63 ± 3.46	0.376

[a] Three specimens at each condition, except at $S = 0.5$, $E(n) = 0.58$, at which two specimens were examined.

[b] 1 in. = 25.4 mm.

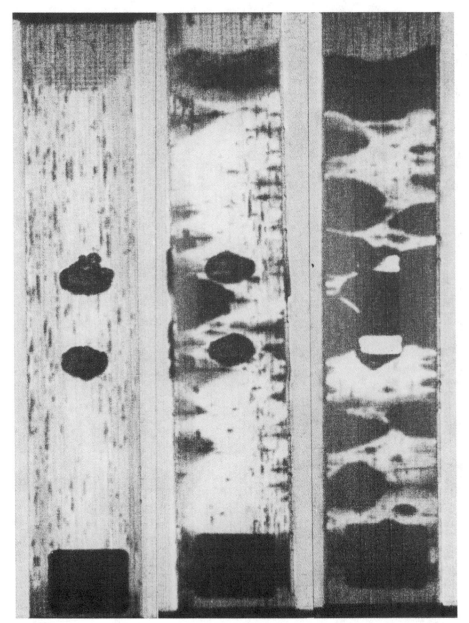

FIG. 6—*C-scans of residual strength specimens, S = 0.5 tabs appear in contrast to background;* (left) *short life fraction,* (middle) *medium life fraction, and* (right) *long life fraction.*

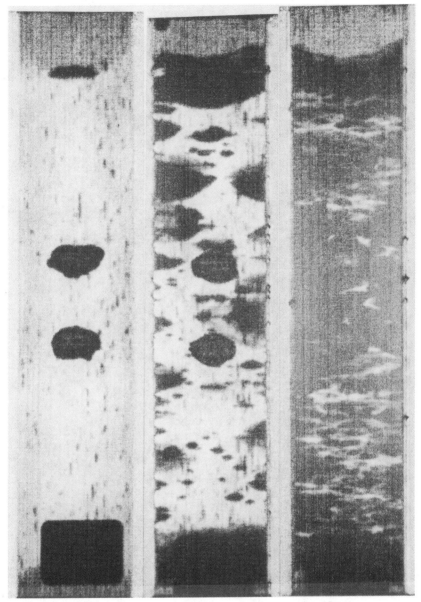

FIG. 7—*C-scans of residual strength specimens, S = 0.45 tabs appear in contrast to background;* (left) *short life fraction,* (middle) *medium life fraction, and* (right) *long life fraction.*

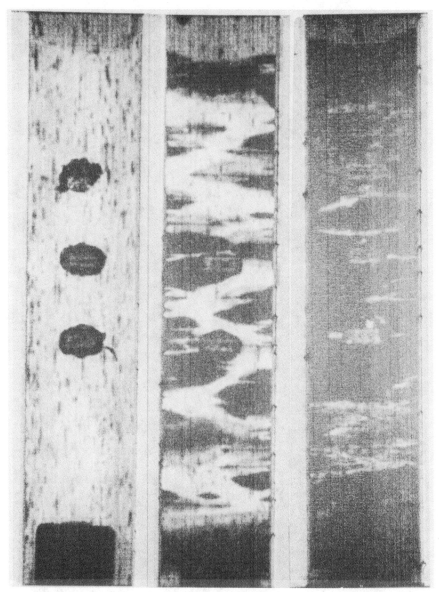

FIG. 8—*C-scans of residual strength specimens,* S *= 0.4 tabs appear in contrast to background;* (left) *short life fraction,* (middle) *medium life fraction, and* (right) *long life fraction.*

Residual Strength

The normalized stress levels chosen for residual tensile strength measurements were: $S = 0.5$, resulting in mean life of 10^5 cycles; $S = 0.45$, resulting in mean life of 650 000 cycles; and

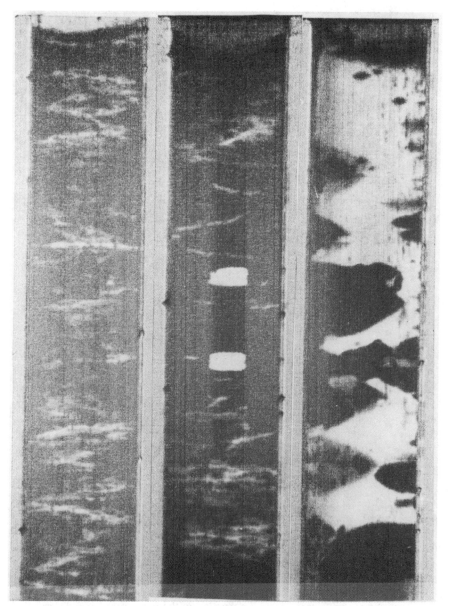

FIG. 9—*C-scans of long life specimens; tabs appear in contrast to background; S = 0.4, 0.45, and 0.5 from left to right.*

$S = 0.4$, resulting in life of over one million cycles. At each stress level, residual strength was determined at $E(n) = 1$ and $E(n) = 0.7$, where n is the fraction of fatigue life, and $E(n)$ is the unloading stiffness at n. To achieve $E(n) = 0.7$ at $S = 0.5$, the life fraction, n, is approximately 0.8; and at $S = 0.45$, the life fraction, n, is approximately 0.3.

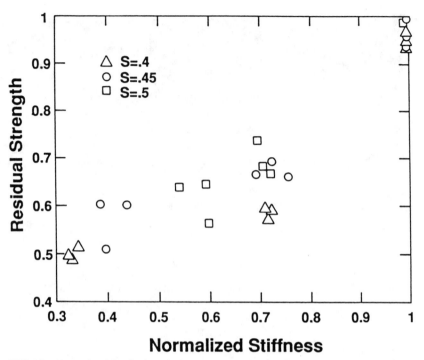

FIG. 10—*Normalized residual strength versus normalized stiffness, S = 0.4, 0.45, and 0.5.*

In addition, residual strength tests were conducted at one other stage of fatigue life, different for each stress level:

1. at $S = 0.5$, for $n > 0.9$;
2. at $S = 0.45$, for $E(n) \cong$ constant, after a large drop in stiffness; and
3. at $S = 0.4$, after 10^6 cycles.

Three specimens were tested at each stress level and life fraction. Residual strength versus stiffness is shown in Fig. 10.

Discussion

Fatigue Response

A new parameter, SR, was defined. It is the ratio of the maximum applied stress S_{max} and the cross sectional area of the K/E layers, divided by the ultimate strength of the fiber plies. Only at stress levels less than $SR = 1$ are the fiber plies able to carry the maximum load by themselves; even if the aluminum plies are cracked and transfer all load to the fiber reinforced plies

$$SR = \frac{S_{max}}{Area_{K/E}} \cdot \frac{1}{\sigma_{ult}^{K/E}} \tag{5}$$

Examining the S/N data in Fig. 2, two distinct fatigue regimes become evident. The SR = 1 (S = 0.6) ratio indicates a transition point in the cyclic damage mode. At high stress levels, $SR \geq 1$, the fatigue response of ARALL 2 is governed by the initiation and propagation of damage in the aluminum plies. Their failure leads to immediate failure of the laminate. At the low stress levels, $SR \leq 0.8$, a stress redistribution process takes place. The progressive failure of the aluminum layers transfers additional load to the K/E layers, whose subsequent response controls laminate failure. A transition zone in the fatigue failure occurs at $1 > SR > 0.8$. In this range, the fatigue life of the K/E plies is too short to have a noticeable effect on the laminate behavior.

Damage Sequence

Based on the available data, a characteristic sequence of events leading to the ultimate failure of the laminate was determined. The focus is on S values of 0.5, 0.45, and 0.4 because of the more complete investigation performed at these stress levels. Five main stages are recognized in the damage sequence.

Stage 1—Stage 1 will be called the initiation stage. During this stage:

1. A strain hardening effect takes place in the aluminum layers and in the fiber layers (fiber straightening).
2. A similar and small residual strength reduction is measured at all stress levels tested, approximately 4 to 5%.
3. No damage is evident in the C-scans and no cracks are detected on the surface.
4. No changes in unloading stiffness are measured.

Obviously, some form of undetected microscopic-level damage occurs as it is reflected in the reduced residual strength.

Stage 2—Stage 2 will be called the damage development stage. This stage begins with the appearance of cracks at the edges of the aluminum layers and is completed when a saturated or characteristic damage state is reached in the aluminum layers. At $E(n) = 0.7$, Stage 2 is completed.

1. The cracks growing in the aluminum plies cause the delaminations to grow inside the resin-rich interface between the K/E and aluminum plies, causing cohesive, not adhesive, failure.
2. Crack spacing slightly decreases with increasing stress level.
3. Stiffness is reduced gradually. Small cracks at the edges do not influence the strain measured by the extensometer, hence a stiffness reduction is noticed only after the cracks have advanced inward by some distance, dependent on location.

Stage 3—Stage 3 will be called the crack growth stage.

1. Edge cracks in the aluminum continue to grow from the edges perpendicular to the load direction. Two cracks growing in opposite directions may interact and join. Cracks originating away from the edges also appear. These seem to join cracks growing from the edges.
2. Delaminations (cohesive debonding) between the outer aluminum and the fiber layers grow parallel to the load direction and may join with those of neighboring cracks.
3. At lower stress levels, cracks grow longer and delaminations are larger.

4. Stiffness is reduced significantly. The rate of reduction is dependent on the stress level. At S values of 0.45 and below, the normalized stiffness will approach $E(n) \cong 0.34$ and remain constant.
5. A significant reduction in residual strength accompanies the stiffness drop. This reduction is also dependent on stress level. Lower S values result in larger strength degradation at a given stiffness reduction.

Stage 4—Stage 4 will be called the fiber ply degradation stage. The extent of this stage is highly dependent on stress level, and it will be addressed separately for each S value and the correspondent SR value.

1. $S = 0.5$, $SR = 0.84$—Though the reduced stiffness approaches a constant value, the fiber layers are unable to carry the total load for a significant number of cycles. Failure occurs shortly after the aluminum layers are cracked across the width of the specimen. Because of the difficulty in monitoring the events in the last few cycles, not enough data are available to permit comment on the process of fiber ply degradation.
2. $S = 0.45$, $SR = 0.76$—Stage 4 becomes a significant part (greater than 50%) of the total life of the specimens. After the reduced stiffness approaches a constant value of approximately 0.34, the measured normalized residual strength is 0.57. During Stage 4, the stiffness and residual strength decrease slowly. Failure occurs when the normalized residual strength reaches the level of applied load of 0.45.
3. $S = 0.4$, $SR = 0.67$—The reported life of a unidirectional K/E laminate at this stress level is $3.5E + 7$ cycles [16], well over the $1E + 6$ cycles limit of this investigation. The reduced stiffness approaches 0.34 and remains nearly constant to the end of the test. The residual strength at the beginning of Stage 4 is half of the original tensile strength.

Stage 5—Stage 5 is the final fracture. Under tensile loading, the final failure event is controlled by the unidirectional fiber layers. Fibers break and cause the rapid growth of a crack across the width of the specimen as in quasi-static tests. Unlike quasi-static tests, this final crack in the fiber layers may occur at some distance from the large cracks on the outer aluminum layers that grew across the width of the specimen in Stage 3. In most cases, the inner aluminum layer and the fiber layers pull away from the outer aluminum layers exposing delaminated areas of varying length.

The sequence of events just described summarizes the damage development observations made during this investigation. Though the portion of life at each stage varies with stress level, the order is maintained. The *SR* ratio emerges as the single most important parameter in determining the tension-tension fatigue behavior of ARALL 2 laminates. At values above one, life ends shortly after the aluminum layers crack. Stages 3 through 5 collapse into one single event. At values below 0.8, life is controlled by the fiber plies, although there is a great degradation in performance; that is, stiffness and residual strength. A failure mode transition range exists for $1 > SR > 0.8$.

The Characteristic Damage State

Reifsnider et al. [18] reported the existence of a unique characteristic damage state (CDS) for laminated composites, consisting of a specific array of matrix cracks parallel to the fibers in each off-axis ply of a composite laminate. The CDS is attained when each ply becomes saturated with cracks that appear at regularly spaced intervals. This equilibrium spacing is identical for quasi-static and tensile fatigue loading, and is controlled by the constraint of the laminate on the cracked plies. Material properties, stacking sequence, etc., may change the specific form of the equations or boundary conditions, but a CDS can be predicted for all laminates. After the CDS has formed, other damage modes may play an important role in the response of the laminate. The manner in which the CDS evolves into the final fracture event is not yet fully understood.

The reader is referred to Reifsnider and Highsmith [18] for a detailed description of the shear-lag mathematical model and a quasi-three-dimensional finite difference treatment is given in Refs 19 and 20. The formation of a CDS, first reported to exist in graphite-epoxy composites, has been observed and confirmed in widely differing material systems such as aramid-epoxy, glass-epoxy, metal-matrix composites, etc., and now in ARALL hybrid laminates.

The Effect on Stiffness, Strength, and Life

Highsmith and Reifsnider [9] and Highsmith [21] comment on the effects of a stable array of cracks forming in the off-axis plies of a laminate. The cracks that develop in the off-axis plies cause a transfer of load to a neighboring unbroken region. The redistribution of stress causes the unbroken plies to extend an additional increment with no additional load being introduced. The laminate becomes less stiff with each increment in crack length or appearance of a new crack. If the cracked layers have a low stiffness, then the additional load introduced into the unbroken plies is small, causing a small stiffness reduction. Each crack will result in a specific stiffness change. Stiffness changes can be directly related to stresses through measured strains and displacements.

The CDS formation results in a corresponding stiffness reduction that can be predicted. Since the CDS is a unique laminate property, this stiffness reduction is independent of the path, that is, load history, temperature, moisture, residual stresses, etc. Residual strength depends on the way the CDS reduces the effective stiffness tensor and redistributes the internal stresses. Accepting the definition of fatigue life as the number of cycles to failure by fracture, residual strength controls the fatigue behavior.

Preliminary Results for ARALL Laminates

A computer code available at VPI&SU [22] was used to perform a preliminary characterization of the CDS of ARALL 2 laminates. The characteristic crack spacing in the surface plies of aluminum was predicted on the basis of information measured in the context of this investigation where possible. Data available from other sources were introduced where necessary.

The computer code calculates the stress state in cracked and uncracked layers of general laminates. It performs a semi-infinite analysis for the case of one crack or a finite analysis between two cracks. From the stress state, the effective stiffness is calculated. The formulation is similar to a finite difference problem, but it is solved exactly [22].

The input to the program includes the elastic properties of the laminae, geometry, applied strain, and boundary conditions (crack(s) location). The interface between layers is characterized by the value of G/b, the ratio of shear modulus and thickness of the interface. This value is generally unknown. The two common assumptions are

1. G is taken as the shear modulus of the resin and b is the thickness of the resin rich area between layers.
2. G is taken as G_{23} of the ply, and b is half of the ply thickness.

All of the preceding assumptions correspond to fiber-reinforced composites, where all the laminae are of the same material or at least the same kind of interface. The load-transferring interface is a part of both adjacent layers and its properties are evaluated on this basis. This certainly is not the case in ARALL 2 laminates and the following assumptions were made.

1. The aluminum layers crack through their whole thickness and are not a part of the interface. This assumption is supported by the evidence that delaminations originating at the cracked aluminum layers grow inside the resin rich area within the K/E plies.
2. Since there is no information regarding the interface, G is taken to be G_{23} of the K/E plies, and b is taken to be half the K/E ply thickness, yielding a value of $G/b = 1.48 \times 10^{13}$ N/m^3 (54.25 $E6$ psi/in.).

A semi-infinite analysis was performed for the case of one crack in the outer aluminum ply. The upper bound on the distance between cracks is evaluated by determining the distance away from an existing crack where the strain approaches the global strain that caused the first crack, Fig. 11.

The available information on the characteristic crack spacing is used to perform a finite stress analysis between two adjacent cracks in the outer aluminum ply. The strain distributions between the two cracks in the aluminum plies and K/E ply between them are shown in Fig. 12. The calculated local stress concentrations in the fiber plies in the vicinity of cracks range from 2.3 for the case of a single crack in the outer aluminum plies to 3.7 for parallel cracks in the inner and outer aluminum plies.

The actual average crack spacing, as given in Table 2, is 8.76 mm (0.345 in.). The value obtained for the lower bound on the crack spacing from the semi-infinite analysis is 5.59 mm (0.22 in.). Considering the approximate nature of the assumptions and the presence of large delaminations, the preliminary analysis shows encouraging predictive capabilities.

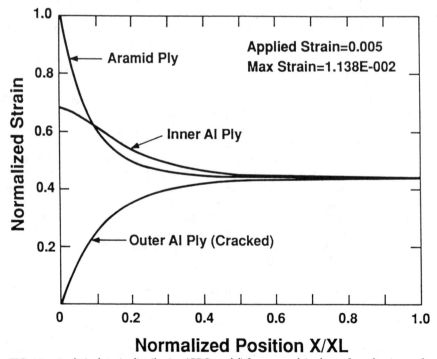

FIG. 11—*Analytical strain distribution (CDS model) for one crack in the surface aluminum ply.*

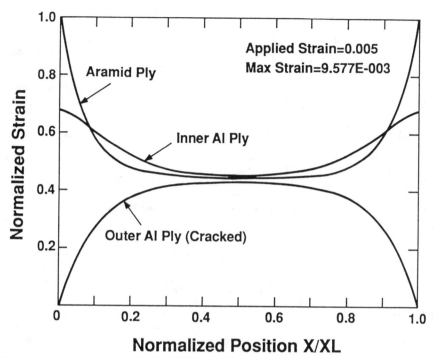

FIG. 12—*Analytical strain distribution (CDS model) between two cracks in the surface aluminum ply.*

Summary and Conclusions

Tension-tension ($R = 0.1$) cyclic loading of ARALL-2 laminates in the fiber direction at normalized maximum stress levels (S) ranging from 0.4 to 0.9 of the ultimate tensile strength yielded lives of 8×10^3 to over one million cycles. At low stress levels ($S \leq 0.45$), life is controlled by the fatigue behavior of the fiber plies. At high stress levels ($S \geq 0.6$), life is determined by the initiation of damage in the aluminum plies. At intermediate stress levels, a transition fatigue failure occurs. The SR ratio was defined and shown to account for this transition in terms of the ultimate tensile strength of the fiber plies.

The reduction in normalized dynamic stiffness correlates well with residual strength and reflects the damage state of the laminate. Five stages were recognized in the damage sequence; their relative length is stress level dependent. Each stage is accompanied by changes in stiffness and strength.

The characteristic damage state model was modified in an attempt to predict a lower bound on the spacing of the stable array of cracks that develops during cyclic loading in the surface aluminum plies.

Acknowledgments

The research presented in this report was conducted under a contract from the Alcoa Technical Center. The authors express their sincere appreciation to Alcoa for their support and to Dr. Richard Stiffler, technical monitor, for his valuable assistance and personal interest in the program.

References

[1] Vogelesang, L. B., Marissen, R., and Schijve J., "A New Fatigue Resistant Material: Aramid Reinforced Aluminum Laminate (Arall)," *Proceedings, Eleventh Symposium, International Committee on Aeronautical Fatigue*, Noordwijkerhout, The Netherlands, 1981.

[2] Bucci, R. J., Mueller, L. N., Vogelesang, L. B., and Gunnick, J. W., "Arall Laminates Properties and Design Update," *Proceedings, Thirty-Third International SAMPE Symposium on Materials—Pathway to the Future*, Anaheim, CA, 1988.

[3] Teply, J. L., "Analytical Modeling of Arall Laminates," Arall Laminates Technical Conference, Seven Springs, PA, 1987.

[4] Yeh, J. R., "Fracture Mechanics of Delamination in Arall Laminates," Arall Laminates Technical Conference, Seven Springs, PA, 1987.

[5] Marissen, R., "Fatigue Crack Growth in Aramid Reinforced Aluminum Laminates (Arall)—Mechanisms and Predictions," DFVLR Institute für Werkstoff-Förschung, DFVLR-FB 84-37, 1984.

[6] Ritchie, R. O., Weikang, Y., and Bucci, R. J., "Fatigue Crack Propagation in Arall Laminates: Measurement of the Effect of Crack-Tip Shielding from Crack Bridging," to be published.

[7] Reifsnider, K. L., *Fatigue of Composite Materials*, Composite Materials Series, Vol. 4, Elsevier, Oxford, 1989.

[8] Henneke, E. G., "Nondestructive Evaluation of Fiber Reinforced Composite Laminates," *Proceedings, Eleventh World Conference on Nondestructive Testing*, Vol. 2, 1985, pp. 1332–1343.

[9] Highsmith, A. L. and Reifsnider, K. L., "Stiffness Reduction Mechanisms in Composite Laminates," *Damage in Composite Materials, ASTM STP 775*, K. L. Reifsnider, Ed., American Society for Testing and Materials, Philadelphia, 1982, pp. 103–117.

[10] O'Brien, T. K., "Stiffness Change as a Nondestructive Damage Measurement," *Mechanics of Nondestructive Testing*, W. W. Stinchcomb, Ed., Plenum Press, New York, 1980, pp. 101–122.

[11] Reifsnider, K. L. and Stinchcomb, W. W., "Stiffness Change as a Fatigue Damage Parameter for Composite Laminates," *Advances in Aerospace Structures, Materials and Dynamics—AD-06*, Yuceoglu, U., Sierakowski, R. L., and Glasgow, D. A., Eds., Book No. H00272, American Society of Mechanical Engineers, New York, 1983.

[12] Osiroff, R. and Stinchcomb, W. W., "Long Term Characterization of ARALL Laminates," CCMS-89-06 Report, Virginia Polytechnic Institute and State University, Blacksburg, VA, 1989.

[13] Reifsnider, K. L. and Stinchcomb, W. W., "A Critical Element Model of the Residual Strength and Life of Fatigue Loaded Composite Coupons," *Composite Materials: Fatigue and Fracture, ASTM STP 907*, H. T. Hahn, Ed., American Society for Testing and Materials, Philadelphia, 1986, pp. 298–313.

[14] *Data Manual for Kevlar 49 Aramid*, E. I. DuPont de Nemours, Wilmington, DE, 1986.

[15] *Military Standard Handbook, MIL-HDBK-5D*, "Metallic Materials and Elements for Aerospace Vehicle Structures," U. S. Department of Defense, revised 1 May 1985, pp. 3–73–3–139.

[16] Johnson, W. S, " Mechanisms of Fatigue Damage in Boron/Aluminum Composites," *Damage in Composite Materials, ASTM STP 775*, K. L. Reifsnider, Ed., American Society for Testing and Materials, Philadelphia, 1982, pp. 83–102.

[17] Osiroff, R., Stinchcomb, W. W., and Reifsnider, K. K., "Fatigue Behavior of Aramid/Aluminum (ARALL) Laminates," *Visco-Plastic Behavior of New Materials*, PVP-Vol. 184, MD-Vol. 17, D. Hui and T. J. Kozik, Eds., American Society of Mechanical Engineers, 1989.

[18] Reifsnider, K. L. and Highsmith, A., "Characteristic Damage States: A New Approach to Representing Fatigue Damage in Composite Laminates," *Materials, Experimentation and Design in Fatigue*, Westbury House, U.K., 1981, pp. 246–260.

[19] Highsmith, A. L. and Reifsnider, K. L., "Internal Load Distribution Effects During Fatigue Loading of Composite Laminates," *Composite Materials: Fatigue and Fracture, ASTM STP 907*, H. T. Hahn, Ed., American Society for Testing and Materials, Philadelphia, 1986, pp. 233–252.

[20] Talug, A., "Analysis of Stress Fields in Composite Laminates with Interior Cracks," Ph.D. thesis, College of Engineering, Virginia Polytechnic Institute and State University, Blacksburg, VA, 1978.

[21] Highsmith, A. L., "Stiffness Reduction from Transverse Cracking in Fiber Reinforced Composite Laminates," Masters thesis, College of Engineering, Virginia Polytechnic Institute and State University, Blacksburg, VA, 1981.

[22] Moore, R. H. and Dillard, D. A., "Elastic and Time Dependent Matrix Cracking in Cross-ply Composite Laminates," CCMS Report CCMS-88-19, Virginia Polytechnic Institute and State University, Blacksburg, VA, April 1988.

Christopher D. Wilson[1] and Dale A. Wilson[2]

Effective Crack Lengths by Compliance Measurement for ARALL-2 Laminates

REFERENCE: Wilson, C. D. and Wilson, D. A., **"Effective Crack Lengths by Compliance Measurement for ARALL-2 Laminates,"** *Composite Materials: Fatigue and Fracture (Third Volume), ASTM STP 1110,* T. K. O'Brien, Ed., American Society for Testing and Materials, Philadelphia, 1991, pp. 791–805.

ABSTRACT: As a means of determining a stress intensity factor solution, the compliance properties of an ARALL-2 laminated-sheet composite were investigated. Fatigue crack growth rate (FCGR) tests were conducted on middle crack tension (MT) specimens fabricated from a layup consisting of three sheets of 2024-T3 aluminum bonded together with unidirectional aramid fibers embedded in epoxy. Excellent fatigue crack growth properties are obtained by the presence of unbroken aramid fibers in the wake of the crack tip. These unbroken fibers act as a bridging mechanism to inhibit further crack growth. To quantify the effect of maximum fatigue load on compliance, a series of FCGR tests were performed. An effective crack length was defined to be the length of a through-the-thickness crack with the same compliance level as a fiber-bridged fatigue crack. A specimen slotted with a jeweler's saw was used to simulate a through-the-thickness crack. As maximum fatigue load increased, the difference between surface measured fatigue crack and sawcut crack length increased. Effective crack lengths were determined to be at least 10 mm shorter than surface measured crack lengths for a 76-mm-wide specimen. The bridging zone was estimated to be at least 5 mm. Compliance and stress intensity factor as functions of effective crack length were determined.

KEY WORDS: composite materials, fracture, fatigue (materials), aramid aluminum laminates, compliance measurement, effective crack length, fatigue crack growth rate testing, fiber bridging, linear elastic fracture mechanics, stress intensity factor

ARALL[3] laminates are a new family of high-strength structural composite materials developed in the early 1980s for tensile fatigue applications in aircraft. ARALL laminates are fabricated by bonding thin aluminum alloy sheets with strong uniaxial aramid fibers embedded in an epoxy matrix. Excellent fatigue properties are obtained by the presence of unbroken aramid fibers behind the crack tip. These unbroken fibers act as a bridging mechanism to inhibit crack growth. As a hybrid composite material, ARALL laminates combine the durability of high-strength aluminum alloys with the strength of aramid fibers. The current investigation involves fatigue crack growth rate tests of the ARALL-2 laminate. As shown in Fig. 1, a layup consisting of three sheets of 0.3 mm (0.012 in.) thick 2024-T3 aluminum alloy with aramid/epoxy layers sandwiched between the sheets was used. This 3/2 layup has a total laminate thickness of 1.3 mm (0.052 in). Other versions of ARALL use 7075-T6, 7475-T76, or 2024-T8 aluminum alloy with 2/1 to 5/4 ply layups [1].

Typical fatigue crack growth behavior of 2024-T3 and ARALL-2 laminates are shown in Fig. 2. The ARALL family has exhibited fatigue lives several orders of magnitude higher than

[1] Aerospace engineer, George C. Marshall Space Flight Center, Huntsville, AL 35812.

[2] Professor, Department of Mechanical Engineering, Tennessee Technological University, Cookeville, TN 38505

[3] ARALL LAMINATE is a registered trademark of the Aluminum Company of America.

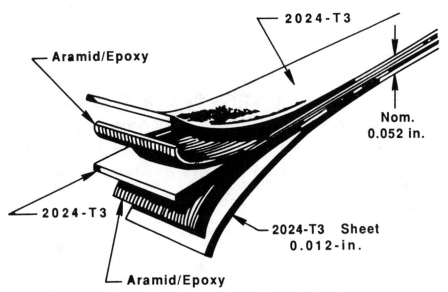

FIG. 1—*The 3/2 ARALL-2 layup (1 in. = 25.4 mm).*

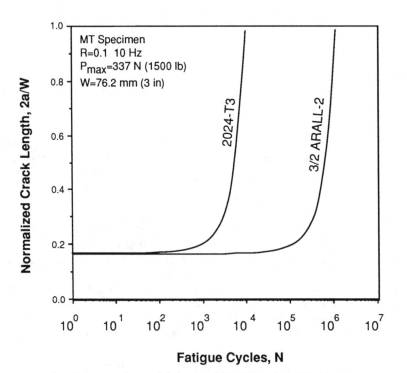

FIG. 2—*Fatigue crack growth behavior of 2024-T3 and ARALL-2 laminates.*

comparable aluminum. In tests where an overload cycle was applied, the crack grew to only a short length before the bridging mechanism of the aramid fibers arrested further crack growth. Crack growth was reinitiated by increasing the maximum fatigue loading. Early work by Marissen characterized the excellent fatigue crack growth properties of ARALL laminates due to the restraint on crack opening caused by unbroken aramid fibers behind the crack tip. This mechanism reduces the stress intensity at the crack tip so that the stress intensity factor decreases with increasing crack length. A decreasing crack growth rate with increasing crack length can result [2].

A detailed study of fiber bridging was performed by Ritchie, Yu, and Bucci [3]. Ritchie developed an experimental procedure to quantify the primary mechanisms of crack-tip shielding from the extensive bridging of unbroken fibers behind the crack tip. For crack extension perpendicular to the tensile loading direction, controlled delamination occurs between the aluminum and the epoxy/fiber interface. This delamination causes a stress redistribution ahead and behind the crack tip that permits aramid fibers to remain intact in the wake of the crack tip. The fiber bridging of a fatigue crack in an ARALL laminate is shown in Fig. 3. Yeh [4] examined the delamination characteristics of ARALL laminates by employing special singular elements in a finite element analysis. He concluded that stress intensity factors of ARALL laminates are independent of the length of delamination. Using a combination of back face strain and d-c potential measurement, Ritchie was able to determine an effective stress intensity factor range that accounted for crack closure (wedging through the contact of crack surfaces) and for crack bridging. Closure was determined to be of secondary importance in governing the effective stress intensity factor range. Longitudinal crack growth rates lost their crack-size and load history dependence when the effective stress intensity factor range developed by Ritchie was used [3].

The mechanical properties of ARALL-2 laminates are shown in comparison with 2024-T3 and a carbon fiber composite in Table 1. The ultimate tensile strength of ARALL-2 laminates falls between the values cited for the other materials. The density of ARALL-2 laminates is over 15% lower than the aluminum, but the density of the carbon fiber composite is almost 50% lower than ARALL-2 laminates. These properties demonstrate that mechanically,

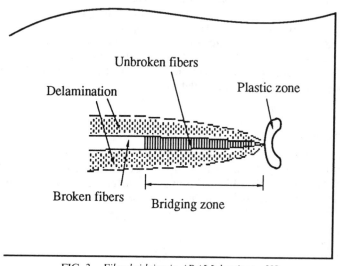

FIG. 3—*Fiber bridging in ARALL laminates* [3].

TABLE 1—*Properties of ARALL-2, 2024-T3, and a unidirectional carbon fiber composite* [3].

Mechanical Properties	Direction	ARALL-2	2024-T3	Angle	Unidirectional Carbon Fiber[a]
σ_{uts}, ksi[b] (tensile)	L[c]	102	66	0°	180
	T[d]	46	65	90°	8
σ_{ys}, ksi (tensile)	L	52	50	0°	NA
	T	34	45	90°	NA
σ_{ys}, ksi	L	NA[e]	43	0°	180
(compressive)	T	NA	40	90°	30
E, 10^6 psi	L	9.5	10.5	0°	NA
	T	6.5	10.5	90°	NA
$\varepsilon_{ultimate}$, μin./in.	L	17 000	25 000	0°	8700
	T	78 000	136 000	90°	4750
Poisson's ratio	L	0.29	0.33	0°	NA
	T	0.22	0.33	90°	NA
Density[f], lb/in.³		0.083	0.101		0.056

[a] *DoD/NASA Advanced Composites Design Guide,* Vol. 1A, generic F180 graphite/epoxy composite, 60% fiber content, for example, AS/3501-6, T300/5208.
[b] 1 kPa = 1 kN/m^2 = 1.45038 × 10^{-4} ksi.
[c] L = longitudinal direction (0°).
[d] T = transverse direction (90°).
[e] Not available.
[f] 1 kg/m^3 = 3.6127 × 10^{-5} lb/in.^{3}.

ARALL-2 laminates represent a material compromise between aluminum alloys and carbon fiber composites.

The ARALL-2 laminate's outstanding fatigue and damage tolerance properties makes it a logical choice for structural components under tensile loading. Additional studies have been conducted that indicate that ARALL laminates have excellent low-velocity impact characteristics, damping characteristics, and durability. To further examine the capability of ARALL laminates, an ARALL-based wing panel for the Fokker F27 was designed, built, and successfully tested. Lugs, winded tubes, and armor plates have been designed using ARALL laminates [5]. A redesigned cargo door for the C-17 military transport was built using ARALL laminates. A weight savings of 23% when compared to aluminum was achieved. An ARALL lower wing for the Fokker 50 is being designed and tested for use in the early 1990s. Additionally, ARALL crack stoppers have been used for the retrofitting of older aircraft. Many other examples of the current and potential use of ARALL laminates exist [6].

Technical Background

Linear elastic fracture mechanics (LEFM) deals with the failure of load bearing structures by fracture before general yielding occurs in the net section. The stress intensity factor, K, is the fundamental parameter of LEFM and is used to characterize crack extension. In general, the stress intensity factor gives the magnitude of the elastic stress field in the region near the crack tip as

$$K = \sigma \sqrt{\pi a}\, f\left(\frac{a}{W}\right) \tag{1}$$

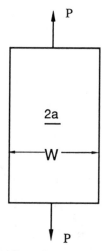

FIG. 4—*Middle crack tension (MT) specimen.*

where σ is the stress at a given location, a is the flaw size, W is the specimen width, and $f(a/W)$ is a function depending on geometry and crack orientation. Stress intensity solutions have been developed for various specimen geometries.

For the middle crack tension (MT) specimen shown in Fig. 4, the Mode I (opening mode) stress intensity factor K_I has the general form given by Eq 1 and can be written

$$K_I = \sigma\sqrt{\pi a} \ \sqrt{\left(\frac{W}{\pi a}\right)\tan\left(\frac{\tau a}{W}\right)} \tag{2}$$

This equation was developed by Irwin as an approximate solution to the periodic crack problem where cracks are evenly spaced in an infinitely long isotropic plate. Irwin's solution has an accuracy within 5% for $2a/W \leq 0.5$. Numerous other solutions based on approximations or guesses have been developed [7].

The energy approach to LEFM is based on the strain energy release rate, G. Also known as the crack driving force, G is the total energy absorbed during cracking per unit increase in crack length and per unit thickness. For the MT specimen shown in Fig. 4 (with thickness B), the strain energy release rate is

$$G = \frac{P^2}{4B}\frac{\partial C}{\partial a} \tag{3}$$

where $\partial C/\partial a$ is the slope of the compliance-crack length curve. For the plane stress case, the relationship between K and G is $K_I^2 = EG$, where E is Young's modulus. Combining this relationship with Eq 3 yields

$$K_I^2 = \frac{P^2 E}{4B}\frac{\partial C}{\partial a} \tag{4}$$

The stress intensity becomes

$$K_I = \frac{\sqrt{2W}}{2}\left(\frac{P}{BW}\right)\sqrt{\frac{\partial(BEC)}{\partial\left(\frac{2a}{W}\right)}} \tag{5}$$

where the compliance slope has been nondimensionalized [8,9].

With the stress intensity solution in a form that relies on knowledge of the specimen's change in compliance with respect to flaw size, experimental compliance measurements are needed for a series of different crack lengths for a given specimen. The derivative $\partial C/\partial a$ can be then calculated using a simple scheme, such as the secant method for determining the slope of a curve at discrete points. Using this method yields $\partial C/\partial a$

$$\left.\frac{\partial C}{\partial a}\right|_i = \frac{C_{i+1} - C_{i-1}}{a_{i+1} - a_{i-1}} \tag{6}$$

Other, more complicated schemes can be used to calculate $\partial C/\partial a$.

By using compliance measurements to determine the stress intensity for a specific flaw size, it is possible to obtain a stress intensity factor solution for an anisotropic material such as fiber-reinforced composites or laminated sheet composites that would be extremely difficult to obtain otherwise. Therefore, compliance determination of a material is an important method in developing stress intensity factor solutions.

Test Program

The objective of this test program was to determine a simple technique for quantifying the fiber-bridging effect on ARALL laminates. All of the fatigue crack growth rate (FCGR) tests were conducted in general accordance with ASTM Test Method for Measurements of Fatigue Crack Growth Rates (E 647-88). Although this standard was established for the testing of monolithic materials, it serves as an appropriate set of guidelines for the testing of laminated sheet composites such as ARALL laminates. Because the crack growth occurs only in the aluminum layers of an ARALL laminate, Marissen assumed that crack growth could be predicted using the same methods used for metallic materials if the load-carrying capability of the aramid fibers could be quantified. Additionally, it was assumed that anistropy of the elastic properties of the ARALL laminate can be neglected in stress intensity calculations [2].

Constant-load-amplitude FCGR tests were conducted on specimens made from ARALL-2 3/2 layup described in the beginning of this paper to determine the stress intensity and FCGR properties as a function of crack length and maximum fatigue load [10]. Middle crack tension (MT) specimens were 76.2 mm (3 in.) wide by 279.4 mm (11 in.) long. Each specimen had a starter hole drilled in the center of the gage length area through which a jeweler's saw was used to cut a 12.7-mm (0.5 in.) starter notch in the specimen. Twelve 6.35-mm-diameter bolts were used to attach a specimen to a pin and clevis fixture. The MT specimen used is shown in Fig. 5.

Four FCGR tests were conducted on ARALL-2 panels with the fatigue loading in the fiber direction (0°). Each test was conducted at room temperature using a load ratio $R = 0.1$. A servohydraulic test machine was used to perform the tests at a frequency of 10 Hz using a sine wave form. To determine the effects of maximum fatigue load on compliance behavior, the tests were conducted using four different loads: 337, 674, 899, and 1124 N (1500, 3000, 4000, and 5000 lb).

FIG. 5—*MT specimen mounted in load train* [10].

At discrete time intervals, each test was halted and the crack length was measured using a ×10 traveling microscope. A computer-controlled compliance test program was developed using HP BASIC on a microcomputer [*11*]. The compliance was calculated by taking the reciprocal of the slope of the load-displacement curve. Several hundred load-displacement pairs were gathered from the load cell of the test machine and from a crack opening displacement (COD) gage with a 12.7 mm (0.5 in.) gage length mounted on the centerline of a specimen. Elapsed time for this operation was minimized to avoid transient crack growth rates due to long interruptions in the test. The fatigue test was continued until relatively large $2a/W$ ratios were achieved.

An additional test examined the compliance in the fiber direction without using mechanical fatigue as a means of increasing the flaw size. A through-cut notch was made with a jeweler's saw, the notch length was measured, and compliance data was taken. The notch was then lengthened using a jeweler's saw and the compliance measurement at the new notch length was made. This process was continued for a number of different notch lengths. This test established a means of quantifying the fiber-bridging effects that take place when the aramid fibers remain behind the crack tip. This sawcut test produced a series of through-cut compliance values that simulated sawing out the fibers behind the crack tip.

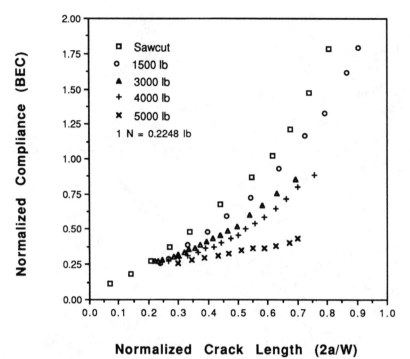

Normalized Crack Length (2a/W)

FIG. 6—*Compliance data for ARALL-2 laminate* [10].

Results

The compliance data obtained by the FCGR tests were normalized using the nominal thickness, B, and Young's modulus, E, to obtain BEC, a nondimensional quantity. The value of E used was the longitudinal modulus from Table 1. The crack length, $2a$, was normalized using the specimen width, W, to obtain $2a/W$. The normalized compliance versus surface crack length data are plotted in Fig. 6 for the specimens fatigue loaded in the fiber direction, as well as for the specimen in which only a sawcut was used. The trend shows that the compliance of ARALL-2 laminates is a function of flaw size and maximum fatigue load. Compliance values decreased for a given flaw size with an increase in fatigue load. Additionally, the compliance data were used to develop a definition for effective crack lengths in ARALL-2 laminates.

In modeling maximum fatigue load effects on the compliance of ARALL-2 laminates, a normalized load variable, P_{max}/P_{ys}, was defined where P_{max} is the maximum fatigue load and P_{ys} is the load corresponding to tensile yielding of an uncracked specimen. Based on the tensile strength given in Table 1, P_{ys} was calculated to be 1824 N (8112 lb). The P_{max}/P_{ys} and $2a/W$ were used as the two independent variables for the model. The compliance-crack length data was fitted using second-order polynomials such that

$$BEC = d_0 + d_1 \left(\frac{2a}{W}\right) + d_2 \left(\frac{2a}{W}\right)^2 \tag{7}$$

To account for the effect of maximum fatigue load on compliance for each specimen, the coefficients, d_0, d_1, and d_0, were fitted to second-order polynomials with the normalized load as the independent variable yielding

TABLE 2—*Compliance fit coefficients (ASTM E 647-88).*

Maximum Fatigue Load, N (lb)	d_0	d_1	d_2
337 (1500)	0.1794	−0.0910	2.0247
674 (3000)	0.2090	−0.0533	1.4353
899 (4000)	0.2942	−0.5337	1.7591
1124 (5000)	0.1695	0.2823	0.1164

$$d_0 = e_0 + e_1\left(\frac{P_{max}}{P_{ys}}\right) + e_2\left(\frac{P_{max}}{P_{ys}}\right)^2$$

$$d_1 = f_0 + f_1\left(\frac{P_{max}}{P_{ys}}\right) + f_2\left(\frac{P_{max}}{P_{ys}}\right)^2 \tag{8}$$

$$d_2 = g_0 + g_1\left(\frac{P_{max}}{P_{ys}}\right) + g_2\left(\frac{P_{max}}{P_{ys}}\right)^2$$

The resulting coefficients, d_n, are tabulated in Table 2. Each d_n coefficient was then fitted as a function of P_{max}/P_{ys} with a second-order polynomial to produce a single compliance function such that $BEC = f(2a/W, P_{max}/P_{ys})$. The coefficients, e_n, f_n, and g_n, are included in Table 3. Note that it was necessary to use a third-order term in fitting the 337 N (1500 lb) data to maintain a standard deviation with the same order of magnitude as the other data sets.

By comparing the compliance data for the sawcut test with the compliance data for a fatigued specimen, the effect of fiber bridging was determined. Since the sawcut simulated a through-the-thickness crack, an effective crack length that corresponds to the same compliance level as the fatigue crack measured from the surface could be found. An example of this method of quantifying the effect of fiber bridging based on compliance levels is shown in Fig. 7a. In this figure, the specimen loaded to 337 N (1500 lb) is compared to the sawcut specimen to determine equivalent flaw sizes for the same level of compliance. A normalized surface flaw size of 0.64 is equivalent to a sawcut through the thickness flaw size of 0.58. In other words, for a fatigue load of 337 N, the effective crack length is 0.06 shorter than the crack length as measured on the surface. Using $W = 76.2$ mm (3.0 in.), this difference translates to 4.57 mm (0.18 in.) at the compliance level shown in Fig. 7a. It can be observed from this figure that the difference between the effective crack length, $(2a/W)_{eff}$, and the surface crack length, $2a/W$, becomes greater as $2a/W$ increases. This increase means that the fiber-bridging mechanism has a greater effect on compliance for larger flaw sizes. The magnitude of the increase may vary with specimen width.

The difference between $(2a/W)_{eff}$ and $2a/W$ becomes greater as the maximum fatigue load was increased. At a maximum fatigue load of 899 N (4000 lb), a surface measured crack length of 0.66 has an effective crack length of 0.46 for an equivalent compliance. At this flaw size, the

TABLE 3—*Compliance model coefficients (ASTM E 647-88).*

n	e_n	f_n	g_n
1	−0.0219	0.8385	1.1008
2	1.3093	−6.1124	6.8483
3	−1.5531	8.0221	−12.9480

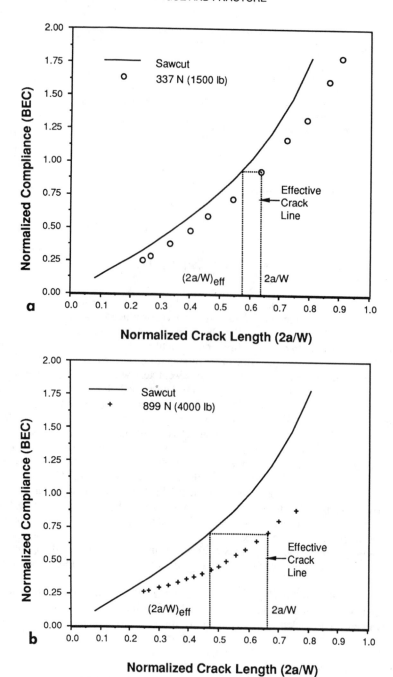

FIG. 7—*Effective crack length determination for ARALL-2 laminate.*

difference is 30% or 15.24 mm (0.6 in.). The difference between $2a/W$ and $(2a/W)_{\mathrm{eff}}$ steadily increases with flaw size as shown in Fig. 7b. The same trend can be discerned for all data sets used in the model by reexamining Fig. 6.

In an effort to collapse the compliance-crack length data for varying fatigue loads to a single curve, effective crack lengths were determined for each specimen. The sawcut compliance data were fitted as a polynomial function of crack length

$$\left(\frac{2a}{W}\right)_{\mathrm{sawcut}} = -0.0188 + 0.8935(BEC) - 0.3320(BEC)^2$$
$$+ 0.0628(BEC)^3 - 0.0068(BEC)^4 \tag{9}$$

By applying Eq 7 to each specimen, the modeled compliance was calculated. At each data point, the modeled compliance was used as the input variable in Eq 9 to calculate an effective crack length based on the sawcut compliance. This process was graphically demonstrated in Fig. 7. The resulting effective crack length versus surface crack length plots are shown in Fig. 8. Effective crack lengths were fitted as functions of surface crack length for each maximum fatigue load. To estimate an effective crack length for a load between 337 N (1500 lb) and 1124 N (5000 lb), a linear interpolation can be used between load curves. The coefficients for these curves are listed in Table 4 and can be used in the following equation

$$\left(\frac{2a}{W}\right)_{\mathrm{eff}} = C_0 + C_1\left(\frac{2a}{W}\right) + C_2\left(\frac{2a}{W}\right)^2 + C_3\left(\frac{2a}{W}\right)^3 \tag{10}$$

Good correlation between the effective crack lengths for the fatigue crack growth rate data and the sawcut compliance-crack length curve was found.

In Fig. 8, a line is drawn where $(2a/W)_{\mathrm{eff}} = 2a/W$ to represent the case where no fiber bridging takes place. For maximum fatigue loads less than 337 N (1500 lb), effective crack lengths

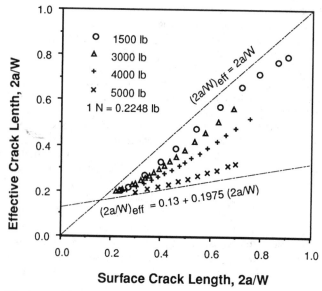

FIG. 8—Effective crack length as a function of surface crack length for ARALL-2 laminate.

TABLE 4—*Effective crack length coefficients.*

Load, N (lb)	C_0	C_1	C_2	C_3
337 (1500)	0.1336	−0.1521	2.1348	−1.2787
674 (3000)	0.1236	0.1742	0.6719	...
899 (4000)	0.1320	0.1388	0.5007	...
1124 (5000)	0.1309	0.1691	0.1424	...

can be estimated by interpolating between this line and the 337 N curve. The intersection of the line where $(2a/W)_{\text{eff}} = 2a/W$ and the effective crack length curves was determined to establish the lower bound of the definition for effective crack lengths. The value for which the effective crack length was equal to the surface crack length was determined to be approximately 0.162. This point is likely notch size dependent. The normalized surface crack length must be greater than 0.162 for the effective crack length definition to be valid. Additionally, the slopes of the effective crack length versus surface crack length curves converge to the same value as the surface crack length approaches zero. The effective crack length for a zero-length surface crack was empirically determined by least-squares curve fits to be approximately equal to 0.13. Since these points are not consistent with the physical definition of the effective crack length, the line connecting them represents the lower bound of the effective crack length definition

$$\left(\frac{2a}{W}\right)_{\text{eff}} = 0.13 + 0.1975\left(\frac{2a}{W}\right) \tag{11}$$

From Eq 11, the minimum bridging zone for the MT specimen was determined to be $0.13W$.[4] For a specimen width of 76.2 mm, the bridging zone length is a minimum of 5 mm. In comparison, Ritchie [3] examined the compact tension (CT) geometry. For a specimen width of 50 mm, Ritchie reported the length of the bridging zone to be 3 to 5 mm. Because the bridging zone measured by Ritchie includes the effect of bending loads (potentially more damaging to the fibers than tension loads alone), it is reasonable to assume that the bridging zone for the CT geometry would be smaller than the bridging zone measured for the MT geometry.

For maximum fatigue loads greater than 1124 N (5000 lb), the value for effective crack length can be estimated by interpolating between the 1124 N curve and the lower bound line defined by Eq 11. The definition for effective crack length is consistent in the region below the line where $(2a/W)_{\text{eff}} = 2a/W$ and above the line defined by Eq 11. If smaller sawcut notches were used, it would be possible to extend the region of valid effective crack lengths to include values smaller than 0.162.

Using effective crack lengths allowed the development of a single compliance versus crack length curve shown in Fig. 9. By definition, the effective crack lengths fall precisely on the sawcut compliance curve. Equation 6 can be applied to this curve to calculate the compliance slope data necessary to calculate stress intensity using Eq 5. This method will yield a single stress intensity versus effective crack length curve that accounts for maximum fatigue load in Fig. 10. The stress intensity factor was normalized by dividing K_I by $\sigma\sqrt{W}$.

The tangent solution of Irwin is shown for comparison. It is noted that the secant method used in determining the compliance slope is a point-to-point method, and small discontinuities in the compliance slope may cause variations in stress intensity solutions. It is also noted that the compliance of ARALL-2 is notch size dependent, as shown in Fig. 11, where the compliance values for small notch sizes in the sawcut test approach a much smaller value than the

---

[4] This value accounts for the two crack tips in the MT specimen, each with a bridging zone.

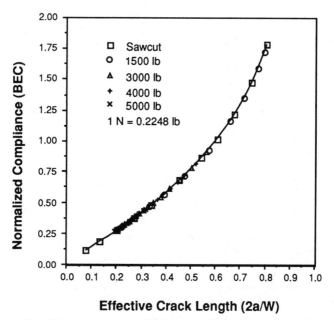

FIG. 9—*Compliance as a function of effective crack length for ARALL-2 laminate.*

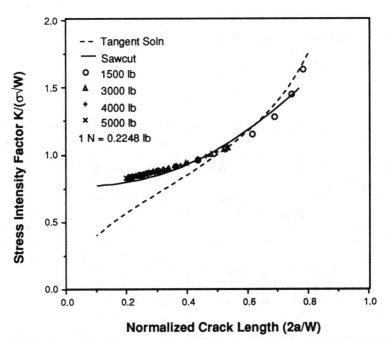

FIG. 10—*Stress intensity as a function of effective crack length for ARALL-2 laminate.*

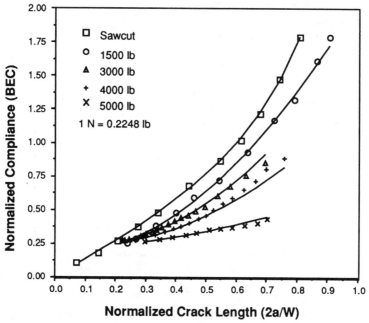

FIG. 11—*Compliance model for ARALL-2 laminate.*

FCGR tests with the larger crack starter notch. Specimen width, test frequency, and load ratio are additional factors that will influence both the crack growth behavior and the compliance behavior.

One specific requirement of ASTM E 647-88 was not satisfied. Section 7.2.2 requires the specimen to be predominantly elastic at all values of applied fatigue load. To ensure that this elastic state exists, the following empirical requirement has been established

$$\sigma_{ys} \geq 1.25 \frac{P_{max}}{(W - 2a)B} \tag{12}$$

where $W - 2a$ is the uncracked width of the specimen and B is the specimen thickness. For larger crack lengths, $W - 2a$ is small enough to create a stress that violates this requirement. This situation occurred for larger crack lengths in the tests conducted with maximum fatigue loads of 899 N (4000 lb) and 1124 N (5000 lb). However, this violation can be considered to be the result of using $W - 2a$ as measured from the surface of the specimen, instead of using a corrected value that accounts for fiber bridging. The fiber-bridging mechanism makes the uncracked width of the specimen appear to be longer than the surface measurement. Additionally, the yield strength used in this calculation is likely to be conservative since the fibers left intact behind the crack tip make the specimen stiffer. Herakovich reported that the longitudinal failure stress of the aramid/epoxy is 1141 MPa (165.5 ksi) [12]. Since the tensile failure strength of the aramid/epoxy ply is over three times greater than that of the average tensile yield of the 3/2 layup, an effective yield strength that accounts for the unbroken fibers behind the crack tip would be higher than the yield strength normally used in determining this requirement.

Visual inspection and limited fractographic examination revealed that the amount of remaining fibers in the wake of the crack tip were greater in specimens with FCGR tests using higher maximum fatigue loads. The number of cycles recorded for the tests with higher maximum fatigue loads was an order of magnitude less than the number of cycles in the test with a maximum fatigue load of 337 N (1500 lb) [10]. The difference in number of fatigue cycles accounts for the increase in the amount of remaining fibers for higher fatigue load tests.

Conclusions

The results clearly show that the maximum fatigue load should be used, along with crack length, as an independent variable in predicting the compliance of ARALL-2 laminates loaded in the fiber direction. Compliance as a function of normalized crack length and normalized load was modeled using second-order polynomials. The model is simple enough to be adapted to include the effects of load ratio and test frequency on the compliance of ARALL-2 laminates as data becomes available.

A through-cut compliance test was conducted on a specimen loaded in the fiber direction. The through-cut notch, made using a jeweler's saw, was used to simulate a through-the-thickness fatigue crack. By comparing the surface crack length of a specimen in which fiber bridging was present, to the flaw size of the sawcut specimen at the same compliance level, an effective crack length for ARALL-2 laminates was defined. The difference between the surface crack length and the effective crack length increased with flaw size and maximum fatigue load. The effect of maximum fatigue load on the enhanced performance was bounded using the effective crack length concept. By defining effective crack lengths, it was possible to quantify the actual effect that fiber bridging has on the compliance of ARALL-2 laminates and establish a single stress intensity-crack length curve. Without using effective crack lengths, the four FCGR tests previously described would produce four distinct stress intensity-crack length curves [10]. Like the compliance model, the effective crack length definition could be easily expanded to include other effects on compliance behavior.

References

[1] Bucci, R. J., Mueller, L. N., Vogelesang, L. B., and Gunnink, J. W., "ARALL Laminates Properties and Design Update," Alcoa Laboratories Technical Report No. 57-87-26, Alcoa Center, PA, 1987.

[2] Marissen, R., "Flight Simulation Behavior of Aramid Reinforced Aluminum Laminates (ARALL)," *Engineering Fracture Mechanics,* Vol. 19, No. 2, 1984, pp. 261–277.

[3] Ritchie, R. O., Yu, Weikang, and Bucci, R. J., "Fatigue Crack Propagation in ARALL Laminates: Measurement of the Effect of Crack-Tip Shielding from Crack Bridging," *Engineering Fracture Mechanics,* Vol. 32, No. 3, 1989, pp. 361–377.

[4] Yeh, J. R., "Fracture Mechanics of Delamination in ARALL Laminates," *Engineering Fracture Mechanics,* Vol. 30, No. 6, 1988, pp. 827–837.

[5] Vogelesang, L. B., Chen, D., Roebroeks, G. H. J. J., and Vlot, A., "ARALL Laminates: Past, Present, and Future," presented at the ARALL Laminates User's Conference, Seven Springs, PA, 26 Oct. 1987.

[6] *ARALLETTER,* Aluminum Company of America, New Kensington, PA, Vol. 2, No. 1, May 1989.

[7] Tada, H., Paris, P., and Irwin, G., *The Stress Analysis of Cracks Handbook,* Paris Productions, St. Louis, MO, 1985.

[8] Broek, D., *Elementary Engineering Fracture Mechanics,* Martinus Nijhoff, Dordrecht, The Netherlands, 1987.

[9] Ewalds, H. L. and Wanhill, R. J. H., *Fracture Mechanics,* Edward Arnold, Baltimore, MD, 1986.

[10] Wilson, C. D., "An Investigation of the Fracture and Fatigue Properties of ARALL-2," Masters thesis, Tennessee Technological University, Cookeville, TN, June 1986.

[11] "Instron Series 2490 Intelligent Interface Examples Package," Instron Corporation, Canton, MA, 1986.

[12] Herakovich, C., "Mechanical and Thermal Characterization of Unidirectional Aramod/Epoxy," presented at the ARALL Laminates Technical Conference, Champion, PA, 25–28 Oct. 1987.

Thomas C. Lee[1] and Dale A. Wilson[2]

An Investigation of the Effects of Temperature on the Impact Behavior and Residual Tensile Strength of an ARamid Aluminum Laminate (ARALL-2 Laminate)

REFERENCE: Lee, T. C. and Wilson, D. A., **"An Investigation of the Effects of Temperature on the Impact Behavior and Residual Tensile Strength of an ARamid Aluminum Laminate (ARALL-2 Laminate),"** *Composite Materials: Fatigue and Fracture (Third Volume), ASTM STP 1110,* T. K. O'Brien, Ed., American Society for Testing and Materials, Philadelphia, 1991, pp. 806–821.

ABSTRACT: The main objective of this investigation was to determine the effects of temperature on the impact behavior of ARALL-2 laminate. Post-impact studies included the assessment of localized damage and the determination of residual tensile strength. Longitudinal specimens (0° fiber orientation) were subjected to impact damage at temperatures of $-29, 24, 66, 93, 121,$ and 138°C ($-20, 75, 150, 200, 250,$ and 280°F) using an instrumented pendulum impact testing system and then loaded in tension to failure. Damage assessment included ultrasonic C-scanning of the impact areas and a scanning electron microscope (SEM) analysis.

The effects of temperature on the impact response of ARALL-2 were made apparent through the behavior of the aluminum layers. Impact damage and, consequently, residual tensile strength were found to vary considerably with temperature. The 24°C (75°F) impacts produced the least amount of internal damage, displayed the longest backsurface cracks, and absorbed the greatest amount of energy. As the impact temperature increased, the internal damage area increased and the energy absorbed decreased. Backsurface cracks decreased with increasing temperature, disappearing altogether at 93°C (200°F) and above.

A maximum tensile strength reduction of 41% occurred for the specimens impacted at 24°C (75°F). Accordingly, the residual tensile strength increased with increasing impact temperature. Since the tensile load was carried primarily by the fibers, this indicated that as temperature increased, fewer fibers experienced breakage. The $-29°C$ ($-20°F$) impact specimens deviated from these trends. This was attributed to the behavior of the metal adhesive. Less localized deformation occurred, and more unbroken fibers remained to bridge the crack, thus contributing to the residual tensile strength.

KEY WORDS: ARamid aluminum laminate, metal matrix composite, ARALL-2, impact damage, temperature effects, energy absorption, matrix plasticity, brittle fracture, residual strength, composite materials, fracture, fatigue (materials)

Within the aerospace industry, there is a constant objective to develop more efficient and more economical aircraft. It is well known that the more prohibitive costs of air travel are associated with fuel consumption and the service losses incurred during downtime maintenance. Obviously, fuel consumption is greatly affected by weight. Thus, there is a search for new materials and construction techniques that offer substantial weight savings. Since down-

---

[1] Engineer, McDonnell Douglas Space Systems Company, Huntsville, AL 35806.
[2] Professor, Department of Mechanical Engineering, Tennessee Technological University, Cookeville, TN 38505.

time is largely associated with aircraft inspections, efforts are aimed at developing more reliable materials that would allow greater intervals between routine inspections, and ultimately, greater service time. As a result, the last three decades have seen increased usage of fiber-reinforced composite materials. Because of high strength-to-weight ratios, composite structures offer considerable weight savings when compared to components based on more traditional engineering materials. Additionally, composites offering good fatigue resistance are especially attractive for cyclicly loaded fuselage skins and wing surfaces.

When utilizing fiber-reinforced composite materials, however, it is necessary for the design engineer to make additional material considerations. Composites are currently finding applications in areas highly susceptible to foreign body impact. Shock loading due to impact is known to cause localized damage and substantial reductions in material strengths. Therefore, investigating impact response and post-impact behavior is essential when characterizing a new composite material. When examining current design procedures, unfortunately, there is typically little or no consideration of impact behavior [1]. The very nature of the impact phenomenon does not lend itself to analytical formulations. As a result, design equations and material guidelines that govern composites in general cannot be applied efficiently. In practice, a new material must be tested explicitly for impact behavior under very specific conditions to determine its suitability for a given engineering application. Therefore, the development of a new fiber-reinforced laminate by the Delft University of Technology, the Netherlands, was the incentive for launching the current investigation.

Developed in the early 1980s, and currently under license for manufacture by the Aluminum Company of America (ALCOA), ARALL[3] laminate (ARamid aluminum laminate) actually represents a family of composite materials. ARALL-2 laminate, the focus material of this study, is a laminate built up of three thin sheets of 2024-T3 aluminum alloy and two layers of strong, unidirectional aramid fibers impregnated with a metal adhesive. This alternating 3/2 layup combines the advantages of aluminum alloys, such as design familiarity and machinability, and the fracture resistance of aramid fibers. ARALL laminate was developed primarily as a fatigue-resistant material, and there is considerable research available documenting its fatigue and fracture properties [2-4]. However, since it is hoped that this material will find extensive usage in future construction of aircraft, there is a need for characterization of the impact behavior of ARALL laminate. With this in mind, the interest of this work is concerned with determining the effects of temperature on the impact response of ARALL-2 laminate. To evaluate the consequences of such damage, post-impact studies will include the determination of residual tensile strength and the assessment of localized damage due to impact by means of the ultrasonic C-scan method and post-mortem fractographic studies.

Preliminary Considerations

There are many parameters involved in the characterization of impact loading. Currently, however, there is no single "most important" parameter to consider. Previous works have characterized impact as hard- or soft-body impact [5]. Hard-body impact, inflicted by hailstones, gravel, and hand tools (dropped during maintenance), primarily produces very localized damage in the form of indentations, axial and lateral cracking, and delamination. Usually limited to the immediate vicinity of the impact, this damage results in substantial strength degradation. Bird strikes are the main cause of soft-body impact. Depending on the particular projectile, this type of impact generally results in overall structural damage due to a large transfer of energy.

Additional efforts have successfully made characterizations of impact behavior based on impact velocity [1,6,7]. The general trend indicates that low-velocity impact results in non-

---

[3] ARALL LAMINATE is a registered trademark of the Aluminum Company of America.

visual or near-nonvisual damage that can substantially reduce residual strength properties. Intermediate-velocity impact (large indentations or severe matrix cracking or both) and high-velocity impact (perforation) are easily detected and thus can be repaired prior to component failure.

Since the impact response of a material can be thought of as its ability to absorb and dissipate energy while under shock loading, impact can be classified as high energy or low energy. History shows that high-energy impact typically results in immediate failure [8]. Damage from low-energy impact, however, may not be detected and may lead to sudden, catastrophic failure. It should be noted that care must be taken when impact energies are used as a means of comparing the impact behavior of different materials or even the same material under different loading conditions. A specific energy level can mask the relative effects of mass and velocity. This is made apparent by the fact that a variety of mass-velocity combinations can produce the same energy level. Earlier work has shown that low mass-high velocity impact and large mass-low velocity impact, both at identical energy levels, did not result in identical damage [9].

In this investigation, all impact loading will be considered low-energy impact. In light of the preceding discussion, however, further characterization of the impact loading presented herein is in order. Thus, the impact loading introduced to the ARALL-2 laminate test coupons may be categorically identified as hard-body, low-velocity, low-energy impact.

When investigating the post-impact behavior of a composite, the interactions of the individual components of the material during impact must be considered. One of the primary factors influencing the ability of a unidirectional composite to absorb energy during impact is the type of fiber used. For example, carbon- and boron-reinforced composites generally absorb low amounts of energy before causing laminate failure due to the low strains to failure of these particular fibers. E-glass fibers, on the other hand, have relatively high strains to failure. Composites reinforced with these fibers are found to absorb more energy before laminate failure occurs [3,10,11].

Another important factor is the ability of the matrix material to absorb energy. If, while under impact loading, the matrix undergoes plastic deformation, it can absorb a relatively large amount of energy. However, if the impact causes the matrix to experience brittle fracture, very little energy will be absorbed. Additionally, the level of adhesion between fiber and matrix affects energy absorption. High levels of adhesion encourage brittle failure and minimal energy absorption. Low levels, while allowing maximum energy to be absorbed, permit excessive delamination and possible premature catastrophic failure. The optimal condition would be some intermediate adhesion level that would allow progressive delamination, while discouraging brittle fracture of the matrix (depending on the specific material), and thereby providing acceptable energy absorption [12].

In recent works, residual tensile strength testing of impact damaged, resin matrix composites has shown strength degradation of up to 50% or more [1,12,13]. Reports indicate that fiber failure or fiber splitting or both on the back surface are major contributors to strength reduction [14]. This information is not necessarily indicative of metal matrix composite behavior, however, unless the effects of matrix plasticity are considered. Specimens of a boron/aluminum laminate impacted at room temperature exhibited strength reductions of up to 30% [5,15]. Although partially attributed to fiber breakage, the tensile strength was reported to drop sharply with the appearance of small transverse cracks on the back surface. Additionally, any moisture absorbed by the fibers or the resin material or both through these backsurface cracks may further weaken the laminate. Reports clearly indicate that moisture absorption reduced the tensile strength of composite panels, caused premature failure of pressure vessel type structures, and shortened the fatigue life of an S-glass epoxy composite [16–18].

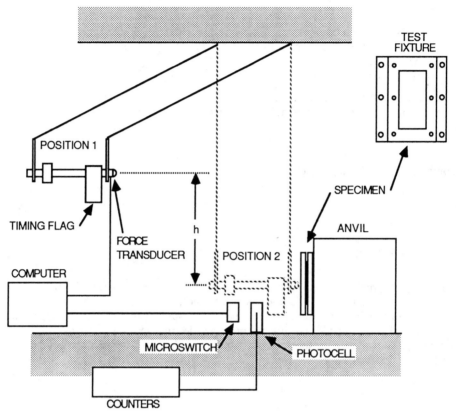

FIG. 1—*Schematic of instrumented pendulum impact testing system. Point of release and point of contact are indicated by Positions 1 and 2, respectively.*

Experimental Program

Theoretical Discussion

An instrumented pendulum impact testing system was used in this investigation to introduce transverse shock loading to the ARALL-2 laminate test coupons, as schematically shown in Fig. 1. Initial considerations of pendulum impact involve the relationship between impactor mass, release point, and velocity at impact. Assuming a conservative system, the energy at any point is governed by the law of conservation of energy, namely

$$T_1 + U_1 = T_2 + U_2 \qquad (1)$$

Subscripts 1 and 2 reflect the position of the pendulum as seen in Fig. 1. T_i and U_i represent the kinetic energy and the potential energy of the system, respectively, and are given by

$$T_i = \tfrac{1}{2}mV^2 \qquad (2)$$

$$U_i = mgh \qquad (3)$$

where

    m = mass of the pendulum,
    V = velocity of the pendulum,
    h = height above reference, and
    g = acceleration due to gravity.

At Position 1, the velocity of the pendulum, V, is zero. Along the reference plane at Position 2, the potential energy in the system is zero. Thus, Eq 1 reduces to

$$U_1 = T_2 \tag{4}$$

Combining Eqs 2 through 4 yields

$$mgh = \tfrac{1}{2}mV^2 \tag{5}$$

or

$$V = \sqrt{2gh} \tag{6}$$

Equation 6 reveals that the impact velocity is independent of the mass of the impactor, and depends solely on the release height, h, above reference.

In order to establish an impact energy history, the instantaneous velocity of the pendulum is necessary. However, additional information is needed to determine the velocity of the pendulum after contact with the specimen is established. Assuming a stepwise reduction in velocity, let the velocity at Time i be given by

$$V(i) = V(i - 1) - \Delta V \tag{7}$$

where, from Eq 6

$$V(i - 1) = \sqrt{2gh} \text{ , for } i > 1 \tag{8}$$

and ΔV is to be determined.

The computerized data acquisition system utilized in this investigation provides a force-time history. From this digitally stored curve, the force at any given time is known. Thus, from Newton's second law

$$F = m\frac{dV}{dt} \tag{9}$$

and solving for dV provides the basis to find ΔV, as

$$\Delta V = \frac{1}{2}[F(i - 1) + F(i)]\frac{\Delta t}{m} \tag{10}$$

where

    $F(i)$ = force at Time i, and
    Δt = time interval between data.

The impact energy history is the product of the impact velocity and the impulse. The impact velocity must be corrected by deceleration during impact, as given by Eqs 8 and 10. The impulse is the area under the force-time curve [7]. Thus, the impact energy is given by

$$E(i) = E(i - 1) + \frac{1}{2}[F(i - 1) + F(i)]\, \Delta V(i)\, \Delta t \qquad (11)$$

where

$E(i)$ = the energy at Time i, and
$E(0) = 0$.

The energy absorbed by the specimen during impact is determined by comparing the kinetic energy of the pendulum before and after impact. Recalling Eq 2

$$E_a = \frac{1}{2}m(V_{in}^2 - V_{out}^2) \qquad (12)$$

where

E_a = the energy absorbed by the specimen,
V_{in} = the impact velocity, and
V_{out} = the rebound velocity.

Impact Testing

The 76.2 by 127.0 by 1.32 mm (3.0 by 5.0 by 0.052 in.) specimens were cut from larger sheets with the fibers running parallel to the long edge. Since the specimens were to be heated prior to impact, an insulating material was needed. Further, the insulation material must not deform under transverse shock loading thereby absorbing impact energy. Ceramic was chosen for its excellent insulating capabilities as well as its extreme rigidity. Ceramic plates, 6.4 mm (0.25 in.) thick, were cut into strips 6.4 mm (0.25 in.) wide using a low-speed, diamond-coated wafering saw. The strips were clamped on either side of the specimen and around all four edges, in an aluminum picture frame type fixture. This provided an insulating border with unobstructed front and rear faces.

Silicon rubber resistance heaters, with a power density of 0.78 W/cm^2 (5.0 W/in.^{2}), were used to heat the specimens. Two heaters, 50.8 by 50.8 mm (2.0 by 2.0 in.), were used on the back surface of the coupon, while two heaters, 25.4 by 50.8 mm (1.0 by 2.0 in.), were used on the front. This arrangement provided a total of 60 W of heating power. Thin panels of insulating fabric were used as blankets to maintain specimen temperature. The front panel had a 25.4-mm (1.0-in.)-diameter hole to allow an unobstructed impact from the indenter. The rear asbestos panel was slit in an "X" fashion to allow for specimen deformation during impact. Heat to the specimen was controlled by regulating the voltage supply across the heaters using a Variac autotransformer.

A hemispherical impactor with a diameter of 12.7 mm (0.5 in.) was supported by four steel cables suspended beneath a large, rigid frame. The lengths of the cables were adjusted such that, at the lowest point of its trajectory, the impactor was level and just touching the center of the specimen to be impacted. A series of preliminary tests were performed at room temperature to establish the desired level of damage. An energy level of 8.81 N·m (6.5 ft·lb), which provided sufficient back-surface cracking to make moisture absorption a threat, was chosen for impact testing. After being released, and just prior to impact, the pendulum passed through the beam of a microswitch that triggered a microcomputer to begin collecting data. A force transducer located behind the impactor nose provided a voltage signal to an analog to digital (A/D) converter card with a sampling rate of 10 000 datum points per second. The data from the A/D card were stored on floppy disk for later analysis.

In order to verify the impact velocity and to determine the rebound velocity, a 50.8-mm (2.0-in.) wide triggering flag was suspended beneath the impactor. As the pendulum approached the bottom of its path, the flag passed between two photoelectric sensors triggering an electronic counter. By knowing the duration that the sensors were interrupted and also the width of the flag, the impact velocity was calculated. Similarly, on rebound, the flag triggered a second counter, and the rebound velocity was calculated.

In this investigation, longitudinal specimens were impacted at -29, 24, 66, 93, 121, and 138°C (-20, 75, 150, 200, 250, and 280°F). Duplicate impact tests were performed at each temperature resulting in 24 damaged specimens. These temperatures were selected to provide information on the impact behavior of ARALL-2 laminate from well below freezing to a point above the maximum operating temperature of the metal adhesive (121°C, 250°F) used in the laminate. The specimens were heated, as discussed previously, until all four thermocouples indicated the desired temperature with ± 2.8°C (± 5°F). To reach -29°C (-20°F), the specimen/fixture assembly was saturated with a stream of liquid Freon-12. The impacted specimens were then visually inspected to thoroughly study the extent of damage under each loading condition. Back surfaces were examined for cracks, and, when present, were measured by means of a traveling microscope. The impacted specimens were then prepared for ultrasonic C-scanning. The two-dimensional view provided by this inspection technique revealed a full-scale representation of the area. By direct measurement of the C-scans, the total internally damaged area was determined.

Residual Tensile Strength Testing

Residual tensile strength tests were performed at room temperature using an Instron advanced testing machine, Model 1332. The specimens in this investigation were 76.2 by 127.0 by 1.32 mm (3.0 by 5.0 by 0.052 in.). This rectangular coupon, while efficient for impact testing, did not lend itself readily to tension testing. Therefore, it was necessary to reduce the gage width sufficiently to control the failure location. This reduction of specimen width introduced stress concentrations near the impact damage region that must be considered. A finite element (FE) model was constructed and used to optimize reduction radii and gage width as well as estimate actual stress in the damage region. Details of the FE model can be found in Ref 19. To ensure uniformity, a machining template was used to modify the rectangular specimens to the geometry shown in Fig. 2.

To establish baseline tensile strengths, undamaged specimens were modified according to the gage width reduction methods outlined previously, and loaded to failure. Following the guidelines set forth by ASTM Test Methods of Tension Testing of Metallic materials (E 8-88), a stress rate of 344.8 kPa/min (50 ksi/min) was chosen. This ensured a total testing time of less than 2 min. Duplicate residual tensile strength tests were then performed on all damaged specimens.

Results and Discussion

Impact Damage Analysis

The total energy absorbed during impact was calculated for each specimen using the methods discussed previously. Impact energy plotted as a function of temperature is shown in Fig. 3. Back surface crack lengths (all occurring perpendicular to the fiber direction) as a function

FRONT   EDGE

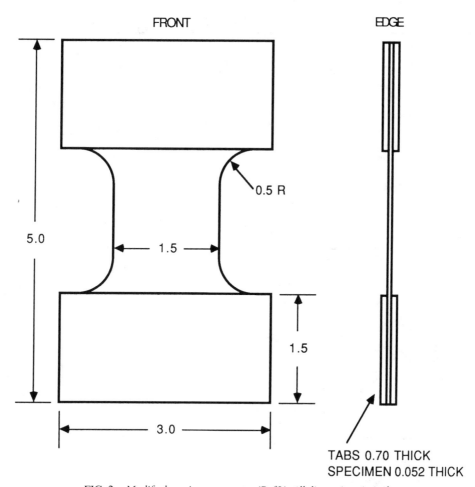

0.5 R

5.0

1.5

1.5

3.0

TABS 0.70 THICK
SPECIMEN 0.052 THICK

FIG. 2—*Modified specimen geometry (Ref 2). All dimensions in inches.*

of temperature are shown in Fig. 4. As can be seen, energy absorption and backsurface cracking vary considerably with temperature. The 24°C (75°F) impacts resulted in the greatest
amount of energy absorbed, at 7.73 N·m (5.7 ft·lb). As temperature increased, energy absorption decreased with a minimum occurring for 138°C (280°F) impacts, at 7.19 N·m (5.3 ft·lb).
Specimens impacted at −29°C (−20°F) deviated from this trend by displaying an energy
absorption of 7.59 N·m (5.6 ft·lb). After studying the backsurface crack data, a correlation to
energy absorption becomes apparent. The impacts at 24°C (75°F), that is, the temperature
which resulted in the greatest amount of energy absorption, produced the largest backsurface
cracks, at 8.13 mm (0.32 in.). As impact temperature increased, consequently decreasing
energy absorption, the backsurface crack lengths decreased until disappearing altogether at
impact temperatures of 93°C (200°F) and above. Reviewing energy absorption data revealed
a corresponding decrease at 93°C (200°F). As expected, the −29°C (−20°F) impacts deviated
from the trend of crack length increasing with decreasing impact temperature by producing a
backsurface crack length of 6.86 mm (0.27 in.).

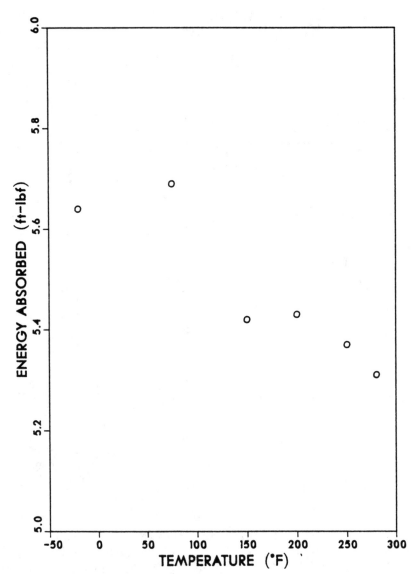

FIG. 3—*Energy absorbed during impact versus temperature.*

An important factor governing the ability of a composite to absorb energy is the ability of the matrix to deform plastically under impact loading [*12*]. The ductility of the matrix material (aluminum layers) in the ARALL-2 laminate should not vary significantly with temperature [*20*]. Thus, the fact that energy absorption did vary with temperature suggests that the interlaminar bonds degrade with increasing temperature due to variations in the properties of the metal adhesive (3M AF163-2). This is especially true at higher impact temperatures considering the maximum service temperature of the adhesive is 93°C (200°F). Reports indicate that composites reinforced with fibers possessing low strains to failure typically are able to absorb less energy [*5,10,11*]. Additionally, the typically low strain to failure of aramid fibers

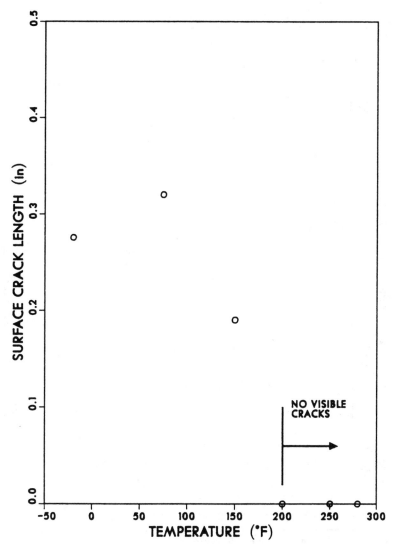

FIG. 4—*Backsurface crack length versus temperature.*

might suggest that energy absorption is governed by fiber breakage as temperature increases [3]. The fact that the energy absorption decreased with increasing temperature indicates that fewer and fewer fibers are actually breaking. The deviation for the general trends at the −29°C (−20°F) impacts, therefore, may be attributed to the increased brittleness of the outer aluminum layers.

To further evaluate the extent of impact damage to the longitudinal specimens, the impact regions were inspected using the ultrasonic C-scan method. This nondestructive examination technique provided a full-size map of the internally damaged area. By measuring the region displaying varying degrees of signal attenuation of each C-scan, the internally damaged areas were determined. These damaged areas are presented as a function of temperature in Fig. 5. The 24°C (75°F) resulted in the least amount of internal damage, with an area of 58.1 mm^2

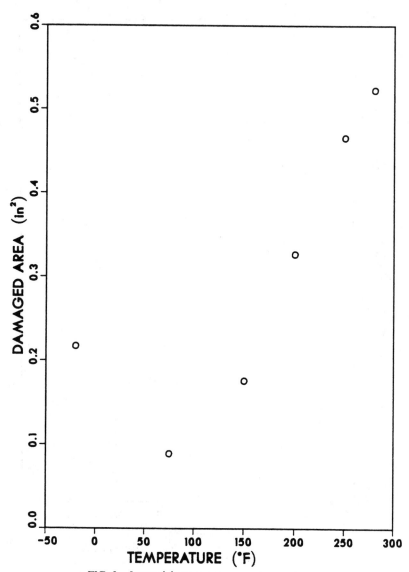

FIG. 5—*Internal damage area versus temperature.*

(0.09 in.²). As the impact temperature increased, the internally damaged area increased, with a maximum area of 335.5 mm² (0.52 in.²) at 138°C (280°F). Visual examination revealed this increasing internal damage area to be associated with significant out-of-plane deformation, and, therefore, ply delamination. Supplementing this information with the energy absorption data and the backsurface crack length data provides interesting results. The 24°C (75°F) impacts produced the least amount of internally damaged area, the longest backsurface cracks, and the greatest amount of absorbed energy. As the temperature increased, internally damaged areas increased, while the energy absorption decreased and the backsurface cracks disappeared

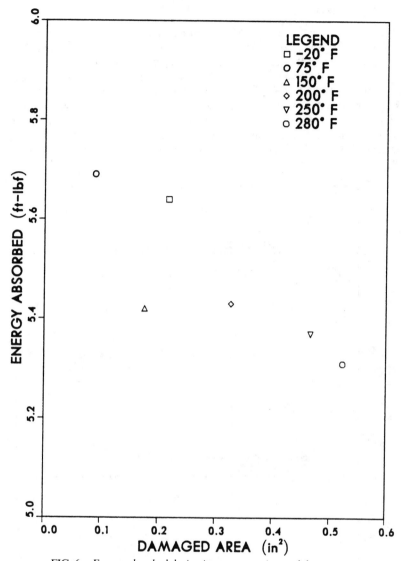

FIG. 6—*Energy absorbed during impact versus internal damage area.*

altogether. This indicates that for impacts above room temperature, delamination governs energy absorption. As the temperature increases, resulting in a decaying internal structural integrity, less energy is required to cause ply separation and delamination. This is supported by recognizing that decreasing levels of adhesion permit excessive delamination and, therefore, increasing ability to absorb energy before failure. Since out-of-plane deformation is not easily measured, energy absorption as a function of internally damaged areas is shown in Fig. 6 to provide insight into the correlation between energy absorption and delamination.

Residual Tensile Strength Analysis

Residual tensile strengths of the ARALL-2 laminate specimens as a function of temperature are shown in Fig. 7. The dotted line at 75.0 ksi represents the ultimate strength of an unimpacted specimen with the reduced gage width as previously discussed. As can be seen, a maximum strength reduction occurred for the 24°C (75°F) impact specimens, with a residual tensile strength of 413.8 kPa (60.0 ksi). As impact temperature increased, the residual tensile strength increased, with a maximum of 517.2 kPa (75.0 ksi) at 138 °C (280°F). Since the load is carried primarily by the fibers in the longitudinal specimens, these results further support the previous conclusion that as temperature increases, fewer and fewer fibers actually experience breakage. Additionally, as impact temperature increases, absorbed energy is consumed

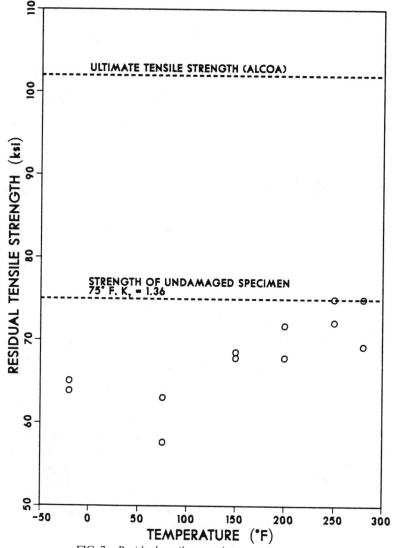

FIG. 7—*Residual tensile strength versus temperature.*

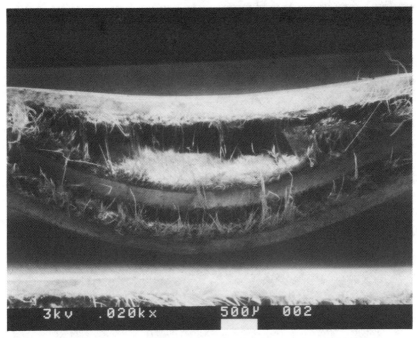

FIG. 8—*Photomicrograph of 24°C (75°F) impact specimen showing highly localized damage. Scale mark represents 500 µin.*

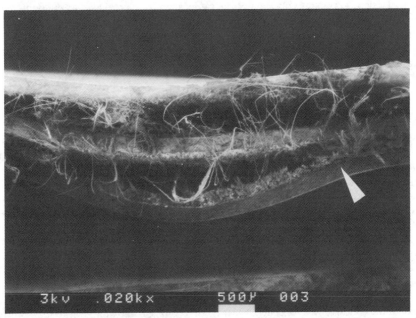

FIG. 9—*Photomicrograph of 66°C (150°F) impact specimen showing transition from region of fiber breakage to region of fiber pullout and delamination, as shown by the arrow. Scale mark represents 500 µin.*

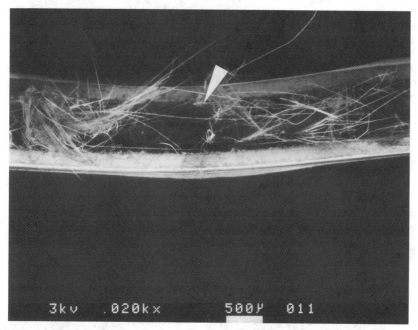

FIG. 10—*Photomicrograph of 121°C (250°F) impact specimen showing large damaged area relative to the contact area of the indentor. Arrow indicates region of excessive delamination and ply separation. Scale mark represents 500 µin.*

by work of delamination and ply separation, rather than fiber breakage. As encountered previously, the −29°C (−20°F) impact deviated from the general trends, showing a residual tensile strength of 461.4 kPa (66.9 ksi). This suggests that the back surface cracks at −29°C (−20°F) may have been enhanced due to adhesive degradation and hinted at more severe fiber breakage than actually occurred. As a result, more unbroken fibers existed to bridge the crack and contribute to the residual tensile strength.

Fractographic Analysis

Photomicrographs taken using an ISI-SR-50 scanning election microscope (SEM) display the varying effects of temperature on the internal damage of ARALL-2 laminate. In all photomicrographs, the loading axis is perpendicular to the plane of the page. At 24°C (75°F), the damage is highly localized, as shown in Fig. 8. Extensive fiber breakage and the well-defined indentation are apparent. The impacted surface (top of photo) separated from the remaining plies during tension testing and was not the result of impact testing. As the temperature increased to 66°C (150°F), as shown in Fig. 9, the damage area increased and is accompanied by less severe indentation damage. Also, the transition from a region of fiber breakage due to impact to a region of fiber pullout and delamination is evident, as indicated by the arrow. At 121°C (250°F), as shown in Fig. 10, the damaged area is large relative to the contact area of the indentor. This is due to the significant reduction in structural integrity associated with the high-impact temperature. This should be expected, however, considering that the service temperature of the adhesive was exceeded. Further, the damage is shown to be predominantly ply separation and delamination plus extensive fiber pullout. This resulted in the center aluminum ply failing on a different plane than the two outer aluminum skins.

Conclusions

In this investigation, temperature was found to have significant effects on the impact behavior of ARALL-2 laminate. The influence of temperature was most apparent in the behavior of the aluminum layers and the metal adhesive used during laminate construction. Since the ARALL laminate "family of materials" is currently in a state of development and improvement, new laminates utilizing different alloy/adhesive combinations would merit a continuation of the investigation presented herein.

An obvious path for further study of the ARALL family of composite materials follows the incorporation of different adhesives. Sufficient study should enable the optimization of a laminate to a particular situation by carefully controlling delamination and fiber breakage by varying the adhesion levels. As seen by this investigation, the fact that the behavior of the ARALL-2 laminate varies greatly should not be viewed as a material flaw, but as a material feature that should be exploited to meet the needs of a given application.

References

[1] Joshi, S. P. and Sun, C. T., *Journal of Composite Materials,* Vol. 19, Jan 1985, pp. 51–66.

[2] Vogelesang, L. B. and Gunnick, J. W., "ARALL: A Materials Challenge for the Next Generation of Aircraft," Department of Aerospace Engineering, Delft University of Technology, The Netherlands, 1986.

[3] Bucci, R. J., Mueller, L. N., Schultz, R. U., and Prohaska, J. L., "ARALL Laminates—Results from a Cooperative Test Program" ALCOA Laboratories, ALCOA Center, PA, 1987.

[4] Wilson, C. D., "An Investigation of Fracture and Fatigue Properties of ARALL-2," Master's thesis, Tennessee Technological University, Cookeville, TN, 1988.

[5] Awerbuch, J. and Hahn, H. T., *Journal of Composite Materials,* Vol. 10, July 1976, pp. 231–257.

[6] Broutman, L. J. and Rotem, A., "Impact Strength and Toughness of Fiber Composite Materials," *Foreign Object Impact Damage to Composites, ASTM STP 568,* American Society for Testing and Materials, Philadelphia, 1973, pp. 114–133.

[7] Winkel, J. D. and Adams, D. F., *Composites,* Vol. 16, No. 4, Oct 1985, pp. 268–278.

[8] Davies, C. K. L., Turner, S., and Williamson, K. H., *Composites,* Vol. 16, No. 4, Oct. 1985, pp. 279–285.

[9] Williams, J. G., O'Brien, T. K., and Chapman, A. J., III, "Comparison of Toughened Composite Laminates using NASA Standard Damage Tolerance Tests," presented at ACEE Composite Structures Technology Conference, Seattle, WA, Aug. 1984.

[10] Grove, R. and Smith B., "Compendium of Post-Failure Analysis Techniques for Composite Materials," Air Force Wright Aeronautical Labs, Wright Patterson AFB, OH, Jan. 1987.

[11] Clark, G. and Van Blaricum, T. J., "Carbon Fiber Composite Coupons—Static and Fatigue Behavior After Impact," Department of Defense, Structures Report ARL-STRUC-R-422, Washington, DC, Aug. 1986.

[12] Mallick, P. K., *Fiber-Reinforced Composites: Materials Manufacturing and Design,* Marcel Dekker, Inc., New York 1988.

[13] Aleszka, J. C., *Journal of Testing and Evaluation,* Vol. 6, No. 3, May 1978, pp. 202–210.

[14] Reed, R. P. and Schuster, D. M., *Journal of Composite Materials,* Vol. 4, Oct. 1970, pp. 514–525.

[15] Caprino, G., Visconti, I. C., and Di Ilio, A., *Composites,* Vol. 15, No. 3, July 1984, pp. 231–234.

[16] Vogelesang, L. B. and Gunnick, J. W., "ARALL: A Material for the Next Generation of Aircraft, the State of the Art," Department of Aerospace Engineering, Delft University of Technology, The Netherlands, 1986.

[17] "ALCOA ARALL Laminate Sheet," ALCOA Aerospace Technical Fact Sheet, Aluminum Company of America, New Kensington, PA, 1985.

[18] Gunnick, J. W., Verbruggen, M. L. C. E., and Vogelesang, L. B., "ARALL, A Light Weight Structural Material for Impact and Fatigue Sensitive Structures," *Vertica,* Vol. 10, No. 2, 1986, pp. 241–254.

[19] Lee, T. C., "An Investigation of the Effects of Temperature on the Impact Behavior and Residual Strength of an Aramid/Aluminum Laminate," Master's thesis, Tennessee Technological University, Cookeville, TN, 1991.

[20] Van Vlack, L. H., *Elements of Materials Science and Engineering,* Addison-Wesley Publishing Company, Reading, MA, 1980, pp. 219–220.

Author Index

A

Allen, D. H., 56
An, D., 659
Armanios, E. A., 269, 340
Ashby, M. F., 617
Avery, W. B., 476

B

Badir, A. M., 269
Bahei-El-Din, Y. A., 696
Bandyopadhyay, S., 667
Bathias, C., 638
Beaumont, P. W. R., 596, 617
Bhatia, N. M., 126
Boniface, L., 9
Bradley, W. L., 393
Bucinell, R. B., 528
Burch, I. A., 667

C

Camponeschi, E. T., Jr., 439, 550
Cantwell, W. J., 70
Chinatambi, N., 187
Coxon, B. R., 476
Crews, J. H., Jr., 169
Curtis, D. C., 581

D

Davies, M., 581
Davies, P., 70
Dost, E. F., 30, 476

G

Ghosn, L., 711
Grande, D. H., 30
Guynn, E. G., 393

H

Harris, C. E., 56
Hashemi, S., 143
Hooper, S. J., 89, 107

I

Ilcewicz, L. B., 30, 476

J

Jar, P-Y., 70
Johnson, W. S., 753

K

Kageyama, K., 210
Kan, H-P., 126
Kantzos, P., 711
Kausch, H-H., 70
Kikuchi, M., 210
Kinloch, A. J., 143
Kortschot, M. T., 596
Koury, J. L., 528
Krempl, E., 659

L

Lai, D., 638
Lam, P. W. K., 686
Lee, J-W., 56
Lee, T. C., 806

M

Madan, R. C., 457
Mahler, M. A., 126
Majumdar, B. S., 732
Martin, R. H., 243
McCool, J. W., 30
Moore, D. R., 581
Murri, G. B., 312

N

Naik, R. A., 169, 753
Neville, D. J., 70
Newaz, G. M., 732
Nuismer, R. J., 528

O

O'Brian, T. K., 1, 269, 312
Ochoa, O. O., 393
Ogin, S. L., 9
Osiroff, R., 772

P

Palazotta, A., 373
Parnas, L., 340
Piggott, M. R., 686
Poe, C. C., Jr., 501
Poursartip, A., 187

R

Reifsnider, K. L., 772
Richard, H., 70
Russell, A. J., 226

S

Salpekar, S. A., 269, 312
Shivakumar, K. N., 169

Slater, B., 581
Smith, P. A., 9
Spearing, M., 617
Sriram, P., 269
Stinchcomb, W. W., 772
Subramanian, R., 89, 107

T

Telesman, J., 711
Toubia, R. F., 89, 107
Tratt, M. D., 359

U

Underwood, J. H., 667

W

Whitcomb, J. D., 393, 417
Wilder, B., 373
Williams, G., 143
Wilson, C. D., 791
Wilson, D. A., 791, 806

Y

Yanagisawa, N., 210

Subject Index

A

Absorption, effect on interlaminar fracture toughness, 107

Adhesive joints, Mode I behavior under mixed-mode loadings, 187

Aligned carbon/epoxy laminates, fatigue failure, 686

Aluminum plies, fatigue properties, 772

Angle-ply laminates, 56

APC-2 composite, fiber microbuckling, 393

ARALL laminates, damage and performance, 772

ARALL-2 laminates, effective crack lengths, 791

Aramid aluminum laminates (*see* ARALL-2 laminates)

AS4/3501-5A graphite fabric-reinforced epoxy, delamination growth, 359

AS4/3501-6 carbon/epoxy composites, compression testing, 439

AS4 fibers, PEEK composites with, 70

AS4/PEEK composites
effects of T-tabs and large deflections, 169
fatigue behavior, 581
fiber microbuckling, 393
Mode II delamination, 226

ASTM standards
D 695-80, 72
D 695-89, 550, 581
D 790-86, 667
D 3410-87, 439, 550
E 399, 211

B

Bending tests, for bulk flexural strength, 667

Biaxial fatigue, interlaminar stress effects, 659

Boron/aluminum composites, fracture, 696

Brittle fracture, ARALL-2 laminates, 806

Buckling, cylindrical composite panels with implanted delaminations, 373

C

Cantilever beams
double, fatigue delamination onset prediction, 312
split, Mode III delamination testing, 243

Carbon/bismaleimide laminates, fracture strength, 667

Carbon/epoxy laminates
fatigue failure, 686
fracture strength, 667
low-velocity impact damage, 457
stabilized end notched flexure test, 210
thick-sections, compression testing, 439

Carbon fiber composites, Mode II delamination, 226

Carbon fiber/PEEK composites, cooling rate effects, 70

Carbon fiber/polymer composites, mixed-mode fracture, 143

Carbon fiber-reinforced plastics, stacking sequence effects, 476

Circular delamination, 373

Compliance method, effective crack lengths for ARALL-2 laminates, 791

Composite laminates (*see also* specific laminate)
matrix cracking, 30
transverse ply cracking, 9

Composite plates, filament-wound, effects of quasi-static impact events, 528

Compression after impact strength, quasi-isotropic laminates, 476

Compression behavior
delamination effects, 359
fiber-reinforced composite materials, review, 550

Compression fatigue
CF/PEEK, 581
hole effect in T300/N5208 composites, 638

Compression strength
edge effects, 393
transverse, composite plates, 528

Compression testing
fiber-reinforced composite materials, review, 550
thick-section composites, 439
Computer-aided testing, for crack length and fracture toughness, 210
Cooling rate effects
on carbon fiber/PEEK composites, 70
on fatigue behavior of CF/PEEK composites, 581
Corner radii, tensile stress, porosity effects, 126
Crack bridging, in metal matrix composites, 711
Cracked-lap shear specimens, Mode I strain energy release rate, 187
Crack growth, interlaminar Mode II, 210
Crack lengths
computer-aided testing system, 210
effective, for ARALL-2 laminates, 791
Crack opening displacement, surface opening displacement and, 187
Crack shear displacements, direct measurement, 210
Creep, carbon fiber/PEEK composites, cooling rate effects, 70
Cross-ply laminates
fatigue damage growth, 617
with matrix cracks, 56
Crystallinity, cooling rate effects in CF/PEEK, 70
Cylindrical composite panels, effects of inserted circular delaminations, 373

D

Damage growth
notched graphite/epoxy laminates, 617
and strain energy release rate, 617
Damage mechanics
ARALL laminates subjected to tensile cyclic loading, 772
fatigue-related, notched graphite/epoxy laminates, 617
notched strength modeling, 596
Damage modeling, damage zone in T300/N5208 composites, 638
Damage tolerance, low-velocity impacts, 457
Deflections, effects on double cantilever beam tests, 169
Delamination
carbon fiber/PEEK composites, cooling rate effects, 70
combined matrix cracking and free edge effects, 287
direct measurement of crack shear displacements, 210
edge, 89, 107
effects of T-tabs and large deflections, 169
effects on
compression behavior, 359
stress distribution in zero degree ply, 596
free-edge, 269
growth in compressively loaded composites, 359
implanted within cylindrical composite panel, 373
interply, in SCS-6/TI-15-3 composites, 732
local, shear deformation model, 269
Mode I under mixed-mode loadings, 187
Mode II in toughened composites, 226
in tapered composite laminates, 312
transverse crack-tip, 269
Discontinuities, in fibrous metal matrix composites, 696
Double cantilever beams
effects of large deflections and T-tabs, 169
fatigue delamination onset prediction, 312
Dropped ply, 312

E

Edge delamination
moisture effects, 89
partially saturated, tension test, 107
Edge effects, in fiber microbuckling, 393
Effective crack lengths
for ARALL-2 laminates, 791
and stress intensity factor, 791
End notched flexure test, stabilized, 210
Environmental effects
jet fuel absorption on delamination, 107
moisture and notch geometry effects on fracture toughness, 667

F

Fatigue behavior
ARALL laminates subjected to tensile cyclic loading, 772
carbon/epoxy laminates, 686
carbon fiber/PEEK composites, cooling rate effects, 70
uniaxial zero-to-tension, of graphite/epoxy tubes, 659

Fatigue cracks
  growth in unidirectional SCS-6/Ti-15-3
    composites, 711
  growth rate testing, ARALL-2 laminates,
    791
  initiation, in notched SCS-6/TI-15-3
    composites, 753
Fatigue damage, growth modeling, 617
Fiber bridging, and crack growth in
    ARALL-2 laminates, 791
Fiber composites
  fatigue failure, 686
  mixed-mode fracture, 143
Fiberglass/epoxy composites, compression
    testing of thick sections, 439
Fiber-matrix debonding, in SCS-6/TI-15-3
    composites, 732, 753
Fiber microbuckling, edge effects, 393
Fiber-reinforced composites
  compressive response, review, 550
  notched, fatigue damage mechanics, 617
  stabilized end notched flexure test, 210
  tapered laminates, delamination analysis,
    340
Fiber-reinforced plies, fatigue properties,
    772
Fiber shear, notched composite laminates,
    393
Fiber stress, zero degree, 732
Fiber type, effects CF/PEEK fatigue
    behavior, 581
Fibrous metal matrix composites, fracture,
    696
Fickian moisture diffusion, 89
Filament-wound composite plates, effects of
    quasi-static impact events, 528
Filament-wound motor cases, nonvisible
    damage and residual tensile strength,
    501
Finite element analyses
  fibrous metal matrix composites with
    discontinuities, 696
  maximum delamination stress in
    thermomechanical fatigue, 732
  splitting and delamination effects on stress
    distribution, 596
  unidirectional tapered laminates with ply
    drops, 312
Finite element analyses, three-dimensional
  combined effect of matrix cracking and
    free edge, 287
  prediction of delamination onset in
    compressively loaded composites,
    359

split cantilever beam Mode III
    delamination testing, 243
weave effects on composite moduli and
    stresses, 417
Finite element modeling
  fatigue damage growth, 617
  plane strain, 373
Flange-web corners, porosity effects,
    126
Flexural strength, effects of notch geometry
    and moisture, 667
Fracture behavior
  carbon fiber/PEEK composites, cooling
    rate effects, 70
  fibrous metal matrix composites with
    discontinuities, 696
Fracture strength, effects of notch geometry
    and moisture, 667
Fracture toughness, interlaminar
  absorption effects, 89, 107
  computer-aided testing system, 210
  effects of
    large deflections, 169
    moisture, 667
    notch geometry, 667
    T-tabs, 169
  fiber composite laminates, 143
  Mode II, 226
  Mode III testing of split cantilever beams,
    243
Fracture toughness, translaminar, effects of
    notch geometry and moisture, 667
Free edge, and matrix cracking, combined
    effects, 287
Free-edge delamination, fracture analysis,
    269

G

Geometric nonlinearity, large deflection-
    and T-tab-related, 169
Graphite/epoxy laminate composites,
    30
  edge delamination, 107
  impacter shape effects, 501
  notched, cross-ply, damage-based strength
    models, 596
Graphite/epoxy tubes, uniaxial zero-to-
    tension fatigue behavior, 659
Graphite/PEEK composites, edge
    delamination, 107
Growth law, notched graphite/epoxy
    laminate damage, 617

H

Hertz's law, 501
Hole effect, in compression fatigue of T300/
    N5208 composites, 638
Holes, effect on fracture strength, 696
Hole size, effects on strength, 596

I

IM6/5245C composites, Mode II
    delamination, 226
IM6 fibers, PEEK composites with, 70
IM6/PEEK composites, fatigue behavior,
    581
IM7/8551-7
    effects of T-tabs and large deflections, 169
    low-velocity impact damage, 457
    matrix cracking, 30
    Mode II delamination, 226
Impact
    carbon fiber/PEEK composites, cooling
        rate effects, 70
    low-velocity
        effect on composite structures, 457
        impacter shape effects, 457
        temperature effects on ARALL-2
            laminates, 806
Impact damage
    ARALL-2 laminates, temperature effects,
        806
    resistance, stacking sequence effects, 476
    in stitched laminates, 457
Impact energy, effect on damage to bonded
    stiffened structures, 457
Impacters, shape effects, 501
Interlaminar fracture toughness, fiber
    composite laminates, 143
Interlaminar layers, resin-rich, 30
Interlaminar stresses, effects on graphite/
    epoxy tube fatigue, 659
Interlaminar tensile strength, porous
    structures, 126
Internal state variable concept, 56
Internal stresses, carbon fiber/PEEK
    composites, cooling rate effects, 70
Interply delamination, in SCS-6/TI-15-3
    composites, 732
Isothermal fatigue, in quasi-isotropic metal
    matrix composites, 732

J

Jet fuel absorption, effect on interlaminar
    fracture toughness, 107

L

Linear elastic fracture mechanics
    damage-based notch strength modeling,
        596
    effective crack lengths for ARALL-2
        laminates, 791
Load-displacement curves, nonlinearity, 226
Loading tabs, effects on double cantilever
    beam tests, 169
Low-velocity impact
    effect on composite structures, 457
    impacter shape effects, 501

M

Matrix cracking
    and free edge, combined effects, 287
    in IM7/8551-7 composites, 30
    in SCS-6/TI-15-3 composites, 753
    upper bounds of reduced axial and shear
        moduli, 56
Matrix cracks, in composite laminates, 9
Matrix crack tip delamination, fracture
    analysis, 269
Matrix plasticity, ARALL-2 laminates, 806
Matrix structure, carbon fiber/PEEK
    composites, cooling rate effects, 70
Metal matrix composites
    damage growth, 753
    fatigue crack growth, 711
    fatigue crack initiation, 753
    temperature effects on impact damage
        and residual tensile strength, 806
Mixed-mode loading
    composite laminates, delamination onset
        of, 359
    delamination, Mode I behavior, 187
Mixed-mode tests, fiber composite
    materials, 143
Mode I strain energy release rate, 187
Mode II delamination, in toughened
    composites, 226
Mode II interlaminar crack growth, 210
Mode III delamination testing, split
    cantilever beams, 243
Moisture, effect on fracture strength, 667
Motor cases, nonvisible damage and
    residual tensile strength, 501

N

Nondestructive inspection
    porosity effects on flange-web corner
        strength, 126

stacking sequence effects on impact
damage and residual strength, 476
Nonlinearity
geometric, double cantilever beams,
169
Mode II delamination in toughened
composites, 226
Nonvisible damage, impacter shape effects,
501
Notched, cross-ply graphite/epoxy
laminates, damage-based strength
models, 596
Notched composites
fatigue crack initiation, 753
fiber microbuckling, 393
Notched strength modeling, damage-based,
596
Notches
effect on fracture strength in fibrous metal
matrix composites, 696
geometry, effect on fracture strength, 667

O

Open-hole/notched composites, fiber
microbuckling, 393

P

Plastic zone, from discontinuities, 696
Ply drops, 312
Ply group thickness, 30
effect on damage resistance and residual
strength, 476
Porosity, effects on flange-web corner
strength, 126
Pulse-echo ultrasonics, 476

Q

Q3DG program, 89, 107
Quasi-isotropic laminates
fatigue damage growth, 617
stacking sequence effects, 476
Quasi-isotropic metal matrix composites,
thermomechanical fatigue, 732
Quasi-static impact, effects on composite
plates, 528

R

Radiography, damage monitoring with,
596
Rayleigh-Ritz energy method, 528
Reduced axial, upper bounds, 56

Residual strength
ARALL-2 laminates, temperature effects,
806
and impact damage, 457
notched graphite/epoxy laminates, 617
stacking sequence effects, 476
thick graphite/epoxy composites,
impacter shape effects, 501
Residual stress, SCS-6/TI-15-3 composites,
732
Resin-rich interlaminar layers, 30

S

S2/3501-6 fiberglass/epoxy composites,
thick-sections, compression testing,
439
S2/SP250 glass/epoxy laminates, tapered,
delamination analysis, 340
Scanning electron microscopy
ARALL-2 impact damage, 806
damage modes to SCS-6/TI-15-3
composites, 732
for damage monitoring, 596
fatigue crack growth in metal matrix
composites, 711
SCS-6/Ti-15-3 composites
fatigue crack growth, 711
notched, fatigue crack initiation and
growth, 753
thermomechanical fatigue, 732
Shear deformation model, for local
delamination, 269
Shear moduli, upper bounds, 56
Split cantilever beams, Mode III
delamination testing, 243
Splitting, effects on stress distribution in
zero degree ply, 596
Stabilized end notched flexure test, 210
Stacking sequence, effect on damage
resistance and residual strength, 476
Stiffness reduction
cross-ply laminates, 56
and fatigue damage, 772
Stitched laminates, impact damage, 457
Strain energy release rate, 9
and damage growth, power law
relationship, 617
mixed mode, 89, 107
Mode I, in cracked-lap shear specimens, 187
Mode III delamination testing, 243
notched graphite/epoxy laminates, 617
tapered laminates, 312
tapered laminates under tensile loading,
340

Strain measurement, optical, 596
Stress analyses, woven composites, 417
Stress distribution in zero ply, splitting and delamination effects, 596
Stress intensity factor
  and effective crack length, 791
  of transverse ply cracking, 9
Subcritical damage, effect on notched strength of cross-ply composites, 596
Surface opening displacement, calculation, 187

T

T300/914C carbon-fiber epoxy composites, fatigue damage mechanics, 617
T300/934 graphite/epoxy laminates, transverse crack-tip and free-edge delamination, 269
T300/5208 composites, effects of T-tabs and large deflections, 169
T300/N5208 composites, hole effect in compression fatigue, 638
Temperature effects
  on ARALL-2 laminates, 806
  on fatigue behavior of CF/PEEK composites, 581
Tensile loading, delamination of tapered laminates under, 340
Tensile residual strength
  ARALL-2 laminates, temperature effects, 806
  impacter shape effects, 501
Tensile strength, interlaminar, 126
Tension cyclic loading, ARALL laminates, 772
Tension fatigue, CF/PEEK, 581
Tension-tension cyclic loading, effects on T300/914C laminates, 617
Thermomechanical fatigue, in quasi-isotropic metal matrix composites, 732
Thermoplastic composites, fatigue behavior, 581

Ti/SCS composites (see SCS-6/Ti-15-3 composites)
Toughened matrices, stacking sequence effects, 476
Transverse microcracking, SCS-6/TI-15-3 composites, 732
Transverse ply cracking, composite laminates, 9
Tropical exposure, effects of notch geometry and moisture, 667
Tubular specimens, interlaminar stress effects on fatigue behavior, 659
Two-degree of freedom model, composite plates subjected to transverse impacts, 528

U

Ultrasonic signals, attenuation loss, 126
Unidirectional boron/aluminum composites, fracture, 696
Unidirectional SCS-6/Ti-15-3 composites, fatigue crack growth, 711
Upper bounds, reduced axial and shear moduli, 56

V

Virtual crack closure technique, 287

W

Water absorption, effect on interlaminar fracture toughness, 107
Wavelength dispersive spectroscopy, thermomechanical fatigue of SCS-6/TI-15-3 composites, 732
Weaves, effect on composite moduli and stresses, 417
Weibull statistics, residual strength calculation with, 617
Woven composites, three-dimensional stress analysis, 417